Lecture Notes in Computer Science 7668

Commenced Publication in 1973
Founding and Former Series Editors:
Gerhard Goos, Juris Hartmanis, and Jan van Leeuwen

Steven Galbraith   Mridul Nandi (Eds.)

Progress in Cryptology - INDOCRYPT 2012

13th International Conference on Cryptology in India
Kolkata, India, December 9-12, 2012
Proceedings

 Springer

Volume Editors

Steven Galbraith
University of Auckland
Department of Mathematics
Private Bag 92019
Auckland 1142, New Zealand
E-mail: s.galbraith@math.auckland.ac.nz

Mridul Nandi
Indian Statistical Institute
Applied Statistics Unit
203 B.T. Road
Kolkata 700108, West Bengal, India
E-mail: mridul@isical.ac.in

ISSN 0302-9743   e-ISSN 1611-3349
ISBN 978-3-642-34930-0   e-ISBN 978-3-642-34931-7
DOI 10.1007/978-3-642-34931-7
Springer Heidelberg Dordrecht London New York

Library of Congress Control Number: 2012951708

CR Subject Classification (1998): E.3, K.6.5, D.4.6, C.2, J.1, G.2.1

LNCS Sublibrary: SL 4 – Security and Cryptology

Typesetting: Camera-ready by author, data conversion by Scientific Publishing Services, Chennai, India

Printed on acid-free paper

Springer is part of Springer Science+Business Media (www.springer.com)

Preface

Indocrypt 2012, the 13th International Conference on Cryptology in India, took place during December 9–12, 2012. It was hosted by the Indian Statistical Institute in Kolkata. The Indocrypt series of conferences began in 2000 under the leadership of Bimal Roy. This series is now well established as an international forum for presenting high-quality cryptography research.

This year 99 papers were submitted for consideration. The authors of the submitted papers were from institutions across 25 countries and five continents.

As in previous years, the submission deadline was split into two: authors were required to register titles and abstracts by July 23, 2012, while final versions of papers had to be submitted by July 28. During that week the titles and abstracts were made available to the Program Committee (PC) to enable them to select their preferred articles for review. Most papers were refereed by three committee members, and papers co-authored by a PC member were refereed by five committee members. We thank Tanja Lange for hosting the ichair system, which was used to manage submissions and online discussions, on her Web server.

The review stage was very tight, with only three weeks for reviewing papers and only two weeks (August 20 to September 1) for the online discussions. It was a difficult challenge for the 33 PC members and 86 sub-reviewers to give every paper a fair assessment in such a short time. A total of 313 referee reports were written, and 314 comments were posted on the online discussions. At the end of the discussion process 28 papers were accepted for the proceedings (five of them conditionally accepted subject to successful revision according to referee suggestions). Authors were notified on September 3 and had around 2 weeks to revise their papers according to the suggestions of the referees.

We would like to thank all the authors of submitted papers for supporting the Indocrypt conference. The best way to support a conference is to submit papers to it and attend it. We also wish to thank the members of the PC and their sub-reviewers (a list is given here) for devoting their time and knowledge to the selection of papers.

The proceedings include the revised versions of the 28 selected papers. Revisions were not checked by the PC and the authors bear the full responsibility for the contents of the respective papers. The proceedings also contain invited papers by Nigel Smart and Vinod Vaikuntanathan, as well as a one-page abstract of the invited lecture by Orr Dunkelman.

The organization of the conference involved many individuals. We express our heart-felt gratitude to the General Chair, Bimal Roy, Director of Indian

Statistical Institute, Kolkata, and Subhamoy Maitra for taking care of the actual hosting of the conference. Members of the Cryptology Research Group at the Indian Statistical Institute provided invaluable secretarial support. Finally, we would like to acknowledge Springer for their active cooperation and timely production of the proceedings.

December 2012

Steven Galbraith
Mridul Nandi

Organization

Indocrypt 2012 was organized by the Indian Statistical Institute, Kolkata, in collaboration with the Cryptology Research Society of India.

General Chair

Bimal K. Roy   Indian Statistical Institute, Kolkata, India

Program Co-chairs

Steven Galbraith   University of Auckland, New Zealand
Mridul Nandi   Indian Statistical Institute, Kolkata, India

Program Committee

Daniel Bernstein   University of Illinois at Chicago, USA
Sanjit Chatterjee   Indian Institute of Science, India
Chen-Mou Cheng   National Taiwan University, Taiwan
Joan Daemen   STMicroelectronics
Philippe Gaborit   Universite de Limoges, France
Steven Galbraith   University of Auckland, New Zealand
Vipul Goyal   Microsoft Research India, India
Shay Gueron   University of Haifa, Israel
Goichiro Hanaoka   National Institute of AIST, Japan
Tetsu Iwata   University of Nagoya, Japan
Stanislaw Jarecki   UC Irvine, USA
Tanja Lange   Technische Universiteit Eindhoven,
                                   The Netherlands
Kristin Lauter   Microsoft, USA
Keith Martin   Royal Holloway, University of London, UK
Willi Meier   FHNW Switzerland
David Naccache   Ecole Normale Superieure, Paris, France
Mridul Nandi   Indian Statistical Institute, Kolkata, India
Jesper Buus Nielsen   Aarhus University, Denmark
Elisabeth Oswald   University of Bristol, UK
Carles Padro   Nanyang Technological University, Singapore
Arpita Patra   ETH Zurich, Switzerland
Goutam Paul   Jadavpur University, India

Manoj M. Prabhakaran	University of Illinois, Urbana-Champaign, USA
Bart Preneel	Katholieke Universiteit Leuven, Belgium
C. Pandu Rangan	Indian Statistical Institute, Chennai, India
Christian Rechberger	Danish Technical University, Denmark
Phillip Rogaway	University of California, Davis, USA
Dipanwita Roy Chowdhury	Indian Institute of Technology, Kharagpur, India
Palash Sarkar	Indian Statistical Institute, Kolkata, India
Nicolas Sendrier	INRIA Rocquencourt, France
Damien Stehle	Ecole Normale Superieure de Lyon, France
Dominique Unruh	University of Tartu, Estonia
Fre Vercauteren	Katholieke Universiteit Leuven, Belgium

External Referees

Mohamed Ahmed
  Abdelraheem
Elena Andreeva
Diego F. Aranha
Subhadeep Banik
Guido Bertoni
Rishiraj Bhattacharyya
Zeeshan Bilal
Begül Bilgin
Praloy Biswas
Julia Borghoff
Joppe Bos
Charles Bouillaguet
Christina Boura
Stanislav Bulygin
Donghoon Chang
Yun-An Chang
Melissa Chase
Anupam Chattopadhyay
Ashish Choudhury
Sherman S.M. Chow
Craig Costello
Abhijit Das
Christophe Doche
Keita Emura
Reza Rezaeian Farashahi
Thomas Fuhr
David Galindo
Sugata Gangopadhyay
Gerhard Hancke

Carmit Hazay
Javier Herranz
Huseyin Hisil
Chethan Kamath
Ferhat Karakoc
Sandip Karmakar
Kouhei Kasamatsu
Simon Knellwolf
Ozgul Kucuk
Fabien Laguillaumie
Gregory Landais
Cédric Lauradoux
Anh Le
Gregor Leander
Benoit Libert
Changshe Ma
Subhamoy Maitra
Preetha Mathews
Takahiro Matsuda
Filippo Melzani
Miodrag Mihaljevic
Nicky Mouha
Sean Murphy
Michael Naehrig
Souradyuti Paul
Roel Peeters
Albrecht Petzoldt
Maria Naya Plasencia
Sukhendu Quila
Dhiman Saha

Yusuke Sakai
Subhabrata Samajdar
Somitra Sanadhya
Yu Sasaki
Jacob C.N. Schuldt
Gautham Sekar
Sharmila Deva Selvi
Sourav Sengupta
Jae Hong Seo
Kyoji Shibutani
Francesco Sica
Alice Silverberg
Shashank Singh
Ben Smith
Valentin Suder
Enrico Thomae
Michael Tunstall
Berkant Ustaoglu
Vinod Vaikuntanathan
Ayineedi Venkateswarlu
Sree Vivek
Colin Walter
Carolyn Whitnall
Christopher Wolf
Marcin Wójcik
Shota Yamada
Naoto Yanai
Bo-Yin Yang
Wei Zhang

Sponsoring Institutions

Defence Research and Developement Organization (D.R.D.O.), India
Google Inc., India
Microsoft Research, Bangalore
National Board of Higher Mathematics (N.B.H.M.), Mumbai
Reserve Bank of India (R.B.I.), Kolkata
Tata Consultancy Services (T.C.S.), Kolkata

Table of Contents

Hash Functions and Stream Cipher

Cryptanalysis of Block Ciphers

Time Memory Trade-Off

Hardware

Elliptic Curve

Digital Signature

Symmetric Key Design and Provable Security

How to Compute on Encrypted Data

Vinod Vaikuntanathan[*]

University of Toronto
vinodv@cs.toronto.edu

Abstract. A fully homomorphic encryption scheme enables computation of arbitrary functions on encrypted data. Fully homomorphic encryption has long been regarded as cryptography's prized "holy grail" – extremely useful yet rather elusive. Starting with the groundbreaking work of Gentry in 2009, the last three years have witnessed numerous constructions of fully homomorphic encryption involving novel mathematical techniques, and a number of exciting applications. We will take the reader through a journey of these developments and provide a glimpse of the exciting research directions that lie ahead.

1 Introduction

Encryption has traditionally been viewed as a mechanism that enables *secure communication*, namely the problem of transmitting a message from Alice to Bob over a public channel while keeping it hidden from an eavesdropper. In particular, Public-key Encryption – conceived in the seminal work of Diffie and Hellman [DH76] and first constructed by Rivest, Shamir and Adleman [RSA83] – provides a way for Alice to encrypt a message into a ciphertext using Bob's public key, and for Bob to decrypt the ciphertext to obtain the message using his secret key. In this view of encryption schemes, *access to encrypted data is all or nothing* – having the secret decryption key enables one to learn the entire message, but without the decryption key, the ciphertext is completely useless.

This state of affairs raises an intriguing question, first posed by Rivest, Adleman and Dertouzos in 1978: *Can we do arbitrary computations on data while it remains encrypted, without ever decrypting it?* This asks for the seemingly fantastical ability to perform computations on encrypted data without being able to "see" the data. Such ability also gives rise to a number of useful applications including the ability to privately outsource arbitrary computations to the "cloud" and the ability to store all data encrypted and perform computations on encrypted data, decrypting only when necessary.

---

[*] Supported by an NSERC Discovery Grant and by DARPA under Agreement number FA8750-11-2-0225. The U.S. Government is authorized to reproduce and distribute reprints for Governmental purposes notwithstanding any copyright notation thereon. The views and conclusions contained herein are those of the author and should not be interpreted as necessarily representing the official policies or endorsements, either expressed or implied, of DARPA or the U.S. Government.

S. Galbraith and M. Nandi (Eds.): INDOCRYPT 2012, LNCS 7668, pp. 1–15, 2012.

Fully Homomorphic encryption (FHE) is a special type of encryption system that permits *arbitrarily complex computation* on encrypted data. Long regarded as a "holy grail" of cryptography, fully homomorphic encryption was first shown to be possible in the recent, breakthrough work of Gentry [Gen09a, Gen09b]. We will take the reader through a journey of the fascinating mathematical techniques underlying these developments, which in turn raise a number of exciting new questions in cryptography.

Organization of this Survey. In this survey, we focus on the recent developments in fully homomorphic encryption, starting from the work of Brakerski and Vaikuntanathan [BV11b]. For more on the early work on homomorphic encryption and its history, we refer the reader to expository writings [Gen10a, Vai11] on the subject as well as the original papers [Gen09a, Gen09b, DGHV10, SV11]. We also describe a few applications of fully homomorphic encryption and a (highly incomplete) list of research directions. Our intention in this short survey is to give the reader a taste of the main ideas and developments in this area, while referring to the original works for detailed expositions.

2   Gentry's Blueprint for FHE

Gentry's work showed not only the first fully homomorphic encryption scheme, but a general method (a "blue-print") to construct such systems. This blue-print has been instantiated with a number of cryptographic assumptions, yielding progressively simpler and more efficient schemes [DGHV10, SV10, SS10, BV11a].

Notwithstanding the elegance and generality of the blue-print, schemes constructed along these lines suffer from a number of deficiencies, including the reliance on a host of non-standard cryptographic assumptions, and severe limitations on efficiency. In Section 3, we describe new developments in fully homomorphic encryption that solve some of these issues.

Gentry's construction has three components.

Step 1: Somewhat Homomorphic Encryption. The first step in Gentry's blueprint is to construct a *somewhat homomorphic encryption (SWHE) scheme*, namely an encryption scheme capable of evaluating "low-degree" polynomials homomorphically. More precisely, letting κ be the security parameter, the scheme is able to evaluate ℓ-variate polynomials for some $\ell = \mathsf{poly}(\kappa)$ where each monomial has degree at most κ^ϵ (for some constant $\epsilon < 1$). Starting with Gentry's scheme, the ciphertexts in all subsequent constructions contain some "noise" that increases during homomorphic operations. Homomorphic multiplication increases the noise significantly more than addition, and thus the limitation on low-degree polynomials.

Step 2: The Bootstrapping Theorem. The somewhat homomorphic encryption unfortunately falls well short of evaluating arbitrary functions on encrypted data. To obtain an FHE, Gentry provided a remarkable *bootstrapping theorem* which states that given an SWHE scheme that can evaluate its own decryption function (plus an additional operation), one can transform it into a "leveled" FHE scheme, in a completely generic way. Such an SWHE scheme is called a bootstrappable encryption scheme. Furthermore, if we are willing to make an additional assumption – namely that it is safe to encrypt the leveled FHE secret key under its own public key, a requirement that is referred to as "circular security" – then the transformation gives us a "pure" (as opposed to "leveled") FHE scheme. Bootstrapping "refreshes" a ciphertext by running the decryption function on it homomorphically, using an encrypted secret key (given in the evaluation key), resulting in a reduced noise.

For a formal definition of a bootstrappable encryption scheme and Gentry's bootstrapping theorem, we refer the reader to [Gen09b, Gen09a, Vai11].

Step 3: Squashing the Decryption Circuit. The final piece in the puzzle is to determine if we can apply the bootstrapping theorem to the known SWHE schemes, namely determine if they are in fact capable of evaluating their own decryption circuits (plus some). Surprisingly, as if by a strange law of nature, this turned out to *not* be the case for all the (then available) SWHE schemes. For example, the decryption circuits in the schemes of [Gen09b, DGHV10, SV11] turn out to have degree $\widetilde{\Omega}(\kappa)$ which is larger than their homomorphic capacity, namely polynomials of degree $O(\kappa^\epsilon)$ for some $\epsilon < 1$.

Thus, the final step is to *squash the decryption circuit* of the SWHE scheme, namely transform the scheme into one with the same homomorphic capacity but a decryption circuit that is simple enough to allow bootstrapping. Gentry [Gen09b] showed a way to do this by adding a "hint" about the secret key to the evaluation key. The hint is a large set of elements that has a secret sparse subset that sums to the original secret key. In order to ensure that the hint does not reveal damaging information about the secret key, the security of this transformation relies on a new "sparse subset sum" assumption. The sparsity pushes the decryption complexity at the cost of the additional assumption. This approach can be adapted to the later schemes [DGHV10, SV10, BV11a] as well, and it crucially utilizes the fact that the decryption equation of these schemes is (almost) a linear equation in the secret key.

2.1   Other Instantiations

A number of other works construct FHE schemes following this framework, essentially by building various instantiations of the SWHE scheme. Gentry's original construction [Gen09b] of an SWHE scheme was based on (a variant of) the bounded distance decoding problem on ideal lattices drawn according to a certain distribution.[1] In a subsequent work, Gentry [Gen10b] showed a worst-case to average-case reduction for this problem, thus basing the security of his scheme on a worst-case problem over ideal lattices. Smart and Vercauteren [SV10] construct an SWHE scheme, following Gentry's construction closely, but basing it on

---

[1] Roughly speaking, ideal lattices correspond to a geometric embedding of ideals in a number field. For a formal definition, see [Gen09b, LPR10].

the average-case hardness of a "small principal ideal problem" (unlike [Gen09b], there is no known worst-case to average-case connection for this problem). Brakerski and Vaikuntanathan [BV11a] showed yet another scheme based on the average-case hardness of the "Ring Learning with Errors" (Ring LWE) problem which, by the results of Lyubashevsky, Peikert and Regev [LPR10], is as hard as various standard worst-case problems on ideal lattices. This scheme, in addition, has very simple and efficient algorithms. See [GH11b, CMNT11, NLV11] for other variants and optimizations.

All these schemes then go through the squashing and bootstrapping transformations to construct FHE, and therefore, they all have to rely on the sparse subset sum assumption.

3   A New Era of Fully Homomorphic Encryption

Gentry's blueprint and its instantiations leave open a number of important questions.

First, all the constructions based on the blueprint rely on multiple complexity assumptions, the most problematic of them being the little-studied "sparse subset sum assumption" used in squashing the decryption circuit. Is this assumption necessary? In addition, the constructions use ideals in various rings either explicitly or implicitly [DGHV10]. Ideals are a natural mathematical object to construct fully homomorphic encryption since they natively support both addition and multiplication operations, but are they necessary for FHE? A final concern is that all the constructions ultimately rely on the hardness of approximating lattice problems to within a subexponential factor (in the dimension n of the lattice). Can we base security on the hardness of approximation in the polynomial range?

Secondly, schemes that follow Gentry's blueprint turn out to have inherent efficiency limitations (see [BGV12] for an argument to this effect). When speaking of efficiency, we are interested in the length of the ciphertext (per bit encrypted) and the keys, and the time it takes to encrypt and decrypt. More importantly, it turns out that the bottleneck in practical deployments of FHE is the per-gate evaluation time, defined as the ratio of the time it takes to evaluate a circuit C homomorphically to the time it takes to evaluate C on plaintext inputs. The schemes that follow Gentry's blue-print [Gen09b, DGHV10, GH11b, SV10, BV11b] have a per-gate evaluation time of $\Omega(\kappa^4)$ (where κ is the security parameter), even by fairly generous estimates.

A series of new works address these concerns. In particular, Brakerski and Vaikuntanathan [BV11b] show that (leveled) FHE can be based on the hardness of the much more standard "learning with error" (LWE) problem introduced by Regev [Reg05] which, by the results of Regev [Reg05] and Peikert [Pei09] is as hard as solving various short vector problems on arbitrary (not ideal) lattices *in the worst case*. In effect, they show how to obtain a direct construction of a bootstrappable encryption scheme without having to squash the decryption circuit and thus, without relying on the non-standard sparse subset sum assumption.

In a concurrent work, Gentry and Halevi [GH11a] show how to get rid of squashing as well, using a completely different technique. Their construction still relies on ideal lattices.

In a recent development, Brakerski, Gentry and Vaikuntanathan [BGV12] build on (a refinement of) the main technique in [BV11b] to construct an FHE scheme with asymptotically linear efficiency (namely, a per-gate computation time of $\widetilde{O}(\kappa)$) under the assumption that short vector problems on arbitrary lattices are hard to approximate to within a slightly super-polynomial factor (more precisely, $n^{O(\log n)}$ where n is the dimension of the lattice) in the worst-case.

We now describe the ideas behind the work of [BGV12] which builds on [BV11b] and, at the time of writing, constitutes the state-of-the-art in fully homomorphic encryption, both in terms of the mildness of assumptions as well as efficiency.

3.1 The BGV Result

The starting point for our description of the scheme is a somewhat homomorphic encryption scheme based on "Ring LWE" [LPR10] (although the scheme can be instantiated with a number of other SWHE schemes in the literature). The scheme works over the rings $R = Z[x]/(f(x))$ where $f(x)$ is an irreducible polynomial of degree n and $R_q = R/qR$ where q is a prime modulus. An additional parameter is an "error distribution" χ over R that outputs polynomials with "small coefficients". As before, we describe a secret-key system with message space $R_2 := R/2R$, for simplicity.

- SH.Keygen(1^κ): Sample $sk := \mathbf{s} \leftarrow R_q$ to be a polynomial with small coefficients chosen from the error distribution χ.
- SH.Enc($sk, \mu \in R_2$): Sample $\mathbf{a} \leftarrow R_q$ at random, and a polynomial $\mathbf{e}$ with small coefficients from an error distribution χ. Output $\mathbf{c} := (\mathbf{a}, \mathbf{as} + 2\mathbf{e} + \mu)$.
- SH.Dec($sk, \mathbf{c} = (\mathbf{a}, \mathbf{b})$): Compute $\widetilde{\mu} := \mathbf{b} - \mathbf{as}$ over R_q and output $\mu := \widetilde{\mu}$ (mod 2).

As before, the success of decryption is contingent on the noise in the ciphertext being "small enough". Homomorphic operations in this scheme increase the noise – multiplication more than addition – and the noise after homomorphically evaluating a multivariate polynomial with degree D and A monomials turns out to be $O(A \cdot n^{O(D)})$. In other words, the noise increases *exponentially in the degree of the polynomial.*

The key contribution in the work of [BGV12] is a new *noise-management technique* that keeps the noise in check by reducing it after homomorphic operations, without bootstrapping. Essentially, the noise will grow to $O(A \cdot n^{O(d)})$, where *d is the depth of the circuit computing the polynomial.* Since the degree D of a circuit is usually exponentially larger than its depth d, we achieve an exponential improvement in the homomorphic capacity.

The key technical tool they use for noise management is the "modulus switching" technique developed by Brakerski and Vaikuntanathan [BV11b], the essence of which is captured in the following lemma.

In words, the lemma says that an evaluator, who does not know the secret key $\mathbf{s}$ but instead only knows a bound on its length, can transform a ciphertext $\mathbf{c}$ modulo q into a different ciphertext $\mathbf{c}'$ modulo p while preserving correctness – namely, $(\mathbf{c}'\mathbf{s} \pmod p) \pmod 2 = (\mathbf{cs} \pmod q) \pmod 2$. The transformation from $\mathbf{c}$ to $\mathbf{c}'$ involves simply scaling by (p/q) and rounding appropriately! Most interestingly, if $\mathbf{s}$ is short and p is sufficiently smaller than q, the "noise" in the ciphertext actually decreases – namely,

$$|\mathbf{c}'\mathbf{s} \pmod p| < |\mathbf{cs} \pmod q|$$

Lemma 1. *Let p and q be two odd moduli, and let $\mathbf{c} = (\mathbf{a}, \mathbf{b})$ be a ciphertext modulo q. Define $\mathbf{c}' = (\mathbf{a}', \mathbf{b}')$ to be the integer vector closest to $(p/q) \cdot \mathbf{c} = ((p/q)\cdot\mathbf{a}, (p/q)\cdot\mathbf{b})$ such that $\mathbf{c}' = \mathbf{c} \bmod 2$. Then, for any $\mathbf{s}$ with $|\mathbf{b} - \mathbf{as} \bmod q| < q/2 - (q/p) \cdot \ell_1(\mathbf{s})$, we have*

$$(\mathbf{b}' - \mathbf{a}'\mathbf{s} \bmod p) \bmod 2 = (\mathbf{b} - \mathbf{as} \bmod q) \bmod 2$$
$$and \ |\mathbf{b}' - \mathbf{a}'\mathbf{s} \bmod p| < (p/q) \cdot |\mathbf{b} - \mathbf{as} \bmod q| + \ell_1(\mathbf{s})$$

where $\ell_1(\mathbf{s})$ is the ℓ_1-norm of (the co-efficient vector corresponding to) $\mathbf{s}$.

Amazingly, this trick permits the evaluator to reduce the magnitude of the noise without knowing the secret key, and without bootstrapping. In other words, modulus switching gives us a very powerful and lightweight way to *manage the noise* in FHE schemes!

The BGV Noise Management Technique. At first, it may look like modulus switching is not a very effective noise management tool. If p is smaller than q, then of course modulus switching may reduce the magnitude of the noise, but it reduces the modulus size by essentially the same amount. In short, the ratio of the noise to the "noise ceiling" (the modulus size) does not decrease at all. Isn't this ratio what dictates the remaining homomorphic capacity of the scheme, and how can potentially worsening (certainly not improving) this ratio do anything useful?

In fact, it's not just the ratio of the noise to the "noise ceiling" that's important. The *absolute magnitude of the noise* is also important, especially in multiplications. Suppose that $q \approx x^k$, and that you have two mod-q SWHE ciphertexts with noise of magnitude x. If you multiply them, the noise becomes x^2. After 4 levels of multiplication, the noise is x^{16}. If you do another multiplication at this point, you reduce the ratio of the noise ceiling (i.e. q) to the noise level by a huge factor of x^{16} – i.e., you reduce this gap very fast. Thus, the actual magnitude of the noise impacts how fast this gap is reduced. After only $\log k$ levels of multiplication, the noise level reaches the ceiling.

Now, consider the following alternative approach. Choose a *ladder of gradually decreasing moduli* $\{q_i \approx q/x^i\}$ for $i < k$. After you multiply the two mod-q

ciphertexts, switch the ciphertext to the smaller modulus $q_1 = q/x$. As the lemma above shows, the noise level of the new ciphertext (now with respect to the modulus q_1) goes from x^2 back down to x. (Let's suppose for now that $\ell_1(\mathbf{s})$ is small in comparison to x so that we can ignore it.) Now, when we multiply two ciphertexts (wrt modulus q_1) that have noise level x, the noise again becomes x^2, but then we switch to modulus q_2 to reduce the noise back to x. In short, each level of multiplication only reduces the ratio (noise ceiling)/(noise level) by a factor of x (not something like x^{16}). With this new approach, we can perform about k (not just $\log k$) levels of multiplication before we reach the noise ceiling. We have just increased (without bootstrapping) the number of multiplicative levels that we can evaluate by an exponential factor!

This exponential improvement is enough to achieve leveled FHE without squashing or bootstrapping. For *any* polynomial k, we can evaluate circuits of depth k. The performance of the scheme degrades with k – e.g., we need to set $q = q_0$ to have bit length proportional to k – but it degrades only polynomially with k.

Performance-wise, this scheme trounces previous (bootstrapping-based) FHE schemes (at least asymptotically; the concrete performance remains to be seen). Instantiated with ring-LWE, it can evaluate L-level arithmetic circuits with per-gate computation $\tilde{O}(\kappa \cdot L^3)$ – i.e., computation *quasi-linear* in the security parameter. Since the ratio of the largest modulus (namely, $q \approx x^L$) to the noise (namely, x) is exponential in L, the scheme relies on the hardness of approximating short vectors to within an exponential in L factor. The performance can be improved further using batching tricks.

In essence, the new noise management technique allows us to evaluate exponentially deeper circuits at the same cost as before. Combining this technique with bootstrapping gives us further performance gains and allows us to base security on better assumptions – namely, the hardness of approximating shortest vector to a quasi-polynomial factor (in the dimension n). See [BGV12] for more details.

Brakerski's Refinement. Brakerski [Bra12] showed a conceptually simpler version of the BGV scheme, together with improvements in the complexity assumptions. His construction does away with the need for the ladder of moduli and instead works with a single modulus throughout. In essence, his construction highlights the essential use of the modulus reduction technique, namely as a better way of doing homomorphic multiplication. For more details, we refer the reader to his work [Bra12].

Multi-key Homomorphism. In a homomorphic encryption scheme, we can compute functions on (many) ciphertexts, all encrypted under the same key. We need to go beyond this when working in a multi-user setting. Assume that Alice and Bob store their encrypted medical records in the cloud. Of course, they use independent public / private key pairs since they do not trust each other. How can the cloud compute a joint function on both their inputs, encrypted under different keys? Lopéz-Alt, Tromer and Vaikuntanathan [LTV12] introduce the notion of multi-key homomorphic encryption where an evaluator can compute any function on inputs encrypted under multiple unrelated public keys. Of course, Alice and Bob have to cooperate in order to decrypt the evaluated ciphertext. They showed a construction of multi-key FHE, based on the NTRU encryption scheme [HPS98] and leveraging the techniques in [BGV12].

4 Applications of Fully Homomorphic Encryption

Aside from the multitude of scenarios where it is beneficial to keep all data encrypted and to perform computations on encrypted data, fully homomorphic encryption has been used to solve a number of other problems in cryptography. We briefly describe a number of such problems.

4.1 Verifiably Outsourcing Computation

While fully homomorphic encryption enables a client to privately outsource the computation of any function to a server, it does not provide any correctness guarantees whatsoever. In particular, the server could potentially compute a different function on the client's encrypted input, without the client noticing it. Indeed, a practical concern is that cloud computing services have the financial incentive to perform a much cheaper (and incorrect!) computation than what the client outsourced, especially if they can get away with it undetected. Verifiable Computation (VC) is the problem of outsourcing computation to a server where the client can verify the correctness of the answer, without spending much computational effort. Ideally, we like this process to be non-interactive, namely, we allow a message from the client to the server communicating his input (and perhaps additional information) and a second message back from the server to communicate the result of the computation together with a "proof" that the computation was performed correctly.

Verifiability and Privacy in the setting of outsourcing computation seem to be orthogonal constraints. Yet, Gennaro, Gentry and Parno [GGP10] and subsequently, Chung, Kalai and Vadhan [CKV10] showed how to use fully homomorphic encryption in an essential way to construct a non-interactive verifiable computation protocol for any function. In their protocol, the verifier is efficient in an amortized sense. Namely, the prover and the verifier run an expensive pre-processing phase which they can then re-use for polynomially many protocol executions. Thus, even though the off-line computational cost of the verifier is large, this is amortized by many executions in which his per-execution on-line cost is small (in particular, independent of the complexity of the function).

We refer the reader to [GGP10, CKV10, AIK10] for details on these solutions, and [Mic00, GKR08, Rot09] for other related work on this problem.

4.2 Short Non-Interactive Zero-Knowledge Proofs

In a non-interactive zero-knowledge (NIZK) proof system [BFM88, FLS99] for an NP language (say, Circuit SAT), a prover and a verifier have a Boolean circuit C on n variables, and the prover wishes to convince the verifier that there is a satisfying input to C, namely an input x such that $C(x) = 1$, without revealing any (more) information about x itself. NIZK proof systems have been extremely useful in constructing various fundamental cryptographic primitives such as secure signatures [BG89] and public-key encryption schemes that resist chosen ciphertext (CCA) attacks [NY90, RS91, DDN00].

In all the known constructions of non-interactive zero-knowledge proof systems [BFM88, FLS99, GOS06b, GOS06a], the size of the proof is $\mathsf{poly}(|C|, |x|)$, which is rather large compared to the size of the (non zero-knowledge) NP proof, namely the witness x itself. A natural question to ask is: can we design NIZK proof systems where the size of the proof depends only on the size of the witness (namely, x) and not on the size of the statement (namely, C)? Fully Homomorphic Encryption provides an affirmative answer to this question, as noted by Gentry [Gen09a].

In a nutshell, the idea is that FHE reduces the problem of constructing a NIZK proof for Circuit SAT to constructing a NIZK proof for a "simpler" language. The prover chooses the public key pk for an FHE scheme, prepares encryptions $\psi_i \leftarrow \mathsf{FHE.Enc}(pk, x_i)$ of the bits of the satisfying assignment x, and constructs $\psi^* = \mathsf{FHE.Eval}(evk, C, \psi_1, \ldots, \psi_n)$. The task of the prover is now reduced to constructing a NIZK proof π^* to show that ψ^* is an encryption of 1 together with proofs π_i that the input encryptions ψ_i are well-formed, all of which are statements whose size is independent of the size of the circuit C. The proof consists of $\psi_1, \ldots, \psi_n$ together with $(\pi_1, \ldots, \pi_n, \pi^*)$. The verifier, upon receipt of the entire proof, checks the NIZK proofs of well-formedness $\pi_1, \ldots, \pi_n$, computes $\psi^* = \mathsf{FHE.Eval}(evk, C, \psi_1, \ldots, \psi_n)$ and then checks the NIZK proof π^*. Note that this procedure crucially relies on the fact that the homomorphic evaluation algorithm $\mathsf{FHE.Eval}$ is deterministic.

4.3   Other Applications

Another important application of FHE schemes is in constructing efficient and minimally interactive secure multi-party computation (MPC) protocols where, in addition, the burden of the computation rests on a designated party (called the "cloud"), and the rest of the parties (called the "clients") do very little by way of computation. Asharov et al. [AJLA$^+$12] showed how to obtain an efficient MPC protocol of this form by leveraging a key homomorphism property of the schemes of [BV11b, BV11a, BGV12].

Other applications of fully homomorphic encryption include Private Information Retrieval (PIR) schemes [OS07, Gen09a, BV11b], Proxy Re-encryption [Gen09a], Key Dependent Message (KDM)-secure encryption schemes [App11], leakage-resilient cryptography [JV10], Oblivious RAMs [GO96] and One-time Programs [GKR08]. We refer the reader to the original papers for more details. Some of these applications do not require the full power of FHE – for PIR, it is sufficient to have a somewhat homomorphic encryption scheme capable of evaluating simple database indexing functions.

5   Open Problems and Future Directions

The study of fully homomorphic encryption has led to a number of new and exciting concepts and questions, as well as a powerful tool-kit to address them. We conclude this survey by describing a number of research directions related to FHE and more generally, the problem of computing on encrypted data.

5.1   Can Fully Homomorphic Encryption Be Practical?

While Gentry's original construction was viewed as being impractical, recent constructions and implementation efforts have drastically improved the efficiency of fully homomorphic encryption. The initial implementation efforts focused on Gentry's original scheme and its variants [SV10, GH11b, SV11, CMNT11] which seemed to pose rather inherent efficiency bottlenecks. Later implementations leverage the recent algorithmic advances [BV11a, BV11b, BGV12] that result in asymptotically better FHE systems, as well as new algebraic techniques to improve the concrete efficiency of these schemes [NLV11, GHS11, SV11].

A significant hurdle to the practicality of fully homomorphic encryption is that it is designed to work with a circuit model of computation and is inherently not suited to work with uniform models such as Turing machines and RAM machines. To illustrate this, assume that you are given an encryption of a positive integer x and you are asked to homomorphically compute the following program on $\mathsf{Enc}(x)$: "while (x< 10^6) x++;". The techniques we have at our disposal require us to flatten this program into a circuit resulting in a huge blow-up in size.

The problem is conceptually even deeper than this. Consider, instead, the program "while (x< 10^6) // do nothing". Depending on the value of x, this computation either terminates immediately or does not terminate at all. The question, then, is – given an encrypted x, how long will you run the computation? The fundamental source of difficulty is the existence of data-dependent loops in general programs. Clearly, the process of running a program with data-dependent loops on an encrypted input in an *effective way* leaks some information about the input. In this case, it is the information on whether $x < 10^6$ or not. The question is, can we define and achieve the minimal information that we have to leak to perform effective homomorphic computations.

5.2   FHE from Other Assumptions

The state of the art (non-leveled) FHE schemes rely on the bootstrapping technique and consequently, some form of the circular security assumption. A central question in this area is whether such an assumption is in fact necessary.

All the constructions of fully homomorphic encryption to date derive their security from the hardness of various lattice problems. Can we build FHE schemes from the traditional number theoretic machinery? Are there FHE schemes whose security can be based on the hardness of factoring? How about the hardness of computing discrete logarithms? At the time of this writing, we do not have any hint as to whether such constructions are possible, or whether there are inherent limitations. We believe this is a very interesting avenue to pursue.

Yet another interesting question along these lines is to determine what sort of FHE schemes can be based on "general assumptions". Private Information Retrieval seems intimately connected to FHE schemes. In particular, PIR schemes can be used to homomorphically evaluate decision trees. A special type of (communication efficient) PIR scheme was used by Ishai and Paskin [IP07] to homomorphically evaluate branching programs. Furthermore, it is easy to see that PIR gives us an *inefficient* homomorphic encryption scheme for arbitrary functions. This is because one could write down the truth table of any function, treat it as a database and use a PIR scheme to retrieve the appropriate entry of the database. These observations point to a deeper connection between these two primitives, which we believe is worth exploring.

5.3   Non-malleability and Homomorphic Encryption

Homomorphism and Non-malleability are antipodal properties of an encryption scheme. Homomorphic encryption schemes permit anyone to transform an encryption of a message m into an encryption of $f(m)$ for non-trivial functions f. Non-malleable encryption, on the other hand, prevents precisely this sort of thing – it requires that no adversary be able to transform an encryption of m into an encryption of any "related" message. In reality, what we need is a combination of both properties that *selectively permit homomorphic computations*. Namely, the evaluator should be able to homomorphically compute any function from some pre-specified class $\mathcal{F}_{\text{hom}}$, yet she should not be able to transform an encryption of m into an encryption of $f(m)$ for any $f \notin \mathcal{F}_{\text{hom}}$. Thus, the question is: *Can we control what is being (homomorphically) computed?*

Formalizing this notion turns out to be tricky. Boneh, Segev and Waters [BSW12] propose the notion of *targeted malleability* – a candidate formalization of such a requirement – as well as constructions of such encryption schemes. Their encryption scheme is based on a strong "knowledge of exponent-type" assumption, and allows iterative evaluation of at most t functions, where t is a pre-specified constant. Improving their construction as well as the underlying complexity assumptions is an important open problem.

Furthermore, it is interesting to extend the definition of non-malleability to allow for chosen ciphertext attacks. Consider, for example, implementing an encrypted targeted advertisement system that generates advertisements depending on the contents of a user's e-mail. Since the e-mail is stored encrypted (with the user's public key), the e-mail server performs a homomorphic evaluation and computes an encrypted advertisement to be sent back to the user. The user decrypts it, and performs an action depending on what she sees. Namely, if the advertisement is relevant, she might choose to click on it, but otherwise, she will ignore it. Now, if the e-mail server is privy to this information, namely whether the user clicked on the ad or not, they can use this as a restricted "decryption oracle" to break the security of the user's encryption scheme and perhaps even recover her secret key. Such attacks are ubiquitous whenever we compute on encrypted data, almost to the point that CCA security seems like a necessity. Yet, it is easy to see that chosen ciphertext secure (CCA2-secure) homomorphic encryption schemes cannot exist. To resolve this conundrum, we need an appropriate definition of security "CCA2-like" security for homomorphic encryption and constructions that achieve the definition.

5.4   FHE and Functional Encryption

Homomorphic encryption schemes permit anyone to evaluate functions on encrypted data, but the evaluators never see any information about the result. Is is possible to construct an encryption scheme where a user can compute $f(m)$ *in the clear* from an encryption of a message m, but she should learn no other information about m (including the intermediate results in the computation of f)? Thus, the question is: *Can we control what the evaluator can see?* Such an encryption scheme is called a functional encryption scheme, first defined by Sahai and Waters [SW05] and explored in a number of works ([KSW08, BSW12, AFV11] and many others). Although these constructions work for several interesting families of functions (such as monotone formulas and inner products), constructing a *fully functional encryption scheme* is wide open.

More generally, what we need is a new, broad vision for encryption systems that provide us with *fine-grained control* over what one can see and what one can compute on data.

Acknowledgments. This survey has benefited immensely from discussions with Boaz Barak, Dan Boneh, Zvika Brakerski, Craig Gentry, Shafi Goldwasser, Shai Halevi, Kristin Lauter, Michael Naehrig, Nigel Smart and Marten van Dijk.

References

[AFV11]   Agrawal, S., Freeman, D.M., Vaikuntanathan, V.: Functional Encryption for Inner Product Predicates from Learning with Errors. In: Lee, D.H., Wang, X. (eds.) ASIACRYPT 2011. LNCS, vol. 7073, pp. 21–40. Springer, Heidelberg (2011)

[AIK10]   Applebaum, B., Ishai, Y., Kushilevitz, E.: From Secrecy to Soundness: Efficient Verification via Secure Computation. In: Abramsky, S., Gavoille, C., Kirchner, C., Meyer auf der Heide, F., Spirakis, P.G. (eds.) ICALP 2010, Part I. LNCS, vol. 6198, pp. 152–163. Springer, Heidelberg (2010)

[AJLA+12]   Asharov, G., Jain, A., López-Alt, A., Tromer, E., Vaikuntanathan, V., Wichs, D.: Multiparty Computation with Low Communication, Computation and Interaction via Threshold FHE. In: Pointcheval, D., Johansson, T. (eds.) EUROCRYPT 2012. LNCS, vol. 7237, pp. 483–501. Springer, Heidelberg (2012)

[App11]   Applebaum, B.: Key-dependent message security: Generic amplification and completeness. In: Paterson [Pat11], pp. 527–546

[BFM88]   Blum, M., Feldman, P., Micali, S.: Non-interactive zero-knowledge and its applications (extended abstract). In: STOC, pp. 103–112 (1988)

[BG89]   Bellare, M., Goldwasser, S.: New Paradigms for Digital Signatures and Message Authentication Based on Non-interactive Zero Knowledge Proofs. In: Brassard, G. (ed.) CRYPTO 1989. LNCS, vol. 435, pp. 194–211. Springer, Heidelberg (1990)

[BGV12]   Brakerski, Z., Gentry, C., Vaikuntanathan, V.: (leveled) fully homomorphic encryption without bootstrapping. In: Goldwasser [Gol12], pp. 309–325

[Bra12]   Brakerski, Z.: Fully Homomorphic Encryption without Modulus Switching from Classical GapSVP. In: Safavi-Naini, R., Canetti, R. (eds.) CRYPTO 2012. LNCS, vol. 7417, pp. 868–886. Springer, Heidelberg (2012)

[BSW12]   Boneh, D., Segev, G., Waters, B.: Targeted malleability: homomorphic encryption for restricted computations. In: Goldwasser [Gol12], pp. 350–366

[BV11a]   Brakerski, Z., Vaikuntanathan, V.: Fully Homomorphic Encryption from Ring-LWE and Security for Key Dependent Messages. In: Rogaway, P. (ed.) CRYPTO 2011. LNCS, vol. 6841, pp. 505–524. Springer, Heidelberg (2011)

[BV11b]   Brakerski, Z., Vaikuntanathan, V.: Efficient fully homomorphic encryption from (standard) lwe. In: Ostrovsky [OST11], pp. 97–106

[CKV10]   Chung, K.-M., Kalai, Y., Vadhan, S.: Improved Delegation of Computation Using Fully Homomorphic Encryption. In: Rabin, T. (ed.) CRYPTO 2010. LNCS, vol. 6223, pp. 483–501. Springer, Heidelberg (2010)

[CMNT11]   Coron, J.-S., Mandal, A., Naccache, D., Tibouchi, M.: Fully Homomorphic Encryption over the Integers with Shorter Public Keys. In: Rogaway, P. (ed.) CRYPTO 2011. LNCS, vol. 6841, pp. 487–504. Springer, Heidelberg (2011)

[DDN00]   Dolev, D., Dwork, C., Naor, M.: Nonmalleable cryptography. SIAM J. Comput. 30(2), 391–437 (2000)

[DGHV10]   van Dijk, M., Gentry, C., Halevi, S., Vaikuntanathan, V.: Fully homomorphic encryption over the integers. In: Gilbert, H. (ed.) EUROCRYPT 2010. LNCS, vol. 6110, pp. 24–43. Springer, Heidelberg (2010), full version in http://eprint.iacr.org/2009/616.pdf

[DH76]   Diffie, W., Hellman, M.: New directions in cryptography. IEEE Transactions on Information Theory IT-22, 644–654 (1976)

[FLS99]   Feige, U., Lapidot, D., Shamir, A.: Multiple noninteractive zero knowledge proofs under general assumptions. SIAM J. Comput. 29(1), 1–28 (1999)

[Gen09a]   Gentry, C.: A fully homomorphic encryption scheme. PhD thesis, Stanford University (2009), http://crypto.stanford.edu/craig

[Gen09b]   Gentry, C.: Fully homomorphic encryption using ideal lattices. In: STOC, pp. 169–178 (2009)

[Gen10a]   Gentry, C.: Computing arbitrary functions of encrypted data. Commun. ACM 53(3), 97–105 (2010)

[Gen10b]   Gentry, C.: Toward Basing Fully Homomorphic Encryption on Worst-Case Hardness. In: Rabin, T. (ed.) CRYPTO 2010. LNCS, vol. 6223, pp. 116–137. Springer, Heidelberg (2010)

[GGP10]   Gennaro, R., Gentry, C., Parno, B.: Non-interactive Verifiable Computing: Outsourcing Computation to Untrusted Workers. In: Rabin, T. (ed.) CRYPTO 2010. LNCS, vol. 6223, pp. 465–482. Springer, Heidelberg (2010)

[GH11a]   Gentry, C., Halevi, S.: Fully homomorphic encryption without squashing using depth-3 arithmetic circuits. In: Ostrovsky [Ost11], pp. 107–109

[GH11b]   Gentry, C., Halevi, S.: Implementing Gentry's fully-homomorphic encryption scheme. In: Paterson [Pat11], pp. 129–148

[GHS11]   Gentry, C., Halevi, S., Smart, N.: Personal communication (2011)

[GKR08]   Goldwasser, S., Kalai, Y.T., Rothblum, G.N.: Delegating computation: interactive proofs for muggles. In: STOC, pp. 113–122 (2008)

[GO96]   Goldreich, O., Ostrovsky, R.: Software protection and simulation on oblivious rams. J. ACM 43(3), 431–473 (1996)

[Gol12]   Goldwasser, S. (ed.): Innovations in Theoretical Computer Science 2012, Cambridge, MA, USA, January 8-10. ACM (2012)

[GOS06a]   Groth, J., Ostrovsky, R., Sahai, A.: Non-interactive Zaps and New Techniques for NIZK. In: Dwork, C. (ed.) CRYPTO 2006. LNCS, vol. 4117, pp. 97–111. Springer, Heidelberg (2006)

[GOS06b]   Groth, J., Ostrovsky, R., Sahai, A.: Perfect Non-interactive Zero Knowledge for NP. In: Vaudenay, S. (ed.) EUROCRYPT 2006. LNCS, vol. 4004, pp. 339–358. Springer, Heidelberg (2006)

[HPS98]   Hoffstein, J., Pipher, J., Silverman, J.H.: NTRU: A Ring-Based Public Key Cryptosystem. In: Buhler, J.P. (ed.) ANTS 1998. LNCS, vol. 1423, pp. 267–288. Springer, Heidelberg (1998)

[IP07]   Ishai, Y., Paskin, A.: Evaluating Branching Programs on Encrypted Data. In: Vadhan, S.P. (ed.) TCC 2007. LNCS, vol. 4392, pp. 575–594. Springer, Heidelberg (2007)

[JV10]   Juma, A., Vahlis, Y.: Protecting Cryptographic Keys against Continual Leakage. In: Rabin, T. (ed.) CRYPTO 2010. LNCS, vol. 6223, pp. 41–58. Springer, Heidelberg (2010)

[KSW08]   Katz, J., Sahai, A., Waters, B.: Predicate Encryption Supporting Disjunctions, Polynomial Equations, and Inner Products. In: Smart, N.P. (ed.) EUROCRYPT 2008. LNCS, vol. 4965, pp. 146–162. Springer, Heidelberg (2008)

[LOS+10]   Lewko, A., Okamoto, T., Sahai, A., Takashima, K., Waters, B.: Fully Secure Functional Encryption: Attribute-Based Encryption and (Hierarchical) Inner Product Encryption. In: Gilbert, H. (ed.) EUROCRYPT 2010. LNCS, vol. 6110, pp. 62–91. Springer, Heidelberg (2010)

[LPR10]   Lyubashevsky, V., Peikert, C., Regev, O.: On Ideal Lattices and Learning with Errors over Rings. In: Gilbert, H. (ed.) EUROCRYPT 2010. LNCS, vol. 6110, pp. 1–23. Springer, Heidelberg (2010)

[LTV12]   López-Alt, A., Tromer, E., Vaikuntanathan, V.: On-the-fly multiparty computation on the cloud via multikey fully homomorphic encryption. In: Karloff, H.J., Pitassi, T. (eds.) STOC, pp. 1219–1234. ACM (2012)

[Mic00]   Micali, S.: Computationally sound proofs. SIAM J. Comput. 30(4), 1253–1298 (2000)

[NLV11]   Naehrig, M., Lauter, K., Vaikuntanathan, V.: Can homomorphic encryption be practical? In: Cachin, C., Ristenpart, T. (eds.) CCSW, pp. 113–124. ACM (2011)

[NY90]   Naor, M., Yung, M.: Public-key cryptosystems provably secure against chosen ciphertext attacks. In: STOC, pp. 427–437 (1990)

[OS07]   Ostrovsky, R., Skeith III, W.E.: A Survey of Single-Database Private
            Information Retrieval: Techniques and Applications. In: Okamoto, T.,
            Wang, X. (eds.) PKC 2007. LNCS, vol. 4450, pp. 393–411. Springer,
            Heidelberg (2007)
[Ost11]   Ostrovsky, R. (ed.): IEEE 52nd Annual Symposium on Foundations of
            Computer Science, FOCS 2011, Palm Springs, CA, USA, October 22-25.
            IEEE (2011)
[Pat11]   Paterson, K.G. (ed.): EUROCRYPT 2011. LNCS, vol. 6632, pp. 2011–
            2030. Springer, Heidelberg (2011)
[Pei09]   Peikert, C.: Public-key cryptosystems from the worst-case shortest vector
            problem: extended abstract. In: STOC, pp. 333–342 (2009)
[Reg05]   Regev, O.: On lattices, learning with errors, random linear codes, and
            cryptography. In: STOC, pp. 84–93 (2005)
[Rot09]   Rothblum, G.: Delegating Computation Reliably: Paradigms and Con-
            structions. PhD thesis, MIT (2009),
            http://dspace.mit.edu/handle/1721.1/54637
[RS91]   Rackoff, C., Simon, D.R.: Non-interactive Zero-Knowledge Proof of
            Knowledge and Chosen Ciphertext Attack. In: Feigenbaum, J. (ed.)
            CRYPTO 1991. LNCS, vol. 576, pp. 433–444. Springer, Heidelberg
            (1992)
[RSA83]   Rivest, R.L., Shamir, A., Adleman, L.M.: A method for obtaining
            digital signatures and public-key cryptosystems (reprint). Commun.
            ACM 26(1), 96–99 (1983)
[SS10]   Stehlé, D., Steinfeld, R.: Faster Fully Homomorphic Encryption. In: Abe,
            M. (ed.) ASIACRYPT 2010. LNCS, vol. 6477, pp. 377–394. Springer,
            Heidelberg (2010)
[SV10]   Smart, N.P., Vercauteren, F.: Fully Homomorphic Encryption with Rel-
            atively Small Key and Ciphertext Sizes. In: Nguyen, P.Q., Pointcheval,
            D. (eds.) PKC 2010. LNCS, vol. 6056, pp. 420–443. Springer, Heidelberg
            (2010)
[SV11]   Smart, N.P., Vercauteren, F.: Fully homomorphic simd operations. Cryp-
            tology ePrint Archive, Report 2011/133 (2011),
            http://eprint.iacr.org/2011/133
[SW05]   Sahai, A., Waters, B.: Fuzzy Identity-Based Encryption. In: Cramer,
            R. (ed.) EUROCRYPT 2005. LNCS, vol. 3494, pp. 457–473. Springer,
            Heidelberg (2005)
[Vai11]   Vaikuntanathan, V.: Computing blindfolded: New developments in fully
            homomorphic encryption. In: Ostrovsky [OST11], pp. 5–16

From Multiple Encryption to Knapsacks – Efficient Dissection of Composite Problems[*]

Orr Dunkelman

Computer Science Department,
University of Haifa,
Haifa 31905,
Israel
orrd@cs.haifa.ac.il

Abstract. In this talk, we show some interesting relations between the problem of attacking multiple encryption schemes and attacking knapsack systems. The underlying relation, of problems of a bicomposite nature, allows introducing a series of algorithms for the *dissection* of these problems, thus offering significantly better time/memory tradeoffs than previously known algorithms.

For the case of finding the keys used in a multiple-encryption scheme with r independent n-bit keys, previous error-free attacks required time T and memory M satisfying $TM = 2^{rn}$. Our new technique yields the first algorithm which never errs and finds all the possible keys with a smaller product of TM (e.g., for 7-encryption schemes in time $T = 2^{4n}$ and memory $M = 2^n$). The improvement ratio we obtain increases in an unbounded way as r increases, and if we allow algorithms which can sometimes miss solutions, we can get even better tradeoffs by combining our dissection technique with parallel collision search (offering better complexities than the parallel collision search variants).

After discussing multiple encryption, we show that exactly the same algorithm can be used to offer attacks on knapsacks, which work for any knapsack, that offer the best known time-memory tradeoff curve. This algorithm can be used to handle also more general types of knapsacks, involving a combination of modular additions, XORs, and any T-functions.

---

[*] This is a joint work with Itai Dinur, Nathan Keller, and Adi Shamir.

S. Galbraith and M. Nandi (Eds.): INDOCRYPT 2012, LNCS 7668, p. 16, 2012.
© Springer-Verlag Berlin Heidelberg 2012

Using the Cloud to Determine Key Strengths

Thorsten Kleinjung[1], Arjen K. Lenstra[1], Dan Page[2], and Nigel P. Smart[2]

[1] EPFL IC LACAL, Station 14, CH-1015 Lausanne, Switzerland
[2] Dept. Computer Science, University of Bristol, Merchant Venturers Building,
Woodland Road, Bristol, BS8 1UB, United Kingdom

Abstract. We develop a new methodology to assess cryptographic key
strength using cloud computing, by calculating the true economic cost
of (symmetric- or private-) key retrieval for the most common cryptographic primitives. Although the present paper gives both the current
(2012) and last year's (2011) costs, more importantly it provides the tools
and infrastructure to derive new data points at any time in the future,
while allowing for improvements such as of new algorithmic approaches.
Over time the resulting data points will provide valuable insight in the
selection of cryptographic key sizes.[1]

1 Introduction

An important task for cryptographers is the analysis and recommendation of parameters, crucially including key size and thus implying key strength, for cryptographic primitives; clearly this is of theoretic and practical interest, relating
to the study and the deployment of said primitives respectively. As a result,
considerable effort has been, and is being, expended with the goal of providing
meaningful data on which such recommendations can be based. Roughly speaking, two main approaches dominate: use of special-purpose hardware designs, including proposals such as [15,16,38,39] (some of which have even been realised),
and use of more software-oriented (or at least less bespoke) record setting computations such as [4,5,13,18]. The resulting data can then be extrapolated using
complexity estimates for the underlying algorithms, and appropriate versions of
Moore's Law, in an attempt to assess the longevity of associated keys. In [20]
this results in key size recommendations for public-key cryptosystems that offer security comparable to popular symmetric cryptosystems; in [24] it leads to
security estimates in terms of hardware cost or execution time. Existing work
for estimating symmetric key strengths (as well as other matters) is discussed in
[40].

It is not hard to highlight disadvantages in these approaches. Some special-
purpose hardware designs are highly optimistic; they all suffer from substantial
upfront costs and, as far as we have been able to observe, are always harder

---

[1] We see this as a living document, which will be updated as algorithms, the Amazon
pricing model, and other factors change. The current version is V1.1 of May 2012;
updates will be posted on http://www.cs.bris.ac.uk/~nigel/Cloud-Keys/

S. Galbraith and M. Nandi (Eds.): INDOCRYPT 2012, LNCS 7668, pp. 17–39, 2012.
© Springer-Verlag Berlin Heidelberg 2012

to get to work and use than expected. On the other hand, even highly speculative designs may be useful in exploring new ideas and providing insightful lower bounds. Despite being more pragmatic, software-oriented estimates do not necessarily cater adequately for the form or performance of future generations of general-purpose processors (although so far straightforward application of Moore's Law is remarkably reliable, with various dire prophecies, such as the "memory wall", not materialising yet). For some algorithms, scaling up effort (e.g., focusing on larger keys) requires no more than organisational skill combined with patience. Thus, record setting computation that does not involve new ideas may have little or no scientific value: for the purposes of assessing key strength, a partial calculation is equally valuable. For some other algorithms, only the full computation adequately prepares for problems one may encounter when scaling up, and overall (in)feasibility may thus yield useful information. Finally, for neither the hardware- nor software-oriented approach, is there a uniform, or even well understood, metric by which "cost" should be estimated. For example, one often overlooks the cost of providing power and cooling, a foremost concern for modern, large-scale installations.

Despite these potential disadvantages and the fact that papers such as [20] and [24] are already a decade old, their results have proved to be quite resilient and are widely used. This can be explained by the fact that standardisation of key size requires some sort of long term extrapolation: there is no choice but to take the inherent uncertainty and potential unreliability for granted. In this paper we propose to complement traditional approaches, using an alternative that avoids reliance on special-purpose hardware, one time experiments, or record calculations, and that adopts a business-driven and thus economically relevant cost model. Although extrapolations can never be avoided, our approach minimises their uncertainty because whenever anyone sees fit, he/she can use commodity hardware to repeat the experiments and verify and update the cost estimates. In addition we can modify our cost estimates as our chosen pricing mechanism alters over time.

The current focus on cloud computing is widely described as a significant long-term shift in how computing services are delivered. In short, cloud computing enables any party to rent a combination of computational and storage resources that exist within managed data centers; the provider we use is the Amazon Elastic Compute Cloud [3], however others are available. The crux of our approach is the use of cloud computing to assess key strength (for a specific cryptographic primitive) in a way that provides a useful relationship to a true economic cost. Crucially, we rely on the fact that cloud computing providers operate as businesses: assuming they behave rationally, the pricing model for each of their services takes into account the associated purchase, maintenance, power and replacement costs. In order to balance reliability of revenue against utilisation, it is common for such a pricing model to incorporate both long-term and supply and demand driven components. We return to the issue of supply and demand below, but for the moment assume this has a negligible effect on the longer-term pricing structure of Amazon in particular. As a result, the Amazon pricing model provides a valid way to attach a monetary cost to key strength. A provider may clearly be expected to update their infrastructures and pricing model as technology and economic conditions dictate; indeed we show the effact of this over the last 18 months or so. However, by ensuring our approach is repeatable, for example using commodity cloud computing services and platform-agnostic software (i.e., processor non-specific), we are able to track results as they evolve over time. We suggest this, and the approach as a whole, should therefore provide a robust understanding of how key size recommendations should be made.

In Section 2 we briefly explain our approach and those aspects of the Amazon Elastic Compute Cloud that we depend on. In Section 3 we describe our analysis applied to a number of cryptographic primitives, namely DES, AES, SHA-2, RSA and ECC. Section 4 and Section 5 contains concluding remarks. Throughout, all monetary quantities are given in US dollars.

2   The Amazon Elastic Compute Cloud

Amazon Elastic Compute Cloud (EC2) is a web-service that provides computational and storage resource "in the cloud" (i.e., on the Internet) to suit the specific needs of each user. In this section we describe the current (per May 2012) EC2 hardware platform and pricing model, and compare it with the previous (per Feb 2011) pricing model, from a point of view that is relevant for our purposes.

EC2 Compute Units. At a high level, EC2 consists of numerous installations (or data centers) each housing numerous processing nodes (a processor core with some memory and storage) which can be rented by users. In an attempt to qualify how powerful a node is, EC2 uses the notion of an *EC2 Compute Unit*, or *ECU* for short. One ECU provides the equivalent computational capacity of a 1.0-1.2 GHz Opteron or Xeon processor circa 2007, which were roughly state of the art when EC2 launched. When new processors are deployed within EC2, they are given an ECU-rating; currently there are at least four different types of core, rated at 1, 2, 2.5 and 3.25 ECUs. The lack of rigour in the definition of an ECU (e.g., identically clocked Xeon and Opteron processors do not have equivalent computational capacity) is not a concern for our purposes.

Instances. An *instance* refers to a specified amount of dedicated compute capacity that can be purchased: it depends on the processor type (e.g., 32- or 64-bit), the number of (virtualised) cores (1, 2, 4 or 8), the ECU-rating per core, memory and storage capacity, and network performance. There are currently thirteen different instances, partitioned into the *instance types* termed "standard" (std), "micro", "high-memory" (hi-m), "high-CPU" (hi-c), "cluster compute" (cl-C), and "cluster GPU" (cl-G). The later two cluster types are intended for High Performance Computing (HPC) applications, and come with 10 Gb ethernet connections; for all other instance types except micro, there are two or three subinstances of different sizes, indicated by "Large (L)", "Extra Large (EL)" and so on. Use of instances can be supported by a variety of operating systems, ranging from different versions of Unix and Linux through to Windows.

Pricing Model. Instances are charged per instance-hour depending on their capacity and the operating system used. In 2007 this was done at a flat rate of $ 0.10 per hour on a 1.7 GHz processor with 1.75 GB of memory [1]; in 2008 the pricing model used ECUs charged at the same flat rate of $ 0.10 per hour per ECU [2]. Since then the pricing model has evolved [3]. Currently, instances can be purchased at three different price bands, "on-demand", "reserved" and "spot", charged differently according to which of four different geographic locations one is using (US east coast, US west coast, Ireland or Singapore).

On-demand pricing allows purchase of instance-hours as and when they are needed. After a fixed annual (or higher three year) payment per instance, re-served pricing is significantly cheaper per hour than on-demand pricing: it is intended for parties that know their requirements, and hence can reserve them, ahead of when they are used. Spot pricing is a short-term, market determined price used by EC2 to sell off unused capacity. In 2012, the "reserved" instances (which is of more interest for our purposes) were further divided into light, medium and heavy utilization bands), and discount of at least 20% is available for those spending at least two million dollars or more.

Table 1. Instance Technical Specifications

Instance	cores	ECUs per core	total ECUs	M2050 GPUs	RAM GB
standard L	2	2	4	0	7.5
high-memory EL	2	3.25	6.5	0	17.1
high-CPU EL	8	2.5	20	0	7
cluster compute	(not specified)		33.5	0	23
cluster GPU			33.5	2	22

For all instances and price bands, Windows usage is more expensive than Linux/Unix and is therefore not considered. Similarly, 32-bit instances are not considered, because for all large computing efforts and price bands 32-bit cores are at least as expensive as 64-bit ones. To enable a longitudinal study we restrict to the five remaining relevant instances which were available in both 2011 and 2012, and which are most suited to our needs. In Table 1 we present the technical specification of these instances.

Table 2 lists the current pricing of the relevant remaining instances in both Feb 2011 and Feb 2012, of which k-fold multiples can be purchased at k times the price listed, for any positive integer k. Although of course there is a clear upper bound on k due to the size of the installation of the cloud service. To simplify our cost estimate we ignore this upper bound, and assume that the provider will provide more capacity (at the same cost) as demand increases. One thing is clear is that prices have dropped over the preceeding twelve months, the question is by how much. Separate charges are made for data transfer and so on, and the table does not take into account the 20% discount for large purchases.

Table 2. February 2011 and 2012 US east coast instance pricing, in US dollars, using 64-bit Linux/Unix

Instance	on-dem.	2011				on-dem.	2012			
		reserved					reserved			
	per	fixed payment		per hour		per	fixed payment		per hour	
	hour	1 yr	3 yr	1 yr	3 yr	hour	1 yr	3 yr	1 yr	3 yr
	δ	α	τ	ϵ_1	ϵ_2	δ	α	τ	ϵ_1	ϵ_2
std L	$0.34	$910	$1400	$0.120	$0.120	$0.32	$780	$1200	$0.060	$0.052
hi-m EL	$0.50	$1325	$2000	$0.170	$0.170	$0.45	$1030	$1550	$0.088	$0.070
hi-c EL	$0.68	$1820	$2800	$0.240	$0.240	$0.66	$2000	$3100	$0.160	$0.140
cl-C	$1.60	$4290	$6590	$0.560	$0.560	$1.30	$4060	$6300	$0.297	$0.297
cl-G	$2.10	$5630	$8650	$0.740	$0.740	$2.10	$6830	$10490	$0.494	$0.494

With, for a given instance, δ, α, τ, ϵ_1, ϵ_3 the four pricing parameters as indicated in Table 2 and using Y for the number of hours per year, it turns out that we have

$$1.962(\tau + 3\epsilon_3 Y) < 3\delta Y < 2.032(\tau + 3\epsilon_3 Y) \qquad \text{(2011 prices)}$$
$$2.351(\tau + 3\epsilon_3 Y) < 3\delta Y < 3.489(\tau + 3\epsilon_3 Y) \qquad \text{(2012 prices)} \qquad (1)$$

That is, for any instance using on-demand pricing continuously for a three year period was approximately two times (now three times) as expensive as using reserved pricing with a three year term for the entire three year period. Furthermore, for all instances reserved pricing for three consecutive (or parallel) annual periods was approximately 1.3 times (now 1.5 times) as expensive as a single three year period:

$$1.292(\tau + 3\epsilon_3 Y) < 3(\alpha + \epsilon_1 Y) < 1.305(\tau + 3\epsilon_3 Y) \qquad \text{(2011 prices)}$$
$$1.416(\tau + 3\epsilon_3 Y) < 3(\alpha + \epsilon_1 Y) < 1.594(\tau + 3\epsilon_3 Y) \qquad \text{(2012 prices)} \qquad (2)$$

There are more pricing similarities between the various instances. Suppose that one copy of a given instance is used for a fixed number of hours h. We assume that h is known in advance and that the prices do not change during this period (cf. remark below on future developments). Which pricing band(s) should be used to obtain the lowest overall price depends on h in the following way. For small h use on-demand pricing, if h is larger than a first cross-over point γ_α but at most one year, reserved pricing with one year term must be used instead. Between one year and a second cross-over point γ_τ one should use reserved pricing with one year term for a year followed by on-demand pricing for the remaining $h - Y$ hours, but for longer periods up to three years one should use just reserved pricing with a three year term. After that the pattern repeats. This holds for all instances, with the cross-over values varying little among them, as shown below.

The first cross-over value γ_α satisfies $\delta\gamma_\alpha = \alpha + \epsilon_1\gamma_\alpha$ and thus $\gamma_\alpha = \frac{\alpha}{\delta - \epsilon_1}$.

The second satisfies $\alpha + \epsilon_1 Y + \delta(\gamma_\tau - Y) = \tau + \epsilon_3\gamma_\tau$ and thus $\gamma_\tau = \frac{(\delta - \epsilon_1)Y + \tau - \alpha}{\delta - \epsilon_3}$.

For the 2011 prices we find $\gamma_\alpha \approx 4100$ and $\gamma_\tau \approx Y + 2200$ in all price instances. But for 2012 there is more variation between the cut-off points.

Our Requirements and Approach. For each cryptographic primitive studied, our approach hinges on the use of EC2 to carry out a negligible yet representative fraction of a certain computation. Said computation, when carried to completion, should result in a (symmetric- or private-) key or a collision depending on the type of primitive. For DES, AES, and SHA-2 this consists of a fraction of the symmetric-key or collision search, for RSA it is small parts of the sieving step and the matrix step of the Number Field Sieve (NFS) integer factorisation method, and for ECC it is a small number of iterations of Pollard's rho method for the calculation of a certain discrete logarithm.

Note that in each case, the full computation would require at least many thousands of years when executed on a single core[2]. With the exception of the NFS matrix step and disregarding details, each case allows embarrassing parallelisation with only occasional communication with a central server (for distribution of inputs and collection of outputs). With the exception of the NFS sieving and matrix steps, memory requirements are negligible. Substantial storage is required only at a single location. As such, storage needs are thus not further discussed. Thus, modulo details and with one exception, anything we could compute on a single core in y years, can be calculated on n such cores in y/n years, for any $n > 0$.

Implementing a software component to perform each partial computation on EC2 requires relatively little upfront cost with respect to development time. In addition, execution of said software also requires relatively little time (compared to the full computation), and thus the partial computation can be performed using EC2's most appropriate pricing band for short-term use; this is the only actual cost incurred. Crucially, it results in a reliable estimate of the number of ECU years, the best instance(s) for the full computation, and least total cost (as charged by EC2) to do so (depending on the desired completion time). Obviously the latter cost(s) will be derived using the most appropriate applicable long-term pricing band.

Minimal Average Price. Let $\mu(h)$ denote the minimal average price per hour for a calculation that requires h hours using a certain fixed instance. Then we have

$$
\mu(h) = \begin{cases}
\delta & \text{for } 0 < h \leq \gamma_\alpha \quad \text{(constant global maximum)} \\
\frac{\alpha}{h} + \epsilon & \text{for } \gamma_\alpha < h \leq Y \quad \text{(with a local minimum at } h = Y) \\
\frac{\alpha + \epsilon Y + (h - Y)\delta}{h} & \text{for } Y < h \leq \gamma_\tau \quad \text{(with a local maximum at } h = \gamma_\tau) \\
\frac{\tau}{h} + \epsilon & \text{for } \gamma_\tau < h \leq 3Y \quad \text{(with the global minimum at } h = 3Y).
\end{cases}
$$

For $h > 3Y$ the pattern repeats, with each three year period consisting of four segments, reaching decreasing local maxima at $h = 3kY + \gamma_\alpha$, decreasing local

---

[2] The processor type can be left unspecified but quantum processors are excluded; to the best of our knowledge and consistent with all estimates of the last two decades, it will always take at least another decade before such processors are operational.

minima at $h = 3kY + Y$, decreasing local maxima at $h = 3kY + \gamma_\tau$, and the global minimum at $h = 3kY + 3Y$, for $k = 1, 2, 3, \ldots$. For all relevant instances, Figure 1 depicts the graphs of the minimal average prices for periods of up to six years, per ECU, in both 2011 and 2012.

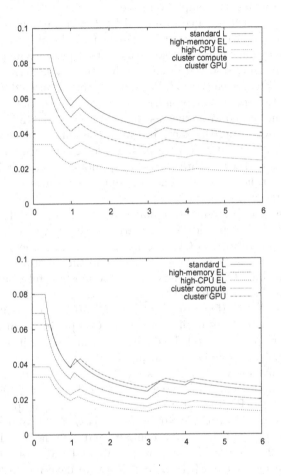

Fig. 1. Feb 2011 (top) and Feb 2012 (bottom) minimal average prices per hour per ECU, in US dollars, as a function of the number of years required by the calculation.

Consequences. Given the embarrassingly parallel nature and huge projected execution time of each full computation, the above implies that we can always reach lowest (projected) cost by settling for a three year (projected) completion time. Faster completion would become gradually more expensive until completion time $\frac{\tau}{\alpha}Y > Y$ is reached[3], at which point one should switch right away to a shorter completion time of one year; faster than one year again becomes gradually more expensive until γ_α is reached, at which point the cost has reached

---

[3] Note that $\frac{\tau}{\alpha} \approx 1.5$, so $\frac{\tau}{\alpha}Y$ is about a year and a half for all instances.

its (constant) maximum and the completion time is only limited (from below) by the number of available on-demand copies of the required instance. We stress yet again that this all refers to projected computation, none of which is actually completed: all we need is an estimate of the cost of completion, possibly depending on the desired completion time.

Thus assume that using a certain fixed EC2-instance, a full-scale security assessment is expected to take y years for a y that will be at least many thousands. The full computation can be completed using that instance by purchasing about $y/3$ copies of it using reserved pricing with a three year term, and to use each copy for the full three year period. Whether or not this (i.e., doing this full computation over the three year period) is also optimal for that instance depends on issues beyond our control (such as changed pricing and inflation) and is commented on below. Using the same instance, the entire calculation can also be completed in one year at ≈ 1.5 (resp. ≈ 1.3 for 2011 prices) times higher cost (cf. Inequalities (2)), or at three times (resp. double) the cost in an arbitrarily short period of time (assuming enough copies of the instance are available in EC2, cf. Inequality (1)).

With all instances behaving so similarly over time for the different price bands, the best instance for a particular calculation is selected by taking the one with the lowest hourly rate per ECU as long as it provides adequate memory and network bandwidth per core. From Table 2 and the description of our needs, it follows that the high-CPU EL instance will be most useful to us: using EC2 for a computation that is estimated to require y ECU years for some large y will cost about $\$\frac{0.66yY}{2.5\cdot20} \approx 115y$ (resp. $\$\frac{0.68yY}{2\cdot20} \approx 150y$) (cf. Inequality (1)) if we can afford to wait three years until completion. Completing it in a year will cost about $\$170y$ (resp. $\$195y$) (cf. Inequality (2)), and doing it as fast as possible will cost $\$290y$ (resp. $\$300y$) (cf. Inequality (1)).

In a similar fashion we derive costs $\$\frac{1.30yY}{2.4\cdot33.5} \approx 140y$, $\$210y$ and $\$340y$ (resp. $\$210y$, $\$272y$ and $\$420y$) for a y ECU year calculation done in three years, one year, and "ASAP", respectively, using EC2 cluster-compute instances, and costs $\$\frac{0.45yY}{3.4\cdot6.5} \approx 180$, $\$270y$ and $\$605y$ (resp. $340y$, $\$440y$ and $\$675y$) for high-memory EL instances. Thus, cluster compute and high-memory EL instances are approximately 20% and 50%-100% (resp. 40% and 120% for 2011 prices) more expensive than high-CPU EL instances. Note we have not in the above taken into account the 20% discount for large usage of reserved instances in the 2012 prices.

Accommodating Future Developments. As one can see prices and pricing models will change over time, and so may security assessment strategies and their interaction with advances in processor design and manufacture. In particular, one could imagine that if a party decided to use the Amazon cloud for key recovery or collision search then the increase in demand would induce Amazon to increase the instance costs. However, we assume that the effect of such supply and demand on the pricing is relatively constant over a long period of time. Thus, we assume the non-spot prices are a relatively accurate reflection of the actual economic cost to Amazon (bar a marginal profit) of providing the service.

The cost estimates produced by our approach are valid at the moment they are calculated (in the way set forth above), but cannot take any future developments into account. However, this problem can be mitigated by adopting an open source model for the software components and using (as far as is sensible) platform-agnostic programming techniques; for example, this permits the software to maintain alignment with the latest algorithmic and processor developments, and to add or remove primitives as and when appropriate (cf. [37]). Almost all our test software used has been made available on a web-site which will be updated, as years pass by, with the latest costs:

http://www.cs.bris.ac.uk/~nigel/Cloud-Keys/

EC2 versus Total Cost of Ownership (TCO). The approach set forth above associates a monetary cost to key strength, but does so at negligible actual cost; this is useful for many purposes. However, no key recovery nor collision is completed. The question remains, if one desires to complete a computation, whether doing so on EC2 is less expensive than acquiring a similar platform and operating it oneself.

TCO includes many costs that are hard to estimate. Nevertheless, the following may be useful. At moderate volume, a dual node server with two processors, each with twelve 1.9 GHz cores, and 32 GB of memory per node can be purchased for approximately $ 8000. This implies that at that fixed cost approximately $2 \cdot 2 \cdot 12 \cdot 1.9 = 91.2$ ECUs with $1\frac{1}{3}$ GB of memory per core can be purchased. At $ \$ \frac{20}{91.2} \cdot 8000 \approx 1750$ per 20 ECUs this compares favourably to the fixed three year payment $\tau = 3100$ for the 20 ECUs of EC2-instance high-CPU EL. Power consumption of the above server is estimated to be bounded by 600 Watts. Doubling this to account for cooling and so on, we arrive at approximately 265 Watts for 20 ECUs, thus about a quarter kWh. At a residential rate of $ \$ 0.25$ per kWh we find that running our own 20 ECUs for three years costs us $ \$ 1750 + 3Y \frac{265}{1000} \cdot 0.25 \approx 3500$, as opposed to $ \$ 3100 + 3Y \cdot 0.16 \approx 7304$ for EC2's high-CPU EL.

Although in low-volume such a server could be supported without additional infrastructure, in high-volume they require a data center and maintenance personnel; this also implies lower electricity rates and quantum discount for acquisition however. Given expected cost and lifespan of infrastructures and salary costs, it is not unreasonable to estimate that TCO is at most one half of the cost of EC2. We conclude that full computations are still best conducted on one's own equipment.

3   Results

In this section we detail the application of our approach to five different cryptographic primitives: the block ciphers DES and AES, the cryptographic hash function SHA-2, and the public-key cryptosystems RSA and ECC. The first is chosen for historic reasons, whilst the others are the primitives of choice in many current applications.

For each of the five primitives, the fastest methods to recover the symmetric-key (DES and AES), to find a collision (SHA-2), or to derive the private-key (RSA and ECC) proceed very similarly, though realised using entirely different algorithms. In each algorithm, a huge number of identical computations are performed on different data; they can be carried out simultaneously on almost any number (and type) of client cores. With one exception (the NFS matrix step, as alluded to above), each client operates independently of all others, as long as there are central servers tasked with distributing inputs to clients and collecting their outputs. Furthermore, for each client the speed at which inputs are processed (and outputs produced, if relevant) is constant over time: a relatively short calculation per type of client along with a sufficiently accurate estimate of the number of inputs to be processed (or outputs to be produced, if relevant) suffices to be able to give a good indication of the total computational effort required.

The client-server approach has been common in a cryptographic context since the late 1980s, originally implemented using a variety of relatively crude application-specific, pre-cloud approaches [9,23] (that continue to the present day [5,19]), and later based on more general web-based services such as [8] that support collaborative compute projects (such as [28]). Thus, for each primitive under consideration, the problem of managing the servers is well understood. Additionally, the computational effort expended by said servers is dwarfed by the total computation required of the clients. As a result, in this section we concentrate on a series of experiments using the client software only: for each primitive, we execute the associated client software for a short yet representative period of time on the most appropriate on-demand EC2 instance, use the results to extrapolate the total key retrieval cost using the corresponding reserved EC2 instance, and relate the result to a discussion of prior work.

We implemented a software component to perform each partial computation on EC2, focusing on the use of platform-agnostic (i.e., processor non-specific) programming techniques via ANSI-C. In particular, we used processor-specific operations (e.g., to cope with carries efficiently) only in situations where an alternative in C was at least possible; in such cases we abstracted the operations into easily replaceable macros or used more portable approaches such as compiler intrinsics where possible. As motivated above, the goal of this was portability and hence repeatability; clearly one could produce incremental improvements by harnessing processor-specific knowledge (e.g., of instruction scheduling or register allocation), but equally clearly these are likely to produce improvement by only a (small) constant factor and may defeat said goal.

3.1 DES

We first examine DES (which is considered broken with current technology) to provide a baseline EC2 cost against which the cost of other key retrieval efforts can be measured.

Prior Work. As one of the oldest public cryptographic primitives, DES has had a considerable amount of literature devoted to its cryptanalysis over the years.

Despite early misgivings and suspicions about its design and the development of several new cryptanalytic techniques, the most efficient way to recover a single DES key is still exhaustive search. This approach was most famously realised by *Deep Crack* developed by the EFF [15]. Designed and built in 1998 at a cost of $ 200,000, this device could find a single DES key in 22 hours. Various designs, often based on FPGAs, have been presented since specifically for DES key search. The most famous of these is COPACOBANA [16], which at a cost of $ 10,000 can perform single DES key search in 6.4 days on average. One can also extrapolate from suitable high-throughput DES implementations. For example [34] presents an FPGA design, which on a Spartan FPGA could perform an exhaustive single DES key search in 9.5 years. Using this design, in [40][p. 19] it is concluded that a DES key search device which can find one key every month can be produced for a cost of $ 750. Alternatively, using a time-memory trade-off [33], one can do a one-time precomputation at cost comparable to exhaustive key search, after which individual keys can be found at much lower cost: according to [32] a DES key can be found in half an hour on a $ 12 FPGA, after a precomputation that takes a week on a $ 12,000 device.

DES Key Search. In software the most efficient way to implement DES key search is to use the bit-sliced implementation method of Biham [6]. In our experiments we used software developed by Matthew Kwan[4] for the RSA-sponsored symmetric-key challenge eventually solved by the DESCHALL project in 1997. Our choice was motivated by the goal of using platform-agnostic software implementations of the best known algorithms.

Given a message/ciphertext pair the program searches through all the keys trying to find the matching key. On average one expects to try 2^{55} keys (i.e., one half of the key space) until a match is found. However in the bit-slice implementation on a 64-bit machine one evaluates DES on the message for 64 keys in parallel. In addition there are techniques, developed by Rocke Verser for the DESCHALL project, which allow one to perform an early abort if one knows a given set of 64 keys will not result in the target ciphertext. For comparison using processor extensions, we implemented the same algorithm on 128 keys in parallel using the SSE instructions on the x86 architecture.

DES Key Search on EC2. Using EC2 we found that y, the expected number of years to complete a DES calculation on a single ECU, was $y = 97$ using a vanilla 64-bit C implementation, and $y = 51$ using the 128-bit SSE implementation. Combined with our prior formulae of $ 115y$, $ 170y$ and $ 290y$ (resp. $ 150y$, $ 195y$ and $ 300y$) for high-CPU EL instances, we estimate the cost of using EC2 to recover a single DES key as in Table 3. The values are so low that the Amazon bulk discount has not been applied, however we see that the cost of obtaining a DES key over a three year period has fallen by 25% in the last year. However, obtaining the DES key almost instantly has only fallen by 5%.

In comparing these to earlier figures for special-purpose hardware, one needs to bear in mind that once a special-purpose hardware device has been designed

---

[4] Available from http://www.darkside.com.au/bitslice/

Table 3. Cryptanalytic Strength of DES on EC2

Implementation Technique	ECU Years	Estimated Key Retrieval Cost					
		2011 Prices			2012 Prices		
		3 Years	1 Year	ASAP	3 Years	1 Year	ASAP
Vanilla C	97	$14,550	$18,915	$29,100	$11,155	$16,490	$28,130
SSE Version	51	$7,650	$9,945	$15,300	$5,865	$8,670	$14,790

and built, the additional cost of finding subsequent keys after the first one is essentially negligible (bar the maintenance and power costs). Thus, unless time-memory trade-off key search is used, the cost-per-key of specialised hardware is lower than using EC2. We repeat that our thesis is that dedicated hardware gives a point estimate, whereas our experiments are repeatable. Thus as long as our costs are scaled by an appropriate factor to take into account the possibility of improving specialised hardware, our estimates (when repeated annually) can form a more robust method of determining the cost of finding a key.

We end with noting that whilst few will admit to using DES in any application, the use of three-key triple DES (or 3DES) is widespread, especially in the financial sector. Because the above cost underestimates the cost of 2^{55} full DES encryptions (due to the early abort technique), multiplying it by 2^{112} lower bounds the cost for a full three-key 3DES key search (where the factor of 3 incurred by the three distinct DES calls can be omitted by properly ordering the search).

3.2   AES

Prior Work. Since its adoption around ten years ago, AES has become the symmetric cipher of choice in many new applications. It comes in three variants, AES-128, AES-192, and AES-256, with 128-, 192-, and 256-bit keys, respectively. The new cipher turned out to have some unexpected properties in relation to software side-channels [30], which in turn triggered the development of AES-specific instruction set extensions [17]. Its strongest variant, AES-256, was shown to have vulnerabilities not shared with the others [7]. These developments notwithstanding, the only known approach to recover an AES key is by exhaustive search. With an AES-128 key space of 2^{128} this is well out of reach and therefore it is not surprising that there seems little work on AES specific hardware to realise a key search. One can of course extrapolate from efficient designs for AES implementation. For example using the FPGA design in [41] which on a single Spartan FPGA can perform the above exhaustive key search in $4.6 \cdot 10^{23}$ years, the authors of [40] estimate that a device can be built for $\$2.8 \cdot 10^{24}$ which will find an AES-128 key in one month.

AES Key Search on EC2. In software one can produce bit-slice versions of AES using the extended instruction sets available on many new processors [25]. However, in keeping with our principle of simple code, which can be run on multiple versions of today's computers as well as future computers, we decided to use a traditional AES implementation in our experiments. We found the estimated number of years y on a single ECU to finish a single AES computation was $y \approx 10^{24}$. Using high-CPU EL instances our costs become (rounding to the nearest order of magnitude) those give in Table 4.

Table 4. Cryptanalytic Strength of AES on EC2

Algorithm	ECU Years	Estimated Key Retrieval Cost 3 Years / 1 Year / ASAP
AES-128	10^{24}	$\approx \$10^{26}$

Again our comments for DES concerning the cost of specialised hardware versus our own estimates apply for the case of AES, although in the present case the estimates are more closely aligned. However, in the case of AES the new Westmere 32nm Intel core has special AES instructions [17]. It may be instructive to perform our analysis on the EC2 service, once such cores are available on this service[5]. Whilst a 3-to-10 fold performance improvement for AES encryption using Westmere has been reported; our own experiments on our local machines only show a two fold increase in performance for key search.

3.3   SHA-2

Prior Work. The term SHA-2 denotes a family of four hash functions; SHA-224, SHA-256, SHA-384 and SHA-512. We shall be concentrating on SHA-256 and SHA-512; the SHA-224 algorithm only being introduced to make an algorithm compatible with 112-bit block ciphers and SHA-384 being just a truncated version of SHA-512. The three variants SHA-256, SHA-384 and SHA-512 were standardised by NIST in 2001, with SHA-224 being added in 2004, as part of FIPS PUB 180-2 [26]. The SHA-2 family of algorithms is of the same algorithmic lineage as MD4, MD5 and SHA-1.

Cryptographic hash functions need to satisfy a number of security properties; for example preimage-resistance, collision resistance, etc. The property which appears easiest to violate for earlier designs, and which generically is the least costly to circumvent, is that of collision resistance. Despite the work on cryptanalysis of the related hash functions MD4, MD5 and SHA-1 [43,44,45,46,42], the best known methods to find collisions for the SHA-2 family still are the generic ones. Being of the Merkle-Damgård family each of the SHA-2 algorithms consists of a compression function, which maps b-bit inputs to h-bit outputs, and a chaining method. The chaining method is needed to allow the hashing of messages of more than b bits in length. The input block size $b = 512$ for SHA-256 but $b = 1024$ for SHA-384 and SHA-512. The output block size h is given by the name of the algorithm, i.e., SHA-h.

---

[5] As of April 2012 none of the 64-bit instances we ran on the EC2 service had the Westmere 32nm Intel core on them.

SHA-2 Collision Search. The best generic algorithm for collision search is the parallel "distinguished points" method of van Oorschot and Wiener [29]. In finding collisions this method can be tailored in various ways; for example one could try to obtain two meaningful messages which produce the same hash collision. In our implementation we settle for the simplest, and least costly, of all possible collision searches; namely to find a collision between two random messages.

The collision search proceeds as follows. Each client generates a random element $x_0 \in \{0,1\}^h$ and then computes the iterates $x_i = H(x_{i-1})$ where H is the hash function. When an iterate x_d meets a given condition, say the last 32-bits are zero, we call the iterate a distinguished point. The tuple (x_d, x_0, d) is returned to a central server and the client now generates a new value $x_0 \in \{0,1\}^h$ and repeats the process. Once the server finds two tuples (x_d, x_0, d) and $(y_{d'}, y_0, d')$ with $x_d = y_{d'}$ a collision in the hash function can be obtained by repeating the two walks from x_0 and y_0.

By the birthday paradox we will find a collision in roughly $\sqrt{\pi \cdot 2^{h-1}}$ applications of H, with n clients providing an n-fold speed up. If $1/p$ of the elements of $\{0,1\}^h$ are defined to be distinguished then the memory requirement of the server becomes $O(\sqrt{\pi \cdot 2^{h-1}}/p)$.

SHA-2 Collision Search on EC2. We implemented the client side of the above distinguished points algorithm for SHA-256 and SHA-512. On a single ECU we found that the expected number of years needed to obtain a collision was given by the "ECU Years" values in Table 5, resulting in the associated collision finding costs, where again we round to the nearest order of magnitude and use high-CPU EL instances. Note, that SHA-256 collision search matches the cost of AES-128 key retrieval, as one hopes would happen for a well designed hash function of output twice the key size of a given well designed block cipher.

Table 5. Cost of SHA-2 Collision on EC2

Algorithm	ECU Years	Estimated Collision Search Cost 3 Years / 1 Year / ASAP
SHA-256	10^{24}	$\approx \$10^{26}$
SHA-512	10^{63}	$\approx \$10^{63}$

3.4   RSA

NFS Background. As the oldest public key algorithm, RSA (and hence integer factorisation) has had a considerable amount of research applied to it over the years. The current best published algorithm for factoring integers is Coppersmith's variant [12] of the Number Field Sieve method (NFS) [22]. Based on loose heuristic arguments and asymptotically for $n \to \infty$, its expected run time to factor n is

$$L(n) = \exp((1.902 + o(1))(\log n)^{1/3}(\log \log n)^{2/3}),$$

where the logarithms are natural. For the basic version, i.e., not including Coppersmith's modifications, the constant "1.902" is replaced by "1.923". To better understand and appreciate our approach to get EC2-cost estimates for NFS, we need to know the main steps of NFS. We restrict ourselves to the basic version.

Polynomial Selection. Depending on the RSA modulus n to be factored, select polynomials that determine the number fields to be used.

Sieving Step. Find elements of the number fields that can be used to derive equations modulo n. Each equation corresponds to a sparse k-dimensional zero-one vector, for some $k \approx \sqrt{L(n)}$, such that each subset of vectors that sums to an all-even vector gives a 50% chance to factor n. Continue until at least $k + t$ equations have been found for a small constant $t > 0$ (hundreds, at most).

Matrix Step. Find at least t independent subsets as above.

Square Root. Try to factor n by processing the subsets (with probability of success $\geq 1 - (\frac{1}{2})^t$).

Because $L(n)$ number field elements in the sieving step have to be considered so as to find $k + t \approx \sqrt{L(n)}$ equations, the run time is attained by the sieving and matrix steps with memory requirements of both steps, and central storage requirements, behaving as $\sqrt{L(n)}$. The first step requires as little or as much time as one desires – see below. The run time of the final step behaves as $\sqrt{L(n)}$ with small memory needs.

The set of number field elements to be sieved can be parcelled out among any number of independent processors. Though each would require the same amount $\sqrt{L(n)}$ of memory, this sieving memory can be optimally shared by any number of threads; smaller memories can be catered-for as well at small efficiency loss. Although all clients combined report a substantial amount of data to the server(s), the volume per client is low. The resulting data transfer expenses are thus not taken into account in our analysis below. The matrix step can be split up in a small number (dozens, at most) of simultaneous and independent calculations. Each of those demands fast inter-processor communication and quite a bit more memory than the sieving step (though the amounts are the same when expressed in terms of the above L-function).

It turns out that more sieving (which is easy, as sieving is done on independent processors) leads to a smaller k (which is advantageous, as it makes the matrix step less cumbersome). It has been repeatedly observed that this effect diminishes, but the trade-off has not been analysed yet.

Unlike DES, AES, or ECC key retrieval or SHA-2 collision finding methods, NFS is a multi-stage method which makes it difficult to estimate its run time. As mentioned, the trade-off between sieving and matrix efforts is as yet unclear and compounded by the different platforms (with different EC2 costs) required for the two calculations. The overall effort is also heavily influenced by the properties of the set of polynomials that one manages to find in the first step. For so-called *special* composites finding the best polynomials is easy: in this case the special number field sieve applies (and the "1.902" or "1.923" above is replaced by "1.526"). For generic composites such as RSA moduli, the situation is not so clear. The polynomials can trivially be selected so that the (heuristic) theoretical NFS run time estimate is met. As it is fairly well understood how to predict a good upper bound for the sieving and matrix efforts given a set of polynomials, the overall NFS-effort can easily be upper bounded. In practice, however, this upper bound is too pessimistic and easily off by an order of magnitude. It turns out that one can quickly recognise if one set of polynomials is "better" than some other set, which makes it possible (combined with smart searching strategies that have been developed) to efficiently conduct a search for a "good" set of polynomials. This invariably leads to substantial savings in the subsequent steps, but it is not yet well understood how much effort needs to be invested in the search to achieve lowest overall run time.

The upshot is that one cannot expect that for any relevant n a shortened polynomial selection step will result in polynomials with properties representative for those one would find after a more extensive search. For the purposes of the present paper we address this issue by simply skipping polynomial selection, and by restricting the experiments reported below to a fixed set of moduli for which good or reasonable polynomials are known – and to offer others the possibility to improve on those choices. The fixed set that we consider consists of the k-bit RSA moduli RSA-k for $k = 512, 768, 896, 1024, 2048$ as originally published on the now obsolete RSA Challenge list [35]. That we consider only these moduli does not affect general applicability of our cost estimates, if we make two assumptions: we assume that for RSA moduli of similar size NFS requires a similar effort, and that cost estimates for modulus sizes between 512 and 2560 bits other than those above follow by judicious application of the L-function. Examples are given below. We find it hazardous to attach significance to the results of extrapolation beyond 2560.

Prior Work. Various special purpose hardware designs have been proposed for factoring most notably TWINKLE [38], TWIRL [39] and SHARK [14]. SHARK is speculated to do the NFS sieving step for RSA-1024 in one year at a total cost of one billion dollars. With the same run time estimate but only ten million dollars to build and twenty million to develop, plus the same costs for the matrix step, TWIRL would be more than two orders of magnitude less expensive. Not everyone agrees, however, that TWIRL can be built and will perform as proposed.

In 1999 NFS was used to factor RSA-512 [10] using software running on commodity hardware. Using much improved software and a better set of polynomials the total effort required for this factorisation would now be about 3 months on a single 2.2GHz Opteron core. In 2009, this same software version of NFS (again running on regular servers) was used to factor RSA-768 [18]. The total effort of this last factorisation was less than 1700 years on a single 2.2GHz Opteron core with 2 GB of memory: about 40 years for polynomial selection, 1500 years for sieving, 155 years for the matrix, and on the order of hours for the final square root step. The matrix step was done on eight disjoint clusters, with its computationally least intensive but most memory demanding central stage done on a single cluster and requiring up to a TB of memory for a relatively brief period of time. According to [18], however, half the amount of sieving would have sufficed. Combined with the rough estimate that this would have doubled the matrix effort and based on our first assumption above, 1100 years on a 2.2GHz core will thus be our estimate for the factoring effort of *any* 768-bit RSA modulus. Note that the ratio $\frac{1100}{0.25} = 4400$ of the efforts for RSA-768 and RSA-512 is of the same order of magnitude as $\frac{L(2^{768})}{L(2^{512})} \approx 6150$ (twice omitting the "$o(1)$"), thus not providing strong evidence against our second assumption above.

Factoring on EC2. Based on the sieving and matrix programs used in [18] we developed two simplified pieces of software that perform the most relevant sieving and matrix calculations without outputting any other results than the time required for the calculations. Sieving parameters (such as polynomials defining the number fields, as described above) are provided for the fixed set of moduli RSA-512, RSA-768, RSA-896, RSA-1024 and RSA-2048. For the smallest two numbers they are identical to the parameters used to derive the timings reported above, for both steps resulting in realistic experiments with threading and memory requirements that can be met by EC2. For RSA-896 our parameters are expected to be reasonable, but can be improved, and RSA-896 is small enough to allow meaningful EC2 sieving experiments. For the largest two numbers our parameter choices allow considerable improvement (at possibly substantial computational effort spent on polynomial selection), but we do not expect that it is possible to find parameters for RSA-1024 or RSA-2048 that allow realistic sieving experiments on the current EC2: for RSA-1024 the best we may hope for would be a sieving experiment requiring several hours using on the order of 100 GB of memory, increasing to several years using petabytes of memory for RSA-2048.

The simplified matrix program uses parameters (such as the k value and the average number of non-zero entries per vector) corresponding to those for the sieving. It produces timing and cost estimates only for the first and third stage of the three stages of the matrix step. The central stage is omitted. For RSA-512 and RSA-768 this results in realistic experiments that can be executed using an EC2 cluster instance (possibly with the exception of storage of the full RSA-768 matrix). For the other three moduli the estimated sizes are beyond the capacity of EC2. It is even the case that at this point it is unclear to us how the central stage of the RSA-2048 matrix step should be performed at all: with the approach

Table 6. RSA Sieving Step

Modulus	Instance	ECU Years	Estimated Sieving Cost		
			3 Years	1 Year	ASAP
RSA-512	high-CPU	0.36	N/A	N/A	\$107
RSA-768	EL	1650	\$190,000	\$280,000	\$480,000
RSA-896		$1.5 \cdot 10^5$	$\$2.2 \cdot 10^7$	$\$3.3 \cdot 10^7$	$\$9.1 \cdot 10^7$
RSA-1024	high-mem EL	$2 \cdot 10^6$	$\$2.9 \cdot 10^8$	$\$4.3 \cdot 10^8$	$\$1.2 \cdot 10^9$
RSA-2048		$2 \cdot 10^{15}$	$\$\approx 10^{17}$		

Table 7. RSA Matrix Step

Modulus	Instance	ECU Years	Estimated Matrix Cost		
			3 Years	1 Year	ASAP
RSA-512	high-CPU EL	0.1	N/A	N/A	$30
RSA-768		680	$95,000	$142,000	$231,000
RSA-896	cluster	$3 \cdot 10^4$	$3.3 \cdot 10^6$	$5.0 \cdot 10^6$	$1.0 \cdot 10^7$
RSA-1024	compute	$8 \cdot 10^5$	$8.9 \cdot 10^7$	$1.3 \cdot 10^8$	$2.7 \cdot 10^8$
RSA-2048		$8 \cdot 10^{14}$	$\approx 10^{17}$		

used for RSA-768 the cost of the central stage would by far dominate the overall cost, whereas for the other three moduli the central stage is known or expected to be negligible compared to the other two.

Table 6 and Table 7 list the most suitable ECU instance for each program and the five moduli under consideration so far, and the resulting timings and EC2 2012 cost estimates (including 20% discount where appropriate, i.e. for one and three year costs of over two million dollars) for RSA-512, RSA-768, and RSA-896. The figures in italics (RSA-896 matrix step, both steps for RSA-1024 and RSA-2048) are just crude L-based extrapolations[6].

The rough cost estimate for factoring a 1024-bit RSA modulus in one year is of the same order of magnitude as the SHARK cost, without incurring the SHARK development cost and while including the cost of the matrix step.

3.5   ECC

Prior Work. The security of ECC (Elliptic Curve Cryptography) relies on the hardness of the Elliptic Curve Discrete Logarithm Problem (EC-DLP). In 1997 Certicom issued a series of challenges of different security levels [11]. Each security level is defined by the number of bits in the group order of the elliptic curve. The Certicom challenges were over binary fields and large prime fields, and ranged from 79-bit to 359-bit curves. The curves are named with the following convention. ECCp-n refers to a curve over a large prime field with a group order of n bits, ECC2-n refers to a similar curve over a binary field, and ECC2K-n refers to a curve with additional structure (a so-called Koblitz curve) with group order of n bits over a binary field.

The smaller "exercise" challenges were solved quickly: in December 1997 and February 1998 ECCp-79 and ECCp-89 were solved using 52 and 716 machine days on a set of 500 MHz DEC Alpha workstations, followed in September 1999 by ECCp-97 in an estimated 6412 machine days on various platforms from different contributors. The first of the main challenges were solved in November 2002 (ECCp-109) and in April 2004 (ECC2-109). Since then no challenges have been solved, despite existing efforts to do so.

---

[6] We note the huge discrepancy between EC2-factoring cost and the monetary awards that used to be offered for the factorizations of these moduli [35].

ECC Key Search. The method for solving EC-DLP is based on Pollard's rho method [31]. Similar to SHA-2 collision search, it searches for a collision between two walks. We first define a deterministic "pseudorandom" walk on the group elements; each client starts such a walk, and then when they reach a distinguished point, the group element and some additional information is sent back to a central server. Once the servers received two identical distinguished points one can solve the EC-DLP using the additional information. We refer to [29] for details.

ECC Key Search on EC2. Because most deployed ECC systems, including the recommended NIST curves, are over prime fields, we focus on elliptic curves defined over a field of prime order p. We took two sample sets: the Certicom challenges [11] which are defined over fields where p is a random prime (ECC-p-X), and the curves over prime fields defined in the NIST/SECG standards [27,36], where p is a so-called generalised Mersenne prime (secpX-r1), thereby allowing more efficient field arithmetic. Of the latter we took the random curves over fields of cryptographically interesting sizes listed in the table below.

All of the curves were analysed with a program which used Montgomery arithmetic for its base field arithmetic. The NIST/SECG curves were also analysed using a program which used specialised arithmetic, saving essentially a factor of two. Table 8 summarises the costs of key retrieval using high-CPU EL EC2 instances in May 2012, rounded to the nearest order of magnitude for the larger p. We present the costs for the small curves for comparison with the effort spent in the initial analysis over a decade ago. Note that general orders of magnitude correlate with what we expect in terms of costs related to AES, SHA-2, etc.

Table 8. Cryptanalytic Strength of ECC on EC2

Curve Name Name	ECU Years	Estimated Key Retrieval Cost 3 Years / 1 Year / ASAP
ECCp-79	3.5 days	N/A / N/A / $ 2
ECCp-89	104 days	N/A / N/A / $82
ECCp-97	5	$580 / $850 / $1,500
ECCp-109	300	$35,000 / $51,000 / $87,000
ECCp-131	10^6	$\approx \$10^8$
ECCp-163	10^{10}	$\approx \$10^{12}$
ECCp-191	10^{15}	$\approx \$10^{17}$
ECCp-239	10^{22}	$\approx \$10^{24}$
ECCp-359	10^{40}	$\approx \$10^{42}$
secp192-r1	10^{15}	$\approx \$10^{17}$
secp224-r1	10^{20}	$\approx \$10^{22}$
secp256-r1	10^{25}	$\approx \$10^{27}$
secp384-r1	10^{44}	$\approx \$10^{46}$

4   Extrapolate

Comparing the current pricing model to EC2's 2008 flat rate of $0.10 per hour per ECU, we find that prices have dropped by a factor of three (short term) to six (three year). This shaves off about two bits of the security of block ciphers over a period of about three years, following closely what one would expect based on Moore's law. The most interesting contribution of this paper is that our approach allows anyone to measure and observe at what rate key erosion continues in the future. A trend that may appear after doing so for a number of years could lead to a variety of useful insights – not just concerning cryptographic key size selection but also with respect to the sanity of cloud computing pricing models.

In future versions of this paper these issues will be elaborated upon in this section. Right now the required data are, unavoidably, still lacking.

5   Can One Do Better?

If by better one means can one reduce the overall costs of breaking each cipher or key size, then the answer is yes. This is for a number of reasons: Firstly one could find a different utility computing service which is cheaper; however we selected Amazon EC2 so as to be able to repeat the experiment each year on roughly the same platform. Any price differences which Amazon introduce due to falling commodity prices, or increased power prices, are then automatically fed into our estimates on a year basis. Since it is unlikely that Amazon will cease to exist in the near future we can with confidence assume that the EC2 service will exist in a year's time.

Secondly, we could improve our code by fine tuning the algorithms and adopting more efficient implementation techniques. We have deliberately tried not to do this. We want the code to be executed next year, and the year after, on the platforms which EC2 provides, therefore highly specialised performance improvements have not been considered. General optimisation of the algorithm can always be performed and to enable this we have made the source code available on a public web site, http://www.cs.bris.ac.uk/~nigel/Cloud-Keys/. However, we have ruled out aggressive optimisations as they would only provide a constant improvement in performance and if costs to break a key are of the order of 10^{20} dollars then reducing this to 10^{18} dollars is unlikely to be that significant in the real world.

Finally, improvements can come from algorithmic breakthroughs. Although for all the algorithms we have discussed algorithmic breakthroughs have been somewhat lacking in the last few years, we intend to incorporate them in our code if they occur.

Acknowledgements. The work of the first two authors was supported by the Swiss National Science Foundation under grant numbers 200021-119776, 206021-128727, and 200020-132160. The work in this paper was partially funded by the European Commission through the ICT Programme under Contract ICT-2007-216676 ECRYPT II. The fourth author's research was also partially supported by a Royal Society Wolfson Merit Award and by the ERC through Advanced Grant ERC-2010-AdG-267188-CRIPTO.

References

1. Amazon Elastic Compute Cloud Limited Beta (July 2007), http://web.archive.org/web/20070705164650rn_2/ www.amazon.com/b?ie=UTF8&node=201590011
2. Amazon Elastic Compute Cloud Beta (May 2008), http://web.archive.org/web/20080501182549rn_2/ www.amazon.com/EC2-AWS-Service-Pricing/b?ie=UTF8&node=201590011
3. Amazon Elastic Compute Cloud (Amazon EC2), http://aws.amazon.com/ec2/
4. Bahr, F., Boehm, M., Franke, J., Kleinjung, T.: Subject: RSA200. Announcement, May 9 (2005)
5. Bailey, D.V., Batina, L., Bernstein, D.J., Birkner, P., Bos, J.W., Chen, H.-C., Cheng, C.-M., van Damme, G., de Meulenaer, G., Perez, L.J.D., Fan, J., Güneysu, T., Gurkaynak, F., Kleinjung, T., Lange, T., Mentens, N., Niederhagen, R., Paar, C., Regazzoni, F., Schwabe, P., Uhsadel, L., Van Herrewege, A., Yang, B.-Y.: Breaking ECC2K-130. Cryptology ePrint Archive, Report 2009/541 (2009), http://eprint.iacr.org/2009/541
6. Biham, E.: A Fast New DES Implementation in Software. In: Biham, E. (ed.) FSE 1997. LNCS, vol. 1267, pp. 260–272. Springer, Heidelberg (1997)
7. Biryukov, A., Khovratovich, D., Nikolić, I.: Distinguisher and Related-Key Attack on the Full AES-256. In: Halevi, S. (ed.) CRYPTO 2009. LNCS, vol. 5677, pp. 231–249. Springer, Heidelberg (2009)
8. The BOINC project, http://boinc.berkeley.edu/
9. Caron, T.R., Silverman, R.D.: Parallel implementation of the quadratic sieve. J. Supercomputing 1, 273–290 (1988)
10. Cavallar, S., Dodson, B., Lenstra, A.K., Lioen, W., Montgomery, P.L., Murphy, B., te Riele, H., Aardal, K., Gilchrist, J., Guillerm, G., Leyland, P., Marchand, J., Morain, F., Muffett, A., Putnam, C., Putnam, C., Zimmermann, P.: Factorization of a 512-Bit RSA Modulus. In: Preneel, B. (ed.) EUROCRYPT 2000. LNCS, vol. 1807, pp. 1–18. Springer, Heidelberg (2000)
11. Certicom Inc. The Certicom ECC Challenge, http://www.certicom.com/index.php/the-certicom-ecc-challenge
12. Coppersmith, D.: Modifications to the number field sieve. J. of Cryptology 6, 169–180 (1993)
13. Hayashi, T., Shimoyama, T., Shinohara, N., Takagi. T.: Breaking pairing-based cryptosystems using η_T pairing over $GF(3^{97})$. Cryptology ePrint Archive, Report 2012/345 (2012), http://eprint.iacr.org/2012/345
14. Franke, J., Kleinjung, T., Paar, C., Pelzl, J., Priplata, C., Stahlke, C.: SHARK: A Realizable Special Hardware Sieving Device for Factoring 1024-Bit Integers. In: Rao, J.R., Sunar, B. (eds.) CHES 2005. LNCS, vol. 3659, pp. 119–130. Springer, Heidelberg (2005)
15. Gilmore, J. (ed.): Cracking DES: Secrets of Encryption Research, Wiretap Politics & Chip Design. Electronic Frontier Foundation. O'Reilly & Associates (1998)
16. Güneysu, T., Kasper, T., Novotný, M., Paar, C., Rupp, A.: Cryptanalysis with COPACOBANA. IEEE Transactions on Computers 57, 1498–1513 (2008)

17. Gueron, S.: Intel's New AES Instructions for Enhanced Performance and Security. In: Dunkelman, O. (ed.) FSE 2009. LNCS, vol. 5665, pp. 51–66. Springer, Heidelberg (2009)
18. Kleinjung, T., Aoki, K., Franke, J., Lenstra, A.K., Thomé, E., Bos, J.W., Gaudry, P., Kruppa, A., Montgomery, P.L., Osvik, D.A., te Riele, H., Timofeev, A., Zimmermann, P.: Factorization of a 768-Bit RSA Modulus. In: Rabin, T. (ed.) CRYPTO 2010. LNCS, vol. 6223, pp. 333–350. Springer, Heidelberg (2010)
19. Kleinjung, I., Bos, J.W., Lenstra, A.K., Osvik, D.A., Aoki, K., Contini, S., Franke, J., Thomé, E., Jermini, P., Thiémard, M., Leyland, P., Montgomery, P.L., Timofeev, A., Stockinger, H.: A heterogeneous computing environment to solve the 768-bit RSA challenge. Cluster Computing 15, 53–68 (2012)
20. Lenstra, A.K.: Unbelievable Security; Matching AES Security Using Public Key Systems. In: Boyd, C. (ed.) ASIACRYPT 2001. LNCS, vol. 2248, pp. 67–86. Springer, Heidelberg (2001)
21. Lenstra, A.K.: Key Lengths. In: The Handbook of Information Security, ch. 114. Wiley (2005)
22. Lenstra, A.K., Lenstra Jr., H.W. (eds.): The development of the number field sieve. Lecture Notes in Math., vol. 1554. Springer (1993)
23. Lenstra, A.K., Manasse, M.S.: Factoring by Electronic Mail. In: Quisquater, J.J., Vandewalle, J. (eds.) EUROCRYPT 1989. LNCS, vol. 434, pp. 355–371. Springer, Heidelberg (1990)
24. Lenstra, A.K., Verheul, E.R.: Selecting Cryptographic Key Sizes. J. of Cryptology 14, 255–293 (2001)
25. Matsui, M., Nakajima, J.: On the Power of Bitslice Implementation on Intel Core2 Processor. In: Paillier, P., Verbauwhede, I. (eds.) CHES 2007. LNCS, vol. 4727, pp. 121–134. Springer, Heidelberg (2007)
26. NIST. Secure Hash Signature Standard (SHS) – FIPS PUB 180-2, http://csrc.nist.gov/publications/fips/fips180-2/fips180-2.pdf
27. NIST. Digital Signature Standard (DSS) – FIPS PUB 186-2, http://csrc.nist.gov/publications/fips/fips186-2/fips186-2-change1.pdf
28. NFS@home, http://escatter11.fullerton.edu/nfs
29. van Oorschot, P.C., Wiener, M.J.: Parallel collision search with cryptanalytic applications. J. of Cryptology 12, 1–28 (1999)
30. Osvik, D.A., Shamir, A., Tromer, E.: Efficient Cache Attacks on AES, and Countermeasures. J. of Cryptology 23, 37–71 (2010)
31. Pollard, J.: Monte Carlo methods for index computation mod p. Math. Comp. 32, 918–924 (1978)
32. Quisquater, J.-J., Standaert, F.: Exhaustive key search of the DES: Updates and refinements. In: SHARCS 2005 (2005)
33. Quisquater, J.-J., Standaert, F.: Time-memory tradeoffs. In: Encyclopedia of Cryptography and Security, pp. 614–616. Springer (2005)
34. Rouvroy, G., Standaert, F.-X., Quisquarter, J.-J., Legat, J.-D.: Design strategies and modified descriptions to optimize cipher FPGA implementations: Fact and compact results for DES and Triple-DES. In: ACM/SIGDA - Symposium on FPGAs, pp. 247–247 (2003)
35. The RSA challenge numbers, formerly on http://www.rsa.com/rsalabs/node.asp?id=2093, now on for instance http://en.wikipedia.org/wiki/RSA_numbers
36. SECG. Standards for Efficient Cryptography Group. SEC2: Recommended Elliptic Curve Domain Parameters version 1.0, http://www.secg.org

37. http://csrc.nist.gov/groups/ST/hash/sha-3/
38. Shamir, A.: Factoring large numbers with the TWINKLE device (2000) (manuscript)
39. Shamir, A., Tromer, E.: Factoring Large Numbers with the TWIRL Device. In: Boneh, D. (ed.) CRYPTO 2003. LNCS, vol. 2729, pp. 1–26. Springer, Heidelberg (2003)
40. Smart, N.P. (ed.): ECRYPT II: Yearly report on algorithms and keysizes (2009-2010), http://www.ecrypt.eu.org/documents/D.SPA.13.pdf
41. Standaert, F.-X., Rouvroy, G., Quisquater, J.-J., Legat, J.-D.: Efficient Implementation of Rijndael Encryption in Reconfigurable Hardware: Improvements and Design Tradeoffs. In: Walter, C.D., Koç, Ç.K., Paar, C. (eds.) CHES 2003. LNCS, vol. 2779, pp. 334–350. Springer, Heidelberg (2003)
42. Stevens, M., Sotirov, A., Appelbaum, J., Lenstra, A., Molnar, D., Osvik, D.A., de Weger, B.: Short Chosen-Prefix Collisions for MD5 and the Creation of a Rogue CA Certificate. In: Halevi, S. (ed.) CRYPTO 2009. LNCS, vol. 5677, pp. 55–69. Springer, Heidelberg (2009)
43. Wang, X., Feng, D., Lai, X., Yu, H.: Collisions for Hash Functions MD4, MD5, HAVAL-128 and RIPEMD. Cryptology ePrint Archive, Report 2004/199 (2004), http://eprint.iacr.org/2004/199
44. Wang, X., Yao, A., Yao, F.: New Collision Search for SHA-1. Crypto 2005 Rump session (2005), http://www.iacr.org/conferences/crypto2005/r/2.pdf
45. Wang, X., Yin, Y.L., Yu, H.: Finding Collisions in the Full SHA-1. In: Shoup, V. (ed.) CRYPTO 2005. LNCS, vol. 3621, pp. 17–36. Springer, Heidelberg (2005)
46. Wang, X., Yu, H.: How to Break MD5 and Other Hash Functions. In: Cramer, R. (ed.) EUROCRYPT 2005. LNCS, vol. 3494, pp. 19–35. Springer, Heidelberg (2005)

A Unified Characterization of Completeness and Triviality for Secure Function Evaluation

Hemanta K. Maji[1,*], Manoj Prabhakaran[2,**], and Mike Rosulek[3,***]

[1] University of California, Los Angeles
hmaji@cs.ucla.edu
[2] University of Illinois, Urbana-Champaign
mmp@illinois.edu
[3] University of Montana
mikero@cs.umt.edu

Abstract. We present unified combinatorial characterizations of completeness for 2-party secure function evaluation (SFE) against passive and active corruptions in the information-theoretic setting, so that all known characterizations appear as special cases.

In doing so we develop new technical concepts. We define several notions of isomorphism of SFE functionalities and define the *"kernel"* of an SFE functionality. An SFE functionality is then said to be *"simple"* if and only if it is strongly isomorphic to its kernel. An SFE functionality $\mathcal{F}'$ is a core of an SFE functionality $\mathcal{F}$ if it is "redundancy free" and is weakly isomorphic to $\mathcal{F}$. Then:

- An SFE functionality is complete for security against passive corruptions if and only if it is not simple.
- A deterministic SFE functionality is complete for security against active corruptions if and only if it has a core that is not simple. We conjecture that this characterization extends to randomized SFE as well.

We further give explicit combinatorial characterizations of simple SFE functionalities.

Finally, we apply our new notions of isomorphism to reduce the problem of characterization of trivial functionalities (i.e., those securely realizable without setups) for the case of general SFE to the same problem for the case of simple symmetric SFE.

1 Introduction

Two party secure function evaluation (SFE) is a fundamental concept in modern cryptography. In a seminal work, Kilian [Kil88] introduced the notion of a *complete* function for SFE: given access to an ideally secure implementation of a complete function, every SFE function can be securely implemented using a

* Supported by NSF CI postdoctoral fellowship.
** Supported by NSF grant CNS 07-47027.
*** Supported by NSF grant CCF-1149647.

S. Galbraith and M. Nandi (Eds.): INDOCRYPT 2012, LNCS 7668, pp. 40–59, 2012.

protocol, without relying on any computational assumptions. He showed that the oblivious transfer functionality (OT) is complete for secuity against active corruption. Earlier results [GMW87,GV87,HM86] already implied that OT is complete for security against passive corruption. Since then several works have *characterized* which functions are complete, under passive and active corruption, in the information-theoretic setting. While complete functionalities are in a sense the "most complex" functionalities, at the other extreme are the *trivial* functionalities which are not useful as setups, because they can be securely realized from scratch.

In this work we develop a unified framework for stating these results in the information-theoretic setting. Unlike previous characterizations, we do not give separate characterizations for SFE with one output or two outputs[1] or for subclasses of deterministic and randomized SFE. We summarize our definitions and characterizations below. For simplicity, we restrict ourselves to "finite" functionalities through out, though the results do extend to functionalities with *polynomial-sized domains* (but not necessarily exponential-sized domains).

Our Results. We define strong and weak isomorphisms among SFE functionalities (Definition 5). We also use the notion of the kernel of an SFE functionality (Definition 3). We define an SFE functionality to be simple if it is strongly isomorphic to its kernel. For characterizing completeness and triviality against active corruption, we define an SFE functionality $\mathcal{F}'$ to be a core of an SFE functionality $\mathcal{F}$ if it is "redundancy free" and is weakly isomorphic to $\mathcal{F}$.

Completeness. For the case of completeness for security against passive adversaries (passive-completeness, for short), we obtain a complete characterization by piecing together and extending known results for symmetric and asymmetric SFE. In the case of (standalone or UC) security against active corruption, we identify a gap in the known characterizations, but our unified presentation gives a natural conjecture to fill this gap.

Our characterizations for completeness are as follows.

- A (possibly randomized) SFE functionality is passive-complete if and only if it is not simple (e.g. Theorem 1).
- A *deterministic* SFE functionality is UC or standalone-complete if and only if it has a core that is not simple (Theorem 2). The same characterization holds for UC/standalone-completeness of "channel" functionalities as well. We conjecture that this characterization holds for UC/standalone-completeness of all SFE functionalities.

Triviality. It has been known that a *deterministic* SSFE functionality is passive-trivial if and only if it is "decomposable" and is active-trivial if and only if it is "saturated" [Kus89,MPR09,KMR09]. These characterizations were extended

---

[1] SFE functionalities which produce only one output have been considered in the literature either in the form of "symmetric" SFE (a.k.a. SSFE, which give the output to both parties) or in the form of "asymmetric" SFE (which give the output to only one party).

to general (not necessarily symmetric) SFE in [KMR09]. Our contribution is in characterizing passive-triviality for general (not necessarily symmetric) SFE in terms of characterization of passive-triviality for SSFE, in a manner that applies to *both deterministic and randomized SFE*. Briefly, we show that:

- An SFE functionality $\mathcal{F}$ is passive-trivial if and only if it is simple and its kernel (which is a simple SSFE functionality) is passive-trivial.
- An SFE functionality is standalone-trivial if and only if it has a simple core whose kernel (which is a simple SSFE functionality) is standalone-trivial.

Interestingly, the characterization of passive-trivial and standalone-trivial *randomized* SFE still remains open. If this characterization is carried out for simple SSFE functionality, then our results show how to extend it to general SFE.

We heavily rely on prior work which gave characterizations of completeness and triviality for various special cases. Our main contribution is perhaps in identifying how the different results can be unified using arguably elegant definitions. For instance, we unify the characterization of UC- and standalone-completeness for deterministic SFE [KM11] and for randomized channels [CMW04]; further our formulation gives a plausible conjecture for extending these characterizations to all SFE functionalities.

Related Work. The first complete functionality that was discovered was Oblivious Transfer. It was shown to be complete in the information-theoretic setting, against passive adversaries in [GMW87,GV87,HM86], and against active adversaries in [Kil88] (and explicitly extended to UC security in [IPS08]). All subsequent completeness results build on this.

Completeness (w.r.t. security against passive and active corruption) was characterized in [Kil91,Kil00,CK88,CMW04,KM11], for subclasses of SFE functionalities. We observe that a result in [MOPR11] implicitly extended the characterization of completeness w.r.t. security against passive corruption to all SFE functionalities. (Our simpler characterization is proven based on these results.)

On the front of triviality, seminal results in secure multi-party computation established that *all* functionalities are trivial for passive and standalone security, either under computational assumptions [Yao86,GMW87] or, in the setting involving more than two parties, under restrictions on how many parties can be corrupted [CCD88,BGW88]. These results are mostly not applicable for the setting we are considering (information-theoretic security for two-party computation). [Kus89] introduced the notion of "decomposability" and used it to characterize passive-trivial functionalities. Initial proofs considered only perfect security [Kus89,Bea89], but later results extended this to the case of statistical security [MPR09,KMR09]. Triviality for UC-security has a relatively simpler characterization and is well-understood [CKL03,PR08].

2   Preliminaries

A two-party *secure function evaluation (SFE)* functionality $\mathcal{F}(f_A, f_B)$ is a trusted party whose behavior is specified using two functions $f_A : X \times Y \times R \to Z_A$

and $f_B : X \times Y \times R \to Z_B$. The trusted party takes inputs $x \in X$ from Alice and $y \in Y$ from Bob, samples a private local randomness $r \xleftarrow{\$} R$ and evaluates $a = f_A(x, y, r)$ and $b = f_B(x, y, r)$, the respective outcomes of Alice and Bob. A *deterministic* SFE is one for which $|R| = 1$.

We consider, in the information-theoretic setting, security against adversaries who may be passive (a.k.a. honest-but-curious) or active (a.k.a. byzantine, or malicious). In the latter case, the adversary could be standalone (does not interact with the environment during the course of the protocol, except via protocol input/output for reactive functionalities), or not. Correspondingly, we have three notions of security: *passive security, standalone security* and *UC security*. In settings involving active adversaries we consider *security with abort* (if either party is corrupt, the functionality delivers the output of the corrupt party first and then delivers the output to the honest party when instructed by the corrupt party); guaranteed output security notions are beyond the scope of this work.

We adopt the following useful categories of SFE from the literature:

- *Symmetric SFE (SSFE)*, for which $f_A = f_B$. That is, both the parties get the same output.
- *Asymmetric SFE*, for which either f_A or f_B is a constant function. In other words, only one party gets output. A special case of an asymmetric SFE is a *channel* in which the party receiving the output has no input (i.e., if f_B is not a constant function, then $|Y| = 1$).
- There are SFE functionalities which fall into neither of these classes. Sometimes we will use the term *general SFE* to stress that we are considering an SFE which is not necessarily of the above two types.

We restrict ourselves to the universe of *finite* SFE functionalities: the input and output spaces are finite — that is, have size $O(1)$, as a function of the security parameter. In particular, the maximum number of bits needed to represent the input to the parties (and, in the case of randomized functionalities, the number of bits in the random tape of the functionality) does not grow with the security parameter. We remark that for simplicity we considered the distribution over R to be uniform (but $|R|$ need not be a power of 2). However, any fixed arbitrary distribution (which does not change with the security parameter) could be considered, without affecting our results.

Security Notions and Hybrids. The *real-ideal* paradigm for security [GMW87] is used to define security for multi-party computation. Informally, a protocol *securely realizes* a functionality if, for every adversary attacking the actual protocol in the real world, there is a corresponding ideal world adversary (called the simulator) which can achieve the same effect in any environment. Depending on the type of the security required, the capabilities of the adversary vary. We consider three kinds of security: security against passive adversaries (passive-security, for short), security against active adversaries (standalone-security) and Universally Composable security (UC-security). In passive-security the adversary follows the protocol exactly and the ideal world adversary (simulator) does not alter the input it sends to the functionality. In standalone security the adversary can actively corrupt the parties, but it does not interact with the outside environment during the course of the protocol execution.

We consider secure realization of functionalities in presence of "setups" as well. A $\mathcal{G}$-hybrid is a world where trusted implementation of the functionality $\mathcal{G}$ is accessible to both parties. In the real world, parties can communicate via private communication channel (like the plain model) as well as invoke evaluations of $\mathcal{G}$. In the ideal world, the simulator pretends to provide the access of such a setup to the adversary. A protocol π in the $\mathcal{G}$-hybrid, represented as $\pi^{\mathcal{G}}$, securely realizes a functionality $\mathcal{F}$ if for every real world adversary, there exists an ideal world simulator which can simulate identical behavior.

If a functionality has a secure protocol (in some security model) without any setup (i.e., the functionality is *realizable* in that security model), then it is of no value as a setup, as the access to such an ideal functionality can be replaced by an implementation. We shall refer to such functionalities as *trivial functionalities* (for the corresponding security model or reduction). In terms of reducibility, a trivial functionality reduces to every functionality. At the other extreme, we can consider functionalities to which every functionality reduces. Such functionalities, any of which can replace any other setup, are called *complete functionalities* (for the corresponding security model or reduction).

For brevity, we shall write "*passive-trivial*," "*UC-complete*," etc. to stand for "trivial w.r.t. reductions that are secure against passive adversaries," "complete w.r.t. reductions that are UC-secure" etc.

3   Definitions: Isomorphism, Kernel, Simple SFE and Core

In this section, we define the terms useful in stating unified completeness results for 2-party SFE in various security notions.

3.1   Graph of an SFE Functionality

Given a 2-party SFE $\mathcal{F}(f_A, f_B)$ we define a bipartite graph $G(\mathcal{F})$ as follows.

Definition 1 (Graph of a 2-party SFE). *Given a SFE functionality $\mathcal{F}(f_A, f_B)$, its corresponding* graph $G(\mathcal{F})$ *is a weighted bipartite graph constructed as follows. Its partite sets are $X \times Z_A$ and $Y \times Z_B$. For every $(x, a) \in X \times Z_A$ and $(y, b) \in Y \times Z_B$, the edge joining these two vertices is assigned weight*

$$\mathsf{wt}\Big((x, a), (y, b)\Big) := \frac{\Pr_{r \overset{\$}{\leftarrow} R}\big[f_A(x, y, r) = a \ \wedge\ f_B(x, y, r) = b\big]}{|X \times Y|}.$$

The choice of the normalizing constant $1/|X \times Y|$ is arbitrary. For this particular choice of constant, we can view the weight of an edge as representing the joint-distribution probability of input-output pairs seen by the two parties when $(x, y, r) \overset{\$}{\leftarrow} X \times Y \times R$.

We remark that such representations have appeared in the literature for a long time. In particular, on squaring the (the unweighted version of) the above bipartite graph, the two parts separate into two *characteristic graphs* as defined by Witsenhausen [Wit76] for the correlated source which samples input-output pairs for the two parties. The bipartite graph itself, but again for correlated sources, has appeared in later works, both in information theory (e.g. [KTRR03]) and in cryptography (e.g. [WW06]). The graph $G(\mathcal{F})$ as defined above for SFE functionalities was considered in [MOPR11] (for proving the result mentioned in Footnote 3).

For a combinatorial characterization of what we shall define as a simple SFE functionality, the following definition will be useful.

Definition 2 (Product Distribution Graph). *A weighted bipartite graph with partite sets U and V and weight function* wt *is a* product distribution graph *if there exist*

1. *non-empty partitions $\{U_1, \ldots, U_n\}$ and $\{V_1, \ldots, V_n\}$ of U and V respectively, and*
2. *probability distributions p over U, q over V, and c over $[n]$,*

such that for all $u \in U$ and $v \in V$, the weight on edge (u, v) is given by

$$\mathsf{wt}(u,v) = \begin{cases} p_u \cdot q_v / c_k & \text{if } \exists k \in [n], \text{ s.t. } c_k > 0, \text{ and } u \in U_k, v \in V_k, \\ 0 & \text{otherwise.} \end{cases}$$

Intuitively, a bipartite graph G is a product distribution graph if sampling an edge of G corresponds to first sampling a *connected component* of G, and then within that component further sampling two nodes *independently* from the two partite sets of the component. Thus when $u \in U_k$ and $v \in V_k$, $\mathsf{wt}(u,v) = c_k \cdot (p_u/c_k) \cdot (q_v/c_k)$, where c_k is the probability of selecting the kth component and p_u/c_k (resp. q_v/c_k) is the probability of sampling u (resp. v) conditioned on selecting the kth component.

Given an SFE functionality $\mathcal{F}$, it is convenient to define an associated SSFE functionality as the "common information" that Alice and Bob both get from $\mathcal{F}$ [MOPR11].

Definition 3 (Kernel— Common-information in a SFE). *The* kernel *of a SSFE $\mathcal{F}$ is a symmetric SFE which takes inputs x and y from the parties, samples $r \xleftarrow{\$} R$ and computes $a = f_A(x, y, r)$ and $b = f_B(x, y, r)$. Then it outputs to both parties the connected component of $G(\mathcal{F})$ which contains the edge $\big((x, a), (y, b)\big)$.*

Note that the kernel of $\mathcal{F}$ is a symmetric functionality and is randomized only for randomized SFE $\mathcal{F}$. For example, let $\mathcal{F}$ be (possibly biased) symmetric coin tossing functionality. Its kernel is a randomized functionality and, incidentally, is identical to $\mathcal{F}$ itself. Further, kernel of a kernel is the kernel itself.

3.2 Isomorphisms

We introduce a couple of notions of isomorphism between SFE functionalities, which we use in all our subsequent definitions. The definitions of isomorphism presented here are refinements of a notion of isomorphism in [MOPR11], which in turn was preceded by similar notions (restricted to deterministic SFE) in [KMR09,MPR09].

Crucial to defining isomorphism is the following notion of "locality" for a protocol:

Definition 4 (Local Protocol). *In a* local protocol *for $\mathcal{F}$ which uses $\mathcal{G}$ as a setup, each party maps her $\mathcal{F}$-input to a $\mathcal{G}$-input, calls $\mathcal{G}$ once with that input and, based on her local view (i.e. her given $\mathcal{F}$-input, the output of $\mathcal{G}$, and possibly local randomness), computes her final output, without any further communication between the parties.*

Definition 5 (Isomorphism). *We say that $\mathcal{F}$ and $\mathcal{G}$ are* strongly isomorphic *to each other if there exist two local protocols π_1 and π_2 such that:*

1. *$\pi_1^{\mathcal{G}}$ UC-securely realizes $\mathcal{F}$ and $\pi_2^{\mathcal{F}}$ UC-securely realizes $\mathcal{G}$;*
2. *$\pi_1^{\mathcal{G}}$ passive-securely realizes $\mathcal{F}$ and $\pi_2^{\mathcal{F}}$ passive-securely realizes $\mathcal{G}$; and*
3. *$\mathcal{F}$ and $\mathcal{G}$ have the same input domains, and in $\pi_1^{\mathcal{G}}$ and $\pi_2^{\mathcal{F}}$, the parties invoke the given functionality with the same input as they get from the environment.*

$\mathcal{F}$ and $\mathcal{G}$ are said to be isomorphic *to each other if conditions 1 and 2 are satisfied. $\mathcal{F}$ and $\mathcal{G}$ are said to be* weakly isomorphic *to each other if condition 1 is satisfied.*

A few remarks on the definition mentioned above are in order.

1. It is not hard to see that Condition 1, which is required by all three definitions of isomorphisms, is equivalent to the (seemingly weaker) condition obtained by replacing UC-security with standalone security. This is because of the nature of a local reduction.

2. Condition 2 might seem weaker than Condition 1 since the former requires security against a weaker adversary. But security against a weaker adversary is not always a weaker requirement, since it requires that the ideal-world adversaries (simulators) are also weaker (passive, in this case).

3. All these notions of isomorphism are equivalence relations. In particular, they are transitive due to secure composition of local reductions (under all three notions of security).

4. Another consequence of secure composition is that isomorphism (and hence strong isomorphism) preserves UC and standalone reducibility, as well as reducibility against passive adversaries, between functionalities, and weak isomorphism preserves UC and standalone reducibility (but not necessarily passive reducibility) between functionalities. For example, if $\mathcal{F}$ UC-securely (resp., standalone-securely or passive-securely) reduces to $\mathcal{G}$, and $\mathcal{F}$ and $\mathcal{F}'$ are isomorphic to each other, and $\mathcal{G}$ and $\mathcal{G}'$ are isomorphic to each other, then $\mathcal{F}'$ UC-securely (resp., standalone-securely or passive-securely) reduces to $\mathcal{G}'$.

As we shall see shortly, an important property of an SFE functionality is whether or not it is (strongly) isomorphic to its kernel.

Definition 6 (Simple SFE). *A (possibly randomized) SFE functionality $\mathcal{F}$ is said to be* simple *if it is strongly isomorphic to its kernel.*

We shall see from our characterizations that the above definition remains unaltered if strong isomorphism is replaced by isomorphism (see Lemma 3). However, for *deriving* the characterizations, it is convenient to use the stricter notion of isomorphism in this definition.

Though Definition 6 is in terms of isomorphism, below we give an explicit combinatorial characterization of simple SFE functionalities. This combinatorial characterization will be useful in seeing how our definition unifies several definitions in the literature for special cases (in Section 3.3).

Lemma 1. *The following statements are equivalent.*

1. $\mathcal{F}$ *is simple.*
2. $G(\mathcal{F})$ *is a product distribution graph.*
3. *For any nodes $u_0, u_1 \in X \times Z_A$ and $v_0, v_1 \in Y \times Z_B$, the weights in $G(\mathcal{F})$ satisfy*

$$\mathsf{wt}(u_0, v_0)\mathsf{wt}(u_1, v_1) = \mathsf{wt}(u_0, v_1)\mathsf{wt}(u_1, v_0)$$

We prove Lemma 1 in Appendix A.

3.3 Special Cases of Simple SFE

The definition of simple functionalities that we presented above unifies several definitions that appeared in the literature for special classes of functionalities.

- **Deterministic Symmetric SFE.** The first instance where simple functionalities were identified was for the special case of deterministic symmetric SFE: in this case a functionality is *not* simple if and only if the matrix representing the function f has an "OR minor" (i.e., $\exists x_0, x_1, y_0, y_1, z_0, z_1$ with $z_0 \neq z_1$ and $f(x_a, y_b) = z_{a \vee b}$, for $a, b \in \{0, 1\}$) [Kil91].
- **Randomized Symmetric SFE.** In [Kil00] this was generalized to randomized symmetric SFE functionality: in this case (as described in Appendix B) a functionality is not simple iff $\exists x_0, x_1, y_0, y_1, z$ such that

$$\Pr[f(x_0, y_0) = z] > 0; \text{ and } \Pr[f(x_0, y_1) = z] > 0; \text{ and}$$
$$\Pr[f(x_0, y_0) = z] \cdot \Pr[f(x_1, y_1) = z] \neq \Pr[f(x_1, y_0) = z] \cdot \Pr[f(x_0, y_1) = z].$$

  It is easy to see that this is a generalization of the previous definition by setting $z = z_1$.
- **Randomized Asymmetric SFE.** In [Kil00], the characterization of simple functionalities, specialized to the case of randomized *asymmetric* SFE too appears. Kilian gives a combinatorial condition for being non-simple, but also notes (the more intuitive characterization) that the condition does not hold (i.e., the functionality *is* simple) if and only if the functionality has a passive-secure protocol which involves a single deterministic message from Alice to Bob. Equivalently, a (possibly randomized) asymmetric SFE is simple if and only if it is strongly isomorphic to a deterministic functionality in which Bob has no input.

– **Deterministic SFE.** Another generalization, this time to deterministic, but general (not necessarily symmetric or asymmetric) SFE, appears in [KM11]: as described in Appendix B, a deterministic SFE functionality is not simple iff it has an "OT-core": i.e., there are inputs x, x' for Alice and y, y' for Bob such that $f_A(x, y) = f_A(x, y')$, $f_B(x, y) = f_B(x', y)$ and $\left(f_A(x', y), f_B(x, y') \right) \neq \left(f_A(x', y'), f_B(x', y') \right)$.

All these special cases of the definition of simple functionalities were identified to characterize complete functionalities (see Theorem 1 and Theorem 2).

3.4   Redundant Inputs and Core of an SFE Functionality

To study security against active adversaries alone (i.e., not also against passive adversaries) it is useful to have a notion of "redundant" inputs of an SFE functionality that will never be needed by an active adversary. Combinatorial definitions of redundancy have appeared in the literature before for special classes of SFE functionalities, but our definition is in terms of weak isomorphism and applies to all SFE functionalities.

To state our definition we use the following notation. Given a function $f : X \times Y \times R \to Z$, and $x \in X$ (resp. $y \in Y$), let $f|_{X \setminus \{x\}}$ (resp. $f|_{Y \setminus \{y\}}$) denote the function obtained by restricting f to the domain $(X \setminus \{x\}) \times Y \times R$ (resp. $X \times (Y \setminus \{y\}) \times R$). For a functionality $\mathcal{F}(f_A, f_B)$, and $x \in X$ (resp. $y \in Y$), let $\mathcal{F}|_{X \setminus \{x\}}$ (resp. $\mathcal{F}|_{Y \setminus \{y\}}$) denote the functionality $\mathcal{F}'(f_A|_{X \setminus \{x\}}, f_B|_{X \setminus \{x\}})$ (resp. $\mathcal{F}'(f_A|_{Y \setminus \{y\}}, f_B|_{Y \setminus \{y\}})$).

Definition 7 (Redundant Inputs). *A functionality $\mathcal{F}$ with input domain $X \times Y$ is said to have a* redundant input *$x \in X$ (resp. $y \in Y$) if $\mathcal{F}$ is weakly isomorphic to $\mathcal{F}|_{X \setminus \{x\}}$ (resp. $\mathcal{F}|_{Y \setminus \{y\}}$). $\mathcal{F}$ is said to be* redundancy-free *if it has no redundant inputs.*

We highlight two special cases:

– For deterministic SFE functionalities, Alice's input x is redundant iff there is an input $x' \neq x$ that *dominates* x: i.e., Alice can substitute x' for x without Bob noticing (i.e., for all inputs y of Bob, $f_B(x, y) = f_B(x', y)$) while still allowing her to calculate her correct output (i.e., there is a deterministic mapping $T_{x,x'}$ such that for all inputs y of Bob, $f_A(x, y) = T_{x,x'}(f_A(x', y))$).
– For (possibly randomized) asymmetric functionalities (in which only Bob receives a non-constant output), Alice's input x is redundant iff Alice could instead send a "convex combination" of other inputs to achieve the same effect for Bob. That is, there exists $x_1, \ldots, x_k \in X$, $r_1, \ldots, r_k \in \mathbb{R}$ with $x \notin \{x_1, \ldots, x_k\}$, $\sum_i r_i = 1$ and for all $y \in Y$, we have $f_B(x, y) \equiv \sum_i r_i f_B(x_i, y)$. In the previous expression, the output of $f_B(\cdot, \cdot)$ is interpreted as a probability distribution over Z_B (equivalently, a stochastic vector in $\mathbb{R}^{|Z_B|}$).

Definition 8 (Core). *An SFE functionality $\mathcal{F}'$ is said to be a* core *of an SFE functionality $\mathcal{F}$ if $\mathcal{F}$ and $\mathcal{F}'$ are weakly isomorphic to each other and $\mathcal{F}'$ is redundancy-free.*

For any SFE functionality, one can find a core by successively removing from its domain, one at a time, inputs that are redundant (based on the set of inputs that have not been removed yet). To see this, note that if $\mathcal{F}'$ is obtained by removing a single redundant input from the domain of $\mathcal{F}$, then by definition of being redundant, $\mathcal{F}$ and $\mathcal{F}'$ are weakly isomorphic to each other. This process must terminate after a constant number of steps, since the domains are finite. By transitivity of weak isomorphism the final redundancy free functionality obtained is indeed weakly isomorphic to $\mathcal{F}$ and hence a core of $\mathcal{F}$.

From this and the transitivity of weak isomorphism, it follows that the core of $\mathcal{F}$ is *unique* up to weak isomorphism. In fact, for the two special cases of deterministic SFE and randomized channel SFE that are required in Theorem 2 and Theorem 4, the core of $\mathcal{F}$ is unique with respect to (plain) isomorphisms whose input mapping is a bijection. For the former case, this was explicitly observed in [KM11] (where a core of $\mathcal{F}$ was called the redundancy-free version of $\mathcal{F}$). For the latter case, consider the set of points in $\mathbb{R}^{|Z_B|}$ denoting the probability distributions of $f_B(x)$; then the inputs in a core correspond to the vertices of the convex-hull of this set of points. (If the points for multiple values of x coincide on a vertex of the convex-hull, a core will retain exactly one of these inputs.)

To characterize active security, one will typically consider only a core of the functionality. Redundancy-free functionalities also have the convenient property that a protocol for a redundancy-free functionality that is secure against active adversaries is also secure against passive adversaries:[2]

Lemma 2. *If $\mathcal{F}$ is redundancy-free, then any protocol for $\mathcal{F}$ that is standalone-secure is also passive-secure.*

Note that this is not true in general (i.e., when $\mathcal{F}$ has redundant inputs). In a general active-secure protocol, the simulator for a passively corrupt adversary may not be a passive ideal adversary.

Proof: Let π be such a protocol for $\mathcal{F}$. It suffices to show that in π, the simulator for a passive adversary is without loss of generality passive itself. By symmetry, consider a passive dummy adversary $\mathcal{A}$ for Alice, which runs the protocol honestly and outputs its entire view. Note that $\mathcal{A}$ receives an actual input x from the environment.

Let $\mathcal{S}$ be the simulator for $\mathcal{A}$. Let x denote the input provided by the environment, and let $\mathcal{E}_x$ denote the event that $\mathcal{S}$ sends something other than x to $\mathcal{F}$. If for all x, $\mathcal{E}_x$ is negligible, then we are essentially done. We can modify $\mathcal{S}$ to always send x to $\mathcal{F}$; the interaction's outcome changes only by a negligible amount and hence the modified $\mathcal{S}$ is a passive ideal adversary and a valid simulator for $\mathcal{A}$.

Otherwise, fix an x such that $\mathcal{E}_x$ is non-negligible. Then there is a way to condition the randomness of $\mathcal{S}$ so that $\mathcal{E}_x$ always occurs, and the outputs reported by both parties is indistinguishable from the correct output, for all possible inputs of Bob. Call the resulting simulator $\mathcal{S}_x$. Then the following is a local

---

[2] Thus redundancy-free functionalities are a special case of what are called "deviation-revealing functionalities" [PR08], a notion that is defined more generally for reactive functionalities.

protocol for $\mathcal{F}$ using $\mathcal{F}|_{X \setminus \{x\}}$: Bob runs the dummy protocol; if Alice's input is not x, then she runs the dummy protocol. If Alice's input is x, she runs $\mathcal{S}_x$ and reports the prescribed output of the simulated adversary. The properties established for $\mathcal{S}_x$ show that this is a secure protocol for $\mathcal{F}$ in which Alice never uses input x. Since there is always a local protocol for $\mathcal{F}|_{X \setminus \{x\}}$ using $\mathcal{F}$, we have that $\mathcal{F}$ and $\mathcal{F}|_{X \setminus \{x\}}$ are weakly isomorphic, so x is redundant in $\mathcal{F}$. This contradicts the redundancy-freeness of $\mathcal{F}$, so this case cannot happen. $\qquad \square$

4   Completeness of Two Party Functionalities

The first complete functionality that was discovered was Oblivious Transfer. It was shown to be complete against passive adversaries in [GMW87,GV87,HM86], and against active adversaries in [Kil88] (and explicitly extended to UC security in [IPS08]). All subsequent completeness results build on this.

We have a full understanding of SFE functionalities that are complete under security against *passive* adversaries.

Theorem 1. *A finite (possibly randomized) 2-party SFE functionality is passive-complete in the information theoretic setting if and only if it is not simple.*

The first step towards such a characterization was taken by Kilian, for the special case of deterministic symmetric SFE [Kil91]. As mentioned before, for this case the complete functionalities are those with an OR minor. Later, Kilian extended it to the setting of randomized, *symmetric* SFE functionalities, and also for randomized *asymmetric* SFE functionalities [Kil00]. [KM11] includes the case of deterministic general SFE. We observe that a result in [MOPR11] can be used to obtain the complete characterization.[3] Our proof below directly uses Kilian's characterization (rather than extending Kilian's protocol as in [MOPR11]) along with Lemma 1.

Proof: [Proof of Theorem 1]

For the first direction, suppose for contradiction that $\mathcal{F}$ is passive-complete and it is simple. This implies that $\mathcal{K}$, the kernel of $\mathcal{F}$, is also passive-complete. Now, we shall invoke the completeness characterization of randomized symmetric functionalities SFE by Kilian [Kil00] to show that $\mathcal{K}$ is not complete. Let $U_{x,k} := (\{x\} \times Z_A) \cap U_k$ be the set of nodes in the k-th connected component of $\mathrm{G}(\mathcal{F})$ which are of the form $u = (x,a)$ for some $a \in Z_A$. The probability that a randomly sampled edge lies in the k-th component and its corresponding Alice and Bob inputs are x and y, respectively, is:

---

[3] [MOPR11] extends the protocol in [Kil00] for asymmetric SFE to show that if an SFE functionality $\mathcal{F}$ is not (strongly) isomorphic to its kernel, then it is complete for security against passive adversaries. (Though the statement in [MOPR11] is not in terms of strong isomorphism, the protocols that establish completeness of $\mathcal{F}$ only uses the condition of $\mathcal{F}$ not being *strongly* isomorphic to its kernel.) On the other hand, a functionality which is (strongly) isomorphic to its kernel is not complete, since the kernel (which is an SSFE) is itself simple and hence not complete by one of the characterizations in [Kil00].

$$\sum_{(u',v')\in U_{x,k}\times V_{x,k}} \mathsf{wt}(u',v').$$

But $\mathsf{wt}(u',v') = p_{u'} \cdot q_{v'}/c_k$, because $G(\mathcal{F})$ is a product distribution graph. So, the previous probability expression can be re-written as:

$$\left(\sum_{u'\in U_{x,k}} p_{u'}\right) \times \left(\sum_{v'\in V_{y,k}} q_{v'}\right) \frac{1}{c_k} = P_{x,k} \times Q_{y,k}.$$

Now it is easy to verify that:

$$\Pr[k|x_0,y_0]\cdot\Pr[k|x_1,y_1] = |X\times Y|^2 P_{x_0,k} P_{x_1,k} Q_{y_0,k} Q_{y_1,k} = \Pr[k|x_0,y_1]\cdot\Pr[k|x_1,y_0],$$

for every $k \in [n]$ and $(x,y) \in X \times Y$. By [Kil00], this implies that $\mathcal{K}$ is not a passive-complete SSFE.

Next, we prove the more interesting direction: if $\mathcal{F}$ is not simple then $\mathcal{F}$ is passive-complete. For this it is enough to show how to use $\mathcal{F}$ to passive-securely realize a channel C in which Alice has two inputs 0 and 1, and the distributions D_0 and D_1 of the output that Bob receives on input 0 and 1 respectively are such that they are not identical, but nor do they have disjoint supports. This is because by a characterization in Kilian [Kil00], such asymmetric non-trivial channels are passive-complete.

First we describe the channel C we shall securely realize using $\mathcal{F}$. By Lemma 1 we know that $G(\mathcal{F})$ is not a product distribution graph. Given $u \in X \times Z_A$, we can consider the following distribution D_u over $Y \times Z_B$. The probability of a node $v \in Y \times Z_B$ induced by D_u is: $\mathsf{wt}(u,v)/\sum_{v'\in Y\times Z_B} \mathsf{wt}(u,v')$. Since, $G(\mathcal{F})$ is not a product distribution graph, there is some connected component with two nodes $u, u' \in X \times Z_A$ such that D_u and $D_{u'}$ are not identical distributions over $Y \times Z_B$. Since u and u' are connected, there is a path $(u = \hat{u}_0, \hat{v}_0, \hat{u}_1, \hat{v}_1, \ldots, \hat{u}_t = u')$ in $G(\mathcal{F})$. Then there must exist $\hat{u}_i, \hat{u}_{i+1}$ such that $D_{\hat{u}_i}$ and $D_{\hat{u}_{i+1}}$ are not identical. Let $D_0 = D_{\hat{u}_i}$ and $D_1 = D_{\hat{u}_{i+1}}$. Then D_0 and D_1 are not identical, but their supports intersect (at $\hat{v}_i$).

Now we describe how to securely realize the channel C by invoking $\mathcal{F}$ several times. For convenience, let $u_0 = \hat{u}_i$ and $u_1 = \hat{u}_{i+1}$ so that $D_b = D_{u_b}$ for $b \in \{0,1\}$. For $b \in \{0,1\}$, let p_b be the probability that when $\mathcal{F}$ is invoked with random inputs $(x,y) \xleftarrow{\$} X \times Y$, Alice sees outcome a and $(x,a) = u_b$. We know that $\min\{p_0, p_1\} \geq 1/|X \times Y \times R| = \Theta(1)$. To implement the channel, Alice and Bob invoke the functionality $\mathcal{F}$ with uniformly drawn inputs κ times, where κ is the security parameter. Let I_0 and I_1 be the set of indices of the executions where Alice's input-output pair is u_0 and u_1 respectively. With probability at least $1 - 2^{-\Omega(\kappa)}$ both these sets are non-empty. To send a bit b via channel C, Alice sends a random index $i \xleftarrow{\$} I_b$ to Bob and Bob interprets his corresponding input-output pair in the i-th invocation of $\mathcal{F}$ as the output of the channel. It is not hard to show that this is a passive-secure realization of C. □

A consequence of the above characterizations is the following lemma which gives an alternate definition for a simple functionality (where strong isomorphism in Definition 6 is replaced by isomorphism).

Lemma 3. *An SFE functionality $\mathcal{F}$ is simple if and only if it is isomorphic to its kernel.*

Proof: Clearly, if $\mathcal{F}$ is simple, i.e. strongly isomorphic to its kernel $\mathcal{K}$, then it is also isomorphic to its kernel. For the converse, assume for contradiction that $\mathcal{F}$ is isomorphic to its kernel $\mathcal{K}$ but $\mathcal{F}$ is not simple. Since $\mathcal{F}$ is not simple, by Theorem 1, $\mathcal{F}$ is passive-complete. $\mathcal{F}$ is isomorphic to $\mathcal{K}$ implies that $\mathcal{K}$ itself is passive-complete. But kernel of $\mathcal{K}$ is identical to $\mathcal{K}$ and, hence, they are strongly isomorphic to each other. This implies that $\mathcal{K}$ is simple and passive complete — a contradiction by Theorem 1.   □

For the case of active corruption, in standalone as well as the UC setting, a characterization of complete functionalities is known for special cases. This was first shown for the special case of deterministic, *asymmetric* SFE (in which f_A is the constant function) by Kilian [Kil00]. The complete characterization for deterministic SFE — including the extension to UC security — is due to Kraschewski and Müller-Quade [KM11], who phrased it in terms of the presence of an OT-core (see Section 3.3). For the case of *channels* (i.e., asymmetric SFE in which only one party has an input and only the other party gets an output), UC and standalone-completeness was characterized in [CK88,CMW04].[4] Our characterization unifies these two results into a common characterization.

Theorem 2. *A finite 2-party SFE functionality that is*

– *deterministic, or*
– *a channel*

is UC or standalone-complete in the information theoretic setting if and only if it has a core that is not simple.

Proof: We rely on the characterizations of [KM11] and [CMW04] to prove this result.

First consider the case of deterministic 2-party SFE. Kraschewski and Müller-Quade [KM11] showed that $\mathcal{F}$ is UC or standalone-complete if and only if the "redundancy-free version" of $\mathcal{F}$ has an OT-core (see Section 3.3). For deterministic SFE $\mathcal{F}$, a redundancy-free version of $\mathcal{F}$ in the sense of [KM11] is the same as a core of $\mathcal{F}$, (and in fact is isomorphic to every core of $\mathcal{F}$). Also, as discussed in Section 3.3, a deterministic SFE has an OT-core if and only if it is not simple. Thus the characterization of [KM11] can be recast as saying that a deterministic SFE $\mathcal{F}$ is UC or standalone-complete if and only if it has a core that is not simple (and equivalently, every core of $\mathcal{F}$ is not simple).

Next we consider the case of channels. Crépeau, Morozov and Wolf [CMW04] showed that complete channels are exactly those channels for which, after removing "redundant inputs," the resulting channel is "non-trivial." As we described after Definition 8, for an asymmteric SFE, and in particular for a channel SFE $\mathcal{F}$,

---

[4] [CMW04] does not explicitly deal with UC-security. However the simulator implicit in the analysis of the protocol in [CMW04] is a straightline simulator, and can be used to argue UC-completeness as well.

redundancy-free version of $\mathcal{F}$ in the sense of [CMW04] is isomorphic to every core of $\mathcal{F}$. Here a trivial channel is what [Kil00] characterized as non-simple (randomized) asymmetric SFE (see Section 3.3). Thus the characterization of [CMW04] too can be recast as saying that a deterministic SFE $\mathcal{F}$ is standalone-complete if and only if it has a core that is not simple (and equivalently, every core of $\mathcal{F}$ is not simple). The proof in [CMW04] can be extended to cover UC-completeness as well. $\qquad\square$

Extending this characterization to cover randomized SFE remains an open problem. We conjecture that the same characterization as in Theorem 2 holds for all SFE (and not just deterministic SFE or channel SFE).

5 Characterizing Trivial SFE

There are three main classes of trivial SFE functions, depending on the type of security. The simplest 2-party functionalities are the ones which are trivial under UC security. The functionalities are equivalent to noiseless channels [CKL03]. A much richer class of functionalities is obtained by considering triviality under information theoretic *passive security* (this section), and triviality under information theoretic *standalone active security* (Section 5.2. We focus on these two low-complexity classes below. These two classes have been characterized only restricted to deterministic functionalities. Our characterization reduces the problem of characterizing triviality of general SFE functionalities to the problem of characterizing triviality of simple SSFE functionalities. We remark that it still remains open to give a combinatorial characterization of trivial SSFE outside of determinsitic SFE.

5.1 Passive Trivial SFE

Theorem 3. *A finite 2-party SFE functionality $\mathcal{F}$ is passive-trivial in the information theoretic setting if and only if it is simple and its kernel (which is a simple SSFE functionality) is passive-trivial.*

Proof: If $\mathcal{F}$ is simple, then it is strongly isomorphic to its kernel. Hence, if the latter is passive-trivial, then so is $\mathcal{F}$.

The other direction is a simple consequence of Theorem 1. If $\mathcal{F}$ is not simple, then by Theorem 1, it is passive-complete. A passive-complete functionality is not passive-trivial (as otherwise, all functionalities will be passive-trivial, which is not the case). $\qquad\square$

An interesting special case of this appeared in [Kil00]: for an asymmetric deterministic SFE, its kernel is simply a constant functionality and is passive-trivial. Hence an asymmetric SFE is passive-trivial if and only if it is simple. That is, any asymmetric SFE is either passive-trivial or is complete.

5.2   Standalone Trivial SFE Functionalities

Theorem 4. *A finite 2-party SFE functionality is UC- or standalone-trivial in the information theoretic setting if and only if it has a simple core whose kernel $\mathcal{K}$ (which is a simple SSFE functionality) is respectively UC or standalone-trivial.*

Proof: We give the proof for standalone-triviality; the argument for UC-triviality is similar.

Suppose a finite 2-party SFE functionality $\mathcal{F}$ has a simple core $\mathcal{F}'$ whose kernel $\mathcal{K}$ is standalone-trivial. Since, $\mathcal{F}'$ is simple, i.e. strongly isomorphic to $\mathcal{K}$, and $\mathcal{K}$ is standalone trivial, we conclude that $\mathcal{F}'$ is also standalone trivial. Since $\mathcal{F}$ is weakly-isomorphic to $\mathcal{F}'$ and weak isomorphism preserves standalone triviality, $\mathcal{F}$ itself is standalone trivial.

To see the converse, suppose $\mathcal{F}$ is a standalone-trivial SFE. Let $\mathcal{F}'$ be a core of $\mathcal{F}$. Standalone triviality of $\mathcal{F}$ implies that $\mathcal{F}'$ is also standalone trivial. Now, Lemma 2 implies that $\mathcal{F}'$ is also passive trivial and, in particular, it is not passive complete. By Theorem 1, $\mathcal{F}'$ is simple. Now, the core $\mathcal{K}$ of $\mathcal{F}'$ is standalone trivial because $\mathcal{F}$ is weakly isomorphic to $\mathcal{F}'$ and $\mathcal{F}'$ is strongly isomorphic to $\mathcal{K}$.

Note that if *any* core of $\mathcal{F}$ is standalone trivial, then so are all cores. Because $\mathcal{F}$ is weakly isomorphic to both cores and standalone triviality of one of them shall entail standalone triviality of the other core.   □

References

Bea89.   Beaver, D.: Perfect privacy for two-party protocols. In: Feigenbaum, J., Merritt, M. (eds.) Proceedings of DIMACS Workshop on Distributed Computing and Cryptography, vol. 2, pp. 65–77. American Mathematical Society (1989)

BGW88.   Ben-Or, M., Goldwasser, S., Wigderson, A.: Completeness theorems for non-cryptographic fault-tolerant distributed computation (extended abstract). In: Simon, J. (ed.) STOC, pp. 1–10. ACM (1988)

CCD88.   Chaum, D., Crépeau, C., Damgård, I.: Multiparty unconditionally secure protocols. In: Simon, J. (ed.) STOC, pp. 11–19. ACM (1988)

CK88.   Crépeau, C., Kilian, J.: Achieving oblivious transfer using weakened security assumptions (extended abstract). In: FOCS, pp. 42–52. IEEE (1988)

CKL03.   Canetti, R., Kushilevitz, E., Lindell, Y.: On the Limitations of Universally Composable Two-Party Computation Without Set-Up Assumptions. In: Biham, E. (ed.) EUROCRYPT 2003. LNCS, vol. 2656, pp. 68–86. Springer, Heidelberg (2003)

CMW04.   Crépeau, C., Morozov, K., Wolf, S.: Efficient Unconditional Oblivious Transfer from Almost Any Noisy Channel. In: Blundo, C., Cimato, S. (eds.) SCN 2004. LNCS, vol. 3352, pp. 47–59. Springer, Heidelberg (2005)

GMW87.   Goldreich, O., Micali, S., Wigderson, A.: How to play ANY mental game. In: Aho, A.V. (ed.) STOC, pp. 218–229. ACM (1987); See [Gol04, ch. 7] for more details

Gol04.   Goldreich, O.: Foundations of Cryptography: Basic Applications. Cambridge University Press (2004)

GV87.   Goldreich, O., Vainish, R.: How to Solve Any Protocol Probleman Efficiency Improvement. In: Pomerance, C. (ed.) CRYPTO 1987. LNCS, vol. 293, pp. 73–86. Springer, Heidelberg (1988)

HM86.   Haber, S., Micali, S.: Unpublished Manuscript (1986)

IPS08.   Ishai, Y., Prabhakaran, M., Sahai, A.: Founding Cryptography on Oblivious Transfer – Efficiently. In: Wagner, D. (ed.) CRYPTO 2008. LNCS, vol. 5157, pp. 572–591. Springer, Heidelberg (2008)

Kil88.   Kilian, J.: Founding cryptography on oblivious transfer. In: Simon, J. (ed.) STOC, pp. 20–31. ACM (1988)

Kil91.   Kilian, J.: A general completeness theorem for two-party games. In: Koutsougeras, C., Vitter, J.S. (eds.) STOC, pp. 553–560. ACM (1991)

Kil00.   Kilian, J.: More general completeness theorems for secure two-party computation. In: Frances Yao, F., Luks, E.M. (eds.) STOC, pp. 316–324. ACM (2000)

KM11.   Kraschewski, D., Müller-Quade, J.: Completeness Theorems with Constructive Proofs for Finite Deterministic 2-Party Functions. In: Ishai, Y. (ed.) TCC 2011. LNCS, vol. 6597, pp. 364–381. Springer, Heidelberg (2011)

KMR09.   Künzler, R., Müller-Quade, J., Raub, D.: Secure Computability of Functions in the IT Setting with Dishonest Majority and Applications to Long-Term Security. In: Reingold, O. (ed.) TCC 2009. LNCS, vol. 5444, pp. 238–255. Springer, Heidelberg (2009)

KTRR03.   Koulgi, P., Tuncel, E., Regunathan, S.L., Rose, K.: On zero-error coding of correlated sources. IEEE Transactions on Information Theory 49(11), 2856–2873 (2003)

Kus89.   Kushilevitz, E.: Privacy and communication complexity. In: FOCS, pp. 416–421. IEEE (1989)

MOPR11.   Maji, H.K., Ouppaphan, P., Prabhakaran, M., Rosulek, M.: Exploring the Limits of Common Coins Using Frontier Analysis of Protocols. In: Ishai, Y. (ed.) TCC 2011. LNCS, vol. 6597, pp. 486–503. Springer, Heidelberg (2011)

MPR09.   Maji, H.K., Prabhakaran, M., Rosulek, M.: Complexity of Multi-party Computation Problems: The Case of 2-Party Symmetric Secure Function Evaluation. In: Reingold, O. (ed.) TCC 2009. LNCS, vol. 5444, pp. 256–273. Springer, Heidelberg (2009)

PR08.   Prabhakaran, M., Rosulek, M.: Cryptographic Complexity of Multi-Party Computation Problems: Classifications and Separations. In: Wagner, D. (ed.) CRYPTO 2008. LNCS, vol. 5157, pp. 262–279. Springer, Heidelberg (2008)

Wit76.   Witsenhausen, H.S.: The zero-error side information problem and chromatic numbers (corresp.). IEEE Transactions on Information Theory 22(5), 592–593 (1976)

WW06.   Wolf, S., Wullschleger, J.: Oblivious Transfer Is Symmetric. In: Vaudenay, S. (ed.) EUROCRYPT 2006. LNCS, vol. 4004, pp. 222–232. Springer, Heidelberg (2006)

Yao86.   Yao, A.C.-C.: How to generate and exchange secrets. In: FOCS, pp. 162–167. IEEE Computer Society (1986)

A   Proof of Lemma 1

Proof: We shall show the following implications: (1) $\Rightarrow$ (3) $\Rightarrow$ (2) $\Rightarrow$ (1). In fact, first we shall show (2) $\Leftrightarrow$ (3) because this result gives a local test to check whether a graph is product distribution graph or not, which could be of independent interest.

Proof of (2) $\Rightarrow$ (3): Let $G(\mathcal{F})$ be the graph of functionality $\mathcal{F}$ and $\{U_1, \ldots, U_n\}$ and $\{V_1, \ldots, V_n\}$ be the partition of the left and right partite sets of the connected components. The only interesting case is when there exists a $k \in [n]$ such that $u_0, u_1 \in U_k$ and $v_0, v_1 \in V_k$; otherwise $\mathsf{wt}(u_0, v_0)\mathsf{wt}(u_1, v_1) = 0 = \mathsf{wt}(u_0, v_1)\mathsf{wt}(u_1, v_0)$ and condition 3 holds trivially. Let p_u and q_v be the distributions over U and V respectively and c_k be the distribution over the connected components of $G(\mathcal{F})$. Now, we have: $\mathsf{wt}(u_0, v_0)\mathsf{wt}(u_1, v_1) = p_{u_0} q_{v_0} \times p_{u_1} q_{v_1} / c_k^2 = p_{u_0} q_{v_1} \times p_{u_1} q_{v_0} / c_k^2 = \mathsf{wt}(u_0, v_1)\mathsf{wt}(u_1, v_0)$.

Proof of (3) $\Rightarrow$ (2): Let U_k and V_k be the partite sets of the k-th component of the graph. Let W be the sum of weights on all edges of the graph and W_k be the sum of weights of edges in the k-th component. We define c_k as the distribution over the partitions such that probability of $k \in [n]$ is W_k/W. We represent the weight of the edge between i and j node as $\mathsf{wt}(i, j)$. The probability distribution over the edges is $\mathsf{wt}^*(i, j) = \mathsf{wt}(i, j)/W$. We define the weight of the node $i \in U$ as $p_i = \sum_{j' \in V} \mathsf{wt}(i, j')/W$. It is easy to observe that p_i is a probability distribution over U. Similarly, define the probability of $j \in V$ as $q_j = \sum_{i' \in U} \mathsf{wt}(i', j)/W$. Now, consider $i \in U_k$ and $j \in V_k$ and evaluate the expression $p_i \times q_j/c_k$:

$$
\begin{aligned}
p_i \times q_j/c_k &= \left(\sum_{j' \in V} \mathsf{wt}(i, j') \right) \times \left(\sum_{i' \in U} \mathsf{wt}(i', j) \right) \times \frac{1}{W_k W} \\
&= \left(\sum_{j' \in V_k} \mathsf{wt}(i, j') \right) \times \left(\sum_{i' \in U_k} \mathsf{wt}(i', j) \right) \cdot \frac{1}{W_k W} \\
&= \sum_{(i', j') \in U_k \times V_k} \mathsf{wt}(i, j')\mathsf{wt}(i', j) \frac{1}{W_k W} \\
&= \sum_{(i', j') \in U_k \times V_k} \mathsf{wt}(i, j)\mathsf{wt}(i', j') \frac{1}{W_k W} \\
&= (\mathsf{wt}(i, j)/W) \times (W_k/W_k) = \mathsf{wt}^*(i, j)
\end{aligned}
$$

Finally, we show the equivalence of $\mathcal{F}$ being simple with the other two statements. In the following, we shall represent the kernel of $\mathcal{F}$ as $\mathcal{K}$.

Proof of (1) $\Rightarrow$ (3): Suppose we are given a local protocol $\pi_{\mathcal{F}}$ which securely realizes $\mathcal{F}$ in the $\mathcal{K}$ hybrid. By definition, parties invoke $\mathcal{K}$ with the same input as their input for $\mathcal{F}$; and this protocol is passive, standalone and UC secure. Consider the experiment where $x \xleftarrow{\$} X$ and $y \xleftarrow{\$} Y$. We shall condition our analysis on the output of $\mathcal{K}$ being k in the real protocol, when inputs to Alice and Bob are x and y. The probability that Alice outputs a is denoted by $\tilde{p}_k(x,a)$, since $\pi_{\mathcal{F}}$ is a local protocol. Similarly, the probability that Bob outputs b is represented by $\tilde{q}_k(y,b)$.

Let c_k be the probability that $\mathcal{K}$ gives k as output to both parties when $x \xleftarrow{\$} X$ and $y \xleftarrow{\$} Y$. Note that c_k is the sum of weights on edges in the k-th component in $G(\mathcal{F})$. The probability of edge joining $u = (x,a)$ and $v = (y,b)$ in the real world execution is $\tilde{\mathsf{wt}}(u,v) = c_k \times \tilde{p}_k(u)\tilde{q}_k(v)$, for every $u \in U_k$ and $v \in V_k$.

By security of the protocol, we can claim that:[5]

$$\left| \tilde{\mathsf{wt}}(u,v) - \mathsf{wt}(u,v) \right| \leq \mathrm{negl}(\kappa)$$
$$\Leftrightarrow \quad \left| c_k \tilde{p}_k(u)\tilde{q}_k(v) - \mathsf{wt}(u,v) \right| \leq \mathrm{negl}(\kappa)$$

Consider drawing an edge from $G(\mathcal{F})$ with probability equal to the weight on the edge. Let $p_k(u)$ be the probability that Alice's node is $u = (x,a)$, conditioned on the event that an edge in the kth component is selected. Formally, $p_k(u) = \sum_{v' \in V_k} \mathsf{wt}(u,v')/c_k$. Similarly, we define $q_k(v) = \sum_{u' \in U_k} \mathsf{wt}(u',v)/c_k$. By security of the protocol, we can claim that $|\tilde{p}_k(u), p_k(u)| \leq \mathrm{negl}(\kappa)$ and $|\tilde{q}_k(v), q_k(v)| \leq \mathrm{negl}(\kappa)$. Here we use union bounds over V_k and U_k respectively.

Thus, we can conclude that

$$\left| c_k p_k(u) q_k(v) - \mathsf{wt}(u,v) \right| \leq \mathrm{negl}(\kappa)$$

Note that the function $\mathsf{wt}(\cdot,\cdot)$ assigns values which are integral multiples of $1/|X \times Y \times R|$. Therefore, c_k, $p_k(u)$ and $q_k(v)$ are also integral multiples of $1/|X \times Y \times R|$. So, if $c_k p_k(u) q_k(v)$ is not equal to $\mathsf{wt}(u,v)$, then

$$\left| c_k p_k(u) q_k(v) - \mathsf{wt}(u,v) \right| \geq \frac{1}{|X \times Y \times R|} \text{ which is non-negligible}$$

Thus, we can conclude that $\mathsf{wt}(u,v) = c_k p_k(u) q_k(v)$ and this trivially satisfies condition 3 (alternately, interpret $p_k(u)c_k$ as the distribution over U and $q_k(v)c_k$ as the distribution over V).

Proof of (2) $\Rightarrow$ (1): Let the distribution over U and V be p_u and q_v respectively; and the distribution over the connected components be c_k. From the product distribution guarantee we have $\mathsf{wt}(u,v) = p_u \cdot q_v / c_k$.

1. Computing $\mathcal{F}$ in $\mathcal{K}$ hybrid (Protocol $\pi_{\mathcal{F}}$): We provide the algorithm for Alice; and Bob's algorithm is symmetrically defined. On input x, Alice sends x to $\mathcal{K}$ setup and receives the connected component k as output. Given x and k there is distribution over her output $d_{x,k}(a)$ as induced by the edges in the k-th component of $G(\mathcal{F})$. She locally samples her outcome according to this distribution. Formally, the probability of her output being a is:

---

[5] Since $\mathcal{F}$ is a finite functionality, the probability of an edge in $G(\mathcal{F})$, when evaluated with random input, is at least $1/|X \times Y \times R|$. Thus, significant fraction of the soundness error in a particular edge propagates as soundness error in the overall experiment where $x \xleftarrow{\$} X$ and $y \xleftarrow{\$} Y$.

$$\frac{p_{(x,a)}}{\sum_{(x,a')\in U_k} p_{(x,a')}}$$

2. Computing $\mathcal{K}$ in $\mathcal{F}$ hybrid (Protocol $\pi_\mathcal{K}$): Again, we provide Alice's algorithm. On input x, Alice sends x to $\mathcal{F}$ setup and receives her output a. Given (x, a) there is a unique k such that the node (x, a) lies in the k-th component of $G(\mathcal{F})$. Alice outputs k.

Below we prove the security of these protocols. For simplicity, we shall assume that in the first protocol (and in the simulation for the second) it is possible to sample an outcome *exactly* according to a requisite distribution. In general this is not true (for instance, when the probabilities involved have binary infinite binary expansions). But this assumption can be removed by carrying out the sampling to within an exponentially small error (using polynomially many coins); this affects the security error only by an exponentially small amount.

Proofs for Protocol $\pi_\mathcal{F}$: Let us argue the correctness of the protocol. Define $W_{x,y}(k)$ as the weight of edges in k-th component of $G(\mathcal{F})$ when Alice and Bob inputs are x and y respectively; and $W_{x,y} = \sum_{k\in[n]} W_{x,y}(k)$. Consider the event that the edge connecting (x, a) and (y, b) in $G(\mathcal{F})$ lies in the k-th connected component. In both real and ideal worlds the probability of this event is $W_{x,y}(k)/W_{x,y}$. Now, we shall analyze the probability of joint distribution of (x, a) and (y, b) conditioned on this event. Let $U_{x,k}$ be the subset of the k-th connected component's left partite set which have Alice input x. Similarly, define $V_{y,k}$. The probability of the edge connecting $u = (x, a)$ and $v = (y, b)$ in the ideal world is:

$$\frac{\mathsf{wt}(u, v)}{W_{x,y}(k)}$$

The probability of the edge connecting u and v in the real world is:

$$\frac{p_u}{\sum_{u'\in U_{x,k}} p_{u'}} \times \frac{q_v}{\sum_{v'\in V_{y,k}} q_{v'}} = \frac{\mathsf{wt}(u, v)}{W_{x,y}(k)}$$

This shows that the protocol $\pi_\mathcal{F}$ is perfectly correct.

For security, we shall construct a simulator for Alice. Malicious Bob's case is analogous. When malicious Alice is invoked with inputs x, she sends $\tilde{x}$ to $\mathcal{K}$. The simulator, who is implementing the setup $\mathcal{K}$, forwards $\tilde{x}$ to the external $\mathcal{F}$ functionality. It receives an outcome $\tilde{a}$ from the external functionality. Next, the simulator sends the connected component in $G(\mathcal{F})$ which contains the vertex $(\tilde{x}, \tilde{a})$. Simulation is perfect because the probability of malicious Alice seeing $\tilde{k}$ in real and ideal work is exactly $W_{\tilde{x},y}(\tilde{k})/W_{\tilde{x},y}$.

Proofs for Protocol $\pi_\mathcal{K}$: The correctness of the protocol is trivial. Both in the real and ideal world, the probability of k being the output when Alice and Bob have inputs x and y respectively is $W_{x,y}(k)/W_{x,y}$.

For security, we shall construct a simulator for malicious Alice. When malicious Alice is invoked with inputs x, she sends $\tilde{x}$ to $\mathcal{F}$. The simulator, who is implementing the setup $\mathcal{F}$, forwards $\tilde{x}$ to the external $\mathcal{K}$ functionality and receives the connected component $\tilde{k}$ from the external functionality. It samples a node $\tilde{u} = (\tilde{x}, \tilde{a})$ from $U_{\tilde{x}, \tilde{k}}$ according to the distribution $p_{\tilde{u}}$ and sends $\tilde{a}$ as the output of $\mathcal{F}$. The simulation is perfect (up to sampling *exactly* with the requisite probabilities) because the probability of malicious Alice seeing $\tilde{a}$ in the real world is:

$$\frac{\sum_{v' \in V_{y,\tilde{k}}} \mathsf{wt}(\tilde{u}, v')}{W_{\tilde{x}, y}}$$

While the probability of the same event in the simulation is:

$$\frac{W_{\tilde{x}, y}(\tilde{k})}{W_{\tilde{x}, y}} \times \frac{p_{\tilde{u}}}{\sum_{u' \in U_{\tilde{x}, \tilde{k}}} p_{u'}} = \frac{W_{\tilde{x}, y}(\tilde{k})}{W_{\tilde{x}, y}} \times \frac{c_k \sum_{v' \in V_{y, \tilde{k}}} \mathsf{wt}(\tilde{u}, v')}{c_k \sum_{(u', v') \in U_{\tilde{x}, \tilde{k}} \times V_{y, \tilde{k}}} \mathsf{wt}(u', v')}$$

$$= \frac{W_{\tilde{x}, y}(\tilde{k})}{W_{\tilde{x}, y}} \times \frac{\sum_{v' \in V_{y, \tilde{k}}} \mathsf{wt}(u, v')}{W_{\tilde{x}, y}(\tilde{k})}$$

This is identical to the previous expression. □

B Special Cases of Simple SFE

For the special case (possibly randomzied) SSFE functionalities, note that each connected component in the graph of an SSFE functionality has the same output value z. So we observe that Kilian's condition that $\exists x_0, x_1, y_0, y_1, z$ such that

$$\Pr[f(x_0, y_0) = z] > 0; \text{ and } \Pr[f(x_0, y_1) = z] > 0; \text{ and}$$
$$\Pr[f(x_0, y_0) = z] \cdot \Pr[f(x_1, y_1) = z] \neq \Pr[f(x_1, y_0) = z] \cdot \Pr[f(x_0, y_1) = z].$$

can be rephrased as follows: *there exists some connected component in the graph of the functionality that is not a product distribution*, or equivalently, *the functionality is not simple*. In terms of the above values x_0, x_1, y_0, y_1, z, this component (with output z) has nodes (x_0, z), (y_0, z), (x_0, z), (x_1, z). They are connected because the edges $((x_0, z), (y_0, z))$ and $((x_0, z), (y_1, z))$ are present, and (x_1, z) is connected with them either by the edge $((x_1, z), (y_0, z))$ or by the edge $((x_1, z), (y_1, z))$ (i.e., it is not the case that $\Pr[f(x_1, y_1) = z] = \Pr[f(x_1, y_0) = z] = 0$). This connected component is not a product distribution, because if it were, then $\Pr[f(x_0, y_0) = z] \Pr[f(x_1, y_1) = z] = p_A(x_0, z) p_A(x_1, z) p_B(y_0, z) p_B(y_1, z) = \Pr[f(x_1, y_0) = z] \Pr[f(x_0, y_1) = z]$, for some functions p_A and p_B.

To see the simplification in the case of deterministic SFE, note that in the graph of a deterministic SFE, a connected component must be a complete bipartite graph to have a product distribution. So, to *not* be a product graph, there must be two distinct nodes (x, a), (x', a') on the left and two nodes (y, b), (y', b') on the right such that there are edges $((x, a), (y, b))$, $((x, a), (y', b'))$, $((x', a'), (y, b))$, but the edge $((x', a'), (y', b'))$ is not present. That is, there are inputs $x \neq x'$ for Alice and $y \neq y'$ for Bob[6] such that $f_A(x, y) = f_A(x, y')$, $f_B(x, y) = f_B(x', y)$ and $(f_A(x', y'), f_B(x', y')) \neq (f_A(x', y), f_B(x, y'))$ (i.e., either $f_A(x', y) \neq f_A(x', y')$ or $f_B(x, y') \neq f_B(x', y')$ or both). That is the tuple (x, x', y, y') is an OT-core.

---

[6] If $x = x'$, then $a = f_A(x, y) = f_A(x', y) = a'$ and (x, a) and (x', a') are not distinct; similarly if $y = y'$ then $b = b'$.

On the Non-malleability
of the Fiat-Shamir Transform

Sebastian Faust[1,*], Markulf Kohlweiss[2],
Giorgia Azzurra Marson[3,**], and Daniele Venturi[1,*]

[1] Aarhus University
[2] Microsoft Research
[3] Technische Universität Darmstadt

Abstract. The Fiat-Shamir transform is a well studied paradigm for removing interaction from public-coin protocols. We investigate whether the resulting non-interactive zero-knowledge (NIZK) proof systems also exhibit non-malleability properties that have up to now only been studied for NIZK proof systems in the common reference string model: first, we formally define simulation soundness and a weak form of simulation extraction in the random oracle model (ROM). Second, we show that in the ROM the Fiat-Shamir transform meets these properties under lenient conditions. A consequence of our result is that, in the ROM, we obtain truly efficient non malleable NIZK proof systems essentially for free. Our definitions are sufficient for instantiating the Naor-Yung paradigm for CCA2-secure encryption, as well as a generic construction for signature schemes from hard relations and simulation-extractable NIZK proof systems. These two constructions are interesting as the former preserves both the leakage resilience and key-dependent message security of the underlying CPA-secure encryption scheme, while the latter lifts the leakage resilience of the hard relation to the leakage resilience of the resulting signature scheme.

1 Introduction

Zero-knowledge proof systems [26] are a powerful tool for designing cryptographic primitives and protocols. They force malicious parties to behave according to specification while allowing honest parties to protect their secrets. Non-interactive zero-knowledge (NIZK) proofs [10] consist of a single proof message passed from the prover to the verifier. They are particularly useful for designing public-key encryption and signature schemes as the proof can be added to the

---

* The authors acknowledge support from the Danish National Research Foundation and The National Science Foundation of China (under the grant 61061130540) for the Sino-Danish Center for the Theory of Interactive Computation, and also from the CFEM research center (supported by the Danish Strategic Research Council) within which part of this work was performed.
** Supported by CASED. Part of the work done while a Master Student in Rome, La Sapienza.

S. Galbraith and M. Nandi (Eds.): INDOCRYPT 2012, LNCS 7668, pp. 60–79, 2012.
© Springer-Verlag Berlin Heidelberg 2012

ciphertext and signature respectively. Understanding the most efficient NIZK proofs that are sufficiently strong, i.e., sufficiently non-malleable, for building signature and encryption schemes with strong security properties is thus of fundamental importance in cryptography. It was shown by Goldreich and Oren [25] that NIZK proofs are unattainable in the standard model. To avoid this impossibility result, one must rely on additional assumptions, such as common reference strings [9] (CRS model) or idealizations of hash functions [7] (random oracle model, ROM).

With the aim of finding the "right" definition in the non-interactive case, several flavors of non-malleability [19] have been introduced for NIZK in the CRS model [37,38,24,31]. The notion of *simulation soundness*, which bridges soundness and zero knowledge, guarantees that soundness holds even after seeing accepting proofs, for both true and *false* statements, produced by the simulator. This strengthened soundness notion was first proposed by Sahai in [37], and later improved by De Santis et al. [38]. The notion of simulation extraction [38,28] in addition requires that accepting proofs allow to extract witnesses. Different variants of simulation extraction have been proposed by [15,18].

Until recently, zero-knowledge in general and NIZK in particular were considered to be primarily of theoretical interest. Significant exceptions being efficient Σ-protocols [16,17] and their non-interactive relatives based on the Fiat-Shamir (FS) transform [21]. A Σ-protocol is a three-move interactive scheme where the prover sends the first message and the verifier sends a random challenge as the second message. In a nutshell, the Fiat-Shamir transform removes the interaction by computing the challenge as the hash value of the first message and the theorem that is being proven. Σ-protocols and the Fiat-Shamir transform are widely used in the construction of efficient identification [21], anonymous credential [14], signature [35,1], e-voting schemes [8], and many other cryptographic constructions [11,6,23].

Most work on the provable security of zero-knowledge has, however, been conducted either on interactive proof systems in the plain model or on NIZK in the CRS model, while practitioners often preferred Fiat-Shamir based NIZK proofs for their simplicity and efficiency. The use of the Fiat-Shamir transform was most thoroughly explored in the security proofs of signature schemes in the random oracle model [35,1], but was otherwise often used heuristically. The question thus arises whether one can lay sound foundations for the FS transform in the light of recent research on CRS-based NIZKs. To this end, we provide non-malleability definitions for NIZK in the random oracle model that closely follow the established CRS-based definitions [28]. An earlier result oriented in the same direction, but concerning a Σ-protocol for a *specific* language,[1] was given by Fouque and Pointcheval [23]. Their proof strategy relies on the forking lemma [35] and (implicitly) on the fact that the Σ-protocol they consider has a particular property called *strong special honest-verifier zero-knowledge* (SS-HVZK). Since there exist Σ-protocols that do not satisfy the SS-HVZK property, Fouque

---

[1] This is the language used in the Naor-Yung transform when the underlying encryption is the ElGamal scheme.

and Pointcheval's proof cannot be immediately extended to the general case. Moreover, we make the random oracle explicit in our definition, which is crucial as definitions in the random oracle model can be brittle [40].

Our first observation is that much less is required to show simulation soundness for any FS-NIZK proof. Namely, in the random oracle model, simulation soundness simply follows from the soundness and the HVZK properties of the underlying interactive protocol. In particular, it is neither necessary to rely on the forking lemma, nor on the strong property of SS-HVZK. We also show that the proof strategy of [23], when generalized properly to any Σ-protocol, yields something more than just simulation soundness. In fact, one gets some form of simulation extractability, which we call *weak simulation extractability*. In a nutshell, *full simulation extractability* requires that even after seeing many simulated proofs, whenever an adversary outputs a new accepted proof, we can build an algorithm to extract a valid witness. Sometimes, such a strong extraction property is called *online* extraction [22] because the extractor outputs a witness directly after receiving the adversary's proof. In comparison, our notion is weaker in that it allows the extractor to fully control the adversary (i.e., rewind it).

Our contribution. Our contributions are threefold. First, we formally define the notions of zero-knowledge (which holds trivially for the Fiat-Shamir transform), simulation soundness, and simulation extractability for NIZKs in the random oracle model. Second, we show that simulation soundness and a weak form of simulation extractability come for free if one uses the FS-transform for turning Σ-protocols into NIZK proof systems. Third, we investigate the consequences of this result by showing that our definitions are sufficient for instantiating the Naor-Yung paradigm for constructing CCA2-secure encryption schemes, and generic construction for signature schemes from hard relations and simulation-extractable NIZK proof systems [18]. These two constructions are particularly interesting as the former preserves both leakage resilience and key-dependent message security of the underlying CPA-secure encryption scheme, while the latter lifts the leakage resilience of the hard relation to the leakage resilience of the resulting signature scheme. To our knowledge, these are the most efficient schemes having such properties, if one is willing to rely on the ROM.[2]

Related work. The only other efficient transform for Σ-protocols yielding simulation soundness (again in the random oracle model) is Fischlin's transform [22] which is designed with the purpose of online extraction and is less efficient than the classical Fiat-Shamir transform. Therefore, it would be interesting to investigate whether Fischlin's transform achieves a stronger form of simulation extractability. We notice that in the interactive case, a general transform from any Σ-protocol to an (unbounded) simulation-sound Σ-protocol using one-time signatures has been proposed [24]. In the common reference string model the most efficient simulation-sound or simulation extractable NIZK proof system are

---

[2] In particular we obtain as a special case the Alwen et al. [4] leakage-resilient signature scheme based on the Okamoto identification scheme.

based on Groth-Sahai proofs [29]. One has however to pay the price of proving a structure-preserving CCA secure encryption [18] (for true-simulation extraction) or a structure-preserving signature scheme [3] (for full simulation extraction).

2   Preliminaries

Notation. Let k be a security parameter. A function ν is called *negligible* if $\nu(k) \leq k^{-c}$ for any $c > 0$ and sufficiently large k. Given two functions f, g, we write $f \approx g$ if there exists a negligible function ν such that $|f(k) - g(k)| < \nu(k)$. Given an algorithm $\mathcal{A}$, $y \leftarrow \mathcal{A}(x)$ means that y is the output of $\mathcal{A}$ on input x; when $\mathcal{A}$ is randomized, then y is a random variable. We write $\mathcal{A}^H$ to denote the fact that $\mathcal{A}$ has oracle access to some function H. PPT stands for probabilistic-polynomial time. A *decision problem* related to a language $\mathcal{L} \subseteq \{0, 1\}^*$ consists in determining whether a string x is in $\mathcal{L}$ or not. Given an instance x, we say that $\mathcal{A}$ decides (or recognizes) $\mathcal{L}$ if, after a finite number of steps, the algorithm halts and outputs $\mathcal{A}(x) = 1$ if $x \in \mathcal{L}$, otherwise $\mathcal{A}(x) = 0$. (Sometimes, we may call "theorem" a string belonging to the language at hand.) We can associate to any NP-language $\mathcal{L}$ a polynomial-time recognizable relation $\mathcal{R}_{\mathcal{L}}$ defining $\mathcal{L}$ itself, that is $\mathcal{L} = \{x : \exists w \text{ s.t. } (x, w) \in \mathcal{R}_{\mathcal{L}}\}$, where $|w| \leq poly(|x|)$. The string w is called a *witness* or *certificate* for membership of $x \in \mathcal{L}$. For NP, w corresponds to the non-deterministic choices made by $\mathcal{A}$.

Interactive protocols. An interactive proof system (IPS) for membership in $\mathcal{L}$ is a two-party protocol, where a prover wants to convince an efficient verifier that a string x belongs to $\mathcal{L}$. In a zero-knowledge interactive proof system, a prover $\mathcal{P}$ can convince a verifier $\mathcal{V}$ that $x \in \mathcal{L}$ without revealing anything beyond the fact that the statement is indeed true. Informally, this means that $\mathcal{V}$ cannot exploit the interaction with $\mathcal{P}$ for gaining extra-knowledge. Such a property is formalized by requiring the existence of an efficient algorithm $\mathcal{S}$, the *zero-knowledge simulator*, which produces messages indistinguishable from conversations between an honest prover $\mathcal{P}$ and a malicious verifier $\mathcal{V}^*$. Besides the zero-knowledge property, any proof system satisfies two standard requirements: proving true statements is always possible, while it should be infeasible to convince the verifier to accept a false statement as correct. These two conditions are called *completeness* and *soundness* respectively. Related to the concept of interactive proof systems, but even more subtle, is the notion of *proof of knowledge*. In a proof of knowledge (PoK), $\mathcal{P}$ wants to convince $\mathcal{V}$ that he *knows* a secret witness which implies the validity of some assertion, and not merely that the assertion is true. To formalize the fact that a prover actually "knows something", we require that there exists an efficient algorithm $\mathcal{E}$, called *knowledge extractor*, that when given complete access to the program of the prover can extract the witness.

An IPS or an interactive PoK is called *public-coin* when the verifier's moves consist merely of tossing coins and sending their outcomes to the prover. (In contrast, in a *private-coin* IPS the verifier does not need to show the outcome of the coins to the prover [27].) We are mainly interested in a specific class of

$$
\begin{array}{lcl}
\text{Prover} \quad \mathcal{P}(x, w; 1^k) & & \text{Verifier} \quad \mathcal{V}(x; 1^k) \\
\hline
\alpha \leftarrow \mathcal{P}_0(x, w; \rho) & \xrightarrow{\alpha} & \\
 & \xleftarrow{\beta} & \beta \xleftarrow{\$} \mathcal{V}_0(x, \alpha) \\
\gamma \leftarrow \mathcal{P}_1(\alpha, \beta, x, w; \rho) & \xrightarrow{\gamma} & \text{Accept iff } \mathcal{V}_1(x, \alpha, \beta, \gamma) = 1
\end{array}
$$

Fig. 1. A Σ-protocol for a language $\mathcal{L}$

public-coin interactive PoK systems for NP-languages, called Σ-protocols. Here, the parties involved share a string x belonging to a language $\mathcal{L} \in$ NP and the prover also holds a witness w for membership of $x \in \mathcal{L}$. Thus, the prover $\mathcal{P}$ wants to convince the verifier $\mathcal{V}$ that it "knows" a witness w for x, i.e. that x is in the language, without revealing the witness itself. Σ-protocols have a 3-move shape where the first message α, called *commitment*, is sent by the prover and then, alternatively, the parties exchange the other messages β and γ, called (respectively) *challenge* and *response*. The interaction is depicted in Figure 1. Besides the standard properties held by any IPS, Σ-protocols satisfy a flavour of zero-knowledge — called *honest-verifier zero knowledge* (HVZK) — saying that an *honest* verifier taking part in the protocol does not learn anything beyond the validity of the theorem being proven.

Definition 1 (Σ-protocols). *A Σ-protocol $\Sigma = (\mathcal{P}, \mathcal{V})$ for an NP-language $\mathcal{L}$ is a three-round public-coin IPS where $\mathcal{P} = (\mathcal{P}_0, \mathcal{P}_1)$ and $\mathcal{V} = (\mathcal{V}_0, \mathcal{V}_1)$ are PPT algorithms, with the following additional proprieties:*

Completeness. *If $x \in \mathcal{L}$, any proper execution of the protocol between $\mathcal{P}$ and $\mathcal{V}$ ends with the verifier accepting $\mathcal{P}$'s proof.*

Honest-Verifier Zero Knowledge (HVZK). *There exists an efficient algorithm $\mathcal{S}$, called zero-knowledge simulator, such that for any PPT distinguisher $\mathcal{D} = (\mathcal{D}_0, \mathcal{D}_1)$ and for any $(x, w) \in \mathcal{R}_\mathcal{L}$, the view of the following two experiments, real and simulated, are computationally indistinguishable:*

$$
\begin{array}{l|l}
\textbf{Experiment } \mathsf{Exp}_{\Sigma, \mathcal{D}}^{\text{REAL}}(1^k) & \textbf{Experiment } \mathsf{Exp}_{\Sigma, \mathcal{D}}^{\text{SIM}}(\mathcal{S}, 1^k) \\
(x, w, \delta) \leftarrow \mathcal{D}_0(1^k) & (x, w, \delta) \leftarrow \mathcal{D}_0(1^k) \\
\pi \leftarrow \langle \mathcal{P}(x, w; 1^k), \mathcal{V}(x; 1^k) \rangle & \pi \leftarrow \mathcal{S}(x, 1^k) \\
\textit{Output } \mathcal{D}_1(\pi, \delta) & \textit{Output } \mathcal{D}_1(\pi, \delta)
\end{array}
$$

where $\langle \mathcal{P}(x, w), \mathcal{V}(x) \rangle$ denotes the verdict returned at the end of the interaction between $\mathcal{P}$ and $\mathcal{V}$ on common input x and private input w.

Soundness. *If $x \notin \mathcal{L}$ then any malicious (even unbounded) prover $\mathcal{P}^*$ is accepted only with negligible probability.*

Special soundness. *There exists an efficient algorithm $\mathcal{E}$, called special extractor, such that given two accepting conversations (α, β, γ) and $(\alpha, \beta', \gamma')$ for a string x, where $\beta \neq \beta'$, then $w \leftarrow \mathcal{E}(\alpha, \beta, \gamma, \beta', \gamma', x)$ is such that $(x, w) \in \mathcal{R}_\mathcal{L}$.*

The special soundness property is strong enough to imply both soundness and that Σ-protocols are PoK [17]. Sometimes Σ-protocols are required to meet stronger notions of HVZK. We discuss these notions and implications and non implications between them in the full version [20].

A non-standard condition that many Σ-protocols satisfy, introduced by Fischlin in [22], requires that responses are quasi unique, i.e. given an accepting proof it should be infeasible to find a new valid response for that proof.

Definition 2 (Quasi unique responses). *A Σ-protocol has* quasi unique responses *if for any PPT $\mathcal{A}$ and for any security parameter k it holds:*

$$\text{Prob}[(x, \alpha, \beta, \gamma, \gamma') \leftarrow \mathcal{A}(1^k) : \mathcal{V}(x, \alpha, \beta, \gamma) = \mathcal{V}(x, \alpha, \beta, \gamma') = 1 \ \wedge \ \gamma \neq \gamma'] \approx 0.$$

A Σ-protocol has *unique responses* if the probability above is zero. The latter condition, defined by Unruh in [39], is also known as *strict soundness*.

Min-entropy of commitments. Following [1,2], we use the concept of min-entropy to measure how likely it is for a commitment to collide with a fixed value.

Definition 3 (Min-entropy of commitment). *Let k be a security parameter and $\mathcal{L}$ be an NP-language with relation $\mathcal{R_L}$. Consider a pair $(x, w) \in \mathcal{R_L}$ and let $(\mathcal{P}, \mathcal{V})$ be an arbitrary three-round IPS. Denote with $\text{Coins}(k)$ the set of coins used by the prover and consider the set $A(x, w) = \{\mathcal{P}_0(x, w; \rho) : \rho \leftarrow \text{Coins}(k)\}$ of all possible commitments associated to w. The min-entropy function associated to $(\mathcal{P}, \mathcal{V})$ is defined as $\varepsilon(k) = \min_{(x,w)}(-\log_2 \mu(x, w))$, where the minimum is taken over all possible (x, w) drawn from $\mathcal{R_L}$ and $\mu(x, w)$ is the maximum probability that a commitment takes on a particular value, i.e., $\mu(x, w) = \max_{\alpha \in A(x,w)}(Prob[\mathcal{P}_0(x, w; \rho) = \alpha : \rho \leftarrow \text{Coins}(k)])$.*

We say that $(\mathcal{P}, \mathcal{V})$ is *non-trivial* if $\varepsilon(k) = \omega(\log(k))$ is super-logarithmic in k. Often, the commitment is drawn uniformly from some set. In order for $(\mathcal{P}, \mathcal{V})$ to be non-trivial, this set must have size exponential in k. Notice that most of natural Σ-protocols meet such a condition and, in fact, non-triviality is quite easy to achieve, e.g. by appending redundant random bits to the commitment.

Forking lemma. To prove our second main result, we make use of the following version of the forking lemma, which appeared in [6].

Lemma 1 (General forking lemma). *Fix an integer Q and a set $\mathcal{H}$ of size $h \geq 2$. Let P be a randomized program that on input $y, h_1, \ldots, h_Q$ returns a pair, the first element of which is an integer in the range $0, \ldots, Q$ and the second element of which we refer to as a side output. Let IG be a randomized algorithm that we call the input generator. The accepting probability of P, denoted acc, is defined as the probability that $J \geq 1$ in the experiment $y \leftarrow \mathsf{IG}; h_1, \ldots, h_Q \leftarrow \mathcal{H}; (J, s) \leftarrow \mathsf{P}(y, h_1, \ldots, h_Q)$.*

The forking algorithm $\mathsf{F_P}$ associated to P is the randomized algorithm that on input y proceeds as follows.

Algorithm $\mathsf{F_P}(y)$
    Pick coins ρ for P at random

$h_1, \ldots, h_Q \leftarrow \mathcal{H}$
$(I, s) \leftarrow \mathsf{P}(y, h_1, \ldots, h_Q; \rho)$
If $I = 0$ *return* $(0, \perp, \perp)$
$h'_I, \ldots, h'_Q \leftarrow \mathcal{H}$
$(I', s') \leftarrow \mathsf{P}(y, h_1, \ldots, h_{I-1}, h'_I, \ldots, h'_Q; \rho)$
If $(I = I') \wedge (h_I \neq h'_I)$ *return* $(1, s, s')$ *else return* $(0, \perp, \perp)$

Let $\mathrm{ext} = \mathrm{Prob}[b = 1 : y \leftarrow \mathsf{IG}; (b, s, s') \leftarrow \mathsf{F_P}(y)]$, *then* $\mathrm{ext} \geq \mathrm{acc}\left(\frac{\mathrm{acc}}{Q} - \frac{1}{h}\right)$.

3   Properties of NIZKs in the Random Oracle Model

Removing interaction. The Fiat-Shamir transform was originally designed to turn three-round identification schemes into efficient signature schemes. As Σ-protocols are an extension of three-round identification schemes, it is not surprising that they can be considered as a starting point for the Fiat-Shamir transform. The Fiat-Shamir paradigm applies to any Σ-protocol (and more generally to any three-round public-coin proof system): We start from an interactive protocol $(\mathcal{P}, \mathcal{V})$ and remove the interaction between $\mathcal{P}$ and $\mathcal{V}$ by replacing the challenge, chosen at random by the verifier, with a hash value $H(\alpha, x)$ computed by the prover, where H is a hash function modeled as a random oracle. Thus, the interactive protocol $(\mathcal{P}, \mathcal{V})$ is turned into a non-interactive one: The resulting protocol, denoted $(\mathcal{P}^H, \mathcal{V}^H)$, is called *Fiat-Shamir proof system*.

Throughout this paper, we refer to the so called *explicitly programmable* random oracle model [40] (EPROM) where the simulator is allowed to program the random oracle explicitly. We model this by defining the zero-knowledge simulator $\mathcal{S}$ of a non-interactive zero-knowledge proof system as a stateful algorithm that can operate in two modes: $(h_i, st) \leftarrow \mathcal{S}(1, st, q_i)$ takes care of answering random oracle queries (usually by lazy sampling) while $(\pi, st) \leftarrow \mathcal{S}(2, st, x)$ simulates the actual proof. Note that calls to $\mathcal{S}(1, \cdots)$ and $\mathcal{S}(2, \cdots)$ share the common state st that is updated after each operation.

Definition 4 (Unbounded non-interactive zero knowledge). *Let $\mathcal{L}$ be a language in* NP. *Denote with $(\mathcal{S}_1, \mathcal{S}_2)$ the oracles such that $\mathcal{S}_1(q_i)$ returns the first output of $(h_i, st) \leftarrow \mathcal{S}(1, st, q_i)$ and $\mathcal{S}_2(x, w)$ returns the first output of $(\pi, st) \leftarrow \mathcal{S}(2, st, x)$ if $(x, w) \in \mathcal{R}_{\mathcal{L}}$. We say a protocol $(\mathcal{P}^H, \mathcal{V}^H)$ is a NIZK proof for language $\mathcal{L}$ in the random oracle model, if there exists a PPT simulator $\mathcal{S}$ such that for all PPT distinguishers $\mathcal{D}$ we have*

$$\mathrm{Prob}[\mathcal{D}^{H(\cdot), \mathcal{P}^H(\cdot, \cdot)}(1^k) = 1] \approx \mathrm{Prob}[\mathcal{D}^{\mathcal{S}_1(\cdot), \mathcal{S}_2(\cdot, \cdot)}(1^k) = 1],$$

where both $\mathcal{P}$ and $\mathcal{S}_2$ oracles output $\perp$ if $(x, w) \notin \mathcal{R}_{\mathcal{L}}$.

A well known fact is that, in the random oracle model, the Fiat-Shamir transform allows to efficiently design digital signature schemes [21] and non-interactive zero-knowledge protocols. In fact, an appealing characteristic of this transform is that many properties of the starting protocol are still valid after applying it. In particular, it has been proven that the Fiat-Shamir transform turns any three-round public-coin zero-knowledge interactive proof system into a NIZK proof system [7]. It is straightforward to prove that the same holds when the starting protocol is ZK only with respect to a honest verifier, as stated in the following Theorem.

Theorem 1 (Fiat-Shamir NIZKs). *Let k be a security parameter. Consider a non-trivial three-round public-coin honest-verifier zero-knowledge interactive proof system $(\mathcal{P}, \mathcal{V})$ for a language $\mathcal{L} \in$ NP. Let H be a function with range equal to the space of the verifier's coins. In the random oracle model the proof system $(\mathcal{P}^H, \mathcal{V}^H)$, derived from $(\mathcal{P}, \mathcal{V})$ by applying the Fiat-Shamir transform, is unbounded non-interactive zero-knowledge.*

Proof (sketch). To prove that the proof system $(\mathcal{P}^H, \mathcal{V}^H)$ is non-interactive zero-knowledge it is sufficient to show that there exists a simulator $\mathcal{S}$ as required in Definition 4. This can be done by invoking the HVZK simulator associated with the underlying interactive proof system. In particular, $\mathcal{S}$ works as follows:

- To answer query $q = (x, \alpha)$ to $\mathcal{S}_1$, $\mathcal{S}(1, st, q)$ lazily samples a lookup table $\mathcal{T}_H$ kept in state st. It checks whether $\mathcal{T}_H(q)$ is already defined. If this is the case, it returns the previously assigned value; otherwise it returns and sets a fresh random value (of the appropriate length).
- To answer query x to $\mathcal{S}_2$ (respectively $\mathcal{S}_2'$), $\mathcal{S}(2, st, x)$ calls the HVZK simulator of $(\mathcal{P}, \mathcal{V})$ on input x to obtain a proof (α, β, γ). Then, it updates $\mathcal{T}_H$ in such a way that $\beta = \mathcal{T}_H(x, \alpha)$. If $\mathcal{T}_H$ happens to be already defined on this input, $\mathcal{S}$ returns failure and aborts.

We call this simulator *canonical*. The main result of Fiat-Shamir [21] (expressed for their particular identification protocol) is that $\mathcal{S}$ is a "good" NIZK simulator. The crucial step in the proof is that the starting protocol $(\mathcal{P}, \mathcal{V})$ is non-trivial (cf. Definition 3), thus the probability of failure in each of the queries to $\mathcal{S}_2'$ is upper-bounded by $\text{Prob[failure]} \leq 2^{-\varepsilon(k)}$, which is negligible in k.   □

Simulation soundness. The soundness property of a proof system ensures that no malicious prover can come up with an accepting proof for a string that does not belong to the language in question (i.e., for a false theorem). However, it is not clear whether this condition still holds *after* the attacker observes valid proofs for adaptively chosen (true or false) statements. The notion of *simulation soundness* deals with this case.

Definition 5 (Unbounded simulation soundness). *Let $\mathcal{L}$ be a language in NP. Consider a proof system $(\mathcal{P}^H, \mathcal{V}^H)$ for $\mathcal{L}$, with zero-knowledge simulator $\mathcal{S}$. Denote with $(\mathcal{S}_1, \mathcal{S}_2')$ the oracles such that $S_1(q_i)$ returns the first output of $(h_i, st) \leftarrow \mathcal{S}(1, st, q_i)$ and $\mathcal{S}_2'(x)$ returns the first output of $(\pi, st) \leftarrow \mathcal{S}(2, st, x)$. We say that $(\mathcal{P}^H, \mathcal{V}^H)$ is simulation sound with respect to $\mathcal{S}$ in the random oracle model, if for all PPT adversaries $\mathcal{A}$ the following holds:*

$$\text{Prob}[(x^\star, \pi^\star) \leftarrow \mathcal{A}^{\mathcal{S}_1(\cdot), \mathcal{S}_2'(\cdot)} : (x^\star, \pi^\star) \notin \mathcal{T} \wedge x^\star \notin \mathcal{L} \wedge \mathcal{V}^{\mathcal{S}_1}(x^\star, \pi^\star) = 1] \approx 0,$$

where $\mathcal{T}$ is the list of pairs (x_i, π_i), i.e., respectively queries asked to and proofs returned by the simulator.

We stress that the above definition relies crucially on the zero-knowledge property of $(\mathcal{P}^H, \mathcal{V}^H)$, as we use a probability experiment that defines a property of $\mathcal{S}$ to define a property about $(\mathcal{P}^H, \mathcal{V}^H)$. In particular the definition is most meaningful for a simulator $\mathcal{S}$ for which the simulation of the random oracle of $\mathcal{S}_1$ is consistent with a truly random oracle H. Also note that $\mathcal{S}'_2$ allows $\mathcal{A}$ to ask for simulated proofs of false statements.

The possibility to request proofs of false statements has an interesting consequence: simulation soundness holds only with respect to specific simulators and not in general for all NIZK simulators. In particular, one can construct a NIZK proof system $(\mathcal{P}^H, \mathcal{V}^H)$ that is simulation sound with respect to a simulator $\mathcal{S}$ but for which there exists a valid NIZK simulator $\hat{\mathcal{S}}$, such that $(\mathcal{P}^H, \mathcal{V}^H)$ cannot be simulation sound with respect to $\hat{\mathcal{S}}$. To see this, consider a $\mathcal{V}^H$ that accepts all proofs if $H(0) = 0$. $\hat{\mathcal{S}}$ simulates a consistent random oracle until it receives a proof of a false statement (one of which could be hard-coded in $\hat{\mathcal{S}}$ or easy to recognize) at which point it sets $\mathcal{T}_H(0) = 0$. Note that a similar counterexample exists for CRS-based NIZK [28]: $\hat{\mathcal{S}}_2$ can simply return the simulation trapdoor when queried on a false statement.

Simulation extractability. Combining simulation soundness and knowledge extraction, we may require that even after seeing (polynomially) many simulated proofs, whenever $\mathcal{A}$ makes a new proof it is possible to extract a witness. This property is called simulation extractability, and implies simulation soundness. Indeed, if we can extract a witness from the adversary's proof even with small probability, then obviously the statement must belong to the language in question. We introduce a weaker flavor of simulation extractability which we call *weak simulation extractability*. The main difference with full simulation extractability is that the extractor $\mathcal{E}_{\mathcal{A}}$ is now given complete control over the adversary $\mathcal{A}$, meaning that it is allowed to rewind $\mathcal{A}$ and gets to see the answers of $(\mathcal{S}_1, \mathcal{S}'_2)$. Moreover, we require that if $\mathcal{A}$ outputs an accepting proof with some probability, then $\mathcal{E}_{\mathcal{A}}$ can extract with almost the same probability.

Definition 6 (Weak simulation extractability). *Let $\mathcal{L}$ be a language in* NP. *Consider a NIZK proof system $(\mathcal{P}^H, \mathcal{V}^H)$ for $\mathcal{L}$ with zero-knowledge simulator $\mathcal{S}$. Let $(\mathcal{S}_1, \mathcal{S}'_2)$ be oracles returning the first output of $(h_i, st) \leftarrow \mathcal{S}(1, st, q_i)$ and $(\pi, st) \leftarrow \mathcal{S}(2, st, x)$ respectively. We say that $(\mathcal{P}^H, \mathcal{V}^H)$ is weakly simulation-extractable with extraction error ν and with respect to $\mathcal{S}$ in the random oracle model, if for all PPT adversaries $\mathcal{A}$ there exists an efficient algorithm $\mathcal{E}_{\mathcal{A}}$ with access to the answers $\mathcal{T}_H, \mathcal{T}$ of $(\mathcal{S}_1, \mathcal{S}'_2)$ respectively such that the following holds. Let:*

$$\mathrm{acc} = \mathrm{Prob}\big[(x^\star, \pi^\star) \leftarrow \mathcal{A}^{\mathcal{S}_1(\cdot), \mathcal{S}'_2(\cdot)}(1^k; \rho) : (x^\star, \pi^\star) \notin \mathcal{T}; \mathcal{V}^{\mathcal{S}_1}(x^\star, \pi^\star) = 1\big]$$

$$\mathrm{ext} = \mathrm{Prob}\big[(x^\star, \pi^\star) \leftarrow \mathcal{A}^{\mathcal{S}_1(\cdot), \mathcal{S}'_2(\cdot)}(1^k; \rho);$$
$$w^\star \leftarrow \mathcal{E}_{\mathcal{A}}(x^\star, \pi^\star; \rho, \mathcal{T}_H, \mathcal{T}) : (x^\star, \pi^\star) \notin \mathcal{T}; (x^\star, w^\star) \in \mathcal{R}_{\mathcal{L}}\big],$$

where the probability space in both cases is over the random choices of S and the adversary's random tape ρ. Then, there exist a constant $d > 0$ and a polynomial p such that whenever $\mathrm{acc} \geq \nu$, *we have* $\mathrm{ext} \geq \frac{1}{p}(\mathrm{acc} - \nu)^d$.

The above definition is inspired by similar notions in the context of proofs of knowledge [5,30,39]. The value ν is called *extraction error* of the proof system. We omit for better readability that values $\mathrm{acc}, \mathrm{ext}, p, \nu$ all depend on the security parameter k. Note that a non-negligible extractor error can be made exponentially small by sequential repetitions (see the full version [20] for a proof).

Proposition 1 (Extraction error amplification). *Let $(\mathcal{P}^H, \mathcal{V}^H)$ be a weakly simulation extractable NIZK proof system with extraction error ν. Then, the proof system $(\mathcal{P}'^H, \mathcal{V}'^H)$ obtained by repeating sequentially $(\mathcal{P}^H, \mathcal{V}^H)$ for a number n of times yields a weakly simulation extractable NIZK proof system with extraction error ν^n.*

It is useful to look at the relation between weak simulation extractability and the following stronger property modeling online-extraction.

Definition 7 (Full Simulation extractability). *Let $\mathcal{L}$ be a language in NP. Consider a NIZK proof system $(\mathcal{P}^H, \mathcal{V}^H)$ for $\mathcal{L}$ with simulator S. Let (S_1, S_2') be oracles returning the first output of $(h_i, st) \leftarrow S(1, st, q_i)$ and $(\pi, st) \leftarrow S(2, st, x)$ respectively. We say that $(\mathcal{P}^H, \mathcal{V}^H)$ is strongly simulation extractable with respect to S in the random oracle model, if there exists an efficient algorithm $\mathcal{E}$ such that for all PPT adversaries $\mathcal{A}$ the following holds. Let:*

$$\mathrm{Prob}\big[w^* \leftarrow \mathcal{E}(st, x^*, \pi^*) \ : \ (x^*, \pi^*) \leftarrow \mathcal{A}^{S_1(\cdot), S_2'(\cdot)}(1^k; \rho);$$
$$(x^*, \pi^*) \notin \mathcal{T}; \quad \mathcal{V}^{S_1}(x^*, \pi^*) = 1; \quad (x^*, w^*) \notin \mathcal{R}_{\mathcal{L}}\big] \approx 0$$

where $\mathcal{T}$ is the list of transcripts (x_i, π_i) returned by the simulator and the probability space is over the random choices of S and the adversary's randomness ρ.

4 On the Non-malleability of the Fiat-Shamir Transform

4.1 Simulation Soundness

We now show that NIZK proofs obtained via the Fiat-Shamir transform from any IPS of the public-coin type additionally satisfying the HVZK property are simulation sound. Since Σ-protocols are a special class of HVZK public-coin IPSs, we get as a corollary that Fiat-Shamir NIZK proofs obtained from Σ-protocols are simulation-sound.

Theorem 2 (Simulation soundness of the Fiat-Shamir transform). *Consider a non-trivial three-round public-coin HVZK interactive proof system $(\mathcal{P}, \mathcal{V})$ for a language $\mathcal{L} \in \mathrm{NP}$, with quasi unique responses. In the random oracle model, the proof system $(\mathcal{P}^H, \mathcal{V}^H)$ derived from $(\mathcal{P}, \mathcal{V})$ via the Fiat-Shamir transform is a simulation-sound NIZK with respect to its canonical simulator S.*

Proof. We assume that $(\mathcal{P}^H, \mathcal{V}^H)$ is a non-interactive zero-knowledge proof system with the simulator $\mathcal{S}$ described in the proof of Theorem 1, and show that $(\mathcal{P}^H, \mathcal{V}^H)$ is simulation sound. We proceed by contradiction. Suppose there exists a PPT adversary $\mathcal{A}$ that breaks the simulation soundness of the non-interactive protocol with non-negligible probability

$$\epsilon := \mathrm{Prob}\Big[(x^\star, \pi^\star) \leftarrow \mathcal{A}^{\mathcal{S}_1(\cdot), \mathcal{S}_2'(\cdot)} : (x^\star, \pi^\star) \notin \mathcal{T} \wedge x^\star \notin \mathcal{L} \wedge \mathcal{V}^{\mathcal{S}_1}(x^\star, \pi^\star) = 1\Big].$$

In such a case, we are able to build two reductions $\hat{\mathcal{P}}$ and $\mathcal{P}^*$ which, by using $\mathcal{A}$ as a black-box, violate either the quasi unique response or the soundness properties of the underlying interactive protocol $(\mathcal{P}, \mathcal{V})$ respectively, contradicting our hypothesis. Recall that $\mathcal{S}_1$ simulates answers to the RO, while $\mathcal{S}_2'$ replies with an accepting proof π. Without loss of generality we assume that whenever adversary $\mathcal{A}$ succeeds and outputs an accepting proof $(\alpha^\star, \gamma^\star)$, she has previously queried the oracle $\mathcal{S}_1$ on input $(x^\star, \alpha^\star)$. The argument for this is that it is straightforward to transform any adversary that violates this condition into an adversary that makes one additional query to $\mathcal{S}_1$ and wins with the same probability.

A simple but crucial observation is that adversary $\mathcal{A}$ may have learned $\alpha^\star$ by querying the oracle $\mathcal{S}_2'$ on input $x^\star$ or might have computed it itself. We denote the first by the event proof, the second by the event $\overline{\mathsf{proof}}$. As these events are mutually exclusive and exhaustive, we have:

$$\mathrm{Prob}[\mathcal{A}\ \mathsf{wins}] \;=\; \mathrm{Prob}[\mathcal{A}\ \mathsf{wins} \wedge \mathsf{proof}] \;+\; \mathrm{Prob}[\mathcal{A}\ \mathsf{wins} \wedge \overline{\mathsf{proof}}].$$

Now we have two different cases to analyze, each of them corresponding to the probability in the expression above.

In the first case (when proof happens), we assume that $x^\star$ is asked to $\mathcal{S}_2'$ and the answer is a proof of the type $(\alpha^\star, -)$. We show how to use an adversary $\mathcal{A}$ that makes use of $(x^\star, \alpha^\star)$ in its fake proof to build a reduction $\hat{\mathcal{P}}$. In this way we bound $\mathrm{Prob}[\mathcal{A}\ \mathsf{wins} \wedge \mathsf{proof}]$ by the probability that $\hat{\mathcal{P}}$ wins in breaking the quasi unique response property.

Consider an algorithm $\hat{\mathcal{P}}$ which runs $\mathcal{A}$ internally as a black-box. Thus, $\hat{\mathcal{P}}$ sees all queries $\mathcal{A}$ makes to the oracles $\mathcal{S}_1$ and $\mathcal{S}_2'$ and produces their answers. The internal description of $\hat{\mathcal{P}}$ follows:

- $\hat{\mathcal{P}}$ answers $\mathcal{S}_1$ and $\mathcal{S}_2'$ and keeps lists $\mathcal{T}_H$ and $\mathcal{T}$ respectively as the real simulator $\mathcal{S}$ would.
- When $\mathcal{A}$ outputs a fake proof $(\alpha^\star, \gamma^\star)$ for $x^\star$, $\hat{\mathcal{P}}$ looks through its lists $\mathcal{T}$ and $\mathcal{T}_H$ until it finds $(x^\star, (\alpha^\star, \gamma))$ and $((x^\star, \alpha^\star), \beta)$ respectively;
- It returns $(x^\star, \alpha^\star, \beta, \gamma^\star, \gamma)$.

We claim that algorithm $\hat{\mathcal{P}}$ breaks the quasi unique response property. Indeed, the proof produced by $\mathcal{A}$ is accepting by $\mathcal{V}^H$ on common input $x^\star$. On the other hand, the proof $(\alpha^\star, \gamma)$ is given by the simulator, therefore it must be accepting for $x^\star$. Given this, it holds $\mathcal{V}^H(x^\star, \alpha^\star, \gamma^\star) = \mathcal{V}^H(x^\star, \alpha^\star, \gamma) = 1$, that means

$$\mathcal{V}(x^\star, \alpha^\star, H(x^\star, \alpha^\star), \gamma^\star) \;=\; \mathcal{V}(x^\star, \alpha^\star, H(x^\star, \alpha^\star), \gamma) = 1,$$

where $H(x^\star, \alpha^\star) = \beta$. The conclusion is that either $\gamma = \gamma^\star$, that is excluded since $\mathcal{A}$ cannot win by printing a simulated proof, or algorithm $\hat{\mathcal{P}}$ succeeds in breaking the quasi unique response property. We obtain:

$$\text{Prob}[\mathcal{A} \text{ wins} \wedge \text{proof}] = \text{Prob}[\hat{\mathcal{P}} \text{ wins}] \leq negl(k).$$

In case proof does not happen, we can use adversary $\mathcal{A}$ that does not query $\mathcal{S}'_2$ with input $x^\star$ to build a reduction $\mathcal{P}^\star$ and bound $\text{Prob}[\mathcal{A} \text{ wins} \wedge \overline{\text{proof}}]$ by the probability $\text{Prob}[\mathcal{P}^\star \text{ wins}] \cdot Q$ of breaking the soundness of the underlying interactive scheme. $\mathcal{P}^\star$ runs $\mathcal{A}$ as a black-box and has to simulate its environment by answering the queries to $\mathcal{S}_1$ and $\mathcal{S}'_2$ in a consistent way. More precisely, $\mathcal{P}^\star$ works as follows. It guesses uniformly at random an index $j \in [Q]$ and replies to queries to $\mathcal{S}_1$ and $\mathcal{S}'_2$ in the following way:

1. Answer query (x_i, α_i) to $\mathcal{S}_1$:
   (a) Query $1 \leq i \leq j - 1$: Returns $H(x_i, \alpha_i)$ if it is already defined; otherwise it samples a random value β_i and sets $H(x_i, \alpha_i) := \beta_i$.
   (b) Query j: Runs the protocol with the honest verifier $\mathcal{V}$ for statement x_j, using as a commitment the value α_j. Obtains challenge β_j from $\mathcal{V}$ and program the oracle as $H(x_j, \alpha_j) := \beta_j$. The answer to $\mathcal{A}$'s query is β_j.
   (c) Query $j + 1 \leq i \leq Q$: Proceed as in Step 1a.
2. Answer query x to $\mathcal{S}'_2$: Run the HVZK simulator of the interactive protocol on input x to obtain an accepting proof (α, β, γ), and program the oracle H in such a way that $H(x, \alpha) := \beta$. If the NIZK simulator returns failure, which happens when $H(x, \alpha)$ is already defined, output failure and abort, otherwise output (α, γ).
3. Answer $\mathcal{V}$'s challenge: Let $x^\star, (\alpha^\star, \gamma^\star)$ be the instance and the proof output by $\mathcal{A}$. Return $\gamma^\star$ to $\mathcal{V}$ as the response to challenge β_j in step 1b.

We need to estimate the probability that $\mathcal{P}^\star$ succeeds in breaking the soundness of the interactive scheme $(\mathcal{P}, \mathcal{V})$ in terms of the probability that $\mathcal{A}$ outputs an accepting proof $(\alpha^\star, \gamma^\star)$ for a false statement $x^\star$. Suppose that $(x^\star, \alpha^\star)$ has been asked to the random oracle as the $j^\star$-th query and we have $j = j^\star$, i.e., $\mathcal{P}^\star$ guesses the correct index for which $\mathcal{A}$ outputs an accepting proof for a false statement $x^\star$. In such a case, $\mathcal{P}^\star$ breaks the soundness of $(\mathcal{P}, \mathcal{V})$. Hence, we get:

$$\text{Prob}[\mathcal{P}^\star \text{ wins}] = \text{Prob}[\mathcal{A} \text{ wins} \wedge j = j^\star \wedge \overline{\text{proof}}]$$
$$= \text{Prob}[\mathcal{A} \text{ wins} \wedge \overline{\text{proof}}] \cdot \text{Prob}[j = j^\star],$$

where the second equality comes from the fact that $\mathcal{P}^\star$ guesses $j^\star$ correctly indipendently of the event that $\mathcal{A}$ is successful and $\overline{\text{proof}}$ happens. Since the index j is chosen at random in $[Q]$, we have $\text{Prob}[\mathcal{P}^\star \text{ wins}] = \frac{1}{Q} \cdot \text{Prob}[\mathcal{A} \text{ wins} \wedge \overline{\text{proof}}]$. Whenever $\mathcal{P}^\star$ wins, it breaks the soundness of the interactive scheme: by hypothesis, this happens only with negligible probability. Therefore:

$$\text{Prob}[\mathcal{A} \text{ wins} \wedge \overline{\text{proof}}] = Q \cdot \text{Prob}[\mathcal{P}^\star \text{ wins}] \leq negl(k).$$

Now we can bound the probability that $\mathcal{A}$ succeeds. As we assume, $\mathcal{A}$ breaks the simulation soundness of the scheme with non-negligible probability ϵ:

$$\text{Prob}[\mathcal{A} \text{ wins}] \leq \text{Prob}[\mathcal{A} \text{ wins} \wedge \text{proof}] + \text{Prob}[\mathcal{A} \text{ wins} \wedge \overline{\text{proof}}] \leq negl(k),$$

thus $\epsilon \leq negl(k)$, that is a contradiction.   □

On the quasi-unique responses condition. We remark that assuming $(\mathcal{P}, \mathcal{V})$ has quasi-unique responses is not an artifact of the proof. In fact, without this property, proofs would be malleable and breaking the simulation soundness would be an easy task. Consider a FS-NIZK proof system for which responses are not quasi-unique. An efficient adversary $\mathcal{A}$ can always query the simulator on input a false statement $x^\star$, obtaining a simulated proof $\mathcal{S}'_2(x^\star) \to \pi^\star = (\alpha^\star, \beta^\star, \gamma^\star)$. Given $\pi^\star$, $\mathcal{A}$ might be able to find, with non-negligible probability, a new response $\gamma^{\star\star} \neq \gamma^\star$ such that $(\alpha^\star, \beta^\star, \gamma^{\star\star})$ is also accepting. Hence, the scheme cannot be simulation sound.

4.2   Weak Simulation Extractability

The argument Fouque and Pointcheval use in [23] to show that the proof system they consider is simulation sound is roughly as follows. Assume there exists an adversary $\mathcal{A}$ which outputs a pair $(x^\star, \pi^\star)$ breaking the simulation soundness, as in the experiment of Definition 5. Then, one can invoke a suitable version of the forking lemma to show that it is possible to "extract" a witness $w^\star$ for $x^\star$ from such an adversary, contradicting the fact that $x^\star$ is false. The reduction simulates the list $\mathcal{T}$ for $\mathcal{A}$ in the simulation soundness experiment, in particular one needs to fake accepting proofs for (adaptively chosen and potentially false) theorems selected by the attacker. In order to do so, Fouque and Pointcheval (implicitly) rely on the SS-HVZK property. The next theorem is a generalization of the above strategy which does not rely on the SS-HVZK property and indeed applies to arbitrary languages. Moreover, we are able to prove a stronger statement, namely that Fiat-Shamir proofs satisfy weak simulation extractability (and not only simulation soundness). For simplicity the following theorem assumes (perfect) unique responses, but could be generalized using the same reduction as for Theorem 2.

Theorem 3 (Weak simulation extractability of the Fiat-Shamir transform). *Let $\Sigma = (\mathcal{P}, \mathcal{V})$ be a non-trivial Σ-protocol with unique responses for a language $\mathcal{L} \in$ NP. In the random oracle model, the NIZK proof system $\Sigma_{FS} = (\mathcal{P}^H, \mathcal{V}^H)$ resulting by applying the Fiat-Shamir transform to Σ is weakly simulation extractable with extraction error $\nu = \frac{Q}{h}$ for the canonical simulator $\mathcal{S}$. Here, Q is the number of random oracle queries and h is the number of elements in the range of H. Furthermore, the extractor $\mathcal{E}_\mathcal{A}$ needs to run $\mathcal{A}^{\mathcal{S}_1(\cdot), \mathcal{S}_2'(\cdot)}$ twice, where $\mathcal{A}$ and $\mathcal{E}_\mathcal{A}$ are both defined in Definition 6.*

Proof. Let $\mathcal{S}$ be the canonical zero-knowledge simulator described in the proof of Theorem 2. Denote with $(x^\star, \alpha^\star, \gamma^\star)$ the pair statement/proof returned by

$\mathcal{A}^{\mathcal{S}_1(\cdot),\mathcal{S}_2'(\cdot)}$; we describe an extractor $\mathcal{E}_\mathcal{A}$, able to compute a witness $w^\star$ by rewinding $\mathcal{A}$ once.

We want to exploit the general forking lemma1. In order to do so, we define program $P(1^k, h_1, \ldots, h_Q; \rho_P)$ as follows: P virtually splits ρ_P into two random tapes ρ and ρ_S (e.g. by using even bits for ρ and odd bits for ρ_S) and runs internally $\mathcal{A}^{\mathcal{S}_1(\cdot),\mathcal{S}_2'(\cdot)}$ with randomness ρ. P uses values $(h_1, \ldots, h_Q)$ to simulate fresh answers of $\mathcal{S}_1$, and ρ_S to simulate answers of $\mathcal{S}_2$. If $\mathcal{A}^{\mathcal{S}_1(\cdot),\mathcal{S}_2'(\cdot)}$ outputs $(x^\star, (\alpha^\star, \gamma^\star))$, P checks that it is a valid proof and not in $\mathcal{T}$ (otherwise it returns $(0, \perp)$). Then, because of the unique response property, $(x^\star, \alpha^\star)$ must correspond to some fresh query to $\mathcal{S}_1$ and P outputs $(J, (x^\star, \alpha^\star, \gamma^\star))$, where $J > 0$ is the index corresponding to the random oracle query $(x^\star, \alpha^\star)$. We say that P is successful whenever $J \geq 1$, and we denote with acc the corresponding probability. Given program P, we consider two related runs of P with the same random tape but different hash values, as specified by the forking algorithm F_P of Lemma 1. Denote with $(I, (x^\star, \alpha^\star, \gamma^\star)) \leftarrow P(1^k, h_1, \ldots, h_Q; \rho)$ and $(I', (x^{\star\star}, \alpha^{\star\star}, \gamma^{\star\star})) \leftarrow P(1^k, h_1, \ldots, h_{I-1}, h_I', \ldots, h_Q'; \rho)$ the two outputs of $\mathcal{A}$ in these runs. By the forking lemma we know that with probability ext $\geq$ acc(acc$/Q - 1/h$) the forking algorithm will return indexes I, I' such that $I = I'$, $I \geq 1$ and $h_I \neq h_I'$.

Notice that since F_P's forgeries are relative to the *same* random oracle query $I = I'$, we must have $x^\star = x^{\star\star}$ and $\alpha^\star = \alpha^{\star\star}$; on the other hand we have $h_I \neq h_I'$. We are thus in a position to invoke the special extractor $\mathcal{E}$ for the underlying proof system, yielding a valid witness $w^\star \leftarrow \mathcal{E}(\alpha^\star, h_I, \gamma^\star, h_I', \gamma^{\star\star}, x^\star)$ such that $(x^\star, w^\star) \in \mathcal{R}_\mathcal{L}$.

Assume now that acc $\geq \nu$. By applying the general forking lemma1 we obtain that ext $\geq$ acc$^2/Q -$ acc$/h$. Since Q is polynomial while h is exponentially large in the security parameter, for sure $\frac{Q}{h} < 1$ (in particular, it is negligible in k). As $\nu := \frac{Q}{h}$, we have:

$$\frac{\text{acc}^2}{Q} - \frac{\text{acc}}{h} = \frac{1}{Q}(\text{acc}^2 - \text{acc} \cdot \nu).$$

Now, since acc $\geq \nu$, we have acc $\cdot \nu \geq \nu^2$, that is $\nu^2 -$ acc $\cdot \nu \leq 0$. Hence,

$$\frac{1}{Q}(\text{acc}^2 - \text{acc} \cdot \nu) \geq \frac{1}{Q}(\text{acc}^2 - 2\text{acc} \cdot \nu + \nu^2) = \frac{1}{Q}(\text{acc} - \nu)^2.$$

The previous inequality matches the definition of weak extractability with values $p = Q$ and $d = 2$. $\qquad\qquad\square$

5   Applications

In the literature there is a large number of applications for simulation-sound or extractable NIZKs. One of the first request for simulation soundness comes from the setting of public key encryption, for the design of encryption schemes with *chosen-ciphertext security* using the Naor-Yung (NY) paradigm [34]. At a high level, the NY works as follows: given two key pairs (sk, pk) and (sk', pk')

for CPA-secure encryption schemes Π and Π' respectively, a ciphertext consists of two encryptions c, c' of the same message m, under different keys pk, pk', and a NIZK proof π that both c and c' encrypt m. In order to achieve security against *adaptive* chosen-ciphertext attacks (CCA2), the underlying NIZK must be simulation-sound. While achieving CCA2 security is probably one of the most prominent application of simulation soundness, simulation-sound or extractable proofs have been used, e.g., to build also leakage-resilient signatures or KDM secure encryption. In this section, we review some important applications of such proof systems and show how our result provides more efficient constructions in the ROM or generalizes earlier results that use the Fiat-Shamir transform.

5.1 Leakage Resilience

Simulation-sound and simulation-extractable NIZK proofs have been very useful in constructing leakage-resilient encryption and signature schemes [18,32,33]. Here, we consider these works and show that our result immediately yields efficient leakage-resilient schemes in the random oracle model.

Leakage-resilient signatures. A signature scheme is leakage resilient if it is hard to forge a signature even given (bounded) leakage from the signing key. Obviously, this requires that the amount of leakage given to the adversary has to be smaller than the length of the secret key, as otherwise the leakage may just reveal such a key, trivially breaking the security of the signature scheme.

We instantiate the generic construction of leakage-resilient signatures based on leakage-resilient hard relations and simulation-extractable NIZKs of [18] using the Fiat-Shamir transform. Let $\mathcal{R}$ be a λ-leakage-resilient hard relation with sampling algorithm $\mathsf{Gen}_{\mathcal{R}}$ (for detailed definitions, see the full version [20]). Let $(\mathcal{P}^H, \mathcal{V}^H)$ be a NIZK argument[3] for relation $\mathcal{R}'$ defined by $\mathcal{R}'((pk, m), sk) \Leftrightarrow \mathcal{R}(pk, sk)$. Consider the following signature scheme:

$\mathsf{KeyGen}(1^k)$: Calls $(pk, sk) \leftarrow \mathsf{Gen}_{\mathcal{R}}(1^k)$ and returns the same output.
$\mathsf{Sign}(sk, m)$: Outputs $\sigma \leftarrow \mathcal{P}^H((pk, m), sk)$.[4]
$\mathsf{Vrf}(pk, m, \sigma)$: Verifies the signature by invoking $\mathcal{V}^H((pk, m), \sigma)$.

Notice that $\sigma \leftarrow \mathcal{P}^H((pk, m), sk)$ is a NIZK proof for the hard relation obtained by applying the Fiat-Shamir transform.

We chose to state the theorem below using an argument system as this is the minimal requirement under which leakage resilience of the scheme can be proven. Since our FS-based protocols are weakly simulation-extractable NIZK *proof* systems, they automatically satisfy the hypothesis of Theorem 4.

Theorem 4. *If $\mathcal{R}$ is a 2λ-leakage-resilient hard relation and $(\mathcal{P}^H, \mathcal{V}^H)$ is a weakly simulation-extractable NIZK argument with negligible extraction error*

---

[3] As opposed to a proof system where soundness needs to hold unconditionally, in an *argument* system it is sufficient that soundness holds with respect to a computationally bounded adversary.

[4] Note that m is part of the instance being proven.

for relation $\mathcal{R}'((pk, m), sk) \Leftrightarrow \mathcal{R}(pk, sk)$, *then the above scheme is a λ-leakage-resilient signature scheme in the random oracle model.*

The proof of the theorem from above follows the one of Theorem 4.3 in [18]. A couple of subtleties arise, though. The main idea of the proof is to build a reduction from an adversary $\mathcal{A}$ breaking λ-leakage resilience of the signature scheme to an adversary $\mathcal{B}$ breaking the hardness of the 2λ-leakage-resilient hard relation $\mathcal{R}$. Roughly speaking, in the reduction $\mathcal{B}$ is given some instance pk and simulates the signing queries of $\mathcal{A}$ by using the zero-knowledge simulator of the NIZK, and the leakage queries by using the leakage oracle for the relation $\mathcal{R}$. At some point $\mathcal{A}$ outputs a forgery $\sigma^\star$ and $\mathcal{B}$ invokes the extractor of Theorem 3 to get $sk^\star \leftarrow \mathcal{E}_\mathcal{A}(pk, \sigma^\star)$. The first issue is that we are only guaranteed *weak* simulation-extractability, whereas the proof of [18] relies on full simulation-extractability. [5] However, this is not a problem because we just need to show that $\mathcal{B}$ outputs a valid witness with non-negligible probability. A second issue involves the extractor of Theorem 3, which needs to rewind $\mathcal{A}$ once and, thus, to simulate twice its environment (including the leakage queries). This causes the loss of a factor 2 in the total amount of tolerated leakage. We refer the reader to the full version [20] for the details.

We emphasize that the leakage-resilient signature scheme of Alwen et al. [4], obtained by applying the Fiat-Shamir transform to the Okamoto identification scheme, follows essentially the above paradigm. Here, one may view the public and secret keys of the Okamoto ID scheme form an instance of a leakage-resilient hard relation, while the NIZK proof corresponds to the Fiat-Shamir transform applied to the Okamoto identification protocol.

Naor-Yung with leakage. The definition of IND-CPA and IND-CCA security of an encryption scheme can be extended to the leakage setting by giving the adversary access to a leakage oracle. Naor and Segev [33] show that the Naor-Yung paradigm instantiated with a simulation-sound NIZK allows to leverage CPA-security to CCA-security even in the presence of leakage. In other words, if Π is CPA-secure against λ-key-leakage attacks, the encryption scheme obtained by applying the Naor-Yung paradigm to (Π, Π), using a simulation-sound NIZK, is CCA2-secure against λ-key-leakage attacks. In the full version [20] we revisit their proof in the ROM, dealing with the issue that the leakage queries can potentially depend on H. We stress that for the proof only *simulation soundness* is needed (i.e., our result from Theorem 2) and not weak simulation extractability.

In what follows, we propose a concrete instantiation of the result above, relying on the BHHO encryption scheme from [12]. Let $\mathbb{G}$ be a group of prime-order q. For randomly selected generators $g_1, \ldots, g_\ell \xleftarrow{\$} \mathbb{G}$, the public key is a tuple $pk = (g_1, \ldots, g_\ell, h)$, where $h = \prod_{i=1}^\ell g_i^{z_i}$ for a secret key $sk = (z_1, \ldots, z_\ell) \in \mathbb{Z}_q^\ell$. To encrypt a message $m \in \mathbb{G}$, choose a random $r \xleftarrow{\$} \mathbb{Z}_q$ and output $c = (c_1, \ldots, c_{\ell+1}) = (g_1^r, \ldots, g_\ell^r, m \cdot h^r)$. The message m can be recovered by computing $m = c_{\ell+1} \cdot (\prod_{i=1}^\ell c_i^{z_i})^{-1}$.

---

[5] Actually, they rely on a weaker property called *true* simulation-extractability [18].

Assuming that the DDH problem is hard in $\mathbb{G}$, Naor and Segev [33] showed that the BHHO encryption scheme is CPA-secure against λ-key-leakage attacks for any $\ell = 2 + \frac{\lambda + \omega(\log k)}{\log q}$, where k is the security parameter. Applying the Naor-Yung paradigm, consider the language:

$$\mathcal{L} = \{(c, pk, c', pk') : \exists r, r' \in \mathbb{Z}_q, m \in \mathbb{G} \text{ s.t.}$$

$$c = (g_1^r, \ldots, g_\ell^r, h^r \cdot m), \ c' = (g_1^{r'}, \ldots, g_\ell^{r'}, h'^{r'} \cdot m)\},$$

where $c = (c_1, \ldots, c_{\ell+1})$ and $c' = (c_1', \ldots, c_{\ell+1}')$ are BHHO encryptions with randomness r and r', using public keys $pk = (g_1, \ldots, g_\ell, h)$ and $pk' = (g_1, \ldots, g_\ell, h')$ respectively. The pair $w = (r, r')$ is a witness for a string $x = (c, pk, c', pk') \in \mathcal{L}$. Consider the following interactive protocol $\Sigma = (\mathcal{P}, \mathcal{V})$ for the above language:

1. $\mathcal{P}$ chooses s, s' at random from $\mathbb{Z}_q$ and computes the commitment:

$$\vec{\alpha} = ((\alpha_1, \ldots, \alpha_\ell), (\alpha_1', \ldots, \alpha_\ell'), \alpha'') = ((g_1^s, \ldots, g_\ell^s), (g_1^{s'}, \ldots, g_\ell^{s'}), h^s \cdot (h')^{s'}).$$

2. The verifier $\mathcal{V}$ chooses a random challenge $\beta \xleftarrow{\$} \mathbb{Z}_q$.
3. The prover computes the response $\vec{\gamma} = (\gamma, \gamma') = (s - \beta \cdot r, s' + \beta \cdot r')$.
4. Given a proof $\pi = (\vec{\alpha}, \beta, \vec{\gamma})$, the verifier $\mathcal{V}$ checks that:

$$(\alpha_1, \ldots, \alpha_\ell) = (g_1^\gamma \cdot c_1^\beta, \ldots, g_\ell^\gamma \cdot c_\ell^\beta)$$

$$(\alpha_1', \ldots, \alpha_\ell') = (g_1^{\gamma'} \cdot (c_1')^{-\beta}, \ldots, g_\ell^{\gamma'} \cdot (c_\ell')^{-\beta})$$

$$\alpha'' = h^\gamma \cdot (h')^{\gamma'} \cdot (c_{\ell+1} \cdot (c_{\ell+1}')^{-1})^\beta.$$

In the full version [20] we prove that the above protocol is a Σ-protocol for the language $\mathcal{L}$. With the Naor-Yung paradigm applied to the BHHO encryption scheme we get a ciphertext (c, c', π) consisting of $4\ell + 3$ elements in $\mathbb{G}$ plus 2 elements in $\mathbb{Z}_q$. Moreover, the fact that the BHHO encryption scheme is CPA-secure against key leakage together with the result of Naor-Segev, show that the above instantiation is CCA-secure against key-leakage attacks.

Corollary 1. *Let k be a security parameter. Assuming that the DDH problem is hard in $\mathbb{G}$, the Naor-Yung paradigm applied to the BHHO encryption scheme yields an encryption scheme that is CCA-secure against λ-key-leakage attacks in the random oracle model for $\lambda = \ell \log q (1 - \frac{2}{\ell} - \frac{\omega(\log k)}{\ell \log q}) = L(1 - o(1))$, where L is the length of the secret key. An encryption consists of $4\ell + 3$ elements in $\mathbb{G}$ plus 2 elements in $\mathbb{Z}_q$.*

5.2 Key-Dependent Message Security

Key-dependent message (KDM) security of a public-key encryption scheme requires that the scheme remains secure even against attackers allowed to see encryptions of the value $f(sk)$, where $f \in \mathcal{F}$ for some class of functions $\mathcal{F}$.

Camenisch, Chandran and Shoup [13] show that a variation of the Naor-Yung paradigm instantiated with a simulation-sound NIZK can still leverage CPA-security to CCA-security, even in the context of KDM security. We revisit their proof in the random oracle model in the full version [20]. Also in this case, only *simulation soundness* is needed for the proof.

Roughly, for some function family $\mathcal{F}$, if Π is KDM$[\mathcal{F}]$-CPA secure and Π' is CPA-secure, the scheme Π'' obtained by applying the Naor-Yung paradigm to (Π, Π') — i.e., an encryption of $m \in \mathcal{M}$ is a tuple $c'' = (c, c', \pi)$ where c encrypts m under Π, c' encrypts m under Π' and π is a simulation-sound NIZK proof that c and c' encrypt the same message — is KDM$[\mathcal{F}]$-CCA secure.

Let $sk_i[j]$ denote the j-th bit of sk_i. The BHHO encryption scheme was the first KDM-CPA secure encryption scheme, with respect to the class of all projection functions $\mathcal{F}_\downarrow = \mathcal{F}_{\text{read}} \cup \mathcal{F}_{\text{flip}}$, where

$$\mathcal{F}_{\text{read}} = \left\{ f_{i,j} : \ \vec{sk} \to sk_i[j] \right\}_{i,j} \quad \text{and} \quad \mathcal{F}_{\text{flip}} = \left\{ f_{i,j} : \ \vec{sk} \to 1 - sk_i[j] \right\}_{i,j}.$$

More generally, when the message space is a linear space over $\mathbb{Z}_q$, we define the function class $\mathcal{PJ}(\mathcal{F}_\downarrow)$ as the class of all affine combinations of elements in $\mathcal{F}_\downarrow$.

Now we can instantiate the general transform of [13] as follows. We choose Π to be BHHO, Π' to be ElGamal (say with $pk' = h' = g_1^{z_1}$) and we build a Σ-protocol Σ' for the Naor-Yung language relative to Π and Π'. Protocol Σ' can be easily derived from protocol Σ of the last section, by just compressing the commitment as in $\vec{\alpha} = ((g_1^s, \ldots, g_\ell^s), g_1^{s'}, h^s \cdot (h')^{s'})$ (and simplifying the verification procedure accordingly). Hence, Theorem 2 yields the following result.

Corollary 2. *Assuming that the DDH problem is hard in* $\mathbb{G}$*, the Naor-Yung paradigm instantiated with BHHO and ElGamal encryption schemes yields a* KDM$[\mathcal{PJ}(\mathcal{F}_\downarrow)]$*-CCA secure encryption scheme in the random oracle model. An encryption consists of* $\ell + 3$ *elements in* $\mathbb{G}$ *plus 3 elements in* $\mathbb{Z}_q$*.*

Beyond Naor-Yung. Another paradigm that yields chosen-ciphertext security from NIZKs, based on *proving knowledge of the plaintext*, was suggested by Rackoff and Simon [36]. Such a construction is somewhat more natural and more efficient than the twin-encryption paradigm: a message m is encrypted (only once) under a CPA-secure encryption scheme, and a NIZK proof of knowledge of the plaintext is attached to the ciphertext. However, (to the best of our knowledge) truly efficient constructions for sufficiently strong NIZK proofs of knowledge are not available even using random oracles. One can hope that using the weaker from of extractability afforded by the Fiat-Shamir transform one could at least obtain NM-CPA secure encyption, and this is indeed what is aimed at in the ongoing work of [8].

Acknowledgments. We thank Marc Fischlin and Ivan Damgård for the useful feedbacks provided on earlier versions of the paper.

References

1. Abdalla, M., An, J.H., Bellare, M., Namprempre, C.: From Identification to Signatures via the Fiat-Shamir Transform: Minimizing Assumptions for Security and Forward-Security. In: Knudsen, L.R. (ed.) EUROCRYPT 2002. LNCS, vol. 2332, pp. 418–433. Springer, Heidelberg (2002)

2. Abdalla, M., An, J.H., Bellare, M., Namprempre, C.: From identification to signatures via the Fiat-Shamir transform: Necessary and sufficient conditions for security and forward-security. IEEE Transactions on Information Theory 54(8), 3631–3646 (2008)
3. Abe, M., Fuchsbauer, G., Groth, J., Haralambiev, K., Ohkubo, M.: Structure-Preserving Signatures and Commitments to Group Elements. In: Rabin, T. (ed.) CRYPTO 2010. LNCS, vol. 6223, pp. 209–236. Springer, Heidelberg (2010)
4. Alwen, J., Dodis, Y., Wichs, D.: Leakage-Resilient Public-Key Cryptography in the Bounded-Retrieval Model. In: Halevi, S. (ed.) CRYPTO 2009. LNCS, vol. 5677, pp. 36–54. Springer, Heidelberg (2009)
5. Bellare, M., Goldreich, O.: On Defining Proofs of Knowledge. In: Brickell, E.F. (ed.) CRYPTO 1992. LNCS, vol. 740, pp. 390–420. Springer, Heidelberg (1993)
6. Bellare, M., Neven, G.: Multi-signatures in the plain public-key model and a general forking lemma. In: ACM Conference on Computer and Communications Security, pp. 390–399 (2006)
7. Bellare, M., Rogaway, P.: Random oracles are practical: A paradigm for designing efficient protocols. In: ACM Conference on Computer and Communications Security, pp. 62–73 (1993)
8. Bernhard, D., Pereira, O., Warinschi, B.: On necessary and sufficient conditions for private ballot submission. Cryptology ePrint Archive, Report 2012/236 (2012), http://eprint.iacr.org/
9. Blum, M., Feldman, P., Micali, S.: Non-interactive zero-knowledge and its applications (extended abstract). In: STOC, pp. 103–112 (1988)
10. Blum, M., De Santis, A., Micali, S., Persiano, G.: Noninteractive zero-knowledge. SIAM J. Comput. 20(6), 1084–1118 (1991)
11. Boneh, D., Boyen, X., Shacham, H.: Short Group Signatures. In: Franklin, M. (ed.) CRYPTO 2004. LNCS, vol. 3152, pp. 41–55. Springer, Heidelberg (2004)
12. Boneh, D., Halevi, S., Hamburg, M., Ostrovsky, R.: Circular-Secure Encryption from Decision Diffie-Hellman. In: Wagner, D. (ed.) CRYPTO 2008. LNCS, vol. 5157, pp. 108–125. Springer, Heidelberg (2008)
13. Camenisch, J., Chandran, N., Shoup, V.: A Public Key Encryption Scheme Secure against Key Dependent Chosen Plaintext and Adaptive Chosen Ciphertext Attacks. In: Joux, A. (ed.) EUROCRYPT 2009. LNCS, vol. 5479, pp. 351–368. Springer, Heidelberg (2009)
14. Camenisch, J., Lysyanskaya, A.: An Efficient System for Non-transferable Anonymous Credentials with Optional Anonymity Revocation. In: Pfitzmann, B. (ed.) EUROCRYPT 2001. LNCS, vol. 2045, pp. 93–118. Springer, Heidelberg (2001)
15. Chase, M., Lysyanskaya, A.: On Signatures of Knowledge. In: Dwork, C. (ed.) CRYPTO 2006. LNCS, vol. 4117, pp. 78–96. Springer, Heidelberg (2006)
16. Cramer, R., Damgård, I., Schoenmakers, B.: Proof of Partial Knowledge and Simplified Design of Witness Hiding Protocols. In: Desmedt, Y.G. (ed.) CRYPTO 1994. LNCS, vol. 839, pp. 174–187. Springer, Heidelberg (1994)
17. Damgård, I.: On Σ-protocols (2002), http://www.daimi.au.dk/~ivan/Sigma.ps
18. Dodis, Y., Haralambiev, K., López-Alt, A., Wichs, D.: Efficient Public-Key Cryptography in the Presence of Key Leakage. In: Abe, M. (ed.) ASIACRYPT 2010. LNCS, vol. 6477, pp. 613–631. Springer, Heidelberg (2010)
19. Dolev, D., Dwork, C., Naor, M.: Nonmalleable cryptography. SIAM J. Comput. 30(2), 391–437 (2000)
20. Faust, S., Kohlweiss, M., Marson, G.A., Venturi, D.: On the non-malleability of the Fiat-Shamir transform. Cryptology ePrint Archive (2012), http://eprint.iacr.org/

21. Fiat, A., Shamir, A.: How to Prove Yourself: Practical Solutions to Identification and Signature Problems. In: Odlyzko, A.M. (ed.) CRYPTO 1986. LNCS, vol. 263, pp. 186–194. Springer, Heidelberg (1987)
22. Fischlin, M.: Communication-Efficient Non-interactive Proofs of Knowledge with Online Extractors. In: Shoup, V. (ed.) CRYPTO 2005. LNCS, vol. 3621, pp. 152–168. Springer, Heidelberg (2005)
23. Fouque, P.-A., Pointcheval, D.: Threshold Cryptosystems Secure against Chosen-Ciphertext Attacks. In: Boyd, C. (ed.) ASIACRYPT 2001. LNCS, vol. 2248, pp. 351–368. Springer, Heidelberg (2001)
24. Garay, J.A., MacKenzie, P.D., Yang, K.: Strengthening zero-knowledge protocols using signatures. J. Cryptology 19(2), 169–209 (2006)
25. Goldreich, O., Oren, Y.: Definitions and properties of zero-knowledge proof systems. J. Cryptology 7(1), 1–32 (1994)
26. Goldwasser, S., Micali, S., Rackoff, C.: The knowledge complexity of interactive proof systems. SIAM J. Comput. 18(1), 186–208 (1989)
27. Goldwasser, S., Sipser, M.: Private coins versus public coins in interactive proof systems. In: STOC, pp. 59–68 (1986)
28. Groth, J.: Simulation-Sound NIZK Proofs for a Practical Language and Constant Size Group Signatures. In: Lai, X., Chen, K. (eds.) ASIACRYPT 2006. LNCS, vol. 4284, pp. 444–459. Springer, Heidelberg (2006)
29. Groth, J., Sahai, A.: Efficient Non-interactive Proof Systems for Bilinear Groups. In: Smart, N.P. (ed.) EUROCRYPT 2008. LNCS, vol. 4965, pp. 415–432. Springer, Heidelberg (2008)
30. Halevi, S., Micali, S.: More on proofs of knowledge. Cryptology ePrint Archive, Report 1998/015 (1998), http://eprint.iacr.org/
31. Jain, A., Pandey, O.: Non-malleable zero knowledge: Black-box constructions and definitional relationships. Cryptology ePrint Archive, Report 2011/513 (2011), http://eprint.iacr.org/
32. Katz, J., Vaikuntanathan, V.: Signature Schemes with Bounded Leakage Resilience. In: Matsui, M. (ed.) ASIACRYPT 2009. LNCS, vol. 5912, pp. 703–720. Springer, Heidelberg (2009)
33. Naor, M., Segev, G.: Public-Key Cryptosystems Resilient to Key Leakage. In: Halevi, S. (ed.) CRYPTO 2009. LNCS, vol. 5677, pp. 18–35. Springer, Heidelberg (2009)
34. Naor, M., Yung, M.: Public-key cryptosystems provably secure against chosen ciphertext attacks. In: STOC, pp. 427–437 (1990)
35. Pointcheval, D., Stern, J.: Security arguments for digital signatures and blind signatures. J. Cryptology 13(3), 361–396 (2000)
36. Rackoff, C., Simon, D.R.: Non-interactive Zero-Knowledge Proof of Knowledge and Chosen Ciphertext Attack. In: Feigenbaum, J. (ed.) CRYPTO 1991. LNCS, vol. 576, pp. 433–444. Springer, Heidelberg (1992)
37. Sahai, A.: Non-malleable non-interactive zero knowledge and adaptive chosen-ciphertext security. In: FOCS, pp. 543–553 (1999)
38. De Santis, A., Di Crescenzo, G., Ostrovsky, R., Persiano, G., Sahai, A.: Robust Non-interactive Zero Knowledge. In: Kilian, J. (ed.) CRYPTO 2001. LNCS, vol. 2139, pp. 566–598. Springer, Heidelberg (2001)
39. Unruh, D.: Quantum proofs of knowledge. To appear in CRYPTO (2012)
40. Wee, H.: Zero Knowledge in the Random Oracle Model, Revisited. In: Matsui, M. (ed.) ASIACRYPT 2009. LNCS, vol. 5912, pp. 417–434. Springer, Heidelberg (2009)

Another Look at Symmetric Incoherent Optimal Eavesdropping against BB84

Arpita Maitra[1,*] and Goutam Paul[2,**]

[1] Applied Statistics Unit, Indian Statistical Institute,
Kolkata 700 108, India
arpita76b@rediffmail.com
[2] Department of Computer Science and Engineering,
Jadavpur University, Kolkata 700 032, India
goutam.paul@ieee.org

Abstract. The BB84 protocol is used by Alice (the sender) and Bob (the receiver) to settle on a secret classical bit-string by communicating qubits over an insecure quantum channel where Eve (the Eavesdropper) can have access. In this paper, we revisit a well known eavesdropping technique against BB84. We claim that there exist certain gaps in understanding the existing eavesdropping strategy in terms of cryptanalytic view and we try to bridge those gaps in this paper.

First we refer to the result where it is shown that in the six-state variant of the BB84 protocol (Bruß, Phys. Rev. Lett., 1998), the mutual information between Alice (the sender) and Eve (the eavesdropper) is higher when two-bit probe is used compared to the one-bit probe and hence the two-bit probe provides a stronger eavesdropping strategy. However, from cryptanalytic point of view, we show that Eve has the same success probability in guessing the bit transmitted by Alice in both the cases of the two-bit and the one-bit probe. Thus, we point out that having higher mutual information may not directly lead to obtaining higher probability in guessing the key bit.

It is also explained in the work of Bruß that the six-state variant of the BB84 protocol is more secure than the traditional four-state BB84. We look into this point in more detail and identify that this advantage is only achieved at the expense of communicating more qubits in the six-state protocol. In fact, we present different scenarios, where given the same number of qubits communicated, the security comparison of the four and six-state protocols is evaluated carefully.

Keywords: Advantage, BB84, Key Distribution, Optimal Eavesdropping, Quantum Cryptography.

* The work of the first author was supported by the WOS-A fellowship of the Department of Science and Technology, Government of India.
** This work was done in part while the second author was visiting RWTH Aachen, Germany as an Alexander von Humboldt Fellow.

S. Galbraith and M. Nandi (Eds.): INDOCRYPT 2012, LNCS 7668, pp. 80–99, 2012.

1   Introduction

Establishing a common secret key between two parties at a distance is a pre-requisite for executing a symmetric key cryptographic protocol between them. The seminal paper by Diffie and Hellman [9] presents a nice idea in this direction using the Discrete Logarithm problem. However, the pioneering work of Shor [17] showed that the key distribution [9] as well as the public key crypto-systems like RSA [16] and ECC [11] are not secure in the quantum computing model. On the other hand, there are lattice and coding theory based public key algorithms [4] that are believed to be secure in the quantum computing model and these are the main focus in the domain of post-quantum cryptography. However, these algorithms are quite complex and considerable works are going on for efficient implementation of such schemes on low end devices. In this regard, it is notable that provably secure quantum key distribution protocols exist and amongst them BB84 [1] is the first and the most cited one. It has not only been verified exper-imentally [3] in laboratory, but now-a-days some companies are manufacturing devices [15] to implement this protocol. In this scenario, it is important to study various eavesdropping models for these protocols and this is the motivation for our current work.

The famous BB84 protocol [1] relies on the conjugate bases $Z = \{|0\rangle, |1\rangle\}$ and $X = \{|+\rangle, |-\rangle\}$, where $|+\rangle = \frac{|0\rangle + |1\rangle}{\sqrt{2}}$ and $|-\rangle = \frac{|0\rangle - |1\rangle}{\sqrt{2}}$. Alice randomly selects one of the two orthogonal bases and encodes 0 and 1 respectively by a qubit prepared in one of the two states in each base. To be specific, Alice encodes 0 to $|0\rangle$ or $|+\rangle$, and 1 to $|1\rangle$ or $|-\rangle$, depending on the chosen basis Z or X respectively. Bob measures the qubits one by one, randomly selecting the basis from the same set of bases. After the measurement, Alice and Bob publicly announce the sequence of bases used by them and discard the bases that do not match. They identify the sequence of bits corresponding to the bases that match and the resulting bit string, followed by error correction and privacy amplification [2], becomes the common secret key.

Fuchs et al. (Phy. Rev. A, 1997) presented an optimal eavesdropping strategy on the four-state BB84 protocol. Later, Bruß (Phys. Rev. Lett., 1998) described the use of the basis $\left\{ \frac{|0\rangle + i|1\rangle}{\sqrt{2}}, \frac{|0\rangle - i|1\rangle}{\sqrt{2}} \right\}$ ($i = \sqrt{-1}$) along with the above two to show that the BB84 protocol with three conjugate bases (six-state protocol) provides improved security. Bruß had also shown that for the six-state protocol, the mutual information between Alice (the sender) and Eve (the eavesdropper) is higher when two-bit probe is used compared to the one-bit probe and hence provides a stronger eavesdropping strategy. In this paper, we revisit the problem towards a critical and concrete analysis in terms of Eve's success probability in guessing the qubits that Alice has sent.

The security of the BB84 protocol is based on the fact that if one wants to dis-tinguish two non-orthogonal quantum states, then obtaining any information is only possible at the expense of introducing disturbance in the state(s). There are several works in the literature, e.g., [6,7,10], that studied the relationship between "the amount of information obtained by Eve" and "the amount of disturbance created on the qubits that Bob receives from Alice". There are also several models for analysis of these problems. As example, Eve can work on each individual qubit as opposed to a set of qubits studied together. While the first one is called the *incoherent attack* [10], the second one is known as the *coherent attack* [7]. In this paper, we study the incoherent attack.

Another interesting issue in specifying the eavesdropping scenario is whether there will be equal error probability at Bob's end corresponding to different bases. If this is indeed equal, then we call it *symmetric* and that is what we concentrate on here. It creates certain constraint on Eve in terms of extracting information from the communicated qubits, as the disturbance created on the qubits that Bob receives should be equal for all the bases. That is, as far as Alice and Bob are concerned, the interference by Eve will produce a binary symmetric channel between them, with an error probability that we will denote by D. There is also another model where this is not equal and then we call the eavesdropping model as *asymmetric*. Different error rates for different bases would be a clear indication to Alice and Bob that an eavesdropper (Eve) is interfering in the communication line. One may refer to [7] for details on this and it has been commented in the same paper that given any asymmetric attack (coherent or incoherent), one can always get a symmetric attack that can match the results of the non-symmetric strategy.

In both [10,6], the security of BB84 is analyzed in terms of the mutual information between Alice and Eve. When measuring her probe, Eve has two choices. One option is that she measures both her qubits - this is referred as a *two-bit probe*. Alternatively, she can either measure only one of her two qubits [5,6] or may interact with one qubit at her disposal - both of these lead to identical results and therefore we refer any one of them as *one-bit probe*. In [6], it was claimed that the eavesdropping using the two-bit probe provides identical information to Eve using the one-bit probe in case of four-state protocol; however, for the six-state protocol, the two-bit probe leaks more information to Eve than the one-bit probe.

1.1 Organization of the Paper

In Section 2, we revisit the background material in detail. Sections 3 and 4 contain our main contributions. We re-examine the security in the light of Eve's *success probability* of guessing what was sent by Alice. In practice, Eve's goal is to determine the secret key bits that Alice sends to Bob. Eve's individual probes and hence individual guesses are independent. After measurement of the i-th probe, Eve makes a guess of the i-th secret key bit, i.e., she has to decide whether the i-th bit was 0 or 1. If her decided bit matches with what Alice has sent, then we call it a *success*, else it is an *error*. Eve's strategy would be to minimize the *error probability* in her guess, i.e., to maximize the *success probability*.

The mutual information between Alice and Eve gives a theoretical measure about the average information contained in the random variable associated with one of them about the random variable associated with the other. However, from the point of view of guessing the secret key established between Alice and Bob, Eve's success probability is a more practical parameter of cryptanalytic interest than the mutual information between Alice and Eve. The difference between the attacker's success probability and the probability of random guess (in this case, the probability of random guess is $\frac{1}{2}$) gives the attacker's *advantage*.

In Section 3, we present an analysis of the success probabilities of the four-state and the six-state protocols and show that there is no extra advantage of the two-bit probe over the one-bit probe in the six-state protocol. We show that these two probes do not differ in terms of success probability of Eve's guess about the bits sent by Alice, though the mutual information is different.

In Section 4, we propose a multi-round version of the BB84 protocol. Using this strategy, Alice and Bob can decrease Eve's advantage. Though the concept is similar to privacy amplification [2], we study the multi-round communication as part of the key distribution steps from a different viewpoint as follows. Both in the traditional 4-state BB84 protocol [1] and in the six-state one [6], Bob measures first and then Alice publishes the bases she used. Thus, while the six-state protocol is more secure than the four-state one, the disadvantage of the six-state scheme is that, on an average, only one-third of the qubits are kept and the rest two-third are discarded, which is worse than in the case of four-state scheme, where half of the received qubits are discarded. Hence, for a fair comparison between our multi-round versions of these two protocols, we must ensure that the same number of qubits communicated between Alice and Bob and in the end, the secret keys established are of the same bit length. In this setting, we critically evaluate the security parameters of both the protocols.

2   Review of Optimal Eavesdropping [6,10]

In this part, we study a generic version of BB84 with the bases $\{|0\rangle, |1\rangle\}$ and $\{|\psi\rangle, |\psi_\perp\rangle\}$, where $|\psi\rangle = a|0\rangle + b|1\rangle$ and $|\psi_\perp\rangle = b^*|0\rangle - a^*|1\rangle$. We characterize the values of a, b based on the eavesdropping model presented in [6,10]. We take each of a, b nonzero, as otherwise both the base will coincide (up to rotation). It is also trivial to see that $|a|^2 + |b|^2 = 1$ from normality condition. Under the symmetric incoherent optimal eavesdropping strategy [6,10], we get certain constraints on a, b as given in Theorem 1 in the next section. If one takes a state $|\psi\rangle$ such that the conditions on a, b as given in Theorem 1 are not admitted, then the symmetric attack of [10] needs to be modified properly.

Following [19], let $\{|\phi_i\rangle|i = 1, \ldots, N\}$ and $\{|\Phi_i\rangle|i = 1, \ldots, N\}$ be two orthonormal bases for an N dimensional Hilbert space. Such a pair of bases will be called *conjugate*, if and only if $|\langle\phi_i|\Phi_j\rangle|^2 = \frac{1}{N}$ for any i, j. Here $\langle\phi_i|\Phi_j\rangle$ is the inner product between $|\phi_i\rangle, |\Phi_j\rangle$. The case $N = 2$ is considered here. The analysis with non-conjugate bases has been presented by Phoenix [13] and it has been shown that the original proposal of [1] using the conjugate bases provides the optimal security.

In the absence of eavesdropper or any channel noise, Bob exactly knows the state that has been sent by Alice, if measured in the correct basis. However,

Eve's interaction does not allow that to happen. Consider the scenario when Alice sends one of two orthogonal states $|\psi\rangle$ and $|\psi_\perp\rangle$ to Bob and Eve has her own initial two-qubit state $|W\rangle$. Eve's interaction with the state being sent from Alice to Bob can be modeled as the action of a unitary operator U on three qubits as follows.

$$U(|\psi\rangle, |W\rangle) = \sqrt{F'}|\psi\rangle|E'_{00}\rangle + \sqrt{D'}|\psi_\perp\rangle|E'_{01}\rangle,$$
$$U(|\psi_\perp\rangle, |W\rangle) = \sqrt{D'}|\psi\rangle|E'_{10}\rangle + \sqrt{F'}|\psi_\perp\rangle|E'_{11}\rangle. \tag{1}$$

Thus, when Alice sends $|\psi\rangle$ (respectively $|\psi_\perp\rangle$), then Bob receives $|\psi\rangle$ (respectively $|\psi_\perp\rangle$) with probability F' (this is called *fidelity*) and receives $|\psi_\perp\rangle$ (respectively $|\psi\rangle$) with probability D' (this is called *disturbance*). One may note that $F' + D' = 1$.

After Bob measures the qubit he receives, Eve tries to obtain information about Bob's qubit. As example, if Eve obtains $|E'_{00}\rangle$ after measurement, she knows that Bob has received $|\psi\rangle$. The problem with Eve is that, if she tries to extract such information with certainty, then $|E'_{00}\rangle$, $|E'_{01}\rangle$, $|E'_{10}\rangle$ and $|E'_{11}\rangle$ need to be orthogonal and in that case the error probability D' at Bob's end will be very high and Bob will abort the protocol. Thus all of $|E'_{00}\rangle$, $|E'_{01}\rangle$, $|E'_{10}\rangle$, $|E'_{11}\rangle$ cannot be orthogonal and Eve has to decide the relationship among these 2-qubit states for optimal eavesdropping strategy.

Now let us consider the case for the $\{|0\rangle, |1\rangle\}$ basis.

$$U(|0\rangle, |W\rangle) = \sqrt{F}|0\rangle|E_{00}\rangle + \sqrt{D}|1\rangle|E_{01}\rangle,$$
$$U(|1\rangle, |W\rangle) = \sqrt{D}|0\rangle|E_{10}\rangle + \sqrt{F}|1\rangle|E_{11}\rangle. \tag{2}$$

The case for the generalized basis $\{|\psi\rangle, |\psi_\perp\rangle\}$ has already been expressed in (1). As we are studying the symmetric attack here, we consider that the fidelity F and the disturbance D are same for all the cases, i.e., $F = F'$ and $D = D'$.

We have considered $|\psi\rangle = a|0\rangle + b|1\rangle$ and $|\psi_\perp\rangle = b^*|0\rangle - a^*|1\rangle$, where a, b are nonzero. Hence, by linearity and then using Equation (2), we get

$$U(|\psi\rangle, |W\rangle) = aU(|0\rangle, |W\rangle) + bU(|1\rangle, |W\rangle)$$
$$= |0\rangle(a\sqrt{F}|E_{00}\rangle + b\sqrt{D}|E_{10}\rangle) + |1\rangle(a\sqrt{D}|E_{01}\rangle + b\sqrt{F}|E_{11}\rangle). \tag{3}$$

Substituting $|\psi\rangle = a|0\rangle + b|1\rangle$ and $|\psi_\perp\rangle = b^*|0\rangle - a^*|1\rangle$ in the first one of Equation (1), we obtain

$$U(|\psi\rangle, |W\rangle) = |0\rangle(a\sqrt{F}|E'_{00}\rangle + b^*\sqrt{D}|E'_{01}\rangle) + |1\rangle(b\sqrt{F}|E'_{00}\rangle - a^*\sqrt{D}|E'_{01}\rangle). \tag{4}$$

Equating the right hand sides of Equations (3) and (4), we get

$$\sqrt{F}|E'_{00}\rangle = \sqrt{F}\left(|a|^2|E_{00}\rangle + |b|^2|E_{11}\rangle\right) + \sqrt{D}\left(ab^*|E_{01}\rangle + a^*b|E_{10}\rangle\right), \tag{5}$$
$$\sqrt{D}|E'_{01}\rangle = ab\sqrt{F}\left(|E_{00}\rangle - |E_{11}\rangle\right) - \sqrt{D}\left(a^2|E_{01}\rangle - b^2|E_{10}\rangle\right). \tag{6}$$

Similarly, comparing two different expressions for $U(|\psi_\perp\rangle, |W\rangle)$, we get

$$\sqrt{D}|E'_{10}\rangle = a^*b^*\sqrt{F}\left(|E_{00}\rangle - |E_{11}\rangle\right) + \sqrt{D}\left(b^{*2}|E_{01}\rangle - a^{*2}|E_{10}\rangle\right), \tag{7}$$
$$\sqrt{F}|E'_{11}\rangle = \sqrt{F}\left(|b|^2|E_{00}\rangle + |a|^2|E_{11}\rangle\right) - \sqrt{D}\left(ab^*|E_{01}\rangle + a^*b|E_{10}\rangle\right). \tag{8}$$

As explained in [10,7], for a symmetric attack, we have the following constraints.

(i) The scalar products $\langle E_{ij}|E_{kl}\rangle$ and $\langle E'_{ij}|E'_{kl}\rangle$, are such that $\langle E_{ij}|E_{kl}\rangle = \langle E_{kl}|E_{ij}\rangle$ and $\langle E'_{ij}|E'_{kl}\rangle = \langle E'_{kl}|E'_{ij}\rangle$, for $i,j,k,l \in \{0,1\}$. This assumption implies that all the inner products must be real.

(ii) Any element of $\{|E_{00}\rangle, |E_{11}\rangle\}$ is orthogonal to any element of $\{|E_{01}\rangle, |E_{10}\rangle\}$. Similar orthogonality condition holds between the pairs $\{|E'_{00}\rangle, |E'_{11}\rangle\}$ and $\{|E'_{01}\rangle, |E'_{10}\rangle\}$.

(iii) Further, we take $\langle E_{00}|E_{11}\rangle = \langle E'_{00}|E'_{11}\rangle = x$, $\langle E_{01}|E_{10}\rangle = \langle E'_{01}|E'_{10}\rangle = y$, where x, y are real. It is evident that all the other inner products are zero due to the orthogonality conditions.

We have $\langle E'_{00}|E'_{01}\rangle = 0$ and replacing them as in (5) and (6), we get

$$ab(|a|^2 - |b|^2)(1 - x) - D\left[ab\left(|a|^2 - |b|^2\right)(2 - x) + \left(a^3 b^* - a^* b^3\right)y\right] = 0. \quad (9)$$

From (9) we get the following

$$D = \frac{ab\left(|a|^2 - |b|^2\right)(1 - x)}{ab\left(|a|^2 - |b|^2\right)(2 - x) + \left(a^3 b^* - a^* b^3\right)y}. \quad (10)$$

The expression of D in (10) is not defined when the denominator is zero. Given $y \neq 0$, the denominator of (10) is 0 if and only if $\left(|a| = |b| = \frac{1}{\sqrt{2}}\right)$ AND $\left(\arg(\frac{a}{b}) \equiv 0 \mod \frac{\pi}{2}\right)$. Under this condition, we get that $a = \pm b$ or $\pm \imath b$.

When $a = \pm b$ or $\pm \imath b$, D cannot be calculated from (10) as the denominator will be zero. However, taking $\langle E'_{01}|E'_{01}\rangle = 1$ and putting there the expression of $|E'_{01}\rangle$ from (6), we get the value of D as follows

$$D = \frac{1-x}{2-x+y}, \quad \text{when } a = \pm b \quad (11)$$

$$= \frac{1-x}{2-x-y}, \quad \text{when } a = \pm \imath b. \quad (12)$$

Now consider the case when denominator of D in (10) is not zero. It has already been considered that $\langle E'_{00}|E'_{10}\rangle = 0$. Now replacing them as in (5) and (7) and plugging in the value of D from (10), we get $(1 - x)y\left(a^2 b^{*2} - a^{*2}b^2\right) = 0$.

We have considered that $\langle E_{00}|E_{11}\rangle = \langle E'_{00}|E'_{11}\rangle = x$, and $\langle E_{01}|E_{10}\rangle = \langle E'_{01}|E'_{10}\rangle = y$, where both x, y are real. Thus, it is natural to consider that $0 < x, y < 1$; otherwise, the vectors will be either orthogonal or the same. In such a situation, from $(1 - x)y\left(a^2 b^{*2} - a^{*2}b^2\right) = 0$, we get $\left(a^2 b^{*2} - a^{*2}b^2\right) = 0$, i.e., $ab^* = \pm a^* b$. This holds if and only if $a = \pm rb, \pm \imath rb$, where $r = \frac{|a|}{|b|} \neq 1$. The $r = 1$ case has already been taken care of.

For $r \neq 1$, when we put $a = \pm rb$ in (10), we get $D = \frac{1-x}{2-x+y}$, as given in (11) already. Now taking the inner product of both sides of (6) and (7) and putting $D = \frac{1-x}{2-x+y}$, we get $\langle E'_{01}|E'_{10}\rangle = \left((b^*)^2 + (a^*)^2\right)^2 y$ which has been assumed to be y. Thus, $\left((b^*)^2 + (a^*)^2\right)^2 = 1$, and given $a = \pm rb$, we obtain either both a, b are real of both a, b are imaginary.

However, for $r \neq 1$, if we put $a = \pm \imath r b$ in (10), we get $D = \frac{1-x}{2-x-y}$ as in (12). Then following the similar manner as before, we get one of a, b is real and the other one is imaginary. Thus we have the following result.

Theorem 1. *Consider symmetric incoherent eavesdropping with $0 < x, y < 1$, on the BB84 protocol with the bases $|0\rangle$, $|1\rangle$ and $|\psi\rangle = a|0\rangle + b|1\rangle$, $|\psi_\perp\rangle = b^*|0\rangle - a^*|1\rangle$. We have (i) $D = \frac{1-x}{2-x+y}$ if and only if a, b are either both real or both imaginary and (ii) $D = \frac{1-x}{2-x-y}$ if and only if one of a, b is real and the other one is imaginary.*

Theorem 1 identifies that for such eavesdropping where BB84 protocol is implemented with the bases $|0\rangle$, $|1\rangle$ and $|\psi\rangle$, $|\psi_\perp\rangle$, the form of $|\psi\rangle$ is restricted given $0 < x, y < 1$. When $r \neq 1$, then the bases $|0\rangle$, $|1\rangle$ and $|\psi\rangle$, $|\psi_\perp\rangle$ cannot be conjugate. To have conjugate bases, one must take $r = 1$, i.e., $|a| = |b| = \frac{1}{\sqrt{2}}$. As the simplest example, it is natural to consider $a = b = \frac{1}{\sqrt{2}}$, which gives $|\psi\rangle = \frac{|0\rangle+|1\rangle}{\sqrt{2}}$ and $|\psi_\perp\rangle = \frac{|0\rangle-|1\rangle}{\sqrt{2}}$ that has indeed been used in BB84 protocol [1]. On such conjugate bases, the eavesdropping idea of [10] works that we discuss in the next section.

In [10], the conjugate bases $|0\rangle$, $|1\rangle$ and $\frac{|0\rangle+|1\rangle}{\sqrt{2}}$, $\frac{|0\rangle-|1\rangle}{\sqrt{2}}$ have been considered. That is in this case, $a = b = \frac{1}{\sqrt{2}}$ and $D = \frac{1-x}{2-x+y}$, as in Equation (11).

In [6], three conjugate bases $|0\rangle$, $|1\rangle$; $\frac{|0\rangle+|1\rangle}{\sqrt{2}}$, $\frac{|0\rangle-|1\rangle}{\sqrt{2}}$ and $\frac{|0\rangle+\imath|1\rangle}{\sqrt{2}}$, $\frac{|0\rangle-\imath|1\rangle}{\sqrt{2}}$ have been exploited for the BB84 protocol. Thus, while considering $a = b = \frac{1}{\sqrt{2}}$ one gets $D = \frac{1-x}{2-x+y}$, but in case of $a = \frac{1}{\sqrt{2}}$, $b = \frac{\imath}{\sqrt{2}}$ we obtain $D = \frac{1-x}{2-x-y}$. To have the symmetric attack possible, we need $\frac{1-x}{2-x+y} = \frac{1-x}{2-x-y}$ and thus $y = 0$. For $y = 0$, both (11) and (12) reduce to

$$D = \frac{1-x}{2-x}. \tag{13}$$

However, there are complex numbers a, b, where $|a| = |b| = \frac{1}{\sqrt{2}}$, but $a \neq \pm b, \pm \imath b$ and in those case a, b are not as given in Theorem 1. As example, one can take, $|\psi\rangle = \frac{1+\imath}{2}|0\rangle + \frac{1}{\sqrt{2}}|1\rangle$ and $|\psi_\perp\rangle = \frac{1}{\sqrt{2}}|0\rangle - \frac{1-\imath}{2}|1\rangle$. Symmetric attack in the attack model of [6,10] is not directly possible in these cases when y is nonzero. However, if Eve uses a phase-covariant cloner or orients her probes appropriately, then she can mount the same attack. Thus, by no choice of a, b, Alice and Bob can avoid the symmetric attack on the four-state protocol.

3   Eavesdropper's Success Probability as a Function of Disturbance at Receiver End

In this part, we critically revisit the attack models of [10] and [6] in the light of success probability of Eve's guess about the qubit that was actually sent by Alice. In the analysis, we require to compute the probabilities of different related events. These probabilities form the components for the mutual information between Alice and Eve as well as the success probability for Eve's guess. First in Section 3.1, we compute these individual probabilities and for the sake of completeness show the calculation of mutual information also. Next in Section 3.2, we derive the success probabilities of Eve's guess for various cases and discuss how they give different insight from mutual information.

We introduce a few notations for the sake of our analysis. Let A, B, V be the random variables corresponding to the bit sent by Alice, the bit received by Bob and the outcome observed by Eve due to her measurement. Eve performs the measurement after Alice and Bob announce their bases. After the announcement, Eve discards the probes corresponding to the qubits for which Alice and Bob's bases do not match and works with the probes corresponding to the bases that match. For one-bit probe, Eve measures her second qubit in the bases Z or X, as used by Alice. Similarly, for two-bit probe, Eve measures in the bases $\{|00\rangle, |01\rangle, |10\rangle, |11\rangle\}$ when Alice and Bob use the Z basis and she measures in the basis $\{|++\rangle, |+-\rangle, |-+\rangle, |--\rangle\}$ when Alice and Bob use the X basis. In this paper, we calculate all probabilities considering the Z basis only. Symmetry gives the same results when the X basis is used. Hence, without loss of generality, V can be assumed to be in $\{0, 1\}$ for one-bit probe, and it can be assumed to be in $\{00, 01, 10, 11\}$ for two-bit probe. In the subsequent discussion, we use the term Eve's *observation* to denote the observed outcome V of her measurement.

3.1   Probability Analysis and Mutual Information

We follow the standard definitions of mutual information and conditional entropy from information theory [8]. The mutual information between Alice and Bob is given by

$$I^{AB} = H(A) - H(A|B), \tag{14}$$

and the mutual information between Alice and Eve is given by

$$I^{AV} = H(A) - H(A|V), \tag{15}$$

where $H(\cdot)$ is the Shannon entropy function.

We assume that Alice randomly generates the bits to be transmitted, so that $P(A = 0) = P(A = 1) = \frac{1}{2}$. Hence $H(A) = -\frac{1}{2}\log_2(\frac{1}{2}) - \frac{1}{2}\log_2(\frac{1}{2}) = 1$. Also,
$P(B = 0 \mid A = 1) = P(B = 1 \mid A = 0) = D$ and
$P(B = 0 \mid A = 0) = P(B = 1 \mid A = 1) = 1 - D$.
Hence, $P(B = 0) = P(B = 1) = \frac{1}{2}$ and the conditionals $P(A \mid B)$ are identical with the conditionals $P(B \mid A)$. Thus,
$H(A \mid B = 0) = H(A \mid B = 1) = -D\log_2(D) - (1 - D)\log_2(1 - D)$ and
$H(A \mid B) = P(B = 0)H(A \mid B = 0) + P(B = 1)H(A \mid B = 1)$
$= -D\log_2(D) - (1 - D)\log_2(1 - D)$. So from Equation (14) we have

$$I^{AB} = 1 + D\log_2(D) + (1 - D)\log_2(1 - D). \tag{16}$$

Recall that (one may refer to Section 2 for details) the general unitary transformation designed by Eve is as follows:

$U(|0\rangle, |W\rangle) = \sqrt{F}|0\rangle|E_{00}\rangle + \sqrt{D}|1\rangle|E_{01}\rangle$, and
$U(|1\rangle, |W\rangle) = \sqrt{D}|0\rangle|E_{10}\rangle + \sqrt{F}|1\rangle|E_{11}\rangle$, where $F = 1 - D$.

If we rewrite the interactions expressed in [10, Equations 50-51] in our notation, we obtain the following expressions for $|E_{ij}\rangle$'s.

$$|E_{00}\rangle = \sqrt{1-D}\frac{|00\rangle+|11\rangle}{\sqrt{2}} + \sqrt{D}\frac{|00\rangle-|11\rangle}{\sqrt{2}}, |E_{01}\rangle = \sqrt{1-D}\frac{|01\rangle+|10\rangle}{\sqrt{2}} - \sqrt{D}\frac{|01\rangle-|10\rangle}{\sqrt{2}},$$

$$|E_{10}\rangle = \sqrt{1-D}\frac{|01\rangle+|10\rangle}{\sqrt{2}} + \sqrt{D}\frac{|01\rangle-|10\rangle}{\sqrt{2}}, |E_{11}\rangle = \sqrt{1-D}\frac{|00\rangle+|11\rangle}{\sqrt{2}} - \sqrt{D}\frac{|00\rangle-|11\rangle}{\sqrt{2}}.$$

For $i \in \{0,1\}$, by Bayes' Theorem, Eve's posterior probability $P(A = i \mid V = v)$ of what Alice sent is given by

$$\frac{P(A = i) \cdot P(V = v \mid A = i)}{P(V = v)} = \frac{P(A = i) \cdot P(V = v \mid A = i)}{\sum\limits_{j=0,1} P(A = j) \cdot P(V = v \mid A = j)}$$

$$= \frac{P(V = v \mid A = i)}{P(V = v \mid A = 0) + P(V = v \mid A = 1)}. \quad (17)$$

Again, the likelihoods $P(V = v \mid A = i)$ are computed as

$$P(B = 0 \mid A = i)P(V = v \mid A = i, B = 0)$$
$$+P(B = 1 \mid A = i)P(V = v \mid A = i, B = 1)$$
$$= P(B = 0 \mid A = i)P(V = v \mid E_{i0}) + P(B = 1 \mid A = i)P(V = v \mid E_{i1}). (18)$$

After the announcement of the bases in the BB84 protocol, Eve measures her qubit in the corresponding bases. The likelihoods for the attack in [10] when computed using Equation (18) turns out to be as shown in Table 1 below.

Table 1. Values of $P(V = v \mid A = i) = P(A = i \mid V = v)$ for the attack model of [10]

	$V = 0$	$V = 1$
$A = 0$	$\frac{1}{2} + \sqrt{D(1-D)}$	$\frac{1}{2} - \sqrt{D(1-D)}$
$A = 1$	$\frac{1}{2} - \sqrt{D(1-D)}$	$\frac{1}{2} + \sqrt{D(1-D)}$
Marginal of V	$\frac{1}{2}$	$\frac{1}{2}$

For example, $P(V = 0 \mid A = 0)$ is given by $P(B = 0 \mid A = 0)P(V = 0 \mid E_{00}) + P(B = 1 \mid A = 0)P(V = 0 \mid E_{01}) = (1-D) \cdot \left(\frac{1}{\sqrt{2}}\left(\sqrt{1-D} + \sqrt{D}\right)\right)^2 + D \cdot \left(\frac{1}{\sqrt{2}}\left(\sqrt{1-D} + \sqrt{D}\right)\right)^2 = \frac{1}{2} + \sqrt{D(1-D)} = f(D)$, say.

Note that since $P(A = 0) = P(A = 1) = \frac{1}{2}$, the half of the sum of each column in Table 1 gives the marginal probability of V for that column. In Equation (17), putting the value of $P(V = v \mid A = i)$ from Table 1, we find that the posteriors are identical with the corresponding likelihoods. Hence $H(A \mid V = 0) = H(A \mid V = 1) = -f(D)\log_2 f(D) - (1 - f(D))\log_2(1 - f(D))$.

Also, from Table 1, we have $P(V = 0) = P(V = 1) = \frac{1}{2}$, giving $H(A|V) = P(V = 0)H(A \mid V = 0) + P(V = 1)H(A \mid V = 1) = -f(D)\log_2 f(D) - (1 - f(D))\log_2 (1 - f(D))$. Substituting in Equation (15), we have

$$I^{AV} = 1 + f(D)\log_2 f(D) + (1 - f(D))\log_2 (1 - f(D)). \tag{19}$$

Note that the above computation is shown assuming a one-bit probe. It is easy to show that, for the four-state protocol, the one-bit and the two-bit probes give identical mutual information between Alice and Eve. The expression for this mutual information is given by Equation (19) which matches with [10, Equation 65].

Next, the interactions of [6, Equations 9-15], when expressed in our notations, become $|E_{00}\rangle = \beta|10\rangle + \sqrt{1 - |\beta|^2}|01\rangle$, $|E_{01}\rangle = |00\rangle$, $|E_{10}\rangle = |11\rangle$, and $|E_{11}\rangle = \sqrt{1 - |\beta|^2}|10\rangle + \beta|01\rangle$. From Equation (13) (Section 2), we obtain, $D = \frac{1-x}{2-x}$, which gives, $x = \frac{1-2D}{1-D}$. Noting that, $x = \langle E_{00}|E_{11}\rangle$, we get

$$|\beta|^2 = \frac{1}{2}\left(1 + \frac{\sqrt{D(2 - 3D)}}{1 - D}\right). \tag{20}$$

Technically, the square-root in Equation (20) should be written with a $\pm$ sign. However, for simplicity, we show all calculation with the $+$ sign here. The calculation with the $-$ sign would be similar.

Table 2. Values of $P(V = v \mid A = i) = P(A = i \mid V = v)$ for one-bit probe of [6]

	$V = 0$	$V = 1$
$A = 0$	$D + (1 - D)|\beta|^2$	$1 - D - (1 - D)|\beta|^2$
$A = 1$	$1 - D - (1 - D)|\beta|^2$	$D + (1 - D)|\beta|^2$
Marginal of V	$\frac{1}{2}$	$\frac{1}{2}$

Table 3. Values of $P(V = v \mid A = i)$ for two-bit probe of [6]

	$V = 00$	$V = 01$	$V = 10$	$V = 11$
$A = 0$	D	$1 - D - (1 - D)|\beta|^2$	$(1 - D)|\beta|^2$	0
$A = 1$	0	$(1 - D)|\beta|^2$	$1 - D - (1 - D)|\beta|^2$	D
Marginal of V	$\frac{D}{2}$	$\frac{1-D}{2}$	$\frac{1-D}{2}$	$\frac{D}{2}$

For one-bit probe, the likelihoods for [6] when computed using Equation (18) turns out to be as shown in Table 2.

From Equation (17), we find that in this case also, the posteriors are identical with the corresponding likelihoods.

Table 4. Values of $P(A = i \mid V = v)$ for two-bit probe of [6]

	$V = 00$	$V = 01$	$V = 10$	$V = 11$
$A = 0$	1	$1 - |\beta|^2$	$|\beta|^2$	0
$A = 1$	0	$|\beta|^2$	$1 - |\beta|^2$	1
Marginal of V	$\frac{D}{2}$	$\frac{1-D}{2}$	$\frac{1-D}{2}$	$\frac{D}{2}$

For ease of calculation, let us denote

$$f_1(D) = D + (1 - D)|\beta|^2 = \frac{1}{2}\left(1 + D + \sqrt{D(2 - 3D)}\right). \qquad (21)$$

Hence $H(A \mid V = 0) = H(A \mid V = 1)$ can be written as $-f_1(D)\log_2 f_1(D) - (1 - f_1(D))\log_2 (1 - f_1(D))$.

Also, from Table 2, we have $P(V = 0) = P(V = 1) = \frac{1}{2}$, giving $H(A|V) = P(V = 0)H(A \mid V = 0) + P(V = 1)H(A \mid V = 1) = -f_1(D)\log_2 f_1(D) - (1 - f_1(D))\log_2 (1 - f_1(D))$. Substituting in Equation (15), we have

$$I_1^{AV} = 1 + f_1(D)\log_2 f_1(D) + (1 - f_1(D))\log_2 (1 - f_1(D)). \qquad (22)$$

This expression matches with [6, Equation 18].

Now, consider the two-bit probe. The likelihoods for [6] when computed using Equation (18) turns out to be as shown in Table 3.

From Equation (17), the posteriors are computed as given in Table 4. Hence $H(A \mid V = 00) = H(A \mid V = 11) = 0$ and $H(A \mid V = 01) = H(A \mid V = 10) = -|\beta|^2\log_2|\beta|^2 - (1 - |\beta|^2)\log_2(1 - |\beta|^2) = h(D)$ (say). Thus, $H(A|V) = P(V = 00)H(A \mid V = 00) + P(V = 01)H(A \mid V = 01) + P(V = 10)H(A \mid V = 10) + P(V = 11)H(A|V = 11) = \frac{D}{2}\cdot 0 + \frac{1-D}{2}\cdot h(D) + \frac{1-D}{2}\cdot h(D) + \frac{D}{2}\cdot 0 = (1-D)\cdot h(D)$. Substituting in Equation (15), we have

$$I_2^{AV} = 1 - (1 - D)h(D). \qquad (23)$$

Again, this matches with [6, Equation 17].

If one plots the curves of I^{AV}, I_1^{AV} and I_2^{AV} against D, one can find that for all values of $D \in (0, \frac{1}{2})$, the relation $I_1^{AV} < I_2^{AV} < I^{AV}$ holds. From this, it is concluded in [6] that the six-state protocol is more secure than the four-state protocol. Moreover, within the six-state protocol, two-bit probe helps Eve in obtaining more mutual information than the one-bit probe. However, we present a different view on both of these claims.

3.2   Optimal Success Probability and Its Implications

We introduce a few relevant definitions first and then proceed with the analysis.

Definition 1. *A strategy S of the Eavesdropper is a function of her observation v such that for each v, it produces a unique guess $S(v)$ about the bit sent by Alice to Bob.*

Definition 2. *For some observation* v, *if the Eavesdropper's guess matches with the bit sent by Alice, i.e., if* $S(v) = A$, *we call this event a* **success**.

Definition 3. *For some observation* v, *if the Eavesdropper's guess does not match with the bit sent by Alice, i.e., if* $S(v) \neq A$, *we call this event a* **failure** *or an* **error**.

Thus, the *conditional error probability* of Eve is given by $P(error \mid V = v) = P(S(v) \neq A \mid V = v)$ and the *error probability* of Eve is given by

$$P(error) = \sum_v P(V = v) P(error \mid V = v)$$

$$= \sum_v P(V = v) P(S(v) \neq A \mid V = v). \tag{24}$$

The *success probability* of Eve is given by $P(success) = 1 - P(error)$.

Definition 4. *If* $P(success)$ *is the success probability of the Eavesdropper in guessing the bit sent by Alice through some strategy* S, *and* $P(prior)$ *is the probability denoting the Eavesdropper's prior knowledge about the bit sent by Alice before applying any strategy, then the* **advantage** *of the Eavesdropper for the particular strategy is defined as* $A(D) = |P(success) - P(prior)|$.

Since Alice chooses the bit to be sent uniformly at random over $\{0, 1\}$, in our case $P(prior) = \frac{1}{2}$ and so $A(D) = \left| P(success) - \frac{1}{2} \right|$.

Maximizing the success probability or the advantage is equivalent to minimizing the error probability. Note that Eve's success or error probability is a feature of the particular strategy devised by Eve. Her goal is to choose the best possible strategy in determining the secret key.

Definition 5. *Out of all possible strategies, the one giving the maximum success probability or the minimum error probability, is called the* **optimal strategy** S_{opt}. *The corresponding success (or error) probability is called the* **optimal success (or error) probability** *of the Eavesdropper and the corresponding advantage is called the* **optimal advantage** *of the Eavesdropper.*

In the result below, we formulate how Eve can decide the optimal strategy.

Theorem 2. *The optimal strategy is given by*

$$S_{opt}(v) = \operatorname*{argmax}_i P(A = i \mid V = v),$$

and the corresponding optimal success probability is given by

$$P_{opt}(success) = \sum_v \max_i P(A = i, V = v),$$

where the notation $\operatorname*{argmax}_i$ *denotes the particular value* i_{opt} *of the argument* i *which maximizes the above conditional probability across all values* i.

Proof. Since $P(V = v)$ is independent of the strategy S, an optimum strategy that minimizes $P(error)$ must minimize $P(S(v) \neq A \mid V = v)$ for each v, as per Equation (24). In other words, for each v, it should maximize $P(S(v) = A \mid V = v)$. This means that $S(v)$ should produce a guess $i \in \{0, 1\}$ for which $P(A = i|V = v)$ is maximum. For the particular observation v, denote this optimal value of i by $i_{opt}(v)$. With this optimal strategy the *optimal error probability* turns out to be $P_{opt}(error) = \sum_v P(V = v)P(A \neq i_{opt}(v) \mid V = v) = \sum_v P(A \neq i_{opt}(v), V = v)$ and the *optimal success probability* becomes

$$P_{opt}(success) = 1 - P_{opt}(error) = 1 - \sum_v P(A \neq i_{opt}(v), V = v)$$
$$= \sum_v P(A = i_{opt}(v), V = v). \text{ Hence the result follows.} \qquad \square$$

Since $P(A = 0) = P(A = 1) = \frac{1}{2}$, if we multiply each likelihood in Tables 1, 2 and 3 by $\frac{1}{2}$, we get the corresponding joint probabilities $P(A = i, V = v)$'s and the optimal success probability is given by summing the maximum joint probability (corresponding to the row $i_{opt}(v)$) for each column v.

Thus, for the attack model of [10], the optimal success probability is computed from Table 1 as

$$P_{opt}^{4\text{-}state}(success) = \frac{1}{2}\left(\frac{1}{2} + \sqrt{D(1-D)}\right) + \frac{1}{2}\left(\frac{1}{2} + \sqrt{D(1-D)}\right)$$
$$= \frac{1}{2} + \sqrt{D(1-D)} = f(D). \tag{25}$$

It can be easily shown that, like the mutual information, the success probabilities are also the same in both the probes (one-bit and two-bit) for the four-state protocol.

Since the six-state protocol [6] has different mutual information between Alice and Eve for the one-bit and the two-bit probes, one may be tempted to conclude that Eve has different success probabilities in these two probes. However, we are going to show that this is not the case. In spite of having different mutual information, both the probes lead to the same success probability for the six-state protocol.

For the one-bit probe of the six-state protocol [6], the optimal success probability is computed from Table 2 as

$$P_{opt1}^{6\text{-}state}(success) = \frac{1}{2}\left(D + (1-D)|\beta^2|\right) + \frac{1}{2}\left(D + (1-D)|\beta|^2\right)$$
$$= D + (1-D)|\beta|^2 = f_1(D). \tag{26}$$

Note that in the above derivation, we have used the fact that $D + (1-D)|\beta|^2 \geq 1 - D - (1-D)|\beta^2|$, which follows from $D + (1-D)|\beta^2| \geq \frac{1}{2}$ as per Equation (21).

For the two-bit probe of [6], the optimal success probability is computed from Table 3 as

$$P_{opt2}^{6\text{-}state}(success) = \frac{1}{2} \cdot D + \frac{1}{2} \cdot (1-D)|\beta|^2 + \frac{1}{2} \cdot (1-D)|\beta|^2 + \frac{1}{2} \cdot D$$
$$= D + (1-D)|\beta|^2 = f_1(D). \tag{27}$$

Note that in the above derivation, we have used the fact that $(1 - D)|\beta|^2 \geq 1 - D - (1 - D)|\beta^2|$, which follows from $|\beta^2| \geq \frac{1}{2}$ as per Equation (20).

Hence, we have the following result.

Theorem 3. *For all $D \in (0, \frac{1}{2})$,*

$$P_{opt1}^{6\text{-}state}(success) = P_{opt2}^{6\text{-}state}(success) < P_{opt}^{4\text{-}state}(success).$$

In Figure 1, we plot (as functions of the disturbance D) the optimal mutual information between Alice and Eve (on the left) and the optimal success probability of Eve's guess (on the right).

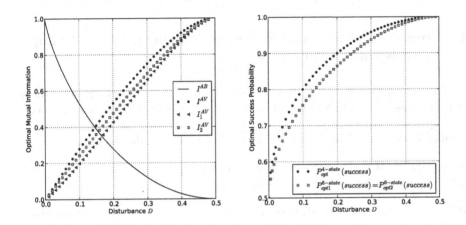

Fig. 1. Optimal mutual information and optimal success probability as a function of disturbance D

As an illustrative example, we show the values of the probabilities for $D = \frac{1}{6}$ in Table 5. The optimal success probability in one-bit probe is given by $\frac{5}{6} \cdot \frac{1}{2} + \frac{5}{6} \cdot \frac{1}{2} =$

Table 5. Values of $P(A = i \mid V = v)$ for $D = \frac{1}{6}$ for both one- and two-bit probes of [6]

	One-bit Probe		Two-bit Probe			
	$V = 0$	$V = 1$	$V = 00$	$V = 01$	$V = 10$	$V = 11$
$A = 0$	$\frac{5}{6}$	$\frac{1}{6}$	1	$\frac{1}{5}$	$\frac{4}{5}$	0
$A = 1$	$\frac{1}{6}$	$\frac{5}{6}$	0	$\frac{4}{5}$	$\frac{1}{5}$	1
Marginal of V	$\frac{1}{2}$	$\frac{1}{2}$	$\frac{1}{12}$	$\frac{5}{12}$	$\frac{5}{12}$	$\frac{1}{12}$

$\frac{5}{6}$ and that in two-bit probe turns out to be the same: $1 \cdot \frac{1}{12} + \frac{4}{5} \cdot \frac{5}{12} + \frac{4}{5} \cdot \frac{5}{12} + 1 \cdot \frac{1}{12} = \frac{5}{6}$. But the mutual information in the first case is $1 + \frac{5}{6} \log_2 \frac{5}{6} + \frac{1}{6} \log_2 \frac{1}{6} = 0.3500$ and in the second case is $1 + \frac{5}{6} \cdot (\frac{4}{5} \log_2 \frac{4}{5} + \frac{1}{5} \log_2 \frac{1}{5}) = 0.3984$.

According to Definition 4, the *optimal advantages* of the eavesdropper in the four-state and in the six-state protocols are respectively given by

$$A_4(D) = \sqrt{D(1-D)}. \tag{28}$$

$$A_6(D) = \frac{D + \sqrt{D(2-3D)}}{2}. \tag{29}$$

Thus, though Eve has more mutual information in the two-bit probe, that does not give any extra cryptographic advantage in guessing the bit sent by Alice. So from the point of view of cryptanalysis, both the one-bit probe and two-bit probe are equivalent even in the six-state BB84.

4   Comparing Four and Six-State Protocols Considering Same Number of Qubits

For BB84 with four states, on average half of the qubits communicated by Alice to Bob is discarded due to mismatch in their bases. For the six-state protocol, the expected number of discarded qubits is two-third of the total number of qubits communicated. So for a fair comparison, we must take the same values of

1. the length of the secret key established, and
2. the total number of qubits communicated

in both the protocols. To establish a secret key of length n bits, the four-state protocol must communicate around $4n$ qubits (in the practical scenario, the exact number is little more than $4n$) and the six-state protocol must communicate around $6n$ qubits (practically little more than that). Therefore, in order to match the total number of bits communicated, the four-state protocol may be repeated $3t$ times and the six-states protocol should be repeated $2t$ times for any positive integer t.

With the above motivation, we define a variant of BB84, called m-BB84 in Table 6. In this protocol, Alice and Bob establish m different keys of the same length by running m independent instances of BB84 and finally establish the actual secret key by bitwise XOR-ing the individual keys together. The main idea behind this scheme is the fact that when several biased bits are XOR-ed together, the bias in the XOR output bit becomes smaller than the bias of each bit. The concept is in the direction to privacy amplification [2]. However, the motivation here is to compare the four-state and six-state protocol under the same footage. Any post-processing including privacy amplification can be performed on the string produced by the multi-round BB84.

The bias in K_j, the j-th bit of the final key K, depends on the biases in the j-th bits of the individual keys. We can use the Piling-up Lemma [18] stated below to compute the bias in K_j. We present the proof also for the sake of completeness.

Table 6. Multi-round BB84 Protocol with parameter (number of rounds) m

---

Protocol m-BB84

1. Alice and Bob run m independent instances of BB84.
(The instances may either be run sequentially,
or they may be run in parallel through separate channels).
2. Suppose they establish m many n-bit secret keys, namely,
$k_1, k_2, \ldots, k_m$. Let $k_{i,j}$ be the j-th bit of the key k_i established
in the i-th instance of BB84, for $1 \leq i \leq m$, $1 \leq j \leq n$.
3. The j-th bit of the final secret key K is given by
$K_j = k_{1,j} \oplus k_{2,j} \oplus \cdots \oplus k_{m,j}$, for $1 \leq j \leq n$.

Lemma 1 (Piling-up Lemma). *Let ϵ_i be the bias in the binary random variable X_i, $i = 1, 2, \ldots, m$, i.e., $P(X_i = 0) = \frac{1}{2} + \epsilon_i$ and $P(X_i = 1) = \frac{1}{2} - \epsilon_i$. Then the bias in the random variable $X_1 \oplus X_2 \oplus \cdots \oplus X_m$ is given by $2^{m-1}\epsilon_1\epsilon_2 \ldots \epsilon_m$, considering the individual random variables as independent.*

Proof. The result trivially holds for $m = 1$. For $m = 2$, we have

$$P(X_1 \oplus X_2 = 0) = P(X_1 = 0, X_2 = 0) + P(X_1 = 1, X_2 = 1)$$
$$= \left(\frac{1}{2} + \epsilon_1\right)\left(\frac{1}{2} + \epsilon_2\right) + \left(\frac{1}{2} - \epsilon_1\right)\left(\frac{1}{2} - \epsilon_2\right) = \frac{1}{2} + 2\epsilon_1\epsilon_2$$

and hence the bias is $2^{2-1}\epsilon_1\epsilon_2$. Assume that the result holds for $m = \ell$, i.e., the bias in XOR of ℓ variables is given by $\delta = 2^{\ell-1}\epsilon_1\epsilon_2 \ldots \epsilon_\ell$. Now, for $k = \ell + 1$, taking $Y = X_1 \oplus X_2 \oplus \cdots \oplus X_\ell$, we can apply the result for $k = 2$ to obtain the bias in $Y \oplus X_{\ell+1}$ as $2\delta\epsilon_{\ell+1} = 2^\ell\epsilon_1\epsilon_2 \ldots \epsilon_{\ell+1}$. Hence, by induction, the result holds for any m. $\qquad\square$

Now, we can formulate the optimal advantage of the adversary for m-BB84 as follows.

Theorem 4. *For a disturbance D in each qubit of the individual instances of BB84, the optimal advantages of the adversary in guessing a bit of the final key of m-BB84 are given by $A_4(D, m) = 2^{m-1}\left(\sqrt{D(1 - D)}\right)^m$, and $A_6(D, m) = \frac{1}{2}\left(D + \sqrt{D(2 - 3D)}\right)^m$ corresponding to the four-state and the six-state protocols respectively.*

Proof. For any bit position j, the computation of the bias follows in the same manner. Hence, without loss of generality, fix a bit position j. Corresponding to this position, there are m key bits, each having the same bias ϵ_i, $1 \leq i \leq m$. The value of this bias is given by Equation (28) for the four-state protocol and by Equation (29) for the six-state protocol. By substituting these expressions for ϵ_i in Lemma 1, the result follows. $\qquad\square$

Note that Equations (28) and (29) can be considered as special cases of Theorem 4 with $m = 1$, i.e., they represent $A_4(D, 1)$ and $A_6(D, 1)$ respectively.

In principle, the higher the value of m, the greater is the reduction of Eve's advantage. However, one should keep in mind that with increasing m, the effective disturbance perceived by Bob also increases. We can formulate this by the following result.

Theorem 5. *For a disturbance D in the channel for each qubit of the individual instances of BB84, the effective disturbance perceived by Bob for each bit of the final key of m-BB84 is given by $\Delta(D,m) = \frac{1}{2} - 2^{m-1}\left(\frac{1}{2} - D\right)^m$.*

Proof. A disturbance D corresponds to a no-error (success) probability of $1 - D = \frac{1}{2} + \left(\frac{1}{2} - D\right)$, i.e., a bias of $\left(\frac{1}{2} - D\right)$ at Bob's end. For any bit position j, the computation follows in the same manner. Hence, without loss of generality, fix a bit position j. Corresponding to this position, there are m key bits, each having the same bias $\epsilon_i = \left(\frac{1}{2} - D\right)$, $1 \leq i \leq m$. By Lemma 1, the equivalent bias (of no-error) for the j-th bit (and so for each bit) of the final key is given by $2^{m-1}\left(\frac{1}{2} - D\right)^m$. Thus, the equivalent no-error probability for each bit of the final key is given by $s = \frac{1}{2} + 2^{m-1}\left(\frac{1}{2} - D\right)^m$. The equivalent disturbance is given by $1 - s$. □

As discussed already, for fair comparison we should always compare four-state $3t$-BB84 with six-state $2t$-BB84 for any fixed integral value of t. Because of Theorem 5, higher t means more error for Alice and Bob. Hence, we would restrict our subsequent discussion for $t = 1$, i.e, we would compare the four-state 3-BB84 with the six-state 2-BB84, though in principle similar comparison holds for any t.

We consider three scenarios for our comparative study. Let D_4 and D_6 denote the disturbances in each qubit of the individual instances of the four and the six-state protocols respectively. For comparison in equal footing, we take $D_6 = D$ and express all the other quantities in terms of D.

4.1 Scenario 1: Equal Disturbance in Each Qubit of the Individual Instances of Four-State and Six-State BB84

Here, $D_4 = D_6 = D$. In Figure 2 (top), we plot the optimal advantages of Eve and the effective disturbances of Bob as a function of the disturbance D for $D \in [0, \frac{1}{2}]$.

As pointed out in [6], one can note that for all $D \in [0, 0.5]$, $A_4(D, 1) > A_6(D, 1)$. That is, the eavesdropper can obtain more information in the traditional 4-state BB84 [1] than the 6-state modification [6]. However, we note that $A_4(D, 3) \leq A_6(D, 2)$ for $D \leq 0.27$ (up to two decimal places). Thus, at the expense of same number of qubits, for the range of disturbance ≤ 0.27, the four-state BB84 is more secure (as eavesdropper obtains less information) than the six-state BB84 in the model we discussed above. But this greater security comes at the cost of greater effective disturbance at Bob's end, as depicted by the plot.

As a numerical example, consider $D = 0.1$. Then $A_4(D, 1) = 0.3$, which is more than $A_6(D, 1) = 0.2562$. Again, $A_4(D, 3) = 0.108$, which is less than $A_6(D, 2) = 0.1312$, implying that the four-state 3-BB84 is more secure. However,

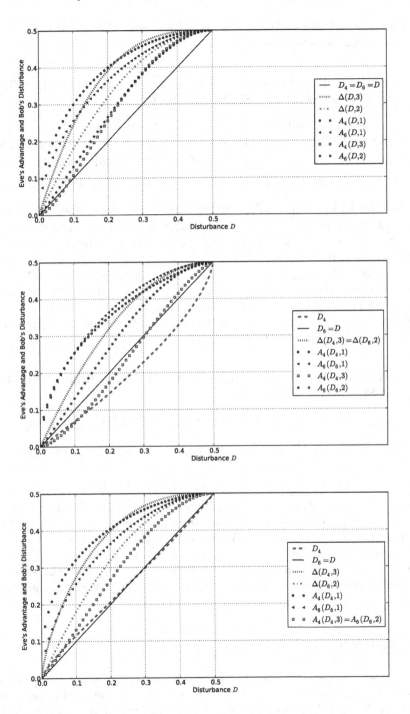

Fig. 2. Eavesdropper's advantages and Bob's disturbances against $D_6 = D$, for three cases: Scenario 1 (top), Scenario 2 (middle) and Scenario 3 (bottom)

its effective disturbance $\Delta(D_4, 3) = 0.244$ is more than that of the six-state 2-BB84 one, which is $\Delta(D_6, 2) = 0.18$.

4.2 Scenario 2: Equal Effective Disturbance in Each Bit of the Final Key of Four-State and Six-State BB84

In this scenario, we consider that Eve chooses different values of D_4 and D_6 so that the effective disturbances $\Delta(D_4, 3)$ and $\Delta(D_6, 2)$ are equal. Using Theorem 5, we can write $\Delta(D_4, 3) = \frac{1}{2} - 2^2 \left(\frac{1}{2} - D_4\right)^3$, and $\Delta(D_6, 2) = \frac{1}{2} - 2 \left(\frac{1}{2} - D_6\right)^2$. Equating the right hand sides and substituting $D_6 = D$, we obtain $D_4 = \frac{1}{2} - \left(\frac{1}{2} \left(\frac{1}{2} - D\right)^2\right)^{\frac{1}{3}}$. Now we plot Eve's optimal advantages $A_4(D_4, 3)$ and $A_6(D_6, 2)$ using Theorem 4 and the quantities for Bob's disturbances in Figure 2 (middle). Note that for the entire range of D, the four-state 3-BB84 is more secure than the six-state 2-BB84.

As a numerical example, consider $D_6 = 0.1$. Then $\Delta(D_6, 2) = 0.18$. For $\Delta(D_4, 3)$ to have the same value, we must have $D_4 = 0.0691$. For the single instance, we have $A_4(D_4, 1) = 0.2536$ to be marginally less than $A_6(D_6, 1) = 0.2562$, but for multiple instances with the same number of qubits, $A_4(D_4, 3) = 0.0653$ is much less than $A_6(D_6, 2) = 0.1312$.

4.3 Scenario 3: Equal Advantages for Eve for Four-State 3-BB84 and Six-State 2-BB84

From Theorem 4, we have $A_4(D_4, 3) = 2^2 \left(\sqrt{D_4(1 - D_4)}\right)^3$, and $A_6(D_6, 2) = \frac{1}{2} \left(D_6 + \sqrt{D_6(2 - 3D_6)}\right)^2$. Equating the right hand sides and substituting $D_6 = D$, we obtain $D_4 = \frac{1}{2} - \frac{1}{2}\sqrt{1 - \left(D + \sqrt{D(2 - 3D)}\right)^{\frac{4}{3}}}$. In Figure 2 (bottom), we plot Bob's effective disturbances $\Delta(D_4, 3)$ and $\Delta(D_6, 2)$ using Theorem 5, along with Eve's advantages. Here also, the four-state protocol offers more (individual as well as effective) disturbance at Bob's end than the six-state one.

As a numerical example, consider $D_6 = 0.1$. Then $A_6(D_6, 2) = 0.1312$. For $A_4(D_4, 3)$ to have the same value, we must have $D_4 = 0.1159$. The effective disturbances are $\Delta(D_4, 3) = 0.2734 > \Delta(D_6, 2) = 0.18$. Also, for the single instances, $A_4(D_4, 1) = 0.3201 > A_6(D_6, 1) = 0.2562$.

5   Conclusion

In this paper, we revisit the symmetric incoherent eavesdropping strategy of Fuchs et al. [10] and Bruß [6] on the four and the six-state BB84 protocols respectively in the light of the success probability of Eve. We show that both the one-bit and the two-bit probes in the six-state have the same success probability for Eve. Further, we critically compare the security issues in the four and the six-state protocols when same number of qubits are used in both the cases. Though the theoretical results of [6] as well as ours are correct, our results are placed from the cryptanalytic viewpoint of optimal eavesdropping and thus the interpretation is different from what claimed in [6].

References

1. Bennett, C.H., Brassard, G.: Quantum Cryptography: Public key distribution and coin tossing. In: Proceedings of the IEEE International Conference on Computers, Systems, and Signal Processing, Bangalore, India, pp. 175–179. IEEE, New York (1984)
2. Bennett, C.H., Brassard, G., Robert, J.M.: Privacy amplification by public discussion. SIAM Journal on Computing 17(2), 210–229 (1988)
3. Bennett, C.H., Bessette, F., Brassard, G., Salvail, L., Smolin, J.: Experimental quantum cryptography. Journal of Cryptology 5(1), 3–28 (1992)
4. Bernstein, D.J., Buchmann, J., Dahmen, E. (eds.): Post-Quantum Cryptography. Springer (2009)
5. Biham, E., Mor, T.: Bounds on Information and the Security of Quantum Cryptography. Phys. Rev. Lett. 79, 4034–4037 (1997)
6. Bruß, D.: Optimal eavesdropping in quantum cryptography with six states. Physical Review Letters 81, 3018–3021 (1998) (quant-ph/9805019)
7. Cirac, J.I., Gisin, N.: Coherent eavesdropping strategies for the 4 state quantum cryptography protocol. Physics Letters A 229(1), 1–7 (1997) (quant-ph/9702002)
8. Cover, T., Thomas, J.: Elements of Information Theory, 1st edn., pp. 16–20. John Wiley & Sons, Inc. (1991)
9. Diffie, W., Hellman, M.E.: New Directions in Cryptography. IEEE Transactions on Information Theory 22, 644–654 (1976)
10. Fuchs, C.A., Gisin, N., Griffiths, R.B., Niu, C.S., Peres, A.: Optimal eavesdropping in quantum cryptography. I. Information bound and optimal strategy. Physical Review A 56(2), 1163–1172 (1997)
11. Miller, V.S.: Use of Elliptic Curves in Cryptography. In: Williams, H.C. (ed.) CRYPTO 1985. LNCS, vol. 218, pp. 417–426. Springer, Heidelberg (1986)
12. Nielsen, M.A., Chuang, I.L.: Quantum Computation and Quantum Information. Cambridge University Press (2002)
13. Phoenix, S.J.D.: Quantum cryptography without conjugate coding. Physical Review A 48(1), 96–102 (1993)
14. Quantum Key Distribution Equipment. ID Quantique (IDQ), http://www.idquantique.com/
15. Quantum Key Distribution System (Q-Box). MagiQ Technologies Inc., http://www.magiqtech.com
16. Rivest, R.L., Shamir, A., Adleman, L.: A Method for Obtaining Digital Signatures and Public Key Cryptosystems. Communications of the ACM 21, 120–126 (1978)
17. Shor, P.: Algorithms for Quantum Computation: Discrete Logarithms and Factoring. In: Foundations of Computer Science (FOCS), pp. 124–134. IEEE Computer Society Press (1994)
18. Stinson, D.: Cryptography Theory and Practice, 3rd edn., pp. 80–81. Chapman & Hall / CRC (2005)
19. Wiesner, S.: Conjugate Coding (1970) (manuscript); subsequently published in SIGACT News 15(1), 78–88 (1983)

On-Line/Off-Line Leakage Resilient Secure Computation Protocols

Chaya Ganesh[1,*], Vipul Goyal[2], and Satya Lokam[2]

[1] Indian Institute of Technology, Madras, India
chaya.ganesh@gmail.com
[2] Microsoft Research, India
{vipul,satya}@microsoft.com

Abstract. We study the question of designing leakage-resilient secure computation protocols. Our model is that of *only computation leaks information with a leak-free input encoding phase*. In more detail, we assume an offline phase called the *input encoding phase* in which each party encodes its input in a specified format. This phase is assumed to be free of any leakage and may or may not depend upon the function that needs to be jointly computed by the parties. Then finally, we have a *secure computation phase* in which the parties exchange messages with each other. In this phase, the adversary gets access to a *leakage oracle* which allows it to download a function of the computation transcript produced by an honest party to compute the next outgoing message.

We present two main constructions of secure computation protocols in the above model. Our first construction is based only on the existence of (semi-honest) oblivious transfer. This construction employs an encoding phase which is dependent of the function to be computed (and the size of the encoded input is dependent on the size of the circuit of the function to be computed). Our second construction has an input encoding phase independent of the function to be computed. Hence in this construction, the parties can simple encode their input and store it as soon as it is received and then later on run secure computation for any function of their choice. Both of the above constructions, tolerate *complete leakage* in the secure computation phase.

Our second construction (with a function independent input encoding phase) makes use of a fully homomorphic encryption scheme. A natural question that arises is "can a leakage-resilient secure computation protocol with function independent input encoding be based on simpler and weaker primitives?". Towards that end, we show that any such construction would imply a secure two-party computation protocol with sub-linear communication complexity (in fact, communication complexity independent of the size of the function being computed).

Finally, we also show how to extend our constructions for the continual leakage case where there is: a one time leak-free input encoding phase, a leaky secure computation phase which could be run multiple times for different functionalities (but the same input vector), and, a leaky refresh phase after each secure computation phase where the input is "re-encoded".

---

* Work done in part while visiting Microsoft Research, India.

S. Galbraith and M. Nandi (Eds.): INDOCRYPT 2012, LNCS 7668, pp. 100–119, 2012.
© Springer-Verlag Berlin Heidelberg 2012

Keywords: leakage, secure computation, protocol, malicious adversary, continual leakage, communication complexity.

1   Introduction

Secure multi-party computation allows a set of n parties to compute a joint function of their inputs while keeping their inputs private. The first general solutions for the problem of secure computation were presented by Yao [Yao86] for the two-party case (with security against semi-honest adversaries) and Goldreich, Micali and Wigderson [GMW87] for the multi-party case (with security against malicious adversaries). These (and subsequent) protocols for secure computation were designed under the assumption that the local computations done by each of the parties on their private data are opaque.

In recent years, a vibrant area of research dealing with "leakage on the local computations" has emerged. This is a well motivated direction since sometimes, the local computations are not fully opaque and real world adversaries can exploit leakage from side channel attacks. There has been rapid progress on developing cryptographic primitives resilient against leakage of internal state in various models. There have been proposals of leakage resilient pseudorandom generators, public key encryption scheme, signature scheme, identity based encryption, etc (see [DP08], [FKPR10], [NS09] and the references therein). While significant progress has been made on designing cryptographic *primitives* secure in the presence of leakage, to our knowledge, there has not been any work on designing cryptographic *protocols* which tries to relax the assumption that the honest party machines are black-boxes with all internal computation hidden from the adversary.[1] In particular, while the protocol is running, what if the adversary manages to obtain a side channel allowing it to peek inside the honest party machines (in addition to completely corrupting some of the parties)? Can one still design secure computation protocols in this setting?

Our Results. We study the question of designing leakage-resilient secure computation protocols in this work. Our model is that of *only computation leaks information with a leak-free input encoding phase*. In more detail, our model has the following two phases.

We first assume an *offline phase* called the *input encoding phase*. This phase can be run in isolation and the parties need not be connected to the network. Hence, this phase is assumed to be free of any leakage. In this phase, each party encodes its input in a specified format. The encoding may or may not depend upon the function that needs to be jointly computed by the parties later on.

Then finally, we have a *secure computation phase* in which the parties exchange messages with each other. In this phase, the adversary gets access to a *leakage oracle* which allows it to download a function of the computation transcript produced by an honest party to compute the next outgoing message (this

---

[1] Please see the end of this section for a discussion of the concurrent independent work.

is, of course in addition to be able to completely corrupt a subset of the parties). If in any given round, the computation transcript "touches" only a subset of the bits of the encoded input (stored as a result of the input encoding phase), the leakage can depend only on that subset of the bits (i.e., only computation leaks information [MR04]).

Note that to hope to be able to have secure computation protocols, an input encoding phase is indeed *necessary*. This is because the security is violated even if a single bit of the initial input of an honest party is leaked to the adversary. Furthermore, if the adversary can later download a function of the entire encoded input, it can at least download a single bit of the initial input. Hence the assumption "only computation leaks information" also appears to be necessary. We emphasize that the security guarantees our protocols should satisfy correspond the standard ideal world (with no leakage allowed in the ideal world).

We present two main constructions of secure computation protocols in the above model. Our first construction is based only on the existence of (semi-honest) oblivious transfer. This construction employs an encoding phase which is dependent of the function to be computed (and the size of the encoded input is dependent on the size of the circuit of the function to be computed). Our second construction has an input encoding phase independent of the function to be computed. Hence in this construction, the parties can simple encode their input and store it as soon as it is received and then later on run secure computation for any function of their choice.

Both of the above constructions, somewhat surprisingly, tolerate *complete leakage* in the secure computation phase. That is, the adversary can fully observe the entire computation transcripts of the honest parties in the secure computation phase (including the bits of the encoded input that were used in running the protocol).

Our second construction (with function independent input encoding) makes use of a fully homomorphic encryption (FHE) scheme [Gen09] (see also [vDGHV10], [Gen10]). A natural question that arises is "can a leakage-resilient secure computation protocol with function independent input encoding phase be based on simpler and weaker primitives?". In fact, can we even have a leakage-resilient secure computation protocol (based on weaker primitives) where the *size* of the encoded input is independent of the size of the circuit of the function to be computed? Towards that end, we show that any such construction would imply a secure two-party computation protocol with sub-linear communication complexity (in fact, communication complexity independent of the size of the function being computed). Note that constructing such a protocol was a central open problem in the field of secure computation (until a construction for FHE was proposed by Gentry [Gen09]). Currently, the only known way to construct a sub-linear communication complexity secure computation protocol is to rely on a FHE scheme.

Finally, we also show how to extend our constructions for the continual leakage case where there is: a one time leak-free input encoding phase, a leaky secure computation phase which could be run multiple times for different functionalities (but the same input vector), and, a leaky refresh phase after each secure computation phase where the input is "re-encoded". As before, the secure computation phase can tolerate complete leakage. However in the refresh phase, the leakage is bounded by a parameter t (which can be any apriori chosen polynomial in the security parameter).

Our Techniques. Our primary tool is to construct and store a number of garbled circuits in the input encoding phase and use them later on in the secure computation phase. For simplicity, we focus on the two-party case; similar ideas are applicable to the multi-party setting as well. The idea of our first construction is as follows. We compile any given (semi-honest) secure computation protocol Π into a leakage-resilient one as follows. In the input encoding phase, every party creates and stores a number of garbled circuits corresponding to the next message function of the underlying protocol. In the secure computation phase, the outgoing message of a party is computed as an output of one of the these garbled circuits. To evaluate such a garbled circuit, only the appropriate wire keys are "read" from the encoded input (one for each input wire of the garbled circuit). Now the computation transcript of the secure computation phase consists primarily of the transcript of evaluation of such garbled circuits (one for each round). However such evaluation transcripts can be simulated and hence the only valuable information that is revealed from leakage is the output of such a garbled circuit, which is just the next output message in the underlying protocol Π (there are caveats like a party needs to keep secret state between the rounds which can be dealt with using standard ideas). This idea is similar in spirit to the ones used to construct one-time programs [GKR08]. In one-time programs, the security is based on the fact that only the appropriate wire keys are read from the given hardware tokens. Hence, one-time programs enable secure evaluation computation of a (non-interactive) function in the only-computation leaks information model.

The above idea "almost works" except for the following (rather subtle) problem. The real adversary $\mathcal{A}$ (given to us) expects to see the leakage along with every message it receives and then may *adaptively decide* the next outgoing message.

- To use the simulator S_Π, we need to construct an adversary $\mathcal{A}'$ for the protocol Π on which S_Π can be run.
- One natural high level idea is to construct another adversary $\mathcal{A}'$ for Π using the real adversary $\mathcal{A}$. To use $\mathcal{A}$, the new adversary $\mathcal{A}'$ will produce all the leakage to be given to $\mathcal{A}$ internally by constructing a *simulated* garbled circuit for each round. Hence now it seems S_Π can be simply run on $\mathcal{A}'$ (which is now a new adversary for the same protocol Π).
- The simulator S_Π of the underlying protocol Π may work by rewinding $\mathcal{A}'$. However once $\mathcal{A}'$ is rewound by S_Π in any given round, for that round, $\mathcal{A}'$ ends up giving to $\mathcal{A}$ *multiple* evaluation transcripts of the same simulated garbled circuit (on different inputs). However since we know how to prove the indistinguishability of a simulated garbled circuit from a real one only if the

garbled circuit is evaluated *once*, our hybrid arguments completely break down. In more detail, once our simulator starts rewinding in the hybrid experiments, we can no longer rely on the indistinguishability of the real garbled circuit execution from a simulated one.[2]

– An initial idea to solve this problem is to have $\mathcal{A}'$ generate the simulated garbled circuit from scratch (to be given to $\mathcal{A}$) every time the protocol in rewound in a given round. However now since the machine $\mathcal{A}'$ needs to be aware of when it is being rewound and be allowed to change its random tape every time it is rewound, the success of the simulator S_Π is no longer guaranteed.

Our final idea to solve this problem continues to be to generate the simulated garbled circuit from scratch every time the protocol in rewound in a given round. Since the adversary of the underlying protocol $\mathcal{A}'$ needs to be aware of when it is being rewound, and cannot be allowed to change its random tape every time it is rewound, the simulator applies a PRF on the current view to choose new random tape for $\mathcal{A}'$ for the rest of the execution after rewinding.

For our second construction, we first use FHE for the evaluation of the next message function and hence end up with the next message to be sent in an encrypted form. We then use a garbled circuit (constructed in the input encoding phase) to get this next message decrypted. This is as opposed to directly using a garbled circuit to perform the entire computation. Hence the garbled circuit generated in the input encoding phase need only be able to decrypt ciphertexts of size dependent only the size of the messages in the protocol Π (and independent of the complexity of the next message function in particular). Finally, our underlying protocol Π has communication complexity (i.e., message sizes) independent of circuit size of the function to be computed (and dependent only the security parameter and the size of the input). Such a construction for Π can be obtained again by using FHE and Kilian's efficient PCP based on zero-knowledge arguments [Kil95]. This construction also suffers from similar caveats relating to adversary choosing the next outgoing message adaptively on the leakage (and the solutions to these problems make use of the same ideas as in the previous construction).

Finally, we prove that any leakage-resilient secure computation protocol C_Π with function independent input encoding implies a secure two-party computation protocol Σ with computation complexity independent of the size of the circuit being computed. The main idea we use to construct such a low communication complexity protocol is for the protocol participant of Σ to internally emulate the parties of C_Π but partition the load of emulating such parties in a highly "unbalanced" way. In more detail, the second party in Σ does much of the work of emulating *both parties* of the protocol C_Π. This leads to the second party knowing about the internal state of both parties in the protocol C_Π; however here we can rely on the leakage-resilience of C_Π to argue security. The only

---

[2] The problem of additional leakage while a simulator rewinds the adversary has also been observed in the context of leakage-resilient zero-knowledge [GJS11]. In [GJS11], such leakage in fact translated to additional leakage in the ideal world. For a comparison of our work to [GJS11], please see the end of this Section.

communication is when the second party needs to read some bits of the encoded input of the first party in C_Π.

Concurrent Independent Work. Independent of our work, leakage resilient secure computation protocols have been proposed in the "plain" model where there is no leakage-free offline phase and the adversary may ask for leakage on the entire memory [GJS11], [DHP11], [BCH11]. These works necessarily relax the security definition by allowing leakage on the input in the ideal world and hence are incomparable to ours.

2 Preliminaries and Model

2.1 Definition - Leakage Resilient Secure Computation

We model the leakage during the computation, by giving the adversary access to a leakage oracle.

Definition 1. *Leakage Oracle:* *Given a computation s (we assume s is given by a Boolean circuit) and λ, $0 \leq \lambda \leq 1$, a leakage oracle $\mathcal{O}_s^\lambda$ is defined as follows. For an input w, oracle $\mathcal{O}_s^\lambda(w)$ evaluates the queried PPT function on the transcript of the computation of s on w, where w includes the input and randomness to the computation. Specifically, let $\tau(s, w)$ denote the transcript of computation of s on w, i.e., it consists of the input, output, and all the intermediate values in the circuit during the computation. Let h be any PPT leakage function whose output length is at most a λ-fraction of its input length. Now, on query h, $\mathcal{O}_s^\lambda(w)$ responds with $h(\tau(s, w))$. We call $\mathcal{O}_s^\lambda = \{\mathcal{O}_s^\lambda(w)\}_w$ the λ-bounded leakage oracle for s, where w ranges over all possible inputs to circuit s.*

When $\lambda = 1$, we call this the complete *leakage oracle for s and omit the superscript.*

We now give a formal definition of security against malicious adversaries in the presence of leakage in the ideal/real simulation paradigm. The execution in the ideal and real world are as follows:

Execution in the Ideal Model

Inputs: Each party obtains an input; the ith party's input is denoted by x_i; we assume that the inputs are of the same length n. The adversary $\mathcal{A}$ receives an auxiliary-input z.

Send Inputs to Trusted Party: The honest party P_j sends its received input x_j to the trusted party. A malicious party may depending on its input (as well as on its coin tosses and the auxiliary input) either abort or send some other input of the same length to the trusted party.

Trusted Party Answers the First Party: In case the trusted party received the input pair, (u_1, u_2), the trusted party first replies to the first party with $f(u_1, u_2)$.

Trusted Party Answers the Second Party: In case the first party is malicious, it may depending on its input and the trusted party's answer, decide to abort the trusted party. In this case, the trusted party sends abort to the second party. Otherwise, the trusted party sends $f(u_1, u_2)$ to the second party.

Outputs: An honest party always outputs the message it obtained from the trusted party. A malicious party may output any arbitrary (probabilistic polynomial-time computable) function of the initial input and the message obtained from the trusted party.

Definition 2. *The output of honest party and the adversary in an execution of the above model is denoted by*
$IDEAL_{f,\mathcal{A}(z)}(x_1, x_2).$

Execution in the Real Model

Let P_1 and P_2 be two parties trying to securely compute functionality $f :$ $(\{0,1\}^*)^2 \rightarrow (\{0,1\}^*)^2$. W.l.o.g., let P_1 be the corrupt party (P_2 is honest). We want to model leakage from the honest party P_2. Note that we only need to consider leakage from honest parties. Let $\mathcal{A}$ be a PPT adversary controlling malicious party P_1. The adversary $\mathcal{A}$ can deviate from the protocol in an arbitrary way. In particular, $\mathcal{A}$ can access leakage from P_2 via its leakage oracle throughout the execution of the protocol. In addition, $\mathcal{A}$'s behavior may depend on some auxiliary input z. More details follow.

Our real world execution proceeds in two phases: a leakage-free *Encoding phase* and a *Secure Computation Phase* in which leakage may occur. In the Encoding phase, each party P_i, locally computes an encoding of its input and any other auxiliary information. The adversary has no access to the leakage oracle during the encoding phase. In the Secure Computation phase, the parties begin to interact, and the adversary gets access to the leakage from the local computation of P_2. Specifically, let s_2 be a circuit describing the internal computation of P_2. The circuit s_2 takes as input the history of protocol execution of the *secure computation* phase so far, and some specified bits of the encoded input, and outputs the next message from P_2 and the new internal state of P_2. Thus, by Definition 1, the leakage is accessible to the adversary via the oracle $\mathcal{O}_{s_2}$, which we abbreviate by $\mathcal{O}_2$. Thus, at the end of round j of the protocol, $\mathcal{A}$ can query $\mathcal{O}_2$ for any PPT function h_j applied on the transcript of the jth round computation of the honest party P_2. We denote this leakage information by l_2^j.

We consider two variants of the above model: one in which the *Encoding phase* can depend on the function to be computed, and another stronger model in which the leakage-free Encoding phase is independent of the function.

Definition 3. *We define* $REAL_{\pi,\mathcal{A}(z)}^{\mathcal{O}^\lambda}(x_1, x_2, \kappa)$ *to be the output pair of the honest party and the adversary $\mathcal{A}$ from the real execution of π as defined above on inputs (x_1, x_2), auxiliary input z to $\mathcal{A}$, with oracle access to $\mathcal{O}_2^\lambda$, and security parameter κ.*

Definition of Security

Definition 4. *Let f, π, be as described above. Protocol π is said to* securely compute f in the presence of $\lambda-$ leakage *if for every non-uniform probabilistic* *polynomial-time pair of algorithms $\mathcal{A} = (A_1, A_2)$ for the real model, there exists a non-uniform probabilistic polynomial-time pair $S = (S_1, S_2)$ for the ideal model such that*

$$IDEAL_{f,S(z)}(x_1, x_2) \stackrel{c}{\equiv} REAL^{\mathcal{O}^\lambda}_{\pi,\mathcal{A}(z)}(x_1, x_2, \kappa)$$

where $x_1, x_2, z \in \{0,1\}^*$, such that $|x_1| = |x_2|$ and $|z| = \mathsf{poly}(|x_1|)$. When $\lambda = 1$, protocol π is said to securely compute f in the presence of complete leakage.

The model for the multi-party setting is analogous to one given above for the two-party case. For lack of space, the details of the definition for the multi-party case are provided in the full version.

3   The Basic Construction

We give a compiler that transforms any semi-honest secure multiparty protocol into a leakage resilient multiparty protocol secure against malicious parties and resilient against complete leakage ($\lambda = 1$) from honest parties. For simplicity, we describe our results for the two party case. They can be naturally extended to the multiparty case using known techniques. We sketch such an extension in the full version.

The high level idea of the compiler is to garble the "Next Message" functions of a two party computation protocol in a leakage-free preprocessing phase. The internal states of the parties are maintained in an encrypted form using a semantically secure public key encryption scheme. The private input of each party is included as part of that party's initial (encrypted) state. All updates to the internal states are also performed on the corresponding ciphertexts. The garbled circuit of the Next Message function of a party acts on its encrypted internal state and the message from the other party to produce its new (encrypted) state and the message to be sent to the other party.

We start with any protocol Π secure against malicious adversaries, which is compiled by the compiler C. Let t be the number of rounds in Π. The compiled protocol C_Π is secure against malicious adversaries in the presence of complete leakage.

3.1   The Compiler

- **Input:** P_1 has input $x_1 \in \{0,1\}^n$ and P_2 has input $x_2 \in \{0,1\}^n$.
- **Encoding phase:** This is done in a leakage-free setting. For concreteness, we assume that in the original protocol Π, P_1 sends the first message. We use the notation m_i^j to denote the jth message sent by party i. Thus jth message from party P_1 followed by jth message from party P_2 constitutes the jth round.

Party P_1 does the following:

1. Initialize the secret state st_1^0 with the private input x_1 and private random tape for the protocol Π.

2. Choose a public key and secret key pair (pk_1, sk_1) of an Encryption scheme E.
   $(pk_1, sk_1) \leftarrow KeyGen(1^\kappa)$
3. Encrypt the initial secret state st_1^0 under pk_1.
   $E[st_1^0] \leftarrow Encrypt(pk_1, st_1^0)$
4. Let $m_1^j \leftarrow \text{NextMsg}_1^j(m_2^{j-1}, st_1^{j-1})$ be the Next Message function of Π to compute the jth message m_1^j that is to be sent to P_2. Let $NextMsgC_1^j$ be the circuit with the following functionality for $j \neq t$. It has the keys pk_1 and sk_1 hardcoded and takes as input $m_2^{j-1}, E[st_1^{j-1}]$. It decrypts $E[st_1^{j-1}]$, executes $\text{NextMsg}_1^j(m_2^{j-1}, st_1^{j-1})$, and hence computes the next message of P_1 as m_1^j. It outputs this next message and the new encrypted state $(E[st_1^j], m_1^j)$ and halts. For $j = t$, the circuit $NextMsgC_1^j$ takes the encrypted state and the message received, computes the message to be sent, updates the state, and outputs the state in the clear.
   $E[st_1^j], m_1^j \leftarrow NextMsgC_1^j(m_2^{j-1}, E[st_1^{j-1}])$.
5. For every round j, using the garbled circuit construction of Yao, garble the circuit $NextMsgC_1^j(m_2^{j-1}, E[st_1^{j-1}])$ to get the garbled circuit NextMsgGC_1^j.

P_2 acts symmetrically.

- **Secure Computation Phase**: This runs the leakage resilient protocol $C_\Pi(x_1, x_2)$ defined below.
  The parties emulate the underlying protocol Π, by replacing every call to the Next Message function by an invocation of the corresponding garbled circuit computed during the preprocessing phase.
  Party P_1 does the following:
  1. In round j, evaluate the garbled circuit NextMsgGC_1^j under the wire keys corresponding to the input $m_2^{j-1}, E[st_1^{j-1}]$. That is, from the encoded input, only the wire keys corresponding to the input $m_2^{j-1}, E[st_1^{j-1}]$ are "touched".

  $$E[st_1^j], m_1^j \leftarrow \text{NextMsgGC}_1^j(m_2^{j-1}, E[st_1^{j-1}])$$

  2. Send m_1^j to the other party.
  3. Update the secret state with $E[st_1^j]$ and wait for the next message from P_2 if it exists; otherwise halt.
  P_2 acts symmetrically.

4   Proof of Security

Theorem 1. *Assuming Π securely computes f in the malicious model, the protocol C_Π securely computes f as in Definition 4 for $\lambda = 1$.*

The proof constructs a simulator (ideal world adversary) whose output is computationally indistinguishable from the view of the real world adversary in an actual run of the protocol. The simulator will access the real world adversary and a trusted party.

4.1 Description of the Simulator

W.l.o.g., we assume P_1 is corrupt. Let A be the adversary controlling the corrupt party. We describe the simulator S.

- Let S_Π be the simulator for the protocol Π. We construct an adversary $\mathcal{A}'$ for the underlying protocol Π, on which S_Π can be run.
- Observe that the behavior of $\mathcal{A}'$ is the same as $\mathcal{A}$, except that $\mathcal{A}$ expects to see the leakage along with every message it receives. The simulator constructs an adversary $\mathcal{A}'$ for Π and runs S_Π on $\mathcal{A}'$. For each round j, S gives $\mathcal{A}$ the round j message m_2^j that S_Π gives $\mathcal{A}'$. Whenever S_Π rewinds $\mathcal{A}'$, S rewinds $\mathcal{A}$. Along with each message, the simulator also gives $\mathcal{A}$, the associated leakage that it computes in the following way.
- The simulator picks a key pair (pk_1, sk_1) of the Encryption scheme E. Let $E[m]$ denote the encryption of a message m under pk_1.
- Throughout, with every message m_2^j that S sends to the adversary, it also sends the associated leakage l_2^j which S computes in the following way: In round j, S picks a random string $rand_j$. It then constructs a fake garbled circuit of the Next Message function of Π that always outputs m_2^j and $E[rand_j]$. S evaluates this fake garbled circuit under the wire keys corresponding to inputs m_1^{j-1} and $E[rand_{j-1}]$, and sets l_2^j to the transcript of this evaluation. The *evaluation transcript* consists of garbled gate tables of every gate in the circuit, output decryption table, and a single garbled value for every wire in the circuit. **Whenever S rewinds A and is required to give leakage for the same round again, it constructs a fresh fake garbled circuit** as opposed to giving the evaluation transcript of the same one on a different input. Now since the adversary of the underlying protocol $\mathcal{A}'$ needs to be aware of when it is being rewound, and cannot be allowed to change its random tape every time it is rewound, S applies a PRF on the current view to choose new random tape for $\mathcal{A}'$ for the rest of the execution after rewinding. In more detail, the random tape for the rest of the execution of $\mathcal{A}'$ after being rewound is set to $PRF(view)$. The view consists of the concatenation of input, random tape, messages received so far. This takes care of the fact that the underlying adversary $\mathcal{A}'$ cannot be allowed to change its random tape every time it is rewound, and guarantees that the randomness is fresh for each execution. This idea is similar to how Resettable protocols are constructed [CGGM00].

The above idea works for any general rewinding strategy that S_Π may employ, and the success of the simulator S is guaranteed.

We prove the indistinguishability of the views going from the real protocol to the simulated one through a series of hybrids. We prove that the real protocol view is indistinguishable from that in the simulated execution by proving indistinguishability between every pair of successive hybrids. The detailed proof appears in Appendix A.1.

An extension to the multi-party case can be found in the full version.

5   Construction with Function Independent Encoding

We now give a construction of another compiler such that the input encoding phase is independent of the function f to be computed. To do this, we start with a generic protocol Π with round complexity and communication complexity independent of the function to be computed. We then use a compiler C in a similar way as before to transform Π into a protocol C_Π that is secure against a malicious adversary in the presence of complete leakage. Recall that the secure computation phase is subject to complete leakage, i.e., all the data and internal states of local computations in the online phase are completely visible to the adversary, whereas the input encoding phase is leakage-free.

5.1   The Generic Protocol Π

Our generic two party computation protocol Π, which is secure against semi-honest adversaries, is constructed based on a Fully Homomorphic Encryption (FHE) scheme E.

1. **Input:** P_1 has input $x_1 \in \{0,1\}^n$ and P_2 has input $x_2 \in \{0,1\}^n$
2. P_1 generates a key pair for FHE scheme E and sends the public key to P_2.
   $(pk, sk) \leftarrow KeyGen(1^\kappa)$
3. P_1 encrypts her input and sends the ciphertext to P_2.
   $E[x_1] \leftarrow Encrypt(x_1, pk)$
4. P_2 encrypts his input and evaluates the circuit $\mathcal{C}$ homomorphically.
   $E[f(x_1, x_2)] \leftarrow Eval(\mathcal{C}, E[x_1], E[x_2], pk)$
   P_2 sends $E[f(x_1, x_2)]$ to P_1.
5. P_1 decrypts to obtain $f(x_1, x_2)$. She sends $f(x_1, x_2)$ to P_2.

It is clear that the communication complexity of the above protocol Π depends only on the size of the input n and the security parameter κ and is independent of the function f (in particular, independent of the size of the circuit $\mathcal{C}$).

5.2   Compiler C

The protocol Π described above is compiled with Kilian's efficient WI arguments based on PCP to enforce honest behavior [Kil95], and this protocol is compiled with our compiler C to produce the compiled protocol C_Π.

- **Input Encoding Phase:** This phase is independent of the function to be computed.
  1. Party i chooses a key pair (pk_i, sk_i) of an FHE $E = (KeyGen, Encrypt, Decrypt, Eval)$.
     $$(pk_i, sk_i) \leftarrow KeyGen(1^\kappa)$$
  2. The initial internal state st_i^0 of Party i is initialized with the private input and randomness of P_i for the execution of Π. It is then encrypted under the chosen key pk_i.
     $$E[st_i^0] \leftarrow Encrypt(pk_i, st_i^0)$$

3. Let $Decrypt(sk_i, c)$ be the decryption circuit of the FHE scheme for Party i. Using the Garbled circuit construction of Yao, garble the decryption circuit to get $DecGC_i$. Let t be the number of rounds in the protocol Π. Construct t independent garbled circuits of $Decrypt(sk_i, c)$. Let $DecGC_i^j$ denote the garbled decryption circuit of Party i to be used in round j. The length of the ciphertext that the garbled circuit should handle as the input is the length of the messages in Π. Thus it is polynomial in the security parameter and the input size and independent of f.

– **Secure Computation Phase:**
P_1 does the following:
1. *Encryption:* Let m_2^{j-1} be the message received. Party P_1 encrypts the received message under its own public key.

$$E[m_2^{j-1}] \leftarrow Encrypt(pk_1, m_2^{j-1})$$

2. *Homomorphic evaluation:* Let $NextMsg_1^j$ be the circuit description of the next message function used by Party P_1 in round j in protocol Π. This circuit is evaluated homomorphically to get an encryption of m_1^j in round j.

$$(E[m_1^j], E[st_1^j]) \leftarrow Eval(E[st_1^{j-1}], E[m_2^{j-1}], NextMsg_1^j)$$

3. *Garbled circuit evaluation:* Party 1 decrypts $E[m_1^j]$ by evaluating the Garbled circuit $DecGC_1^j$ under the wire keys corresponding to input $E[m_1^j]$.
$$m_1^j \leftarrow DecGC_1^j(sk_1, E[m_1^j])$$
4. Party P_1 sends m_1^j to P_2.
5. Party P_1 updates the secret state to $E[st_1^j]$.
Party P_2 acts symmetrically.

The security guarantees of the protocol C_Π are stated in the following theorem.

Theorem 2. *The protocol C_Π is secure against complete leakage, as in Definition 4, with an encoding phase independent of the function to be computed.*

The proof of this theorem follows from ideas similar to that in theorem 1 and due to space limitations, we give the proof in the full version.

5.3 Extension to Multiparty Case

We now sketch an extension of C_Π to the multi-party case. The underlying semi-honest secure protocol Π which we compile is as follows.

– The parties run a multi-party protocol for the following functionality: The functionality takes as input, the vector of inputs of all the parties. It generates the key pair for a fully homomorphic encryption scheme E. The functionality encrypts the input of each party and generates n-out-of-n secret

shares of the decryption key. Each party gets as output, encryption of the inputs of all parties, the public key, and a share each of the secret key.
- The first party now evaluates the circuit homomorphically to get the encrypted output.

$$E[C(x_1, \ldots, x_n)] \leftarrow Eval(E[x_1], \ldots, E[x_n], C)$$

- The parties now run a multi-party protocol for the following functionality. The functionality takes as input the secret shares of the decryption key of each party, and outputs the decrypted output.

The above protocol is compiled as in the two-party case with Kilian's efficient WI arguments based on PCP to enforce honest behavior [Kil95]. This is compiled with compiler C as discussed in 5.2 to get a leakage resilient multi-party protocol C_Π.

6   Is FHE Necessary for a Construction with Function Independent Encoding?

Our protocol C_Π in the previous section uses Fully Homomorphic Encryption as a component to realize function-independent encoding. In this section, we address the question if this is necessary. We show that leakage resilience with function independent encoding implies two-party secure computation with a communication complexity that is independent of the function to be computed. In particular, given a leakage resilient protocol C_Π that securely computes f in the presence of complete leakage and uses a function-independent encoding, we construct a secure two party computation protocol Σ whose communication complexity is independent of f. Realizing the latter task was a major open question and is currently known to be possible only if one uses FHE. Thus, being able to avoid FHE in building a leakage resilient protocol with function-independent encoding would have an alternative solution to this open question that does not rely on FHE.

In what follows, we denote by P_1 and P_2 the players in the leakage resilient protocol C_Π and by p_1 and p_2 the players in the communication-efficient protocol Σ. Given C_Π we show how to build Σ. The protocol Σ is secure against semi-honest adversaries which do not deviate from the protocol, but they may only try to get more information than they are authorized to. This can be extended to the malicious case by compiling Σ using PCP-based efficient zero knowledge arguments to prove honest behavior ([GMW87], [Kil95]). (The communication complexity of Kilian's argument system is independent of the complexity of verifying the statement in question.)

Protocol Σ: Parties p_1 and p_2 with respective inputs x_1 and x_2 wish to jointly compute $f(x_1, x_2)$.

- *Input Encoding Phase:* The parties encode their input (and random tape) locally exactly as per the instructions of the protocol C_Π (see encoding phase of section 5.2). Let m be the size of the encoded input.

– *Secure Computation Phase:* Party p_2 runs the programs of *both* parties P_1 and P_2 in protocol C_Π. That is, p_2 internally runs the parties P_1 and P_2 of C_Π. Whenever party p_2 needs to read a bit of the encoded input of Party 1, it queries p_1 with the index, and gets the encoded input bit. Party p_2 computes the output $f(x_1, x_2)$, and sends the output to party p_1.

Security of Σ: Here we only provide a proof sketch of the security of the protocol Σ. The details are straightforward.

Party p_1 is corrupt: We first consider the case when p_1 is corrupt. All p_1 learns in the execution of Σ is the indices queried by p_2. We observe that a dishonest P_1 learns this *and the leakage* in the protocol C_Π. Thus, the view of dishonest p_1 in Σ is a strict subset of the view of dishonest P_1 in C_Π. By the security of C_Π, protocol Σ is secure.

Party p_2 is corrupt: In the case when p_2 is dishonest, the view of p_2 consists of the entire view of P_2 in C_Π and the view of P_1 in C_Π in the secure computation phase. This is exactly the same as the view of a dishonest P_2 in the presence of the leakage oracle. By the leakage resilience of C_Π, protocol Σ is secure when p_2 is dishonest.

Thus, we conclude that Σ is secure two-party computation protocol.

Communication complexity of Σ: The communication complexity is the maximum number of queries from p_2 to p_1. If the size of the encoded input is m, then the communication complexity of the protocol Σ is $O(m \log m)$, independent of the size of the circuit being evaluated for f. Even the protocol secure against malicious adversaries has communication complexity independent of f, due to Kilian's PCP based construction of zero knowledge argument. Thus, we have

Theorem 3. *For two party computation, if there exists a leakage-resilient secure protocol with function-independent encoding, then there exists a secure protocol (against malicious adversaries) with communication complexity independent of the function being computed.*

7 Continual Leakage Resilient Protocol

We now construct a protocol which runs the input encoding phase just once, and runs the computation phase of the protocol more than once without having to run the leak-free phase again.

The Model. The ideal and the real model remain similar to the one time computation case (see Section 2). The main difference in the ideal world is that now there are k interactions: the parties interact k times, and compute a different functionality of the same inputs x_i. That is $f_j(x_1, x_2)$ is the functionality computed in the jth interaction with the trusted party. In the real model, apart from the leakage-free encoding phase and a leaky secure computation phase, there is a *refresh* phase as well. The adversary is allowed bounded leakage (of up to t bits) in *each* refresh phase. Between every secure computation phase, there is a refresh phase. However the leakage free encoding phase is only one time. More details are provided in appendix B.

Our compiler starts with the protocol Π and produces the compiled protocol C_Π.

- **Input Encoding phase:** This phase is independent of the function to be computed. The first three steps of this phase remain the same as before.
  1. Party i chooses a key pair (pk_i, sk_i) of an FHE $E = (KeyGen, Encrypt, Decrypt, Eval)$.

  $$(pk_i, sk_i) \leftarrow KeyGen(1^\kappa)$$

  2. The initial internal state st_i^0 of Party i is initialized with the private input and randomness of P_i for the execution of Π_{INT}. It is then encrypted under the chosen key pk_i.

  $$E[st_i^0] \leftarrow Encrypt(pk_i, st_i^0)$$

  3. Let $Decrypt(sk_i, c)$ be the decryption circuit of the FHE scheme for Party i. Using the Garbled circuit construction of Yao, garble the decryption circuit to get $DecGC_i$. Let r be the number of rounds in the protocol Π_{INT}. Construct r independent garbled circuits of $Decrypt(sk_i, c)$. Let $DecGC_i^j$ denote the garbled decryption circuit of Party i to be used in round j. The length of the ciphertext that the garbled circuit should handle as the input is the length of the messages in Π_{INT}. Thus it is polynomial in the security parameter and the input size and independent of f.
  4. **Virtual Player Initialization.** Each party i initializes $3t + 1$ virtual players (where t is the leakage bound of the refresh step). The virtual player j holds a share $sk_i[j]$ of the secret key sk_i as input. In other words, at this stage, the party simply divides the secret key sk_i into $3t + 1$ shares $\{sk_i[1], \ldots, sk_i[3t + 1]\}$ (using an additive secret sharing scheme) and stores these shares.
- **Secure Computation phase:**
  This phase remains the same as before. P_1 does the following:
  1. *Encryption:* Let m_2^{j-1} be the message received. Party P_1 encrypts the received message under its own public key.

  $$E[m_2^{j-1}] \leftarrow Encrypt(pk_1, m_2^{j-1})$$

  2. *Homomorphic evaluation:* Let $NextMsg_1^j$ be the circuit description of the next message function used by Party P_1 in round j in protocol Π_{INT}. This circuit is evaluated homomorphically to get an encryption of m_1^j in round j.

  $$(E[m_1^j], E[st_1^j]) \leftarrow Eval(E[st_1^{j-1}], E[m_2^{j-1}], NextMsg_1^j)$$

  3. *Garbled circuit evaluation:* Party 1 decrypts $E[m_1^j]$ by evaluating the Garbled circuit $DecGC_1^j$ under the wire keys corresponding to input $E[m_1^j]$.

  $$m_1^j \leftarrow DecGC_1^j(sk_1, E[m_1^j])$$

4. Party P_1 sends m_1^j to P_2.

5. Party P_1 updates the secret state to $E[st_1^j]$.

Party P_2 acts symmetrically.

– **Refresh Phase:**

Let $\mathcal{F}$ be the following $3t + 1$-party (randomized) functionality:

- It takes as input $sk_i[j]$ from player j for $j \in [3t + 1]$. It reconstructs the secret key sk_i.
- It computes $3t+1$ shares of sk_i using fresh randomness (using an additive secret sharing scheme as before). Denote the j-th share by $sk_i'[j]$. It gives $sk_i'[j]$ as output to the j-th player.
- It also computes, for every round, a garbled circuit for decryption of a ciphertext using sk_i (as constructed in the leakage-free encoding phase). More precisely, it constructs r independent garbled circuits for $Decrypt(sk_i, c)$ and outputs that to each player.

Party P_1 does the following. It internally runs the BGW protocol (guaranteeing security for semi-honest players) [BOGW88] among the $3t + 1$ players. The players hold inputs $sk_i[1], \ldots, sk_i[3t+1]$ and run the BGW protocol for the functionality $\mathcal{F}$. *The internal computation of each player is modeled as a separate sub-computation.* Hence, the adversary is allowed to ask for leakage individually on each of the $3t + 1$ subcomputation as well as on the protocol transcript generated by the (virtual) interaction.

Note that The j virtual players started with shares $sk_i[1], \ldots, sk_i[3t + 1]$ and ended with new shares $sk_i'[1], \ldots, sk_i'[3t + 1]$ using which the refresh phase can be run again. Furthermore, at the end of Refresh phase, P_1 has r independent garbled circuits of $Decrypt(sk_i, c)$ (obtained as output by each player), and hence the Secure Computation phase can be run again.

Sketch of Proof of security. The security guarantees of the protocol C_Π are stated in the following theorem:

Theorem 4. *The protocol C_Π is secure against λ-continual leakage as in Definition 7 tolerating complete leakage in the secure computation phase, and t bits of leakage in the refresh phase, where t can be any apriori fixed polynomial in the security parameter κ.*

We construct a simulator and prove that the output is computationally indistinguishable from the view of a real world adversary in an actual run of the protocol. The simulator proceeds by simulating the Secure computation and the Refresh phase of each execution.

Simulation in the Secure Computation phase: The description of the simulator in this phase is the same as described in section 2.

Simulation during the Refresh phase: Since the adversary is restricted to t-bits of leakage, it can request leakage from at most t of the $3t + 1$ subcomputations. Invoke the BGW simulator to simulate the view of the adversary by corrupting these (at most) t players and constructing their view.

The indistinguishability of the views follows from theorem 2 and the security of the BGW construction [BOGW88].

References

BCH11.   Bitansky, N., Canetti, R., Halevi, S.: Leakage tolerant interactive protocols. Cryptology ePrint Archive, Report 2011/204 (2011)

BOGW88.   Ben-Or, M., Goldwasser, S., Wigderson, A.: Completeness theorems for non-cryptographic fault-tolerant distributed computation (extended abstract). In: STOC, pp. 1–10. ACM (1988)

CGGM00.   Canetti, R., Goldreich, O., Goldwasser, S., Micali, S.: Resettable zero-knowledge (extended abstract). In: Proceedings of 32rd Annual ACM Symposium on Theory of Computing (STOC), pp. 235–244. ACM Press (2000)

DHP11.   Damgaard, I., Hazay, C., Patra, A.: Leakage resilient secure two-party computation. Cryptology ePrint Archive, Report 2011/256 (2011)

DP08.   Dziembowski, S., Pietrzak, K.: Leakage-resilient cryptography. In: FOCS, pp. 293–302. IEEE Computer Society (2008)

FKPR10.   Faust, S., Kiltz, E., Pietrzak, K., Rothblum, G.N.: Leakage-Resilient Signatures. In: Micciancio, D. (ed.) TCC 2010. LNCS, vol. 5978, pp. 343–360. Springer, Heidelberg (2010)

Gen09.   Gentry, C.: Fully homomorphic encryption using ideal lattices. In: Mitzenmacher, M. (ed.) STOC, pp. 169–178. ACM (2009)

Gen10.   Gentry, C.: Toward Basing Fully Homomorphic Encryption on Worst-Case Hardness. In: Rabin, T. (ed.) CRYPTO 2010. LNCS, vol. 6223, pp. 116–137. Springer, Heidelberg (2010)

GJS11.   Garg, S., Jain, A., Sahai, A.: Leakage-Resilient Zero Knowledge. In: Rogaway, P. (ed.) CRYPTO 2011. LNCS, vol. 6841, pp. 297–315. Springer, Heidelberg (2011)

GKR08.   Goldwasser, S., Kalai, Y.T., Rothblum, G.N.: One-Time Programs. In: Wagner, D. (ed.) CRYPTO 2008. LNCS, vol. 5157, pp. 39–56. Springer, Heidelberg (2008)

GMW87.   Goldreich, O., Micali, S., Wigderson, A.: How to play any mental game or a completeness theorem for protocols with honest majority. In: Proceedings of 19th Annual ACM Symposium on Theory of Computing, pp. 218–229 (1987)

Kil95.   Kilian, J.: Improved Efficient Arguments. In: Coppersmith, D. (ed.) CRYPTO 1995. LNCS, vol. 963, pp. 311–324. Springer, Heidelberg (1995)

MR04.   Micali, S., Reyzin, L.: Physically Observable Cryptography. In: Naor, M. (ed.) TCC 2004. LNCS, vol. 2951, pp. 278–296. Springer, Heidelberg (2004)

NS09.   Naor, M., Segev, G.: Public-Key Cryptosystems Resilient to Key Leakage. In: Halevi, S. (ed.) CRYPTO 2009. LNCS, vol. 5677, pp. 18–35. Springer, Heidelberg (2009)

vDGHV10. van Dijk, M., Gentry, C., Halevi, S., Vaikuntanathan, V.: Fully Homomorphic Encryption over the Integers. In: Gilbert, H. (ed.) EUROCRYPT 2010. LNCS, vol. 6110, pp. 24–43. Springer, Heidelberg (2010)

Yao86.   Yao, A.C.: How to generate and exchange secrets. In: FOCS 1986: Proceedings of 27th Annual Symposium on Foundations of Computer Science, pp. 162–167 (1986)

A   The Basic Construction: Missing Details

A.1   Indistinguishability Argument for Theorem 1

We prove the indistinguishability of the views going from the real protocol to the simulated one through a series of hybrids. We prove that the real protocol view is indistinguishable from that in the simulated execution by proving indistinguishability between every pair of successive hybrids.

- **Hybrid H_0:** This is the output distribution of a real execution of the protocol. Clearly, H_0 is identical to
  $REAL^{\mathcal{O}_2}_{C_\Pi, \mathcal{A}(z)}(\bar{x}, \kappa)$.
- **Hybrid H_1:** This is identical to H_0 except in the leakages. The simulator replaces the leakage in the following way: A fake garbled circuit, $\widehat{GC}$ of the Next message function of the honest party in protocol Π is constructed that always outputs the correct next message and the encrypted state as in the previous hybrid. The leakage is set to the evaluation transcript of this fake circuit. The leakage in all rounds is replaced by this simulated leakage.

  In distribution $H_{1,i}$, the first i leakages are evaluation of fake circuits, and the rest are real. We have the sub hybrids $H_{1,0}, \ldots, H_{1,t}$ with $H_{1,0} = H_0$ and $H_{1,t} = H_1$.

  Indistinguishability from H_0, $H_0 \stackrel{c}{\equiv} H_1$: The two hybrids differ only in the leakage. H_0 consists of the evaluation of correct garbled circuits whereas, in H_1, fake garbled circuits are evaluated. For contradiction, assume there is a distinguisher D and a polynomial p, such that

  $$|Pr[D(H_{1,0}) = 1] - Pr[D(H_{1,t}) = 1]| > \frac{1}{p(\kappa)}$$

  it follows that, $\exists i$ such that,

  $$|Pr[D(H_{1,i}) = 1] - Pr[D(H_{1,i+1}) = 1]| > \frac{1}{tp(\kappa)}$$

  Two neighbouring hybrids $H_{1,i}$ and $H_{1,i+1}$ differ only in the $i+1$st transcript. The $i+1$st transcript is the evaluation of real Garbled circuit in $H_{1,i}$, whereas it is simulated in $H_{1,i+1}$. A distinguisher can be constructed that can distinguish between the distribution ensemble consisting of $\widehat{GC}$ and a single garbled value for each input wire, and the distribution ensemble consisting of a real garbled version of C, together with garbled values corresponding to the real input. This contradicts the security of Yao's garbled circuit protocol.
- **Hybrid H_2:** This is the same as H_1, except that in the leakages: the simulator changes the output of the fake garbled circuit to the encryption of a random string instead of the correct encrypted state. S internally maintains the correct state to honestly run Π, but only in the leakage, the fake garbled circuit outputs the encryption of a random string. In distribution $H_{2,i}$, the fake garbled circuits of the first i leakages output the encryption of a random string, and the rest are real. $H_{2,0}$ is the same as H_1, and hybrid $H_{2,t}$ is H_2.

Indistinguishability from H_1, $H_1 \stackrel{c}{\equiv} H_2$: Assume the existence of a distinguisher D that can distinguish between $H_{2,0}$ and $H_{2,t}$ it follows that, $\exists i$ such that, D that can distinguish between $H_{2,i}$ and $H_{2,i+1}$ Two neighbouring hybrids $H_{2,i}$ and $H_{2,i+1}$ differ only in the $i+1$st transcript. The $i+1$st transcript is the evaluation of a fake garbled circuit that outputs the correct encrypted state in $H_{2,i}$ and encryption of a random string in $H_{2,i+1}$. D can be used to distinguish two ciphertexts under the encryption scheme E. This contradicts the semantic security of E.

- **Hybrid H_3:** In this experiment, S simulates the execution of Π. Let S_Π be the simulator for the underlying protocol Π. S runs S_Π which rewinds $\mathcal{A}'$ and extracts the input. S forwards to $\mathcal{A}$, the protocol messages that S_Π simulates and gives to $\mathcal{A}'$. The output of the fake garbled circuits in the leakage are also changed from the correct next message to the ones output by S_Π.

  Indistinguishability from H_2, $H_2 \stackrel{c}{\equiv} H_3$: The hybrids H_3 and H_2 differ in the protocol messages. In H_2, the protocol Π is run honestly whereas in H_3, the protocol messages as simulated by S_Π is used. If there is is a PPT distinguisher that can distinguish between H_3 and H_2, then the distinguisher D can be used to distinguish between a real execution of protocol Π and a simulated one, which contradicts the security of Π. Indistinguishability therefore follows from the security of Π.

Therefore $H_3 \stackrel{c}{\equiv} H_0$. The hybrid H_3 is the same as the execution of the simulator S, and we have seen that H_0 is the real execution of C_Π. It follows that C_Π is secure as per Definition 4.

$$IDEAL_{f,S(z)}(\bar{x}, \kappa) \stackrel{c}{\equiv} REAL^{\mathcal{O}_2}_{C_\Pi, \mathcal{A}(z)}(\bar{x}, \kappa)$$

B Continual Leakage Resilient Protocol: The Model

Execution in the Ideal Model
k **interactions:** The parties interact k times, and compute a different functionality of the same inputs x_i. That is $f_j(x_1, x_2)$ is the functionality computed in the jth interaction with the trusted party.

Definition 5. *The outputs of honest party and the adversary in k executions of the above model is denoted by*

$$IDEAL_{k,f,\mathcal{A}(z)}(x_1, x_2).$$

Execution in the Real Model. Our real world execution proceeds in three phases: a leakage-free *Encoding phase* and a *Secure Computation Phase* and a *Refresh Phase* in which leakage may occur. In the Encoding phase, each party P_i, locally computes an encoding of its input and any other auxiliary information. The adversary has no access to the leakage oracle during the encoding phase. In the Secure Computation phase, the parties begin to interact, and the adversary gets access to the leakage from the local computation of P_2. Specifically, let s_2 be a circuit describing the internal computation of P_2. The circuit s_2 takes as input the history of protocol execution of the *secure computation* phase so far, and some specified bits of the encoded input, and outputs the next message from P_2 and the new internal state of P_2. Thus, by Definition 1, the leakage is accessible to the adversary via the oracle $\mathcal{O}_{s_2}$, which we abbreviate by $\mathcal{O}_2$. Thus, at the end of round j of the protocol, $\mathcal{A}$ can query $\mathcal{O}_2$ for any PPT function h_j applied on the transcript of the jth round computation of the honest party P_2. We denote this leakage information by l_2^j. After each execution, the parties run a *Refresh* phase. The adversary continues to have access to the leakage oracle during the Refresh phase. After Refresh, the protocol, that is the Secure Computation phase can be run again on the same input. Specifically, the *Encoding phase* is run just once in the beginning for k executions of the protocol. The *Refresh phase* is run after each execution of the *Secure Computation phase*.

Definition 6. *We define $REAL^{\mathcal{O}^\lambda}_{k,\pi,\mathcal{A}(z)}(x_1, x_2, \kappa)$ to be the output pairs of the honest party and the adversary $\mathcal{A}$ from k real executions of π as defined above on inputs (x_1, x_2), auxiliary input z to $\mathcal{A}$, with oracle access to $\mathcal{O}_2^\lambda$, and security parameter κ.*

Definition of Security

Definition 7. *Let f, π, be the executions in the ideal world and the real world as described above. Protocol π is said to* securely compute f in the presence of $\lambda-$ continual leakage *if for every non-uniform probabilistic polynomial-time pair of algorithms $\mathcal{A} = (A_1, A_2)$ for the real model, there exists a non-uniform probabilistic polynomial-time pair $S = (S_1, S_2)$ for the ideal model such that*

$$IDEAL_{k,f,S(z)}(x_1, x_2) \stackrel{c}{\equiv} REAL^{\mathcal{O}^\lambda}_{k,\pi,\mathcal{A}(z)}(x_1, x_2, \kappa)$$

for a polynomial number of executions k, where $x_1, x_2, z \in \{0,1\}^*$, such that $|x_1| = |x_2|$ and $|z| = \text{poly}(|x_1|)$. When $\lambda = 1$ in the Secure Computation phase, protocol π is said to securely compute f in the presence of complete leakage.

Leakage Squeezing of Order Two

Claude Carlet[1], Jean-Luc Danger[2,3],
Sylvain Guilley[2,3], and Houssem Maghrebi[2]

[1] LAGA, UMR 7539, CNRS, Department of Mathematics,
University of Paris XIII and University of Paris VIII,
2 rue de la liberté, 93 526 Saint-Denis Cedex, France
[2] TELECOM-ParisTech, Crypto Group,
37/39 rue Dareau, 75 634 Paris Cedex 13, France
[3] Secure-IC S.A.S., 80 avenue des Buttes de Coësmes,
35 700 Rennes, France
claude.carlet@univ-paris8.fr,
{jean-luc.danger,sylvain.guilley,houssem.maghrebi}@telecom-paristech.fr

Abstract. In masking schemes, *leakage squeezing* is the study of the optimal shares' representation, that maximizes the resistance order against high-order side-channel attacks. Squeezing the leakage of first-order Boolean masking has been problematized and solved previously in [8]. The solution consists in finding a bijection F that modifies the mask, in such a way that its graph, seen as a code, be of greatest dual distance. This paper studies second-order leakage squeezing, *i.e.* leakage squeezing with two independent random masks. It is proved that, compared to first-order leakage squeezing, second-order leakage squeezing at least increments (by one unit) the resistance against high-order attacks, such as high-order correlation power analyses (HO-CPA). Now, better improvements over first-order leakage squeezing are possible by relevant constructions of squeezing bijections. We provide with linear bijections that improve by strictly more than one (instead of one) the resistance order. Specifically, when the masking is applied on bytes (which suits AES), resistance against 1st-order (resp. 2nd-order) attacks is possible with one (resp. two) masks. Optimal leakage squeezing with one mask resists HO-CPA of orders up to 5. In this paper, with two masks, we provide resistance against HO-CPA not only of order $5 + 1 = 6$, but also of order 7.

Keywords: High-order side-channel attacks, leakage squeezing, Boolean logic, rate 1/3 linear codes with 3 disjoint information sets, AES.

1 Introduction

Masking is an implementation-level strategy to thwart side-channel attacks. A dth-order masking scheme consists in replacing the manipulation of one sensitive variable X by the manipulation of a vector of $d + 1$ variables $S_0, \cdots, S_d$ called shares, in such a way that:

- X can be deterministically reconstructed from all the shares, while
- no information on X can be retrieved knowing strictly less than $d+1$ shares.

S. Galbraith and M. Nandi (Eds.): INDOCRYPT 2012, LNCS 7668, pp. 120–139, 2012.
© Springer-Verlag Berlin Heidelberg 2012

In this case, sometimes referred to as *perfect masking*, it has been shown that:

- arbitrary computations can be carried out (see for instance [18]), and that
- the leaked information is nonzero, but decreases exponentially as $\mathcal{O}\left(\sigma^{-2 \times d}\right)$, where σ^2 is the variance of the noise that characterizes the measurement process [7].

Besides, it has been often reported that the cost overhead of masking, in terms of program executable file size or running time for software applications and in terms of implementation area for hardware applications, is too high for its adoption in real-world products. Therefore, the optimization of masking is of great practical importance.

The typical behavior of computing devices is to leak a non-injective and noisy function of the shares. It is usually modeled as a deterministic function of the shares plus an additive white Gaussian noise (AWGN). This model is justified by the fact that an attacker can only measure an aggregated function of each computing element's leakage, such as the total current drawn by the circuit. This means that the measurement indeed consists in the sum of the individual leakages of each processed bit, that can be partitioned into:

- the sum of the individual leakages of the bits of the sensitive variable X (which is obviously non-injective, as it projects words of identical Hamming weight onto the same image), and
- the sum of the individual leakages of each non-sensitive variable bits (that obeys a multinomial distribution, well approximated by a normal law).

Depending on the execution platform, the leakage of one bit can be modelled according to:

- its activity (the leakage is observed when the bit changes values), or
- its value (the leakage differs according to the bit's state).

Without loss of generality, we assume the first kind of leakage, which corresponds to the behavior of CMOS logic. The second kind of leakage is a particular case where the previous value is constant and null. Additionally, it can be assumed that every bit of a sensitive variable leaks an identical amount, irrespective of its neighbors. These assumptions lead to the so-called Hamming distance leakage model, *i.e.* a model in which the attacker records the noisy version of the sum of bitflips occurring in X.

This model might not comply exactly with the actual real-world leakage. One research direction is to study the impact of imperfections in the model (because of chip's design variability), that can be quantified for instance with the "perceived information" [17] metric. Another research direction is to do the most of "off-the-shelf" imperfect hardware. For instance, in the case when the countermeasure designer can influence the chip's manufacturing, he can ask that the indistinguishability of the bits and their non-interference explicitly figure in the product specifications. Technically, these requirements can be met; as a matter of fact, the gates that hold the bits of X:

- can be different instances of the same register flip-flop, constrained to have an identical fanout, and
- can be placed far away one from each other, with their output routing wires adequately shielded from nearby aggressors, so as to reduce their cross-talk.

Experimental feedback (from *in silico* measurements) indicates that those constraints are realistic [22]; for instance, in dual-rail logics, such constraints are enforced [21], with varied efficiency in a "static setup" (*i.e.* the only entropy comes from the data). However, in a "dynamic setup", such as masking, these constraints can definitely improve the trustworthiness on the accuracy of the leakage model.

Glitching is another flaw that limits the efficiency of the masking countermeasures; it is a "logical" coupling (as opposed to the "technological" nature of the cross-talk) that produces a higher-order leakage, not captured in the model. For example, it is reported in [11] a glitch that combines the two shares of a first-order countermeasure, thereby unintentionally disclosing one bit of X through the leakage function $\mathscr{L}$. The designer can opt to hide the computations in a synchronous memory table, that evaluates the output at once [20]. In such condition, no glitch is possible, since glitches stem from a race between two signals that converge to the inputs of a gate. However, tables are expensive. Nonetheless, it is possible to break the tables into smaller elements, provided each of them remain glitch-free. This is possible if every computing element receives its inputs simultaneously. Such strategy can be implemented at the gate-level if every gate is clocked and the combinational logic behaves like a very fast pipeline, as explained in [12]. Also, the designer can take advantage of recent works about "threshold implementations" [14] or "multi-party computation" [16], that both aim at securing masked combinational logic against insidious leakage conveyed by glitches. Their principle is to partition the combination functions into non-interfering submodules that compute on d shares or less, which denies all possibility of glitchy recombination that could disclose (all or part of) the sensitive variable X. In the other case when the countermeasure designer must use an already hard-wired circuit, then profiling can be used to characterize to which extent the leakage conditions are satisfied. The stochastic method [19] allows to precisely assess the leakage model. Notably, first-order coefficients should be checked to be as equal one to each other as possible, and second-order coefficients as small as possible [5] with respect to first-order coefficients. Eventually, it is known that in implementations with combinational logics (*e.g.* the sbox of the DPA contest v2 [3]), the Hamming distance 0 signs much less than the others. The reason is that the combinational nets of the sbox are already prepositioned, and thus the next identical computation does not require to recompute them. On tables, this "memory effect", also termed "clockwise collision" [6], is less visible, as all accesses, even identical consecutive ones, draw some current due to the dynamic character of lookups.

In the sequel, we assume that the leakage $\mathscr{L}$ is equal, or close enough, to the assumed model. In this case, the security of the masking countermeasure can be greatly enhanced. Notably, the indiscernibility and the non-interference of the bits can be taken advantage of to reach $(d+1)$th-order security with strictly less than $d+1$ shares. This strategy is called "leakage squeezing" [9]. It can be seen as a constructive combination of *masking* (through the splitting of X into shares) and of *hiding* (through the leakage function $\mathscr{L}$ properties, namely the leakage in Hamming distance). The figure 1, whose layout is inspired from [10, p. 12], illustrates the symbiosis of the *masking* and *hiding* countermeasures tactics in the leakage squeezing. Roughly speaking, masking is a "software" countermeasure, in that it is implemented by the designer (in assembly language or hardware description language), whereas hiding is a "hardware" countermeasure, in that it is a native property of the device.

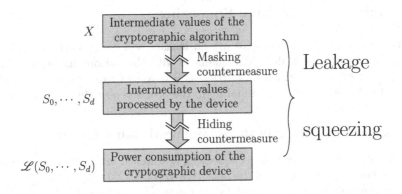

Fig. 1. Principle of the leakage squeezing, that takes on attributes from both the "masking" and the "hiding" strategies

A masking scheme involves a group $(\mathcal{X}, \perp)$, where $\mathcal{X}$ is the support of the sensitive variable X and $\perp$ an internal composition law. By definition of a group, the zero element 0 is neutral, *i.e.* $\forall X \in \mathcal{X}, X \perp 0 = 0 \perp X = X$, and for all element $X \in \mathcal{X}$, there is an opposite element denoted by $-X \in \mathcal{X}$ that satisfies $X \perp -X = -X \perp X = 0$. Several conventions can be adopted; in the most commonly encountered one, the sensitive variable is obtained as $X = S_0 \perp \cdots \perp S_d$. Under this assumption, $S_1, \cdots, S_d$ are independent uniformly distributed random variables on $\mathcal{X}$, and $S_0 \doteq X \perp \perp_{i=1}^{d} -S_i$. In digital circuits, the set $\mathcal{X}$ is made up of vectors of n bits. For example, $n = 8$ in AES, that manipulates bytes; also, $n = 4$ for DES, since it is usually the output of the sboxes that are targeted. Classical examples of masking are:

- Boolean masking, with $(\mathbb{F}_2^n, \oplus)$, or
- arithmetic masking, with $(\mathbb{Z}_{2^n}, \boxplus)$, where $\boxplus$ represents the modular addition.

In this paper, we will be making use of Boolean masking, as it lessens the degradation of performances in the context of hardware implementations: bits are masked one by one, hence the impact of the masking on the critical path is lowered (in particular, we avoid the carry propagation inherent to the arithmetic masking). Also, in $\mathbb{F}_2^n$, the opposite of a share $-S_i$ is the share itself ($-S_i = S_i$), hence the masking and demasking hardware can be factored.

The rest of this paper is structured as follows. In Sec. 2, we explain briefly the rationale about first-order leakage squeezing. Its extension to second-order leakage squeezing is tackled with in Sec. 3. In this section, we show that this generalization is not trivial. Nonetheless, we manage to characterize the adequate bijections and present some interesting solutions. Eventually, conclusions and perspectives are in Sec. 4. A case study on linear second-order leakage squeezing is given in Appendix A for $n = 8$. This last section details some practicalities: the article remains self-contained even without reading it.

2   Reminder on Leakage Squeezing

In this section, we recall the prior art on leakage squeezing at the destination of the reader who is not already acquainted with the notions introduced in [8] by Maghrebi *et al.* The gist of the article is Sec. 3; so this section can be safely skipped by the reader interested mainly in the progress over the state-of-the-art.

2.1   Leakage Squeezing in the Hamming Distance Model

The principle of first-order leakage squeezing is sketched in Fig. 2. The functional computation is carried out on the sensitive variable X, that is mixed with a random uniformly distributed mask (also of n-bit size) denoted by M. The shares are $(S_0, S_1) = (X \oplus M, M)$. As opposed to straightforward first-order masking, the shares are not held as such in registers. Instead, the two registers contain $X \oplus M$ (*i.e.* S_0) and $F(M)$ (*i.e.* $F(S_1)$). The function F must be a bijection, so that the mask value can be recovered from $F(S_0)$. In Fig. 2, the computational logic is concealed in memory tables (to ensure a glitch-free computation). However, any other "more optimized" (tables with $2n$-bit addresses are expensive) logic would also be suitable. The computation is conducted in such a way to respect the invariant:

$$X = \underbrace{S_0}_{\text{Masked data path}} \oplus \underbrace{S_1}_{\text{Mask path}} . \tag{1}$$

The scheme presented in Fig. 2 allows to compute (X', M') from (X, M) in one clock cycle:

- $X' = C(X)$ is a combinational function of X, where $C : \mathbb{F}_2^n \to \mathbb{F}_2^n$ is the expected functionality,
- $M' = R(M)$ is the mask refresh function. Two options are possible: either the mask M' is derived from M deterministically through $R : \mathbb{F}_2^n \to \mathbb{F}_2^n$, or it disregards M and it is drawn fresh from a true random number entropic source.

After one iteration, the invariant condition of Eqn. (1) is still met: $X' = S_0' \oplus S_1' = (X' \oplus M') \oplus F^{-1} \circ F(M')$.

In a hardware setup, the shares leak in the Hamming distance model. The leakage is thus equal to $\mathscr{L} = \mathsf{HW}((X \oplus M) \oplus (X' \oplus M')) + \mathsf{HW}(F(M) \oplus F(M'))$, that can be rewritten as $\mathscr{L} = \mathsf{HW}(Z \oplus M'') + \mathsf{HW}(F(M) \oplus F(M \oplus M'')) = \mathsf{HW}(Z \oplus M'', D_{M''}F(M))$. In this expression:

- HW is the Hamming weight function,
- Z is the difference between two consecutive values of the masked data ($Z \doteq X \oplus X'$),
- M'' is the difference between two consecutive values of the mask ($M'' \doteq M \oplus M'$) and
- $D_Y F(X)$ is the Boolean derivative of F in direction Y taken at point X.

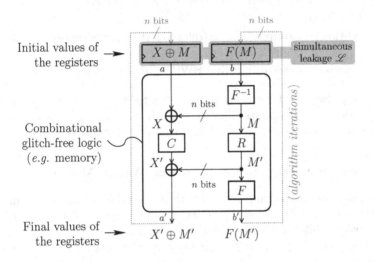

Fig. 2. Setup of the first-order masking countermeasure with bijection F

In the rest of this section, we recapitulate in one single page the key steps described extensively in [8] to find the first-order optimal leakage squeezing. The paper [8] is thus hereafter only surveyed, to highlight the reasoning. The section 3 will conduct step-by-step an accurate and self-contained analysis of the two-mask case.

It is shown in [8] that this leakage function is unexploitable by a dth-order correlation power analysis if all the terms $\mathbb{E}[\mathsf{HW}(Z \oplus M'')^p \times \mathsf{HW}(D_{M''}F(M))^q \mid Z = z]$, whatever p, q such as $p + q \leq d$ do not depend on z. In this expression, the capital letters represent random variables, and $\mathbb{E}$ is the expectation operator. The condition on F is equivalent to finding a bijection $F : \mathbb{F}_2^n \to \mathbb{F}_2^n$ that satisfies:

$$\forall a \in \mathbb{F}_2^{n*}, \quad \widehat{\mathsf{HW}^p}(a) = 0 \quad \text{or} \quad \mathbb{E}[\widehat{\mathsf{HW}^q \circ D_{(\cdot)}F(M)}](a) = 0 . \tag{2}$$

The term HW^p (resp. HW^q) represents the Hamming weight function raised at the power of $p \in \mathbb{N}$ (resp. $q \in \mathbb{N}$). The "hat" symbol represents the Fourier transform, that turns a function $f : \mathbb{F}_2^n \to \mathbb{Z}$ into $\hat{f} : \mathbb{F}_2^n \to \mathbb{Z}, x \mapsto \sum_{y \in \mathbb{F}_2^n} f(y)(-1)^{y \cdot x}$.

Eventually this expression, $\mathbb{E}[\mathsf{HW}^q \circ D_{(.)}F(M)]$ designates the function:

$$\mathbb{E}[\mathsf{HW}^q \circ D_{(.)}F(M)] : \mathbb{F}_2^n \;\to\; \mathbb{Z}$$
$$m'' \;\mapsto\; \mathbb{E}[\mathsf{HW}^q \circ D_{m''}F(M)] = \tfrac{1}{2^n} \textstyle\sum_m \mathsf{HW}^q(D_{m''}F(m)) \ .$$

The Eqn. (2) can be simplified, as Theorem 1 below is proved in [8, Appendix A.1].

Theorem 1. $\forall a \in \mathbb{F}_2^n, \forall p \in \mathbb{N}, \quad \widehat{\mathsf{HW}^p}(a) = 0 \iff \mathsf{HW}(a) > p$.

So the condition for the leakage squeezing to reach order d is simply to have: for all $a \in \mathbb{F}_2^{n*}$ and for all p such that $\mathsf{HW}(a) \le p$ and for all q such as $q \le d - p$, $\widehat{\mathbb{E}[\mathsf{HW}^q \circ D_{(.)}F(M)]}(a) = 0$.

This condition is also equivalent to (refer to forthcoming Lemma 1 at page 130):

$$\forall p, \forall (a, b) \text{ such that } \mathsf{HW}(a) \le p \text{ and } \mathsf{HW}(b) \le d - q, \text{ we have } \widehat{(b \cdot F)}(a) = 0 \ .$$

As shown in [8, Sec. 4], this condition can be related to "complementary information set" codes (*also known as* CIS codes [1]). It is equivalent that the indicator of the graph $\{(x, F(x)); x \in \mathbb{F}_2^n\}$ of F is d-th order correlation immune.

2.2   Leakage Squeezing in the Hamming Weight Model

If the device leaks in Hamming weight, then the relations are still valid if we replace the derivative $D_{(.)}F$ of F by F itself. Such an analysis is conducted in [7]. It is also worthwhile mentioning that if F is linear, the two problems are the same, because $D_m F(x) = F(x \oplus m) \oplus F(x) = F(x \oplus m \oplus x) = F(m)$, irrespective of x. This property is important, as a recent scholar work has shown empirically that on FPGAs, both Hamming distance and Hamming weight leakage models should be envisioned [13].

3   Second-Order Leakage Squeezing

3.1   Goal

In this section, an improvement of the leakage squeezing where two masks are used is studied. More precisely,

- the masked data ($X \oplus M_1 \oplus M_2$, also noted $X \oplus M$, where $M \doteq M_1 \oplus M_2$) is processed as is, *i.e.* through a bijection that is the identity (denoted by Id),
- the first mask (M_1) is processed through bijection F_1 and
- the second mask (M_2) is processed through bijection F_2.

This second-order masking scheme is illustrated in Fig. 3. With respect to the first-order masking scheme (Fig. 2 – described in § 2.1), the processing of the masked sensitive data is unchanged, and only the masks processing differs: each

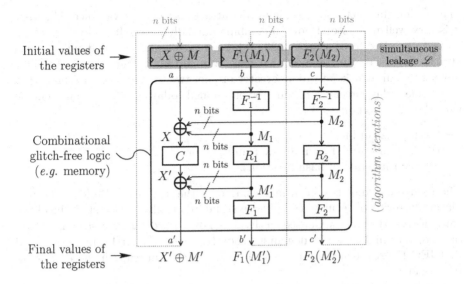

Fig. 3. Setup of the second-order leakage squeezing masking countermeasure with bijections F_1 and F_2

mask can be seeded independently and evolves from a different diversification function (noted R_1 and R_2).

The leakage function is thus:

$$\mathscr{L} = \mathsf{HW}((X{\oplus}M_1{\oplus}M_2){\oplus}(X'{\oplus}M_1'{\oplus}M_2'), F_1(M_1){\oplus}F_1(M_1'), F_2(M_2){\oplus}F_2(M_2')) \ .$$

As previously, $Z \doteq X \oplus X'$, and furthermore we denote: $M_1'' \doteq M_1 \oplus M_1'$ and $M_2'' \doteq M_2 \oplus M_2'$. Hence the leakage:

$$\begin{aligned}\mathscr{L} &= \mathsf{HW}(Z \oplus M_1'' \oplus M_2'', F_1(M_1) \oplus F_1(M_1 \oplus M_1''), F_2(M_2) \oplus F_2(M_2 \oplus M_2'')) \\ &= \mathsf{HW}(Z \oplus M_1'' \oplus M_2'', D_{M_1''}F_1(M_1), D_{M_2''}F_2(M_2)) \ .\end{aligned} \quad (3)$$

3.2   Motivation

It could be argued that the security brought by first-order leakage squeezing is already high enough, and resisting at still higher orders is a superfluous refinement. Admittedly, it has seldom been question of high-order attacks of order strictly greater than two in the abundant public literature.

However, searching for greater security can be motivated by "forward security" concerns. Secure elements (*e.g.* smartcards, RFID chips, hardware security modules, *etc.*) contain high-value secrets, and cannot be upgraded. Therefore, one can imagine buying one of these today, and having it attacked with tomorrow's know-how. For instance, with the advance of science, measurements apparati will have a lower noise figure and a greater vertical resolution in the future, thereby reducing the noise in side-channel acquisitions. Now, it is known that the limiting factor for the high-order attacks is the noise variance [24]. Also, it is now well understood how to combine partially successful side-channel attacks with brute force search [2,23]. Therefore, computer-assisted side-channel attacks might greatly enhance what can be done today. Thus, to avoid tomorrow successful attacks of orders greater of one, two, or more orders than what is possible today, precautions must be envisioned today. A parallel can be made with the evolution of:

- the key size of block ciphers,
- the modulus size of asymmetric primitive, or
- the internal state of hash functions.

Those have continuously been increasing over the last years. Besides, the regulation in terms of security compliance standards is always one step ahead the state-of-the-art attacks. Consequently, it is not absurd that side-channel resistance of very high order be demanded soon (*e.g.* with the forthcoming standard ISO/IEC 17 825), hence an incentive for the research in really high-order countermeasures.

Finally, some products supporting second-order countermeasures are already deployed in the field. The second-order leakage squeezing can be mapped in the devices of this installed base at virtually no extra cost, and so the application of this method in real products does not require further architectural development costs. The sole modification is the entry of the masking material in the (F_1, F_2) bijections, and their leaving at the end of the cryptographic application.

3.3   Formalization of Second-Order Leakage Squeezing

The attack fails at order d if $\forall i \leq d, \mathbb{E}\left((\mathscr{L} \mid Z = z)^i\right) = \mathbb{E}\left(\mathscr{L}^i \mid Z = z\right)$ does not depend on z. Indeed, the attacker has thus no bias to relate the leakage at order $i \geq 1$ to the (predictable and key-dependent) sensitive variable Z. Now, the goal of the attacker is to exhibit a bias in $\mathbb{E}\left(\mathscr{L}^d \mid Z = z\right)$ for an exponent d as small as possible, because the noise in $\mathscr{L}^d$ evolves as $\left(\sigma^2\right)^d$ [24], where σ^2 is the variance of the noise (for $d = 1$). Taking into account the formula of $\mathscr{L}$ from Eqn. (3), we have the following expression for $\mathbb{E}\left(\mathscr{L}^i \mid Z = z\right)$:

$$\mathbb{E}\left(\left(\mathsf{HW}(Z \oplus M_1'' \oplus M_2'', D_{M_1''}F_1(M_1), D_{M_2''}F_2(M_2))\right)^i \mid Z = z\right)$$

$$= \frac{1}{2^{4n}} \sum_{m_1'', m_2''} \sum_{m_1, m_2} \left(\mathsf{HW}(z \oplus m_1'' \oplus m_2'', D_{m_1''}F_1(m_1), D_{m_2''}F_2(m_2))\right)^i$$

$$= \frac{1}{2^{4n}} \sum_{\substack{m_1'', m_2'' \\ m_1, m_2}} \left(\underbrace{\mathsf{HW}(z \oplus m_1'' \oplus m_2'')}_{\text{Term } \#0} + \underbrace{\mathsf{HW}(D_{m_1''}F_1(m_1))}_{\text{Term } \#1} + \underbrace{\mathsf{HW}(D_{m_2''}F_2(m_2))}_{\text{Term } \#2}\right)^i .$$

This equation can be developed, to yield a sum of products of the three terms. Let us denote by p, q and r the degrees of each term, that satisfy $p + q + r = i$.

So attacks fail at order d if for all p, q and r such as $p + q + r \leq d$, the function

$$z \mapsto f(z)$$

$$\doteq \sum_{m_1'',m_2''} \sum_{m_1,m_2} \mathsf{HW}^p (z \oplus m_1'' \oplus m_2'') \cdot \mathsf{HW}^q (D_{m_1''} F_1(m_1)) \cdot \mathsf{HW}^r (D_{m_2''} F_2(m_2))$$

$$= \sum_{m_1'',m_2''} \mathsf{HW}^p (z \oplus m_1'' \oplus m_2'') \cdot \sum_{m_1} \mathsf{HW}^q (D_{m_1''} F_1(m_1)) \cdot \sum_{m_2} \mathsf{HW}^r (D_{m_2''} F_2(m_2))$$

$$= \sum_{m_1'',m_2''} \mathsf{HW}^p (z \oplus m_1'' \oplus m_2'') \cdot \mathbb{E}[\mathsf{HW}^q (D_{m_1''} F_1(M_1))] \cdot \mathbb{E}[\mathsf{HW}^r (D_{m_2''} F_2(M_2))]$$

$$= \left\{ \mathsf{HW}^p \otimes \mathbb{E}[\mathsf{HW}^q \circ D_{(.)} F_1(M_1)] \otimes \mathbb{E}[\mathsf{HW}^r \circ D_{(.)} F_2(M_2)] \right\} (z) \qquad (4)$$

is constant. From Eqn. (4), we see that every term to be kept constant is a double convolution product.

Keeping f constant is equivalent to having the Fourier transform $\hat{f}$ of f null everywhere but in zero. The Fourier transform turns a convolution product into a product; therefore,

$$\hat{f} = \widehat{\mathsf{HW}^p} \cdot \overline{\mathbb{E}[\mathsf{HW}^q \circ D_{(.)} F_1(M_1)]} \cdot \overline{\mathbb{E}[\mathsf{HW}^r \circ D_{(.)} F_2(M_2)]} \ .$$

In summary, to resist at order d, we are attempting to find two bijections F_1 and F_2 such as:

$$\forall a \in \mathbb{F}_2^{n*}, \quad \widehat{\mathsf{HW}^p}(a) = 0 \ \text{ or } \ \mathbb{E}[\widehat{\mathsf{HW}^q \circ D_{(.)} F_1(M)}](a) = 0$$

$$\text{or } \ \mathbb{E}[\widehat{\mathsf{HW}^r \circ D_{(.)} F_2(M)}](a) = 0 \ , \qquad (5)$$

for every triple of integers p, q and r such as $p + q + r \leq d$, d being the targeted protection order.

The Fourier support of a function $\psi : \mathbb{F}_2^n \to \mathbb{Z}$ is the set $\left\{ a \in \mathbb{F}_2^n ; \hat{\psi}(a) \neq 0 \right\}$. The equation (5) expresses the fact that the Fourier supports of HW^p, $\mathbb{E}[\mathsf{HW}^q \circ D_{(.)} F_1(M)]$ and $\mathbb{E}[\mathsf{HW}^r \circ D_{(.)} F_2(M)]$ intersect only in the singleton $\{0\}$.

3.4   Gaining at Least One Order with Two Masks Instead of One

It is a well known property that adding one mask increases the security by one order [24]. We here prove that the same benefit can be expected from the leakage squeezing.

Proposition 1. *Let F_1 be a bijection such that the security is reached at order d with one mask. Then, by introducing a second mask processed through whatever bijection F_2, the security is reached at order at least $d + 1$.*

Proof. Let (p, q, r) be any triple of integers such as $p + q + r \leq d + 1$. Then:

- if $r = 0$, $\widehat{\mathsf{HW}^r \circ D_{(.)} F_2} = \widehat{1 \circ D_{(.)} F_2} = \hat{1} = \delta$ is a Kronecker symbol function, hence null for all $a \neq 0$,

- otherwise, $r > 0$ and for all p, q, we have $p + q \leq d + 1 - r$ (by hypothesis), and so $p + q \leq d$. Thus, we have $\widehat{\mathsf{HW}^p}(a) \cdot \widehat{\mathsf{HW}^q \circ F_1}(a) = 0$, which implies that either $\widehat{\mathsf{HW}^p}(a) = 0$ or $\widehat{\mathsf{HW}^q \circ F_1}(a) = 0$ for $a \neq 0$. $\qquad\square$

3.5 Problem Equivalent Formulation in Terms of Boolean Theory

We shall need the next lemma, which was already more or less explicit in [8].

Lemma 1. *Let* $F : \mathbb{F}_2^n \to \mathbb{F}_2^n$ *be any function, let* q *be an integer such that* $0 < q < n$ *and let* $a \in \mathbb{F}_2^n$ *be nonzero. We have* $\sum_{z,m} \mathsf{HW}^{q'}(F(m) \oplus F(m \oplus z))(-1)^{a \cdot z} = 0$ *for every* $0 < q' \leq q$ *if and only if* $\widehat{b \cdot F}(a) = 0$ *for every* $b \in \mathbb{F}_2^n$ *such that* $\mathsf{HW}(b) \leq q$.

Proof. According to [8, Eqn. (15)], we have:

$$\sum_{z,m} \mathsf{HW}^{q'}(F(m) \oplus F(m \oplus z))(-1)^{a \cdot z} = \tag{6}$$

$$\frac{1}{2^{q'}} \sum_{j=0}^{q'} \binom{q'}{j} n^{q'-j} (-1)^j \sum_{k_1 + \cdots + k_n = j} \binom{j}{k_1, \cdots, k_n} \left(\sum_{x \in \mathbb{F}_2^n} (-1)^{(\oplus_{i=1}^n k_i e_i) \cdot F(x) + a \cdot x} \right)^2.$$

Since, for $b = \oplus_{i=1}^n k_i e_i$, we have $\sum_{x \in \mathbb{F}_2^n} (-1)^{(\oplus_{i=1}^n k_i e_i) \cdot F(x) + a \cdot x} = -2\widehat{b \cdot F}(a)$, the condition "$\widehat{b \cdot F}(a) = 0$ for every $b \in \mathbb{F}_2^n$ such that $\mathsf{HW}(b) \leq q$" is then clearly sufficient. Conversely, let the condition "$\widehat{b \cdot F}(a) = 0$ for every $b \in \mathbb{F}_2^n$ such that $\mathsf{HW}(b) \leq k$" be denoted by $P(k)$. We prove $P(k)$ by induction on $k \in \mathbb{N}$. $P(0)$ is clearly satisfied since $a \neq 0$. Assume that $P(k)$ is satisfied for some $0 \leq k \leq q-1$, then applying the hypothesis with $q' = k + 1$ implies that $\widehat{b \cdot F}(a) = 0$ for every b such that $\mathsf{HW}(b) = k + 1$ since we have only squares in (6) multiplied by coefficients which are all of the same sign and $P(k + 1)$ is then satisfied. This completes the proof by induction. $\qquad\square$

Incidentally, we remark that the Theorem 1 of previous Sec. 2.1 is also an immediate consequence of Lemma 1 with $F = \mathsf{Id}$.

We characterize now Eqn. (5) in terms of Fourier transform.

Proposition 2. *Let* F_1 *and* F_2 *be two permutations of* $\mathbb{F}_2^n$ *and* d *an integer smaller than* n. *The condition:*

$$\forall a \neq 0, \forall (p, q, r), \tag{7}$$

$$(p + q + r \leq d) \Longrightarrow \begin{cases} \widehat{\mathsf{HW}^p}(a) = 0 \ or \\ \sum_{z,m} \mathsf{HW}^q(F_1(m) \oplus F_1(m \oplus z))(-1)^{a \cdot z} = 0 \ or \\ \sum_{z,m} \mathsf{HW}^r(F_2(m) \oplus F_2(m \oplus z))(-1)^{a \cdot z} = 0 \ . \end{cases}$$

is satisfied if and only if:

$$\forall a \in \mathbb{F}_2^n, a \neq 0, \exists q, r / \begin{cases} \mathsf{HW}(a) + q + r = d - 1, \\ \forall b \in \mathbb{F}_2^n, \mathsf{HW}(b) \leq q \implies \widehat{b \cdot F_1}(a) = 0, \\ \forall c \in \mathbb{F}_2^n, \mathsf{HW}(c) \leq r \implies \widehat{c \cdot F_2}(a) = 0. \end{cases} \qquad (8)$$

Proof. Condition (7) is satisfied for every (p, q, r) such that $p + q + r \leq d$ if and only if it is satisfied when p is minimal such that $\widehat{\mathsf{HW}^p}(a) \neq 0$, r is minimal such that $\sum_{z,m} \mathsf{HW}^r(F_2(m) \oplus F_2(m \oplus z))(-1)^{a \cdot z} \neq 0$ and $p + q + r \leq d$. We know that the minimum value of p such that $\widehat{\mathsf{HW}^p}(a) \neq 0$ equals $\mathsf{HW}(a)$. Let r be the minimal element defined above. Condition (7) implies then:

$$\forall q \leq d - \mathsf{HW}(a) - r, \quad \sum_{z,m} \mathsf{HW}^q(F_1(m) \oplus F_1(m \oplus z))(-1)^{a \cdot z} = 0.$$

According to Lemma 1, this latter condition is equivalent to $\forall b, \mathsf{HW}(b) \leq d - \mathsf{HW}(a) - r \implies \widehat{b \cdot F_1}(a) = 0$ and we obtain the condition:

$$\forall a \neq 0, \exists r / \begin{cases} \forall b, \mathsf{HW}(b) \leq d - \mathsf{HW}(a) - r \implies \widehat{b \cdot F_1}(a) = 0, \\ \forall c, \mathsf{HW}(c) < r \qquad\qquad \implies \widehat{c \cdot F_2}(a) = 0. \end{cases}$$

Now, let us replace r by $r' \doteq r - 1$. Thus $\mathsf{HW}(c) < r$ is equivalent to $\mathsf{HW}(c) \leq r'$, and condition $\mathsf{HW}(a) + q + r = d$ is equivalent to $\mathsf{HW}(a) + q + r' = d - 1$. This shows that Eqn. (8) is necessary. Clearly it is also sufficient. □

It is clear from Proposition 2 that any choice of F_2 allows to increase by one the resistance order provided by F_1 (this has already been mentioned in Sec. 3.4). Indeed, let us denote by d_1 the maximal order of resistance of F_1 in the one mask situation. Then, $\forall a \neq 0, \forall p, q, p + q \leq d_1 \implies \widehat{b \cdot F_1}(a) = 0$. By reference to Eqn. (8), for a given $a \neq 0$, we choose:

- $q = d_1 - \mathsf{HW}(a)$, thus $\forall b \in \mathbb{F}_2^n, \mathsf{HW}(b) \leq q \implies \widehat{b \cdot F_1}(a) = 0$ (by definition of d_1).
- $r = 0$, thus $\forall c \in \mathbb{F}_2^n, \mathsf{HW}(c) \leq r \implies \widehat{c \cdot F_2}(a) = 0$ (indeed, $c = 0$, hence $\widehat{c \cdot F_2}(a) = \delta(a) = 0$ since $a \neq 0$).

Consequently, Eqn. (8) is met with $d = d_1 + 1$.

So, one strategy can be to start from F_1, the optimal solution with one mask (this solution is known from [8]), and then to choose F_2 so as to increase as much as possible the resistance degree. Another strategy is to find F_1 and F_2 jointly. This problem seems not to be a classical one in the general case. In the next section, we show however that it becomes a problem of coding theory when F_1 and F_2 are linear.

3.6   Solutions When F_1 and F_2 Are Linear

In this section, F_1 and F_2 are assumed to be linear. For every $b, x \in \mathbb{F}_2^n$, we have $b \cdot F_1(x) = F_1^t(b) \cdot x$, where F_1^t is the so-called adjoint operator of F_1, that is, the linear mapping whose matrix is the transpose of the matrix of F_1. Then, for every nonzero $a \in \mathbb{F}_2^n$, we have $\widehat{b \cdot F_1}(a) = -\frac{1}{2}\sum_{x\in\mathbb{F}_2^n}(-1)^{(F_1^t(b)\oplus a)\cdot x}$, which equals $-2^{n-1} \neq 0$ if $F_1^t(b) = a$ and is null otherwise. Let us denote by L_1 (resp. L_2) the inverse of mapping F_1^t (resp. F_2^t). Then $\widehat{b \cdot F_1}(a)$ (resp. $\widehat{c \cdot F_2}(a)$) equals $-2^{n-1} \neq 0$ if $b = L_1(a)$ (resp. if $c = L_2(a)$) and is null otherwise.

Let also $a \neq 0$. From Eqn. (8) of Proposition 2, we can choose:

- $q = \mathsf{HW}(L_1(a)) - 1$ and
- $r = \mathsf{HW}(L_2(a)) - 1$.

Thus $d = \min\{\mathsf{HW}(a) + \mathsf{HW}(L_1(a)) + \mathsf{HW}(L_2(a)) - 1; a \neq 0\}$, which is exactly the minimal distance of the code $\{(x, L_1^t(x), L_2^t(x)); x \in \mathbb{F}_2^n\}$ (of rate 1/3 and with three disjoint information sets) minus the number 1.

Example for Linear F_1 and F_2 for $n = 8$. The optimal linear codes of length $8 \times 3 = 24$ and of dimension 8 have minimal distance 8. For instance, code $[24, 8, 8]$ is a subcode of code $[24, 12, 8]$, that is itself the extension of the quadratic-residue (QR) code of length 23.

By properly arranging the bits in the codewords, the generator matrix can write $(I_8 \quad L_1^t \quad L_2^t)$, where:

$$L_1^t = \begin{pmatrix} 1\,0\,0\,0\,0\,1\,0\,1 \\ 1\,0\,0\,0\,0\,1\,1\,1 \\ 1\,0\,1\,1\,1\,0\,0\,1 \\ 1\,0\,1\,1\,1\,0\,1\,0 \\ 1\,0\,1\,1\,1\,1\,1\,0 \\ 0\,1\,1\,0\,0\,1\,1\,1 \\ 0\,1\,0\,1\,0\,1\,1\,1 \\ 0\,1\,0\,0\,1\,0\,0\,0 \end{pmatrix}, \quad L_2^t = \begin{pmatrix} 0\,1\,1\,0\,0\,1\,1\,0 \\ 0\,0\,0\,1\,1\,0\,1\,0 \\ 1\,1\,1\,1\,0\,1\,1\,1 \\ 1\,0\,1\,0\,0\,0\,0\,1 \\ 1\,1\,0\,1\,1\,0\,1\,0 \\ 1\,1\,1\,1\,0\,0\,0\,1 \\ 1\,1\,1\,1\,1\,1\,1\,0 \\ 1\,0\,0\,0\,1\,1\,1\,1 \end{pmatrix}. \tag{9}$$

Those matrices are of full rank, namely 8, and their inverses are:

$$(L_1^t)^{-1} = (L_1^{-1})^t = \begin{pmatrix} 0\,1\,1\,0\,1\,0\,0\,0 \\ 1\,0\,1\,1\,1\,1\,1\,1 \\ 1\,0\,0\,1\,0\,0\,1\,1 \\ 1\,0\,0\,1\,0\,1\,0\,1 \\ 1\,0\,1\,1\,1\,1\,1\,0 \\ 0\,0\,0\,1\,1\,0\,0\,0 \\ 1\,1\,0\,0\,0\,0\,0\,0 \\ 1\,1\,1\,1\,0\,0\,0\,0 \end{pmatrix}, \quad (L_2^t)^{-1} = (L_2^{-1})^t = \begin{pmatrix} 0\,0\,0\,0\,0\,1\,1\,1 \\ 0\,1\,0\,0\,1\,1\,1\,1 \\ 1\,1\,1\,0\,1\,0\,1\,1 \\ 0\,1\,0\,1\,1\,0\,1\,1 \\ 1\,1\,0\,1\,1\,1\,1\,0 \\ 1\,1\,1\,0\,0\,0\,0\,1 \\ 1\,1\,0\,0\,0\,1\,0\,1 \\ 1\,1\,1\,1\,1\,1\,0\,0 \end{pmatrix}.$$

The technique to find those matrices is described in Appendix A.

We note that the binary linear code $\{(x, F_1(x)); x \in \mathbb{F}_2^8\}$ has minimal distance 3, and that the binary linear code $\{(x, F_2(x)); x \in \mathbb{F}_2^8\}$ has minimal distance 4. So those two codes are non-optimal, because the best linear code of length 16 and dimension 8 has minimal distance 5 [8, Tab. 1 in §4.1]. This noting justifies that it is indeed relevant to search for the bijections doublet (F_1, F_2) together instead of one after the other, independently. It also suggests that non-linear codes might still achieve better.

Example for Linear F_1 and F_2 for $n = 4$. It is also possible to construct a rate $1/3$ linear code of dimension 4 with three distinct information sets, which is suitable to protect DES. This solution generates two bijections, that jointly allow to resist high-order attacks of order 1 to 5 inclusive.

Security Validation. An implementation with a leakage function $\mathscr{L}$ is vulnerable at order d if $\mathbb{E}[(\mathscr{L} - \mathbb{E}[\mathscr{L}])^d \mid Z = z]$ depends on z, *i.e.* if the variance of this random variable is strictly positive $(\mathsf{Var}[\mathbb{E}[(\mathscr{L} - \mathbb{E}[\mathscr{L}])^d \mid Z]] > 0)$. In this case, the asymptotic HO-CPA correlation coefficient $\rho_{\mathrm{opt}}^{(d)}$, equal to [15, Eqn. (15) at page 802]:

$$\rho_{\mathrm{opt}}^{(d)} \doteq \sqrt{\frac{\mathsf{Var}[\mathbb{E}[(\mathscr{L} - \mathbb{E}[\mathscr{L}])^d \mid Z]]}{\mathsf{Var}[(\mathscr{L} - \mathbb{E}[\mathscr{L}])^d]}} \quad , \tag{10}$$

is non-zero. The table 1 gives the values of $\rho_{\mathrm{opt}}^{(d)}$ for the second-order Boolean masking of bytes ($n = 8$), without and with leakage squeezing. The two variances involved in Eqn. (10) were computed using a multiprecision integer library; therefore, when $\rho_{\mathrm{opt}}^{(d)}$ is reported as 0 (integer zero, not the approximated floating number 0.000000), we really mean that $\mathbb{E}[(\mathscr{L} - \mathbb{E}[\mathscr{L}])^d \mid Z = z]$ does not depend on z.

For the sake of comparison, we also report in this table the results obtained with one mask. In such case, both the best linear and non-linear squeezing bijection F can be characterized. It is relevant to consider the linear bijections F as they allow an efficient protection against HO-CPA, whether the device leaks in Hamming weight or distance. The best linear F for leakage squeezing with one mask is secure against attacks of orders up to 4. It can be used with two masks, thereby granting a security up to order $4 + 1 = 5$. Our results, that are not based on the extension of a single mask solution, is a security against HO-CPA of orders up to 7. Therefore, our method provides a free advantage of two orders. Now, with one mask, the best achievable security is gotten by the use of a non-linear F. This function does only protect against attacks that exploit the Hamming distance (and not the Hamming weight), but allows to reach a resistance up to HO-CPA of order 5. here also, our linear solution with two masks is better than merely this code used with one mask extended with another mask: it protected at order up to $7 > 5 + 1$. Besides, it is interesting to compare the first nonzero correlation coefficients with and without leakage squeezing:

- with one mask, $\rho_{\mathrm{opt}}^{(d=2)}(\text{no LS})/\rho_{\mathrm{opt}}^{(d=5)}(\text{LS}) = 0.258199/0.023258 \approx 11$, and
- with two masks, $\rho_{\mathrm{opt}}^{(d=3)}(\text{no LS})/\rho_{\mathrm{opt}}^{(d=8)}(\text{LS}) = 0.038886/0.000446 \approx 87$.

Table 1. Optimal 2O-CPA correlation coefficient for zero-offset attacks at order d, without and with leakage squeezing (LS) on $n = 8$ bits. Results are rounded at the sixth decimal

| Order | One mask (see [8]) | | | Two masks (this paper) | | | |
| | Without LS | With LS | | Without LS | With "partial" LS | | With LS |
d	$F = \mathrm{Id}$	Optimal linear ([8, App. B])	Optimal non-linear ([8, Sec. 4.2])	$F_1 = F_2 = \mathrm{Id}$	$F_1 = \mathrm{Id}$, but F_2 as [8, App. B]	$F_2 = \mathrm{Id}$, but F_2 as [8, Sec. 4.2]	(Non-optimal) linear (cf Sec. 3.6)
1	0	0	0	0	0	0	0
2	0.258199	0	0	0	0	0	0
3	0	0	0	0.038886	0	0	0
4	0.235341	0	0	0	0	0	0
5	0	0.023231	0	0.049669	0	0	0
6	0.197908	0.016173	0.023258	0.003403	0.001286	0	0
7	0	0.042217	0	0.045585	0.000868	0.000726	0
8	0.164595	0.032796	0.046721	0.006820	0.002644	0.000682	0.000446

Table 2. Optimal 2O-CPA correlation coefficient for zero-offset attacks at order d, without and with leakage squeezing (LS) on $n = 4$ bits. Results are rounded at the sixth decimal

| Order | One mask (see [8]) | | Two masks (this paper) | | |
| | Without LS | With LS | Without LS | With "partial" LS | With LS |
d	$F = \text{Id}$	Optimal linear ([7, Eqn. (13)])	$F_1 = F_2 = \text{Id}$	$F_1 = \text{Id}$, and $F_2 \neq \text{Id}$ is [7, Eqn. (13)]	(Non-optimal) linear (cf Sec. 3.6)
1	0	0	0	0	0
2	0.377964	0	0	0	0
3	0	0	0.081289	0	0
4	0.363815	0.191663	0	0	0
5	0	0	0.105175	0.021035	0
6	0.346246	0.283546	0.015973	0.015973	0.022590

So, in front of leakage squeezing, not only the attacker shall conduct an attack of much higher order, but also she will get a very degraded distinguisher value.

On $n = 4$ bits, the optimal first-order leakage squeezing is linear and allows to reach resistance order of 3. The used optimal code is $[8, 4, 4]$. For the second-order leakage squeezing, we can resort to the linear code $[12, 4, 6]$, that improves by two $(6 - 4 = 2)$ orders (with only one additional mask) the resistance against HO-CPA. By the trivial construct of Sec. 3.4, only one additional order of resistance would have been gained. A summary of the results is shown in Tab. 2. The improvement from the "straightforward" to the "squeezed" masking is of two orders with one mask and three orders with two masks.

4   Conclusions and Perspectives

Leakage squeezing has been thoroughly studied in [8] in the context where one sole mask is used. This paper investigates the potential of leakage squeezing extension to second-order leakage squeezing, *i.e.* using two independent masks instead of only one. Trivially, the addition of one mask increases the resistance against HO-CPA by one order. Our analysis allows to characterize (in Proposition 2) the conditions to reach higher resistance. The optimal solutions are not as easy to find as in the case with one mask. Nonetheless, for the special case of linear bijections, we find that one solution (probably not optimal) consists in finding a rate $1/3$ linear code of maximal minimal distance with three disjoint information sets. The optimal $[24, 8, 8]$ linear code fulfills this condition, and makes it possible to resist attacks of all orders from 1 to 7 included. Concretely speaking, this result means that the same security level as a 7th-order attack is attainable with 2 instead of 7 masks, thus at a much lower implementation cost.

Finding better, for instance non-linear, bijections, could allow to further improve on top of these results. In particular, a thorough study of rate $1/2$ codes with two complementary information sets exists [1]. However, such work is missing in general for rate $1/d$ codes with $d > 2$ distinct information sets.

Another perspective is to integrate the second-order leakage squeezing with "hyperpipelined" designs [12], "threshold implementations" [14] or "multi-party computation" [16] masking schemes, so as to improve their order of resistance while at the same time removing the latent leakage by glitches (if the logic is not concealed in memories).

References

1. Carlet, C., Gaborit, P., Kim, J.-L., Solé, P.: A new class of codes for Boolean masking of cryptographic computations, October 6 (2011), http://dblp.uni-trier.de/rec/bibtex/journals/tit/CarletGKS12
2. Dichtl, M.: A new method of black box power analysis and a fast algorithm for optimal key search. J. Cryptographic Engineering 1(4), 255–264 (2011)
3. DPA Contest (2nd edition) (2009-2010), http://www.DPAcontest.org/v2/
4. Grassl, M.: Bounds on the minimum distance of linear codes and quantum codes (2007), http://www.codetables.de/ (accessed on July 23, 2012)

5. Heuser, A., Schindler, W., Stöttinger, M.: Revealing side-channel issues of complex circuits by enhanced leakage models. In: Rosenstiel, W., Thiele, L. (eds.) DATE, pp. 1179–1184. IEEE (2012)
6. Li, Y., Nakatsu, D., Li, Q., Ohta, K., Sakiyama, K.: Clockwise Collision Analysis – Overlooked Side-Channel Leakage Inside Your Measurements. Cryptology ePrint Archive, Report 2011/579 (October 2011), http://eprint.iacr.org/2011/579
7. Maghebi, H., Guilley, S., Carlet, C., Danger, J.-L.: Classification of High-Order Boolean Masking Schemes and Improvements of their Efficiency. Cryptology ePrint Archive, Report 2011/520 (September 2011), http://eprint.iacr.org/2011/520
8. Maghrebi, H., Carlet, C., Guilley, S., Danger, J.-L.: Optimal First-Order Masking with Linear and Non-linear Bijections. In: Mitrokotsa, A., Vaudenay, S. (eds.) AFRICACRYPT 2012. LNCS, vol. 7374, pp. 360–377. Springer, Heidelberg (2012)
9. Maghrebi, H., Guilley, S., Danger, J.-L.: Leakage Squeezing Countermeasure against High-Order Attacks. In: Ardagna, C.A., Zhou, J. (eds.) WISTP 2011. LNCS, vol. 6633, pp. 208–223. Springer, Heidelberg (2011)
10. Mangard, S., Oswald, E., Popp, T.: Power Analysis Attacks: Revealing the Secrets of Smart Cards (December 2006) ISBN 0-387-30857-1
11. Mangard, S., Schramm, K.: Pinpointing the Side-Channel Leakage of Masked AES Hardware Implementations. In: Goubin, L., Matsui, M. (eds.) CHES 2006. LNCS, vol. 4249, pp. 76–90. Springer, Heidelberg (2006)
12. Moradi, A., Mischke, O.: Glitch-free Implementation of Masking in Modern FPGAs. In: HOST, June 2-3, pp. 89–95. IEEE Computer Society, Moscone Center, San Francisco, CA, USA (2012), doi:10.1109/HST.2012.6224326
13. Moradi, A., Mischke, O.: How Far Should Theory Be from Practice? In: Prouff, E., Schaumont, P. (eds.) CHES 2012. LNCS, vol. 7428, pp. 92–106. Springer, Heidelberg (2012)
14. Nikova, S., Rijmen, V., Schläffer, M.: Secure hardware implementation of nonlinear functions in the presence of glitches. J. Cryptology 24(2), 292–321 (2011)
15. Prouff, E., Rivain, M., Bevan, R.: Statistical Analysis of Second Order Differential Power Analysis. IEEE Trans. Computers 58(6), 799–811 (2009)
16. Prouff, E., Roche, T.: Higher-Order Glitches Free Implementation of the AES Using Secure Multi-party Computation Protocols. In: Preneel, B., Takagi, T. (eds.) CHES 2011. LNCS, vol. 6917, pp. 63–78. Springer, Heidelberg (2011)
17. Renauld, M., Standaert, F.-X., Veyrat-Charvillon, N., Kamel, D., Flandre, D.: A Formal Study of Power Variability Issues and Side-Channel Attacks for Nanoscale Devices. In: Paterson, K.G. (ed.) EUROCRYPT 2011. LNCS, vol. 6632, pp. 109–128. Springer, Heidelberg (2011)
18. Rivain, M., Prouff, E.: Provably Secure Higher-Order Masking of AES. In: Mangard, S., Standaert, F.-X. (eds.) CHES 2010. LNCS, vol. 6225, pp. 413–427. Springer, Heidelberg (2010)
19. Schindler, W., Lemke, K., Paar, C.: A Stochastic Model for Differential Side Channel Cryptanalysis. In: Rao, J.R., Sunar, B. (eds.) CHES 2005. LNCS, vol. 3659, pp. 30–46. Springer, Heidelberg (2005)

20. Shah, S., Velegalati, R., Kaps, J.-P., Hwang, D.: Investigation of DPA Resistance of Block RAMs in Cryptographic Implementations on FPGAs. In: Prasanna, V.K., Becker, J., Cumplido, R. (eds.) ReConFig, pp. 274–279. IEEE Computer Society (2010)
21. Tiri, K., Hwang, D., Hodjat, A., Lai, B.-C., Yang, S., Schaumont, P., Verbauwhede, I.: Prototype IC with WDDL and Differential Routing – DPA Resistance Assessment. In: Rao, J.R., Sunar, B. (eds.) CHES 2005. LNCS, vol. 3659, pp. 354–365. Springer, Heidelberg (2005)
22. Tiri, K., Verbauwhede, I.: A VLSI Design Flow for Secure Side-Channel Attack Resistant ICs. In: DATE, pp. 58–63. IEEE Computer Society (2005), http://dx.doi.org/10.1109/DATE.2005.44
23. Veyrat-Charvillon, N., Gérard, B., Renauld, M., Standaert, F.-X.: An optimal Key Enumeration Algorithm and its Application to Side-Channel Attacks. Cryptology ePrint Archive, Report 2011/610 (2011), http://eprint.iacr.org/2011/610/
24. Waddle, J., Wagner, D.: Towards Efficient Second-Order Power Analysis. In: Joye, M., Quisquater, J.-J. (eds.) CHES 2004. LNCS, vol. 3156, pp. 1–15. Springer, Heidelberg (2004)

A Isolation of Three Information Sets from Code $[24, 8, 8]$

If F_1 and F_2 are two linear bijections then the linear code $\{(x, F_1(x), F_2(x)); x \in \mathbb{F}_2^n\}$ has $\{1, \cdots, n\}$, $\{n + 1, \cdots, 2n\}$ and $\{2n + 1, \cdots, 3n\}$ for information sets, since the restriction of the generator matrix of this code to the columns indexed in each of these three sets is invertible. Conversely, if a $[3n, n, d]$ code C is known with three disjoint information sets, then after rearranging the columns of its generator matrix so that these three information sets are $\{1, \cdots, n\}$, $\{n + 1, \cdots, 2n\}$ and $\{2n + 1, \cdots, 3n\}$, we have $C = \{(\phi_0(x), \phi_1(x), \phi_2(x)); x \in \mathbb{F}_2^n\}$ where ϕ_0, ϕ_1 and ϕ_2 are bijective. Then, by trading the dummy variable x for $y = \phi_0(x)$ through one-to-one function ϕ_0, we get $C = \{(y, \phi_1 \circ \phi_0^{-1}(y), \phi_2 \circ \phi_0^{-1}(y)); y \in \mathbb{F}_2^n\}$ and we can take $F_1 = \phi_1 \circ \phi_0^{-1}$ and $F_2 = \phi_2 \circ \phi_0^{-1}$.

One generator matrix for the $[24, 8, 8]$ code can be obtained as a submatrix of extended QR-code of length 23[1], such as:

$$
\begin{array}{c}
\begin{smallmatrix} 1 & 2 & 3 & 4 & 5 & 6 & 7 & 8 & 9 & 10 & 11 & 12 & 13 & 14 & 15 & 16 & 17 & 18 & 19 & 20 & 21 & 22 & 23 & 24 \end{smallmatrix} \\
\downarrow \\
\begin{pmatrix}
0 & 0 & 0 & 0 & 0 & 0 & 0 & 1 & 0 & 1 & 0 & 1 & 0 & 1 & 1 & 1 & 1 & 1 & 1 & 1 & 1 & 1 & 1 & 0 \\
0 & 0 & 0 & 0 & 0 & 0 & 1 & 0 & 1 & 0 & 1 & 0 & 1 & 0 & 1 & 1 & 1 & 1 & 1 & 1 & 1 & 1 & 1 & 0 \\
0 & 0 & 1 & 1 & 1 & 1 & 0 & 0 & 0 & 0 & 1 & 1 & 1 & 1 & 0 & 0 & 1 & 0 & 0 & 1 & 0 & 1 & 1 & 0 \\
0 & 1 & 0 & 1 & 1 & 1 & 0 & 0 & 0 & 0 & 1 & 1 & 1 & 1 & 0 & 1 & 0 & 0 & 1 & 0 & 1 & 0 & 1 & 0 \\
1 & 0 & 0 & 1 & 1 & 1 & 0 & 0 & 0 & 0 & 1 & 1 & 1 & 1 & 1 & 0 & 1 & 0 & 0 & 1 & 0 & 0 & 0 & 0 \\
1 & 1 & 0 & 0 & 1 & 1 & 0 & 0 & 1 & 1 & 0 & 0 & 1 & 1 & 0 & 0 & 0 & 1 & 0 & 1 & 0 & 0 & 1 & 0 \\
1 & 0 & 1 & 1 & 0 & 1 & 0 & 0 & 1 & 1 & 0 & 0 & 1 & 1 & 0 & 0 & 0 & 0 & 1 & 1 & 1 & 0 & 0 & 0 \\
0 & 1 & 1 & 1 & 1 & 0 & 0 & 0 & 1 & 1 & 0 & 0 & 1 & 1 & 0 & 0 & 0 & 1 & 1 & 0 & 0 & 1 & 0 & 1
\end{pmatrix}
\end{array}.
$$

---

[1] See: http://www.mathe2.uni-bayreuth.de/cgi-bin/axel/codedb?extensioncode id+39649+2+8 [4].

The goal is to rearrange the columns of this matrix to get a form:

$$\left(M_0^t \quad M_1^t \quad M_2^t\right) , \tag{11}$$

where M_0, M_1 and M_2 are 8×8 invertible matrices with elements in $\mathbb{F}_2$. The research algorithm is as follows: first, an invertible 8×8 matrix (M_0) is searched. There are $\binom{24}{8} = 735,471$ of them[2]. We find one M_0^t by considering the columns $[\![2,9]\!]$. Second, the $\binom{16}{8} = 12,870$ permutations of columns $\{1\} \cup [\![10,24]\!]$ are tested for a partitioning into two invertible matrices $\left(M_1^t \quad M_2^t\right)$. For instance, M_1^t can be the columns $\{1, 10, 11, 12, 13, 15, 17, 18\}$ and M_2^t the columns $\{14, 16\} \cup [\![19,24]\!]$. Those define the three bijections $\phi_i : \mathbb{F}_2^8 \to \mathbb{F}_2^8, x \mapsto M_i \times x$, for $i \in \{0,1,2\}$. After that, we get a generating matrix in systematic form $\left(I_8 \quad L_1^t \quad L_2^t\right)$; The matrices L_1^t and L_2^t are defined as $L_1^t = M_1 \times M_0^{-1} = \left((M_0^t)^{-1} \times M_1^t\right)^t$ and $L_2^t = M_2 \times M_0^{-1} = \left((M_0^t)^{-1} \times M_2^t\right)^t$.

A priori, it was not clear whether or not the $[24, 8, 8]$ code could be cut into three disjoint information sets. However, in this case, it is, as just described, and in a non-unique way. For instance, the same shape as Eqn. (11) can be obtained by selecting for M_0^t the columns of index 0 modulo 3, for M_1^t the columns of index 1 modulo 3, and for M_2^t the columns of index 2 modulo 3. This partitioning is not equivalent to the previous one, as the columns for the new matrices pick up columns from all three previous ones. However, results in terms of correlation coefficient (*c.f.* Eqn. (10)) are the same.

---

[2] This amount of tries is still manageable on a standard desktop personal computer; all the more so as, in practice, we find very quickly a solution as the number of partitionings that yield an invertible 8×8 matrix is $310,400$ (which represents around 42% of the possible partitionings).

ROSETTA for Single Trace Analysis

Recovery Of Secret Exponent
by Triangular Trace Analysis

Christophe Clavier[1], Benoit Feix[1,2], Georges Gagnerot[1,2], Christophe Giraud[3],
Mylène Roussellet[2], and Vincent Verneuil[2]

[1] XLIM-CNRS, Université de Limoges, France
firstname.familyname@unilim.fr
[2] INSIDE Secure, Aix-en-Provence, France
firstname-first-letterfamilyname@insidefr.com
[3] Oberthur Technologies, Pessac, France
c.giraud@oberthur.com

Abstract. In most efficient exponentiation implementations, recovering the secret exponent is equivalent to disclosing the sequence of squaring and multiplication operations. Some known attacks on the RSA exponentiation apply this strategy, but cannot be used against classical blinding countermeasures. In this paper, we propose new attacks distinguishing squaring from multiplications using a single side-channel trace. It makes our attacks more robust against blinding countermeasures than previous methods even if both exponent and message are randomized, whatever the quality and length of random masks. We demonstrate the efficiency of our new techniques using simulations in different noise configurations.

Keywords: Exponentiation, Side-Channel Analysis, Collision, Correlation, Blinding.

1 Introduction

Although crypto-systems are proven secure against theoretical cryptanalysis, they can be easily broken if straightforwardly implemented on embedded devices such as smart cards. Indeed, the so-called *Side-Channel Analysis* (SCA) takes advantage of physical interactions between the embedded device and its environment during the crypto-system execution to recover information on the corresponding secret key. Examples of such interactions are the device power consumption [16] or its electromagnetic radiation [10]. SCA can be mainly divided into two kinds: *Simple Side-Channel Analysis* (SSCA) and *Differential Side-Channel Analysis* (DSCA). The first kind aims at recovering information on the secret key by using only one execution of the algorithm whereas DSCA uses several executions of the algorithm and applies statistical analysis to the corresponding measurements to exhibit information on the secret key.

Amongst crypto-systems threatened by SCA, RSA [20] is on the front line since it is the most widely used public key crypto-system, especially in embedded

S. Galbraith and M. Nandi (Eds.): INDOCRYPT 2012, LNCS 7668, pp. 140–155, 2012.

environment. Therefore, many researchers have published efficient side-channel attacks and countermeasures specific to RSA over the last decade. Due to the constraints of the embedded environment, countermeasures must not only resist each and every SCA known so far but must also have the smallest impact in terms of performance and memory consumption. Nowadays, the most common countermeasure to prevent SSCA on RSA consists in using an exponentiation algorithm where the sequence of modular operations leaks no information on the secret exponent. Examples of such exponentiation are the square-and-multiply-always [8], the Montgomery ladder [13], the Joye ladder [12], the square-always [7] or the atomic multiply-always exponentiation [4]. The latter is generally favorite due to its very good performance compared to the other non-atomic methods. Regarding DSCA prevention, most common countermeasures consist in blinding the modulus and/or the message, and the exponent [14,8]. Their effect is to randomize the intermediate values manipulated during the exponentiation as well as the sequence of squarings and multiplications. In this paper we denote by *blinded exponentiation* an exponentiation using the atomic implementation presented in [4] where modulus, message and exponent are blinded.

Today blinded exponentiation remain resistant to most SCA techniques. Only the Big Mac attack presented by Walter [24] theoretically threatens this implementation, although no practical result has been ever published. Other attacks introduced later partially threaten this implementation. First, Amiel et al. [1] show how to exploit the average Hamming weight difference between squaring and multiplication operations to recover the secret exponent. Their technique is efficient when the modulus and the message are blinded. However it requires many exponentiation traces using a fixed exponent, so this attack can be thwarted by the randomization of the exponent. To circumvent the blinded exponentiation, they suggested to apply their attack on a single trace but did not try it in practice. Clavier et al. present in [5] the so-called *Horizontal Correlation Analysis*. They apply DSCA using the Pearson correlation coefficient [3] on a single exponentiation side-channel trace. The exponent randomization has no effect against this attack. Modulus and message blinding are efficient only if random masks are large enough (32 bits or more).

Other attacks on the RSA exponentiation are not mentioned in our study as they do not apply to the blinded exponentiation.

In this paper we propose new attacks on the blinded exponentiation which make use of a single execution trace. We achieve this by introducing two new distinguishers — the Euclidean distance and the collision correlation applied to the long-integer multiplication — which allow to efficiently distinguish a squaring from a multiplication operation without the knowledge of the message or the modulus.

Roadmap. In Section 2, we recall some basics on RSA implementations on embedded devices. In particular, we describe the attacks presented in [1,5,24] and we show that one of them can be extended using the collision-correlation technique. In Section 3, we present the principle of the so-called *Rosetta* analysis using two different distinguishers. In Section 4, we put into practice our attack and we demonstrate its efficiency using simulated side-channel traces of long-integer operations using a 32×32-bit multiplier. Moreover, we also compare our technique with previous attacks and show that it is more efficient especially on noisy measurements. We discuss in Section 5 possible methods to counteract Rosetta analysis. Finally, Section 6 concludes this paper.

2   Background

In this section, after presenting some generalities on RSA implementation in the context of embedded environment, we present three of the most efficient side-channel attacks published so far on RSA: the *Big Mac attack* published by Walter at CHES 2001 [24], the one published by Amiel et al. at SAC 2008 [1] and the *Horizontal Correlation Analysis* published at ICICS 2010 by Clavier et al. [5]. Also, we explain how the latter can be extended using a collision-correlation technique.

2.1   RSA Implementation

The standard way of computing an RSA signature S of a message m consists of a modular exponentiation with the private exponent: $S = m^d \bmod N$. The corresponding signature is verified by comparing the message m with the signature S raised to the power of the public exponent: $m \stackrel{?}{=} S^e \bmod N$.

In order to improve its efficiency, the signature is often computed using the Chinese Remainder Theorem (CRT). Let us denote by d_p (resp. d_q) the residue $d \bmod p - 1$ (resp. $d \bmod q - 1$). To compute the signature, the message is raised to the power of d_p modulo p then to the power of d_q modulo q. The corresponding results S_p and S_q are then combined using Garner's formula [11] to obtain the signature: $S = S_q + q(q^{-1}(S_p - S_q) \bmod p)$.

If used exactly as described above, RSA is subject to multiple attacks from a theoretical point of view. Indeed, it is possible under some assumptions to recover some information on the plaintext from the ciphertext or to forge fake signatures. To ensure its security, RSA must be used according to a protocol which mainly consists in formatting the message. Examples of such protocols are the encryption protocol OAEP and the signature protocol PSS, both of them being proven secure and included in the standard PKCS #1 V2.1 [19]. Note that, as they do not require the knowledge of the exponentiated value, the new attacks described in this paper also apply when either OAEP or PSS scheme is used.

From a practical point of view, the RSA exponentiation is also subject to many attacks if straightforwardly implemented. For instance, SSCA, DSCA or collision analysis can be used to recover the RSA private key. SSCA aims at distinguishing a difference of behavior when an exponent bit is a 0 or a 1.

DSCA allows a deeper analysis than SSCA by exploiting the dependency which exists between side-channel measurements and manipulated data values [2]. To this end, thousands of measurements are generally combined using a statistical distinguisher to recover the secret exponent value. Nowadays, the most widespread distinguisher is the Pearson linear correlation coefficient [3].

Finally, collision analysis aims at identifying when a value is manipulated twice during the execution of an algorithm.

Algorithm 1 presents the classical atomic exponentiation which is one of the fastest exponentiation algorithms protected against the SPA.

Algorithm 1. Atomic Multiply-Always Exponentiation

> **Input:** $x, n \in \mathbb{N}$, $d = (d_{v-1}d_{v-2}\ldots d_0)_2$
> **Output:** $x^d \bmod n$

1: $R_0 \leftarrow 1$
2: $R_1 \leftarrow x$
3: $i \leftarrow v - 1$
4: $k \leftarrow 0$
5: **while** $i \geq 0$ **do**
6: $\quad R_0 \leftarrow R_0 \times R_k \bmod n$
7: $\quad k \leftarrow k \oplus d_i$ $\hfill$ [$\oplus$ stands for bitwise X-or]
8: $\quad i \leftarrow i - \neg k$ $\hfill$ [$\neg$ stands for bitwise negation]
9: **return** R_0

When correctly implemented, Alg. 1 defeats SSCA since squarings cannot be distinguished from other multiplications on a side-channel trace, as depicted by Fig. 1.

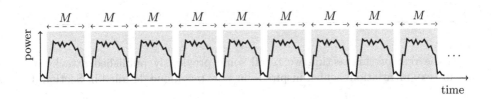

Fig. 1. Atomic multiply-always side-channel leakage

To prevent the implementation of RSA exponentiation from DSCA, the two main countermeasures published so far are based on message and exponent blinding [8,14]. Instead of computing straightforwardly $S = m^d \bmod n$, one rather computes $\tilde{S} = (m + k_0 \cdot n)^{d+k_1 \cdot \varphi(n)} \bmod 2^\lambda \cdot n$ where φ denotes the Euler's totient and k_0 and k_1 are two λ-bit random values, then finally reduce $\tilde{S}$ modulo N to obtain S. Using such a blinding scheme with a large enough λ (32 bits are generally considered as a good compromise between security and cost overhead), the relationship between the side-channel leakages occurring during an exponentiation and the original message and exponent values is hidden to an adversary, therefore circumventing DSCA.

As the modular exponentiation consists of a series of modular multiplications, it relies on the efficiency of the modular multiplication. Many methods have been published so far to improve the efficiency of this crucial operation. Amongst these methods, the most popular are the Montgomery, Knuth, Barrett, Sedlack or Quisquater modular multiplications [17,9]. Most of them have in common that the long-integer multiplication is internally computed by repeatedly calling a smaller multiplier operating on t-bit words. A classic example is given in Alg. 2 which performs the schoolbook long-integer multiplication using a t-bit internal multiplier giving a $2t$-bit result. The decomposition of an integer x in t-bit words is given by $x = (x_{\ell-1}x_{\ell-2}\ldots x_0)_b$ with $b = 2^t$ and $\ell = \lfloor \log_b(x) \rfloor + 1$.

Algorithm 2. Schoolbook Long-Integer Multiplication

    Input: $x = (x_{\ell-1}x_{\ell-2}\ldots x_0)_b, y = (y_{\ell-1}y_{\ell-2}\ldots y_0)_b$
    Output: $x \times y$
1: **for** $i = 0$ **to** $2\ell - 1$ **do**
2:     $z_i \leftarrow 0$
3: **for** $i = 0$ **to** $\ell - 1$ **do**
4:     $R_0 \leftarrow 0$
5:     $R_1 \leftarrow x_i$
6:     **for** $j = 0$ **to** $\ell - 1$ **do**
7:         $R_2 \leftarrow y_j$
8:         $R_3 \leftarrow z_{i+j}$
9:         $(R_5 R_4)_b \leftarrow R_3 + R_2 \times R_1 + R_0$
10:       $z_{i+j} \leftarrow R_4$
11:       $R_0 \leftarrow R_5$
12:     $z_{i+\ell} \leftarrow R_5$
13: **return** z

In the rest of this section we recall some previously published attacks on atomic exponentiations which inspired our new technique detailed in Section 3.

2.2 Attacks Background

Distinguishing Squarings from Multiplications in Atomic Exponentiation. In [1] Amiel et al. present a specific DSCA aimed at distinguishing squaring from other multiplications in the atomic exponentiation. They observe that the average Hamming weight of the output of a multiplication $x \times y$ has a different distribution whether:

- the operation is a squaring performed using the multiplication routine, i.e. $x = y$, with x uniformly distributed in $[0, 2^{\ell t} - 1]$;
- or the operation is an actual multiplication, with x and y independent and uniformly distributed in $[0, 2^{\ell t} - 1]$.

Thus, considering a device with a single long-integer multiplication routine used to perform either $x \times x$ or $x \times y$, a set of N side-channel traces computing multiplications with random operands can be distinguished from a set of N traces computing squarings, provided that N is sufficiently large to make the two distribution averages separable. This attack can thus target an atomic exponentiation such as Alg. 1 even in the case of message and modulus blinding. Regarding the exponent blinding, authors suggest that their attack should be extended to success on a single trace but do not give evidence of its feasibility. We thus study this point in the following of the paper.

Horizontal Correlation Analysis. Correlation analysis on a single atomic exponentiation side-channel trace has been published in [5] where the message is known to the attacker but the exponent is blinded. This attack called *horizontal correlation analysis* requires only one exponentiation trace to recover the full RSA private exponent.

Instead of considering the whole k-th long-integer multiplication side-channel trace T^k as a block, the authors consider each inner side-channel trace segment corresponding to a single-precision multiplication on t-bit words. For instance, if the long-integer multiplication is performed using Alg. 2 on a device provided with a t-bit multiplier, then the trace T^k of the k-th long-integer multiplication $x \times y$ can be split into ℓ^2 trace segments $T^k_{i,j}$, $0 \le i, j < \ell$, each of them representing a single-precision multiplication $x_i \times y_j$. More precisely, for each word y_j of the multiplicand y, the attacker obtains ℓ trace segments $T^k_{i,j}$, $0 \le i, j < \ell$, corresponding to a multiplication by y_j. The slicing of T^k into trace segments $T^k_{i,j}$ is illustrated on Fig. 2.

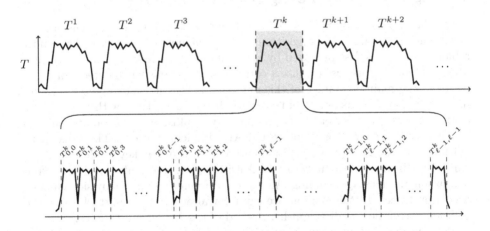

Fig. 2. Horizontal side-channel analysis on exponentiation

In the horizontal correlation analysis the attacker is able to identify whether the k-th long-integer operation T^k is a squaring or a multiplication by computing the correlation factor between the series of Hamming weights of each t-bit word m_j of the message m and the series of corresponding sets of ℓ trace segments $T_{i,j}^k$, $0 \leq i, j < \ell$. This correlation factor is expected to be much smaller when the long-integer operation is a squaring (i.e. $R_0 \leftarrow R_0 \times R_0$ in Alg. 1) than when it is a multiplication by m (i.e. $R_0 \leftarrow R_0 \times R_1$). The correlation factor can be computed by using the Pearson correlation coefficient $\rho(H, T^k)$ where $H = (H_0, \ldots, H_{\ell-1})$, with $H_j = (\mathrm{HW}(m_j), \ldots, \mathrm{HW}(m_j))$, $\mathrm{HW}(m_j)$ standing for the Hamming weight of m_j and $T^k = (T_0^k, \ldots, T_{\ell-1}^k)$ with $T_j^k = (T_{0,j}^k, \ldots, T_{\ell-1,j}^k)$.

Big Mac Attack. Walter's attack needs, as our technique, a single exponentiation side-channel trace to recover the secret exponent. For each long-integer multiplication, the Big Mac attack detects if the operation performed is either $R_0 \times R_0$ or $R_0 \times m$. The multiplications $x_i \times y_j$ — and corresponding trace segments $T_{i,j}^k$ — can be easily identified on the side-channel trace from their specific pattern which is repeated ℓ^2 times in the long-integer multiplication loop. A template side-channel trace is computed (either from the precomputations or from the first squaring operation) to characterize the manipulation of the message during the long-integer multiplication. The Euclidean distance between the template trace and each long-integer multiplication trace T^k is then computed. If it exceeds a threshold then the attack concludes that the operation is a squaring, or a multiplication by m otherwise.

Walter uses the Euclidean distance but we noticed that other distinguisher could be used. In the following section, we extend the Big Mac attack using a collision-correlation technique.

2.3   Big Mac Extension Using Collision Correlation

A specific approach for SCA uses information leakages to detect collisions between data manipulated in algorithms. A side-channel collision attacks against a block cipher was first proposed by Schramm et al. in 2003 [22]. More recently Moradi et al. [18] proposed to use a correlation distinguisher to detect collisions in AES. The main advantage of this approach is that it is not necessary to define a leakage model as points of traces are directly correlated with other points of traces. Later, Clavier et al. [6] presented two collision-correlation techniques defeating different first order protected AES implementations. The same year, Witteman et al. [25] applied collision correlation to public key implementation. They describe an efficient attack on RSA using square-and-multiply-always exponentiation and message blinding. All these techniques require many side-channel execution traces. In this section, we extend Walter's Big Mac attack using the collision correlation as distinguisher instead of the Euclidean distance.

We consider a blinded exponentiation and use the fact that the second and third modular operations in an atomic exponentiation are respectively $1 * \tilde{m}$ and $\tilde{m} * \tilde{m}$, where $\tilde{m}$ is the blinded message. The trace of the second long-integer multiplication yields ℓ multiplication segments for each word $\tilde{m}_j$ of the blinded message. Considering the k-th long-integer multiplication, $k > 3$, we compute the correlation factor between the series of ℓ trace segments T_j^2 — each one being composed of the ℓ trace segments $T_{i,j}^2$ involved in the multiplication by $\tilde{m}_j$ — and the series of ℓ trace segments T_j^k. Since the blinded value of the message does not change during the exponentiation, a high correlation occurs if the k-th long-integer operation is a multiplication, and a low correlation otherwise. Once the sequence of squarings and multiplications is found, the blinded exponent value is straightforwardly recovered. Notice that recovering the blinded value of the secret exponent is not an issue as it can be used to forge signature as well as its non-blinded value.

This attack also works if we use the trace segments T_j^3 of the third long-integer operation instead of the trace segments T_j^2. One can also combine the information provided by the second and third long-integer operations to improve the attack.

Remark. As the original Big Mac, this attack also applies to the CRT RSA exponentiation since no information is required on either the message or the modulus. This is of the utmost importance since, to the best of our knowledge, this is the first practical attack on a CRT RSA fully blinded (message, modulus and exponent) atomic exponentiation.

3 ROSETTA: Recovery Of Secret Exponent by Triangular Trace Analysis

3.1 Attack Principle

The long-integer multiplication $\mathrm{LIM}(x, y)$ in base $b = 2^t$ is given by the classical schoolbook formula:

$$x \times y = \sum_{i=0}^{\ell-1} \sum_{j=0}^{\ell-1} x_i y_j b^{i+j}$$

and illustrated, with for instance $\ell = 4$ by the following matrix M:

$$M = \begin{pmatrix} x_0 y_0 & x_0 y_1 & x_0 y_2 & x_0 y_3 \\ x_1 y_0 & x_1 y_1 & x_1 y_2 & x_1 y_3 \\ x_2 y_0 & x_2 y_1 & x_2 y_2 & x_2 y_3 \\ x_3 y_0 & x_3 y_1 & x_3 y_2 & x_3 y_3 \end{pmatrix}$$

In the case of a squaring, then $x = y$ and the inner multiplications become:

$$S = \begin{pmatrix} x_0 x_0 & x_0 x_1 & x_0 x_2 & x_0 x_3 \\ x_1 x_0 & x_1 x_1 & x_1 x_2 & x_1 x_3 \\ x_2 x_0 & x_2 x_1 & x_2 x_2 & x_2 x_3 \\ x_3 x_0 & x_3 x_1 & x_3 x_2 & x_3 x_3 \end{pmatrix}$$

We consider four observations to design our new attacks, assuming a large enough multiplier size $t \geq 16$:

(Ω_0) LIM(x,y) s.t. $x = y \Rightarrow$ Prob$(x_i \times y_i$ are squaring operations$) = 1$ $\forall i$
(Ω_1) LIM(x,y) s.t. $x \neq y \Rightarrow$ Prob$(x_i \times y_i$ are squaring operations$) \approx 0$ $\forall i$
(Ω_2) LIM(x,y) s.t. $x = y \Rightarrow$ Prob$(x_i \times y_j = x_j \times y_i) = 1$ $\forall i \neq j$.
(Ω_3) LIM(x,y) s.t. $x \neq y \Rightarrow$ Prob$(x_i \times y_j = x_j \times y_i) \approx 0$ $\forall i \neq j$.

From observations (Ω_0) and (Ω_1) one can apply the attack presented in [1] on a single trace as suggested by the authors. The main drawback is that only ℓ such operations are performed during a LIM which represents a small number of trace segments. It is likely to make the attack inefficient for small modulus lengths (with respect to the multiplier size t).

From observations (Ω_2) and (Ω_3) we notice that collisions between $x_i \times y_j$ and $x_j \times y_i$ for $i \neq j$ can be used to identify squarings from other multiplications. Moreover, LIM(x,y) provides $\ell^2 - \ell$ operations $x_i \times y_j$, $i \neq j$, thus $(\ell^2 - \ell)/2$ couples of potential collisions. This represents a fairly large number of trace segments. The principle of our new attack consists in detecting those internal collisions in a single long-integer operation to determine whether it is a squaring or not. Visually, we split the matrix M into an upper-right and a lower-left triangles of terms, thus we call this technique a *triangle trace analysis*.

We present in the following two techniques to identify these collisions on a single long-integer multiplication trace. The first analysis uses the Euclidean distance distinguisher and the second one relies on a collision-correlation technique.

3.2   Euclidean Distance Distinguisher

We use as distinguisher the Euclidean distance between two sets of points on a trace as Walter [24] in the Big Mac analysis. In order to exploit properties (Ω_2) and (Ω_3) we proceed as follows. For each LIM(x,y) operation we compute the following differential side-channel trace:

$$T_{\mathrm{ED}} = \frac{2}{\ell^2 - \ell} \sum_{0 \leq i < j < \ell} \sqrt{(T_{i,j} - T_{j,i})^2}$$

If the operation performed is a squaring then the single-precision multiplications $x_i \times y_j$ and $x_j \times y_i$ store the same value in the result register (or in the memory) at the end of the operation. The side-channel leakage of the result storage of both operations should thus be similar. On the other hand, if $x \neq y$, products differ and the side-channel leakage should present less similarities. Assuming a side-channel leakage function linear in the Hamming weight of the data manipulated, a squaring should result in $E(T_{\mathrm{ED}}) \approx 0$, whereas we should expect a significantly higher value (about $t/2$ for each of the product halves) in the case of a multiplication.

3.3   Collision-Correlation Distinguisher

We define the two following series of trace segments, where the ordering of couples (i,j) is the same for the two series:

$$\Theta_0 = \{T_{i,j} \text{ s.t. } 0 \le i < j \le \ell - 1\}$$

$$\Theta_1 = \{T_{j,i} \text{ s.t. } 0 \le i < j \le \ell - 1\}$$

Each set includes $N = (\ell^2 - \ell)/2$ trace segments of base b multiplications.

In order to determine the operation performed by the LIM we compute the Pearson correlation factor between the two series Θ_0 and Θ_1 as described in [6]:

$$\hat{\rho}_{\Theta_0,\Theta_1}(t) = \frac{\mathrm{Cov}(\Theta_0(t), \Theta_1(t))}{\sigma_{\Theta_0}(t)\sigma_{\Theta_1}(t)}$$

$$= \frac{N \sum (T_{i,j}(t)T_{j,i}(t)) - \sum T_{i,j}(t) \sum T_{j,i}(t)}{\sqrt{N \sum (T_{i,j}(t))^2 - (\sum T_{i,j}(t))^2}\sqrt{N \sum (T_{j,i}(t))^2 - (\sum T_{j,i}(t))^2}}$$

where summations are taken over all couples $0 \le i < j \le \ell - 1$.

In case of a squaring operation, a much higher correlation value $\hat{\rho}_{\Theta_0,\Theta_1}$ is expected than in case of a multiplication. Computing this correlation value for each LIM operation allows to determine its nature and to recover the sequence of exponent bits.

Remark. Contrary to differential analysis on symmetric ciphers, each exponent bit requires to distinguish one hypothesis out of only two, instead of for instance 256 considering a differential attack on AES. Thus fixing a decision threshold is easier when dealing with the exponentiation. This has already been observed when applying DPA or CPA on RSA [2,15] compared to DES or AES.

4   Comparison of the Different Attacks

In order to validate these two techniques, we generated simulated side-channel traces for a classical 32×32-bit multiplier. As generally considered in the literature, we assume a side-channel leakage model linear in the Hamming weight of the manipulated data — here x_i, y_j, and $x_i \times y_j$ — and add a white Gaussian noise of mean $\mu = 0$ and standard deviation σ. We build simulated side-channel traces based on the Hamming weight of the data manipulated in the multiplication operation such that each processed single-precision multiplication generates four leakage points $\mathrm{HW}(x_i)$, $\mathrm{HW}(y_j)$, $\mathrm{HW}(x_i \times y_j \bmod b)$, and $\mathrm{HW}(x_i \times y_j \div b)$, where $\div$ stands for the Euclidean quotient.

Besides validating our two Rosetta variants — the Euclidean distance distinguisher (Rosetta ED) and the collision-correlation one (Rosetta CoCo) — we compare Rosetta with other techniques discussed previously, namely the classical Big Mac, the Big Mac using collision correlation (Big Mac CoCo), and the single trace variant of the Amiel et al. attack presented at SAC 2008.

We proceed in the following way: we randomly select two ℓ-bit integers x and y. Then we generate the side-channel traces of the multiplication $\mathrm{LIM}(x, y)$ and of the squaring $\mathrm{LIM}(x, x)$.

Each different attack is eventually applied and we keep trace of their success or failure to distinguish the squaring from the multiplication. Finally, we estimate the success rate of each technique by running 1 000 such experiments. These tests are performed for three different noise standard deviation values[1]: from no noise ($\sigma = 0$) to a strong one ($\sigma = 7$).

Characterisation and Threshold. A threshold for the attack must be selected for each technique to determine whether the targeted operation is a multiplication or a squaring. Using simulated side-channel traces, it was possible to determine the best threshold value for each technique. Without any knowledge on the component, it is more difficult to fix those threshold values. The attacks could be processed with guess on these thresholds, for instance selecting 0.5 for the collision correlation, but it could not reach optimal efficiency or fail. It is then preferable to determine the best threshold values through a characterization phase of the multiplier, either with an access to an open sample or using the public exponentiation calculation as suggested in [2].

Results. We obtain the success rates given in tables 1 ($\sigma = 0$), 2 ($\sigma = 2$) and 3 ($\sigma = 7$) for different key lengths ranging from 512 bits to 2048 bits. Figures 3 and 4 present a graphic comparison of these results for $\sigma = 0$ and $\sigma = 7$.

Table 1. Success rate with a null noise, $\sigma = 0$

Technique	512 bits	768 bits	1024 bits	1536 bits	2048 bits
Big Mac [24]	0.986	0.990	0.993	0.994	0.995
SAC 2008 [1]	0.533	0.618	0.734	0.858	0.897
Big Mac CoCo (§2.3)	0.999	1.00	1.00	1.00	1.00
Rosetta ED (§3.2)	1.00	1.00	1.00	1.00	1.00
Rosetta CoCo (§3.3)	1.00	1.00	1.00	1.00	1.00

Table 2. Success rate with a moderate noise, $\sigma = 2$

Technique	512 bits	768 bits	1024 bits	1536 bits	2048 bits
Big Mac [24]	0.767	0.775	0.807	0.816	0.818
SAC 2008 [1]	0.546	0.629	0.717	0.805	0.855
Big Mac CoCo (§2.3)	0.981	0.998	0.999	1.00	1.00
Rosetta ED (§3.2)	1.00	1.00	1.00	1.00	1.00
Rosetta CoCo (§3.3)	1.00	1.00	1.00	1.00	1.00

---

[1] Regarding the standard deviation of the noise, a unit corresponds to the side-channel difference related to a one bit difference in the Hamming weight.

Table 3. Success rate with a strong noise, $\sigma = 7$

Technique	512 bits	768 bits	1024 bits	1536 bits	2048 bits
Big Mac [24]	0.557	0.577	0.621	0.614	0.632
SAC 2008 [1]	0.551	0.577	0.623	0.662	0.702
Big Mac CoCo (§2.3)	0.737	0.855	0.909	0.963	0.981
Rosetta ED (§3.2)	0.711	0.821	0.878	0.953	0.992
Rosetta CoCo (§3.3)	0.685	0.816	0.906	0.992	0.997

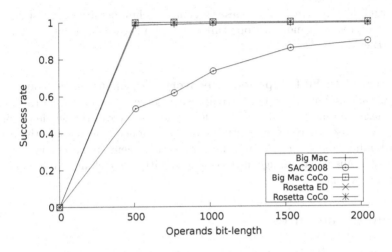

Fig. 3. Success rate of the different attacks with no noise

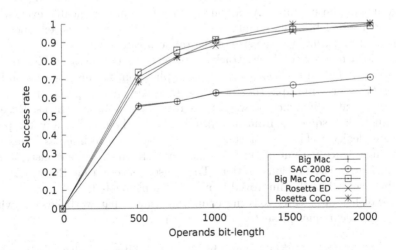

Fig. 4. Success rate of the different attacks with a strong noise, $\sigma = 7$

Results Interpretation. We observe that with no noise (cf. Table 1) all techniques are efficient when applied to large modulus bit lengths (1536 bits or more). For smaller modulus lengths, the SAC 2008 technique is inefficient (probability of success close to 0.5) as expected since the number of useful operations in that case is too small.

In case of a noisy component, we observe that the original Big Mac and the attack from SAC 2008 are not efficient, their probability of success is about 0.5–0.7. Big Mac analysis using collision correlation, and both Rosetta techniques start to be efficient from 1024-bit operands and are very efficient for 1536-bit and 2048-bit operands.

Our study demonstrates that these three last techniques are the most efficient ones and represent a more serious threat for blinded exponentiation than the original Big Mac.

From Partial to Full Exponent Recovery. Depending on the component, on the leakage and noise level of the chip, we observe that the success rate of the attack varies and may reveal too few information to recover the whole exponent value. In the case where uncertainty remains on some exponent bits, the attack from Schindler and Itoh [21] may help to reveal them. If necessary, Rosetta analysis can thus be advantageously combined with this technique to completely recover the exponent.

5   Countermeasures

As for the other attacks considered in this paper, both Rosetta techniques we introduced present the following interesting properties: (i) they make use of a single side-channel trace and, (ii) they do not require the knowledge of the message nor of the modulus. As a consequence they are applicable even when the classical set of blinding countermeasures (message, modulus, exponent) is implemented and whatever the size of the random values used.

A first idea to prevent these attacks is to improve the message blinding by randomizing it before each long-integer multiplication, for instance by adding the modulus n or a multiple thereof to the message. At this point, it is worth noticing a specific difference between both Rosetta and other attacks. Rosetta can distinguish a squaring from a multiplication without using any template or previous leakage. This is not the case with the other techniques — except for the single trace variant of the SAC 2008 attack which we demonstrate not to be efficient in the previous section. The consequence is that Rosetta is still applicable even when this improved blinding is implemented.

We recall hereafter three existing countermeasures that we believe to withstand all the techniques presented in this paper.

Shuffled Long-Integer Multiplication. In [5], a long integer multiplication algorithm with internal single-precision multiplications randomly permuted is presented. More details are given in [23, Sec. 2.7]. This countermeasure makes

Rosetta analysis virtually infeasible as indices i, j of multiplication $x_i \times y_j$ are not known anymore.

Always True Multiplication. This solution consists in ensuring that multiplication operands are always different (or different with high probability). To achieve this objective, before each multiplication $\text{LIM}(x, y)$, both operands x and y are randomized by $x^* = x + r_1.n$ and $y^* = y + r_2.n$. If $r_1 \neq r_2$, two equal operands x and y are traded for x^* and y^* with $x^* \neq y^*$ and the operation $\text{LIM}(x^*, y^*)$ is not a squaring.

Square-Always algorithm. The square-always algorithm presented in [7] processes any multiplication using two squarings. As for the solution of using multiplications of different terms only, Rosetta does not apply. Regular atomic square always algorithms can be used to prevent SSCA. Exponent blinding countermeasure must be associated with this solution.

6   Conclusion

We present in this study new side-channel methods — the Big Mac using collision correlation and the two Rosetta techniques — allowing to distinguish a squaring from a multiplication when the same long-integer multiplication algorithm is used for both operations. They can be used to recover an RSA secret exponent — both in standard or CRT mode — with a single execution side-channel trace. We compare our new techniques with other single trace side-channel analyses and demonstrate that they are more efficient than previous ones, especially on noisy measurements. We show that classical combination of message, modulus and exponent blindings is not sufficient to counteract our analysis and we suggest more advanced countermeasures. As a conclusion, we quote Colin Walter to recall the very interesting property of these attacks: "The longer the key length, the easier the attacks.".

References

1. Amiel, F., Feix, B., Tunstall, M., Whelan, C., Marnane, W.P.: Distinguishing Multiplications from Squaring Operations. In: Avanzi, R.M., Keliher, L., Sica, F. (eds.) SAC 2008. LNCS, vol. 5381, pp. 346–360. Springer, Heidelberg (2009)
2. Amiel, F., Feix, B., Villegas, K.: Power Analysis for Secret Recovering and Reverse Engineering of Public Key Algorithms. In: Adams, C., Miri, A., Wiener, M. (eds.) SAC 2007. LNCS, vol. 4876, pp. 110–125. Springer, Heidelberg (2007)
3. Brier, E., Clavier, C., Olivier, F.: Correlation Power Analysis with a Leakage Model. In: Joye, M., Quisquater, J.-J. (eds.) CHES 2004. LNCS, vol. 3156, pp. 16–29. Springer, Heidelberg (2004)

4. Chevallier-Mames, B., Ciet, M., Joye, M.: Low-cost Solutions for Preventing Simple Side-Channel Analysis: Side-Channel Atomicity. IEEE Transactions on Computers 53(6), 760–768 (2004)
5. Clavier, C., Feix, B., Gagnerot, G., Roussellet, M., Verneuil, V.: Horizontal Correlation Analysis on Exponentiation. In: Soriano, M., Qing, S., López, J. (eds.) ICICS 2010. LNCS, vol. 6476, pp. 46–61. Springer, Heidelberg (2010)
6. Clavier, C., Feix, B., Gagnerot, G., Roussellet, M., Verneuil, V.: Improved Collision-Correlation Power Analysis on First Order Protected AES. In: Preneel, B., Takagi, T. (eds.) CHES 2011. LNCS, vol. 6917, pp. 49–62. Springer, Heidelberg (2011)
7. Clavier, C., Feix, B., Gagnerot, G., Roussellet, M., Verneuil, V.: Square Always Exponentiation. In: Bernstein, D.J., Chatterjee, S. (eds.) INDOCRYPT 2011. LNCS, vol. 7107, pp. 40–57. Springer, Heidelberg (2011)
8. Coron, J.-S.: Resistance against Differential Power Analysis for Elliptic Curve Cryptosystems. In: Koç, Ç.K., Paar, C. (eds.) CHES 1999. LNCS, vol. 1717, pp. 292–302. Springer, Heidelberg (1999)
9. Dhem, J.-F.: Design of an efficient public-key cryptographic library for RISC-based smart cards. PhD thesis, Université catholique de Louvain, Louvain (1998)
10. Gandolfi, K., Mourtel, C., Olivier, F.: Electromagnetic Analysis: Concrete Results. In: Koç, Ç.K., Naccache, D., Paar, C. (eds.) CHES 2001. LNCS, vol. 2162, pp. 251–261. Springer, Heidelberg (2001)
11. Garner, H.: The Residue Number System. IRE Transactions on Electronic Computers 8(6), 140–147 (1959)
12. Joye, M.: Highly Regular m-Ary Powering Ladders. In: Jacobson Jr., M.J., Rijmen, V., Safavi-Naini, R. (eds.) SAC 2009. LNCS, vol. 5867, pp. 350–363. Springer, Heidelberg (2009)
13. Joye, M., Yen, S.-M.: The Montgomery Powering Ladder. In: Kaliski Jr., B.S., Koç, Ç.K., Paar, C. (eds.) CHES 2002. LNCS, vol. 2523, pp. 291–302. Springer, Heidelberg (2003)
14. Kocher, P., Jaffe, J., Jun, B.: Differential Power Analysis. In: Wiener, M. (ed.) CRYPTO 1999. LNCS, vol. 1666, pp. 388–397. Springer, Heidelberg (1999)
15. Messerges, T.S.: Using Second-order Power Analysis to Attack DPA Resistant Software. In: Koç, Ç., Paar, C. (eds.) CHES 2000. LNCS, vol. 1965, pp. 238–251. Springer, Heidelberg (2000)
16. Messerges, T., Dabbish, E., Sloan, R.: Investigations of Power Analysis Attacks on Smartcards. In: The USENIX Workshop on Smartcard Technology (Smartcard 1999), pp. 151–161 (1999)
17. Montgomery, P.: Modular multiplication without trial division. Math. Comp. 44(170), 519–521 (1985)
18. Moradi, A., Mischke, O., Eisenbarth, T.: Correlation-Enhanced Power Analysis Collision Attack. In: Mangard, S., Standaert, F.-X. (eds.) CHES 2010. LNCS, vol. 6225, pp. 125–139. Springer, Heidelberg (2010)
19. PKCS #1. RSA Cryptography Specifications Version 2.1. RSA Laboratories (2003)
20. Rivest, R., Shamir, A., Adleman, L.: A Method for Obtaining Digital Signatures and Public-Key Cryptosystems. Communications of the ACM 21(2), 120–126 (1978)
21. Schindler, W., Itoh, K.: Exponent Blinding Does Not Always Lift (Partial) Spa Resistance to Higher-Level Security. In: Lopez, J., Tsudik, G. (eds.) ACNS 2011. LNCS, vol. 6715, pp. 73–90. Springer, Heidelberg (2011)

22. Schramm, K., Wollinger, T., Paar, C.: A New Class of Collision Attacks and its Application to DES. In: Johansson, T. (ed.) FSE 2003. LNCS, vol. 2887, pp. 206–222. Springer, Heidelberg (2003)

23. Verneuil, V.: Elliptic Curve Cryptography and Security of Embedded Devices. PhD thesis, Université de Bordeaux, Bordeaux (2012)

24. Walter, C.D.: Sliding Windows Succumbs to Big Mac Attack. In: Koç, Ç.K., Naccache, D., Paar, C. (eds.) CHES 2001. LNCS, vol. 2162, pp. 286–299. Springer, Heidelberg (2001)

25. Witteman, M., van Woudenberg, J., Menarini, F.: Defeating RSA Multiply-Always and Message Blinding Countermeasures. In: Kiayias, A. (ed.) CT-RSA 2011. LNCS, vol. 6558, pp. 77–88. Springer, Heidelberg (2011)

Collision Attack on the Hamsi-256 Compression Function

Mario Lamberger[1], Florian Mendel[2], and Vincent Rijmen[2]

[1] NXP Semiconductors, Austria
[2] Katholieke Universiteit Leuven, ESAT/COSIC and IBBT, Belgium

Abstract. Hamsi-256 is a cryptographic hash functions submitted by Küçük to the NIST SHA-3 competition in 2008. It was selected by NIST as one of the 14 round 2 candidates in 2009. Even though Hamsi-256 did not make it to the final round in 2010 it is still an interesting target for cryptanalysts. Since Hamsi-256 has been proposed, it received a great deal of cryptanalysis. Besides the second-preimage attacks on the hash function, most cryptanalysis focused on non-random properties of the compression function or output transformation of Hamsi-256. Interestingly, the collision resistance of the hash or compression function got much less attention. In this paper, we present a collision attack on the Hamsi-256 compression function with a complexity of about $2^{124.1}$.

Keywords: hash function, differential cryptanalysis, collision attack.

1 Introduction

In recent years, significant advances in the field of hash function research have been made which had a formative influence on the landscape of hash functions. Especially the work on MD5 and SHA-1 [19,20] has convinced many cryptographers that these widely deployed hash functions can no longer be considered secure. As a consequence, researchers are evaluating alternative hash functions in the SHA-3 initiative organized by NIST [16]. The goal is to find a hash function which is fast and still secure within the next few decades.

Many new and interesting hash functions have been proposed. Hamsi-256 [12], proposed by Küçük, was one of the 64 submissions to the SHA-3 competition from which 51 submissions were selected for the first round in 2008 and 14 of them advanced to the second round in 2009. Hamsi-256 was one of them. Even though Hamsi-256 was not selected as one of the five finalists in 2010, mainly because of the second-preimage attacks, it is still an interesting target for cryptanalysts. In this work, we focus on the collision resistance on the Hamsi-256 compression function, which in turn gives new insights in the collision resistance of the hash function.

Previous Analysis. Hamsi-256 received a great deal of cryptanalysis during the ongoing SHA-3 competition. However, the only analysis of the Hamsi-256 hash function itself is due to Dinur and Shamir [8,9] and Fuhr [10]. Both attacks

S. Galbraith and M. Nandi (Eds.): INDOCRYPT 2012, LNCS 7668, pp. 156–171, 2012.

are algebraic attacks targeting the second-preimage security of the hash function. The attacks are based on the observation that it is sufficient to show that one of the hashed output bits is wrong to discard a possible second-preimage. Since the output bits of the compression function of Hamsi-256 can be described by low degree polynomials, it is faster to compute a small number of output bits by a fast polynomial evaluation technique than with the original algorithm. The results are second-preimage attacks on Hamsi-256 with a complexity of about 2^{247} and $2^{251.3}$, respectively. But then again, one still needs to test 2^{256} inputs to find a second preimage as in the generic case. In other words, the attacks are a clever way to speed up brute force search. We want to note that in a similar way also the complexity of a generic collision search for the compression function of Hamsi-256 can be improved, resulting in an attack complexity of 2^{125} [8]. Moreover, in [13] Küçük showed a collision attack for a simplified version of the Hamsi-256 compression function, ignoring the message expansion.

Most other attacks published so far are differential attacks targeting the compression function or output transformation of Hamsi-256. Practical near-collisions for the compression function have been shown in [1,14,17,18] and a distinguisher for the compression function has been presented in [6]. Furthermore, non-random properties for the underlying permutation of the Hamsi-256 compression function and output transformation have been demonstrated in [1,4] and a distinguisher for the output transformation of Hamsi-256 has been described in [1].

Our Contribution. In this work, we present a collision attack for the Hamsi-256 compression function. Our collision attack is based on the attack of Çalik and Turan [6] and has a complexity of about $2^{124.1}$ compression function evaluations. The main idea of the attack is very simple. We extend the approach of Çalik and Turan, which was originally used to show non-random properties in the compression function. This is then used to fix some output bits of the compression function to a predefined value faster than in the generic case. Finally, we use a birthday attack on the remaining bits to construct a collision for the compression function of Hamsi-256. Even though the complexity of the attack is very high, namely $2^{124.1}$ compression function evaluations, it demonstrates that the compression function of Hamsi-256 is not collision resistant and gives new insights in the security of Hamsi-256. However, it has to be noted that the attack cannot be extended to the hash function.

Outline. The remainder of the paper is organized as follows. In Section 2, we give a short description of the Hamsi-256 compression function. In Section 3, we describe the distinguishing attack of Çalik and Turan on Hamsi-256, since it is the basis for our collision attack. We present our new attack strategy in Section 4 and apply it to the Hamsi-256 compression function in Section 5. Finally, we conclude in Section 6.

2   Description of Hamsi-256

Hamsi-256 is a cryptographic hash function proposed by Küçük [12] which has been submitted to the SHA-3 competition in 2008. It is an iterated hash function based on the Merkle-Damgård design principle [7,15] and produces a 256-bit hash value. Like most hash functions, Hamsi-256 iterates a compression function f to compute the hash value. It takes a 32-bit message block M_i and a 256-bit chaining value h_{i-1} as input and outputs a 256-bit chaining value h_i. In the following, we give a brief overview of the compression function of Hamsi-256 (see Figure 1).

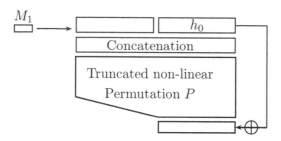

Fig. 1. The compression function of Hamsi-256 [12]

Message Expansion. The message expansion of Hamsi-256 uses a linear code to expand the 32-bit message word M_i into eight 32-bit words $m_0, m_1, \ldots, m_7$, i.e. 256 bits. The $(128, 16, 70)$ linear code defined over $\mathbb{F}_4$ used in the message expansion ensures that any difference in the message word M_i will lead to differences in at least 70 of the 128 columns of the initial state. For a detailed description of the message expansion we refer to [12].

Concatenation. The 256-bit expanded message $(m_0, \cdots, m_7)$ and the 256-bit chaining value $h_{i-1} = (c_0, \ldots, c_7)$ are concatenated to form the 512-bit initial state (see Figure 2). We want to note that the initial state can be considered as both a 4×4 matrix of 32-bit words and 128 columns each consisting of 4 bits.

m_0	m_1	c_0	c_1
c_2	c_3	m_2	m_3
m_4	m_5	c_4	c_5
c_6	c_7	m_6	m_7

Fig. 2. The initial state of the Hamsi-256 compression function after concatenation

Non-linear Permutation. The non-linear permutation P used in the Hamsi-256 compression function is composed of 3 rounds. In each round the round transformation updates the state by means of a sequence of transformations:

Addition of Constants. Predefined constants and a round counter is xored to the whole state. For the value of the constants we refer to [12].

Substitution. Each of the 128 columns of the state (each 4 bits) is updated by the 4-bit s-box of the block cipher Serpent [2].

Diffusion. The linear transformation L of the block cipher Serpent, which accepts four 32-bit input words, and outputs four 32-bit words is applied to the four independent diagonals of the state.

A detailed description of the s-box and the linear transformation L is given in [12] and the differential properties of the s-box and linear transformation of Hamsi-256 have been studied for instance in [1] among others.

Truncation and Feed-Forward. Finally, after the application of the permutation P the second and fourth rows of the state are discarded and the initial chaining value is xored to the truncated state resulting in the initial chaining value for the next iteration or the input to the output transformation to compute the final hash value. For the description of the output transformation of Hamsi-256 we refer to [12].

3   Attack of Çalik and Turan

In this section, we briefly describe the distinguishing attack of Çalik and Turan [6] on the Hamsi-256 compression function, since our collision attack builds upon it. The attack is a differential attack exploiting the fact that for a given input difference not all the output bits of the compression function are affected. This results in a distinguisher for the compression function. Since the message expansion of Hamsi-256 uses a $(128, 16, 70)$ linear code, any difference in the message will lead to differences in at least 70 of the 128 columns of the initial state. In other words, at least 70 s-boxes will be active in the initial state. Therefore, Çalik and Turan consider only differences in the chaining value in their analysis. Furthermore, they restrict themselves to differences in only one column, i.e. one active s-box, in the initial state. Since each column of the initial state contains two bits of the chaining value (see Figure 2), three non-zero differences can be injected to a column: these differences can be $2_x, 8_x$ or a_x for columns 0-63 and $1_x, 4_x$ or 5_x for columns 64-127.

To find the output bits that are not affected by one of the $3 \cdot 128$ possible input differences the authors trace the differences through the round transformations of Hamsi-256 and mark all the bits of the internal state that could have a difference. If there are any unmarked bits in the state after 3 rounds, then one knows that these bits are not affected by the initial difference. However, due to the feed-forward, some of these unaffected bit positions may coincide with the difference in the initial state, resulting in bits that will always change. However, in the paper the authors do not make a distinction between these two cases (and neither will we) and use for both cases the term unaffected bits.

As noted by Çalik and Turan the number of unaffected output bits depends on the Hamming weight of the difference in the initial state. For the differences with Hamming weight 2, i.e. 5_x and a_x, the number of unaffected bits is higher. The reason for this is that differences with Hamming weight 1, i.e. $1_x, 2_x, 4_x$ and 8_x, will lead to a difference with Hamming weight at least 2 at the output of the s-box in round 1, whereas a difference of Hamming weight 2 can lead to a difference with Hamming weight 1, resulting in a sparser difference at the output of round 1. Hence, it is not surprising that one could find 64 solutions with at least one unaffected output bit using an initial difference with Hamming weight 2, cf. [6, Table 4], while no solution could be found using an initial difference with Hamming weight 1. An example with two unaffected output bits and an initial difference with Hamming weight 2 is given in Figure 3.

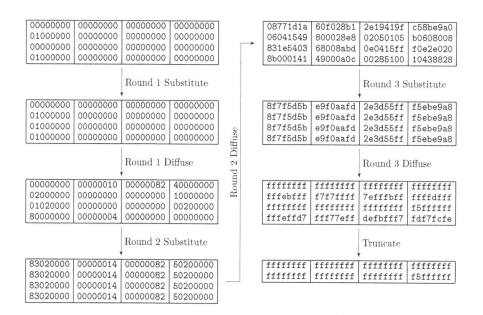

Fig. 3. Propagation of affected bits for the initial difference a_x inserted to column 7 of the initial state [6]

Another factor that influences the number of unaffected output bits are the values of the message bits in the column with the initial difference. By imposing some additional restrictions on the message bits one could observe more unaffected output bits. The reason for this is that for certain choices of the message bits the Hamming weight of the difference at the output of the s-box might be smaller. This results in a sparser difference at the output of round 1. By choosing the message bits carefully, Çalik and Turan could find 8 solutions with two unaffected bits at the output using an initial difference with Hamming weight 1, cf. [6, Table 3]. Furthermore, also for an initial difference with Hamming weight

2 the number of solutions as well as the number of unaffected bits can be increased by choosing the message bits accordingly. For instance, in the example given in Figure 3 the number of unaffected output bits can be increased to 9 bits by setting the message bits in column 7 to 0_x. By additionally also imposing conditions on the chaining bits in column 7, i.e. 0_x or 3_x, the number of unaffected bits can be increased to 62.

Based on these differential properties of the compression function Çalik and Turan describe several attacks on Hamsi-256 in the original paper. First, they show a distinguishing attack on the compression function that needs only a few compression function evaluations. Then, they present a message-recovery attack for the compression function with a complexity of $2^{10.5}$ and a pseudo-preimage attack with complexity of about $2^{254.25}$. For a detailed description of these attacks we refer to [6].

In this work, we will use these differential properties of the compression function to show a collision attack for the Hamsi-256 compression function. We describe the basic idea of the attack in the next section.

4   Basic Attack Strategy

In this section, we present the basic attack strategy employed by our new attack to construct collisions in the compression function of Hamsi-256. It is based on the concept of neutral bits [3] and auxiliary differentials [11], which were originally used to speed up differential collision attacks on hash functions.

The main idea of our attack is quite simple. Assume we can find a distinct input differences (in the following referred to as auxiliary differentials), where each difference only affects a few output bits of the compression function. Further, we assume that there exists at least b output bits (for the sake of simplicity say bits $0, 1, \ldots, b-1$) that are not affected by any of these a auxiliary differences. Then this can be used to find 2^a partial preimages for these b output bits of the compression function with a complexity of about $2^a + 2^b$, compared to the generic case of 2^{a+b} compression function evaluations.

Let $v_1, \ldots, v_a$ denote the a auxiliary differentials not affecting the first b output bits of the compression function and assume for the sake of simplicity that we want to fix these b output bits to 0. Then the attack can be summarized as follows.

1. Choose random values for the chaining input h_{i-1} and the message input m_i. If necessary fulfilling all conditions imposed by the a auxiliary differentials $v_1, \ldots, v_a$.
2. Compute the output of the compression function h_i and check if the b output bits of h_i are correct
   - If all b bits are 0 then continue with step 3
   - else go back to step 1

3. Use the a auxiliary differentials $v_1, \ldots, v_a$ to generate 2^a additional solutions where the first b output bits of the compression function are also 0

$$h_i^{(\underline{d})} = h_i \oplus \bigoplus_{j=1}^{a} d_j \cdot v_j$$

with $\underline{d} = (d_1, \ldots, d_a) \in \{0, 1\}^a$.

Note that we need to repeat step 1-2 about 2^b times to find a correct h_i, resulting in a complexity of about 2^b compression function evaluations to finish step 1 and 2. Since step 3 has a complexity of 2^a, the final complexity of the attack is $2^a + 2^b$ compression function evaluations. But then again, we found $1 + 2^a$ partial preimages where the first b output bits $(0, 1, \ldots, b-1)$ are 0 with complexity of $2^a + 2^b$. This can now be used to construct a collision for the compression function faster than in the generic case.

Since we can find about 2^a partial preimages for the first b with a complexity of $2^a + 2^b$ (instead of 2^{a+b}) we can combine this with a birthday attack to find a collision for the compression function faster than in the generic case. By repeating the attack about t times with $t = 2^{(n-b)/2-a}$ we get $t \cdot 2^a = 2^{(n-b)/2}$ outputs where the first b bits collide and due to the birthday paradox we expect to find at least one pair of outputs where also the remaining $n-b$ bits collide. The result is a collision attack on the compression function with a complexity of about $t \cdot (2^a + 2^b) = 2^{n/2} \cdot (2^{-b/2} + 2^{b/2-a})$ compression function evaluations.

Clearly the complexity of the attack depends the value of a and b. For the above computed complexity we can easily observe that the value

$$2^{-b/2} + 2^{b/2-a} = \frac{2^{b-a} + 1}{2^{b/2}}$$

is minimized if in the numerator we have $a \geq b$, and in the denominator we have b as large as possible, so basically $a = b$. The main question is now which values of a and b we can expect in our attack. On the one hand this number depends on the size of the set S containing all auxiliary differentials, and on the other hand on the number of affected output bits of each of these auxiliary differentials in the set. Moreover, also the number of conditions imposed on the message bits by the auxiliary differentials might be a limiting factor, since this is only 32 bits in the case of the Hamsi-256 compression function.

4.1   Probabilistic Considerations

In the following, we denote by $[n]$ the integer interval $\{1, 2, \ldots, n\}$, by $2^{[n]}$ we mean the set of all subsets of $[n]$ and by $\binom{[n]}{k}$ all subsets of $[n]$ of size k. We are looking at subsets of $\binom{[n]}{k}$ because on average, the number of unaffected output bits is $\approx k$ in our applications. Furthermore, we assume that the auxiliary differentials are independent and that the unaffected bits of each auxiliary differential are randomly distributed.

We want to investigate the probability $P(N, n, k, b)$ that a set $S \subseteq \binom{[n]}{k}$ of size N of auxiliary differentials contains a subset S' of size a such that the elements $s_i \in S'$ satisfying

$$\left| \bigcap_{i=1}^{a} s_i \right| \geq b. \tag{1}$$

Now we have

$$P(N, n, k, b) \leq \binom{N}{a} \cdot \sum_{j=b}^{k} P_r(n, k, j) \tag{2}$$

where $P_r(n, k, j)$ denotes the probability that a randomly chosen subsets $s_i \in \binom{[n]}{k}$ satisfy $|\bigcap_{i=1}^{a} s_i| = j$. The exact distribution is hard to compute, however we can come up with the following approximation. Since each set s_i has size k, we assume that a randomly chosen element $e \in [n]$ is contained in s_i with probability k/n. Thus, the probability for a randomly chosen element $e \in [n]$ to be contained in $\bigcap_{i=1}^{a} s_i$ is $p = (k/n)^a$. From this we deduce that

$$\sum_{j=b}^{k} P_r(n, k, j) = \sum_{j=b}^{k} \binom{n}{j} p^j (1 - p)^{n-j}$$

which in turn leads to

$$P(N, n, k, r) \leq \binom{N}{a} \cdot \sum_{j=b}^{k} \binom{n}{j} p^j (1 - p)^{n-j} \tag{3}$$

5   Application to Hamsi-256

In this section, we will discuss the application of the attack strategy described in the previous section to the Hamsi-256 compression function. Therefore, we first need to find a set S of auxiliary differentials that only affect a few output bits of the compression function of Hamsi-256. Then we need to find a subset of a auxiliary differentials not affecting the same b output bits (for large values of a and b). To construct the set S it seems natural to use the same differentials as Çalik and Turan in their distinguishing attack described in Section 3. However, since we are aiming for a large value of a and b, in order to increase the effectiveness of the attack, we are only interested in auxiliary differentials where the number of unaffected output bits is large. This already rules out all auxiliary differentials with an initial difference of Hamming weight 1. For these cases the maximum number of unaffected output bits is at most 2 (see Section 3). However, auxiliary differentials with an initial difference of Hamming weight 2 might be a good choice, in particular since the number of unaffected output bits can be increased to up to 62 bits by imposing some additional conditions on the chaining and message bits (see Section 3).

In total we found 198 auxiliary differentials with an initial difference of Hamming weight 2. As shown in Figure 5 in the appendix the number of unaffected output bits is on average 30. Note that since we are interested in large values of a and b, we only considered auxiliary differentials where the number of unaffected output bits is at least 10. Now assuming that these 198 auxiliary differentials are independent and randomly distributed, we can use (3) with $k = 30$ to estimate $a, b \approx 4$ that can be used in our collision attack on the Hamsi-256 compression function.

However, the auxiliary differentials in the set S are not independent nor randomly distributed. By doing a brute-force search we found a solution with $a = b = 6$ resulting in a collision attack on the Hamsi-256 compression function with complexity of about 2^{126} compression function evaluations. The six auxiliary differentials including the necessary conditions on the chaining and message bits are given in Table 1 and the affected output bits for these six auxiliary differentials are shown in Table 4 in the appendix. Note that the 10 conditions imposed by the six auxiliary differentials on the message bits can be fulfilled by solving a set of linear equations. This is due to the fact that the message expansion of Hamsi-256 is linear.

Table 1. The six auxiliary differentials used in our attack including all conditions on the chaining and message bits. Note that none of them affects the output bits 131, 200, 201, 202, 237, 238.

	i					
	1	2	3	4	5	6
column	1	2	38	39	110	111
difference	a_x	a_x	a_x	a_x	5_x	5_x
message bits	2_x	2_x	3_x	1_x	$2_x, 3_x$	$2_x, 3_x$
chaining bits	$1_x, 2_x$	$1_x, 2_x$	$1_x, 2_x$	$1_x, 2_x$	$0_x, 3_x$	$1_x, 2_x$

5.1   Improving the Attack

To improve the attack described above we need to find a subset of size a of auxiliary differentials not affecting the same b output bits for larger values of a and b. Therefore, we need to find a set S of auxiliary differentials, where the number of unaffected output bits is larger than 30 on average. To find such auxiliary differentials we need to consider initial differences with more than only one active column at the input of the first round. This significantly increases the search space, but also the complexity to generate the set S. However, since we are interested in auxiliary differentials which are affecting only a few output bits, the search space and hence the complexity can be reduced by only considering initial differences leading to a sparse difference at the input of round 2. To be more precise, we restrict ourselves to initial differences resulting in a single bit difference (one active column) at the input of round 2. One example of such an auxiliary differential with only a single bit difference in column 96 at the input of round 2 is given in Figure 4.

Considering only auxiliary differentials resulting in a singe bit difference at the input of round 2 has several advantages. First of all since only one column

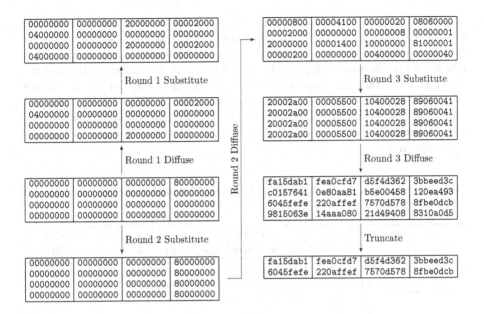

Fig. 4. Propagation of affected bits for the initial difference 1_x inserted to column 96 at the input of round 2. Note that some conditions on the chaining and message bits in the column 5, 66 and 114 have to be fulfilled in order to guarantee that there will be only a single bit difference in the column 96 at the input of round 2.

is active at the input of round 2, this results in sparse auxiliary differentials affecting only a few output bits. Second, due to the fact that the difference at the input of round 2 has Hamming weight 1, this results in at most 7 active columns at the input of round 1. Note that a difference with a higher Hamming weight would result in more active columns, complicating the attack. Moreover, a disadvantage is that the number of conditions on the chaining and message bits, that need to be fulfilled to guarantee that the auxiliary differential holds, would increase. For example the auxiliary differential given in Figure 4 needs 3 conditions on the chaining bits and 5 conditions on the message bits (see Table 2).

Table 2. Detailed information for the auxiliary differential used in the example given in Figure 4

column	5	66	114
difference	a_x	5_x	5_x
message bits	0_x	0_x	$0_x, 1_x$
chaining bits	$1_x, 2_x$	$1_x, 2_x$	$1_x, 2_x$

The best way to find the auxiliary differentials is to start with a single bit difference at the input of round 2 and then compute backward. Due to the properties of the linear layer, one will get 2 to 7 active columns at the input of round 1, resulting in 2-7 conditions on the chaining bits and 3-13 conditions on the message bits. We want to note that since the message input of Hamsi-256 is only 32 bits the increased number of message bit conditions might be the limiting factor for the attack. In Table 3 we list all the auxiliary differentials that result in a single bit difference at the input of round 2. As can be seen in the table the number of unaffected output bits for all the 192 auxiliary differentials is in the range between 100 and 130 bits with an average of 110 bits as shown in Figure 6 in the appendix.

Table 3. List of all possible auxiliary differentials with only a single bit difference at the input of round 2

input round 2		output	input	message
difference	column	# unaffected bits	# active columns	# conditions
1_x	$98, \ldots, 100$	102–103	2	3
	$96, 97; 101, \ldots, 127$	100–125	3	5
2_x	$29, \ldots, 31$	111–116	3	5
	$25, \ldots, 28$	113–114	4	7
	$0, \ldots, 24$	107–129	7	13
4_x	$32, \ldots, 95$	100–125	3	6
8_x	$61, \ldots, 63; 93, \ldots, 95$	111–116	4	7
	$32, \ldots, 60; 64, \ldots, 62$	107–129	5	9

Assuming again that the 192 auxiliary differentials are independent and randomly distributed and the number of unaffected output bits is 110. Then, as above from (3) we would expect to find a solution with $a, b \approx 10$.

This is very close to our result, by using a brute-force search we found $a = 10$ auxiliary differentials not affecting the same $b = 9$ output bits resulting in an attack complexity of about $2^{124.1}$ compression function evaluations. The 10 auxiliary differentials and the list of the affected output bits for each auxiliary differential are given in Table 5 and Table 6 in the appendix.

However, the results of the attack described in the previous section would suggest that there might solutions for larger values of a and b than estimated by (3), since the auxiliary differentials are not independent nor random. Indeed, if we ignore the conditions imposed by the auxiliary differentials on the message bits, then we could find $a = 14$ auxiliary differentials not affecting the same $b = 14$ output bits, however we were not able to find a confirming message input. The reason for this is that we need to fulfill on average about 7 conditions on the message bits per auxiliary differential, while the message input of Hamsi-256 is only 32 bits. Note that for the attack described in the previous section, we had in total only 10 message bit conditions for the 6 auxiliary differentials all together.

6   Conclusion

In this work, we have analyzed the Hamsi-256 compression function with respect to its collision resistance. By exploiting non-random properties of the compression function we could show a collision attack with a complexity of about $2^{124.1}$. The attack is an extension of the distinguishing attack of Çalik and Turan combined with the idea of neutral bits and auxiliary differentials originally used to speed up existing differential collision attacks. Even though the complexity of our attack is very high and close the the generic case it gives some new insights in the security of Hamsi-256.

Acknowledgments. This work was supported in part by the IAP Programme P6/26 BCRYPT of the Belgian State (Belgian Science Policy) and by the European Commission through the ICT programme under contract ICT-2007-216676 ECRYPT II. In addition, this work was supported by the Research Fund KU Leuven, OT/08/027.

References

1. Aumasson, J.-P., Käsper, E., Knudsen, L.R., Matusiewicz, K., Ødegård, R., Peyrin, T., Schläffer, M.: Distinguishers for the Compression Function and Output Transformation of Hamsi-256. In: Steinfeld, R., Hawkes, P. (eds.) ACISP 2010. LNCS, vol. 6168, pp. 87–103. Springer, Heidelberg (2010)
2. Biham, E., Anderson, R., Knudsen, L.: Serpent: A New Block Cipher Proposal. In: Vaudenay, S. (ed.) FSE 1998. LNCS, vol. 1372, pp. 222–238. Springer, Heidelberg (1998)
3. Biham, E., Chen, R.: Near-Collisions of SHA-0. In: Franklin, M. (ed.) CRYPTO 2004. LNCS, vol. 3152, pp. 290–305. Springer, Heidelberg (2004)
4. Boura, C., Canteaut, A.: Zero-Sum Distinguishers for Iterated Permutations and Application to KECCAK-f and Hamsi-256. In: Biryukov, A., Gong, G., Stinson, D.R. (eds.) SAC 2010. LNCS, vol. 6544, pp. 1–17. Springer, Heidelberg (2011)
5. Brassard, G. (ed.): CRYPTO 1989. LNCS, vol. 435. Springer, Heidelberg (1990)
6. Çalık, Ç., Turan, M.S.: Message Recovery and Pseudo-preimage Attacks on the Compression Function of Hamsi-256. In: Abdalla, M., Barreto, P.S.L.M. (eds.) LATINCRYPT 2010. LNCS, vol. 6212, pp. 205–221. Springer, Heidelberg (2010)
7. Damgård, I.: A Design Principle for Hash Functions. In: Brassard [5], pp. 416–427
8. Dinur, I., Shamir, A.: An Improved Algebraic Attack on Hamsi-256. Cryptology ePrint Archive, Report 2010/602 (2010), http://eprint.iacr.org/
9. Dinur, I., Shamir, A.: An Improved Algebraic Attack on Hamsi-256. In: Joux, A. (ed.) FSE 2011. LNCS, vol. 6733, pp. 88–106. Springer, Heidelberg (2011)
10. Fuhr, T.: Finding Second Preimages of Short Messages for Hamsi-256. In: Abe, M. (ed.) ASIACRYPT 2010. LNCS, vol. 6477, pp. 20–37. Springer, Heidelberg (2010)
11. Joux, A., Peyrin, T.: Hash Functions and the (Amplified) Boomerang Attack. In: Menezes, A. (ed.) CRYPTO 2007. LNCS, vol. 4622, pp. 244–263. Springer, Heidelberg (2007)
12. Küçük, Ö.: The Hash Function Hamsi. Submission to NIST (updated) (2009)
13. Küçük, Ö.: Design and Analysis of Cryptographic Hash Functions. Ph.D. thesis, KU Leuven (April 2012)

14. Li, Y., Wang, A.: Using genetic algorithm to find near collisions for the compress function of Hamsi-256. In: BIC-TA, pp. 826–829. IEEE (2010)
15. Merkle, R.C.: One Way Hash Functions and DES. In: Brassard [5], pp. 428–446
16. National Institute of Standards and Technology: Announcing Request for Candidate Algorithm Nominations for a New Cryptographic Hash Algorithm (SHA-3) Family. Federal Register 27(212), 62212–62220 (November 2007), http://csrc.nist.gov/groups/ST/hash/documents/FR_Notice_Nov07.pdf
17. Nikolic, I.: Near Collisions for the Compression Function of Hamsi-256. CRYPTO rump session (2009)
18. Wang, M., Wang, X., Jia, K., Wang, W.: New Pseudo-Near-Collision Attack on Reduced-Round of Hamsi-256. Cryptology ePrint Archive, Report 2009/484 (2009), http://eprint.iacr.org/
19. Wang, X., Yin, Y.L., Yu, H.: Finding Collisions in the Full SHA-1. In: Shoup, V. (ed.) CRYPTO 2005. LNCS, vol. 3621, pp. 17–36. Springer, Heidelberg (2005)
20. Wang, X., Yu, H.: How to Break MD5 and Other Hash Functions. In: Cramer, R. (ed.) EUROCRYPT 2005. LNCS, vol. 3494, pp. 19–35. Springer, Heidelberg (2005)

A   Supporting Material

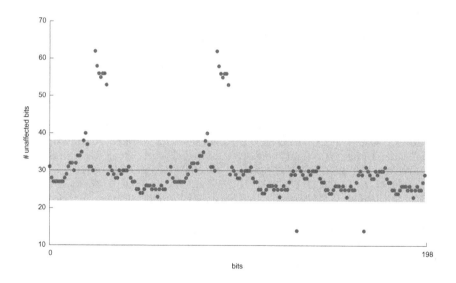

Fig. 5. Number of unaffected bits at the output of the compression function with an initial difference of Hamming weight 2. Note that this does not include auxiliary differentials where the number of unaffected output bits is smaller than 10, since they are not useful for our attack.

Table 4. Affected output bits for the 6 auxiliary differentials used in our collision attack on the Hamsi-256 compression function

i	output	
	affected bits	# unaffected bits
1	7bbfff3f fe1dfaff febaefdf f7fdf7e7 cfbfdfcf 615fffff 2e1affef f7f1fd7b	53
2	bddfff9f fe0efd7f ff5d77ef fbfefbf3 e5dfefe7 a0afffff 970d7ff7 fbf8febd	56
3	3fbfefbf fbddfff9 ffe0efd7 fff5d77e cfbf8feb 6c7dfefe a21affff 7f70d7ff	58
4	9fdff7df fceebffc fff077eb 7ffaebbf e5dfc7f5 263eff7f d10d7fff bfb86bff	62
5	fbffff3f fe1dffff ffffffff fffff7e7 effffef e1ffffff 3e1fffef fff1fdfb	29
6	fdffff9f ff0effff ffffffff fffffbf3 e7fffff7 f0ffffff 9f0fffff fff8fefd	30
	ffffffff ffffffff ffffffff ffffffff effffff ffffffff ff1fffff fff9ffff	6

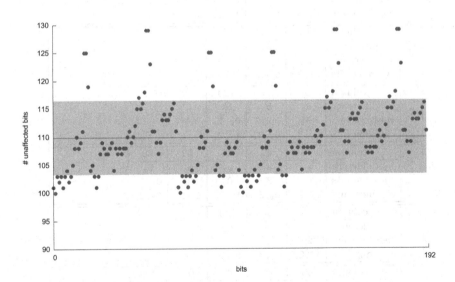

Fig. 6. Number of unaffected bits at the output of the compression function with a single bit difference, i.e. $1_x, 2_x, 4_x$ or 8_x, at the input of round 2

Table 5. List of auxiliary differentials used in our collision attack on the Hamsi-256 compression function including all conditions on the message and chaining bits. Note that all conditions on the message bits can be fulfilled be setting the 32-bit message block $M_i = \{12, 51, \mathtt{aa}, 05\}$ prior to the application of the message expansion.

<table>
<tr><td rowspan="4">1</td><td>column</td><td>3</td><td>4</td><td>64</td><td>112</td><td></td><td></td></tr>
<tr><td>difference</td><td>a_x</td><td>a_x</td><td>5_x</td><td>5_x</td><td></td><td></td></tr>
<tr><td>message bits</td><td>2_x</td><td>0_x</td><td>2_x</td><td>$0_x, 1_x$</td><td></td><td></td></tr>
<tr><td>chaining bits</td><td>$0_x, 3_x$</td><td>$0_x, 3_x$</td><td>$0_x, 3_x$</td><td>$1_x, 2_x$</td><td></td><td></td></tr>

<tr><td rowspan="4">2</td><td>column</td><td>3</td><td>10</td><td>11</td><td>38</td><td>67</td><td>71</td><td>119</td></tr>
<tr><td>difference</td><td>a_x</td><td>a_x</td><td>a_x</td><td>a_x</td><td>5_x</td><td>5_x</td><td>5_x</td></tr>
<tr><td>message bits</td><td>2_x</td><td>3_x</td><td>3_x</td><td>1_x</td><td>1_x</td><td>1_x</td><td>$2_x, 3_x$</td></tr>
<tr><td>chaining bits</td><td>$0_x, 3_x$</td><td>$1_x, 2_x$</td><td>$0_x, 3_x$</td><td>$0_x, 3_x$</td><td>$1_x, 2_x$</td><td>$0_x, 3_x$</td><td>$1_x, 2_x$</td></tr>

<tr><td rowspan="4">3</td><td>column</td><td>3</td><td>38</td><td>67</td><td></td><td></td><td></td></tr>
<tr><td>difference</td><td>a_x</td><td>a_x</td><td>5_x</td><td></td><td></td><td></td></tr>
<tr><td>message bits</td><td>2_x</td><td>1_x</td><td>1_x</td><td></td><td></td><td></td></tr>
<tr><td>chaining bits</td><td>$0_x, 3_x$</td><td>$0_x, 3_x$</td><td>$1_x, 2_x$</td><td></td><td></td><td></td></tr>

<tr><td rowspan="4">4</td><td>column</td><td>3</td><td>64</td><td>112</td><td></td><td></td><td></td></tr>
<tr><td>difference</td><td>a_x</td><td>5_x</td><td>5_x</td><td></td><td></td><td></td></tr>
<tr><td>message bits</td><td>2_x</td><td>2_x</td><td>$0_x, 1_x$</td><td></td><td></td><td></td></tr>
<tr><td>chaining bits</td><td>$0_x, 3_x$</td><td>$0_x, 3_x$</td><td>$1_x, 2_x$</td><td></td><td></td><td></td></tr>

<tr><td rowspan="4">5</td><td>column</td><td>4</td><td>39</td><td>68</td><td></td><td></td><td></td></tr>
<tr><td>difference</td><td>a_x</td><td>a_x</td><td>5_x</td><td></td><td></td><td></td></tr>
<tr><td>message bits</td><td>0_x</td><td>3_x</td><td>2_x</td><td></td><td></td><td></td></tr>
<tr><td>chaining bits</td><td>$0_x, 3_x$</td><td>$0_x, 3_x$</td><td>$1_x, 2_x$</td><td></td><td></td><td></td></tr>

<tr><td rowspan="4">6</td><td>column</td><td>10</td><td>71</td><td>119</td><td></td><td></td><td></td></tr>
<tr><td>difference</td><td>a_x</td><td>5_x</td><td>5_x</td><td></td><td></td><td></td></tr>
<tr><td>message bits</td><td>3_x</td><td>1_x</td><td>$2_x, 3_x$</td><td></td><td></td><td></td></tr>
<tr><td>chaining bits</td><td>$1_x, 2_x$</td><td>$0_x, 3_x$</td><td>$1_x, 2_x$</td><td></td><td></td><td></td></tr>

<tr><td rowspan="4">7</td><td>column</td><td>11</td><td>72</td><td>120</td><td></td><td></td><td></td></tr>
<tr><td>difference</td><td>a_x</td><td>5_x</td><td>5_x</td><td></td><td></td><td></td></tr>
<tr><td>message bits</td><td>3_x</td><td>2_x</td><td>$0_x, 1_x$</td><td></td><td></td><td></td></tr>
<tr><td>chaining bits</td><td>$0_x, 3_x$</td><td>$0_x, 3_x$</td><td>$1_x, 2_x$</td><td></td><td></td><td></td></tr>

<tr><td rowspan="4">8</td><td>column</td><td>16</td><td>42</td><td>45</td><td>52</td><td>90</td><td></td></tr>
<tr><td>difference</td><td>a_x</td><td>a_x</td><td>a_x</td><td>a_x</td><td>5_x</td><td></td></tr>
<tr><td>message bits</td><td>0_x</td><td>3_x</td><td>0_x</td><td>2_x</td><td>$0_x, 1_x$</td><td></td></tr>
<tr><td>chaining bits</td><td>$1_x, 2_x$</td><td>$0_x, 3_x$</td><td>$1_x, 2_x$</td><td>$1_x, 2_x$</td><td>$1_x, 2_x$</td><td></td></tr>

<tr><td rowspan="4">9</td><td>column</td><td>50</td><td>85</td><td>114</td><td></td><td></td><td></td></tr>
<tr><td>difference</td><td>a_x</td><td>5_x</td><td>5_x</td><td></td><td></td><td></td></tr>
<tr><td>message bits</td><td>2_x</td><td>1_x</td><td>2_x</td><td></td><td></td><td></td></tr>
<tr><td>chaining bits</td><td>$1_x, 2_x$</td><td>$0_x, 3_x$</td><td>$0_x, 3_x$</td><td></td><td></td><td></td></tr>

<tr><td rowspan="4">10</td><td>column</td><td>63</td><td>67</td><td>89</td><td>92</td><td>105</td><td></td></tr>
<tr><td>difference</td><td>a_x</td><td>5_x</td><td>5_x</td><td>5_x</td><td>5_x</td><td></td></tr>
<tr><td>message bits</td><td>3_x</td><td>1_x</td><td>2_x</td><td>0_x</td><td>$2_x, 3_x$</td><td></td></tr>
<tr><td>chaining bits</td><td>$1_x, 2_x$</td><td>$1_x, 2_x$</td><td>$1_x, 2_x$</td><td>$1_x, 2_x$</td><td>$1_x, 2_x$</td><td></td></tr>
</table>

Table 6. Affected output bits for the 10 auxiliary differentials used in our improved collision attack on the Hamsi-256 compression function

i	output	
	affected bits	# unaffected bits
1	8777dda7 1f02ab16 ffd419fa 5abe9a0c 7177c1b9 8008bfde f8415ffd 0e3e1aaf	114
2	4f0eefbb 2c3e8576 f5ffa833 18b57d34 7067ef83 bd18917f ebfe82bf 4a1c7c35	109
3	8316afa6 69e19cf7 d58fc0ae 7eb9f506 8143ab86 6a1c7d70 f7a3022f 6d7fd057	119
4	e8576ac7 fa833f5f 57d34d8b eefbb4f0 8117fbf9 e82bffbf d58355e1 7ef8372e	100
5	418b57d3 b4f08e7b 6ac7e857 3e5c3a83 50a1d5c3 352e3cb8 e9d18117 26bfe82b	125
6	8fc0aad5 bff5067e 16afa69b e1ddf769 a0122ff7 7e1057ff c3af86ab 4c7df06e	108
7	c7e0556a 5ffa833f 8b57d34d f0eebbb4 d00917fb bf082bff e1d7c355 263ef837	109
8	8316afa6 69e19cf7 c583c0ae 7eb9f506 81438b86 6a1c7d70 f7a3022f 6d7fd057	123
9	ea0cfd7f 5f4d362d bbeed3c3 a15d8b1f a0affeff 160d5787 fb60dcb9 245fefe6	102
10	ea0cfd7f 5f4d062d bbeed3c3 a15d8b0f a0affeff 160d5787 fb60dcb9 245fef46	107
	efffffff ffffbfff ffffffff ffffffff f1ffffff ff3fffff ffffdfff 7fffffff	9

Generalized Iterated Hash Fuctions Revisited: New Complexity Bounds for Multicollision Attacks

Tuomas Kortelainen[1], Ari Vesanen[2], and Juha Kortelainen[2]

[1] Mathematics Division, Department of Electrical and Information Engineering, University of Oulu
[2] Department of Information Processing Science, University of Oulu

Abstract. We study the complexity of multicollision attacks on generalized iterated hash functions. In 2004 A. Joux showed that the size of a multicollision on any iterated hash function can be increased exponentially while the amount of work (or, equivalently, the length of the collision messages) grows only linearly. In Joux's considerations it was essential that each message block was used only once when computing the hash value. In 2005 M. Nandi and D. Stinson generalized Joux's method to iterated hash functions where each message block could be employed at most twice and in an arbitrary order. In the following year J. Hoch and A. Shamir further extended Joux's ideas, this time to so called ICE hash functions that scan the input message any fixed number of times in an arbitrary order. It was proved that by increasing the work polynomially, exponentially large multicollision sets could be created. The informal attack algorithm of Hoch and Shamir was more rigorously described in [8] where also the amount of work of the attack algorithm (and, as well, the length of the multicollision messages) was more precisely evaluated. In [10] new combinatorial results were proved which allowed a considerably more efficient collision set construction. In this paper we introduce a new set of tools for the combinatorial analysis of long words in which the number of occurrences of any symbol is restricted by a fixed constant. By applying these tools we are able to further shorten the length of the collison messages in an any fixed size collision set leading to a good deal smaller attack complexity. Finally, we study the structure of efficient rules for compression in bounded generalized iterated hash functions (called ICE hash functions in [4]).

1 Introduction

The design principles for message hashing proposed by Merkle and Damgård [2, 11] are applied in most of the contemporary hash functions widely used in practise. However, insecurities have been found and verified in many of these functions [3, 6, 7, 13, 15–17]. Several of the flaws come from the weaknesses in the underlying finite compression apparatus. In recent years, rigorous theoretical study has also found some weaknesses in the iterative structure itself [1].

S. Galbraith and M. Nandi (Eds.): INDOCRYPT 2012, LNCS 7668, pp. 172–190, 2012.
© Springer-Verlag Berlin Heidelberg 2012

In the following we shall describe the attack procedure on bounded generalized iterated hash functions; as the method is fairly complicated, we skip the exact characterization and illustrate the algorithm with simple but nontrivial examples.

Suppose first that we wish to create a usual Joux's multicollision of 2^r messages in an iterated hash function f^+ based on the finite compression function $f : \{0,1\}^n \times \{0,1\}^m \to \{0,1\}^n$ of length n and block size m, $m > n$. Then the compression rule can be described by the sequence $1 \cdot 2 \cdots r$ which tells us that the hash value of the message $x = x_1 x_2 \cdots x_r$ of r blocks is calculated so that

- each block x_i is deployed exactly once in the process; and
- the blocks are used in the order $x_1, x_2, \ldots, x_r$.

Certainly the hash value of x with initial value h_0 is $f^+(h_0, x)$. A 2^r–collision on f^+ can now be found as follows [5]. We start from the initial value h_0 and search two distinct message blocks b_1, b_1' such that $f(h_0, b_1) = f(h_0, b_1')$ and denote $h_1 = f(h_0, b_1)$. By the birthday paradox, the expected number of queries on f is $\tilde{a}\, 2^{\frac{n}{2}}$, where $\tilde{a}$ is approximately 2.5. Then, for each $i = 2, 3, \ldots, r$, we continue by searching message blocks b_i and b_i' such that $b_i \neq b_i'$ and $f(h_{i-1}, b_i) = f(h_{i-1}, b_i')$ and stating $h_i = f(h_{i-1}, b_i)$. Now the set $C = \{b_1, b_1'\} \times \{b_2, b_2'\} \times \cdots \times \{b_r, b_r'\}$ is 2^r-collision in f^+. The expected number of queries on f is clearly $\tilde{a}\, r 2^{\frac{n}{2}}$, i.e., the work the attacker is expected to do is only r times greater than the work she or he has to do to find a single 2-collision. The size of the multicollision set grows exponentially while the work increases only linearly.

More complex compression rules can be developed. Consider messages whose length is l blocks. Let $\alpha = i_1 i_2 \cdots i_s$ where each i_j is in $\mathbb{N}_l = \{1, 2, \ldots, l\}$. Then the compression rule α tells us that, given an l block message $x = x_1 x_2 \cdots x_l$, the hash value of x is calculated by compressing the message blocks in the order $x_{i_1}, x_{i_2}, \ldots, x_{i_s}$. More accurately, the hash value of x (under the rule α and with initial value h_0) is $f_\alpha(h_0, x_1 x_2 \cdots x_l) = f^+(h_0, x_{i_1} x_{i_2} \cdots x_{i_s})$.

Suppose now that in our compression function the parameter $n = 3$ and the hash values of messages of length $l = 18$ blocks are to be calculated. Certainly the length 3 of the hash value is ridiculously small, but the stress is on the principles of the attack method. Assume furthermore that our hash function is 3–bounded (i.e., each block can be deployed at most three times) and that the compression rule is given by:

$$\begin{aligned}
\alpha = {}& 15 \cdot 2 \cdot 7 \cdot 8 \cdot 4 \cdot 18 \cdot 13 \cdot 1 \cdot 17 \cdot 10 \cdot 14 \cdot 9 \cdot 11 \cdot 6 \cdot 3 \cdot 5 \cdot 12 \cdot 16 \cdot \\
& 11 \cdot 4 \cdot 18 \cdot 1 \cdot 5 \cdot 3 \cdot 16 \cdot 8 \cdot 2 \cdot 14 \cdot 17 \cdot 6 \cdot 9 \cdot 10 \cdot 7 \cdot 12 \cdot 13 \cdot 15 \cdot \\
& 16 \cdot 7 \cdot 9 \cdot 1 \cdot 8 \cdot 2 \cdot 10 \cdot 5 \cdot 3 \cdot 17 \cdot 18 \cdot 15 \cdot 11 \cdot 12 \cdot 6 \cdot 14 \cdot 4 \cdot 13
\end{aligned}$$

We wish to create a 4-collision in f_α. Let

$$\begin{aligned}
\alpha_1 = {}& 15 \cdot 2 \cdot 7 \cdot 8 \cdot 4 \cdot 18 \cdot 13 \cdot 1 \cdot 17 \cdot 10 \cdot 14 \cdot 9 \cdot 11 \cdot 6 \cdot 3 \cdot 5 \cdot 12 \cdot 16, \\
\alpha_2 = {}& 11 \cdot 4 \cdot 18 \cdot 1 \cdot 5 \cdot 3 \cdot 16 \cdot 8 \cdot 2 \cdot 14 \cdot 17 \cdot 6 \cdot 9 \cdot 10 \cdot 7 \cdot 12 \cdot 13 \cdot 15, \text{ and} \\
\alpha_3 = {}& 16 \cdot 7 \cdot 9 \cdot 1 \cdot 8 \cdot 2 \cdot 10 \cdot 5 \cdot 3 \cdot 17 \cdot 18 \cdot 15 \cdot 11 \cdot 12 \cdot 6 \cdot 14 \cdot 4 \cdot 13.
\end{aligned}$$

Note that $\alpha = \alpha_1 \alpha_2 \alpha_3$ and that each symbol of $\mathbb{N}_{18}$ occurs in each of the words α_1, α_2, and α_3 exactly once. We shall proceed as follows. As in the basic Joux's

attack, we create a 2^{18}–collision in f_{α_1} with initial value h_0. Let $b_1, b_1', b_2, b_2',$
..., b_{18}, b_{18}' be pairs of message blocks such that $b_i \neq b_i'$ for $i = 1, 2, \ldots, 18$ and
$f_{\alpha_1}(h_0, z_1) = f_{\alpha_1}(h_0, z_2)$ for all $z_1, z_2 \in C_0 := \{b_1, b_1'\} \times \{b_2, b_2'\} \times \cdots \times \{b_{18}, b_{18}'\}$.
The expected number of calls of the compression function f is certainly at most
$\tilde{a} \cdot 18 \cdot 2^{\frac{3}{2}}$. Let

$$\beta_1 = 11 \cdot 4 \cdot 18, \beta_2 = 1 \cdot 5 \cdot 3, \quad \beta_3 = 16 \cdot 8 \cdot 2$$
$$\beta_4 = 14 \cdot 17 \cdot 6, \beta_5 = 9 \cdot 10 \cdot 7, \quad \beta_6 = 12 \cdot 13 \cdot 15.$$

The collision set C_0 is large enough to induce (with initial value h_0) a 2–collision
in $f_{\alpha_1 \beta_1}$, a 2^2–collision in $f_{\alpha_1 \beta_1 \beta_2}$, ..., a 2^6–collision in $f_{\alpha_1 \beta_1 \beta_2 \cdots \beta_6} = f_{\alpha_1 \alpha_2}$. Let
$C_1 \subseteq C_0$ be the 2^6–collision in $f_{\alpha_1 \alpha_2}$ (with initial value h_0).
　Let $\gamma_1 = 16 \cdot 7 \cdot 9 \cdot 1 \cdot 8 \cdot 2 \cdot 10 \cdot 5 \cdot 3$ and $\gamma_2 = 17 \cdot 18 \cdot 15 \cdot 11 \cdot 12 \cdot 6 \cdot 14 \cdot 4 \cdot 13$. Then each
symbol in the words β_1, β_4, and β_6 occurs only in γ_2 and, as well, each symbol
in β_2, β_3, and β_5 occurs only in γ_1. This, and the fact that C_1 is sufficiently
large, induce a 2–collision in $f_{\alpha_1 \alpha_2 \gamma_1}$ and a 2^2–collision in $f_{\alpha_1 \alpha_2 \gamma_1 \gamma_2} = f_\alpha$. The
expected number of compression function calls to create a 4–collision on f_α is at
most $3 \cdot \tilde{a} \cdot 18 \cdot 2^{\frac{3}{2}}$.
　The nested attack method described in the previous example can be generalized and applied to any bounded iterated hash function. Note that the respective
compression rules (one rule for each message length in blocks) are generally not
of such a favorable form as in our example. However, as our generalized iterated
hash function is bounded, we can, by choosing a sufficiently long compression
rule α, always find an arbitrarily large subalphabet A of alph(α) such that the
projection $\pi_A(\alpha)$ is of the required favorable form and induces a collision set of
arbitrary size. The longer is α, the longer are messages in the collision set, and
the more complex is our attack. So it is in our best interests to be able to choose
the compression rule α as short as possible. By carefully analyzing the combinatorial properties of long compression rules in bounded generalized iterated hash
functions, we can do this and thus decrease the length of the colliding messages.
　One more remark is still to the purpose. When constructing the collision set,
our attack method focuses only on message blocks corresponding to the elements
of the active alphabet A. All message blocks linked to symbols in alph(α) \ A
are replaced by a fixed constant message block.
　We shall now recall the earlier work on this subject.
　In [12] Nandi and Stinson considered 2-bounded iterated hash functions. They
were able to show that the attacker can create 2^k-collision with

$$O(k^2 \cdot \ln k \cdot [n + \ln \ln(2k)] \cdot 2^{\frac{n}{2}})$$

compression function calls.
　In the article [4] the results of [12] were further generalized. It was shown
that it is possible to create 2^k-collision in any q-bounded generalized iterated
hash function with $O(g(n, q, k) 2^{\frac{n}{2}})$ compression function calls, where $g(n, q, k)$
is function of n, q and k which is polynomial with respect to n and k.
　The results of [4] were studied more rigorously in [8], where it was shown,
that the approach offered in [4] meant that the expected number of compression

function calls needed to create 2^k-collision for q–bounded generalized iterated hash function would be at most

$$\tilde{a} \cdot q \cdot 2^{2^{2q-3}} k^{(2q-3)2^{2q-1}} n^{(q-1)^2 2^{2q-1}} \cdot 2^{\frac{n}{2}}.$$

A fresh approach and novel tools were created in [10]. It was shown that it is possible to create 2^k-collision in any q-bounded generalized iterated hash function with expected number of compression function calls at most

$$\tilde{a} \cdot q \cdot k^{(2q-3)2^q} n^{(q-1)^2 2^q} \cdot 2^{\frac{n}{2}}.$$

Finally, in this work we will once again improve the methodology and we are able to prove that the expected number of compression function calls needed to create a 2^k-collision is at most

$$\tilde{a} \cdot q \cdot 5^{q-2} \cdot 2^{\lceil \log_2 n \rceil \frac{(q+4)q(q-1)}{6} + \lceil \log_2 k \rceil q} \cdot 2^{\frac{n}{2}}.$$

Note that above the expected numbers of compression function calls all are of the form $\tilde{a} \cdot q \cdot g(n, q, k) \cdot 2^{\frac{n}{2}}$; here $g(n, q, k)$ is, in fact, the length of collision messages.

This paper is organized in the following way. In the second section some preliminaries are given. Then the attack algorithm is a bit more rigorously described and our main result concerning the message complexity is given. The fourth section contains our combinatorial toolbox needed for the complexity considerations. The paper ends with a short conclusion.

2   Basic Concepts and Notation

This section contains helpful background information: Preliminaries on finite strings and hash functions are briefly gone through.

2.1   Prerequisites on Alphabets and Words

Let $\mathbb{Z}$ the set of all integers, $\mathbb{N}$ the set of all natural numbers, and $\mathbb{N}_+ = \mathbb{N} \setminus \{0\}$. For each $i \in \mathbb{N}_+$, denote $\mathbb{N}_i = \{1, 2, \ldots, l\}$. For any finite set S, let $|S|$ be the *cardinality* of S that is to say, the number of elements in S. Given a nonnegative real number x, denote by $\lfloor x \rfloor$ the integer part of x.

An *alphabet* is any finite, nonempty set of abstract symbols called *letters* or *symbols*. Let A be an alphabet. A *word* (over A) is any finite sequence of symbols (in A). Thus, assuming that w is a word over A, we can write $w = x_1 x_2 \cdots x_n$, where n is a nonnegative integer and $x_i \in A$ for $i = 1, 2, \ldots, n$. Here n is the *length* $|w|$ of w. Notice that n may be equal to zero; then w is the *empty word*, often denoted by ϵ, which contains no letters. For each $a \in A$, let $|w|_a$ be the number of occurrences of the letter a in w, and let $\text{alph}(w)$ denote the set of all letters occurring in w at least once. The *powers* of the word w are defined recursively as: $w^0 = \epsilon$, $w^1 = u$, and $w^{i+1} = w^i w$ for $i \in \mathbb{N}_+$. Let $w^+ = \{w^i \mid i \in \mathbb{N}_+\}$. Given $q \in \mathbb{N}_+$, the word w is *q–restricted* if $|w|_a \leq q$ for all $a \in \mathrm{alph}(w)$. The word w is a *permutation* of A, if $\mathrm{alph}(w) = A$ and $|w|_a = 1$ for each $a \in A$ (i.e., the alphabet of w is A and w is 1–restricted).

For each $n \in \mathbb{N}$, denote by A^n the set of all words of length n over A. Let A^* be the set of all words over A and $A^+ = A^* \setminus \{\epsilon\}$.

The *projection morphism* π_B from A^* into B^*, where $B \subseteq A$ is nonempty, is defined by $\pi_B(c) = c$ if $c \in B$ and $\pi_B(c) = \epsilon$ if $c \in A \setminus B$. Given a word w over the alphabet A, define the word $(w)_B$ as follows: $(w)_B = \epsilon$ if $\pi_B(w) = \epsilon$ and $(w)_B = a_1 a_2 \cdots a_s$ if $\pi_B(w) \in a_1^+ a_2^+ \cdots a_s^+$, where $s \in \mathbb{N}_+$, $a_1, a_2, \ldots, a_s \in B$, and $a_i \neq a_{i+1}$ for $i = 1, 2, \ldots, s - 1$.

2.2   Prerequisites in Hash Functions and Their Security Properties

Let us first contemplate traditional hash functions. Call any word over the binary alphabet $\{0, 1\}$ a *message*. A *hash function* (of length n, where $n \in \mathbb{N}_+$) is a mapping $\mathtt{H} : \{0, 1\}^* \to \{0, 1\}^n$. An ideal hash function $\mathtt{H} : \{0, 1\}^* \to \{0, 1\}^n$ is a *(variable input length) random oracle*: for each $x \in \{0, 1\}^*$, the value $\mathtt{H}(x) \in \{0, 1\}^n$ is chosen uniformly at random.

Let $k \geq 2$ be an integer. A k-*collision* in the hash function $\mathtt{H}$ is a k-element subset C of set $\{0, 1\}^*$ such that $\mathtt{H}(x) = \mathtt{H}(y)$ for all $x, y \in C$. Any 2-collision is also called a collision.

There are three basic security properties of hash functions: *collision resistance*, *preimage resistance* and *second preimage resistance*. Historically, the properties are defined as follows.

Collision Resistance. It is computationally infeasible to find $x, x' \in \{0, 1\}^*$, $x \neq x'$, such that $\mathtt{H}(x) = \mathtt{H}(x')$.

Preimage Resistance. Given any $y \in \{0, 1\}^n$, it is computationally infeasible to find $x \in \{0, 1\}^*$ such that $\mathtt{H}(x) = y$.

Second Preimage Resistance. Given any $x \in \{0, 1\}^*$, it is computationally infeasible to find $x' \in \{0, 1\}^*$, $x \neq x'$, such that $\mathtt{H}(x) = \mathtt{H}(x')$.

We wish to have more rigorous definitions to the security properties above, especially to collision resistance, and employ the random oracle model. According to the (generalized) *birthday paradox*, given any hash function $\mathtt{H}$ of length n (random oracle hash functions included), a k-collision can in $\mathtt{H}$ be found (with probability approx. $\frac{1}{2}$) by hashing $(k!)^{\frac{1}{k}} 2^{\frac{n(k-1)}{k}}$ messages [14].

Two remarks can be made immediately:

- In the case $k = 2$ approximately $\sqrt{2} \cdot 2^{\frac{n}{2}}$ hashings (queries on f) are needed; intuitively most of us would expect the number to be around 2^{n-1}.
- Given an integer $k \geq 2$, when n is sufficiently large, finding a $(k + 1)$-collision consumes much more resources than finding a k-collision.

We finally state a new security property for hash functions. The characterization is not yet totally rigorous. It is, however, important for our further considerations.

Multicollision Resistance. The hash function H is k-*collision resistant*, $k \geq 2$ an integer, if to find a k-collision in H is (approximately) as difficult as to find an k-collision in any random oracle hash function G of length n.

2.3 Generalized Iterated Hash Functions

All practical hash functions use iterative structures.

Let $m, n \in \mathbb{N}_+$ be such that $m > n$. Then $H = \{0, 1\}^n$ is the set of *hash values* (of length n) and $B = \{0, 1\}^m$ is the set of *message blocks* (of length m). Any $w \in B^+$ is a *message*. Given a mapping $f : H \times B \to H$, call f a *compression function* (of length n and block size m).

Define the function $f^+ : H \times B^+ \to H$ inductively as follows. For each $h \in H$, $b \in B$ and $x \in B^+$, let $f^+(h, b) = f(h, b)$ and $f^+(h, b\,x) = f^+(f(h, b), x)$. Note that f^+ is nothing but an iterative generalization of the compression function f.

Let $l \in \mathbb{N}_+$ and α be a word such that $\mathrm{alph}(\alpha) \subseteq \mathbb{N}_l = \{1, 2, \ldots, l\}$. Then $\alpha = i_1 i_2 \cdots i_s$, where $s \in \mathbb{N}_+$, $s \geq l$, and $i_j \in \mathbb{N}_l$ for $j = 1, 2, \ldots, s$. Define the *iterated compression function* $f_\alpha : H \times B^l \to H$ (based on α and f) by

$$f_\alpha(h, b_1 b_2 \cdots b_l) = f^+(h, b_{i_1} b_{i_2} \cdots b_{i_s})$$

for each $h \in H$ and $b_1, b_2, \ldots, b_l \in B$. Here clearly α only declares how many times and in which order the message blocks $b_1, b_2, \ldots, b_l$ are used when creating the (hash) value $f_\alpha(h, b_1 b_2 \ldots b_l)$ of the message $b_1 b_2 \cdots b_l$.

Given an integer $k \geq 2$ and $h_0 \in H$, a k-*collision (with initial value h_0) in the iterated compression function* f_α is a set $C \subseteq B^l$ such that the following holds:

1. The cardinality of C is k;
2. For all $u, v \in C$ we have $f_\alpha(h_0, u) = f_\alpha(h_0, v)$; and
3. For any pair of distinct messages $u = u_1 u_2 \cdots u_l$ and $v = v_1 v_2 \cdots v_l$ in C such that $u_i, v_i \in B$ for $i = 1, 2, \ldots, l$, there exists $j \in \mathrm{alph}(\alpha)$ for which $u_j \neq v_j$.

For each $j \in \mathbb{N}_+$, let now $\alpha_j \in \mathbb{N}_j^+$ be such that $\mathrm{alph}(\alpha_j) = \mathbb{N}_j$. Denote $\hat{\alpha} = (\alpha_1, \alpha_2, \ldots)$. Define the *generalized iterated hash function* (gihf, for short) $\mathrm{H}_{\hat{\alpha}, f} : H \times B^+ \to H$ (based on $\hat{\alpha}$ and f) as follows: Given the initial value $h_0 \in H$ and the message $x \in B^j$, $j \in \mathbb{N}_+$, let

$$\mathrm{H}_{\hat{\alpha}, f}(h_0, x) = f_{\alpha_j}(h_0, x) .$$

Thus, given any message x of j blocks and hash value h_0, to obtain the value $\mathrm{H}_{\hat{\alpha}, f}(h_0, x)$, we just pick the word α_j from the sequence $\hat{\alpha}$ and compute $f_{\alpha_j}(h_0, x)$. For more details, see [8] and [4].

The gihf $\mathrm{H}_{\hat{\alpha}, f}$ is q-*bounded*, $q \in \mathbb{N}_+$, if for each $l \in \mathbb{N}_+$, no letter $\mathrm{alph}(\alpha_l)$ occurs in α_l more than q times. A generalized iterated hash function is *bounded* if it is q-bounded for some $q \in \mathbb{N}_+$.

Remark 1. A traditional iterated hash function $\mathrm{H} : B^+ \to H$ based on f (with initial value $h_0 \in H$) can of course be defined by $\mathrm{H}(u) = f^+(h_0, u)$ for each

$u \in B^+$. On the other hand H is a generalized iterated hash function $\mathrm{H}_{\hat{\alpha},f}$: $H \times B^+ \to H$ based on $\hat{\alpha}$ and f where $\hat{\alpha} = (1, 1 \cdot 2, 1 \cdot 2 \cdot 3, \ldots)$ and the initial value is fixed as h_0. Certainly H (as a gihf) is 1-bounded.

Given $k \in \mathbb{N}_+$ and $h_0 \in H$, a *k-collision in the generalized iterated hash function* $\mathrm{H}_{\hat{\alpha},f}$ *(with initial value h_0)* is a set C of k messages such that for all $u, v \in C$, $|u| = |v|$ and $\mathrm{H}_{\hat{\alpha},f}(h_0, u) = \mathrm{H}_{\hat{\alpha},f}(h_0, v)$. Now suppose that C is a k-collision in $\mathrm{H}_{\hat{\alpha},f}$ with initial value h_0. Let $l \in \mathbb{N}_+$ be such that $C \subseteq B^l$, i.e., the length of each message in C is l. Then, by definition, for each $u, v \in C$, the equality $f_{\alpha_l}(h_0, u) = f_{\alpha_l}(h_0, v)$ holds. Since $\mathrm{alph}(\alpha_l) = \mathbb{N}_l$ (and thus each symbol in $\mathbb{N}_l$ occurs in $\mathrm{alph}(\alpha)$), the set C is a k-collision in f_{α_l} with initial value h_0. Thus, a k-collision in the generalized iterated hash function $\mathrm{H}_{\hat{\alpha},f}$ necessarily by definition, is a k-collision in the iterated compression function f_{α_l} for some $l \in \mathbb{N}_+$.

Now, in our security model, the *attacker* tries to find a k-collision in $\mathrm{H}_{\hat{\alpha},f}$. We assume that the attacker

- knows how $\mathrm{H}_{\hat{\alpha},f}$ depends on the compression function f (i.e., she/he knows $\hat{\alpha}$);
- sees the compression function f as a black box (i.e., she/he does not know anything about the internal structure of f); and
- can make (any number of) *queries* (pairs $(h, b) \in H \times B$) on f and get the respective *responses* (values $f(h, b) \in H$).

What do we mean by an attack? A *k-collision attack on* $\mathrm{H}_{\hat{\alpha},f}$ is a probabilistic procedure (based on the birthday paradox) that finds a k-collision in $\mathrm{H}_{\hat{\alpha},f}$ with probability equal to one for any initial value h_0.

The *(message) complexity of a k-collision attack on* $\mathrm{H}_{\hat{\alpha},f}$ is the expected number of queries on f required to get a k-collision $\mathrm{H}_{\hat{\alpha},f}$.

3   Attacking Bounded Generalized Iterated Hash Functions

Below we describe a general attack procedure on generalized iterated hash functions informally, for details we refer to [8].

3.1   Nested Multicollision Attack Schema ($\mathcal{N}MCAS$)

Recall that in a generalized iterated hash function $\mathrm{H}_{\hat{\alpha},f}$ the mapping $f : \{0,1\}^n \times \{0,1\}^m \to \{0,1\}^n$ is a compression function and $\hat{\alpha} = (\alpha_1, \alpha_2, \ldots)$ is a sequence of words such that $\mathrm{alph}(\alpha_l) = \mathbb{N}_l$ for each $l \in \mathbb{N}_+$.

Procedure Schema $\mathcal{N}MCAS$

Input: A generalized iterated hash function $H_{\hat{\alpha},f}$, an initial value $h_0 \in \{0,1\}^n$, a positive integer k.
Output: A 2^k-collision in $H_{\hat{\alpha},f}$.

Step 1: Choose (a large) $l \in \mathbb{N}_+$. Consider the lth element α_l of the sequence $\hat{\alpha}$. Let $\alpha_l = i_1 i_2 \cdots i_s$, where $s \in \mathbb{N}_+$ and $i_j \in \mathbb{N}_l$ for $j = 1, 2, \ldots, s$.

Step 2: Fix a (large) set of *active indices* $Act \subseteq \mathbb{N}_l = \{1, 2, \ldots, l\}$.

Step 3: Factorize the word α_l into nonempty strings appropriately, i.e., find $p \in \{1, 2, \ldots, s\}$ and $\beta_i \in \mathbb{N}_l^+$ such that $\alpha_l = \beta_1 \beta_2 \cdots \beta_p$.

Step 4: Based upon the active indices, create a large multicollision in f_{β_1}. More exactly, find message block sets $M_1, M_2, \ldots, M_l$ satisfying the following properties.

(i) If $i \in \mathbb{N}_l \setminus Act$, then the set M_i consists of one constant message block ω.

(ii) If $i \in Act$, then the set M_i consists of two different message blocks m_{i1} and m_{i2}.

(iii) The set $M = M_1 M_2 \cdots M_l = \{u_1 u_2 \cdots u_l \mid u_i \in M_i \text{ for } i = 1, 2, \ldots, l\}$ is a $2^{|Act|}$-collision in f_{β_1} with initial value h_0.

Step 5: Based on the set $C_1 = M$, find message sets $C_2, C_3, \ldots, C_p$ such that

(iv) $C_p \subseteq C_{p-1} \subseteq \cdots \subseteq C_1 = M$.

(v) For each $j \in \{1, 2, \ldots, p\}$ the set C_j is a (large) multicollision in $f_{\beta_1 \beta_2 \cdots \beta_j}$ with initial value h_0.

(vi) $|C_p| = 2^k$.

Step 6: Output C_p.

Cosider a q–bounded generalized iterated hash function $\mathsf{H}_{\hat{\alpha}, f}$ where the compression function f is of length n and block size m. Suppose that $l \in \mathbb{N}_+$, $p \in \{1, 2, \ldots, q\}$ and $A \subseteq \mathrm{alph}(\alpha_l)$, $|A| = n^{p-1}$ are such that the following properties are satisfied:

(Q1) α_l possesses a factorization $\alpha_l = \beta_1 \beta_2 \cdots \beta_p$ where $A \subseteq \mathrm{alph}(\beta_i)$ for $i = 1, 2, \ldots, p$; and

(Q2) for any $i \in \{1, 2, \ldots, p-1\}$, if $(\beta_i)_A = z_1 z_2 \cdots z_{n^{p-i}k}$ is a factorization of $(\beta_i)_A$ such that $|\mathrm{alph}(z_j)| = n^{i-1}$ for $j = 1, 2, \ldots n^{p-i}k$ and $(\beta_{i+1})_A = u_1 u_2 \cdots u_{n^{p-i+1}k}$ is a factorization of $(\beta_{i+1})_A$ such that $|\mathrm{alph}(u_j)| = n^i$ for $j = 1, 2, \ldots n^{p-i+1}k$, then for each $j_1 \in \{1, 2, \ldots, n^{p-i}k\}$, there exists $j_2 \in \{1, 2, \ldots, n^{p-i-1}k\}$ such that $\mathrm{alph}(z_{j_1}) \subseteq \mathrm{alph}(u_{j_2})$.

Then, following the $\mathcal{NMCAS}$–procedure (and the attack construction described more rigorously in in [8]), the property (Q1) allows the attacker to construct a $2^{|A|}$-collision C_1 in f_{β_1} with any initial value h_0 so that the expected number of queries on f is at most $\tilde{a}|\beta_1| 2^{\frac{n}{2}}$. The property (Q2) ensures that based on the multicollision guaranteed by (Q1), the attacker can proceed and, for $i = 2, 3, \ldots, p$, create a $2^{n^{p-i}k}$-collision C_i in $f_{\beta_1 \beta_2 \cdots \beta_i}$ so that the expected number of queries on f is at most $\tilde{a}|\beta_1 \beta_2 \cdots \beta_i| 2^{\frac{n}{2}}$. Thus finally a 2^k-collision of complexity $\tilde{a}|\alpha| 2^{\frac{n}{2}}$ in $\mathsf{H}_{\hat{\alpha}, f}$ is generated. We refer to Section 5 in [8] for the details of the attack as a statistical experiment.

On the basis of our combinatorial results (culminating to Theorem 6) the following can be proved:

Theorem 1. *Let m, n and q be positive integers such that $m > n$ and $q \geq 2$, $f : \{0,1\}^n \times \{0,1\}^m \to \{0,1\}^n$ a compression function, and $\hat{\alpha} = (\alpha_1, \alpha_2, \ldots)$ a q-bounded sequence of words such that $\mathrm{alph}(\alpha_l) = \mathbb{N}_l$ for each $l \in \mathbb{N}_+$. Then, for each $k \in \mathbb{N}_+$, there exists a 2^k-collision attack on the generalized iterated hash function $\mathsf{H}_{\hat{\alpha},f}$ such that the expected number of queries on f is at most $\tilde{a} \cdot q \cdot 5^{q-2} \cdot 2^{\lceil \log_2 n \rceil \frac{(q+4)q(q-1)}{6} + \lceil \log_2 k \rceil q} \cdot 2^{\frac{n}{2}}$.*

Proof. Choose $d_i = \lceil \log_2 n \rceil$ for $i = 1, 2, \ldots, q-1$, and $d_q = \lceil \log_2 k \rceil$. Then

$$\sum_{i=1}^{q}(q - i + 1)\, i\, d_i = \sum_{i=1}^{q-1}(q - i + 1)\, i\, \lceil \log_2 n \rceil + q \lceil \log_2 k \rceil$$
$$= \frac{(q+4)q(q-1)}{6} \lceil \log_2 n \rceil + q \lceil \log_2 k \rceil .$$

Let $l = 5^{q-2} \cdot 2^{\lceil \log_2 n \rceil \frac{(q+4)q(q-1)}{6} + \lceil \log_2 k \rceil q} \cdot 2^{\frac{n}{2}}$. Now Corollary 3 (and Theorem 6) imply that the properties (Q1) and (Q2) hold. $\qquad\square$

The previous theorem implies that, given $q, k \in \mathbb{N}_+$, then for sufficiently large n, no q-bounded generalized iterated hash function is 2^k-collision resistant.

3.2   Attacks in Practice

We can now shortly compare the upper bounds offered in the different articles. Let us assume, that we want to create 2^k−collision against q−bounded hash function. Articles [4, 8] prove that the total complexity of the attack will be less than $2.5 \cdot q \cdot 2^{2^{2^q-3}} k^{(2q-3)2^{2^q-1}} n^{(q-1)^2 2^{2^q-1}} 2^{\frac{n}{2}}$.

This means that if we assume that for example $n = 256$, $k = 4$ and $q = 3$ we get the expected number of compression function calls is less than

$$2.5 \cdot 3 \cdot 2^{2^{2^3-3}} 4^{(2^3-3)2^{2^3-1}} 256^{(3-1)^2 2^{2^3-1}} 2^{\frac{256}{2}} = 7.5 \cdot 2^{8630}.$$

It is immediately clear that if complexity is this high, the attack has no practical significance. The more efficient way to create the permutations offered by [9] lowers the complexity of the attack to $2.5 \cdot q \cdot k^{(2q-3)2^q} n^{(q-1)^2 2^q} 2^{\frac{n}{2}}$ and thus for the values $n = 256$, $k = 4$ and $q = 3$ the complexity drops dramatically to $2.5 \cdot 3 \cdot 4^{3 \cdot 2^3} 256^{2^2 2^3} 2^{\frac{256}{2}} = 7.5 \cdot 2^{432}$. However it is easy to see that even this is well beyond the complexity required to complete the preimage attack.

In this article we have however once again lowered the upper bound of the attack significantly to less than $2.5 \cdot q \cdot 5^{q-2} 2^{\lceil \log_2 n \rceil \frac{(q+4)q(q-1)}{6} + \lceil log_2 k \rceil q} \cdot 2^{\frac{n}{2}}$.

Our new upper bound gives us (for $n = 256$, $k = 4$ and $q = 3$) the complexity of $7.5 \cdot 5 \cdot 2^{190}$. This is clearly well below the complexity $\sqrt[16]{16!} \cdot 2^{240}$ offered by the brute force attack (see article [14]).

4   Combinatorial Considerations

We now derive the main combinatorial results needed to prove that our multic-collision attack construction can be realized efficiently. Note that, when studying unavoidable regularities in long words with bounded number of symbol occurrences, we give up the use of classical combinatorial results such as Dilworth's Theorem and (generalized) Hall's matching Theorem applied in [4] and in [8, 10, 12]. Instead, we develop a completely new approach which is based purely on concepts of combinatorics on words. The properties of so called nonseparable words appear to be of utmost importance.

4.1   On Nonseparable Words

Let α be a word, $A \subseteq \text{alph}(\alpha)$, and $s \in \mathbb{N}_+$. We say that the word α is (s, A) –*separable* if there exists a factorization $\alpha = \alpha_1\alpha_2$ of α such that $|\text{alph}(\alpha_1) \cap \text{alph}(\alpha_2) \cap A| \geq s$. The word α is (s, A)–*nonseparable* if $A \subseteq \text{alph}(\alpha)$ and α is not (s, A)–separable.

The word α is s–*separable* (s–*nonseparable*, respectively) if α is $(s, \text{alph}(\alpha))$–separable $((s, \text{alph}(\alpha))$–nonseparable, resp.) Let us list three simple properties of nonseparable words.

Lemma 1. *Let s be a positive integer, A an alphabet and α an (s, A)–nonseparable word. Then the following holds.*

(a) *Each subword β of α is (t, B)–nonseparable for any integer $t \geq s$ and set $B \subseteq A$.*

(b) *There exists a factorization $\alpha = \beta\gamma$ of α such that $|[\text{alph}(\beta)\backslash\text{alph}(\gamma)] \cap A| \geq \lfloor\frac{|A|}{2}\rfloor - s$ and $|[\text{alph}(\gamma) \backslash \text{alph}(\beta)] \cap A| \geq \lfloor\frac{|A|}{2}\rfloor - s$.*

(c) *Suppose that $|A| \geq d\,s$ where $d \in \mathbb{N}_+$. Then there exists a factorization $\alpha = \alpha_1\alpha_2 \cdots \alpha_d$ of α such that for each $i \in \{1, 2, \ldots, d\}$ there exists a symbol $a_i \in A$ such that $a_i \in \text{alph}(\alpha_i)$ but $a_i \notin \text{alph}(\alpha_1\alpha_2 \cdots \alpha_{i-1})$ and $a_i \notin \text{alph}(\alpha_{i+1}\alpha_{i+2} \cdots \alpha_d)$.*

Proof. The claim of (a) follows directly from the definition.

To prove (b), let $\alpha = \beta\gamma$ be a factorization of α such that $\text{alph}(\beta)$ contains occurrences of exactly $\lfloor\frac{|A|}{2}\rfloor$ symbols in A. Then γ contains occurrences of at least $\lfloor\frac{|A|}{2}\rfloor$ symbols in A. Since α is (s, A)–nonseparable, $|\text{alph}(\beta)\backslash\text{alph}(\gamma)| \geq \lfloor\frac{|A|}{2}\rfloor - s$ and $|\text{alph}(\gamma) \backslash \text{alph}(\beta)| \geq \lfloor\frac{|A|}{2}\rfloor - s$.

Consider finally (c). Let $\alpha = \alpha_1\alpha_2 \cdots \alpha_d$ be a factorization of α such that for all $i \in \{1, 2, \ldots, d\}$, α_i contains at least s symbols that do not occur in the word $\alpha_1\alpha_2 \cdots \alpha_{i-1}$. Since $|A| \geq d\,s$, the factorization can always be found. Since α is (s, A)–nonseparable, each factor of it is as well. Thus for $i \in \{1, 2, \ldots, d\}$, there exists $a \in \text{alph}(\alpha_i) \backslash \text{alph}(\alpha_1\alpha_2 \cdots \alpha_{i-1})$ such that a does not occur in $\alpha_{i+1}\alpha_{i+2} \cdots \alpha_d$. The claim follows. $\square$

Let now s and p be positive integers. Define the function $T_s^p : \mathbb{Z} \to \mathbb{Z}$ by

$$T_s^p(x) = \lfloor\frac{x}{2^p}\rfloor - 2s .$$

We then define, for each $k \in \mathbb{N}_+$, the function $T_s^{(p,k)} : \mathbb{N} \to \mathbb{N}$ by

$$T_s^{(p,k)}(x) = T_s^p(T_s^p(\cdots (T_s^p(x)) \cdots)) .$$

Here on the right side the function T_s^p is applied k times, i.e.,

$$T_s^{(p,k)} = \underbrace{T_s^p \circ T_s^p \circ \cdots \circ T_s^p}_{k \text{ times}} .$$

What follows now is a series of quite technical lemmata in which the structural properties of (a sequence) of nonseparable words are gradually fortified. We apologize the numerous new notations and strenuous and detailed treatment; they allow us, however, to give the proofs in an accurate form contrary to the state of affairs in many security papers. For space reasons, proofs of some results are omitted and later presented in the full version of the paper.

The first result is a generalization of item b of Lemma 1 for a sequence of nonseparable words.

Lemma 2. *Let p and s be positive integers, A an alphabet, and $\alpha_1, \alpha_2, \ldots, \alpha_p$ a sequence of (s, A)–nonseparable words. Then there exist sets $A_1, A_2 \subseteq A$ and, for each $i \in \{1, 2, \ldots, p\}$, a factorization $\alpha_i = \alpha_{i,1}\alpha_{i,2}$ of α_i as well as a permutation σ_i of $1, 2$ such that*

1. $A_1 \subseteq \mathrm{alph}(\alpha_{1,\sigma_i(1)})$ *and* $A_2 \subseteq \mathrm{alph}(\alpha_{1,\sigma_i(2)})$;
2. $A_1 \cap \mathrm{alph}(\alpha_{i,\sigma_1(2)}) = \emptyset$ *and* $A_2 \cap \mathrm{alph}(\alpha_{i,\sigma_i(1)}) = \emptyset$; *and*
3. $|A_1| \geq T_s^p(|A|)$ *and* $|A_2| \geq T_s^p(|A|)$.

Proof. We proceed by induction on p. Consider the case $p = 1$. Let $\alpha_{1,1}\alpha_{1,2}$ be a factorization of α_1 such that $|\mathrm{alph}(\alpha_{1,1}) \cap A| \geq \lfloor \frac{|A|}{2} \rfloor$ and $|\mathrm{alph}(\alpha_{1,2}) \cap A| \geq \lfloor \frac{|A|}{2} \rfloor$. Then, since α_1 is (s, A)–nonseparable, we have $|\mathrm{alph}(\alpha_{1,1}) \cap \mathrm{alph}(\alpha_{1,2}) \cap A| < s$. This implies that $|[\mathrm{alph}(\alpha_{1,1}) \setminus \mathrm{alph}(\alpha_{1,2})] \cap A| \geq \lfloor \frac{|A|}{2} \rfloor - s$ and $|[\mathrm{alph}(\alpha_{1,2}) \setminus \mathrm{alph}(\alpha_{1,1})] \cap A| \geq \lfloor \frac{|A|}{2} \rfloor - s$. Choose $A_1 = [\mathrm{alph}(\alpha_{1,1}) \setminus \mathrm{alph}(\alpha_{1,2})] \cap A$ and $A_2 = [\mathrm{alph}(\alpha_{1,2}) \setminus \mathrm{alph}(\alpha_{1,1})] \cap A$. Let furthermore σ_1 be the identity permutation of $1, 2$. Since $\lfloor \frac{|A|}{2} \rfloor - s \geq T_s^1(|A|)$, the claims 1, 2, and 3 are satisfied.

Suppose then that $p = k + 1$, $k \in \mathbb{N}_+$. Consider the sequence of words $\alpha_1, \alpha_2, \ldots, \alpha_k$. By the induction hypothesis, there exist $A_1', A_2' \subseteq A$ and, for each $i \in \{1, 2, \ldots, k\}$, a factorization $\alpha_i = \alpha_{i,1}\alpha_{i,2}$ of α_i and a permutaion σ_i of $1, 2$ such that A_1', A_2' and σ_i (for $i = 1, 2, \ldots, k$) satisfy the properties 1, 2, and 3.

Let $\alpha_{k+1} = \alpha_{k+1,1}\alpha_{k+1,2}$ be a factorization of α_{k+1} and i_1, i_2 distinct elements in $\{1, 2\}$ such that

$$|\mathrm{alph}(\alpha_{k+1,i_1}) \cap A_1'| \geq \left\lfloor \frac{1}{2} \lfloor \frac{|A|}{2^k} \rfloor \right\rfloor - s \text{ and } |\mathrm{alph}(\alpha_{k+1,i_2}) \cap A_2'| \geq \left\lfloor \frac{1}{2} \lfloor \frac{|A|}{2^k} \rfloor \right\rfloor - s .$$

Let σ_{k+1} be the permutation of $1, 2$ such that $\sigma_{k+1}(1) = i_1$ and $\sigma_{k+1}(2) = i_2$. Denote

$$A_1 = [\mathrm{alph}(\alpha_{k+1,i_1}) \setminus \mathrm{alph}(\alpha_{k+1,i_2})] \cap A_1' \text{ and } A_2 = [\mathrm{alph}(\alpha_{k+1,i_2}) \setminus \mathrm{alph}(\alpha_{k+1,i_1})] \cap A_2' .$$

Clearly A_1 and A_2 as well as the permutations $\sigma_1, \sigma_2, \ldots, \sigma_{k+1}$ satisfy the claims 1 and 2. Since α_{k+1} is (s, A)-nonseparable, we have

$$|A_1| \geq \left\lfloor \frac{1}{2} \lfloor \frac{|A|}{2^k} \rfloor \right\rfloor - 2s \text{ and } |A_2| \geq \left\lfloor \frac{1}{2} \lfloor \frac{|A|}{2^k} \rfloor \right\rfloor - 2s .$$

Since $|A| \geq 2^{k+1} \lfloor \frac{|A|}{2^{k+1}} \rfloor$, the relations

$$\left\lfloor \frac{1}{2} \lfloor \frac{|A|}{2^k} \rfloor \right\rfloor \geq \left\lfloor \frac{1}{2} \cdot 2 \cdot \lfloor \frac{|A|}{2^{k+1}} \rfloor \right\rfloor = \lfloor \frac{|A|}{2^{k+1}} \rfloor$$

hold, so we deduce that $|A_1| \geq \lfloor \frac{|A|}{2^{k+1}} \rfloor - 2s$ and $|A_2| \geq \lfloor \frac{|A|}{2^{k+1}} \rfloor - 2s$. Since $T_s^p(|A|) = \lfloor \frac{|A|}{2^{k+1}} \rfloor - 2s$, also the condition 3 holds and we are done. $\square$

In our future considerations we need a much stronger tool than the previous lemma can provide. The factorization of each nonseparable word $\alpha_1, \alpha_2, \ldots, \alpha_p$ in Lemma 2 should be refined and the common alphabets for refinements created.

Theorem 2. *Let p and s be positive integers, A an alphabet, and $\alpha_1, \alpha_2, \ldots, \alpha_p$ a sequence of (s, A)-nonseparable words. Then, given $d \in \mathbb{N}_+$, there exist alphabets $A_1, A_2, \ldots, A_{2^d} \subseteq A$ and, for each $i \in \{1, 2, \ldots, p\}$, a factorization $\alpha_i = \alpha_{i,1} \alpha_{i,2} \cdots \alpha_{i,2^d}$ of α_i, and a permutation σ_i of $1, 2, \ldots, 2^d$ such that for each $j \in \{1, 2, \ldots, 2^d\}$*

1. $A_j \subseteq \text{alph}(\alpha_{i, \sigma(j)})$
2. $A_j \cap \text{alph}(\alpha_{i, j'}) = \emptyset$ *for each $j' \in \{1, 2, \ldots, 2^d\}$, $j' \neq \sigma(j)$; and*
3. $|A_j| \geq T_s^{(p,d)}(|A|)$.

Proof. Proceed by induction on d. In the case $d = 1$ our theorem restates the result of Lemma 2.

Consider the case $d = k + 1$, where $k \in \mathbb{N}_+$. Apply Lemma 2 to the words $\alpha_1, \alpha_2, \ldots, \alpha_p$ to obtain the alphabets $A_1, A_2 \subseteq A$, and, for each $i \in \{1, 2, \ldots, p\}$ a permutation δ_i of $1, 2$ and a factorization $\alpha_i = \omega_{i,1} \omega_{i,2}$ of α_i such that the conditions 1, 2, and 3 of the lemma hold (when σ_i is replaced by δ_i and the words $\alpha_{i,1}, \alpha_{i,2}$ by words $\omega_{i,1}, \omega_{i,2}$, respectively). Certainly $|A_1|, |A_2| \geq T_s^p(|A|)$.

Apply the induction hypothesis to the words $\omega_{1, \delta_1(1)}, \omega_{2, \delta_2(1)}, \ldots, \omega_{p, \delta_p(1)}$ to obtain alphabets $B_1, B_2, \ldots, B_{2^k} \subseteq A_1$, and for each $i \in \{1, 2, \ldots, p\}$ a factorization $\omega_{i, \delta_i(1)} = \beta_{i,1} \beta_{i,2} \cdots \beta_{i,2^k}$ of $\alpha_{i, \delta_i(1)}$ and a permutation μ_i of $1, 2, \ldots, 2^k$ such that for each $j \in \{1, 2, \ldots, 2^k\}$

$B_j \subseteq \text{alph}(\beta_{i, \mu(j)})$

$B_j \cap \text{alph}(\beta_{i, j'}) = \emptyset$ for each $j' \in \{1, 2, \ldots, 2^k\}$, $j' \neq \mu_i(j)$; and

$|B_j| \geq T_s^{(p,k)}(|A_1|)$.

Apply the induction hypothesis once more, now to $\omega_{1, \delta_1(2)}, \omega_{2, \delta_2(2)}, \ldots, \omega_{p, \delta_p(2)}$, to obtain alphabets $C_1, C_2, \ldots, C_{2^k} \subseteq C$, and, for each $i \in \{1, 2, \ldots, p\}$, a factorization $\omega_{i, \delta_i(2)} = \gamma_{i,1} \gamma_{i,2} \cdots \gamma_{i,2^k}$ of γ_i and a permutation ρ_i of $1, 2, \ldots, 2^k$ such that for each $j \in \{1, 2, \ldots, 2^k\}$

$C_j \subseteq \mathrm{alph}(\gamma_{i,\rho(j)})$

$C_j \cap \mathrm{alph}(\gamma_{i,j'}) = \emptyset$ for each $j' \in \{1, 2, \ldots, 2^k\}$, $j' \neq \rho_i(j)$; and

$|C_j| \geq T_s^{(p,k)}(|A_2|)$.

Note that $|B_j|, |C_j| \geq T_s^{(p,k)}(T_s^1(|A|)) = T_s^{(p,k+1)}(|A|)$ for $j = 1, 2, \ldots, 2^k$. Let $A_j = B_j$ and $A_{2^k+j} = C_j$ for $j = 1, 2, \ldots, 2^k$. Let $i \in \{1, 2, \ldots, p\}$. Define the factorization $\alpha_i = \alpha_{i,1}\alpha_{i,2} \cdots \alpha_{i,2^{k+1}}$ of α_1 and the permutation σ_i of $1, 2, \ldots, 2^{k+1}$ as follows. If $\delta_i(1) = 1$ and $\delta_i(2) = 2$, then $\alpha_{i,j} = \beta_{i,j}$, $\alpha_{i,2^k+j} = \gamma_{i,j}$ and $\sigma_i(j) = \mu_i(j)$ and $\sigma_i(2^k + j) = \rho_i(j)$ for $j = 1, 2, \ldots, 2^k$. If $\delta_i(1) = 2$ and $\delta_i(2) = 1$, then $\alpha_{i,j} = \gamma_{i,j}$, $\alpha_{i,2^k+j} = \beta_{i,j}$ and $\sigma_i(j) = \rho_i(j)$ and $\sigma_i(2^k+j) = \mu_i(j)$ for $j = 1, 2, \ldots, 2^k$.

Clearly the claims 1, 2, and 3 of our theorem are satisfied with these choices when $p = k + 1$. The induction is thus extended and the proof is complete. $\qquad \square$

Note that in the above theorem properties 1 and 2 imply that the sets $A_1, A_2, \ldots, A_{2^d}$ are pairwise disjoint.

Let p and s be positive integers. For each $d \in \mathbb{N}_+$, denote $D_s(p, d) = 2^p d + 2s \sum_{i=1}^d 2^{p\,i}$. Then obviously $D_s(p, 1) = 2^p + 2\,s\,2^p$ and $D(p, d+1) = 2^p [D(p, d) + 2\,s] = 2^p D(p, d) + 2^{p+1}s$ for any $d \in \mathbb{N}_+$.

It is quite straightforward to see that

$$T_s^{(p,d)}(D_s(p, d)) = \underbrace{(T_s^p \circ T_s^p \circ \cdots \circ T_s^p)}_{k \text{ times}}(D_s(p, d)) = 1 .$$

Obviously $D_s(p, d)$ is the smallest positive integer x such that $T_s^{(p,d)} \in \mathbb{N}_+$.

Corollary 1. *Let p, s and d be positive integers and A an alphabet such that $|A| \geq D_s(p, d)$. Assume furthermore that $\alpha_1, \alpha_2, \ldots, \alpha_p$ are (s, A)–nonseparable words. Then there exists a subalphabet B of A such that $|B| = 2^d$ and $(\alpha_i)_B$ is a permutation of B for all $i \in \{1, 2, \ldots, p\}$.*

Proof. Omitted. $\qquad \square$

Let $s, r \in \mathbb{N}_+$ and $(p_1, d_i) \in \mathbb{N}_+^2$ for $i = 1, 2, \ldots, r$. Define the function $T_s^{\otimes_{i=1}^r (p_i, d_i)}$: $\mathbb{N} \to \mathbb{N}$ by $T_s^{\otimes_{i=1}^r (p_i, d_i)} = T_s^{(p_1, d_1)} \circ T_s^{(p_2, d_2)} \cdots T_s^{(p_r, d_r)}$. Then, for each $x \in \mathbb{N}_+$, the following holds:

$$T_s^{\otimes_{i=1}^r (p_i, d_i)}(x) = T_s^{(p_1, d_1)}(T_s^{(p_2, d_2)}(\cdots (T_s^{(p_r, d_r)}(x)) \cdots)) .$$

It goes without saying that $T_s^{\otimes_{i=1}^r (p_i, d_i)}$ is an increasing function. Note furthermore that $T_s^{\otimes_{i=1}^r (p_i, d_i)} = T_s^{(p_1, d_1)}$ if $r = 1$.

The main result of this section further generalizes Theorem 2 and captures the stepwise refinement of factorizations for i nonseparable words $\alpha_1, \alpha_2, \ldots, \alpha_i$, $i = p, p - 1, \ldots, 2, 1$.

Theorem 3. *Let p and s be positive integers, A an alphabet, and $\alpha_1, \alpha_2, \ldots, \alpha_p$ a sequence of (s, A)–nonseparable words. Given $d_1, d_2, \ldots, d_p \in \mathbb{N}_+$, there exists sets $A_{i,j} \subseteq A$ $(i = 1, 2, \ldots, p; \ j = 1, 2, \ldots, 2^{\sum_{l=i}^p d_l})$ such that*

1. *for each $i \in \{1, 2, \ldots, p\}$ and $j \in \{1, 2, \ldots, 2^{\sum_{l=i}^{p} d_l}\}$*

$$|A_{i,j}| \geq T_s^{\otimes_{l=i}^{p}(i,d_i)}(|A|) \ ;$$

2. *for each $i \in \{2, 3, \ldots, p\}$ and $j \in \{1, 2, \ldots, 2^{\sum_{l=i}^{p} d_l}\}$ there exist exactly $2^{d_{i-1}}$ indices $j' \in \{1, 2, \ldots, 2^{\sum_{l=i-1}^{p} d_l}\}$ such that $A_{i-1,j'} \subseteq A_{i,j}$; and*

3. *for each $i \in \{1, 2, \ldots, p\}$ there exists a factorization*

$$\alpha_i = \alpha_{i,1}\alpha_{i,2} \cdots \alpha_{i,2^{\sum_{l=i}^{p} d_l}}$$

of α_i and a permutation σ_i of $1, 2, \ldots, 2^{\sum_{l=i}^{p} d_l}$ such that for all $j \in \{1, 2, \ldots, 2^{\sum_{l=i}^{p} d_l}\}$

(i) $A_{i,j} \subseteq \mathrm{alph}(\alpha_{i,\sigma_i(j)})$; and

(ii) $A_{i,j} \cap \mathrm{alph}(\alpha_{i,r}) = \emptyset$ for all $r \in \{1, 2, \ldots, 2^{\sum_{l=1}^{p} d_l}\}$ such that $r \neq \sigma_i(j)$.

Proof. Omitted.   □

Let $s, r \in \mathbb{N}_+$ and $(p_i, d_i) \in \mathbb{N}_+^2$ for $i = 1, 2, \ldots, r$. We shall next study those values $x \in \mathbb{N}_+$ for which $T_s^{\otimes_{i=1}^{r}(p_i,d_i)}(x)$ is a positive integer.

Define inductively

$$D_s(\otimes_{i=1}^{r}(p_i, d_i)) = 1 \quad \text{if } r = 0;$$
$$D_s(\otimes_{i=1}^{r}(p_i, d_i)) = D_s(p_r, d_r) \quad \text{if } r = 1;$$
$$D_s(\otimes_{i=1}^{r}(p_i, d_i)) = 2^{p_r d_r} D_s(\otimes_{i=1}^{r-1}(p_i, d_i)) + 2\,s \sum_{i=1}^{d_r} 2^{p_r i} \quad \text{if } r > 1 \ .$$

Oviously

$$D_s(\otimes_{i=1}^{r}(p_i, d_i)) = 2^{\sum_{i=1}^{r} p_i d_i} + 2\,s \Big[2^{\sum_{i=2}^{r} p_i d_i} \sum_{i=1}^{d_1} 2^{p_1 i} + 2^{\sum_{i=3}^{r} p_i d_i} \sum_{i=1}^{d_2} 2^{p_2 i} + \\ + \cdots + 2^{p_r d_r} \sum_{i=1}^{d_{r-1}} 2^{p_{r-1} i} + \sum_{i=1}^{d_r} 2^{p_r i} \Big] \ .$$

Theorem 4. *Let $s, r \in \mathbb{N}_+$ and $(p_i, d_i) \in \mathbb{N}_+^2$ for $i = 1, 2, \ldots, r$. Then*

$$T_s^{\otimes_{i=r-k+1}^{r}(p_i,d_i)}(D_s(\otimes_{i=1}^{r}(p_i, d_i))) = D_s(\otimes_{i=1}^{k}(p_i, d_i))$$

for each $k \in \{0, 1, \ldots, r-1\}$.

Proof. Omitted.   □

Corollary 2. *Let $s, r \in \mathbb{N}_+$ and $(p_1, d_i) \in \mathbb{N}_+^2$ for $i = 1, 2, \ldots, r$. Then*

$$T_s^{\otimes_{i=1}^{r}(p_i,d_i)}(D_s(\otimes_{i=1}^{r}(p_i, d_i))) = 1 \ .$$

On the basis of the facts above it should be clear that $D_s(\otimes_{i=1}^{r}(p_i, d_i))$ is the smallest number $x \in \mathbb{N}_+$ such that $T_s^{\otimes_{i=1}^{r}(p_i,d_i)}(x)$ is a positive integer.

4.2 The Main Combinatorial Results

For practical reasons (in fact, to prove Corollary 3) we need to evaluate the size of the number $D_s(\otimes_{i=1}^p (i, d_i))$.

Theorem 5. *Let p, s and $d_1, d_2, \ldots, d_p$ be positive integers. Then*

$$D_s(\otimes_{i=1}^p (i, d_i)) \leq (4s+1) \, 2^{\sum_{i=1}^p i \, d_i} - 4s \, .$$

Proof. Omitted.   □

Recall that for each $q \in \mathbb{N}_+$, a word α is q-*restricted* if $|\alpha|_a \leq q$ for all $q \in \text{alph}(\alpha)$. Our main combinatorial result for attacking purposes can now be stated.

Theorem 6. *Let $q \geq 2$ and $d_1, d_2, \ldots, d_q \geq 1$ be integers and $s_1, s_2, \ldots, s_q$ parameters defined as follows: $s_q = 2^{\sum_{j=1}^q j \, d_j}$, $s_k = D_{s_{k+1}}(\otimes_{i=1}^k (i, d_i))$, for $k = 2, 3, \ldots, q-1$, and $s_1 = 2^{d_1} s_2$. Let furthermore α be a q-restricted word such that $|\text{alph}(\alpha)| \geq s_1$. Then there exist $p \in \{1, 2, \ldots, q\}$, a factorization $\alpha = \alpha_1 \alpha_2 \cdots \alpha_p$ of α, and alphabets $A_{i,j}$ ($i = 1, 2, \ldots p; \; j = 1, 2, \ldots, 2^{\sum_{k=1}^p d_k}$) such that*

1. *for each $i \in \{1, 2, \ldots p\}$ and $j \in \{1, 2, \ldots, 2^{\sum_{k=i}^p d_k}\}$:*

$$|A_{i,j}| \geq 2^{\sum_{k=1}^{i-1} k \, d_k} \; ;$$

2. *for each $i \in \{2, 3, \ldots, p\}$ and $j \in \{1, 2, \ldots, 2^{\sum_{k=i}^p d_k}\}$ there exist exactly 2^{d_i-1} indices $l \in \{1, 2, \ldots, 2^{\sum_{k=i-1}^p d_k}\}$ such that $A_{i-1,l} \subseteq A_{i,j}$; and*

3. *for each $i \in \{1, 2, \ldots, p\}$ there exists a factorization*

$$\alpha_i = \alpha_{i,1} \alpha_{i,2} \cdots \alpha_{i,2^{\sum_{k=i}^p d_k}}$$

of α_i and a permutation σ_i of $1, 2, \ldots, 2^{\sum_{k=i}^p d_k}$ such that for all $j \in \{1, 2, \ldots, 2^{\sum_{k=i}^p d_k}\}$
(i) $A_{i,j} \subseteq \text{alph}(\alpha_{i,\sigma_i(j)})$; and
(ii) $A_{i,j} \cap \text{alph}(\alpha_{i,l}) = \emptyset$ for all $l \in \{1, 2, \ldots, 2^{\sum_{k=i}^p d_k}\}$ such that $l \neq \sigma(j)$.

Proof. We proceed stepwise as follows.

In the first step we ask whether or not the word α is s_2-nonseparable, i.e., whether or not for each factorization $\alpha = \beta_1 \beta_2$, we have $|\text{alph}(\beta_1) \cap \text{alph}(\beta_2)| < s_2$. Suppose that α is s_2-nonseparable. Let then $\alpha = \alpha_1 \alpha_2 \cdots \alpha_{2^{d_1}}$ be a factorization of α such that for each $i \in \{1, 2, \ldots, 2^{d_1}\}$, the word α_i contains (at least) s_2 different symbols that do not occur in $\alpha_1 \alpha_2 \cdots \alpha_{i-1}$. Since $|\text{alph}(\alpha)| \geq 2^{d_1} s_2$, the factorization always can be found. Let $A_{1,i}$ be the set of all symbols in $\text{alph}(\alpha_i)$ that do not occur in the word $\text{alph}(\alpha_1 \cdots \alpha_{i-1} \alpha_{i+1} \cdots \alpha_{2^{d_1}})$. Since α is s_2-nonseparable, each of the sets $A_{1,1}$, $A_{1,1}$, $A_{1,2}$, $\ldots$, $A_{1,2^{d_1}}$ is nonempty. By choosing $p = 1$, we note that the claims of the theorem hold. We use here convention $\sum_{k=1}^{p-1} k \, d_k = \sum_{k=1}^0 k \, d_k = 0$, i.e., $2^{\sum_{k=1}^{p-1} k \, d_k} = 1$.

Suppose then that the word α is s_2–separable. Let then $\alpha = \alpha_1 \alpha_2$ be a factorization of α such that $|\mathrm{alph}(\alpha_1) \cap \mathrm{alph}(\alpha_2)| \geq s_2$. Recall that, by definition, $s_2 = D_{s_3}(\otimes_{i=1}^{2}(i, d_i))$.

In the second step, we ask whether or not there exists an alphabet $B \subseteq \mathrm{alph}(\alpha_1) \cap \mathrm{alph}(\alpha_2)$ such that one of the words α_1 and α_2 is (s_3, B)–separable. Suppose that this is not the case. Then each of the words α_1 and α_2 is (s_3, A)–nonseparable where $A = \mathrm{alph}(\alpha_1) \cap \mathrm{alph}(\alpha_2)$. Since $|A| \geq D_{s_3}(\otimes_{i=1}^{2}(i, d_i))$, the claims hold by Theorem 3 and Theorem 4.

Assume that $B \subseteq \mathrm{alph}(\alpha_1) \cap \mathrm{alph}(\alpha_2)$ is an alphabet such that one of the words α_1, α_2, say α_2, is (s_3, B)–separable. Let $\alpha_2 = \gamma_1 \gamma_2$ be a factorization of α_2 such that $|\mathrm{alph}(\gamma_1) \cap \mathrm{alph}(\gamma_2) \cap B| \geq s_3$. Redenoting $\alpha_2 := \gamma_1$ and $\alpha_3 := \gamma_2$, we have $\alpha = \alpha_1 \alpha_2 \alpha_3$ and $|\cap_{i=1}^{3} \mathrm{alph}(\alpha_i)| \geq s_3$.

Continuing like this, we describe the general step k, $k \in \{1, 2, \ldots, q\}$ as follows. Let $\alpha = \alpha_1 \alpha_2 \cdots \alpha_k$ be a factorization of α such that $|\cap_{i=1}^{k} \mathrm{alph}(\alpha_i)| \geq s_k$. Assume first that $k < q$. Recall that $s_k = D_{s_{k+1}}(\otimes_{i=1}^{k}(i, d_i))$. We pose the question whether or not there exists an alphabet $B \subseteq \cap_{i=1}^{k} \mathrm{alph}(\alpha_i)$ such that one of the words $\alpha_1, \alpha_2, \ldots, \alpha_k$ is (s_{k+1}, B)–separable. Suppose that this is not the case. Then each of the words $\alpha_1, \alpha_2, \ldots, \alpha_k$ is (s_{k+1}, A)–nonseparable where $A = \cap_{i=1}^{k} \mathrm{alph}(\alpha_i)$. Since $|A| \geq D_{s_{k+1}}(\otimes_{i=1}^{k}(i, d_i))$, the claims hold by Theorem 3 and Theorem 4.

Assume that $B \subseteq \cap_{i=1}^{k} \mathrm{alph}(\alpha_i)$ is an alphabet such that one of the words $\alpha_1, \alpha_2, \ldots, \alpha_k$, say α_k, is (s_{k+1}, B)–separable. Let $\alpha_k = \omega_1 \omega_2$ be a factorization of α_k such that $|\mathrm{alph}(\omega_1) \cap \mathrm{alph}(\omega_2) \cap B| \geq s_{k+1}$. Redenoting $\alpha_k := \omega_1$ and $\alpha_{k+1} := \omega_2$, we have $\alpha = \alpha_1 \alpha_2 \cdots \alpha_{k+1}$ and $|\cap_{i=1}^{k+1} \mathrm{alph}(\alpha_i)| \geq s_{k+1}$.

Suppose that in the general step we have $k = q$. Then, since α is q-restricted, we know that all the words $\alpha_1, \alpha_2, \ldots, \alpha_k$ are $(1, A)$–nonseparable where $A = \cap_{i=1}^{q} \mathrm{alph}(\alpha_i)$. Since $s_q = 2^{\sum_{i=1}^{q} i \, s_i}$, we are again through by Theorems 3 and 4. $\qquad \square$

Corollary 3. *Let $q \geq 2$ and $d_1, d_2, \ldots, d_q \geq 1$ be integers and α a q–restricted word such that $|\mathrm{alph}(\alpha)| \geq 5^{q-2} 2^{\sum_{i=1}^{q}(q-i+1) i \, d_i}$. Let furthermore $s_1, s_2, \ldots, s_q$ be as in Theorem 6. Then $|\mathrm{alph}(\alpha)| \geq s_1$ and all the claims of Theorem 6 hold.*

Proof. Omitted. $\qquad \square$

Let us further study permutations inside q–restricted words. The results are not needed in our attack construction, but have independent combinatorial interest. We wish to find a (good) upper and lower bound for the number $N(m, q)$ defined in the following theorem which is borrowed from [10].

Theorem 7. *For all positive integers m and q there exists a (minimal) positive integer $N(m, q)$ such that the following is true. Let α be a q–restricted word such that $|\mathrm{alph}(\alpha)| \geq N(m, q)$. Then there exist $A \subseteq \mathrm{alph}(\alpha)$ with $|A| = m$ and $p \in \{1, 2, \ldots, q\}$ as well as words $\alpha_1, \alpha_2, \ldots, \alpha_p$ such that $\alpha = \alpha_1 \alpha_2 \cdots \alpha_p$ and for all $i \in \{1, 2, \ldots, p\}$, the word $(\alpha_i)_A$ is a permutation of A.*

In [10] the upper bound $N(m, q+1) \leq N(m^2 - m + 1, q) \leq m^{2^q}$ was attained. The existence of $N(m, q)$ was first (implicitly) proved in [4] where also the first (very large) upper bound for it also was evolved.

Lemma 3. *Let p, s and d be integers such that $d, p \geq 2$ and $s \geq 3$. Then $D_s(p, d) = 2^{pd} + 2s \sum_{i=1}^{d} 2^{pi} \leq 3s\, 2^{pd}$.*

Proof. Obviously $2^{pd} + 2s \sum_{i=1}^{d} 2^{pi} = 2^{pd} + 2s \frac{2^{p(d+1)} - 2^p}{2^p - 1} = 2^{pd}[1 + 2s\frac{1 - 2^{-pd}}{1 - 2^{-p}}]$. Since $2s\frac{1 - 2^{-pd}}{1 - 2^{-p}} \leq \frac{8s}{3}$, the claim holds. $\square$

Theorem 8. *Let $m \geq 2$ and $q \geq 2$ be integers and α be a q–restricted word such that $|alph(\alpha)| \geq 3^{q-2} 2^{\lceil \log_2 m \rceil \cdot \frac{q^2 - q + 2}{2}}$. Then there exist $p \in \{1, 2, \cdots q\}$, a factorization $\alpha = \alpha_1 \alpha_2 \cdots \alpha_p$ of α and set $A \subseteq alph(\alpha)$ such that $|A| = m$ and $(\alpha_i)_A$ is a permutation of A for all $i \in \{1, 2, \cdots, p\}$.*

Proof. Analogous (although simpler) to that of Theorem 6. $\square$

As a consequence from Theorem 8, we get the following upper bound.

Corollary 4. *For all integers $m \geq 2$ and $q \geq 2$ the inequality*

$$N(m, q) \leq 3^{q-2} \cdot 2^{\lceil \log_2 m \rceil \cdot \frac{q^2 - q + 2}{2}}$$

holds.

In the following we shall search a lower bound for $N(m, q)$.

Given $p, q \in \mathbb{N}_+$, call a word α a $P(m, q)$-word if α is q-restricted and there exists an alphabet $A \subseteq alph(\alpha)$, $|A| = m$, integer $p \in \{1, 2, \ldots, q\}$, and permutations $\sigma_1, \sigma_2 \ldots, \sigma_p$ of $1, 2, \ldots, m$ such that

$$\pi_A(\alpha) \in a_{\sigma_1(1)}^+ a_{\sigma_1(2)}^+ \cdots a_{\sigma_1(m)}^+ \cdots a_{\sigma_p(1)}^+ a_{\sigma_p(2)}^+ \cdots a_{\sigma_p(m)}^+$$

We have shown that there exists a smallest positive integer $N(m, q)$ such that if α is q-restricted and $|alph(\alpha)| \geq N(m, q)$, then α is a $P(m, q)$-word. Let $T(m, q) = N(m, q) - 1$. Then there exists a word β such that β is q-restricted and $|alph(\beta)| = T(m, q)$, and β is not a $P(m, q)$-word.

Lemma 4. *Let $m \geq 2$ and q be positive integers and α a q-restricted word such that α is not a $P(m, q)$-word. Assume furthermore that $alph(\alpha) = \{a_1, a_2, \ldots, a_n\}$ where, for all $i, j \in \{1, 2, \ldots, n\}$, $i < j$, the first occurrence of a_i happens before the first occurrence of a_j in α. Let*

$$w = w(a_1, a_2, \ldots, a_n) = a_1 a_2 \ldots a_n a_{n+1} \alpha$$

where a_{n+1} is a new symbol. Then

1. *w is not a $P(m, q + 1)$-word; and*

2. *for all $i, j \in \{1, 2, \ldots, n\}$, we have $\pi_{a_i, a_j}(w) \notin \{a_i a_j, a_j a_i\}$.*

Proof. Omitted. $\square$

We can immediately make the following conclusion.

Corollary 5. *Let* $n, q \in \mathbb{N}_+$ *and*

$$a_{1,1}, a_{1,2}, \ldots, a_{1,n+1}, \ldots, a_{n-1,1}, a_{n-1,2} \ldots, a_{n-1,n+1}$$

distinct symbols. Let $w(\cdot, \cdot, \cdots, \cdot)$ *be the consruction of the previous lemma. Then the word*

$$w(a_{1,1}, a_{1,2}, \ldots, a_{1,n+1})w(a_{2,1}, a_{2,2}, \cdots, a_{2,n+1}) \cdots w(a_{n-1,1}, a_{n-1,2}, \ldots, a_{n-11,n+1})$$

is not a $P(m, q+1)$ *word.*

Corollary 6. *For all* $m, q \in \mathbb{N}_+$, *the inequality*

$$T(m, q+1) \geq (m-1)(T(m, q) + 1)$$

holds.

Proof. In the construction of the lemma we add one letter to the word and the (in the corollary) we make $m - 1$ copies of the word. □

Theorem 9. *For all* $m, q \in \mathbb{N}_+$, $m \geq 2$, *thefollowing inequality holds:*

$$N(m, q) \geq m(m-1)^{q-1} + (m-1)^{q-2} + \cdots + (m-1) + 1$$

Proof. Let us prove by induction on q that

$$T(m, q) \geq [m(m-1)^{q-1} + (m-1)^{q-2} + \cdots + (m-1) + 1] - 1 \ .$$

If $q = 1$, then $T(m, q) = m - 1 = m(m-1)^0 - 1$. Suppose that $T(m, q) \geq [m(m-1)^{q-1} + (m-1)^{q-2} + \cdots + (m-1) + 1] - 1$. Then

$$T(m, q+1) \geq (m-1)[m(m-1)^{q-1} + (m-1)^{q-2} + \cdots + (m-1) + 1]$$
$$= m(m-1)^q + (m-1)^{q-1} + \cdots + (m-1) + 1 - 1 \ .$$

Thus the induction is extended and $N(m, q) \geq m(m-1)^{q-1} + (m-1)^{q-2} + \cdots + (m-1) + 1$ for all positive integers m and q such that $m \geq 2$. □

Corollary 7. *For all integers* $m \geq 2$ *and* $q \geq 2$ *the inequalities*

$$m(m-1)^{q-1} + (m-1)^{q-2} + \cdots + (m-1) + 1 \leq N(m, q) \leq 3^{q-2} \cdot 2^{\lceil \log_2 m \rceil \cdot \frac{q^2 - q + 2}{2}}$$

hold. Furthermore $N(m, 2) = m^2 - m + 1$ *for each integer* $m \geq 1$.

5   Conclusion

We have stated reasons to consider combinatorics on words from a fresh viewpoint and taken some small steps in the new research frame. The results imply more efficient attacks on generalized iterated hash functions and, from their part, confirm the fact that the iterative structure possesses certain security weaknesses.

References

1. Coron, J.-S., Dodis, Y., Malinaud, C., Puniya, P.: Merkle-Damgård Revisited: How to Construct a Hash Function. In: Shoup, V. (ed.) CRYPTO 2005. LNCS, vol. 3621, pp. 430–448. Springer, Heidelberg (2005)
2. Damgård, I.B.: A Design Principle for Hash Functions. In: Brassard, G. (ed.) CRYPTO 1989. LNCS, vol. 435, pp. 416–427. Springer, Heidelberg (1990)
3. Dobbertin, H.: Cryptanalysis of MD4. Journal of Cryptology 11(4), 253–271 (1998)
4. Hoch, J.J., Shamir, A.: Breaking the ICE - Finding Multicollisions in Iterated Concatenated and Expanded (ICE) Hash Functions. In: Robshaw, M. (ed.) FSE 2006. LNCS, vol. 4047, pp. 179–194. Springer, Heidelberg (2006)
5. Joux, A.: Multicollisions in Iterated Hash Functions. Application to Cascaded Constructions. In: Franklin, M. (ed.) CRYPTO 2004. LNCS, vol. 3152, pp. 306–316. Springer, Heidelberg (2004)
6. Klima, V.: Finding MD5 collisions on a notebook PC using multi-message modifications, Cryptology ePrint Archive, Report 2005/102 (2005), http://eprint.iacr.org/2005/102
7. Klima, V.: Huge multicollisions and multipreimages of hash functions BLENDER-n, Cryptology ePrint Archive, Report 2009/006 (2009), http://eprint.iacr.org/2009/006
8. Kortelainen, J., Halunen, K., Kortelainen, T.: Multicollision Attacks and Generalized Iterated Hash Functions. J. Math. Cryptol. 4, 239–270 (2010)
9. Kortelainen, J., Kortelainen, T., Vesanen, A.: Unavoidable Regularities in Long Words with Bounded Number of Symbol Occurrences. In: Fu, B., Du, D.-Z. (eds.) COCOON 2011. LNCS, vol. 6842, pp. 519–530. Springer, Heidelberg (2011)
10. Kortelainen, J., Kortelainen, T., Vesanen, A.: Unavoidable regularities in long words with bounded number of symbol occurrences. J. Comp. Optim. (in print)
11. Merkle, R.C.: A Certified Digital Signature. In: Brassard, G. (ed.) CRYPTO 1989. LNCS, vol. 435, pp. 218–238. Springer, Heidelberg (1990)
12. Nandi, M., Stinson, D.: Multicollision attacks on some generalized sequential hash functions. IEEE Trans. Inform. Theory 53, 759–767 (2007)
13. Stevens, M.: Fast collision attack on MD5. Cryptology ePrint Archive, Report 2006/104 (2006), http://eprint.iacr.org/2006/104
14. Suzuki, K., Tonien, D., Kurosawa, K., Toyota, K.: Birthday paradox for multicollisions, IEICE Transactions 91-A(1), 39–45 (2008)
15. Wang, X., Yu, H.: How to Break MD5 and Other Hash Functions. In: Cramer, R. (ed.) EUROCRYPT 2005. LNCS, vol. 3494, pp. 19–35. Springer, Heidelberg (2005)
16. Wang, X., Yin, Y.L., Yu, H.: Finding Collisions in the Full SHA-1. In: Shoup, V. (ed.) CRYPTO 2005. LNCS, vol. 3621, pp. 17–36. Springer, Heidelberg (2005)
17. Yu, H., Wang, X.: Multi-collision Attack on the Compression Functions of MD4 and 3-Pass HAVAL. In: Nam, K.-H., Rhee, G. (eds.) ICISC 2007. LNCS, vol. 4817, pp. 206–226. Springer, Heidelberg (2007)

A Differential Fault Attack on the Grain Family under Reasonable Assumptions

Subhadeep Banik, Subhamoy Maitra, and Santanu Sarkar

Applied Statistics Unit, Indian Statistical Institute,
203 B T Road, Kolkata 700 108, India
{s.banik_r,subho}@isical.ac.in, sarkar.santanu.bir@gmail.com

Abstract. In this paper we study a differential fault attack against ciphers having the same physical structure as in the Grain family. In particular we demonstrate our attack against Grain v1, Grain-128 and Grain-128a. The existing attacks by Berzati et al. (HOST 2009), Karmakar et al. (Africacrypt 2011) and Banik et al. (CHES 2012) assume a fault model that allows them to reproduce a fault at a particular register location more than once. However, we assume a realistic fault model in which the above assumption is no longer necessary, i.e., re-injecting the fault in the same location more than once is not required. In addition, towards a more practical framework, we also consider the situation in which more than one consecutive locations of the LFSR are flipped as result of a single fault injection.

Keywords: Differential fault attacks, Grain v1, Grain-128, Grain-128a, LFSR, NFSR, Stream Cipher.

1 Introduction

Fault attacks have received serious attention in cryptographic literature for more than a decade [1,2]. Such attacks on stream ciphers have gained momentum ever since the work of Hoch and Shamir [10] and this model of cryptanalysis, though optimistic, could successfully be employed against a number of proposals. Fault attacks study the mathematical robustness of a cryptosystem in a setting that is weaker than its original or expected mode of operation. A typical attack scenario [10] consists of an adversary who injects a random fault (using laser shots/clock glitches [14,15]) in a cryptographic device as a result of which one or more bits of its internal state are altered. The faulty output from this altered device is then used to deduce information about its internal state/secret key. In order to perform the attack, the adversary requires certain privileges like the ability to re-key the device, control the timing of the fault etc. The more privileges the adversary is granted, the more the attack becomes impractical and unrealistic.

The Grain family of stream ciphers [4,8,9] has received a lot of attention and it is in the hardware profile of eStream [3]. In all the fault attacks reported so far [5,6,12] on this cipher, the adversary is granted far too many privileges

S. Galbraith and M. Nandi (Eds.): INDOCRYPT 2012, LNCS 7668, pp. 191–208, 2012.

to make the attacks practical. In fact the designers of Grain themselves have underlined a set of reasonable privileges that may be granted to the adversary while performing the fault attack. Unfortunately, all the existing fault attacks on the Grain family exploited additional assumptions. In this regard, let us refer to the following quote from the designers of Grain [8, Section 6.5] (see also similar comments in [9, Section IV D] and [4, Section 4.4]):

"If an attacker is able to reset the device and to induce a single bit fault many times and at different positions that he can correctly guess from the output difference, we cannot preclude that he will get information about a subset of the state bits in the LFSR. Such an attack seems more difficult under the (more realistic) assumption that the fault induced affects several state bits at (partially) unknown positions, since in this case it is more difficult to determine the induced difference from output differences."

	[6]	[12]	[5]	This work
Multiple fault at same location	Required	Required	Required	Not Required
Multiple IV Initialization	No	Yes	No	No
Single bit Fault	Yes	Yes	Yes	Upto 3-bit toggle allowed
Control over Fault Timing	Required	Required	Required	Required
Multiple Re-Keying	Allowed	Allowed	Allowed	Allowed

Fig. 1. Differential Fault Attack on Grain: Survey of Fault Models

In the published fault attacks [5, 6, 12] on Grain, it has been assumed that attacker has the ability to inject a single bit fault in the same register location over and over again. This is clearly rather optimistic and does not follow the fault model prescribed by the designers. In our work, we have assumed that the adversary has only those privileges that have been allowed by the designers, i.e., we follow the exact fault model provided by the designers and demonstrate that in such a scenario too, the adversary can not only recover "a subset of the LFSR state bits" but also recover the secret key. Furthermore, we consider a situation in which the adversary is unable to induce a single bit toggle every time he injects a fault. The best he can do is to 'influence' upto k bits in random but consecutive LFSR locations without knowing the exact number of bits the injected fault has altered or their locations. We show that for certain small values of k, even under this added constraint the secret key can be recovered. The idea used here is that, with very high confidence, the adversary should be able to identify the situations when the injected fault alters the binary value in only a single register location. He can then use the algorithms described for a single location identification and proceed with the attack.

In this work we assume that the adversary has the following privileges: **(a)** he can reset the cipher with the original Key-IV and restart cipher operations as many times he wishes; this is not a problem if the device ouputs different faulty ciphertexts of the same or known messages and such a model has been used in [5, 6, 11, 12] (actually the attack model requires the original and several faulty key-streams), **(b)** he has full control over the timing of fault injection, and **(c)** he can inject a fault that may affect upto k consecutive LFSR locations but he is unaware of the exact number of bits altered or their locations. In this work we have concentrated on the cases till $k = 3$. As pointed out earlier, these assumptions about the fault model are far more realistic and practical than the ones assumed in [5, 6, 12].

ORGANIZATION OF THIS PAPER. In this section, we continue with a detailed description of the Grain family. In Section 2, we introduce certain tools and definitions that will help us launch the attack. To demonstrate the attack, initially we assume that the attacker is able to induce a single bit toggle at any random LFSR location. The fault location identification routine is explained in Section 3. The general strategy to attack a cipher with the physical structure of Grain is outlined in Section 4. Section 5 demonstrates the attacks on Grain v1, Grain-128 and Grain-128a. In Section 6, we explore a stricter fault model in which the attacker is able to flip the binary values of upto 3 consecutive LFSR locations. Section 7 concludes the paper.

1.1   Brief Description of Grain Family

Any cipher in the Grain family consists of an n-bit LFSR and an n-bit NFSR (see Figure 2). The update function of the LFSR is given by the equation $y_{t+n} = f(Y_t) = y_t \oplus y_{t+f_1} \oplus y_{t+f_2} \oplus \cdots \oplus y_{t+f_a}$, where $Y_t = [y_t, y_{t+1}, \ldots, y_{t+n-1}]$ is an n-bit vector that denotes the LFSR state at the tth clock interval and f is a linear function on the LFSR state bits obtained from a primitive polynomial in $GF(2)$ of degree n. The NFSR state is updated as $x_{t+n} = y_t \oplus g(X_t) = y_t \oplus g(x_t, x_{t+\tau_1}, x_{t+\tau_2}, \ldots, x_{t+\tau_b})$. Here, $X_t = [x_t, x_{t+1}, \ldots, x_{t+n-1}]$ is an n-bit

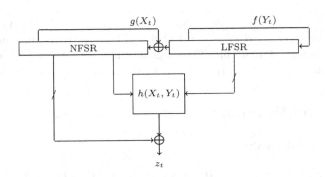

Fig. 2. Structure of Stream Cipher in Grain Family

Table 1. Grain at a glance

	Grain v1	Grain-128	Grain-128a
n	80	128	128
m	64	96	96
Pad	FFFF	FFFFFFFF	FFFFFFFE
$f(\cdot)$	$y_{t+62} \oplus y_{t+51} \oplus y_{t+38}$ $\oplus y_{t+23} \oplus y_{t+13} \oplus y_t$	$y_{t+96} \oplus y_{t+81} \oplus y_{t+70}$ $\oplus y_{t+38} \oplus y_{t+7} \oplus y_t$	$y_{t+96} \oplus y_{t+81} \oplus y_{t+70}$ $\oplus y_{t+38} \oplus y_{t+7} \oplus y_t$
$g(\cdot)$	$x_{t+62} \oplus x_{t+60} \oplus x_{t+52}$ $\oplus x_{t+45} \oplus x_{t+37} \oplus x_{t+33}$ $x_{t+28} \oplus x_{t+21} \oplus x_{t+14}$ $x_{t+9} \oplus x_t \oplus x_{t+63}x_{t+60} \oplus$ $x_{t+37}x_{t+33} \oplus x_{t+15}x_{t+9}$ $x_{t+60}x_{t+52}x_{t+45} \oplus x_{t+33}$ $x_{t+28}x_{t+21} \oplus x_{t+63}x_{t+60}$ $x_{t+21}x_{t+15} \oplus x_{t+63}x_{t+60}$ $x_{t+52}x_{t+45}x_{t+37} \oplus x_{t+33}$ $x_{t+28}x_{t+21}x_{t+15}x_{t+9} \oplus$ $x_{t+52}x_{t+45}x_{t+37}x_{t+33}$ $x_{t+28}x_{t+21}$	$y_t \oplus x_t \oplus x_{t+26} \oplus$ $x_{t+56} \oplus x_{t+91} \oplus x_{t+96} \oplus$ $x_{t+3}x_{t+67} \oplus x_{t+11}x_{t+13}$ $\oplus x_{t+17}x_{t+18} \oplus x_{t+27}x_{t+59}$ $\oplus x_{t+40}x_{t+48} \oplus x_{t+61}$ $x_{t+65} \oplus x_{t+68}x_{t+84}$	$y_t \oplus x_t \oplus x_{t+26} \oplus$ $x_{t+56} \oplus x_{t+91} \oplus x_{t+96} \oplus$ $x_{t+3}x_{t+67} \oplus x_{t+11}x_{t+13}$ $\oplus x_{t+17}x_{t+18} \oplus x_{t+27}x_{t+59}$ $\oplus x_{t+40}x_{t+48} \oplus x_{t+61}$ $x_{t+65} \oplus x_{t+68}x_{t+84}$ $\oplus x_{t+88}x_{t+92}x_{t+93}x_{t+95}$ $\oplus x_{t+22}x_{t+24}x_{t+25} \oplus$ $x_{t+70}x_{t+78}x_{t+82}$
$h(\cdot)$	$y_{t+3}y_{t+25}y_{t+46} \oplus y_{t+3}$ $y_{t+46}y_{t+64} \oplus y_{t+3}y_{t+46}$ $x_{t+63} \oplus y_{t+25}y_{t+46}x_{t+63} \oplus$ $y_{t+46}y_{t+64}x_{t+63} \oplus y_{t+3}$ $y_{t+64} \oplus y_{t+46}y_{t+64} \oplus y_{t+64}$ $x_{t+63} \oplus y_{t+25} \oplus x_{t+63}$	$x_{t+12}x_{t+95}y_{t+95} \oplus x_{t+12}$ $y_{t+8} \oplus y_{t+13}y_{t+20} \oplus x_{t+95}$ $y_{t+42} \oplus y_{t+60}y_{t+79}$	$x_{t+12}x_{t+95}y_{t+94} \oplus x_{t+12}$ $y_{t+8} \oplus y_{t+13}y_{t+20} \oplus x_{t+95}$ $y_{t+42} \oplus y_{t+60}y_{t+79}$
z_t	$x_{t+1} \oplus x_{t+2} \oplus x_{t+4} \oplus$ $x_{t+10} \oplus x_{t+31} \oplus x_{t+43} \oplus$ $x_{t+56} \oplus h$	$x_{t+2} \oplus x_{t+15} \oplus x_{t+36} \oplus$ $x_{t+45} \oplus x_{t+64} \oplus x_{t+73}$ $\oplus x_{t+89} \oplus y_{t+93} \oplus h$	$x_{t+2} \oplus x_{t+15} \oplus x_{t+36} \oplus$ $x_{t+45} \oplus x_{t+64} \oplus x_{t+73}$ $\oplus x_{t+89} \oplus y_{t+93} \oplus h$

vector that denotes the NFSR state at the tth clock interval and g is a non-linear function of the NFSR state bits. The output key-stream is produced by combining the LFSR and NFSR bits as $z_t = x_{t+l_1} \oplus x_{t+l_2} \oplus \cdots \oplus x_{t+l_c} \oplus y_{t+i_1} \oplus y_{t+i_2} \oplus \cdots \oplus y_{t+i_d} \oplus h(y_{t+h_1}, y_{t+h_2}, \ldots, y_{t+h_e}, x_{t+j_1}, x_{t+j_2}, \ldots, x_{t+j_w})$. Here h is a non-linear Boolean function.

KEY SCHEDULING ALGORITHM (KSA). The Grain family uses an n-bit key K, and an m-bit initialization vector IV, with $m < n$. The key is loaded in the NFSR and the IV is loaded in the 0th to the $(m-1)^{th}$ bits of the LFSR. The remaining m^{th} to $(n-1)^{th}$ bits of the LFSR are loaded with some fixed pad $P \in \{0,1\}^{n-m}$. Then, for the first $2n$ clocks, the key-stream bit z_t is XOR-ed to both the LFSR and NFSR update functions.

PSEUDO-RANDOM KEY-STREAM GENERATION ALGORITHM (PRGA). After the KSA, z_t is no longer XOR-ed to the LFSR and the NFSR but it is used as the Pseudo-Random key-stream bit. Thus, during this phase, the LFSR and NFSR are updated as $y_{t+n} = f(Y_t), x_{t+n} = y_t \oplus g(X_t)$.

2 Tools and Definitions

2.1 Differential Grain

Let $S_0 = [X_0 || Y_0] \in \{0,1\}^{2n}$ be the initial state of the Grain family PRGA and S_{0,Δ_ϕ} be the initial state which differs from S_0 in an LFSR location $\phi \in [0, n-1]$. The task is to ascertain how the corresponding internal states in the tth round

S_t and S_{t,Δ_ϕ} will differ from each other, for some integer $t > 0$. One such tool appeared in [6], but our approach is improved and more involved. We present the following algorithm which we will call D-GRAIN that takes as input the difference location $\phi \in [0, n-1]$ and the round r, and returns (i) a set of r integer arrays χ_t, for $0 \leq t < r$, each of length $c + d$, (ii) a set of r integer arrays Υ_t, for $0 \leq t < r$, each of length $e + w$ and (iii) an integer array ΔZ of length r. Note that as defined in Section 1.1, c, d are the number of NFSR, LFSR bits (respectively) which are linearly added to the output function h. Further, w, e are the number of NFSR, LFSR bits (respectively) that are input to the function h.

Now consider the corresponding generalized differential engine Δ_ϕ-Grain with an n-cell LFSR ΔL and an n-cell NFSR ΔN. All the elements of ΔL and ΔN are integers. We denote the tth round state of ΔL as $\Delta L_t = [u_t, u_{t+1}, \ldots, u_{t+n-1}]$ and that of ΔN as $\Delta N_t = [v_t, v_{t+1}, \ldots, v_{t+n-1}]$. Initially all the elements of $\Delta N, \Delta L$ are set to 0, with the only exception that the cell numbered $\phi \in [0, n-1]$ of ΔL is set to 1. The initial states $\Delta N_0, \Delta L_0$ are indicative of the difference between S_0 and S_{0,Δ_ϕ} and we will show that the tth states $\Delta N_t, \Delta L_t$ are indicative of the difference between S_t and S_{t,Δ_ϕ}. ΔL updates itself as $u_{t+n} = u_t + u_{t+f_1} + u_{t+f_2} + \cdots + u_{t+f_a} \mod 2$ and ΔN updates itself as $v_{t+n} = u_t + 2 \cdot OR(v_t, v_{t+\tau_1}, v_{t+\tau_2}, \ldots, v_{t+\tau_b})$. The rationale behind the update functions will be explained later. Here OR is a map from $\mathbb{Z}^{b+1} \to \{0, 1\}$ which roughly represents the logical 'or' operation and is defined as

$$OR(k_0, k_1, \ldots, k_b) = \begin{cases} 0, & \text{if } k_0 = k_1 = k_2 = \cdots = k_b = 0, \\ 1, & \text{otherwise.} \end{cases}$$

Let $\chi_t = [v_{t+l_1}, v_{t+l_2}, \ldots, v_{t+l_c}, u_{t+i_1}, u_{t+i_2}, \ldots, u_{t+i_d}]$ and $\Upsilon_t = [u_{t+h_1}, u_{t+h_2}, \ldots, u_{t+h_e}, v_{t+j_1}, v_{t+j_2}, \ldots, v_{t+j_w}]$. Note that χ_t (respectively Υ_t) is the set of cells in Δ_ϕ-Grain which corresponds to the bits that are linearly added to the output function h (respectively, input to h) in the tth PRGA stage of the actual cipher.

If $\mathbf{V}$ is a vector having non-negative integral elements, then $\mathbf{V} \sqsubseteq \beta$, (for some positive integer β), implies that all elements of $\mathbf{V}$ are less than or equal to β. The tth key-stream element Δz_t produced by this engine is given as

$$\Delta z_t = \begin{cases} 0, & \text{if } \Upsilon_t = \mathbf{0} \text{ AND } \chi_t \sqsubseteq 1 \text{ AND } |\chi_t| \text{ is even} \\ 1, & \text{if } \Upsilon_t = \mathbf{0} \text{ AND } \chi_t \sqsubseteq 1 \text{ AND } |\chi_t| \text{ is odd} \\ 2, & \text{otherwise.} \end{cases}$$

Here $\mathbf{0}$ denotes the all zero vector, and $|\cdot|$ denotes the number of non-zero elements in a vector. Initially $\Delta N_0, \Delta L_0$ represent the difference of S_0 and S_{0,Δ_ϕ}. As the PRGA evolves, the only non-zero element (having value 1) of ΔL propagates and so does the difference between S_t and S_{t,Δ_ϕ}. Since the LFSR in Grain is updated by a linear function, whenever the difference between S_t and S_{t,Δ_ϕ} is fed back via the update function, a 1 is fed back in ΔL. Now when the difference between S_t and S_{t,Δ_ϕ} propagates to some NFSR tap location τ_i (for some value of t), then this difference may or may not be fed back, depending on the nature of the Boolean function g and the current state S_t. Hence in such a case the propagation of the differential is probabilistic. Note that in all such situations, either the integer 2 or 3 is fed back in ΔN as is apparent from the update equation for v_{t+n}. Therefore if

1. some cell in ΔL_t or ΔN_t is 0, then the corresponding bits are equal in S_t and S_{t,Δ_ϕ} with probability 1;
2. some cell in ΔL_t or ΔN_t is 1, then the corresponding bits are different in S_t and S_{t,Δ_ϕ} with probability 1;
3. some cell in ΔL_t or ΔN_t is 2 or 3, then that the corresponding bits are different in S_t and S_{t,Δ_ϕ} with some probability $0 < p_d < 1$.

Also, note that if Υ_t is $\mathbf{0}$, then all the bits of S_t and S_{t,Δ_ϕ} that provide inputs to the non-linear function h are the same (for all choices of S_0). If all elements of χ_t are less than or equal to 1, then each one of the elements of S_t and S_{t,Δ_ϕ} which linearly adds on to the output function h to produce the output key-stream bit is either equal or different with probability 1. When both these events occur, the key-stream bits produced by S_t and S_{t,Δ_ϕ} are definitely the same if $|\chi_t|$ is an even number, as an even number of differences cancel out in GF(2). When this happens, Δ_ϕ-Grain outputs $\Delta z_t = 0$. If $|\chi_t|$ is an odd number, then the key-stream bits produced by S_t and S_{t,Δ_ϕ} are different with probability 1. In this case $\Delta z_t = 1$. In all other cases, the difference of the key-stream bits produced by S_t and S_{t,Δ_ϕ} is equal to 0 or 1 with some probability, and then $\Delta z_t = 2$. We describe the routine D-GRAIN(ϕ, r) in Algorithm 1 which returns the arrays χ_t, Υ_t for $0 \le t < r$ and $\Delta Z = [\Delta z_0, \ldots, \Delta z_{r-1}]$.

Input: ϕ: An LFSR location $\in [0, n-1]$, an integer $r(> 0)$;
Output: An integer array ΔZ of r elements;
Output: Two integer arrays χ_t, Υ_t for $0 \le t < r$;
$[u_0, u_1, \ldots, u_{n-1}] \leftarrow \mathbf{0}, [v_0, v_1, \ldots, v_{n-1}] \leftarrow \mathbf{0}, u_\phi \leftarrow 1, t \leftarrow 0$;
while $t < r$ **do**
    $\Upsilon_t \leftarrow [u_{h_1}, u_{h_2}, \ldots, u_{h_e}, v_{j_1}, v_{j_2}, \ldots, v_{j_w}]$;
    $\chi_t \leftarrow [v_{l_1}, v_{l_2}, \ldots, v_{l_c}, u_{i_1}, u_{i_2}, \ldots, u_{i_d}]$;
    if $\Upsilon_t = \mathbf{0}$ AND $\chi_t \sqsubseteq 1$ **then**
        if $|\chi_t|$ *is EVEN* **then**
          | $\Delta z_t \leftarrow 0$;
        end
        if $|\chi_t|$ *is ODD* **then**
          | $\Delta z_t \leftarrow 1$;
        end
    end
    else
    | $\Delta z_t \leftarrow 2$;
    end
    $t_1 \leftarrow u_0 + u_{f_1} + u_{f_2} + \ldots + u_{f_a} \bmod 2$;
    $t_2 \leftarrow u_0 + 2 \cdot OR(v_0, v_{\tau_1}, v_{\tau_2}, \ldots, v_{\tau_b})$;
    $[u_0, u_1, \ldots, u_{n-2}, u_{n-1}] \leftarrow [u_1, u_2, \ldots, u_{n-1}, t_1]$;
    $[v_0, v_1, \ldots, v_{n-2}, v_{n-1}] \leftarrow [v_1, v_2, \ldots, v_{n-1}, t_2]$;
    $t = t + 1$;
end
$\Delta Z = [\Delta z_0, \Delta z_1, \ldots, \Delta z_{r-1}]$;
Return $[\chi_0, \chi_1, \ldots, \chi_{r-1}]$, $[\Upsilon_0, \Upsilon_1, \ldots, \Upsilon_{r-1}]$, ΔZ

Algorithm 1. D-GRAIN(ϕ, r)

2.2   Derivative of a Boolean Function

Certain properties of Boolean functions have been exploited for the fault attack described in [5]. We use some other properties of the Boolean functions to mount our attack and these are described here. A q-variable Boolean function is a mapping from the set $\{0,1\}^q$ to the set $\{0,1\}$. One important way to represent a Boolean function is by its Algebraic Normal Form (ANF). A q-variable Boolean function $h(x_1,\ldots,x_q)$ can be considered to be a multivariate polynomial over $GF(2)$. This polynomial can be expressed as a sum of products representation of all distinct k-th order products $(0 \leq k \leq q)$ of the variables. The number of variables in the highest order product term with nonzero coefficient is called the *algebraic degree*, or simply the degree of h and denoted by $deg(h)$. Functions of degree at most one are called affine functions.

Definition 1. *Consider a q-variable Boolean function F and any vector $\alpha \in \{0,1\}^q$. We refer to the function $F(\mathbf{x}+\alpha)$ as a translation of F. The set of all possible translations of a given function F is denoted by the term 'Translation Set' and by the symbol $\mathcal{A}_F$. Since a q-variable function can have at most 2^q translations, the cardinality of $\mathcal{A}_F$ is atmost 2^q.*

Definition 2. *Consider a q-variable Boolean function F and its translation set $\mathcal{A}_F$. Any GF(2) linear combination $\hat{F}$ of the functions in $\mathcal{A}_F$, i.e., $\hat{F}(\mathbf{x}) = c_1 F(\mathbf{x} \oplus \alpha_1) \oplus c_2 F(\mathbf{x} \oplus \alpha_2) \oplus \cdots \oplus c_i F(\mathbf{x} \oplus \alpha_i)$, where $c_1, c_2, \ldots, c_i \in \{0,1\}$ is said to be a derivative of F. If $\hat{F}$ happens to be an affine Boolean function and $c_1 = c_2 = \cdots = c_i = 1$ then the set of vectors $\pi = [\alpha_1, \alpha_2, \ldots, \alpha_i]$ is said to be an affine differential tuple of F. If none of the vectors in π is $\mathbf{0}$ then π is said to be a weight i affine differential tuple of F otherwise π is said to be a weight $(i-1)$ affine differential tuple.*

3   Differential Fault Analysis of the Grain Family

In this section, we assume that the attacker has the ability to induce exactly a single bit toggle at a random LFSR location by applying a fault. Later, in Section 6, we will consider a more practical fault model in which an injected fault toggles more than one consecutive bits in LFSR locations.

3.1   Obtaining the Location of the Fault

Let $S_0 \in \{0,1\}^{2n}$ be the initial state of the Grain family PRGA described in Section 1.1 and S_{0,Δ_ϕ} be the initial state resulting after injecting a single bit fault in LFSR location $\phi \in [0, n-1]$. Let $Z = [z_0, z_1, \ldots, z_{2n-1}]$ and $Z^\phi = [z_0^\phi, z_1^\phi, \ldots, z_{2n-1}^\phi]$ be the first $2n$ key-stream bits produced by S_0 and S_{0,Δ_ϕ} respectively. The task for the fault location identification routine is to determine the fault location ϕ by analyzing the difference between Z and Z^ϕ. Of course, in Grain-128a the entire Z and Z_ϕ are not available to the attacker. Thus, we will deal with Grain-128a separately.

Our approach to determine the fault location is an improvement over the one presented in [5]. The basic idea used in [5] was the fact that if a fault is injected in single LFSR location at the beginning of the PRGA, then at certain specific PRGA rounds the key-stream bits are guaranteed to be equal. However, this technique requires multiple fault injections at the same LFSR bit to conclusively identify the fault location. In our work we utilize the added fact that due to a single bit fault at the beginning of the PRGA, the key-stream bits at certain other PRGA rounds are guaranteed to be different as well. This removes the requirement for multiple single-bit fault injection at the same LFSR location.

Grain v1 and Grain-128. As in [5], we define a $2n$ bit vector E_ϕ over $GF(2)$ defined as follows. Let E_ϕ be the bitwise logical XNOR (complement of XOR) of Z and Z_ϕ, i.e., $E_\phi = 1 \oplus Z \oplus Z^\phi$. Similarly we define $\overline{E}_\phi = 1 \oplus E_\phi$. Since S_0 can have 2^{n+m} values (each arising from a different combination of the n bit key and m bit IV, the remaining $n - m$ padding bits are fixed), each of these choices of S_0 may lead to different patterns of E_ϕ. The bitwise logical AND of all such vectors E_ϕ is denoted as the First Signature vector Sgn^1_ϕ for the fault location ϕ. Similarly the bitwise logical AND of all such vectors $\overline{E}_\phi$ is denoted as the Second Signature vector Sgn^2_ϕ for the fault location ϕ. Note that if $Sgn^1_\phi(i)$ $(Sgn^2_\phi(i))$ is 1 for any $i \in [0, 2n - 1]$ then the ith key-stream bit produced by S_0 and S_{0,Δ_ϕ} is equal (different) for all choices of S_0.

This implies that if $\Delta_\phi Z = [\Delta_\phi z_0, \Delta_\phi z_1, \ldots, \Delta_\phi z_{2n-1}]$ is the third output of the routine D-GRAIN$(\phi, 2n)$, then

$$Sgn^1_\phi(i) = \begin{cases} 1, \text{if } \Delta_\phi z_i = 0, \\ 0, \text{otherwise.} \end{cases} \qquad Sgn^2_\phi(i) = \begin{cases} 1, \text{if } \Delta_\phi z_i = 1, \\ 0, \text{otherwise.} \end{cases}$$

Grain-128a. Grain-128a has a different encryption strategy in which the first 64 key-stream bits and every alternate key-stream bit thereof is used to construct the message authentication code and therefore unavailable to the attacker. To circumvent this problem, in Grain-128a every re-keying is followed by a fault injection at the beginning of round 64 of the PRGA instead of round 0. Hence the vectors Z, Z^ϕ are defined as $Z = [z_{64}, z_{66}, \ldots, z_{318}]$ and $Z^\phi = [z^\phi_{64}, z^\phi_{66}, \ldots, z^\phi_{318}]$. As before, we define $E(\phi) = 1 \oplus Z \oplus Z^\phi$ and $\overline{E}(\phi) = 1 \oplus E(\phi)$ and Sgn^1_ϕ, Sgn^2_ϕ are defined as the bitwise AND of all possible $E(\phi), \overline{E}(\phi)$ respectively. Note that if a fault is applied at a random LFSR location ϕ at the 64th PRGA round, then the tth state of Δ_ϕ-Grain will align itself with the $(64+t)^{th}$ state of Grain-128a. This implies that if $\Delta_\phi Z = [\Delta_\phi z_0, \Delta_\phi z_1, \ldots, \Delta_\phi z_{255}]$ is the third output of the routine D-GRAIN$(\phi, 256)$, then

$$Sgn^1_\phi(i) = \begin{cases} 1, \text{if } \Delta_\phi z_{2i} = 0, \\ 0, \text{otherwise.} \end{cases} \qquad Sgn^2_\phi(i) = \begin{cases} 1, \text{if } \Delta_\phi z_{2i} = 1, \\ 0, \text{otherwise.} \end{cases}$$

3.2   Steps for Location Identification

The task for the fault identification routine is to determine the value of ϕ given the vector E_ϕ. For any element $V \in \{0,1\}^l$, define the set $B_V = \{i : 0 \le i <$

l, $V(i) = 1\}$ i.e. B_V is the support of of V. Now define a relation $\preceq$ in $\{0,1\}^l$ such that for any two elements $V_1, V_2 \in \{0,1\}^l$, we will have $V_1 \preceq V_2$ if $B_{V_1} \subseteq B_{V_2}$. Now we check the elements in B_{E_ϕ} and $B_{\overline{E}_\phi}$. By definition, these are the PRGA rounds i during which $z_i = z_i^\phi$ and $z_i \neq z_i^\phi$ respectively. By the definition of the first and second Signature vector proposed above, we know that for the correct value of ϕ, $B_{Sgn_\phi^1} \subseteq B_{E_\phi}, B_{Sgn_\phi^2} \subseteq B_{\overline{E}_\phi}$ and hence $Sgn_\phi^1 \preceq E_\phi, Sgn_\phi^2 \preceq \overline{E}_\phi$. So our strategy would be to search all the n many first Signature vectors and formulate the first candidate set $\Psi_{0,\phi} = \{\psi : 0 \leq \psi \leq n-1, \ Sgn_\psi^1 \preceq E_\phi\}$. If $|\Psi_{0,\phi}|$ is 1, then the single element in $\Psi_{0,\phi}$ will give us the fault location ϕ. If not, we then formulate the second candidate set $\Psi_{1,\phi} = \{\psi : \psi \in \Psi_{0,\phi}, \ Sgn_\psi^2 \preceq \overline{E}_\phi\}$. If $|\Psi_{1,\phi}|$ is 1, then the single element in $\Psi_{1,\phi}$ will give us the fault location ϕ. If $\Psi_{1,\phi}$ has more than one element, we will be unable to decide conclusively at this stage.

However, the task is made simpler if we can access the faulty key-streams Z^ϕ and hence get E_ϕ for all $\phi \in [0, n-1]$. This is possible since our fault model allows multiple re-keying with the same but unknown Key-IV. We need to reset the cipher each time and then inject a fault at any random unknown LFSR location at the beginning of the PRGA. By performing this process around $n \cdot \sum_{i=1}^n \frac{1}{i} \approx n \ln n$, we can expect to hit all the LFSR locations in $[0, n-1]$ and obtained n different faulty key-streams Z^ϕ.

The remaining task is to label each Z^ϕ with a unique $\phi \in [0, n-1]$. Using the techniques outlined above for all the faulty key-streams, we will be able to uniquely label them if (i) for all $\phi \in [0, n-1]$, $|\Psi_{1,\phi}| = 1$, i.e., all the faulty key-streams were assigned unique labels, or (ii) for all $\phi_i \in W = \{\phi_1, \phi_2, \ldots, \phi_j\}$, $|\Psi_{1,\phi_i}| > 1$ AND $|\Psi_{1,\phi_i} - \overline{W}| = 1$, where $\overline{W} = [0, n-1] - W$, i.e., $\overline{W}$ is the set of labels that have already been assigned. The second condition states that even if some faulty key-stream Z^{ϕ_i} has not been labelled uniquely, its second candidate set Ψ_{1,ϕ_i} (along with the element ϕ_i) must contain only those labels that have already been uniquely assigned. Given a random key $K \in_R \{0,1\}^n$ and a random Initial Vector $IV \in_R \{0,1\}^m$ the probability that all n faulty key-streams can be labelled uniquely has been experimentally found to be 1 for all the 3 ciphers Grain v1, Grain-128 and Grain-128a. The experiments were performed over a set of 2^{20} randomly chosen Key-IV pairs. We sum up the fault location identification routine in the following steps.

A. Reset the cipher with the unknown key K and Initial Vector IV and record the first $2n$ fault-free key-stream bits Z.

B. Reset the cipher again with K, IV, and inject a single bit fault in a random LFSR location ϕ, $0 \leq \phi \leq n-1$ at the beginning of the PRGA. Record the faulty key-stream bits Z^ϕ, calculate E_ϕ and $\Psi_{1,\phi}$

C. Repeat Step [**B.**] around $n \ln n$ times so that n different faulty key-stream vectors corresponding to all LFSR locations $\phi \in [0, n-1]$ are obtained and calculate the corresponding $\Psi_{1,\phi}$ vector.

D. Once all the faulty key-stream vectors have been labelled we proceed to the next stage of the attack.

4   Beginning the Attack

Let us first describe some notations that we will henceforth use.

1. $S_t = [x_0^t, x_1^t, \ldots, x_{n-1}^t \;\; y_0^t, y_1^t, \ldots, y_{n-1}^t]$ is used to denote the internal state of the cipher at the beginning of round t of the PRGA. Thus x_i^t (y_i^t) denotes the ith NFSR (LFSR) bit at the start of round t of the PRGA. When $t = 0$, we use $S_0 = [x_0, x_1, \ldots, x_{n-1} \;\; y_0, y_1, \ldots, y_{n-1}]$ to denote the internal state for convenience.

2. S_{t,Δ_ϕ} is used to denote the internal state of the cipher at the beginning of round t of the PRGA, when a fault has been injected in LFSR location ϕ at the beginning of the PRGA round.

3. z_i^ϕ denotes the key-stream bit produced in the ith PRGA round, after faults have been injected in LFSR location ϕ at the beginning of the PRGA round. z_i is the fault-free i^{th} key-stream bit.

4. $\eta_t = [x_{l_1}^t, x_{l_2}^t \ldots, x_{l_c}^t, y_{i_1}^t, y_{i_2}^t \ldots, y_{i_d}^t]$ is the set of elements in S_t which contribute to the output key-stream bit function linearly and $\theta_t = [y_{h_1}^t, y_{h_2}^t, \ldots, y_{h_e}^t, x_{j_1}^t, x_{j_2}^t, \ldots, x_{j_w}^t]$ be the subset of S_t which forms the input to the combining function h.

5. If $\mathbf{v}$ is an integer vector all elements of which are either 0 or 1, then we express $\mathbf{v}$ as a vector over GF(2) and denote it by the symbol $\widetilde{\mathbf{v}}$.

6. If $\mathbf{w}$ is a vector over GF(2) then $\mathcal{P}(\mathbf{w})$ denotes the GF(2) sum of the elements of $\mathbf{w}$.

Determining the LFSR. During PRGA, the LFSR evolves linearly and independent of the NFSR. Hence, y_i^t for any $i \in [0, n-1]$ and $t \geq 0$ is a linear function of $y_0, y_1, \ldots, y_{n-1}$. Let S_0 and S_{0,Δ_ϕ} be two initial states of the Grain PRGA (as described in Section 1.1) that differ in only the LFSR location $\phi \in [0, n-1]$. Let $[\chi_{0,\phi}, \chi_{1,\phi}, \ldots, \chi_{2n-1,\phi}], [\Upsilon_{0,\phi}, \Upsilon_{1,\phi}, \ldots, \Upsilon_{2n-1,\phi}], \Delta_\phi Z$ be the outputs of D-GRAIN($\phi, 2n$).

Let $[\mathbf{0}, \alpha_1]$ be a weight 1 affine differential tuple of h, such that $h(\mathbf{x}) \oplus h(\mathbf{x} \oplus \alpha_1) = h_{01}(\mathbf{x})$ is a function of variables that takes input from LFSR locations only. If, for some round t of the PRGA, we have $\chi_{t,\phi} \sqsubseteq 1, \Upsilon_{t,\phi} \sqsubseteq 1$ and $\widetilde{\Upsilon}_{t,\phi} = \alpha_1$, then we can conclude that the tth round fault-free and faulty internal states S_t and S_{t,Δ_ϕ} differ deterministically in the bit locations that contribute to producing the output key-stream bit at round t. In such a scenario, the GF(2) sum of the fault-free and faulty key-stream bit at round t is given by $z_t \oplus z_t^\phi = \mathcal{P}(\eta_t) \oplus h(\theta_t) \oplus \mathcal{P}(\eta_t \oplus \widetilde{\chi}_{t,\phi}) \oplus h(\theta_t \oplus \widetilde{\Upsilon}_{t,\phi}) = \mathcal{P}(\widetilde{\chi}_{t,\phi}) \oplus h(\theta_t) \oplus h(\theta_t \oplus \alpha_1) = \mathcal{P}(\widetilde{\chi}_{t,\phi}) \oplus h_{01}(\theta_t)$.

Note that in the above equation $\mathcal{P}(\widetilde{\chi}_{t,\phi}) \oplus h_{01}(\theta_t)$ is an affine Boolean function in the LFSR state bits of $S_t = [y_0^t, y_1^t, \ldots, y_{n-1}^t]$ and hence $[y_0, y_1, \ldots, y_{n-1}]$. Since $z_t \oplus z_t^\phi$ is already known to us, this gives us one linear equation in $[y_0, y_1, \ldots, y_{n-1}]$. The trick is to get n such linear equations which are linearly independent by searching over all possible values of ϕ and affine differential tuples of h. Of course h may not have an affine differential tuple $[\mathbf{0}, \alpha_1]$ of weight 1 or even if it does $\widetilde{\Upsilon}_{t,\phi} = \alpha_1$ and $\chi_{t,\phi} \sqsubseteq 1$ may not hold for any t or ϕ. In such situations, one can look at other higher weight affine differential tuples.

Exploring Higher Weight Affine Differential Tuples. Consider λ many fault locations $\phi_i \in [0, n-1]$. Let $[\chi_{0,\phi_i}, \ldots, \chi_{2n-1,\phi_i}], [\Upsilon_{0,\phi_i}, \ldots, \Upsilon_{2n-1,\phi_i}], \Delta_{\phi_i} Z$ be the λ many outputs of D-GRAIN$(\phi_i, 2n)$ for $i \in [1, \lambda]$. Let $[\mathbf{0}, \alpha_1, \alpha_2, \ldots, \ldots, \alpha_\lambda]$ be a weight λ (where λ is an odd number) affine differential tuple of h, such that $h(\mathbf{x}) \oplus \bigoplus_{i=1}^{\lambda} h(\mathbf{x} \oplus \alpha_i) = H_1(\mathbf{x})$ is a function of variables that takes input from LFSR locations only. If for some round t of the PRGA, $\chi_{t,\phi_i} \sqsubseteq 1$, $\Upsilon_{t,\phi_i} \sqsubseteq 1$ and $\widetilde{\Upsilon}_{t,\phi_i} = \alpha_i$ for all $i \in [1, \lambda]$, then by the arguments outlined in the previous subsection, we conclude $z_t \oplus \bigoplus_{i=1}^{\lambda} z_t^{\phi_i} = \mathcal{P}(\eta_t) \oplus h(\theta_t) \oplus \bigoplus_{i=1}^{\lambda} \left(\mathcal{P}(\eta_t \oplus \widetilde{\chi}_{t,\phi_i}) \oplus h(\theta_t \oplus \widetilde{\Upsilon}_{t,\phi_i}) \right) = \bigoplus_{i=1}^{\lambda} \mathcal{P}(\widetilde{\chi}_{t,\phi_i}) \oplus H_1(\theta_t)$.

Note that if λ is odd then we can not exploit differential tuples of the form $[\alpha_1, \alpha_2, \ldots, \ldots, \alpha_\lambda]$ where all $\alpha_i \neq \mathbf{0}$ as an odd number of terms do not cancel out in GF(2). Instead, if $[\alpha_1, \alpha_2, \ldots, \ldots, \alpha_\lambda]$ is a weight λ (λ is an even number) affine differential tuple of h, such that $\bigoplus_{i=1}^{\lambda} h(\mathbf{x} \oplus \alpha_i) = H_2(\mathbf{x})$ is a function of variables, that takes inputs from LFSR locations only, then by the previous arguments we have $\bigoplus_{i=1}^{\lambda} z_t^{\phi_i} = \bigoplus_{i=1}^{\lambda} \left(\mathcal{P}(\eta_t \oplus \widetilde{\chi}_{t,\phi_i}) \oplus h(\theta_t \oplus \widetilde{\Upsilon}_{t,\phi_i}) \right) = \bigoplus_{i=1}^{\lambda} \mathcal{P}(\widetilde{\chi}_{t,\phi_i}) \oplus H_2(\theta_t)$.

Note that each of the above cases gives us one linear equation in $[y_0, y_1, \ldots, y_{n-1}]$. We formally state the routine $\text{FLE}_L(\lambda)$ in Algorithm 2 that attempts to find such linear equations by investigating weight λ affine differential tuples.

Solving the System. Ideally we should get n linearly independent equations in $[y_0, y_1, \ldots, y_{n-1}]$ to solve the LFSR. If a call to $\text{FLE}_L(1)$ does not give us the requisite number of equations then we must call $\text{FLE}_L(2)$ and if required $\text{FLE}_L(3)$ to obtain the required number of equations. Note that the number of iterations in the outer most '**for**' loop is of $\text{FLE}_L(\lambda)$ is $\binom{n}{\lambda} \approx O(n^\lambda)$, so beyond a certain value of λ, it may not be practically feasible to call $\text{FLE}_L(\lambda)$. Assuming that we have n outputs from the successive $\text{FLE}_L(\lambda)$ routines of the form

$$t_i, \quad [\phi_{1,i}, \phi_{2,i}, \ldots, \phi_{\lambda_i,i}], \quad \gamma_i \oplus \bigoplus_{j=0}^{n-1} c_{i,j} y_j, \quad [\alpha_{1,i}, \alpha_{2,i}, \ldots, \alpha_{\lambda_i,i}],$$

$\forall i \in [0, n-1]$, if λ_i is even. Else the last output will be of the form $[\mathbf{0}, \alpha_{1,i}, \alpha_{2,i}, \ldots, \alpha_{\lambda_i,i}]$. Then we can write the equations so obtained in matrix form $LY = W$, where L is the $n \times n$ coefficient matrix $\{c_{i,j}\}$ over GF(2), Y is the column vector $[y_0, y_1, \ldots, y_{n-1}]^t$ and W is a column vector. The ith element $W(i)$ is given by $\gamma_i \oplus z_{t_i} \oplus \bigoplus_{j=1}^{\lambda_i} z_{t_i}^{\phi_{j,i}}$, if λ_i is odd, and $\gamma_i \oplus \bigoplus_{j=1}^{\lambda_i} z_{t_i}^{\phi_{j,i}}$ if λ_i is even, $\gamma_i \in \{0,1\}$.

If the equations are linearly independent then L is invertible. Thus, the solution Y of the above system are obtained by computing $L^{-1}W$. Both L and its inverse may be precomputed and hence the solution can be obtained immediately after recording the faulty bits.

Determining the NFSR. Once the LFSR state has been determined, we proceed to finding the NFSR state. Since the NFSR updates itself non-linearly, the method used to determine the NFSR initial state will be slightly different from the LFSR. If λ is odd, let $[\mathbf{0}, \alpha_1, \alpha_2, \ldots, \alpha_\lambda]$ be a weight λ (where λ is an odd number) tuple of h (not necessarily affine differential), such that

---

Input: λ: An integer > 0;
Output: Set of Rounds t, locations $[\phi_1, \phi_2, \ldots, \phi_\lambda]$, Affine expression in $[y_0, y_1, \ldots, y_{n-1}]$;
    Tuples $[\alpha_1, \ldots, \alpha_\lambda]$

for $\phi_1 = 0$ to $n-1$, $\phi_2 = 0$ to $n-1$, $\ldots$, $\phi_\lambda = 0$ to $n-1$ **do**
   if *All ϕ_j's are pairwise unequal* **then**
      for $i = 1$ to λ **do**
         $([\chi_{0,\phi_i}, \ldots, \chi_{2n-1,\phi_i}], [\Upsilon_{0,\phi_i}, \ldots, \Upsilon_{2n-1,\phi_i}], \Delta_{\phi_i} Z) = \text{D-Grain}(\phi_i, 2n)$
      end
      for $t = 0$ to $2n-1$ **do**
         if $\chi_{t,\phi_i} \sqsubseteq 1$ AND $\Upsilon_{t,\phi_i} \sqsubseteq 1$, $\forall i \in [1, \lambda]$ **then**
           if *λ is odd* **then**

1.1
              $H_1(\mathbf{x}) = h(\mathbf{x}) \oplus_{i=0}^{\lambda} h(\mathbf{x} \oplus \widetilde{\Upsilon}_{t,\phi_i})$;
              if *H_1 is a function only on LFSR bits* **then**
                 Output Round t, Locations $[\phi_1, \phi_2, \ldots, \phi_\lambda]$, Expression
                 $\oplus_{i=1}^{\lambda} \mathcal{P}(\widetilde{\chi}_{t,\phi_i}) \oplus H_1(\theta_t)$;
                 Output Tuple $[0, \widetilde{\Upsilon}_{t,\phi_1}, \ldots, \widetilde{\Upsilon}_{t,\phi_\lambda}]$
              end
           end
           else

1.2
              $H_2(\mathbf{x}) = \oplus_{i=0}^{\lambda} h(\mathbf{x} \oplus \widetilde{\Upsilon}_{t,\phi_i})$;
              if *H_2 is a function only on LFSR bits* **then**
                 Output Round t, Locations $[\phi_1, \phi_2, \ldots, \phi_\lambda]$, Expression
                 $\oplus_{i=1}^{\lambda} \mathcal{P}(\widetilde{\chi}_{t,\phi_i}) \oplus H_2(\theta_t)$;
                 Output Tuple $[\widetilde{\Upsilon}_{t,\phi_1}, \ldots, \widetilde{\Upsilon}_{t,\phi_\lambda}]$
              end
           end
         end
      end
   end
end

Algorithm 2. $\mathrm{FLE}_L(\lambda)$

$h(\mathbf{x}) + \bigoplus_{i=1}^{\lambda} h(\mathbf{x} \oplus \alpha_i) = H_1(\mathbf{x}) = x' \oplus H_{11}(\mathbf{x})$ where x' is a variable that takes input from an NFSR location and $H_{11}(\mathbf{x})$ is a function only on the LFSR variables. If for some round t of the PRGA $\chi_{t,\phi_i} \sqsubseteq 1$ and $\Upsilon_{t,\phi_i} \sqsubseteq 1$ and $\widetilde{\Upsilon}_{t,\phi_i} = \alpha_i$ for all $i \in [1, \lambda]$, then by the arguments outlined in the previous subsection we conclude $z_t \oplus \bigoplus_{i=1}^{\lambda} z_t^{\phi_i} = \mathcal{P}(\eta_t) \oplus h(\theta_t) \oplus \bigoplus_{i=1}^{\lambda} \left(\mathcal{P}(\eta_t \oplus \widetilde{\chi}_{t,\phi_i}) \oplus h(\theta_t \oplus \widetilde{\Upsilon}_{t,\phi_i}) \right) = \bigoplus_{i=1}^{\lambda} \mathcal{P}(\widetilde{\chi}_{t,\phi_i}) \oplus H_1(\theta_t) = \bigoplus_{i=1}^{\lambda} \mathcal{P}(\widetilde{\chi}_{t,\phi_i}) \oplus H_{11}(\theta_t) \oplus x_{j_r}^t$, for some $r \in [1, w]$. Since, the LFSR is already known, $H_{11}(\theta_t)$ can be calculated and that leaves $x_{j_r}^t$ as the only unknown in the equation, whose value is also calculated immediately after recording the faulty bits and solving the LFSR.

The λ even case can be dealt with similarly. We can describe another routine $\mathrm{FLE}_N(\lambda)$ which will help in determining the NFSR state. This routine is similar to the $\mathrm{FLE}_L(\lambda)$ routine described in Algorithm 2. The only differences are that line **1.1** will change to

if $H_1(\mathbf{x}) = x' \oplus H_{11}(\mathbf{x})$ where x' is an NFSR term and $H_{11}(\mathbf{x})$ depends on LFSR variables only.

Line **1.2** of Algorithm 2 will also change accordingly. With the help of $\mathrm{FLE}_N(\lambda)$ routine, we can obtain specific NFSR state bits at various rounds of operation of the PRGA. Due to the shifting property of shift registers, the following equation holds $x_i^t = x_{i-1}^{t+1}$. For example, calculating x_{46}^{30} and x_{50}^{32} is the same as determining the two NFSR state bits of the internal state S_{30}: x_{46}^{30} and x_{52}^{30}.

Hence by using the $\text{FLE}_N(\lambda)$ for successive values of λ, one can obtain all the n NFSR state bits of S_t for some $t \geq 0$. Since the LFSR initial state of S_0 is already known and due to the fact that the LFSR operates independent of the NFSR in the PRGA, the attacker can compute the LFSR state bits of S_t by simply running the Grain PRGA forward for t rounds and thus compute the entire of S_t.

4.1   Finding the Secret Key and Complexity of the Attack

It is known that the KSA, PRGA routines in the Grain family are invertible (see [6, 12]). Once we have all the bits of S_t, by running the PRGA^{-1} (inverse PRGA) routine for t rounds one can recover S_0. Thereafter the KSA^{-1} (inverse KSA) routine can be used to find the secret key.

The attack complexity directly depends on the number of re-keyings to be performed such that all of locations in $[0, n-1]$ of the LFSR are covered. Since each re-keying is followed by exactly one fault injection, the expected number of fault injection is $n \cdot \sum_{i=1}^{n} \frac{1}{i} \approx n \cdot \ln n$. Thereafter, the attack requires one matrix multiplication between an $n \times n$ matrix and an $n \times 1$ vector to recover the LFSR, and solving a few equations to get the NFSR state. After this, t invocations of the PRGA^{-1} and a single invocation of the KSA^{-1} gives us the secret key.

Note that, construction of the matrix L and running the $\text{FLE}_L(\lambda)$ and $\text{FLE}_N(\lambda)$ can be done beforehand and thus do not add to the attack complexity. However, these routines are a part of the pre-processing phase, the exact runtime of which will depend on the nature of the functions g, h and also the choice of taps used in the cipher design.

5   Attacking the Actual Ciphers

Now we will provide the details of the actual attack on Grain v1, Grain-128 and Grain-128a.

Grain v1. In Grain v1 the non linear combining function is of the form $h(s_0, s_1, s_2, s_3, s_4) = s_1 \oplus s_4 \oplus s_0 s_3 \oplus s_2 s_3 \oplus s_3 s_4 \oplus s_0 s_1 s_2 \oplus s_0 s_2 s_3 \oplus s_0 s_2 s_4 \oplus s_1 s_2 s_4 \oplus s_2 s_3 s_4$. Here only s_4 corresponds to an NFSR variable. This function has 4 affine differential tuples of weight 1, only one of which ($[0, \alpha = 11001]$) leads to a derivative which is a function of only LFSR variables. However, $\widehat{\Upsilon}_{t,\phi} = \alpha$ and $\chi_{t,\phi} \sqsubseteq 1$ does not hold for any t or ϕ. Hence one needs to look at higher weight tuples.

A call to $\text{FLE}_L(3)$ returns 78 linearly independent equations. The result is given in Table 2. A call to $\text{FLE}_L(2)$ gives us the 2 other equations required to solve the system. The result is shown in Table 3. One can verify that the linear equations so obtained are linearly independent and thus LFSR can be solved

Table 2. Output of $FLE_L(3)$ for Grain v1 (ADT implies Affine Differential Tuple)

t	ϕ_1	ϕ_2	ϕ_3	Range	Expr.	ADT
$45+i$	$62+i$	$24+i$	$70+i$	$i \in [0,9]$		
$55+i$	$72+i$	$16+i$	$51+i$	$i \in [0,7]$		00000,
$63+i$	$13+i$	$24+i$	$59+i$	$i \in [0,9]$		00100,
$73+i$	$33+i$	$26+i$	$51+i$	$i \in [0,10]$	v_{46}^t	00110,
$84+i$	$44+i$	$37+i$	$38+i$	$i \in [0,6]$		01000
$91+i$	$53+i$	$44+i$	$41+i$	$i \in [0,8]$		
$100+i$	$70+i$	$53+i$	$60+i$	$i \in [0,8]$		
109	79	71	69			
$77+i$	$45+i$	$51+i$	$38+i$	$i \in [0,5]$		00000,
$83+i$	$72+i$	$57+i$	$44+i$	$i \in [0,4]$	$v_3^t \oplus v_{25}^t \oplus v_{64}^t$	01100,
94	62	79	55			10000, 10110
95	78	63	56		$v_3^t \oplus v_{25}^t \oplus v_{46}^t \oplus v_{64}^t$	00000, 01001, 01100, 10110

Table 3. Output of $FLE_L(2)$ for Grain v1   **Table 4.** Output of $FLE_N(1)$ for Grain v1

t	ϕ_1	ϕ_2	Range	Expr.	ADT
$110+i$	$64+i$	$77+i$	$i \in [0,1]$	v_{46}^t	00001, 11000

t	ϕ_1	Range	Expr.	ADT
$55+i$	$23+i$	$i \in [0,14]$		00000,
$70+i$	$77+i$	$i \in [0,2]$	$1 \oplus v_3^t \oplus v_{46}^t \oplus x_{63}^t$	01010
$91+i$	$62+i$	$i \in [0,5]$		

Table 5. Output of $FLE_N(3)$ for Grain v1

t	ϕ_1	ϕ_2	ϕ_3	Range	Expr.	ADT
$17+i$	i	$1+i$	$20+i$	$i \in [0,27]$	$1 \oplus v_3^t \oplus v_{46}^t \oplus x_{63}^t$	00000, 00001, 00010, 10000
$45+i$	$28+i$	$13+i$	$48+i$	$i \in [0,9]$		
$73+i$	$53+i$	$33+i$	$26+i$	$i \in [0,17]$	$1 \oplus v_3^t \oplus x_{63}^t$	00000, 00010, 00100, 00110

readily. A call each to $FLE_N(1)$ and $FLE_N(3)$ gives us all the NFSR bits of S_{80}. The output of these routines are given as in Tables 4 and 5. A look at these tables shows that the attacker can calculate the values of x_{63}^t for all $t \in [17, 96]$. This is equivalent to calculating x_i^{80} for all $i \in [0, 79]$. Thereafter, S_0 and the secret key may be obtained as per the techniques outlined in Section 4.1.

Grain-128. In Grain-128 the non linear combining function is of the form $h(s_0, s_1, \ldots, s_8) = s_0 s_1 \oplus s_2 s_3 \oplus s_4 s_5 \oplus s_6 s_7 \oplus s_0 s_4 s_8$. Only s_0, s_4 correspond to the NFSR variables. This function has 4 affine differential tuples of weight 1 which produce derivatives on LFSR variables. A call to $FLE_L(1)$ produces all the 128 equations needed to solve the LFSR. The output of this routine is given in Table 6.

A call to $FLE_N(1)$ gives us all the NFSR bits of S_{12}. The output of this routine is in Table 7. Thus, $FLE_N(1)$ gives us x_{12}^t for all $t \in [0, 115]$, and x_{95}^t for all $t \in [0, 11]$. This is equivalent to all the NFSR state bits of S_{12}. Thereafter, S_0 and the secret key may be obtained as per the techniques outlined in Section 4.1.

Table 6. Output of $\mathrm{FLE}_L(1)$ for Grain-128

t	ϕ_1	Range	Expr.	ADT
i	$20+i$	$i \in [0, 107]$	v^t_{13}	000 000 000, 000 100 000
$61+i$	$50+i$	$i \in [0, 19]$	v^t_{60}	000 000 000, 000 000 010

Table 7. Output of $\mathrm{FLE}_N(1)$ for Grain-128

t	ϕ_1	Range	Expr.	ADT
i	$8+i$	$i \in [0, 115]$	x^t_{12}	000 000 000, 010 000 000
$33+i$	$75+i$	$i \in [0, 11]$	x^t_{95}	000 000 000, 000 001 000

Table 8. Output of $\mathrm{FLE}_L(1)$ for Grain-128a

t	ϕ_1	Range	Expr.	ADT
$6+2i$	$26+2i$	$i \in [0, 50]$	v^t_{13}	000 000 000, 000 100 000
$108+2i$	$70+2i$	$i \in [0, 12]$		
$2i$	$13+2i$	$i \in [0, 33]$	v^t_{20}	000 000 000, 001 000 000
$28+2i$	$107+2i$	$i \in [0, 10]$	v^t_{60}	000 000 000, 000 000 010
$50+2i$	$1+2i$	$i \in [0, 18]$		

Table 9. Output of $\mathrm{FLE}_N(1)$ for Grain-128a

t	ϕ_1	Range	Expr.	ADT
$50+2i$	$58+2i$	$i \in [0, 34]$		000 000 000, 010 000 000
$120+2i$	$96+2i$	$i \in [0, 15]$	x^t_{12}	
$152+2i$	$102+2i$	$i \in [0, 12]$		000 000 000, 000 001 000
$2i$	$42+2i$	$i \in [0, 42]$	x^t_{95}	
$86+2i$	$38+2i$	$i \in [0, 4]$		

Grain-128a. In Grain-128a, the first 64 key-stream bits and every alternate key-stream bit thereof are used to construct the message authentication code and therefore unavailable to the attacker. To resolve this problem, in Grain-128a every re-keying is followed by a fault injection at the beginning round 64 of the PRGA instead of round 0 and the goal of the attacker is to reconstruct the internal state at the 64th instead of the 0th PRGA round. Note that if a fault is applied at a random LFSR location ϕ at the 64th PRGA round, then the tth state of Δ_ϕ-Grain will align itself with the $(64 + t)^{th}$ state of the actual cipher. Hence, in a slight departure from the notation introduced in the previous section we will call the 64th PRGA state S_0 and all other notations are shifted with respect to t accordingly (e.g., S_t refers to the $(64 + t)^{th}$ PRGA state etc).

The key-stream bit at every odd numbered round (after round 64 of the PRGA) is used for making the MAC and is unavailable to the attacker. Hence after calling $\mathrm{FLE}_L(1)$ the attacker must reject all outputs with an odd value of t. Even then the attacker obtains all the equations required to solve the LFSR. The output is presented in Table 8. Similarly a call to $\mathrm{FLE}_N(1)$ after rejecting outputs with odd values of t, gives us 112 NFSR bits of S_{62}. The output is given in Table 9.

At this point, the attacker could simply guess the remaining 16 bits of S_{62} or give a call to $\mathrm{FLE}_N(2)$ and thus increase the complexity of the preprocessing stage. As it turns out, the attacker can do even better without going for these two options. The 16 NFSR bits not determined at this point are x^{62}_{2i+1}, for $0 \le i \le 15$. Let us now look at the equations for the key-stream bits z_{62+2j} for $j \in [0, 8]$,

$$z_{62+2j} = \bigoplus_{i \in \mathcal{B}} x^{62}_{i+2j} \oplus x^{62}_{15+2j} \oplus y^{62}_{93+2j} \oplus h(\theta_{62+2j}),$$

where $\mathcal{B} = \{2, 36, 45, 64, 73, 89\}$. Now, x^{62}_{15+2j}, $j \in [0, 8]$ is the only unknown in each of these equations and so its value can be calculated immediately. This leaves us with the 7 unknown bits $x^{62}_1, x^{62}_3, \ldots, x^{62}_{13}$. In addition to the entries in Table 9, $\mathrm{FLE}_N(1)$ also gives the output

$$t = 96 + 2i, \phi_1 = 48 + 2i, x_{95}^t, [\mathbf{0}, 000\ 001\ 000], \ \forall i \in [0, 6].$$

This gives us the bits x_{95}^{96+2i} or equivalently x_{127}^{64+2i} for $i \in [0, 6]$. Let us write the NFSR update function g in the form $g(X) = x' \oplus g'(X)$, where x' corresponds to the variable that taps the 0th NFSR location. Then looking at the NFSR update rule for Grain-128a, we have

$$x_{127}^{64+2i} = y_0^{63+2i} \oplus x_0^{63+2i} \oplus g'(X_{63+2i}) = y_{1+2i}^{62} \oplus x_{1+2i}^{62} \oplus g'(X_{63+2i}),$$

$\forall i \in [0, 6]$. Again, $x_{1+2i}^{62}, \ i \in [0, 6]$ is the only unknown in these equations and so its value can be calculated immediately. This gives us all the NFSR bits of S_{62}. Using the techniques in Section 4.1, S_0 can be calculated. Since this state corresponds to the 64th PRGA state, the PRGA^{-1} routine needs to be run 64 more times before invoking the KSA^{-1} routine which would then reveal the secret key.

6   When a Fault Injection Affects More Than One Locations: Some Preliminary Observations

So far we have discussed an attack scenario where an injected fault flips exactly one bit value at a random LFSR location. We now relax the requirements of the attack, and assume a fault model that allows the user to inject a fault that affects more than one locations. Our strategy is that, if the fault injection affects more than one location, we will be able to identify that scenario, and will not use those cases for further processing.

We consider the case when at most three consecutive locations can be disturbed by a single fault injection. Thus, four cases are possible: (a) exactly one LFSR bit is flipped (n cases), (b) 2 consecutive locations $i, i + 1$ of the LFSR are flipped ($n - 1$ cases), (c) 3 consecutive locations $i, i + 1, i + 2$ of the LFSR are flipped ($n - 2$ cases) and (d) locations $i, i + 2$ are flipped but not $i + 1$ ($n - 2$ cases). Studying such a model makes sense if we attack an implementation of Grain where the LFSR register cells are physically positioned linearly one after the other.

It is clear that such a fault model allows a total of $n + n - 1 + 2(n - 2) = 4n - 5$ types of faults out of which only n are single bit-flips. We assume that each of these $4n - 5$ cases are equally probable. The success of our attack that we have described in Section 4 will depend on the ability of the attacker to deduce whether a given faulty key-stream vector has been produced as a result of a single bit toggling of any LFSR location or a multiple-bit toggle. Thus, we need to design a fault location identification algorithm that analyzes a faulty key-stream and (i) if the faulty key-stream has been produced due to a single bit toggling of any LFSR location, the algorithm should output that particular position, and (ii) if the faulty key-stream has been produced due to multiple-bit toggling of LFSR locations, the algorithm should infer that the faulty key-stream could not have been produced due to a single bit toggle.

To solve the problem, will use the same fault location identification technique used in Section 3.2. For the method to be a success, the routine would return the fault location numbers for all possible cases when a single LFSR location is toggled (n out of $4n - 5$ cases), and the empty set $\emptyset$ for all the other $3n - 5$ cases. Let p_s be the probability that the fault location identification technique has succeeded (theoretically, the probability is defined over all possible Key-IV pairs). By performing computer simulations over 2^{20} randomly chosen Key-IV pairs, the value of p_s was found to be 0.99994 for Grain v1, 1.00 for Grain-128 and 0.993 for Grain-128a. Note that assuming this fault model increases the number of re-keyings and hence fault injections to $(4n - 5) \cdot \ln(4n - 5)$.

As the experiments show, the probability of the location identification technique failing is very small. In case the method fails for some particular Key-IV pair, we reset the cipher with the same Key-IV and repeat the fault identification routine and this time inject the fault at PRGA round 1 instead of 0 (round 66 for Grain-128a) and then try to reconstruct this first (66th for Grain-128a) PRGA state using the methods outlined in Section 4. The probability that the location identification routine will fail for both PRGA round 0 and 1 is $(1 - p_s)^2$ (assuming independence) and is thus even smaller. In case the method fails for both round 0 and 1, we repeat the routine on PRGA round 2 and so on.

7   Conclusion

In this paper we outline a general strategy to perform differential fault attack on ciphers with the physical structure of Grain. In particular, the attack is demonstrated on Grain v1, Grain-128 and Grain-128a. The attack also uses a much more practical and realistic fault model compared to the fault attacks on the Grain family reported in literature [5, 6, 12].

References

1. Biham, E., Shamir, A.: Differential Fault Analysis of Secret Key Cryptosystems. In: Kaliski Jr., B.S. (ed.) CRYPTO 1997. LNCS, vol. 1294, pp. 513–525. Springer, Heidelberg (1997)
2. Boneh, D., DeMillo, R.A., Lipton, R.J.: On the Importance of Checking Cryptographic Protocols for Faults. In: Fumy, W. (ed.) EUROCRYPT 1997. LNCS, vol. 1233, pp. 37–51. Springer, Heidelberg (1997)
3. The ECRYPT Stream Cipher Project. eSTREAM Portfolio of Stream Ciphers (revised on September 8, 2008)
4. Ågren, M., Hell, M., Johansson, T., Meier, W.: A New Version of Grain-128 with Authentication. In: Symmetric Key Encryption Workshop. DTU, Denmark (2011)
5. Banik, S., Maitra, S., Sarkar, S.: A Differential Fault Attack on the Grain Family of Stream Ciphers. In: Prouff, E., Schaumont, P. (eds.) CHES 2012. LNCS, vol. 7428, pp. 122–139. Springer, Heidelberg (2012)
6. Berzati, A., Canovas, C., Castagnos, G., Debraize, B., Goubin, L., Gouget, A., Paillier, P., Salgado, S.: Fault Analysis of Grain-128. In: IEEE International Workshop on Hardware-Oriented Security and Trust 2009, pp. 7–14 (2009)

7. Dinur, I., Güneysu, T., Paar, C., Shamir, A., Zimmermann, R.: An Experimentally Verified Attack on Full Grain-128 Using Dedicated Reconfigurable Hardware. In: Lee, D.H. (ed.) ASIACRYPT 2011. LNCS, vol. 7073, pp. 327–343. Springer, Heidelberg (2011)
8. Hell, M., Johansson, T., Meier, W.: Grain - A Stream Cipher for Constrained Environments. ECRYPT Stream Cipher Project Report 2005/001 (2005), http://www.ecrypt.eu.org/stream
9. Hell, M., Johansson, T., Maximov, A., Meier, W.: A Stream Cipher Proposal: Grain-128. In: IEEE International Symposium on Information Theory, ISIT 2006 (2006)
10. Hoch, J.J., Shamir, A.: Fault Analysis of Stream Ciphers. In: Joye, M., Quisquater, J.-J. (eds.) CHES 2004. LNCS, vol. 3156, pp. 240–253. Springer, Heidelberg (2004)
11. Hojsík, M., Rudolf, B.: Differential Fault Analysis of Trivium. In: Nyberg, K. (ed.) FSE 2008. LNCS, vol. 5086, pp. 158–172. Springer, Heidelberg (2008)
12. Karmakar, S., Roy Chowdhury, D.: Fault analysis of Grain-128 by targeting NFSR. In: Nitaj, A., Pointcheval, D. (eds.) AFRICACRYPT 2011. LNCS, vol. 6737, pp. 298–315. Springer, Heidelberg (2011)
13. Knellwolf, S., Meier, W., Naya-Plasencia, M.: Conditional Differential Cryptanalysis of NLFSR-based Cryptosystems. In: Abe, M. (ed.) ASIACRYPT 2010. LNCS, vol. 6477, pp. 130–145. Springer, Heidelberg (2010)
14. Skorobogatov, S.P.: Optically Enhanced Position-Locked Power Analysis. In: Goubin, L., Matsui, M. (eds.) CHES 2006. LNCS, vol. 4249, pp. 61–75. Springer, Heidelberg (2006)
15. Skorobogatov, S.P., Anderson, R.J.: Optical Fault Induction Attacks. In: Kaliski Jr., B.S., Koç, Ç.K., Paar, C. (eds.) CHES 2002. LNCS, vol. 2523, pp. 2–12. Springer, Heidelberg (2003)

Cryptanalysis of Pseudo-random Generators Based on Vectorial FCSRs*

Thierry P. Berger[1] and Marine Minier[2]

[1] XLIM (UMR CNRS 7252), Université de Limoges
123 avenue Albert Thomas, 87060 Limoges Cedex, France
thierry.berger@unilim.fr
[2] Université de Lyon, INRIA
INSA-Lyon, CITI-INRIA, F-69621, Villeurbanne, France
marine.minier@insa-lyon.fr

Abstract. Feedback with Carry Shift Registers (FCSRs) have been first proposed in 2005 by F. Arnault and T. Berger as a promising alternative to LFSRs for the design of stream ciphers. The original proposal called F-FCSR simply filters the content of a FCSR in Galois mode using a linear function. In 2008, Hell and Johannson attacked this version using a method called LFSRization of F-FCSR. This attack is based on the fact that a single feedback bit controls the values of all the carry cells. Thus, a trail of 0 in the feedback bit annihilates the content of the carry register, leading to transform the FCSR into an LFSR during a sufficient amount of time.

Following this attack, a new version of F-FCSR was proposed based on a new ring FCSR representation that guarantees that each carry cell depends on a distinct cell of the main register. This new proposal prevents the LFSRization from happening and remains unbroken since 2009. In parallel, Alaillou, Marjane and Mokrane proposed to replace the FCSR in Galois mode of the original proposal by a Vectorial FCSR (V-FCSR) in Galois mode.

In this paper, we first introduce a general theoretical framework to show that Vectorial FCSRs could be seen as a particular case of classical FCSRs. Then, we show that Vectorial FCSRs used in Galois mode stay sensitive to the LFSRization of FCSRs. Finally, we demonstrate that hardware implementations of V-FCSRs in Galois mode are less efficient than those based on FCSRs in ring mode.

Keywords: stream cipher, FCSRs, n-adic numbers.

Introduction

Since 2003 and the appearance of algebraic attacks, it is well known that LFSRs could no more be used as the transition function of a stream cipher. In 2005,

---

* This work was partially supported by the French National Agency of Research: ANR-11-INS-011.

F. Arnault and T. Berger proposed to replace the classical LFSRs by a FCSR in Galois mode [5] to design a new stream cipher called F-FCSR. The content of the FCSR is then linearly filtered to produce the output stream. Unfortunately, in 2008, Hell and Johannson [13] found an attack called LFSRization of FCSRs against the F-FCSR family of ciphers. This attack exploits the fact that a single feedback bit controls all the carry bits. Then, F. Arnault et al. [8] proposed in 2009 a new FCSR representation called ring representation that prevents this attack. In the same research direction, Alaillou et al. in [1] proposed also an alternative based on particular FCSRs called Vectorial FCSRs. They then designed a stream cipher called Q-SIFR that uses a V-FCSR in Galois mode linearly filtered to produce the output stream.

This paper is devoted to the cryptanalysis of the Q-SIFR stream cipher proposed in [1] and shows that Q-SIFR is sensitive to LFSRization attack. This stream cipher is directly inspired from the family of F-FCSR stream ciphers, but it is based on Vectorial FCSRs. We first present a general theoretical framework for FCSRs that includes V-FCSRs. We then show that practical implementations of V-FCSRs lead to classical FCSRs in Galois mode. So, as classical FCSRs, Q-SIFR suffers from LFSRization attack. Moreover, compared with the latest version of F-FCSR [8], the proposed V-FCSR is not so efficient when looking at hardware implementations.

This paper is organized as follows: in Section 1, we introduce some theoretical background on n-adic topology. In Section 2, we present the family of Algebraic Finite State Machines (AFSMs) and show that V-FCSRs could be seen as a particular case of AFSMs. Section 3 is devoted to the practical implementations of AFSMs and of V-FCSRs which are, in fact, general FCSRs as the ones described in [3]. In Section 4, we recall the design of F-FCSR stream ciphers and the LFSRization attack. Finally, in Section 5, we present our cryptanalysis of Q-SIFR and compare the hardware performances of the latest version of FCSRs and of V-FCSRs.

1 n-adic Topology in Some Polynomial Rings

Let $p(X) \in \mathbb{Z}[X]$ be a monic polynomial of degree d. Let $\mathcal{A}$ be the algebra $\mathcal{A} = \mathbb{Z}[X]/(p(X))$. We identify the $\mathbb{Z}$-module $\mathcal{A}$ with the set of polynomials of degree less than d. Let $n > 0$ be a positive integer.

1.1 n-adic Expansion in $\mathcal{A}$

Let n be a positive integer. Let $\mod n$ and $\operatorname{div} n$ be the maps defined over $\mathbb{Z}$ by the Euclidean division: $a = (a \mod n)n + (a \operatorname{div} n)$, $0 \le (a \mod n) < n$. These maps can be naturally extended to $\mathbb{Z}[X]$: the two maps $\mod_n$ and div_n are defined by:

If $q(X) = \sum_{i=0}^r q_i X^i$, then $\mod_n(q(X)) = \sum_{i=0}^r (q_i \mod n)X^i$ and $\operatorname{div}_n(q(X)) = \sum_{i=0}^r (q_i \operatorname{div} n)X^i$.

Clearly, $q(X) = \mod_n(q(X)) + n \operatorname{div}_n(q(X))$. Under the choice of canonical representatives of degree less than d for $\mathcal{A} = \mathbb{Z}[X]/(p(X))$, we have $\mod_n(\mathcal{A}) \subset$

A and $\mathrm{div}_n(\mathcal{A}) = \mathcal{A}$. So, the restrictions of mod_n and div_n to $\mathcal{A}$ define two applications of $\mathcal{A}$ on $\mathcal{A}$. Note that $\mathcal{S} = \mathrm{mod}_n(\mathcal{A})$ is a finite set with n^d elements, and is a set of representatives of the quotient ring $\mathcal{A}/n\mathcal{A}$.

The ideal $(n) = n\mathcal{A}$ can be used to define a distance d in $\mathcal{A}$:

$$d(q(X), r(X)) = \begin{cases} 0 & \text{if } q(X) = r(X), \\ 2^{-k} & \text{if } q(X) \neq r(X), \end{cases}$$

were k is the largest integer such that $q(X) - r(X) \in (n)^k$ (i.e. $\mathrm{mod}_{n^k}(q(X) - r(X)) = 0$).

Since the intersection of ideals $\cap_{i \in \mathbb{N}^*}((n)^i)$ is zero in $\mathcal{A}$, this distance is a distance in the mathematical sense of this word. In particular, it induces a topology on $\mathcal{A}$, the so-called n-adic topology.

Moreover, as usual, it is possible to define n-adic expansion of an element $q(X) \in \mathcal{A}$: this is the sequence $\mathrm{seq}_n(q(X)) = (s_i(X))_{i \in \mathbb{N}}$ where $s_i(X) = \mathrm{mod}_n(\mathrm{div}_{n^i}(q(X)))$. For the n-adic topology, the series associated to the n-adic expansion of $q(X) \in \mathcal{A}$ converges to $q(X) : q(X) = \sum_{i=0}^{\infty} s_i(X)n^i$.

Finally, we can define $\mathcal{A}_n$ as the completion of $\mathcal{A}$ for the n-adic topology. It is the set of formal series $\left\{\sum_{i=0}^{\infty} s_i(X)n^i \mid s_i(X) \in \mathcal{S}\right\}$. Note that, if $\mathbb{Z}_n$ denotes the set of n-adic numbers, an element $s(X)$ of $\mathcal{A}_n$ can be written in a unique way $s(X) = \sum_{j=0}^{d-1} z_i X^i$, $z_i \in \mathbb{Z}_n$. In particular, $\mathbb{Z}_n^d$ is isomorphic to $\mathcal{A}_n$ as a $\mathbb{Z}$-module, but multiplications of elements of $\mathcal{A}_n$ must be computed modulo $p(X)$.

1.2   Rational Expansions in $\mathcal{A}_n$

Lemma 1. *An element $s(X) = \sum_{i=0}^{\infty} s_i(X)n^i \in \mathcal{A}_n$ is invertible if and only if $s_0(X)$ is invertible modulo $p(X)$.*

Proof. Suppose first that $s_0(X)$ is invertible modulo $p(X)$. Let $a(X) \in \mathcal{A}$ such that $s_0(X)a(X) = 1$ in $\mathcal{A}$. Set $s'(X) = a(X)s(X) = 1 - nt(X)$, $t(X) \in \mathcal{A}_n$. Clearly, $s'(X)$ is invertible and $s'^{-1} = \sum_{i=0}^{\infty} t(X)^i n^i \in \mathcal{A}_n$. So, $s(X)$ is invertible in $\mathcal{A}_n$.

Conversely, if $s(X)$ is invertible in $\mathcal{A}_n$, set $t(X) = s(X)^{-1}$. It is easy to verify, with obvious notations, that $s_0(X)t_0(X) = 1$ in $\mathcal{A}$; i.e. $s_0(X)$ is invertible modulo $p(X)$.

Definition 1. *We denote by $\mathcal{Q}$ the ring of fractions of $\mathcal{A}$, i.e. the elements of $\mathcal{A}_n$ of the form u/v, u, $v \in \mathcal{A}$, v invertible in $\mathcal{A}_n$.*

As a direct consequence of Lemma 1, we obtain the following result:

Proposition 1. *$\mathcal{Q}$ is the set of elements of the form $u/(1 - nv)$, u, $v \in \mathcal{A}$.*

The periodic elements of $\mathcal{A}_n$ are in $\mathcal{Q}$. Indeed, if $s(X) = r(X) \sum_{i=0}^{\infty} n^{Ti}$, $r(X) = \sum_{j=0}^{T-1} s_j(X)n^j$ for some $T \in \mathbb{N}$, then $s(X) = r(X)/(1 - n^T)$. In fact, one can show that $\mathcal{Q}$ is exactly the set of ultimately periodic elements of $\mathcal{A}_n$. It follows from the fact that $\mathcal{A}/n\mathcal{A}$ is finite.

2   Algebraic Finite State Machine and Vectorial FCSRs

2.1   Algebraic Automata in $\mathcal{A}$

Following the notion of Algebraic Feedback Shift Registers introduced in [11] and the matrix approach for the theory of FCSRs presented in [3], we can generalize these two approaches to define a matrix theory of algebraic automata in order to generate the rational elements of $\mathcal{A}_n$, i.e. the elements of $\mathcal{Q}$.

Remember that $\mathcal{S} = \mathrm{mod}_n(\mathcal{A})$ is the finite set of polynomials of degree strictly less than $d = \deg(p(X))$ and with coefficients in $\{0, ..., n-1\}$.

Definition 2. *An algebraic finite state machine (*AFSM*) of size $k \in \mathbb{N}^*$ is an automaton composed of*

- *A set of states $(m, c) \in \mathcal{S}^k \times \mathcal{A}^k$*
- *A transition function defined by a $k \times k$ matrix T with coefficients in $\mathcal{S}$. If the automaton is in state $(m(t), c(t))$ at time t, the transition function is given by*

$$\begin{cases} z(t+1) &= Tm(t) + c(t) \\ m(t+1) &= \mathrm{mod}_n(z(t+1)) \\ c(t+1) &= \mathrm{div}_n(z(t+1)) \end{cases} \tag{1}$$

where $\mathrm{mod}_n(z(t))$ (resp. $\mathrm{div}_n(z(t))$) is the k-tuple obtained by applying mod_n (resp. div_n) to each coordinate of $z(t)$.

To simplify notations, we omitted the indeterminate X in formulas, for example $m_0(t)$ is in fact a polynomial $(m_0(t))(X) \in \mathcal{A}$. Note that in this definition, $m(t)$ and $c(t)$ are column vectors. By abuse of notation, $m(t)$ and $c(t)$ will refer either to a column vector or a row vector, depending on the context.

By analogy with the FCSR automata explained in [3], the register $m(t) = (m_0(t), ..., m_{k-1}(t))$ is called the main register and the register $c(t) = (c_0(t), ..., c_{k-1}(t))$ is called the carry register.

By a method similar to the one developed in [3] Section 5, it is possible to show that AFSMs are finite states machines. More precisely, it is possible to determine a finite parallelogram $V \subset \mathcal{A}^k$, such that if $c(t) \in V$ then $c(t+1) \in V$.

The content of the cells of the main register is in $\mathcal{S}$. We are interested by the sequences observed in each cell of this register. These sequences are naturally interpreted as n-adic elements, i.e. as elements of $\mathcal{A}_n$.

Notations: we denote by $M(t_0) = (M_0(t_0), ..., M_{k-1}(t_0)) \in \mathcal{A}_n^k$ the k-tuple of elements of $\mathcal{A}_n$ observed from time t_0 in the main register. More precisely, $M_i(t_0) = \sum_{t \geq t_0} m_i(t) n^{t-t_0}$.

The main result of this section is the following theorem which is a straightforward generalization of Theorem 2 of [3].

Theorem 1. *If the automaton is in the state $(m(t_0), c(t_0))$ at time t_0, then*

$$M(t_0) = \frac{\mathrm{adj}(I - nT)}{\det(I - nT)} \cdot (m(t_0) + nc(t_0))$$

where $\mathrm{adj}(U)$ denotes the adjoint of a matrix U, i.e. the transpose of cofactors matrix of U, and I denotes the $k \times k$ identity matrix.

In fact, if we set $U = \frac{\mathrm{adj}(I-nT)}{\det(I-nT)}$, U is the inverse of $(I - nT)$. However, the matrix $\mathrm{adj}(I - nT)$ has its coefficients in $\mathcal{A}$. Consequently, this theorem shows that the sequences $M_i = M_i(0)$ observed in each cell m_i of the main register of the automaton from an initial time $t = 0$ are all in $\mathcal{Q}$. More precisely, if we set $q(X) = \det(I - nT)$, they are of the form $M_i = p_i(X)/q(X)$ for some $p_i(X) \in \mathcal{A}$. Moreover, $\mathrm{mod}_n(q(X)) = 1$ and the polynomials $p_i(X)$ are explicitly given by Theorem 1.

The classical binary FCSRs correspond to the particular case $p(X) = X$ and $n = 2$. The polynomial $q(X) = \det(I - nT)$ is called the connection polynomial of the AFSM.

2.2 Galois and Fibonnaci Modes for AFSMs

In the theories of LFSRs, FCSRs and AFSRs, there are two well-known particular modes: the Fibonnaci mode, which refers to the linear recurrence sequences, and the Galois mode, which corresponds to the rational development of series.

These automata are constituted of k cells which are linked by a circular shift. This induces a "ring" structure of the matrix, i.e. $T_{i,i+1} = 1$ for $i = 1$ to $k - 1$ and $T_{k,1} = 1$. For the Fibonnaci mode, the other non-zero entries are on the last row and for the Galois mode, the other non-zero entries are on the first column.

$$
T_G = \begin{pmatrix} q_1 & 1 & & & \\ q_2 & & 1 & (0) & \\ \vdots & & (0) & & \ddots \\ q_{n-1} & & & & 1 \\ 1 & 0 & 0 & \cdots & 0 \end{pmatrix}
\qquad
T_F = \begin{pmatrix} 0 & 1 & & & \\ 0 & & 1 & (0) & \\ \vdots & & (0) & & \ddots \\ 0 & & & & 1 \\ 1 & q_2 & \cdots & q_{n-1} & q_n \end{pmatrix}
$$

2.3 Vectorial FCSR

In [15], A. Marjane and B. Allailou introduced the notion of Vectorial FCSRs (V-FCSRs). The presentation in [15] is different from the one adopted here. However, it is easy to verify that they correspond to the particular case of AFSMs over $\mathcal{A}$ with:

- $n = 2$.
- $p(X) = X^d - \sum_{i=0}^{d-1} \epsilon_i X^i$, with $\epsilon_i \in \{0, 1\}$. Moreover, the polynomial $p(X)$ must be irreducible over $GF(2)$.
- The matrix T is the companion matrix of $q(X) = \det(I - nT)$ (i.e. the automaton is in the so-called Fibonacci form, see [3] for more details).

The fact that $p(X)$ is irreducible over $GF(2)$ has absolutely no consequence on the theory and the results obtained in [15]. The particular form for $p(X)$ is motivated by the fact that coefficients of T are in $\{0, 1\} \subset \mathbb{Z}$, which leads to simple calculus and implementations.

3   Reduction of AFSM over $\mathcal{A}$ to Classical FCSRs

3.1   Arithmetic on $\mathcal{A}$

As a $\mathbb{Z}$-module, under the hypothesis that $p(X)$ is unitary, $\mathcal{A} = \mathbb{Z}[X]/(p(X))$ is isomorphic to $\mathbb{Z}^d$ by the identification of an element $r(X) = \sum_{i=0}^{d-1} r_i X^i$, $r_i \in \mathbb{Z}$ to the list $r = (r_0, ..., r_{d-1}) \in \mathbb{Z}^d$ of its coefficients. Moreover, the multiplication in $\mathcal{A}$ by a fixed polynomial $r(X)$ is $\mathbb{Z}$-linear, so it admits a $d \times d$ matrix representation.

For example, the multiplication by X in $\mathcal{A}$ corresponds to the companion matrix M_X of the polynomial $p(X)$ in $\mathbb{Z}^d$.

$$M_X = \begin{pmatrix} 0 & 1 & 0 & 0 & \cdots\cdots & & 0 \\ \vdots & & 0 & 1 & 0 & \ddots\ddots & \vdots \\ \vdots & & \ddots & \ddots & \ddots & \ddots & \vdots \\ \vdots & & \ddots & \ddots & \ddots & \ddots & 0 \\ 0 & & \cdots & \cdots\cdots & 0 & 0 & 1 \\ -p_0 & -p_1 & \cdots & \cdots\cdots\cdots & & & -p_{d-1} \end{pmatrix}$$

More generally, if $r(X) = \sum_{i=0}^{d-1} r_i X^i$ in $\mathcal{A}$, the corresponding multiplication matrix in $\mathbb{Z}^d$ is $M_{r(X)} = \sum_{i=0}^{d-1} r_i M_X^i$.

The norm of a polynomial $r(X) \in \mathcal{A}$ is the determinant of the matrix $M_{r(X)}$: $\mathrm{norm}(r(X)) = \det(M_{r(X)})$.

3.2   Practical Implementation of AFSM over $\mathcal{A}$

The simplest way to implement an AFSM over $\mathcal{A}$ is to use the representation with the d-tuples of elements of $\mathbb{Z}$ described in the previous section.

So, a k-tuple of elements of $\mathcal{A}$ becomes a k-tuple of d-tuple over $\mathbb{Z}$, which can be identified to an element of $\mathbb{Z}^{kd}$. To any $k \times k$ matrix T over $\mathcal{A}$, we associate a $kd \times kd$ matrix $\mathcal{T}$ over $\mathbb{Z}$ by replacing any entry $r(X)$ of T by the matrix $M_{r(X)}$.

Using this representation of AFSM, Definition 2 becomes

Definition 3. *Let $\mathcal{S}_n$ be the set $\{0, ..., n-1\}$. An algebraic finite state machine (AFSM) of size $k \in \mathbb{N}^*$ is an automaton composed of*

- *A set of states $(m, c) \in \mathcal{S}_n^{kd} \times \mathbb{Z}^{kd}$*
- *A transition function defined by a $k \times k$ matrix T with coefficients in $\mathcal{S}$. Let $\mathcal{T}$ be its corresponding $kd \times kd$ matrix over $\mathbb{Z}$. If the automaton is in the state $(m(t), c(t))$ at time t the transition function is given by*

$$\begin{cases} z(t+1) & = \mathcal{T}m(t) + c(t) \\ m(t+1) = z(t+1) \bmod n \\ c(t+1) & = z(t+1) \mathrm{\ div\ } n \end{cases} \qquad (2)$$

where $z(t) \bmod n$ (resp. $z(t)(\mathrm{div}\ n)$) is the kd-tuple obtained by applying mod n (resp. (div n)) to each coordinate of $z(t)$.

3.3 Reduction of AFSMs to FCSRs

Definition 3 is a particular case of the well known matrix approach of FCSRs presented in [3]. As explained in [3], the generalization from $n = 2$ to any integer n is immediate.

So AFSMs of size k are nothing else than FCSRs of size kd, with the additional constraint that the transition matrix $\mathcal{T}$ will be the image of a $k \times k$ matrix over $\mathcal{A}$.

As explained in [17], Theorem 1, if $\mathcal{T}$ is a matrix over $\mathbb{Z}$ which is the image of a matrix T over $\mathcal{A}$, the link between their determinants is the following:

$$\det(\mathcal{T}) = \text{norm}(\det(T)).$$

As a consequence, if the connection polynomial of an AFSM is $q(X) = \det(I - nT)$, the connection integer $q' = \det(I - n\mathcal{T})$ of the associated FCSR is $q' = \text{norm}(q(X)) = \det(M_{q(X)})$.

If one can observe a rational series $s(X) = p(X)/q(X)$ in a cell of an AFSM with connection polynomial $q(X)$, then the series $s(X)$ can be decomposed as $s(X) = \sum_{j=0}^{d-1} s^{(j)} X^j$, where $s^{(j)}$ is a rational element of $\mathbb{Z}_n$ (the set of n-adic numbers) of the form $s^{(j)} = p^{(j)}/q'$.

4 Pseudo-Random Generators Based on FCSRs

4.1 The Family of F-FCSRs Stream Ciphers

The first F-FCSR stream cipher was presented in 2005 [5]. Other versions have been proposed later [4, 6–8, 10]. They are based on the iterations of an FCSR automaton, which is filtered by a linear function in order to produce the pseudo-random sequence. The resistance against most attacks is provided by the intrinsic non-linearity of the transition function which is quadratic. The resistance to dedicated attacks (i.e. related to 2-adic properties) is insured by the linear filter, which is also the best function to break the correlation attacks. Figure 1 gives a small example of a F-FCSR stream cipher based on a FCSR in Galois mode.

F-FCSR versions until 2008 were based on FCSRs in Galois mode. This mode corresponds to the automata with transition matrices of the form T_G given in Section 2.2. The version F-FCSR-H v2 [7] was one of the height finalists of the eSTREAM European project [16]. Moreover, it was one of the four hardware oriented stream ciphers proposed in the portfolio at the end of the project in September 2008. Unfortunately, it was broken 3 months later [13]. We describe this attack in more details in Subsection 4.2.

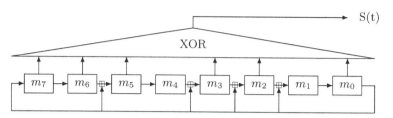

Fig. 1. A toy example of a filtered FCSR (F-FCSR) in Galois mode

The transition matrix of this FCSR in Galois mode is

$$
T_G = \begin{pmatrix}
0 & 1 & 0 & 0 & 0 & 0 & 0 & 0 \\
1 & 0 & 1 & 0 & 0 & 0 & 0 & 0 \\
1 & 0 & 0 & 1 & 0 & 0 & 0 & 0 \\
1 & 0 & 0 & 0 & 1 & 0 & 0 & 0 \\
0 & 0 & 0 & 0 & 0 & 1 & 0 & 0 \\
1 & 0 & 0 & 0 & 0 & 0 & 1 & 0 \\
1 & 0 & 0 & 0 & 0 & 0 & 0 & 1 \\
1 & 0 & 0 & 0 & 0 & 0 & 0 & 0
\end{pmatrix}
$$

4.2 Weakness of F-FCSR Stream Ciphers Based on FCSRs in Galois Mode

In Galois mode, there is a single cell m_0 that controls all the feedbacks. If this cell takes the value 0 during r clocks, most of the carry cells become null after few iterations. If all the carry cells and the feedback cell are null, then the transition function becomes linear. It is not possible that all the carry cells are simultaneously null (see [9]), however they become null except one with a non negligible probability. The cryptanalysis presented in [13] is based on this observation.

4.3 FCSRs in Diversified Mode

Following this attack, a new version of FCSRs was presented in [8]. This is the first introduction of a matrix representation of FCSRs as presented in Section 2. A survey of this approach can be found in [3]. The main idea is to use matrices with more random structure than those of Galois or Fibonnaci modes. For new versions of F-FCSR [8, 10], the authors propose to use matrices with "ring" structure: they have a 1 in the upper-diagonal and at the first position of the last row, and at most two 1 by row and by column. In this way, it is impossible to find two carry cells depending on a same feedback cell. This fact completely discards the attack presented in [13], and new versions remain unbroken since 2009. Figure 2 presents a small example of the new FCSR version.

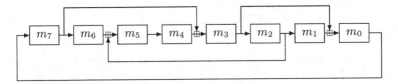

Fig. 2. A toy example of FCSR in ring (or diversified) mode

The transition matrix of this FCSR is

$$T_R = \begin{pmatrix} 0\,1\,0\,1\,0\,0\,0\,0 \\ 0\,0\,1\,0\,0\,0\,0\,0 \\ 0\,0\,0\,1\,0\,0\,0\,0 \\ 0\,0\,0\,0\,1\,0\,0\,1 \\ 0\,0\,0\,0\,0\,1\,0\,0 \\ 0\,0\,1\,0\,0\,0\,1\,0 \\ 0\,0\,0\,0\,0\,0\,0\,1 \\ 1\,0\,0\,0\,0\,0\,0\,0 \end{pmatrix}$$

5   Cryptanalysis of Pseudo Random Generators Based on Vectorial FCSRs

In [1] the authors propose to use a Vectorial FCSR in Galois mode to replace the FCSR in Galois mode of F-FCSR v2 broken in [13]. In this section, we describe this proposal. We then apply the reduction to FCSR automata presented in Section 3. Using this new representation, we show that the stream cipher Q-SIFR based on V-FCSRs in Galois mode remains sensitive to the attack of [13], contrary to the new version of F-FCSR, F-FCSR-H v3 proposed in [8]. We finally compare the hardware performances of Q-SIFR and of F-FCSR-16 v3.

5.1   Description of Q-SIFR

The stream cipher Q-SIFR presented in [1] is based on a V-FCSR. As an AFSM, the parameters of the automaton are: $n = 2$, $p(X) = X^2 - X - 1$, so $d = 2$, and the matrix M_X is

$$M_X = \begin{pmatrix} 0 & 1 \\ 1 & 1 \end{pmatrix}$$

The size of the matrix over $\mathcal{A}$ is $k = 160$. Let $q(X) = u + vX$ with

$u = -19935245913182750153280416113442150364601400879633$ and

$v = -19935245913182750153280416113442150364601400878600$.

The automaton is in Galois form. More precisely, set $d(X) = (1 - q(X))/2$ and $d(X) = \sum_{i=0}^{159} d_i(X)2^i$, $d_i(X) \in \mathcal{S} = \{0, 1, X, X+1\}$. The columns of the transition matrix T of the VFCSR are

- $T^t(1) = (d_{159}(X), d_{158}(X), ..., d_0(X))$,
- $T^t(2) = (10...0)$, $T^t(3) = (010...0)$, ...
- $T^t(160) = (00...010)$.

Since this matrix is the transpose of the companion matrix of $q(X)$, one can verify that $\det(I - 2T) = q(X)$, so $q(X)$ is the connection polynomial of the AFSM.

A practical implementation of such a V-FCSR needs the construction of the corresponding 320×320 matrix $\mathcal{T}$ with coefficients in $\mathbb{Z}$.

Since T is in Galois form (i.e. corresponds to a companion matrix), the first two columns of $\mathcal{T}$ are obtained from the first column of T by replacing each polynomial $d_i(X)$ by the 2×2 matrix $M_{d_i(X)}$. The following columns are $\mathcal{T}^t(3) = (1000..00)$, $\mathcal{T}^t(4) = (0100...00)$, $\mathcal{T}^t(319) = (00...01000)$ and $\mathcal{T}^t(320) = (00...00100)$.

The values of $T^t(1)$, $\mathcal{T}^t(1)$ and $\mathcal{T}^t(2)$ are given in Appendix. A first remark is the fact that the second column contains not only 0 and 1, but also a lot of 2. Consequently, the potential values of carry cells are not only 0 or 1 but also 2 or 3.

One can verify that

$$\tilde{q} = \det(I - 2\mathcal{T}) = 39741402961906954206160047535539796042005214340820825272689327902761723128526374726419918065389490$$

corresponds to the value given in [15].

At each iteration, a linear filter extracts 16 bits of stream.

5.2   LFSRization Attack against Q-SIFR

We will describe in this section how we could apply the LFSRization technique described in [13] against Q-SIFR.

First of all, as the transition matrix $\mathcal{T}$ previously described is in its Galois form and because $d = 2$, all the carry cells of the practical implementation of Q-SIFR only depend on the two least significant bits of the main register $m_0(t)$ and $m_1(t)$. In constrate with the classical FCSRs in Galois mode and due to the matrix form, we must take into account three possible kinds of carries as described in [1]:

- **Classical Carries:** As in the classical FCSR case in Galois mode, V-FCSRs could contain classical carries corresponding with the case where the two values of the first two columns of $\mathcal{T}$ are of the form $(0, 1)$ or $(1, 0)$, i.e. $(\mathcal{T}^t(i, 1), \mathcal{T}^t(i, 2) = (0, 1)$ or $(1, 0))$. In this case, the corresponding updating equations for the bits $m_i(t)$ and $c_i(t)$ are:

$$m_i(t + 1) = m_{i+1}(t) + c_i(t) + m_0(t) \mod 2$$
$$\text{or } m_{i+1}(t) + c_i(t) + m_1(t) \mod 2$$
$$c_i(t + 1) = m_{i+1}(t) + c_i(t) + m_0(t) \operatorname{div} 2$$
$$\text{or } m_{i+1}(t) + c_i(t) + m_1(t) \operatorname{div} 2$$

according the feedback bit and with $m_i(t) \in \{0,1\}$ and $c_i(t) \in \{0,1\}$. There are 2 carries of this form for m_0 and 3 carries of this form for m_1 when looking at $\mathcal{T}$. Thus, we deduce the transition probability matrix between the time t and $t+1$ for the content of $c_i(t)$ seen as a Markov chain under the conditions that during k consecutive clocks, the $m_{i+1}(t)$ values are uniformly distributed and that $m_0(t)$ and $m_1(t)$ are null (this happens with probability $2^{-2 \cdot k}$). The transition probability matrix is then for the two possible values of $c_i(t)$:

$$\mathcal{M}_1 = \begin{pmatrix} 1 & 0 \\ 1/2 & 1/2 \end{pmatrix}$$

- **Double Carries:** If the first two columns of $\mathcal{T}$ are of the form $(1,1)$ i.e. $(\mathcal{T}^t(i,1), \mathcal{T}^t(i,2) = (1,1))$, then the updating equations for the bits $m_i(t)$ and $c_i(t)$ are:

$$m_i(t+1) = m_{i+1}(t) + c_i(t) + m_0(t) + m_1(t) \quad \bmod 2$$
$$c_i(t+1) = m_{i+1}(t) + c_i(t) + m_0(t) + m_1(t) \, \mathrm{div} \, 2$$

with $m_i(t) \in \{0,1\}$ and $c_i(t) \in \{0,1,2\}$. In the same way, we deduce the transition probability matrix of $c_i(t)$ between t and $t+1$ always considering that $m_0(t)$ and $m_1(t)$ are null. The transition probability matrix is:

$$\mathcal{M}_2 = \begin{pmatrix} 1 & 0 & 0 \\ 1/2 & 1/2 & 0 \\ 0 & 1 & 0 \end{pmatrix}$$

There are 82 carries of this form according to $\mathcal{T}$.

- **Triple Carries:** If the first two columns of $\mathcal{T}$ are of the form $(1,2)$ i.e. $(\mathcal{T}^t(i,1), \mathcal{T}^t(i,2) = (1,2))$, then the updating equations for the bits $m_i(t)$ and $c_i(t)$ are:

$$m_i(t+1) = m_{i+1}(t) + c_i(t) + m_0(t) + 2m_1(t) \quad \bmod 2$$
$$c_i(t+1) = m_{i+1}(t) + c_i(t) + m_0(t) + 2m_1(t) \, \mathrm{div} \, 2$$

with $m_i(t) \in \{0,1\}$ and $c_i(t) \in \{0,1,2,3\}$. In the same way, we deduce the transition probability matrix of $c_i(t)$ between t and $t+1$ always considering that $m_0(t)$ and $m_1(t)$ are null. The transition probability matrix is:

$$\mathcal{M}_3 = \begin{pmatrix} 1 & 0 & 0 & 0 \\ 1/2 & 1/2 & 0 & 0 \\ 0 & 1 & 0 & 0 \\ 0 & 1/2 & 1/2 & 0 \end{pmatrix}$$

There are 81 carries of this form according to $\mathcal{T}$.

Now, we are interested in computing the required number of clocks that will cancel the content of the carry register under the hypothesis that $m_0(t)$ and $m_1(t)$ are both null during k consecutive clocks. As noticed in [9], we could not completely cancel the content of the carry register but we could reach a carry register of the form $c^t = (001100...00) \in \mathbb{Z}^{320}$ that will make the content of the main register affine.

To do so, we need to detail the behavior of each type of carry. For the classical carries, we can see that when the two feedback bits $m_0(t)$ and $m_1(t)$ are 0, if the carry bit $c_i(t)$ is 0, it will remain 0 and if the carry bit is 1, it becomes 0 with probability $1/2$. Thus, if $m_0(t)$ and $m_1(t)$ are null, we expect that the general weight of the 5 classical carries will be divided by two after one clock and so on. So, the required number of clocks to cancel the 5 classical carries is $\log_2(5) \approx 3$.

For the double carries, the same reasoning holds: when the two feedback bits $m_0(t)$ and $m_1(t)$ are 0, the value of the carry is either divided by two or the same (only for the case $c_i(t) = 1$ with probability $1/2$). So, the required number of clocks to cancel the 82 classical carries is two times the one for a classical carry $\log_2(2 \times 82) \approx 8$. For the triple carries, when the two feedback bits $m_0(t)$ and $m_1(t)$ are 0, the value of the carry is divided by two or the same. So, the required number of clocks to cancel the 81 classical carries is three times the one required for the classical carry case, so we need $\log_2(3 \cdot 81) \approx 8$ clocks to cancel the triple carries.

Thus, the content of the carry register will be $(001100...00)$ after $\max(3, 8, 8) = 8$ iterations. This happens under the condition "the two feedback bits $m_0(t)$ and $m_1(t)$ are 0" which happens with a probability of $2^{-2\cdot8} = 2^{-16}$.

To complete the LFSRization attack, we need to maintain during a certain number of clocks the state $(001100...00)$ in the carry register to be able to solve the system generated by the outputs. As Q-SIFR linearly outputs 16 bits of the main register at each clock and as the number of unknowns is the size of the main register (i.e. equal to 320 bits), we obtain a linear system of 320 equations and 320 unknowns as soon as 19 iterations under the condition "the two feedback bits $m_0(t)$ and $m_1(t)$ are 0" are performed. This happens with probability $2^{-2*19} = 2^{-38}$. The resolution of the system requires a time complexity of about $(320)^{2.807}$ operations using a smart Gaussian elimination. So the attack requires the observation of $2^{38+16} = 2^{54}$ outputs and has a complexity of $2^{54} \times (320)^{2.807} \approx 2^{77.35}$ operations.

In [13], Hell and Johansson proposed an other method that stays valid in our context to improve the complexity of the attack. This method relies on the structure of the extracting filter. Indeed, the filter has a particular form that allows subsystems resolution. More precisely, at each clock, the 16 output bits are computed in the following way:

$$Z(t) = (z_0(t), z_1(t), \cdots, z_{15}(t)) = (m_0(t) \oplus m_{16}(t) \oplus m_{48}(t) \oplus \cdots \oplus m_{304}(t),$$
$$m_1(t) \oplus m_{49}(t) \oplus \cdots, \cdots, m_{15}(t) \oplus \cdots)$$

Thus, we could pack the equations according to their byte structures: there are 20 equations that only include the particular 20 bits $m_0, m_{16}, \cdots, m_{312}$, there are 20 equations that only include the particular 20 bits $m_1, m_{17}, \cdots, m_{313}$, etc. So, we could build subsytems for those equations from the values $z_0(t), z_7(t+1), \cdots, z_5(t+19)$ and from $z_1(t), z_0(t+1), \cdots, z_6(t+19)$, etc.

Then the system resolution step could be simplified into the resolution of 16 subsystems of 20 linear equations leading to a complexity of the resolution step of $16 \times (20^{2.807}) \approx 2^{16.13}$ operations instead of $(320)^{2.807} \approx 2^{23.35}$ operations. So, the overall complexity of the attack is then $2^{70.13}$ operations.

In [13], the authors proposed a second improvement of their attack: they remark that only 17 clocks under the condition "the two feedback bits $m_0(t)$ and $m_1(t)$ are 0" are required instead of 19. This comes from the fact that at $t+18$, if the feedback bits are not zeros, the content of the carry register at $t+19$ will no more be $(001100...00)$ but something unknown. However, $m(t+19)$ is only required to compute $z(t+19)$: the knowledge of the carry register content at $t+19$ is not necessary except for computing $m(t+20)$ that is useless in our case. This technique remains valid in our case, so we only need to maintain the content of the carry register equal to $(001100...00)$ during 17 clocks instead of 19. So, considering this last optimization, the complexity of the attack becomes $2^{2.8+17.2} \times 16 \cdot (20)^{2.807} \approx 2^{66.13}$ operations and we need to observe 2^{50} outputs.

This attack allows us to recover the internal state of the V-FCSR at a particular time, say t. Using the method proposed in [13], it is also possible from the state at time t denoted by $p(t)$ to recover the key before the key/IV set up especially if the IV is equal to 0. The idea is first to recover the initial state $p(0)$ at the end of the key/IV setup using the relation $p(t) = p(0) \cdot 2^t \mod \tilde{q}$ and then to solve a system of equations depending on the key bits. We think that this method also works against Q-SIFR but as the key/IV schedule of Q-SIFR is not completely specified in [1], we could not completely precise the system to solve.

As we could not simulate the whole attack due to its complexity, we have tested the veracity of the LFSRization of the carry register. We have observed that in average, with a frequency of $2^{-19.93}$, the carry register remains equal to $(001100...00)$ during 11 consecutive clocks. Clearly, it seems that the LFSRization attack proposed here has a sufficient probability to be mounted.

Consequently, contrary to the Q-SIFR authors claim, we have shown that the LFSRization attack still works on Q-SIFR even if this attack is less efficient and less impressive than in the case of F-FCSR-H v2 [7].

5.3 Comparing Hardware Implementations of Q-SIFR and of F-FCSR-16 v3

In [8], the authors repaired not only F-FCSR-H (leading to F-FCSR-H v3) but also F-FCSR-16 v3 which is a version of F-FCSR that outputs 16 bits at each iteration. In this subsection, we are going to compare the hardware performances of Q-SIFR and of this particular version of F-FCSR that have comparable throughput.

The size of the matrix $\mathcal{T}$ of Q-SIFR is 320. The numbers of carry adders is 168. Among these carry cells, 5 are classical adders, 82 are double adders and 81 are triple adders. So, it is necessary to implement $5 + 2 \cdot 82 + 2 \cdot 81 = 331$ simple adders. Leading to a total of 651 binary cells to implement the V-FCSR of Q-SIFR.

In comparison, the size of the ring matrix of F-FCSR-16 v3 is 256 with 130 carry cells. The full implementation of F-FCSR-16 v3 requires 386 binary cells.

Moreover, and when looking at other important hardware parameters such as the critical path (i.e. the maximal number of gates crossed by a bit) and the fan-out (i.e. the maximal number of connections at the output of a cell), F-FCSR-16 v3 is more efficient than Q-SIFR. Indeed, the length of the critical path is 3 for Q-SIFR whereas it is equal to 2 for F-FCSR-16 v3 and the fan-out of QSIFR is 165 whereas it is equal to 2 for F-FCSR-16 v3.

So, in summary, F-FCSR-16 is not only more resistant to the LFSRization attack than Q-SIFR but it has also a better hardware implementation.

6   Conclusion

In this paper, we have shown that the underlying V-FCSR used in the stream cipher proposal Q-SIFR is nothing else than a particular FCSR used in Galois mode. As already noticed in [13], Galois mode could not be used for stream cipher design because of LFSRization attacks that stay efficient against Q-SIFR. Moreover, practical hardware implementations of F-FCSR using V-FCSRs are less efficient than the ones using a FCSR in ring mode.

So, we do not find in this paper an advantage to emphasize V-FCSRs rather than ring FCSRs even when V-FCSRs are used in ring mode.

References

1. Allailou, B., Marjane, A., Mokrane, A.: Design of a Novel Pseudo-Random Generator Based on Vectorial FCSRs. In: Chung, Y., Yung, M. (eds.) WISA 2010. LNCS, vol. 6513, pp. 76–91. Springer, Heidelberg (2011)
2. Arnault, F., Berger, T.P., Pousse, B., Minier, M.: Revisiting LFSRs for cryptographic applications. IEEE Transactions on Information Theory 57(12), 8095–8113 (2011)
3. Arnault, F., Berger, T.P., Pousse, B.: A matrix approach for FCSR automata. Cryptography and Communications 3(2), 109–139 (2011)
4. Arnault, F., Berger, T.P., Lauradoux, C.: The F-FCSR primitive specification and supporting documentation. In: ECRYPT - Network of Excellence in Cryptology, Call for stream Cipher Primitives (2005), http://www.ecrypt.eu.org/stream/
5. Arnault, F., Berger, T.P.: F-FCSR: Design of a New Class of Stream Ciphers. In: Gilbert, H., Handschuh, H. (eds.) FSE 2005. LNCS, vol. 3557, pp. 83–97. Springer, Heidelberg (2005)
6. Arnault, F., Berger, T.P., Lauradoux, C., Minier, M.: X-FCSR – A New Software Oriented Stream Cipher Based Upon FCSRs. In: Srinathan, K., Rangan, C.P., Yung, M. (eds.) INDOCRYPT 2007. LNCS, vol. 4859, pp. 341–350. Springer, Heidelberg (2007)
7. Arnault, F., Berger, T., Lauradoux, C.: F-FCSR Stream Ciphers. In: Robshaw, M., Billet, O. (eds.) New Stream Cipher Designs. LNCS, vol. 4986, pp. 170–178. Springer, Heidelberg (2008)

8. Arnault, F., Berger, T., Lauradoux, C., Minier, M., Pousse, B.: A New Approach for FCSRs. In: Jacobson Jr., M.J., Rijmen, V., Safavi-Naini, R. (eds.) SAC 2009. LNCS, vol. 5867, pp. 433–448. Springer, Heidelberg (2009)
9. Arnault, F., Berger, T.P., Minier, M.: Some Results on FCSR Automata With Applications to the Security of FCSR-Based Pseudorandom Generators. IEEE Transactions on Information Theory 54(2), 836–840 (2008)
10. Berger, T.P., Minier, M., Pousse, B.: Software Oriented Stream Ciphers Based upon FCSRs in Diversified Mode. In: Roy, B., Sendrier, N. (eds.) INDOCRYPT 2009. LNCS, vol. 5922, pp. 119–135. Springer, Heidelberg (2009)
11. Klapper, A., Xu, J.: Algebraic Feedback Shift Registers. Theoretical Computer Science 226, 61–93 (1999)
12. Goresky, M., Klapper, A.: Fibonacci and Galois representations of feedback-with-carry shift registers. IEEE Transactions on Information Theory 48(11), 2826–2836 (2002)
13. Hell, M., Johansson, T.: Breaking the F-FCSR-H Stream Cipher in Real Time. In: Pieprzyk, J. (ed.) ASIACRYPT 2008. LNCS, vol. 5350, pp. 557–569. Springer, Heidelberg (2008)
14. Klapper, A., Goresky, M.: 2-Adic Shift Registers. In: Anderson, R. (ed.) FSE 1993. LNCS, vol. 809, pp. 174–178. Springer, Heidelberg (1994)
15. Marjane, A., Allailou, B.: Vectorial Conception of FCSR. In: Carlet, C., Pott, A. (eds.) SETA 2010. LNCS, vol. 6338, pp. 240–252. Springer, Heidelberg (2010)
16. eSTREAM, ECRYPT Stream Cipher Project,
   http://www.ecrypt.eu.org/stream
17. Silvester, J.R.: Determinants of Blocks Matrices. The Mathematical Gazette 84(501), 460–467 (2000)

Appendix

Value of the First Column of the Matrix T over $\mathcal{A}$:
$T^t(1) = (0, x+1, 1, x+1, x, 0, 1, 0, x+1, 0, x+1, 0, 0, x+1, 0, 0, 0, 0, x+1, 0, x+1, 0, x+1, x+1, x+1, 0, x+1, x+1, x+1, x+1, 0, 0, 0, 0, 0, 0, x+1, 0, 0, x+1, x+1, 0, x+1, 0, x+1, 0, x+1, x+1, 0, x+1, x+1, 0, 0, 0, x+1, 0, x+1, x+1, x+1, x+1, 0, x+1, 0, x+1, 0, x+1, 0, x+1, 0, 0, 0, x+1, 0, 0, x+1, x+1, x+1, 0, x+1, x+1, x+1, x+1, 0, 0, 0, x+1, 0, 0, 0, x+1, x+1, 0, 0, 0, 0, x+1, x+1, 0, x+1, 0, 0, 0, x+1, x+1, x+1, x+1, x+1, x+1, 0, 0, x+1, x+1, 0, 0, 0, 0, x+1, x+1, 0, 0, x+1, x+1, 0, 0, x+1, 0, 0, x+1, 0, 0, x+1, x+1, x+1, x+1, x+1, x+1, x+1, x+1, x+1, 0, x+1, x+1, 0, x+1, 0, 0, 0, 0, x+1, x+1, 0, 0, x+1, 0, x+1, x+1, x+1, 0, x+1, 0, x+1)$

Value of the Columns 1 and 2 of the Corresponding Matrix $\mathcal{T}$ over $\mathbb{Z}$:
$\mathcal{T}^t(1) =$
$(0, 0, 1, 1, 1, 0, 1, 1, 0, 1, 0, 0, 1, 0, 0, 0, 1, 1, 0, 0, 1, 1, 0, 0, 0, 0, 1, 1, 0, 0, 0,$
$0, 0, 0, 0, 0, 1, 1, 0, 0, 1, 1, 0, 0, 1, 1, 1, 1, 1, 1, 0, 0, 1, 1, 1, 1, 1, 1, 1, 1, 0, 0, 0, 0, 0, 0,$
$0, 0, 0, 0, 0, 0, 1, 1, 0, 0, 0, 0, 1, 1, 1, 1, 0, 0, 1, 1, 0, 0, 1, 1, 0, 0, 1, 1, 1, 1, 0, 0, 1, 1, 1,$
$1, 0, 0, 0, 0, 0, 0, 1, 1, 0, 0, 1, 1, 1, 1, 1, 1, 1, 1, 0, 0, 1, 1, 0, 0, 1, 1, 0, 0, 1, 1, 0, 0, 1, 1,$
$0, 0, 0, 0, 0, 0, 1, 1, 0, 0, 0, 0, 1, 1, 1, 1, 1, 1, 0, 0, 1, 1, 1, 1, 1, 1, 1, 1, 0, 0, 0, 0, 0, 0, 1,$
$1, 0, 0, 0, 0, 0, 0, 1, 1, 1, 1, 0, 0, 0, 0, 0, 0, 0, 0, 1, 1, 1, 1, 0, 0, 1, 1, 0, 0, 0, 0, 0, 0, 1, 1,$
$1, 1, 1, 1, 1, 1, 1, 1, 1, 1, 1, 0, 0, 0, 0, 1, 1, 1, 1, 0, 0, 0, 0, 0, 0, 0, 0, 1, 1, 1, 1, 0, 0, 0, 0,$
$0, 1, 1, 1, 1, 0, 0, 0, 0, 1, 1, 0, 0, 0, 0, 1, 1, 1, 1, 1, 1, 1, 1, 1, 1, 1, 1, 1, 1, 1, 1, 1, 1, 0, 0,$
$1, 1, 1, 1, 1, 1, 0, 0, 1, 1, 0, 0, 0, 0, 0, 0, 0, 1, 1, 1, 1, 0, 0, 0, 0, 1, 1, 0, 0, 1, 1, 1, 1, 1, 1,$
$1, 0, 0, 1, 1, 0, 0, 1, 1)$

$\mathcal{T}^t(2) =$
$(0, 0, 1, 2, 0, 1, 1, 2, 1, 1, 0, 0, 0, 1, 0, 0, 1, 2, 0, 0, 1, 2, 0, 0, 0, 0, 1, 2, 0, 0, 0, 0, 0, 0,$
$0, 0, 1, 2, 0, 0, 1, 2, 0, 0, 1, 2, 1, 2, 1, 2, 0, 0, 1, 2, 1, 2, 1, 2, 1, 2, 0, 0, 0, 0, 0, 0, 0, 0, 0, 0,$
$0, 0, 1, 2, 0, 0, 0, 0, 1, 2, 1, 2, 0, 0, 1, 2, 0, 0, 1, 2, 0, 0, 1, 2, 1, 2, 0, 0, 1, 2, 1, 2, 0, 0, 0, 0,$
$0, 0, 1, 2, 0, 0, 1, 2, 1, 2, 1, 2, 1, 2, 0, 0, 1, 2, 0, 0, 1, 2, 0, 0, 1, 2, 0, 0, 1, 2, 0, 0, 0, 0, 0, 0,$
$1, 2, 0, 0, 0, 0, 1, 2, 1, 2, 1, 2, 0, 0, 1, 2, 1, 2, 1, 2, 1, 2, 0, 0, 0, 0, 0, 0, 0, 1, 2, 0, 0, 0, 0, 0, 0,$
$1, 2, 1, 2, 0, 0, 0, 0, 0, 0, 0, 0, 1, 2, 1, 2, 0, 0, 1, 2, 0, 0, 0, 0, 0, 0, 1, 2, 1, 2, 1, 2, 1, 2, 1, 2,$
$1, 2, 1, 2, 0, 0, 0, 0, 1, 2, 1, 2, 0, 0, 0, 0, 0, 0, 0, 0, 1, 2, 1, 2, 0, 0, 0, 0, 1, 2, 1, 2, 0, 0, 0, 0,$
$1, 2, 0, 0, 0, 0, 1, 2, 1, 2, 1, 2, 1, 2, 1, 2, 1, 2, 1, 2, 1, 2, 0, 0, 1, 2, 1, 2, 1, 2, 0, 0, 1, 2,$
$0, 0, 0, 0, 0, 0, 0, 0, 1, 2, 1, 2, 0, 0, 0, 0, 1, 2, 0, 0, 1, 2, 1, 2, 1, 2, 0, 0, 1, 2, 0, 0, 1, 2)$

Faster Chosen-Key Distinguishers on Reduced-Round AES

Patrick Derbez, Pierre-Alain Fouque, and Jérémy Jean

École Normale Supérieure, 45 Rue d'Ulm, 75005 Paris, France
{Patrick.Derbez,Pierre-Alain.Fouque,Jeremy.Jean}@ens.fr

Abstract. In this paper, we study the AES block cipher in the chosen-key setting. The adversary's goal of this security model is to find triplets (m, m', k) satisfying some properties more efficiently for the AES scheme than generic attacks. It is a restriction of the classical chosen-key model, since as it has been defined originally, differences in the keys are possible. This model is related to the known-key setting, where the adversary receives a key k, and tries to find a pair of messages (m, m') that has some property more efficiently than generic attacks. Both models have been called open-key model in the literature and are interesting for the security of AES-based hash functions.

Here, we show that in the chosen-key setting, attacking seven rounds (resp. eight rounds) of AES-128 can be done in time and memory 2^8 (resp. 2^{24}) while the generic attack would require 2^{64} computations as a variant of the birthday paradox can be used to predict the generic complexity. We have checked our results experimentally and we extend them to distinguisers of AES-256.

Keywords: AES, Open-key Model, Chosen-key Distinguisher, Practical Complexities.

1 Introduction

The Advanced Encryption Standard (AES) [16] is nowadays the subject of many attention since attacks coming from hash function cryptanalysis have put its security into question. Related-key attacks and meet-in-the-middle attacks that begin in the middle of the cipher (also known as splice-and-cut attacks) have been proposed to attack the full number of rounds for each AES versions [1,2,4], while other techniques exist for smaller version [5]. This interesting connection between hash functions and block ciphers shows that any improvement on hash function cryptanalysis can be useful for attacking block ciphers and vice-versa.

In this work, we study another model that has been suggested to study the security of hash functions based on AES components. Knudsen and Rijmen [9] have proposed to consider *known-key* attacks since in the hash function domain, the key is usually known and the goal is to find two input messages that satisfy some interesting relations. In some setting, a part of the key can also be chosen (for instance when salt is added to the hash function) and therefore, cryptanalysts have also consider the model where the key is under the control of the

S. Galbraith and M. Nandi (Eds.): INDOCRYPT 2012, LNCS 7668, pp. 225–243, 2012.
© Springer-Verlag Berlin Heidelberg 2012

adversary. The latter model has been called *chosen-key* model and both models belong to the *open-key* model. The chosen-key model has been popularized by Biryukov et al. in [2], since a distinguisher in this model has been extended to a related-key attack on the full AES-256 version.

Related Work. Knudsen and Rijmen in [9] have been the firsts to consider known-key distinguishers on AES and Feistel schemes. The main motivations for this model are the following:

- if there is no distinguisher when the key is known, then there will also be no distinguisher when the key is secret,
- if it is possible to find an efficient distinguisher, finding partial collision on the output of the cipher more efficiently than birthday paradox would predict even though the key is known, then the authors would not recommend the use of such cipher,
- finally, such model where the key is known or chosen can be interesting to study the use of cipher in a compression function for a hash function.

In the same work, they present some results on Feistel schemes and on the AES. Following this work, Minier et al. in [14] extend the results on AES on the Rijndael scheme with larger block-size.

In [2], Biryukov et al. have been the firsts to consider the chosen-key distinguisher for the full 256-bit key AES. They show that in time $q \cdot 2^{67}$, it is possible to construct q-multicollision on Davies-Meyer compression function using AES-256, whereas for an ideal cipher, it would require on average $q \cdot 2^{\frac{q-1}{q+1}128}$ time complexity. In these chosen-key distinguishers, the adversary is allowed to put difference also in the key. Later, Nikolic et al. in [15], describe known-key and chosen-key distinguishers on Feistel and Substitution-Permutation Networks (SPN). The notion of chosen-key distinguisher is more general than the model that we use: here, we let the adversary choose the key, but it has to be the same for the input and output relations we are looking for. We do not consider related-keys in this article. Then in [12], rebound attacks have been used to improve known-key distinguishers on AES by Mendel et al. and in [8], Gilbert and Peyrin have used both the **SuperSBox** and the rebound techniques to get a known-key distinguisher on 8-round AES-128. Last year at FSE, Sasaki and Yasuda show in [18] an attack on 11 Feistel rounds and collision attacks in hashing mode also using rebound techniques, and more recently, Sasaki et al. studied the known-key scenario for Feistel ciphers like Camellia in [17].

Our Results. In this paper, we study 128- and 256-bit reduced versions of AES in the (single) chosen-key model where the attacker is challenged to find a key k and a pair of messages (m, m') such that $m \oplus m' \in E$ and $\text{AES}_k(m) \oplus \text{AES}_k(m') \in F$, where E and F are two known subspaces. On AES-128, we describe in that model a way to distinguish the 7-round AES in time 2^8 and the 8-round AES in time 2^{24}. In the case of the 7-round distinguisher, our technique improves the 2^{16} time complexity of a regular rebound technique [13] on the **SubBytes** layer by computing intersections of small lists. The 8-round distinguisher introduces a problem related the **SuperSBox** construction where the key parameter is under the control of the adversary. As for AES-256, the distinguishers are the natural extensions of the ones on AES-128. Our results are reported in Table 1. We have experimentally checked our results and examples are provided in the appendices. We believe our practical distinguishers can be useful to construct non-trivial inputs for the AES block cipher to be able to check the validity of some theoretical attacks, for instance [7].

Outline of the Paper. The paper is organized as follows. We begin in Section 2 by recalling the AES and the concept of **SuperSBox**. Then in Section 3.1, we precise the chosen-key model in the ideal case to be able to compare our distinguishers to the ideal scenario. Section 3.1 describes the main results of the AES-128 and Section 4 shows how to apply similar results to the AES-256.

Table 1. Comparison of our results to previous ones on reduced-round distinguishers of the AES-128 in the open-key model. Results from [1] are not mentioned since we do not consider related-keys in this paper.

Target	Model	Rounds	Time	Memory	Ideal	Reference
	Known-key	7	2^{56}	-	2^{58} *	[9]
	Known-key	7	2^{24}	2^{16}	2^{64}	[12]
	Single-chosen-key	7	2^{22}	-	2^{64}	[3]
AES-128	**Single-chosen-key**	**7**	2^{8}	2^{8}	2^{64}	**Section 3.2**
	Known-key	8	2^{48}	2^{32}	2^{64}	[8]
	Single-chosen-key	8	2^{44}	-	2^{64}	[3]
	Single-chosen-key	**8**	2^{24}	2^{16}	2^{64}	**Section 3.3**
	Single-chosen-key	**7**	2^{8}	2^{8}	2^{64}	**Section 4.1**
AES-256	**Single-chosen-key**	**8**	2^{8}	2^{8}	2^{64}	**Section 4.2**
	Single-chosen-key	**9**	2^{24}	2^{16}	2^{64}	**Section 4.3**

* Claimed by the authors as a *very inaccurate estimation of the [ideal] complexity*.

2 Description of the AES

The Advanced Encryption Standard [16] is a Substitution-Permutation Network that can be instantiated using three different key bit-lengths: 128, 192, and 256. The 128-bit plaintext initializes the internal state viewed as a 4×4 matrix of bytes as values in the finite field $GF(2^8)$, which is defined via the irreducible polynomial $x^8 + x^4 + x^3 + x + 1$ over $GF(2)$. Depending on the version of the AES, N_r rounds are applied to that state: $N_r = 10$ for AES-128, $N_r = 12$ for AES-192 and $N_r = 14$ for AES-256. Each of the N_r AES round (Figure 1) applies four operations to the state matrix (except the last one where we omit the **MixColumns**):

- **AddRoundKey** (AK) adds a 128-bit subkey to the state.
- **SubBytes** (SB) applies the same 8-bit to 8-bit invertible S-Box S 16 times in parallel on each byte of the state,
- **ShiftRows** (SR) shifts the i-th row left by i positions,
- **MixColumns** (MC) replaces each of the four column C of the state by $M \times C$ where M is a constant 4×4 maximum distance separable circulant matrix over the field $GF(2^8)$, $M = \mathrm{circ}(2,3,1,1)$.

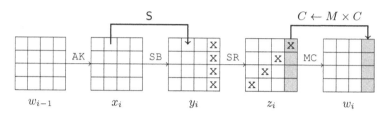

Fig. 1. An AES round applies MC∘SR∘SB∘AK to the state. There are $N_r = 10$ rounds in AES-128.

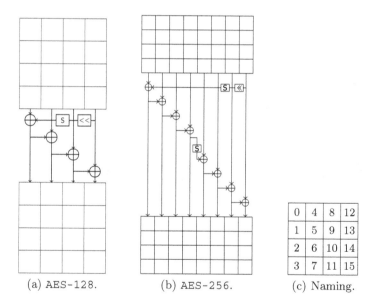

Fig. 2. Key schedules of the variants of the AES (AES-128 and AES-256) – Naming of bytes in a state

After the N_r-th rounds has been applied, a final subkey is added to the internal state to produce the ciphertext. The key expansion algorithm to produce the $N_r + 1$ subkeys for AES-128 is described in Figure 2(a), and in Figure 2(b) for the AES-256. We refer to the official specification document [16] for further details.

SuperSBox. In [6], Rijmen and Daemen introduced the concept of **SuperSBox** to study two rounds of AES. This transformation sees the composition SB ∘ AK(k) ∘ MC ∘ SB as four parallel applications of a 32-bit S-Box, and has been useful for several cryptanalysis works, see for instance [8,10]. Abusing notations, in the sequel, we call **SuperSBox** keyed by the key k the transformation that applies this composition to a single AES-column. In that context, the key k which parameterized the **SuperSBox** is also a 32-bit AES-column. We denote that operation by **SuperSBox**$_k$.

Notations. In this paper, we count the AES rounds from 0 and we refer to a particular byte of an internal state x by $x[i]$, as depicted in Figure 2(c). Moreover, as shown in Figure 1, in the ith round, we denote the internal state after **AddRoundKey** by x_i, after **SubBytes** by y_i, after **ShiftRows** by z_i and after **MixColumns** by w_i. To refer to the difference in a state x, we use the notation Δx.

3   Chosen-Key Distinguishers

3.1   Limited Birthday Distinguishers

In this section, we precise the distinguishers we are using. Our first goal is to distinguish the AES-128 from an ideal keyed-permutation in the chosen-key model. We will derive distinguishers for AES-256 afterwards. We are interested in the kind of distinguishers where the attacker is asked to find a key and a pair of plaintext whose difference is constrained in a predefined input subspace such that the ciphertext difference lies in an other predefined subspace.

Property 1. *Given two subspaces E_{in} and E_{out}, a key k and a pair of messages (x, y) verify the property on a permutation P if $x + y \in E_{in}$ and $P(x) + P(y) \in E_{out}$.*

This type of distinguisher looks like the limited birthday distinguishers introduced by Gilbert and Peyrin in [8] with a very close lower bound proved in [15], except that we allow the attacker more freedom; namely, in the choice of the key bits. To determine how hard this problem is, we need to compare the real-world case to the ideal scenario. In the latter, the attacker faces a family[1] of pseudo-random permutations $\mathcal{F} : \mathcal{K} \times \mathcal{D} \longrightarrow \mathcal{D}$, and would run a limited birthday distinguisher on a particular random permutation F_k to find a pair of messages

---

[1] Where both $\mathcal{K}$ and $\mathcal{D}$ are $\{0,1\}^{128}$ in the case of AES-128.

that conforms to the subspace restrictions of Property 1. The additional freedom of this setting does not help the attacker to find the actual pair of messages that verifies the required property, because the permutation F_k has to be chosen beforehand. Put it another way, the birthday paradox is as constrained as if the key were known since no difference can be introduced in the key bits.

Therefore, even if we let the key to be chosen by the attacker, the limited birthday distinguisher from [8] applies in the same way. For known E_{in} and E_{out}, we denote $n_i = \dim(E_{in})$ and $n_o = \dim(E_{out})$. In terms of truncated differences, n_i (resp. n_o) represents the number of independent active truncated differences in the input (resp. output) of a random permutation $F_k \in \mathcal{F}$ (see Figure 3). Both n_i and n_o range in the interval between 0 and n, where $n = 16$ in the case of AES. Without loss of generality, we assume that $n_i \leq n_o$: the attacker thus considers F_k rather than its inverse, as it is easier to collide on $n - n_o$ differences than on $n - n_i$. The attacker continues by constructing two lists L and L' of 2^{8n_i} plaintexts each by choosing a random value for the $n - n_i$ inactive bytes of the input and considering all the n_i active ones in E_{in}. With a birthday paradox on the two lists L and L', she expects a collision on at most $2n_i$ bytes of the ciphertexts. In the event that $n - n_o \geq 2n_i$, then $n - 2n_i$ bytes have not a zero-difference in the ciphertext. Hence, we need to restart the birthday paradox process about $2^{8(n-n_o-2n_i)}$ times, which costs $2^{8(n-n_o-n_i)}$ in total. Otherwise, if $n - n_o < 2n_i$, then a single birthday paradox with lists of size $2^{8(n-n_o)/2}$ is sufficient to get a collision on the $n - n_o$ required bytes in time $2^{8(n-n_o)/2}$.

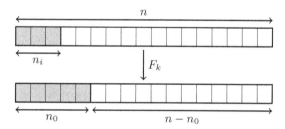

Fig. 3. Assuming $n_i \leq n_o$, the attacker searches for a pair of input to the random permutation F_k differing in n_i known byte positions such that the output differs in n_o known byte positions. A gray cell indicates a byte with a truncated difference.

3.2   Distinguisher for 7-Round AES-128

We consider the 7-round truncated differential characteristic of Figure 4, where the differences in both the plaintext and the ciphertext lie in subspaces of dimension four. Indeed, the output difference lies in a subspace of dimension four since all the operations after the last **SubBytes** layer are linear. With respect to the description of the distinguisher (Section 3.1), the time complexity to find a pair of messages that conforms to those patterns in a family of pseudo-random permutations is 2^{64} basic operations.

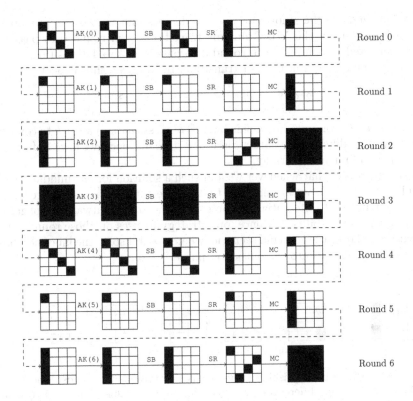

Fig. 4. The 7-round truncated differential characteristic used to distinguish the AES-128 from a random permutation. Black bytes are active, white bytes are not.

The following of this section describes a way to build a key and a pair of messages that conform to the restrictions in time 2^8 basic operations using a memory complexity of 2^8 bytes. This complexity has to be compared to 2^{16} operations, which is the time complexity expected for a straightforward application of the rebound attack [13] on the **SubBytes** layer of the AES. In that case, there are 16 random differential transitions around the AES S-Box, which happens to be all compatible[2] with probability 2^{-16}. Repeating with random differences 2^{16} times, we expect to find a pair of internal states that conforms to the randomized differences. In the following, we proceed slightly differently to reach a solution in time 2^8.

In terms of freedom degrees, we begin by estimating the number of solutions that we expect to verify the truncated differential characteristic. There are 16 bytes in the first message, 4 more independent ones in the second message and 16 others in the key: that makes 36 freedom degrees at the input. On a random input, the probability that the truncated differential characteristic being

---

[2] By compatible, we mean that we can find at least a pair of values that conforms to the differential transition. In the case of the AES S-Box, for a random differential transition $\delta \to \delta'$, this is known to be possible with probability close to $1/2$.

followed depends on the amount of freedom degrees that we loose in probabilistic transitions within the **MixColumns** transitions: 3 in round 0 to pass one $4 \rightarrow 1$ truncated transition, 12 in round 3 to pass four $4 \rightarrow 1$ transitions and 3 again in round 4 for the last $4 \rightarrow 1$ transition. In total, we thus expect

$$2^{8 \times (16+4+16)} \, 2^{-8 \times (3+12+3)} = 2^{8 \times 18}$$

triplets (m, m', k) composed by a pair (m, m') of messages and a key k to conform to the truncated differential characteristic of Figure 4. Hence, we have 18 freedom degrees left to find such a triplet.

First, we observe that whenever we find such a solution for the middle rounds (round 1 to round 4), we are ensured that all the rounds will be covered as in the whole truncated differential characteristic due to an outward propagation occurring with probability 1. Hence, our strategy focuses on those rounds. The context is similar to the rebound scenario, where we first solve the inbound phase and then propagate it into the outbound phase.

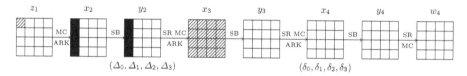

Fig. 5. The 7-round distinguishing attack focuses of the middle rounds. Black bytes have known values and differences, gray bytes have known values, hatched bytes have known differences and white bytes have unknown values and/or differences.

To reduce the number of valid solutions, we begin by fixing some bytes (Figure 5) to a random value: Δz_1 and $x_2[0..3]$. Therefore, we can deduce the values and differences in the first column of x_2 and y_2, as well as the difference Δx_3 by linearity. Let $[\Delta_0, \Delta_1, \Delta_2, \Delta_3]^{\mathrm{T}}$ be the column-vector of deduced differences in Δy_2 and $\mathrm{diag}(\delta_0, \delta_1, \delta_2, \delta_3)$ the differences in the diagonal of Δx_4. Linearly, we can express the differences around the **SubBytes** layer of round 3 (see Figure 6). As a consequence, from the differential properties of the AES S-Box, for $i, j \in \{0, \ldots, 3\}$, Δ_j suggests 2^7 different values for δ_i: we store them in the list $L_{i,j}$.

$$L_{i,j} = \left\{ \delta_i \,\middle/\, \Delta_j \rightarrow \delta_i \text{ is possible} \right\}. \tag{1}$$

Once done, we build the list L_i, for $i \in \{0, \ldots, 3\}$:

$$L_i = \bigcap_{j=0}^{3} L_{i,j} = \left\{ \delta_i \,\middle/\, \forall j \in \{0, \ldots, 3\}, \Delta_j \rightarrow \delta_i \text{ is possible} \right\}. \tag{2}$$

Each $L_{i,j}$ being of size 2^7, we expect each L_i to contain 2^4 elements.

Δx_3

$2\Delta_0$	Δ_3	Δ_2	$3\Delta_1$
Δ_0	Δ_3	$3\Delta_2$	$2\Delta_1$
Δ_0	$3\Delta_3$	$2\Delta_2$	Δ_1
$3\Delta_0$	$2\Delta_3$	Δ_2	Δ_1

SB $\longrightarrow$

Δy_3

$14\delta_0$	$11\delta_1$	$13\delta_2$	$9\delta_3$
$13\delta_3$	$9\delta_0$	$14\delta_1$	$11\delta_2$
$14\delta_2$	$11\delta_3$	$14\delta_0$	$9\delta_1$
$13\delta_1$	$9\delta_2$	$14\delta_3$	$11\delta_0$

Fig. 6. Differences around the **SubBytes** layer of round 3: each Δ_j is fixed, whereas the δ_i are yet to be determined

We continue by setting $\Delta x_4[0]$ to random value in L_0 and $x_4[0]$ to a random value, which allow to determine the value and difference in $y_4[0]$. Since the difference Δy_4 can only take 2^8 values due to the **MixColumns** transition of round 4, we also deduce Δw_4 and the remaining differences in Δy_4. The knowledge of Δy_4 suggests 2^7 possible values for δ_i. As before, we store them in lists called T_i, and we select a value for δ_i in $L_i \cap T_i$ (Figure 7). We expect each intersection to contain about 2^3 elements. More rigorously, if we assume that the lists $L_{i,j}$ and T_i are uniformly distributed, then the probability that L_0, $L_1 \cap T_1$, $L_2 \cap T_2$ and $L_3 \cap T_3$ are not empty is higher than 99.96% (see proof in Appendix C). Finally, we compute the values in x_3 and in the diagonal of x_4.

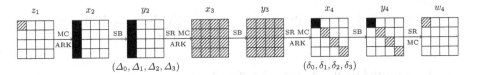

z_1 x_2 y_2 x_3 y_3 x_4 y_4 w_4

$(\Delta_0, \Delta_1, \Delta_2, \Delta_3)$ $(\delta_0, \delta_1, \delta_2, \delta_3)$

Fig. 7. The 7-round distinguishing attack focuses of the middle rounds. Black bytes have known values and differences, gray bytes have known values, hatched bytes have known differences and white bytes have unknown values and/or differences.

We now need to find a key that matches the previous solving in the internal states: we build a partial pair of internal states that conforms to the middle rounds, but that sets 8 bytes on constraints in the key. Namely, if we denote k_i the subkey introduced in round i and $u_i = \mathrm{MC}^{-1}(k_i)$, then both u_3 and k_4 have four known bytes (see Figure 8). We start by fixing all the bytes marked by 1 in u_3 to random values: this allows to compute the values of all 2's in the two last columns of k_3. By the column-wise operations of AES key schedule, we can get the values of all bytes marked by 3. As for the 4's, we get them since there are four known bytes among the eight in the first columns of u_3 and k_3. Again, the key schedule gives the 5's and 6's, and the **MixColumns** the 7's. Finally, we determine values for all the byte tagged by 8 from the key schedule equations. By inverting the key schedule, we are thus able to compute the master key k.

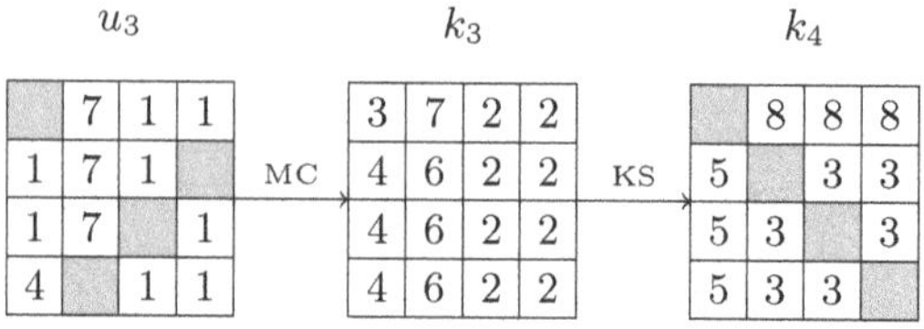

Fig. 8. Generating a compatible key: gray bytes are known, and numbers indicate the order in which we guess or determine the bytes

All in all, we start by getting a partial pair of internal states that conforms to the middle rounds, continue by deriving a valid key that matches the partial known bytes and determine the rest of the middle internal states to get the pair on input messages. The bottleneck of the time and memory complexity occurs when handling the lists of size at most 2^8 elements to compute intersections. Note that those intersections can be done in roughly 2^8 operations by representing lists by 256-bit numbers and then perform a logical AND.

In the end, we build a pair of messages (m, m') and a key k that conforms to the truncated differential characteristic of Figure 4 in time 2^8 basic operations, where it costs 2^{64} in the generic scenario. We note that among the 18 freedom degrees left for the attack, we used only 10 by setting 10 bytes to random values, such that we expect $2^{8 \times 8} = 2^{64}$ solutions in total. All those solutions could be generated in time 2^{64} by iterating over all the possibilities of the bytes marked by 1 in Figure 8.

We implemented the described algorithm to verify that it indeed works, and we found for instance the triplet (m, m', k) reported in Appendix A.

3.3 Distinguisher for 8-Round AES-128

We consider the 8-round truncated differential characteristic of Figure 9, where the matrices of differences in both the plaintext and the ciphertext lie in the same matrix subspaces of dimension four as before. Indeed, the output difference lies in a subspace of dimension four since all the operations after the last **SubBytes** layer are linear. Again, the distinguisher previously described (Section 3.1) claims that the time complexity to find a pair of messages that conforms to those patterns in a family of pseudo-random permutations runs in time 2^{64} operations.

The following of this section describes a way to build a key and a pair of messages that conform to the restrictions in time and memory complexity 2^{24}. We note that it is possible to optimize the memory requirement to 2^{16}. As in the previous section, there are 36 freedom degrees at the input, which shrink to 18 after the consideration of the truncated differential characteristic. Therefore, we also expect $2^{8 \times 18}$ solutions in the end.

First of all, we observe that finding 2^{24} triplets (m, m', k) composed by a key and a pair of internal states that conform to the rounds 2 to 5 is sufficient since

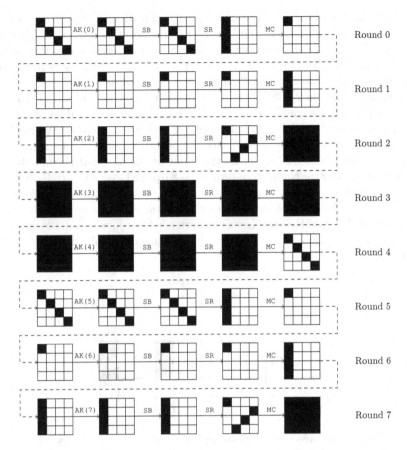

Fig. 9. The 8-round truncated differential characteristic used to distinguish the AES-128. Black bytes are active, white bytes are not.

the propagation in the outward rounds is done with probability 2^{-24} due to the **MixColumns** transition of round 1. The following analysis consequently focuses of those four middle rounds.

We now describe an instance of a problem that we use as a building block in our algorithm, which is related to the keyed **SuperSBox** construction.

Problem 1. *Let a and b two bytes. Given a 32-bit input and output differences Δ_{in} and Δ_{out} of a* **SuperSBox**$_k$ *for a unknown k, find all the pairs of* AES-*columns (c, c') and keys k such that:*

    i. $c + c' = \Delta_{in}$,
    ii. **SuperSBox**$_k(c)$ + **SuperSBox**$_k(c') = \Delta_{out}$,
    iii. **SuperSBox**$_k(c) = [a, b, \star, \star]^{\mathrm{T}}$.

Considering the key k known and the case where there is no restriction on the output bytes (*iii*), we would expect this problem to have one solution on average. Finding it would naively require 2^{32} computations by iterating over the 2^{32} possible inputs and check whether the output has the correct Δ_{out} known difference. The additional constraints on the two output bytes reduce the success of finding a pair (c, c') of input to 2^{-16}, but if we allow the four bytes in the key k to be chosen, then we expect 2^{16} solutions to this problem.

To find all of them in 2^{16} simple operations, we proceed as follows (Figure 10): the two output bytes a and b being known, we can deduce the values of the two associated bytes before the last **SubBytes**, $\tilde{a}$ and $\tilde{b}$ respectively. We can also deduce the difference in those bytes since the output difference is known. Then, we guess the two unset differences at the input of the last **SubBytes**: the differences then propagate completely inside the **SuperSBox**. At both **SubBytes** layers, by the differential properties of the AES S-Box, we expect to find one value on average for each of the six unset transitions. Consequently, the input and output of the **AddRoundKey** operation are known, which determines the four bytes of k. In the end, we find the 2^{16} solutions of Problem 1 in time 2^{16} operations.

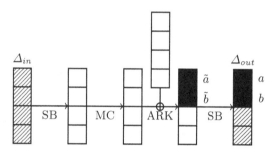

Fig. 10. Black bytes have known values and differences, hatched bytes have known differences and white bytes have unknown values and/or differences

To apply this strategy to the 8-round truncated differential characteristic of Figure 9, we start by randomizing the difference Δy_2, the difference Δw_5 and the values in the first column of w_5. Due to the linear operations involved, we deduce $\Delta x_3 = \Delta w_2$ from Δy_2 and Δy_4 from Δw_4. To use the previous algorithm, we randomize the values of the two first columns of w_4 (situation in Figure 11). Doing so, the four columns of y_4 are constrained on two bytes each and have fixed differences. Consequently, the four **SuperSBoxes** between x_3 and y_4 keyed by the four corresponding columns of k_4 conforms to the requirements[3] of Problem 1. In time and memory complexity 2^{16}, for $i \in \{0, 1, 2, 3\}$, we store the 2^{16} solutions for the ith **SuperSBox** associated to the ith column of x_4 in the list L_i.

We continue by observing that the randomization of the bytes in w_4 actually sets the value of two diagonal bytes in k_5, $k_5[0]$ and $k_5[5]$, which imposes constraints of the elements in the lists L_i. We start by considering the 2^{16} elements

---

[3] The positions of the known output bytes differ, but the strategy applies in the same way.

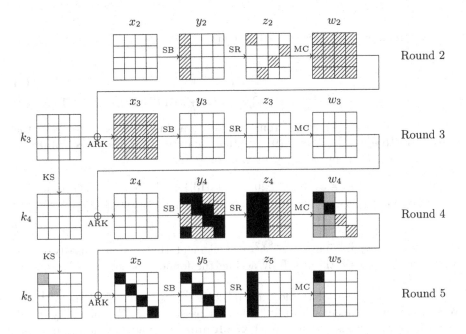

Fig. 11. Black bytes have known values and differences, gray bytes have known values, hatched bytes have known differences and white bytes have unknown values and/or differences

of L_3, and for each of them, we learn the values $x_4[12..15]$ and $k_4[12..15]$. Due to the column-wise operations in the key schedule, we also deduce the value of $k_4[0]$. Filtering the elements of L_0 which share that value of $k_4[0]$, we are left with 2^8 elements for bytes $x_4[0..3]$ and $k_4[0..3]$. At this point, we constructed $2^{16+8} = 2^{24}$ solutions in time 2^{24} that we store in a list $L_{0,3}$.

As $k_5[5]$ has been previously determined, we can deduce $k_4[5] = k_5[5] + k_5[1]$ from the AES key schedule for each of the entry of $L_{0,3}$. Again, this adds an 8-bit constraint on the elements of L_1: we expect 2^8 of them to match the condition on $k_4[5]$. In total, we could construct a list $L_{0,1,3}$ of size $2^{24+8} = 2^{32}$, whose elements would be the columns 0, 1 and 3 of x_4 and k_4, but as soon as we get 2^{24} elements in that list, we stop and discard the remaining possibilities.

Finally, to ensure the correctness of the choice in the remaining column 2, we need to consider the **MixColumns** operation in round 4 and the subkey k_5. Indeed, as soon as we choose an element in both $L_{0,1,3}$ and L_2, x_4, k_4 and k_5 become fully determined, but we need to ensure that the values $x_5[10]$ and $x_5[15]$ equal to the known ones. In particular, for $x_5[10]$, we have:

$$k_4[10] + k_5[6] = k_5[10] \tag{3}$$
$$= w_4[10] + x_5[10] \tag{4}$$
$$= z_4[8] + z_4[9] + 2z_4[10] + 3z_4[11] + x_5[10], \tag{5}$$

and for $x_5[15]$:

$$k_4[11] + k_5[7] + k_4[15] = k_5[11] + k_4[15] \tag{6}$$
$$= k_5[15] \tag{7}$$
$$= w_4[15] + x_5[15] \tag{8}$$
$$= 3z_4[12] + z_4[13] + z_4[14] + 2z_4[15] + x_5[15], \tag{9}$$

where (3), (6) and (7) come from the key schedule, (4) and (8) from the **AddRoundKey** and (5) and (9) use the equations from the **MixColumns**. Hence, for each element of $L_{0,1,3}$, we can compute:

$$S(x_4[8]) + k_4[10] := x_5[10] + k_5[6] + S(x_4[13]) + 2\,S(x_4[2]) + 3\,S(x_4[7]), \tag{10}$$
$$k_4[11] + 2\,S(x_4[11]) := k_5[7] + k_4[15] + 3\,S(x_4[12]) + S(x_4[1]) + S(x_4[6]) + x_5[15] \tag{11}$$

and lookup in L_2 to find $2^{16}\, 2^{-8\times 2} = 1$ element that match those two byte conditions. We create the list L by adding the found element from L_2 to each entry of $L_{0,1,3}$.

All in all, in time and memory complexity 2^{24}, we build L of size 2^{24} and we now exhaust its elements to find one that passes the 2^{-24} probability of the $4 \to 1$ backward transition in the **MixColumns** of round 1. Indeed, an $a \to b$ transition in the **MixColumns** layer cancels $4 - b$ output bytes, so that it would happen with probability $2^{-8(4-b)}$ for a random input a. Consequently, we expect to find a pair (m, m') of messages and a key k that conforms to the 8-round truncated differential characteristic of Figure 9 in time 2^{24} when it requires 2^{64} computations in the ideal case.

Among the 18 available freedom degrees available to mount the attack, we uses 17 of them, which means that we expect to have 2^8 solutions. We could have them in time 2^{32}, but since we discarded 2^8 elements in the algorithm described, we get only 1 in time 2^{24}. We note that it is possible to gain a factor 2^8 in the memory requirements of our attack since we can implement the algorithm without storing the lists L_0, $L_{0,3}$ and $L_{0,1,3}$, by using hash tables for L_1, L_2 and L_3.

We also implemented the described algorithm to verify that it indeed works, and we found for instance the triplet (m, m', k) reported in Appendix B.

4   Extention to AES-256

The two distinguishers described in the previous section can be easily extended in distinguishers on the AES-256. The main idea is to use the 16 additional freedom degrees in the key to extend the truncated differential characteristics by introducing a new fully active round in the middle.

4.1   Distinguisher for 7-Round AES-256

The first step of the attack described in the 7-round distinguisher on AES-128 (Section 3.2) still applies in the case of AES-256 since it does not involve the key schedule. Then, we can generate a compatible key easily since there are only two subkeys involved: we can just choose bytes of k_3 and k_4 as we want, except the imposed ones, and deduce the master key afterwards. This yields to a distinguisher with time and memory complexities around 2^8.

4.2 Distinguisher for 8-Round AES-256

We use a similar approach as the 7-round distinguisher on AES-128 of Section 3.2, but the truncated differential characteristic has one more fully active round in the middle[4].

We begin by choosing values for Δz_1 and $x_2[0..3]$. This allows to deduce Δx_2, Δy_2, and Δx_3. Then, we also set random values for Δw_5 and for the diagonal of x_5 to obtain both Δx_5 and Δy_4. Now, we find a value for Δx_4, which is compatible with Δx_3 and Δy_4. Indeed, we can not take an arbitrary value for Δx_4 because the probability that it fits is very close to 2^{-32}. However, we can find a correct value with the following steps:

1. Store the 2^7 possible values for $\Delta x_4[0]$ in a list L_0.
2. In a similar way, make lists L_1 with $\Delta x_4[1]$, L_2 with $\Delta x_4[2]$ and L_3 with $\Delta x_4[3]$.
3. Choose a value for $(x_3[0], x_3[5], x_3[10], x_3[15])$ and compute $\Delta x_4[0..3]$.
4. If $\Delta x_4[0..3]$ is not in $L_0 \times L_1 \times L_2 \times L_3$, then go back to step 3.

On average, we go back to the step 3 only $\left(2^{8-7}\right)^4 = 2^4$ times since lists are of size 2^7. In the same way, we can obtain values for the other columns of x_4.

At this point, we computed actual values in all those internal states, and we need to generate a compatible key. Finding one can be done using the procedure described in Figure Figure 12. Bytes tagged by 1 are chosen at random, odd steps use the key schedule equations and even steps the properties of **MixColumns**.

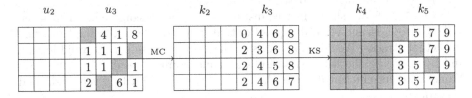

Fig. 12. Generating a compatible key: gray bytes are known, and numbers indicate the order in which we guess or determine the bytes

4.3 Distinguisher for 9-Round AES-256

We begin as in Section 3.3 by choosing the difference Δy_2, the difference Δw_6 and the values in the first column of w_6. Then, we deduce $\Delta w_2 = \Delta x_3$ from Δy_2 and Δy_5 from Δw_5. In addition, we set x_3 to a random value, which allows to

---

[4] In that case, the truncated differential characteristic is thus the one from Figure 9.

determine Δx_4. In order to apply the result from Problem 1 again, we set the values in two first columns of w_5 to random values.

As before, for $i \in \{0, 1, 2, 3\}$, we store in the list L_i the 2^{16} possible values of the i-th column of x_5 and the i-th column of k_5. Unlike previously, we also obtain values of the i-th column of $\mathsf{SR}(k_4)$, but the scenario of the attack still applies. We start by observing that bytes of L_0 allow to compute $k_4[1]$ and $k_4[13]$, which are bytes of L_3. Thus, we can merge L_0 and L_3 in a list $L_{0,3}$ containing 2^{16} elements. Then, we construct the list $L_{0,2,3}$ containing 2^{24} elements of $L_{0,3} \times L_2$. Finally, from bytes of $L_{0,2,3}$, we can compute:

$$3z_5[11] := k_4[2] + \mathsf{S}(k_5[15]) + k_4[6] + k_4[10] + z_5[8] + z_5[9] + 2z_5[10] + x_6[10], \tag{12}$$

$$z_5[14] + k_4[3] := \mathsf{S}(k_5[12]) + k_4[7] + k_4[11] + k_4[15] + 3z_5[12] + z_5[13] + 2z_5[15] + x_6[15]. \tag{13}$$

As a consequence, we expect only one element of L_1 to satisfy those two byte conditions and so, we obtain 2^{24} solutions for the middle rounds. All in all, this yields to a distinguisher with a time complexity around 2^{24} and a memory requirement around 2^{16} using the same trick given in Section 3.3.

5   Conclusion

In this paper, we study the Advanced Encryption Standard and show how to find a pair of messages and a key that satisfy some property a lot more efficiently than a generic attack based on the birthday paradox for both AES-128 and AES-256. Our new results improve the previous claimed ones by reaching very practical complexities, and give new insights of the open-key model for block ciphers, and hash functions based on block ciphers.

On AES-128, we show efficient distinguishers for versions reduced to seven and eight rounds, and verified in practice that they indeed work by implementing the actual attacks. We describe precisely the algorithms to get the valid inputs, and by applying the same strategy, we deduce similar results for AES-256. Namely, we get efficient distinguishers on versions reduced to seven, eight and nine rounds.

References

1. Biryukov, A., Khovratovich, D.: Related-Key Cryptanalysis of the Full AES-192 and AES-256. In: [11], pp. 1-18
2. Biryukov, A., Khovratovich, D., Nikolić, I.: Distinguisher and Related-Key Attack on the Full AES-256. In: Halevi, S. (ed.) CRYPTO 2009. LNCS, vol. 5677, pp. 231–249. Springer, Heidelberg (2009)
3. Biryukov, A., Nikolic, I.: A New Security Analysis of AES-128. In: CRYPTO 2009 rump session, slides only (2009)

4. Bogdanov, A., Khovratovich, D., Rechberger, C.: Biclique Cryptanalysis of the Full AES. In: Lee, D.H., Wang, X. (eds.) ASIACRYPT 2011. LNCS, vol. 7073, pp. 344–371. Springer, Heidelberg (2011)
5. Bouillaguet, C., Derbez, P., Fouque, P.-A.: Automatic Search of Attacks on Round-Reduced AES and Applications. In: Rogaway, P. (ed.) CRYPTO 2011. LNCS, vol. 6841, pp. 169–187. Springer, Heidelberg (2011)
6. Daemen, J., Rijmen, V.: Understanding Two-Round Differentials in AES. In: De Prisco, R., Yung, M. (eds.) SCN 2006. LNCS, vol. 4116, pp. 78–94. Springer, Heidelberg (2006)
7. Dunkelman, O., Keller, N., Shamir, A.: Improved Single-Key Attacks on 8-Round AES-192 and AES-256. In: Abe, M. (ed.) ASIACRYPT 2010. LNCS, vol. 6477, pp. 158–176. Springer, Heidelberg (2010)
8. Gilbert, H., Peyrin, T.: Super-Sbox Cryptanalysis: Improved Attacks for AES-Like Permutations. In: Hong, S., Iwata, T. (eds.) FSE 2010. LNCS, vol. 6147, pp. 365–383. Springer, Heidelberg (2010)
9. Knudsen, L.R., Rijmen, V.: Known-Key Distinguishers for Some Block Ciphers. In: Kurosawa, K. (ed.) ASIACRYPT 2007. LNCS, vol. 4833, pp. 315–324. Springer, Heidelberg (2007)
10. Lamberger, M., Mendel, F., Rechberger, C., Rijmen, V., Schläffer, M.: Rebound Distinguishers: Results on the Full Whirlpool Compression Function. In: [11], pp. 126-143
11. Matsui, M. (ed.): ASIACRYPT 2009. LNCS, vol. 5912. Springer, Heidelberg (2009)
12. Mendel, F., Peyrin, T., Rechberger, C., Schläffer, M.: Improved Cryptanalysis of the Reduced Grøstl Compression Function, ECHO Permutation and AES Block Cipher. In: Jacobson Jr., M.J., Rijmen, V., Safavi-Naini, R. (eds.) SAC 2009. LNCS, vol. 5867, pp. 16–35. Springer, Heidelberg (2009)
13. Mendel, F., Rechberger, C., Schläffer, M., Thomsen, S.S.: The Rebound Attack: Cryptanalysis of Reduced Whirlpool and Grøstl. In: Dunkelman, O. (ed.) FSE 2009. LNCS, vol. 5665, pp. 260–276. Springer, Heidelberg (2009)
14. Minier, M., Phan, R.C.-W., Pousse, B.: Distinguishers for Ciphers and Known Key Attack against Rijndael with Large Blocks. In: Preneel, B. (ed.) AFRICACRYPT 2009. LNCS, vol. 5580, pp. 60–76. Springer, Heidelberg (2009)
15. Nikolić, I., Pieprzyk, J., Sokołowski, P., Steinfeld, R.: Known and Chosen Key Differential Distinguishers for Block Ciphers. In: Rhee, K.-H., Nyang, D. (eds.) ICISC 2010. LNCS, vol. 6829, pp. 29–48. Springer, Heidelberg (2011)
16. NIST: Advanced Encryption Standard (AES), FIPS 197. Technical report, NIST (November 2001)
17. Sasaki, Y., Emami, S., Hong, D., Kumar, A.: Improved Known-Key Distinguishers on Feistel-SP Ciphers and Application to Camellia. In: Susilo, W., Mu, Y., Seberry, J. (eds.) ACISP 2012. LNCS, vol. 7372, pp. 87–100. Springer, Heidelberg (2012)
18. Sasaki, Y., Yasuda, K.: Known-Key Distinguishers on 11-Round Feistel and Collision Attacks on Its Hashing Modes. In: Joux, A. (ed.) FSE 2011. LNCS, vol. 6733, pp. 397–415. Springer, Heidelberg (2011)

A   Solution for the 7-Round Truncated Differential Characteristic on AES-128

Table 2. Example of a pair of messages (m, m') that conforms to the 7-rounds truncated differential characteristic for AES-128 of Section 3.2. The master key found by the attack is: 93CA1344 10A7EBDF B659C8AF ECC59699. The lines in this array contains the values of two internal states before entering the corresponding round, as well as their difference.

Round	m	m'	$m \oplus m'$
Init.	E5FC5DFE 79A851F7 7EB9E366 51C3D9C5	F8FC5DFE 79C951F7 7EB96566 51C3D96E	1D000000 00610000 00008600 000000AB
0	76364EBA 690FBA28 C8E02BC9 BD064F5C	6B364EBA 696EBA28 C8E0ADC9 BD064FF7	1D000000 00610000 00008600 000000AB
1	65CC94D1 85BE1AD3 F3D75BF1 ACCBB8BD	8DCC94D1 85BE1AD3 F3D75BF1 ACCBB8BD	E8000000 00000000 00000000 00000000
2	E93319CD 88F41390 10623230 F66BFBAD	C92309FD 88F41390 10623230 F66BFBAD	20101030 00000000 00000000 00000000
3	89C79074 E09E6F44 F1DBAB2F F984FCC4	1404532A 09774F8D 24BF1AFA CD551921	9DC3C35E E9E920C9 D564B1D5 34D1E5E5
4	867A12E6 BF19139C 1C848362 400030D3	047A12E6 BF5B139C 1C847C62 400030D7	82000000 00420000 0000FF00 00000004
5	84606BEA 0E22D904 3BF29061 9F454807	4B606BEA 0E22D904 3BF29061 9F454807	CF000000 00000000 00000000 00000000
6	FF867544 274436AF 75ECC287 A6BF72F6	3C6A996B 274436AF 75ECC287 A6BF72F6	C3ECEC2F 00000000 00000000 00000000
End	C49E4CB3 0C944043 D5ED6D3B 247E3843	2563B1AF 68F0EC8B A6788B48 EEF27E05	E1FDFD1C 6464ACC8 7395E673 CA8C4646

B   Solution for the 8-Round Truncated Differential Characteristic on AES-128

Table 3. Example of a pair of messages (m, m') that conforms to the 8-round truncated differential characteristic for AES-128 of Section 3.3. The master key found by the attack is: 98C45623 6CA00686 301E836D 614DFAB0. The lines in this array contains the values of two internal states before entering the corresponding round, as well as their difference.

Round	m	m'	$m \oplus m'$
Init.	9588B342 D43D04D4 AB298AE1 E43687DB	0B88B342 D46904D4 AB29D0E1 E4368728	9E000000 00540000 00005A00 000000F3
0	0D4CE561 B89D0252 9B37098C 857B7D6B	934CE561 B8C90252 9B37538C 857B7D98	9E000000 00540000 00005A00 000000F3
1	53FEBB0F 6BFF8E5E B471A8E3 1A2232A3	0EFEBB0F 6BFF8E5E B471A8E3 1A2232A3	5D000000 00000000 00000000 00000000
2	E9F44380 991A8ECB F7B18344 2C936CEB	65B2054A 991A8ECB F7B18344 2C936CEB	8C4646CA 00000000 00000000 00000000
3	2977F65C 3883EDEF 615D3C9E 5CE5384B	8F24A5A9 2398C0D9 10CEDEEF DFEEB0C3	A65353F5 1B1B2D36 7193E271 830B8888
4	BB1DB144 2BE947C3 5FCD89DF DF1CA0EB	82188658 42FFCAAE B337F0CA 09AB1513	3905371C 69168D6D ECFA7915 D6B7B5F8
5	C3E1961D 02A9713E 770A20D4 5470FA8F	8DE1961D 029B713E 770A3AD4 5470FA27	4E000000 00320000 00001A00 000000A8
6	D79D534C 33CC3861 76635DCD 548870C9	EB9D534C 33CC3861 76635DCD 548870C9	3C000000 00000000 00000000 00000000
7	D7F645C6 89358035 09847940 D831EFDE	0211A2F4 89358035 09847940 D831EFDE	D5E7E732 00000000 00000000 00000000
End	16E58308 DFD78F11 A8B05B9D C0A0363E	E49CFA83 D4DC9207 FC4CF3C9 9B3BF6FE	F279798B 0B0B1D16 54FCA854 5B9BC0C0

C   Probability of Success

We are interested in the probability that the intersection of four or five subsets of $\{1, \ldots, 255\}$ each of size 128 being empty.

To evaluate it, let $\mathcal{P}$ denote the set of subsets $X \subset \{1, \ldots, 255\}$ such that $|X| = 128$. We also define:

$$T(n, k) := \{(X_1, \ldots, X_n) \in \mathcal{P}^n \mid |X_1 \cap \ldots \cap X_n| = k\} \qquad \text{for } n \geq 1, k \geq 0.$$

In others words, $|T(n, k)|/|\mathcal{P}^n|$ is the probability that the intersection of n elements from $\mathcal{P}$ has a size equal to k.

Property 2. *The cardinality of* $T(n, k)$ *satisfies the following recurrence relation:*

$$\begin{cases} |T(1, k)| = |\mathcal{P}| \text{ if } k = 128, 0 \text{ otherwise} \\ |T(n+1, k)| = \sum_{l=k}^{128} |T(n, l)| \binom{l}{k} \binom{255-l}{128-k} \end{cases} \qquad \text{for } n \geq 1, k \geq 0.$$

Proof. First, we note that we can partition $\mathcal{P}^n$ by the sets:

$$T(n, Y) := \{(X_1, \ldots, X_n) \in \mathcal{P}^n \mid X_1 \cap \ldots \cap X_n = Y\} \qquad \text{for any subset } Y \subset \{1, \ldots, 255\}.$$

Then, we have:

$$|T(n+1, k)| = \sum_Y |\{(X_1, \ldots, X_{n+1}) \in T(n, Y) \times \mathcal{P} \mid |Y \cap X_{n+1}| = k\}|$$

$$= \sum_Y |T(n, Y)| \times |\{X \in \mathcal{P} \mid |Y \cap X| = k\}|$$

If we fix a set $Y \subset \{1, \ldots, 255\}$, then a set $X \in \mathcal{P}$ such that $|X \cap Y| = k$ is obtained by choosing k elements in Y and $128 - k$ elements in Y^c. As a consequence, we obtain:

$$|T(n+1, k)| = \sum_Y |T(n, Y)| \binom{|Y|}{k} \binom{255 - |Y|}{128 - k}$$

$$= \sum_{l=0}^{255} \binom{l}{k} \binom{255 - l}{128 - k} \sum_{|Y| = l} |T(n, Y)|$$

Finally, we remark that $\{T(n, Y)\}_{|Y| = l}$ is a partition of $T(n, l)$ and thus:

$$|T(n+1, k)| = \sum_{l=0}^{255} \binom{l}{k} \binom{255 - l}{128 - k} |T(n, l)|.$$

$\square$

Using Maple, we found that the probability of failure of the distinguisher described in Section 3.2 is:

$$\frac{T(4, 0)}{|\mathcal{P}|^4} \times \left(\frac{T(5, 0)}{|\mathcal{P}|^5}\right)^3 \approx 0.04\%.$$

The Higher-Order Meet-in-the-Middle Attack and Its Application to the Camellia Block Cipher*
(Extended Abstract)

Jiqiang Lu[1,**], Yongzhuang Wei[2,3], Jongsung Kim[4], and Enes Pasalic[5]

[1] Institute for Infocomm Research, Agency for Science, Technology and Research
1 Fusionopolis Way, #19-01 Connexis, Singapore 138632
lvjiqiang@hotmail.com, jlu@i2r.a-star.edu.sg
[2] Guilin University of Electronic Technology,
Guilin City, Guangxi Province 541004, P.R. China
[3] State Key Lab of Information Security, Institute of Software,
Chinese Academy of Sciences, Beijing 100190, China
walker_wei@msn.com
[4] Department of e-Business, Kyungnam University,
449 Wolyoung-dong, Masan, Kyungnam, Korea
jongsung.k@gmail.com
[5] University of Primorska FAMNIT, Koper, Slovenia
enespasalic@yahoo.se

Abstract. The meet-in-the-middle (MitM) attack is a technique for analysing the security of a block cipher. In this paper, we propose an extension of the MitM attack, which we call the higher-order meet-in-the-middle (HO-MitM) attack; the core idea of the HO-MitM attack is to use multiple plaintexts to cancel some key-dependent component(s) or parameter(s) when constructing a basic unit of "value-in-the-middle". We introduce a novel approach, which combines integral cryptanalysis with the MitM attack, to construct HO-MitM attacks on 10-round Camellia under 128 key bits, 11-round Camellia under 192 key bits and 12-round Camellia under 256 key bits, all of which include FL/FL^{-1} functions. Finally, we apply an existing approach to construct HO-MitM attacks on 14-round Camellia without FL/FL^{-1} functions under 192 key bits and 16-round Camellia without FL/FL^{-1} functions under 256 key bits.

* This paper was presented in part in an invited talk given by J. Lu at the First Asian Workshop on Symmetric Key Cryptography (ASK 2011), Singapore, August 2011. The work was supported by the French ANR project SAPHIR II (No. ANR-08-VERS-014), the Natural Science Foundation of China (No. 61100185), Guangxi Natural Science Foundation (No. 2011GXNSFB018071), the Foundation of Guangxi Key Lab of Wireless Wideband Communication and Signal Processing (No. 11101), China Postdoctoral Science Foundation funded project, and the Basic Science Research Program through the National Research Foundation of Korea funded by Ministry of Education, Science and Technology (No. 2012-0003556).

** The author was with École Normale Supérieure (France) when this work was done.

S. Galbraith and M. Nandi (Eds.): INDOCRYPT 2012, LNCS 7668, pp. 244–264, 2012.
© Springer-Verlag Berlin Heidelberg 2012

Keywords: Block cipher, Camellia, Meet-in-the-middle attack, Integral cryptanalysis.

1   Introduction

The Camellia [1] block cipher has a 128-bit block length, a user key of 128, 192 or 256 bits long, and a total of 18 rounds when used with a 128-bit key and 24 rounds when used with a 192/256-bit key. It has a Feistel structure with key-dependent logical functions FL/FL^{-1} inserted after every six rounds, plus four additional whitening operations at both ends. Camellia is a CRYPTREC e-government recommended cipher [8], a European NESSIE selected block cipher [31], and an ISO international standard [19]. For simplicity, we denote by Camellia-128/192/256 the three versions of Camellia that use 128, 192 and 256 key bits, respectively.

The security of Camellia has been analysed against a variety of cryptanalytic techniques, including differential cryptanalysis [5], higher-order differential cryptanalysis [20, 23], truncated differential cryptanalysis [20], impossible differential cryptanalysis [3, 21], linear cryptanalysis [30], integral (square [9]) cryptanalysis [18, 22], collision attack [33], boomerang attack [34], and rectangle attack [4]; and many cryptanalytic results on Camellia have been obtained. In summary, in terms of the numbers of attacked rounds, the best currently known cryptanalytic results on Camellia with FL/FL^{-1} functions are the impossible differential attacks on 11-round Camellia-128, 12-round Camellia-192 and 14-round Camellia-256 [2, 24], presented recently at FSE 2012 and ISPEC 2012; and the best currently known cryptanalytic results on Camellia without FL/FL^{-1} functions are the impossible differential attacks on 12-round Camellia-128 [28], 14-round Camellia-192 [26] and 16-round Camellia-256 [26, 29].[1]

The meet-in-the-middle (MitM) attack [12] is a technique for analysing the security of a block cipher. In this paper, we propose an extension of the MitM attack, which we call the higher-order meet-in-the-middle (HO-MitM) attack. The core idea of the HO-MitM attack is to use multiple plaintexts to cancel some key-dependent component(s) or parameter(s) when constructing a basic unit of so-called value-in-the-middle. Then we introduce a novel approach, that combines integral cryptanalysis [18, 22] with the MitM attack, to construct a few HO-MitM properties for 5 and 6-round Camellia with FL/FL^{-1} functions, and finally apply these properties to conduct HO-MitM attacks on 10-round Camellia-128 with FL/FL^{-1} functions, 11-round Camellia-192 with FL/FL^{-1} functions and

---

[1] When our work was completed, the best previously published cryptanalytic results on Camellia with FL/FL^{-1} functions were square attack on 9-round Camellia-128 [14], impossible differential attack on 10-round Camellia-192 [7], and higher-order differential and impossible differential attacks on 11-round Camellia-256 [7, 16]; and the best previously published cryptanalytic results on Camellia without FL/FL^{-1} functions were impossible differential attacks on 12-round Camellia-128 [28], 12-round Camellia-192 [25] and 15-round Camellia-256 [7]. We incorporate the newly emerging main results in this revised version.

Table 1. Main cryptanalytic results on Camellia

Cipher	FL/FL⁻¹	Attack Type	Rounds	Data	Memory	Time	Source
Camellia-128	yes	Integral(square)	9	2^{48}CP	2^{53}Bytes	2^{122}Enc.	[14]
		Impossible diff.	10	2^{118}CP	2^{93}Bytes	2^{118}Enc.	[26]
			11	$2^{120.5}$CP	$2^{115.5}$Bytes	$2^{123.8}$Enc.	[2]§
			11†	2^{122}CP	2^{102}Bytes	2^{122}Enc.	[24]§
		HO-MitM	10	2^{93}CP	2^{109}Bytes	$2^{118.6}$Enc.	Sect. 4.2
	no	Impossible diff.	12	$2^{116.3}$CP	2^{73}Bytes	$2^{116.6}$Enc.	[28]
Camellia-192	yes	Impossible diff.	10	2^{121}CP	$2^{155.2}$Bytes	2^{144}Enc.	[7]
			10†	2^{121}CP	$2^{155.2}$Bytes	$2^{175.3}$Enc.	[7]
			11	2^{118}CP	2^{141}Bytes	$2^{163.1}$Enc.	[26]
			12	$2^{120.6}$CP	$2^{171.6}$Bytes	$2^{171.4}$Enc.	[2]§
			12†	2^{123}CP	2^{160}Bytes	$2^{187.2}$Enc.	[24]§
		HO-MitM	11	2^{78}CP	2^{174}Bytes	$2^{187.4}$Enc.	Sect. 4.3
			11	2^{94}CP	2^{174}Bytes	$2^{180.2}$Enc.	Sect. 4.3
	no	Impossible diff.	12‡	2^{119}CP	2^{124}Bytes	$2^{147.3}$Enc.	[25]
			14	2^{117}CP	$2^{122.1}$Bytes	$2^{182.2}$Enc.	[26]
		HO-MitM	14	2^{118}CP	2^{166}Bytes	$2^{164.6}$Enc.	Sect. 5.2
Camellia-256	yes	Integral	10	$2^{60.5}$CP	2^{63}Bytes	$2^{254.3}$Enc.	[26,35]
		Higher-order diff.	11‡	2^{93}CP	2^{98}Bytes	$2^{255.6}$Enc.	[16,26]
		Impossible diff.	11†	2^{121}CP	2^{166}Bytes	$2^{206.8}$Enc.	[7]
			13†	2^{123}CP	2^{208}Bytes	$2^{251.1}$Enc.	[24]§
			14	$2^{121.2}$CP	$2^{180.2}$Bytes	$2^{238.3}$Enc.	[2]§
			14	2^{120}CC	2^{125}Bytes	$2^{250.5}$Enc.	[24]§
		HO-MitM	12	2^{94}CP	2^{174}Bytes	$2^{237.3}$Enc.	Sect. 4.3
	no	Impossible diff.	15	$2^{122.5}$KP	2^{233}Bytes	$2^{236.1}$Enc.	[7]
			16	2^{123}KP	2^{129}Bytes	2^{249}Enc.	[26]
		HO-MitM	16	2^{120}CP	2^{230}Bytes	2^{252}Enc.	Sect. 5.3

§: Newly emerging results; †: Include whitening operations; ‡: Can include whitening operations by making use of an equivalent structure of Camellia.

12-round Camellia-256 with FL/FL⁻¹ functions, all of which do not include the whitening operations. At last, we use an existing approach to construct a few HO-MitM properties for 7 and 8-round Camellia without FL/FL⁻¹ functions, and describe HO-MitM attacks on 14-round Camellia-192 without FL/FL⁻¹ functions and 16-round Camellia-256 without FL/FL⁻¹ functions, both of which do not include the whitening operations. Our HO-MitM results on Camellia-128/192/256 with FL/FL⁻¹ functions, which were among the first to achieve the amounts of attacked rounds of the Camellia versions, show that as far as the numbers of attacked rounds of Camellia with the FL/FL⁻¹ functions are concerned, the HO-MitM attack technique is more efficient than the advanced cryptanalytic techniques studied, except impossible differential cryptanalysis; in this latter case the HO-MitM attacks are now one or two rounds inferior to the best newly emerging impossible differential cryptanalysis results from [2,24].

Our HO-MitM results on Camellia-192/256 without FL/FL^{-1} functions, which were among the first to achieve the amounts of attacked rounds of the Camellia versions as well, match the best currently known cryptanalytic results for the versions of Camellia. Table 1 summarises previous, our and the newly emerging main cryptanalytic results on Camellia, where CP, CC and KP refer respectively to the numbers of chosen plaintexts, chosen ciphertexts and known plaintexts, Enc. refers to the required number of encryption operations of the relevant reduced version of Camellia, "yes" means "with FL/FL^{-1} functions", and "no" means "without FL/FL^{-1} functions".

The remainder of the paper is organised as follows. In the next section, we describe the notation and the Camellia block cipher. We define the HO-MitM attack in Section 3 and present our HO-MitM attack results on Camellia in Sections 4 and 5. Section 6 concludes this paper.

2   Preliminaries

In this section we give the notation used throughout this paper, and briefly describe the Camellia block cipher.

2.1   Notation

The bits of a value are numbered from left to right, starting with 1. We use the following notation throughout this paper.

$\oplus$	bitwise logical exclusive OR (XOR) of two bit strings of the same length
$\cap$	bitwise logical AND of two bit strings of the same length
$\cup$	bitwise logical OR of two bit strings of the same length
$\lll$	left rotation of a bit string
$\|$	bit string concatenation
$\circ$	functional composition. When composing functions X and Y, X $\circ$ Y denotes the function obtained by first applying X and then Y
$\|X\|$	the number of bits in a bit string X
$X[i_1, \cdots, i_j]$	a value made up of bits $(i_1, \cdots, i_j)$ of a bit string X

2.2   The Camellia Block Cipher

Camellia [1] employs a Feistel structure with a 128-bit block length and a variable key length of 128, 192 or 256 bits. It uses the following five functions:

- $\mathbf{S} : \{0,1\}^{64} \rightarrow \{0,1\}^{64}$ is a non-linear substitution constructed by applying eight 8×8-bit S-boxes $S_1, S_2, S_3, S_4, S_5, S_6, S_7$ and S_8 in parallel to the input, where S_1 and S_8 are identical, S_2 and S_5 are identical, S_3 and S_6 are identical, and S_4 and S_7 are identical.

– $\mathbf{P} : GF(2^8)^8 \to GF(2^8)^8$ is a linear permutation equivalent to multiplication by a 8×8 byte matrix P; the matrix P and its reverse P^{-1} are as follows.

$$
P = \begin{pmatrix} 1&0&1&1&0&1&1&1 \\ 1&1&0&1&1&0&1&1 \\ 1&1&1&0&1&1&0&1 \\ 0&1&1&1&1&1&1&0 \\ 1&1&0&0&0&1&1&1 \\ 0&1&1&0&1&0&1&1 \\ 0&0&1&1&1&1&0&1 \\ 1&0&0&1&1&1&1&0 \end{pmatrix}, \quad P^{-1} = \begin{pmatrix} 0&1&1&1&0&1&1&1 \\ 1&0&1&1&1&0&1&1 \\ 1&1&0&1&1&1&0&1 \\ 1&1&1&0&1&1&1&0 \\ 1&1&0&0&1&0&1&1 \\ 0&1&1&0&1&1&0&1 \\ 0&0&1&1&1&1&1&0 \\ 1&0&0&1&0&1&1&1 \end{pmatrix}.
$$

– $\mathbf{F} : \{0,1\}^{64} \times \{0,1\}^{64} \to \{0,1\}^{64}$ is a Feistel function. If X and Y are 64-bit blocks, $\mathbf{F}(X,Y) = \mathbf{P}(\mathbf{S}(X \oplus Y))$.
– $\mathbf{FL}/\mathbf{FL}^{-1} : \{0,1\}^{64} \times \{0,1\}^{64} \to \{0,1\}^{64}$ are key-dependent linear functions. If $X = (X_L \| X_R)$ and $Y = (Y_L \| Y_R)$ are 64-bit blocks, then $\mathbf{FL}(X,Y) = (((((X_L \cap Y_L) \lll 1 \oplus X_R) \cup Y_R) \oplus X_L) \| ((X_L \cap Y_L) \lll 1 \oplus X_R)$, and $\mathbf{FL}^{-1}(X,Y) = (X_L \oplus (X_R \cup Y_R)) \| (((X_L \oplus (X_R \cup Y_R)) \cap Y_L) \lll 1 \oplus X_R)$.

Camellia uses a total of four 64-bit whitening subkeys KW_j, $\frac{N_r-6}{3}$ 64-bit subkeys KI_l for the $\mathbf{FL}$ and $\mathbf{FL}^{-1}$ functions, and N_r 64-bit round subkeys K_i, $(1 \leqslant j \leqslant 4, 1 \leqslant l \leqslant \frac{N_r-6}{3}, 1 \leqslant i \leqslant N_r)$, all derived from an N_k-bit key K, where N_r denotes the number of rounds which is 18 for Camellia-128 and 24 for Camellia-192/256, N_k denotes the key length which is 128 for Camellia-128, 192 for Camellia-192 and 256 for Camellia-256. The key schedule is as follows. First, two 128-bit strings K_L and K_R are generated from K in the following way: For Camellia-128, K_L is the 128-bit key K, and K_R is zero; for Camellia-192, K_L is the left 128 bits of K, and K_R is the concatenation of the right 64 bits of K and the complement of the right 64 bits of K; and for Camellia-256, K_L is the left 128 bits of K, and K_R is the right 128 bits of K. Secondly, depending on the key size, generate one or two 128-bit strings K_A and K_B from (K_L, K_R) by a non-linear transformation; see [1] for detail. Finally, the subkeys are as follows.[2]

– For Camellia-128: $K_2 = (K_A \lll 0)[65 \sim 128], K_3 = (K_L \lll 15)[1 \sim 64], K_9 = (K_A \lll 45)[1 \sim 64], K_{10} = (K_L \lll 60)[65 \sim 128], K_{11} = (K_A \lll 60)[1 \sim 64], \cdots$.
– For Camellia-192/256: $K_1 = (K_B \lll 0)[1 \sim 64], K_2 = (K_B \lll 0)[65 \sim 128], K_3 = (K_R \lll 15)[1 \sim 64], K_4 = (K_R \lll 15)[65 \sim 128], K_7 = (K_B \lll 30)[1 \sim 64], K_8 = (K_B \lll 30)[65 \sim 128], K_{12} = (K_A \lll 45)[65 \sim 128], K_{13} = (K_R \lll 60)[1 \sim 64], K_{14} = (K_R \lll 60)[65 \sim 128], K_{15} = (K_B \lll 60)[1 \sim 64], K_{16} = (K_B \lll 60)[65 \sim 128], K_{17} = (K_L \lll 77)[1 \sim 64], K_{18} = (K_L \lll 77)[65 \sim 128], K_{21} = (K_A \lll 94)[1 \sim 64], K_{22} = (K_A \lll 94)[65 \sim 128], K_{23} = (K_L \lll 111)[1 \sim 64], \cdots$.

Below is the encryption procedure of Camellia, where P is a 128-bit plaintext, represented as 16 bytes, and L_0, R_0, L_i, R_i, $\widehat{L}_i$ and $\widehat{R}_i$ are 64-bit variables.

1. $L_0 \| R_0 = P \oplus (KW_1 \| KW_2)$

---

[2] Here we give only the subkeys concerned in this paper, $(K_A \lll 0)[65 \sim 128]$ represents bits $(65, 66, \cdots, 128)$ of $(K_A \lll 0)$, and so on.

2. For $i = 1$ to N_r:

   if $i = 6$ or 12 (or 18 for Camellia-192/256),

$$\widehat{L}_i = \mathbf{F}(L_{i-1}, K_i) \oplus R_{i-1}, \ \widehat{R}_i = L_{i-1};$$
$$L_i = \mathbf{FL}(\widehat{L}_i, KI_{\frac{i}{3}-1}), \ R_i = \mathbf{FL}^{-1}(\widehat{R}_i, KI_{\frac{i}{3}});$$

   else

$$L_i = \mathbf{F}(L_{i-1}, K_i) \oplus R_{i-1}, \ R_i = L_{i-1};$$

3. Ciphertext $C = (R_{N_r} \oplus KW_3)\|(L_{N_r} \oplus KW_4)$.

We refer to the ith iteration of Step 2 in the above description as Round i, and write $K_{i,j}$ for the j-th byte of K_i, $(1 \leqslant j \leqslant 8)$.

3 The Higher-Order Meet-in-the-Middle Attack

In this section, we first briefly recall the meet-in-the-middle (MitM) attack, and then define the higher-order meet-in-the-middle (HO-MitM) attack.

3.1 The Meet-in-the-Middle Attack

The meet-in-the-middle (MitM) attack was introduced in 1977 by Diffie and Hellman [12]. It usually treats a block cipher $\mathbf{E} : \{0,1\}^n \times \{0,1\}^k \rightarrow \{0,1\}^n$ as a cascade of two sub-ciphers $\mathbf{E} = \mathbf{E}^a \circ \mathbf{E}^b$. The basic unit of input for the MitM attack is a known-plaintext. Given a guess for the subkeys used in $\mathbf{E}^a$ and $\mathbf{E}^b$, if a plaintext produces just after $\mathbf{E}^a$ the same value as the corresponding ciphertext produces just before $\mathbf{E}^b$, then this guess for the subkeys is likely to be correct; otherwise, this guess must be incorrect. Thus, we can find the correct subkey, given a sufficient number of matching plaintext-ciphertext pairs. (The concerned value-in-the-middle can be a truncated one in some circumstances.)

Suppose (P, C) is a known plaintext-ciphertext pair, and let K_a denote the subkeys used in $\mathbf{E}^a$, K_b denote the subkeys used in $\mathbf{E}^b$, and K denote the subkeys used in $\mathbf{E}^a$ and $\mathbf{E}^b$. Obviously, $max\{|K_a|, |K_b|\} \leqslant |K| \leqslant |K_a| + |K_b|$. When checking whether P produces the same value just after $\mathbf{E}^a$ as C produces just before $\mathbf{E}^b$, a straightforward approach is to guess K_a to partially encrypt P through $\mathbf{E}^a$, then guess K_b to partially decrypt C through $\mathbf{E}^b$, and finally check whether the two intermediate values match. This approach requires negligible memory, and has a total time complexity of $2^{|K|}$ partial encryptions/decryptions. However, if the $2^{|K|}$ partial encryptions/decryptions are greater than 2^k full encryptions, then this approach is slower than an exhaustive key search and thus is not effective. Instead, a precomputation table may be helpful, just as in [12], as we now describe.

We precompute $\mathbf{E}^a_{K_a}(P)$ for all possible candidates for K_a and store these values in a hash table indexed by the values (and the overlapping bits between K_a and K_b if any). Then, guess K_b to partially decrypt C through $\mathbf{E}^b$, and check whether the intermediate value matches a value in the precomputation table. If so, the guess for K_b and the corresponding value for K_a are likely to be correct; otherwise, the guess for K_b must be incorrect and we repeat the same process with a different guess for K_b. The off-line precomputation requires a memory of $n \times 2^{|K_a|}$ bits and has a time complexity of $2^{|K_a|}$ partial encryptions. Thus, this approach has a total time complexity of $2^{|K_a|} + 2^{|K_b|}$ partial encryptions/decryptions.[3] Therefore, the approach using a precomputation table is efficient if the $2^{|K_a|} + 2^{|K_b|}$ partial encryptions/decryptions are smaller than 2^k full encryptions.

Both the approaches described above work in a known-plaintext attack scenario. Nevertheless, things may get better under a chosen-plaintext attack scenario. In such an attack scenario, as used in [10], we are able to choose a structure of plaintexts with a particular property, (e.g., a specific byte position takes all the possible values in $\{0,1\}^8$ and the other 15 bytes are fixed); a desirable consequence is that the matched (truncated) value-in-the-middle may be expressed as a function of plaintext and a smaller number of unknown one-bit constants than the number of possible candidates for K_a. As a result, we may generate a precomputation table with a smaller memory and time complexity, and thus give a more efficient attack.

The terminology "the meet-in-the-middle attack" has been abused somewhat to mean a broader type of similar attacks where the matched (truncated) "value-in-the-middle" can be not from the middle or any place of encryption/decryption but is abstracted as the output of some function of plaintext and/or intermediate values, though something like "the meet-in-the-middle-style attack" is more appropriate to term this type of attacks. This is the case for our attacks presented in this paper.

3.2   The HO-MitM Attack

Typically, in the MitM attack a basic unit of value-in-the-middle is obtained from a known-plaintext. We note that we can use multiple plaintexts to construct a basic unit of value-in-the-middle in a MitM attack; we call such an attack a higher-order meet-in-the-middle (HO-MitM) attack. Specifically, the basic idea of the HO-MitM attack can be described as follows, which is an extended version of the basic idea of the MitM attack: It involves treating a block cipher $\mathbf{E}$: $\{0,1\}^n \times \{0,1\}^k \rightarrow \{0,1\}^n$ as a cascade of two sub-ciphers $\mathbf{E} = \mathbf{E}^a \circ \mathbf{E}^b$ for some $\mathbf{E}^a$ and $\mathbf{E}^b$. Given a guess for the subkeys used in $\mathbf{E}^a$ and $\mathbf{E}^b$, if the output of some function[4] (e.g., a truncated XOR sum) of the values that a set of chosen plaintexts produces just after $\mathbf{E}^a$ is equal to the output of the same function of the values that the corresponding ciphertexts produce just before $\mathbf{E}^b$, then this guess for the subkeys is likely to be correct; otherwise, this guess must be incorrect. More

---

[3] When being checked with a plaintext-ciphertext pair, a wrong guess for K will survive with a probability of 2^{-n} in the first approach, and a wrong guess for K_b will survive with a probability of about $\frac{2^{|K_a|}}{2^{|K_a|+|K_b|-|K|}} \times 2^{-n} = 2^{|K|-|K_b|-n}$ in the approach using a precomputation table. Usually, one or more additional plaintext-ciphertext pairs are required to filter out the right subkey, but generally the time complexity associated with these additional plaintext-ciphertext pairs is negligible.

[4] The function should have a distinguishing property.

formally, suppose $\{P_1, P_2, \cdots, P_l\}$ is a set of l chosen plaintexts, $\{C_1, C_2, \cdots, C_l\}$ is the set of the corresponding ciphertexts, and $\mathbf{f} : \{0,1\}^{n \times l} \to \{0,1\}^m$ (for a specific value of m) is some function of l variables of n bits long each. Then, given a guess (K_a^*, K_b^*) for the subkeys (K_a, K_b) used respectively in $\mathbf{E}^a$ and $\mathbf{E}^b$, if $\mathbf{f}(E_{K_a^*}^a(P_1), E_{K_a^*}^a(P_2), \cdots, E_{K_a^*}^a(P_l)) = \mathbf{f}((E_{K_b^*}^b)^{-1}(C_1), (E_{K_b^*}^b)^{-1}(C_2), \cdots, (E_{K_b^*}^b)^{-1}(C_l))$, then the subkey guess (K_a^*, K_b^*) is likely to be correct; otherwise, this subkey guess must be incorrect. This is easy to prove: If (K_a^*, K_b^*) is the correct guess for (K_a, K_b), then $E_{K_a^*}^a(P_i) = E_{K_a}^a(P_i) = (E_{K_b}^b)^{-1}(C_i) = (E_{K_b^*}^b)^{-1}(C_i)$ must hold for all $i = 1, 2, \cdots, l$. Thus, given a sufficient number of sets of chosen plaintexts, we can find the correct subkey in a similar approach as described for the MitM attack in Section 3.1. In particular, it resembles the approach based on the use of a precomputation table in a chosen-plaintext attack scenario. (The definition also works under a known-plaintext attack scenario.)

From the above descriptions, it is easy to see that the fundamental distinction between the basic ideas of the HO-MitM attack and the MitM attack lies in the number of plaintexts used to construct a basic unit of value-in-the-middle: The basic value-in-the-middle concerned in the MitM attack is obtained from a plaintext (we note that it is obtained from two plaintexts in some previously published MitM attacks, as discussed in Section 3.3), whiles the basic value-in-the-middle concerned in the HO-MitM attack is obtained from multiple plaintexts; in other words, while the basic input unit for the MitM attack is a known-plaintext, the basic input unit of the HO-MitM attack is a set of chosen plaintexts.

At first glance, the HO-MitM attack might appear to be a trivial extension of the MitM attack. Generally, we can easily convert a MitM attack to a HO-MitM attack, if we do not consider the consequence caused by the increase of the number of plaintexts in the basic input unit; however, the MitM attack would outperform the HO-MitM attack, for it seems not necessary to use a basic input unit with multiple plaintexts. But we observe that this is not always the case and the HO-MitM attack can be advantageous in some circumstances, because some key-dependent (or sometimes, not necessarily key-dependent but unknown) component(s) or parameter(s) can be canceled when using more than one plaintexts, depending on the cipher being attacked and how to choose these plaintexts. Thus, we may reduce the number of subkeys required when computing the concerned value-in-the-middle, or reduce the number of unknown parameters in the approach using a precomputation table; this is the core of the HO-MitM attack. As a consequence, the HO-MitM attack may have smaller computational workload than the MitM attack, and even more significantly we may break more rounds of a cipher, as shown by its application to Camellia in the following sections.

How to construct a HO-MitM attack (which is equivalent to constructing the $\mathbf{f}$ function to some extent) depends on the design of the cipher to be attacked. In this paper, when constructing HO-MitM attacks for Camellia we use two approaches to cancel some key-dependent component(s)/parameter(s). The first approach, as described in Section 4.1, is to use an integral [18, 22] property, and the HO-MitM attack obtained by this approach is actually a combination of integral cryptanalysis and the MitM attack (thus it is entitled to an alias — the integral-meet-in-the-middle attack),[5] and it is particularly applicable to Camellia-like Feistel ciphers (i.e., Feistel ciphers with some function inserted after some round). The second approach, as described in Section 5.1, is to use a general differential [5] property, and it has a broader applicability in block ciphers with different structures, say substitution-permutation networks and Feistel networks. Notice that the second approach is not novel and has appeared under the name of MitM attacks in the cryptanalytic literature as to be discussed in Section 3.3. Anyway, the basic idea of the HO-MitM attack gives us more flexibility to use a broader property, just provided that it allows us to use multiple plaintexts to cancel some key-dependent parameters somehow, like those potentially useful properties from higher-order differential cryptanalysis [20, 23], structural cryptanalysis [6], etc.

Though we can call a HO-MitM attack with a basic input unit of l plaintexts an l-th order MitM attack, we will not distinguish HO-MitM attacks with different orders in this paper, and we only distinguish between the HO-MitM attack and the MitM attack. The MitM attack corresponds to the special case $l = 1$ under our definition of the HO-MitM attack.

3.3   Related Work

We note that some previously published MitM attacks used a basic input unit of two plaintexts, for example, in [11, 13, 32] the matched "value-in-the-middle" was defined to be a difference between two (truncated) intermediate values with respect to a chosen-plaintext pair, that is the basic input unit is a pair of chosen plaintexts. Thus, by our definition, these attacks can be categorized as HO-MitM attacks (with a basic input unit of two plaintexts). Some collision attacks, like those in [15], are based on checking whether a pair of plaintexts produces the same (truncated) intermediate value in an approach similar to one used for the MitM attack in Section 3.1, and can be seen as a special case of HO-MitM attacks with a basic input unit of two plaintexts, where the matched value-in-the-middle is 0. Thus, the HO-MitM attack with a basic input unit of two plaintexts is not novel, however, these attacks do not take full advantage of possible approaches to cancel key-dependent parameters, and we use a basic input unit of 256 plaintexts to cancel key-dependent parameters in Section 4.

Broadly speaking, integral cryptanalysis [18, 22] and higher-order differential cryptanalysis [20, 23] are based on an idea which is similar to the basic idea of the HO-MitM attack, but a distinction is that in these cryptanalyses we do not need to guess any secret parameter when going through the rounds covered by an integral distinguisher or a higher-order differential.

---

[5] One may treat this combination as an extension of integral cryptanalysis, but we think it is closer to the MitM attack in spirit, because at a high level it is based on an attack principle similar to that for the MitM attack.

4   HO-MitM Attacks on Reduced Camellia-128/192/256 with FL/FL^{-1} Functions

In this section, we describe 5 and 6-round HO-MitM properties of Camellia with **FL/FL**$^{-1}$ functions, and then present HO-MitM attacks on 10-round Camellia-128 with **FL/FL**$^{-1}$ functions, 11-round Camellia-192 with **FL/FL**$^{-1}$ functions and 12-round Camellia-256 with **FL/FL**$^{-1}$ functions, all of which do not include the whitening operations.

4.1   HO-MitM Properties for 5 and 6-Round Camellia with FL/FL^{-1} Functions

We assume the 5-round Camellia is from Rounds 4 to 8 (including the **FL/FL**$^{-1}$ functions between Rounds 6 and 7), and the 6-round Camellia is from Rounds 3 to 8; see Fig. 1-(a).

Proposition 1. *Suppose a set of 2^{16} sixteen-byte values $X^{(i,j)} = (X_L^{(i,j)}||X_R^{(i,j)})$ $= (m_1, m_2, m_3, m_4, m_5, m_6, m_7, m_8, x^{(i)}, y^{(j)}, m_9, m_{10}, m_{11}, m_{12}, m_{13}, m_{14})$ with $x^{(i)}$ and $y^{(j)}$ taking all the possible values in $\{0,1\}^8$ and the other 14 bytes $m_1, m_2, \cdots, m_{14}$ fixed to arbitrary values, $(i, j = 1, \cdots, 256)$. Then:*

1. *If $Z^{(i,j)} = (Z_L^{(i,j)}||Z_R^{(i,j)})$ is the result of encrypting $X^{(i,j)}$ using Rounds 4 to 8 with the **FL/FL**$^{-1}$ functions between Rounds 6 and 7, then the 8-bit value $\mathbf{P}^{-1}(\bigoplus_{j=1}^{256} Z_R^{(i,j)})[49 \sim 56]$ can be expressed as a function of $x^{(i)}$ and 13 constant 8-bit parameters $c_1, c_2, \cdots, c_{13}$, written $\Phi_{c_1, c_2, \cdots, c_{13}}(x^{(i)})$.*

2. *If $Z^{(i,j)} = (Z_L^{(i,j)}||Z_R^{(i,j)})$ is the result of encrypting $X^{(i,j)}$ using Rounds 3 to 8 with the **FL/FL**$^{-1}$ functions between Rounds 6 and 7, then the 8-bit value $\mathbf{P}^{-1}(\bigoplus_{j=1}^{256} Z_R^{(i,j)})[41 \sim 48]$ can be expressed as a function of $x^{(i)}$ and 21 constant 8-bit parameters $c_1', c_2', \cdots, c_{21}'$, written $\Theta_{c_1', c_2', \cdots, c_{21}'}(x^{(i)})$.*

These HO-MitM properties are obtained by using an integral property of Camellia to cancel some key-dependent components **FL**$^{-1}$, and the basic "value-in-the-middle" is obtained from 256 plaintexts. Below we briefly describe where the advantage comes from in the case of the HO-MitM attacks.

For expediency, when encrypting $X^{(i,j)}$, we denote by $Y_t^{(i,j)}$ the value immediately after the **S** operation of Round t, and by $W_t^{(i,j)}$ the value immediately after the **P** operation of Round t, $(3 \leqslant t \leqslant 8)$.

From [35] we know the following integral property holds for Rounds 3 or 4 to 6 with **FL/FL**$^{-1}$:

$$\bigoplus_{j=1}^{256} \mathbf{FL}^{-1}(\widehat{R}_6^{(i,j)}, KI_2) = 0. \tag{1}$$

By the structure of the 5-round Camellia, we have

$$Z_R^{(i,j)} = \mathbf{FL}^{-1}(X_L^{(i,j)} \oplus W_5^{(i,j)}, KI_2) \oplus W_7^{(i,j)}. \tag{2}$$

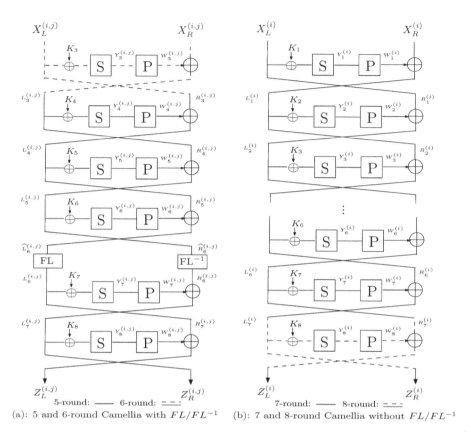

Fig. 1. 5 and 6-round Camellia with $\mathbf{FL}/\mathbf{FL}^{-1}$ and 7 and 8-round Camellia without $\mathbf{FL}/\mathbf{FL}^{-1}$

After applying the $\mathbf{P}^{-1}$ operation to Eq. (2) we get the following equation:

$$\mathbf{P}^{-1}(Z_R^{(i,j)}) = \mathbf{P}^{-1}(\mathbf{FL}^{-1}(X_L^{(i,j)} \oplus W_5^{(i,j)}, KI_2)) \oplus Y_7^{(i,j)}. \tag{3}$$

Observe that $X_L^{(i,j)} \oplus W_5^{(i,j)} = \widehat{R}_6^{(i,j)}$. Thus, by Eqs. (1) and (3) we have

$$\bigoplus_{j=1}^{256} \mathbf{P}^{-1}(Z_R^{(i,j)}) = (\bigoplus_{j=1}^{256} \mathbf{P}^{-1}(\mathbf{FL}^{-1}(X_L^{(i,j)} \oplus W_5^{(i,j)}, KI_2))) \oplus (\bigoplus_{j=1}^{256} Y_7^{(i,j)})$$

$$= \bigoplus_{j=1}^{256} Y_7^{(i,j)}. \tag{4}$$

For the 6-round Camellia, we have

$$Z_R^{(i,j)} = \mathbf{FL}^{-1}(X_R^{(i,j)} \oplus W_3^{(i,j)} \oplus W_5^{(i,j)}, KI_2) \oplus W_7^{(i,j)}. \tag{5}$$

After applying the $\mathbf{P}^{-1}$ operation to Eq. (5) and then by Eq. (1) we have

$$\bigoplus_{j=1}^{256} \mathbf{P}^{-1}(Z_R^{(i,j)}) = (\bigoplus_{j=1}^{256} \mathbf{P}^{-1}(\mathbf{FL}^{-1}(X_R^{(i,j)} \oplus W_3^{(i,j)} \oplus W_5^{(i,j)}, KI_2))) \oplus (\bigoplus_{j=1}^{256} Y_7^{(i,j)})$$

$$= \bigoplus_{j=1}^{256} Y_7^{(i,j)}. \tag{6}$$

Now, observe that the key components $\mathbf{FL}^{-1}(X_L^{(i,j)} \oplus W_5^{(i,j)}, KI_2)$ cancel out in Eqs. (4) and (6). Thus we can compute $\bigoplus_{j=1}^{256} \mathbf{P}^{-1}(Z_R^{(i,j)})$ from the structure of chosen inputs, without guessing the subkeys used in the $\mathbf{FL}^{-1}$ function. This is the origin of the advantage of our HO-MitM attacks. Further, (as given in the full version of this paper), a trivial but complex analysis shows that $\bigoplus_{j=1}^{256} Y_{7,7}^{(i,j)}$ can be expressed as a function of $x^{(i)}$ and 13 constant 8-bit parameters in the 5-round HO-MitM property, and $\bigoplus_{j=1}^{256} Y_{7,6}^{(i,j)}$ can be expressed as a function of $x^{(i)}$ and 21 constant 8-bit parameters in the 6-round HO-MitM property.

In these 5 and 6-round HO-MitM properties, we can regard $x^{(i)}$ as a principle variable and $y^{(j)}$ as a co-variable (note that $y^{(j)}$ is not really a variable, as we use 256 specific values for it), where the co-variable $y^{(j)}$ is used mainly to cancel the key-dependent component $\mathbf{FL}^{-1}$ under the integral property of Camellia.

4.2   Attacking 10-Round Camellia-128 with FL/FL^{-1} Functions

The 5-round HO-MitM property in Proposition 1-1 enables us to break 10-round Camellia-128 with $\mathbf{FL}/\mathbf{FL}^{-1}$ functions. The attacked rounds are from Rounds 2 to 11, and the procedure is as follows. Observe that $\mathbf{P}^{-1}(R_i) = \mathbf{P}^{-1}(L_{i+1}) \oplus S(R_{i+1} \oplus K_{i+1})$.

1. For each of 2^{104} possible values of the 13 constant 8-bit parameters $c_1, c_2, \cdots, c_{13}$, precompute $\Phi_{c_1, c_2, \cdots, c_{13}}(z)$ sequentially for $z = 0, 1, \cdots, 31$. Store the 2^{104} 32-byte sequences in a hash table $\mathcal{L}_\Phi$.

2. Guess a value for $(K_2, K_{3,1}, K_{3,2})$, and we denote the guessed value by $(K_2^*, K_{3,1}^*, K_{3,2}^*)$. Then for $x = 0, 1, \cdots, 31$ and $y = 0, 1, \cdots, 255$, choose plaintext $P^{(x,y)} = (P_L^{(x,y)}, P_R^{(x,y)})$ in the following way, where $\alpha_1, \alpha_2, \cdots, \alpha_8$, $\beta_1, \beta_2, \cdots, \beta_6$ are randomly chosen 8-bit constants:

$$P_L^{(x,y)} = \begin{pmatrix} S_1(x \oplus K_{3,1}^*) \oplus \alpha_1 \\ S_1(x \oplus K_{3,1}^*) \oplus S_2(y \oplus K_{3,2}^*) \oplus \alpha_2 \\ S_1(x \oplus K_{3,1}^*) \oplus S_2(y \oplus K_{3,2}^*) \oplus \alpha_3 \\ S_2(y \oplus K_{3,2}^*) \oplus \alpha_4 \\ S_1(x \oplus K_{3,1}^*) \oplus S_2(y \oplus K_{3,2}^*) \oplus \alpha_5 \\ S_2(y \oplus K_{3,2}^*) \oplus \alpha_6 \\ \alpha_7 \\ S_1(x \oplus K_{3,1}^*) \oplus \alpha_8 \end{pmatrix}^T ,$$

$$P_R^{(x,y)} = \mathbf{P} \begin{pmatrix} S_1(S_1(x \oplus K_{3,1}^*) \oplus \alpha_1 \oplus K_{2,1}^*) \\ S_2(S_1(x \oplus K_{3,1}^*) \oplus S_2(y \oplus K_{3,2}^*) \oplus \alpha_2 \oplus K_{2,2}^*) \\ S_3(S_1(x \oplus K_{3,1}^*) \oplus S_2(y \oplus K_{3,2}^*) \oplus \alpha_3 \oplus K_{2,3}^*) \\ S_4(S_2(y \oplus K_{3,2}^*) \oplus \alpha_4 \oplus K_{2,4}^*) \\ S_5(S_1(x \oplus K_{3,1}^*) \oplus S_2(y \oplus K_{3,2}^*) \oplus \alpha_5 \oplus K_{2,5}^*) \\ S_6(S_2(y \oplus K_{3,2}^*) \oplus \alpha_6 \oplus K_{2,6}^*) \\ S_7(\alpha_7 \oplus K_{2,7}^*) \\ S_8(S_1(x \oplus K_{3,1}^*) \oplus \alpha_8 \oplus K_{2,8}^*) \end{pmatrix}^{\mathrm{T}} \oplus \begin{pmatrix} x \\ y \\ \beta_1 \\ \beta_2 \\ \beta_3 \\ \beta_4 \\ \beta_5 \\ \beta_6 \end{pmatrix}^{\mathrm{T}}.$$

In a chosen-plaintext attack scenario, obtain the ciphertexts for the plaintexts; we denote by $C^{(x,y)}$ the ciphertext for plaintext $P^{(x,y)}$.

3. Guess a value for $(K_{9,7}, K_{10,3}, K_{10,4}, K_{10,5}, K_{10,6}, K_{10,8}, K_{11})$, and we denote the guessed value by $(K_{9,7}^*, K_{10,3}^*, K_{10,4}^*, K_{10,5}^*, K_{10,6}^*, K_{10,8}^*, K_{11}^*)$. Partially decrypt every ciphertext $C^{(x,y)}$ with $(K_{10,3}^*, K_{10,4}^*, K_{10,5}^*, K_{10,6}^*, K_{10,8}^*, K_{11}^*)$ to get the corresponding value for bytes $(1, 2, \cdots, 8, 15)$ just before Round 10; we denote it by $(L_9^{(x,y)}, R_{9,7}^{(x,y)})$. Compute $T^{(x)} = \bigoplus_{y=0}^{255} (\mathbf{P}^{-1}(L_9^{(x,y)})[49 \sim 56] \oplus S_7(R_{9,7}^{(x,y)} \oplus K_{9,7}^*))$. Finally, check whether the sequence $(T^{(0)}, T^{(1)}, \cdots, T^{(31)})$ matches a sequence in $\mathcal{L}_\Phi$; if so, record the guessed value $(K_2^*, K_{3,1}^*, K_{3,2}^*, K_{9,7}^*, K_{10,3}^*, K_{10,4}^*, K_{10,5}^*, K_{10,6}^*, K_{10,8}^*, K_{11}^*)$ and execute Step 4; otherwise, repeat Step 3 with another subkey guess (if all the subkey possibilities are tested in Step 3, repeat Step 2 with another subkey guess).

4. For every recorded value for $(K_{10,3}, K_{10,4}, K_{10,5}, K_{10,6}, K_{10,8})$, exhaustively search the remaining 11 key bytes.

The attack requires $32 \times 256 \times 2^{80} = 2^{93}$ chosen plaintexts. The one-off (i.e., one-time) precomputation requires a memory of $2^{104} \times 32 = 2^{109}$ bytes, and has a time complexity of $2^{104} \times 32 \times 256 \times 2 \times \frac{1}{10} \approx 2^{114.7}$ 10-round Camellia-128 encryptions under the rough estimate that a computation of $\Phi_{c_1,c_2,\cdots,c_{13}}(z)$ equals $256 \times 2 = 512$ one-round Camellia encryptions in terms of time. If the guessed value $(K_2^*, K_{3,1}^*, K_{3,2}^*)$ is correct, the input to Round 4 must have the form $(m_1, m_2, m_3, m_4, m_5, m_6, m_7, m_8, x, y, \beta_1, \beta_2, \beta_3, \beta_4, \beta_5, \beta_6)$, where $m_1, m_2, \cdots, m_8$ are indeterminate constants.

Step 2 has a time complexity of $2^{80} \times 32 \times 256 \times \frac{2+8}{8 \times 10} = 2^{90}$ 10-round Camellia-128 encryptions. Given $(K_2, K_{3,1}, K_{3,2})$, there are only 28 unknown bits for $(K_{9,7}, K_{10,3}, K_{10,4}, K_{10,5}, K_{10,6}, K_{10,8}, K_{11})$, thus Step 3 has a time complexity of about $2^{80+28} \times 32 \times 256 \times \frac{8+5+1}{8 \times 10} \approx 2^{118.5}$ 10-round Camellia-128 encryptions.

In Step 3, if the guessed value $(K_2^*, K_{3,1}^*, K_{3,2}^*, K_{9,7}^*, K_{10,3}^*, K_{10,4}^*, K_{10,5}^*, K_{10,6}^*, K_{10,8}^*, K_{11}^*)$ is correct, the sequence $(T^{(0)}, T^{(1)}, \cdots, T^{(31)})$ must match a sequence in $\mathcal{L}_\Phi$; if the guessed value $(K_2^*, K_{3,1}^*, K_{3,2}^*, K_{9,7}^*, K_{10,3}^*, K_{10,4}^*, K_{10,5}^*, K_{10,6}^*, K_{10,8}^*, K_{11}^*)$ is wrong, the sequence $(T^{(0)}, T^{(1)}, \cdots, T^{(31)})$ matches a sequence in $\mathcal{L}_\Phi$ with a probability of approximately $1 - \binom{2^{104}}{0}(2^{-32 \times 8})^0(1 - 2^{-32 \times 8})^{2^{104}} \approx 2^{-32 \times 8} \times 2^{104} = 2^{-152}$, (assuming the event has a binomial distribution). Consequently, it is expected that about $2^{80+28} \times 2^{-152} = 2^{-44}$ values for $(K_2, K_{3,1}, K_{3,2}, K_{9,7}, K_{10,3}, K_{10,4}, K_{10,5}, K_{10,6}, K_{10,8}, K_{11})$ are recorded in Step 3, meaning only the correct subkey guess will be recorded. Since a total of 40 bits of K_L can be known from $(K_{10,3}, K_{10,4}, K_{10,5}, K_{10,6}, K_{10,8})$, Step 4 takes at most 2^{88} 10-round Camellia-128 encryptions to find the correct 128-bit user key.

Therefore, the attack has a memory complexity of 2^{109} bytes and a total time complexity of approximately $2^{118.6}$ 10-round Camellia-128 encryptions.

4.3   Attacking 11-Round Camellia-192 with FL/FL^{-1} Functions and 12-Round Camellia-256 with FL/FL^{-1} Functions

Similarly, we can use the 6-round HO-MitM property given in Proposition 1-2 to break Rounds 7 to 17 or Rounds 13 to 23 of Camellia-192 with FL/FL^{-1} functions and to break Rounds 7 to 18 of Camellia-256 with FL/FL^{-1} functions. The first 11-round Camellia-192 attack requires 2^{94} chosen plaintexts and a memory of 2^{174} bytes and has a time complexity of approximately $2^{180.2}$ 11-round Camellia-192 encryptions; the second 11-round Camellia-192 attack requires 2^{78} chosen plaintexts and a memory of 2^{174} bytes and has a time complexity of approximately $2^{187.4}$ 11-round Camellia-192 encryptions; and the 12-round Camellia-256 attack requires 2^{94} chosen plaintexts and a memory of 2^{174} bytes and has a time complexity of approximately $2^{237.3}$ 12-round Camellia-256 encryptions. (The details are given in the full version of this paper.)

4.4   A Comparison

We have checked the corresponding MitM properties for the 5 and 6-round Camellia with the FL/FL^{-1} functions, and our result is as follows. For a set of 256 sixteen-byte values $X^{(i)} = (m_1, m_2, m_3, m_4, m_5, m_6, m_7, m_8, x^{(i)}, m_9, m_{10}, m_{11}, m_{12}, m_{13}, m_{14}, m_{15})$ with $x^{(i)}$ taking all the possible values in $\{0,1\}^8$ and the other 15 bytes $m_1, m_2, \cdots, m_{15}$ fixed to arbitrary values, $(i = 1, \cdots, 256)$, then: If $Z^{(i)} = (Z_L^{(i)} \| Z_R^{(i)})$ is the result of encrypting $X^{(i)}$ using Rounds 4 to 8, then $\mathbf{P}^{-1}(Z_R^{(i)})[49 \sim 56]$ is a function of $x^{(i)}$ and 198 constant 1-bit parameters; if $Z^{(i)} = (Z_L^{(i)} \| Z_R^{(i)})$ is the result of encrypting $X^{(i)}$ using Rounds 3 to 8, then $\mathbf{P}^{-1}(Z_R^{(i)})[41 \sim 48]$ is a function of $x^{(i)}$ and 264 constant 1-bit parameters.

Obviously, the numbers of constant 1-bit parameters involved in these MitM properties are much larger than the numbers of constant 1-bit parameters involved in the corresponding HO-MitM properties. Since they are even larger than the key length of Camellia-192/256, it is not preferable to directly use these MitM properties; otherwise, we would like to guess the key bits involved, which are less than the numbers of constant 1-bit parameters involved in the MitM properties. Nevertheless, the MitM properties may potentially become useful in the case we consider only a portion of possible values for the constant 1-bit parameters under a data–memory–time tradeoff [17]; we have checked this direction, and our results are as follows.

Suppose we only consider $\frac{1}{2^{N_1}}$ of the 2^{264} (or 2^{198}) possible values for the 264 (respectively, 198) constant 1-bit parameters in the 6-round (respectively, 5-round) MitM property. For each of the 2^{264-N_1} (respectively, 2^{198-N_1}) possible values for the 264 (respectively, 198) constant 1-bit parameters, we precompute for N_2 chosen inputs $X^{(i)}$. Then, we find we can use the 6-round MitM property to break Rounds 7 to 18 of Camellia-256 with FL/FL^{-1} functions, where we use the 6-round MitM property from Rounds 9 to 14 and guess $(K_{7,1}, K_{7,2}, K_{7,3}, K_{7,5}, K_{7,8}, K_{8,1}, K_{15,6}, K_{16,2}, K_{16,3}, K_{16,5}, K_{16,7}, K_{16,8}, K_{17}, K_{18})$ and a secret 8-bit parameter δ (it has a similar meaning as the δ defined in Section 5.2). The required plaintexts are chosen in a similar approach as in the 14-round Camellia-192 attack in Section 5.2, and the attack procedure is similar to the HO-MitM attack described in Section 4.2, except a major difference: In this 12-round Camellia-256 attack, for every guess of $(K_{7,1}, K_{7,2}, K_{7,3}, K_{7,5}, K_{7,8}, K_{8,1}, \delta)$ we use 2^{N_1+2} structures of N_2 plaintexts $P^{(x)}$ to have a high success probability of 98%. After a similar analysis to that for the HO-MitM attack in Section 4.2, we know that the off-line precomputation phase requires a memory of $N_2 \times 2^{264-N_1} \times \frac{1}{8} = N_2 \times 2^{261-N_1}$ bytes and takes $N_2 \times 2^{264-N_1} \times 3 \times \frac{1}{12} = N_2 \times 2^{262-N_1}$ 12-round Camellia-256 encryptions, and the key-recovery phase requires $2^{N_1+2} \times 2^{56} = 2^{58+N_1}$ chosen plaintexts and takes $N_2 \times 2^{N_1+2} \times 2^{56+158} \times \frac{8+5+1}{8 \times 12} \approx N_2 \times 2^{213.3+N_1}$ 12-round Camellia-256 encryptions (There are only 158 unknown bits for $(K_{15,6}, K_{16,2}, K_{16,3}, K_{16,5}, K_{16,7}, K_{16,8}, K_{17}, K_{18})$ given $(K_{7,1}, K_{7,2}, K_{7,3}, K_{7,5}, K_{7,8}, K_{8,1})$). Therefore, when $N_1 = 24.35$ and $N_2 = 64$, the attack requires $2^{82.35}$ chosen plaintexts and a memory complexity of $2^{242.65}$ bytes, and has a minimum time complexity of $2^{244.65}$ 12-round Camellia-256 encryptions. This MitM attack is slower than the HO-MitM attack on 12-round Camellia-256 mentioned in Section 4.3 which is based on the corresponding 6-round HO-MitM property, and particularly its memory complexity is significantly larger than that for the 12-round HO-MitM attack ($2^{242.65}$ versus 2^{174}).

The 6-round MitM property cannot lead to break 11-round Camellia-192 effectively. The 11-round Camellia-192 that the 5-round MitM property seems to most possibly break are from Rounds 13 to 23, where we use the 5-round MitM property from Rounds 16 to 20 and guess $(K_{13}, K_{14}, K_{15,1}, K_{21,7}, K_{22,3}, K_{22,4}, K_{22,5}, K_{22,6}, K_{22,8}, K_{23})$. There are only 2^{64} possible values for (K_{13}, K_{14}). For every guess of $(K_{13}, K_{14}, K_{15,1})$ we also use 2^{N_1+2} structures of N_2 plaintexts $P^{(x)}$ to have a high success probability 98%. Similarly, the precomputation phase requires a memory of $N_2 \times 2^{198-N_1} \times \frac{1}{8} = N_2 \times 2^{195-N_1}$ bytes and takes $N_2 \times 2^{198-N_1} \times 2 \times \frac{1}{11} = N_2 \times 2^{196.6-N_1}$ 11-round Camellia-192 encryptions, and the key-recovery phase requires $N_2 \times 2^{N_1+2} \times 2^{72} = N_2 \times 2^{74+N_1}$ chosen plaintexts and takes $N_2 \times 2^{N_1+2} \times 2^{72+112} \times \frac{8+5+1}{8 \times 11} \approx N_2 \times 2^{183.4+N_1}$ 11-round Camellia-192 encryptions. Therefore, the smallest total time complexity happens when $N_1 = 6.6$, which is $N_2 \times 2^{191}$ 11-round Camellia-192 encryptions, and under this circumstance the data complexity is $N_2 \times 2^{80.6}$ chosen plaintexts and the memory complexity is $N_2 \times 2^{188.4}$ bytes. However, N_2 should be far larger than 2 to filter out a reasonable number of wrong candidates for $(K_{13}, K_{14}, K_{15,1}, K_{21,7}, K_{22,3}, K_{22,4}, K_{22,5}, K_{22,6}, K_{22,8}, K_{23})$. This means the 5 or 6-round MitM property cannot be used to break 11-round Camellia-192 with $\mathbf{FL}/\mathbf{FL}^{-1}$ functions faster than exhaustive key search (unless some auxiliary trick can be found to improve the attack), but anyway the corresponding 6-round HO-MitM property can easily do so as briefed.

By any means the 5-round MitM property cannot be used to break 10-round Camellia-128 with $\mathbf{FL}/\mathbf{FL}^{-1}$ functions, not to mention the 6-round MitM property, but the corresponding 5-round HO-MitM property does so as presented in Section 4.2.

This comparison shows that the HO-MitM attack technique can achieve some advantages over the MitM attack technique in some circumstances. Besides, we learn from Table 1 that the HO-MitM attack technique works better than integral cryptanalysis (including square cryptanalysis) for Camellia. That is, the HO-MitM attack technique with the alias of the integral-meet-in-the-middle attack can work better than either of its two constituents — integral cryptanalysis and the MitM attack — in some circumstances. (Most recently, we observed that there are 127 and 199 constant 1-bit parameters respectively for the 5 and 6-round HO-MitM properties obtained from the above 5 and 6-round MitM properties by taking XOR between two inputs to cancel some constant parameters, which can be used to break 11-round Camellia-192 and 12-round Camellia-256 but marginally break 10-round Camellia-128.) Anyway, a property of the $\mathbf{FL}^{-1}$ function can be exploited to obtain different 5 and 6-round MitM properties with a smaller number of 1-bit constant parameters, that can be used to devise MitM attacks on the same numbers of attacked rounds of the Camellia versions [27].

5 HO-MitM Attacks on Reduced Camellia-192/256 without $\mathbf{FL/FL}^{-1}$ Functions

In this section we give 7 and 8-round HO-MitM properties of Camellia without $\mathbf{FL/FL}^{-1}$ functions, and then describe HO-MitM attacks on 14-round Camellia-192 without $\mathbf{FL/FL}^{-1}$ functions and 16-round Camellia-256 without $\mathbf{FL/FL}^{-1}$ functions, both of which do not include the whitening operations.

5.1 HO-MitM Properties for 7 and 8-Round Camellia without $\mathbf{FL/FL}^{-1}$ Functions

We construct these 7 and 8-round HO-MitM properties by using a general differential property to cancel some constant parameters, where the basic concerned "value-in-the-middle" is obtained from two plaintexts. See Fig. 1-(b).

Proposition 2. *Suppose a set of 256 sixteen-byte values* $X^{(i)} = (X_L^{(i)} \| X_R^{(i)}) = (m_1, m_2, m_3, m_4, m_5, m_6, m_7, m_8, x^{(i)}, m_9, m_{10}, m_{11}, m_{12}, m_{13}, m_{14}, m_{15})$ *with* $x^{(i)}$ *taking all the possible values in* $\{0,1\}^8$ *and the other 15 bytes* $m_1, m_2, \cdots, m_{15}$ *fixed to arbitrary values,* $(i = 1, \cdots, 256)$. *Let* $i_1, i_2 \in \{1, 2, \cdots, 256\}$ *and* $i_1 \neq i_2$, *then:*

1. *If* $Z^{(i)} = (Z_L^{(i)} \| Z_R^{(i)})$ *is the result of encrypting* $X^{(i)}$ *using 7-round Camellia without* $\mathbf{FL/FL}^{-1}$ *functions, then* $\mathbf{P}^{-1}(Z_R^{(i_1)} \oplus Z_R^{(i_2)})[41 \sim 48]$ *can be expressed as a function of* $x^{(i_1)}, x^{(i_2)}$ *and 20 constant 8-bit parameters* $c_1, c_2, \cdots,$ c_{20}, *written* $\Gamma_{c_1, c_2, \cdots, c_{20}}(x^{(i_1)}, x^{(i_2)})$.

2. *If* $Z^{(i)} = (Z_L^{(i)} \| Z_R^{(i)})$ *is the result of encrypting* $X^{(i)}$ *using 8-round Camellia without* $\mathbf{FL/FL}^{-1}$ *functions, then* $\mathbf{P}^{-1}(Z_R^{(i_1)} \oplus Z_R^{(i_2)})[41 \sim 48]$ *can be expressed as a function of* $x^{(i_1)}, x^{(i_2)}$ *and 28 constant 8-bit parameters* $c_1', c_2', \cdots,$ c_{28}', *written* $\Psi_{c_1', c_2', \cdots, c_{28}'}(x^{(i_1)}, x^{(i_2)})$.

5.2   Attacking 14-Round Camellia-192 without FL/FL^{-1} Functions

We first remind the reader that compared with the above attacks, this attack as well as the attack described in the next subsection uses a different approach to choose plaintexts, that is, there is an additional secret parameter denoted by δ. This approach to choose plaintexts/ciphertexts was introduced in [26].

The 7-round HO-MitM property in Proposition 2-1 can be used to attack 14-round Camellia-192 without **FL/FL**$^{-1}$ functions. We attack Rounds 2 to 15 and use the 7-round HO-MitM property from Rounds 5 to 11, where we guess $(K_2, K_{3,1}, K_{3,2}, K_{3,3}, K_{3,5}, K_{3,8}, K_{4,1}, K_{12,6}, K_{13,2}, K_{13,3}, K_{13,5}, K_{13,7}, K_{13,8}, K_{14}, K_{15})$, plus an additional secret 8-bit parameter δ which is defined to be $\delta = \gamma_1 \oplus \gamma_2 \oplus \gamma_3 \oplus S_4(\gamma_4 \oplus K_{3,4}) \oplus S_6(\gamma_5 \oplus K_{3,6}) \oplus S_7(\gamma_6 \oplus K_{3,7})$, with $\gamma_1, \gamma_2, \cdots, \gamma_6$ being 6 randomly chosen 8-bit constants. Here, δ is used below to allow us to have qualified inputs to Round 5 and know the values at byte (9) of the inputs to Round 5, so that we can sort the computed sequences in the key-recovery phase.

For each possible value of the 20 one-byte parameters $c_1, c_2, \cdots, c_{20}$, precompute $\Gamma_{c_1,c_2,\cdots,c_{20}}(0, z)$ for $z = 1, 2, \cdots, 63$ sequentially. Then for every guess of $(K_2, K_{3,1}, K_{3,2}, K_{3,3}, K_{3,5}, K_{3,8}, K_{4,1}, \delta)$, denoted by $(K_2^*, K_{3,1}^*, K_{3,2}^*, K_{3,3}^*, K_{3,5}^*, K_{3,8}^*, K_{4,1}^*, \delta^*)$, choose 64 plaintexts $P^{(x)} = (P_L^{(x)}, P_R^{(x)})$ in the following way $(x = 0, 1, \cdots, 63)$, where $\alpha_1, \alpha_2, \cdots, \alpha_5, \beta_1, \beta_2, \cdots, \beta_7$ are randomly chosen 8-bit constants:

$$
P_L^{(x)} = \mathbf{P}
\begin{pmatrix}
S_1(S_1(x \oplus K_{4,1}^*) \oplus \alpha_1 \oplus K_{3,1}^*) \\
S_2(S_1(x \oplus K_{4,1}^*) \oplus \alpha_2 \oplus K_{3,2}^*) \\
S_3(S_1(x \oplus K_{4,1}^*) \oplus \alpha_3 \oplus K_{3,3}^*) \\
\gamma_1 \\
S_5(S_1(x \oplus K_{4,1}^*) \oplus \alpha_4 \oplus K_{3,5}^*) \\
\gamma_2 \\
\gamma_3 \\
S_8(S_1(x \oplus K_{4,1}^*) \oplus \alpha_5 \oplus K_{3,8}^*)
\end{pmatrix}^T
\oplus
\begin{pmatrix}
x \oplus \delta^* \\
\beta_1 \\
\beta_2 \\
\beta_3 \\
\beta_4 \\
\beta_5 \\
\beta_6 \\
\beta_7
\end{pmatrix}^T,
$$

$$
P_R^{(x)} = \mathbf{F}(P_L^{(x)}, K_2^*) \oplus
\begin{pmatrix}
S_1(x \oplus K_{4,1}^*) \oplus \alpha_1 \\
S_1(x \oplus K_{4,1}^*) \oplus \alpha_2 \\
S_1(x \oplus K_{4,1}^*) \oplus \alpha_3 \\
\gamma_4 \\
S_1(x \oplus K_{4,1}^*) \oplus \alpha_4 \\
\gamma_5 \\
\gamma_6 \\
S_1(x \oplus K_{4,1}^*) \oplus \alpha_5
\end{pmatrix}^T.
$$

If the guessed value for $(K_2, K_{3,1}, K_{3,2}, K_{3,3}, K_{3,5}, K_{3,8}, K_{4,1}, \delta)$ is correct, the input to Round 5 must have the form $(m_1, m_2, m_3, m_4, m_5, m_6, m_7, m_8, x, m_9, m_{10}, m_{11}, m_{12}, m_{13}, m_{14}, m_{15})$, where $m_1, m_2, \cdots, m_{15}$ are indeterminate constants. The remaining steps are similar to the 10-round Camellia-128 attack.

There are $2^{64+40} = 2^{104}$ possible values for $(K_2, K_{3,1}, K_{3,2}, K_{3,3}, K_{3,5}, K_{3,8}, K_{4,1})$ by the key schedule of Camellia-192, thus the attack requires $64 \times 2^{104+8} = 2^{118}$ chosen plaintexts. Given $(K_2, K_{3,1}, K_{3,2}, K_{3,3}, K_{3,5}, K_{3,8}, K_{4,1})$, there are only 36 unknown bits for $(K_{12,6}, K_{13,2}, K_{13,3}, K_{13,5}, K_{13,7}, K_{13,8}, K_{14}, K_{15})$, so the time complexity in the key recovery phase is approximately $2^{104+8+36} \times 64 \times \frac{8+8+5+1}{8 \times 14} \approx 2^{151.7}$ 14-round Camellia-192 encryptions. As a result, the attack requires a memory of $2^{160} \times 63 \approx 2^{166}$ bytes, and its time complexity is dominated by the time complexity of a one-off precomputation of $\Gamma_{c_1,c_2,\cdots,c_{20}}(0,z)$, which is approximately $2^{160} \times 64 \times 5 \times \frac{1}{14} \approx 2^{164.6}$ 14-round Camellia-192 encryptions under the rough estimate that a computation of $\Gamma_{c_1,c_2,\cdots,c_{20}}(0,z)$ equals 5 one-round Camellia-192 encryptions in terms of time except a one-off computation with connection to the value 0 for each $(c_1, c_2, \cdots, c_{20})$.

Since the attack's time complexity is dominated by the time complexity of the one-off precomputation $\Gamma_{c_1,c_2,\cdots,c_{20}}(0,z)$, we can use a data–time–memory tradeoff to slightly reduce the memory and time complexity by precomputing only for a proportion of the 20 constant 8-bit parameters $c_1, c_2, \cdots, c_{20}$ and then using more data to achieve a reasonable success probability: Such an attack requires 2^{125} chosen plaintexts and a memory of 2^{161} bytes, and has a total time complexity of $2^{160.3}$ 14-round Camellia-192 encryptions, with a success probability of 98%.

5.3 Attacking 16-Round Camellia-256 without FL/FL^{-1} Functions

Similarly, we can use the 8-round HO-MitM property given in Proposition 2-2 to break the first 16 rounds of Camellia-256 without **FL/FL**$^{-1}$ functions, where the 8-round HO-MitM property is used from Rounds 4 to 11, and we guess $(K_1, K_{2,1}, K_{2,2}, K_{2,3}, K_{2,5}, K_{2,8}, K_{3,1}, \delta, K_{12,6}, K_{13,2}, K_{13,3}, K_{13,5}, K_{13,7}, K_{13,8}, K_{14}, K_{15}, K_{16})$, here δ is similar to the δ defined in Section 5.2. For each possible value of the 28 one-byte parameters $c'_1, c'_2, \cdots, c'_{28}$, precompute $\Psi_{c'_1,c'_2,\cdots,c'_{28}}(0,z)$ for $z = 1, 2, \cdots, 63$ sequentially. The one-off precomputation requires a memory of $2^{224} \times 63 \approx 2^{230}$ bytes, and has a time complexity of $2^{224} \times 64 \times 5 \times \frac{1}{16} \approx 2^{228.4}$ 16-round Camellia-256 encryptions under the rough estimate that a computation of $\Psi_{c'_1,c'_2,\cdots,c'_{28}}$ equals 5 one-round Camellia-256 encryptions in terms of time plus a one-off computation with connection to the value 0 for each $(c'_1, c'_2, \cdots, c'_{28})$. Given $(K_1, K_{2,1}, K_{2,2}, K_{2,3}, K_{2,5}, K_{2,8}, K_{3,1})$, there are only 128 unknown bits for $(K_{12,6}, K_{13,2}, K_{13,3}, K_{13,5}, K_{13,7}, K_{13,8}, K_{14}, K_{15}, K_{16})$. After a similar analysis, we learn that the attack requires at most $2^{64+48+8} = 2^{120}$ chosen plaintexts and has a total time complexity of approximately $2^{120+128} \times 64 \times \frac{8+8+8+5+1}{8\times16} \approx 2^{252}$ 16-round Camellia-256 encryptions.

5.4 A Comparison

When constructing the 7 and 8-round HO-MitM properties, we first obtain the corresponding 7 and 8-round MitM properties: The value-in-the-middle $\mathbf{P}^{-1}(X_L^{(i)} \oplus Z_R^{(i)})[41 \sim 48] = Y_{2,6}^{(i)} \oplus Y_{4,6}^{(i)} \oplus Y_{6,6}^{(i)}$ in the 7-round MitM property can be expressed as a function of $x^{(i)}$ and 21 constant 8-bit parameters; and the value-in-the-middle $\mathbf{P}^{-1}(X_R^{(i)} \oplus Z_R^{(i)})[41 \sim 48] = Y_{1,6}^{(i)} \oplus Y_{3,6}^{(i)} \oplus Y_{5,6}^{(i)} \oplus Y_{7,6}^{(i)}$ in the 8-round MitM property can be expressed as a function of $x^{(i)}$ and 30 constant 8-bit parameters, (see Fig. 1-(b) for the undefined notation). Then, by taking XOR under two plaintexts $X^{(i_1)}$ and $X^{(i_2)}$, we cancel the two constant terms $\mathbf{P}^{-1}(X_L^{(i)})[41 \sim 48]$ and $Y_{2,6}^{(i)}$ in the 7-round MitM property, and cancel the three constant terms $\mathbf{P}^{-1}(X_R^{(i)})[41 \sim 48]$, $Y_{1,6}^{(i)}$ and $Y_{3,6}^{(i)}$ in the 8-round MitM property. (The details are given in the full version of this paper.)

The 7 and 8-round MitM properties can be respectively used to break 14-round Camellia-192 without $\mathbf{FL}/\mathbf{FL}^{-1}$ functions and 16-round Camellia-256 without $\mathbf{FL}/\mathbf{FL}^{-1}$ functions; the attacked rounds are the same as in the HO-MitM attacks given in Sections 5.2 and 5.3, and the attack procedures are rather similar as well, except that we use the following way to deal with the unknown 8-bit parameter $\mathbf{P}^{-1}(X_R^{(i)})[41 \sim 48]$ or $\mathbf{P}^{-1}(X_L^{(i)})[41 \sim 48]$: For a 64-byte sequence obtained in the key-recovery phase, we XOR a possible value of $\mathbf{P}^{-1}(X_R^{(i)})[41 \sim 48]$ or $\mathbf{P}^{-1}(X_L^{(i)})[41 \sim 48]$ to all 64 basic units of value-in-the-middle in the sequence and then check the resulting sequence, and repeat this process for all the 256 possible values of $\mathbf{P}^{-1}(X_R^{(i)})[41 \sim 48]$ or $\mathbf{P}^{-1}(X_L^{(i)})[41 \sim 48]$.

Similarly, the MitM attack on 14-round Camellia-192 without $\mathbf{FL}/\mathbf{FL}^{-1}$ functions has a data complexity of $64 \times 2^{104+8} = 2^{118}$ chosen plaintexts, a memory complexity of $64 \times 2^{21 \times 8} = 2^{174}$ bytes and a time complexity of $64 \times 2^{21 \times 8} \times 5 \times \frac{1}{14} + 64 \times 2^{112+36} \times \frac{8+8+5+1}{8 \times 14} \approx 2^{172.6}$ 14-round Camellia-192 encryptions. The time complexity is dominated by the one-off precomputation, and we can use a data–memory–time tradeoff to obtain a 14-round Camellia-192 attack with a data complexity of $2^{118+7} = 2^{125}$ chosen plaintexts, a memory complexity of $2^{174-5} = 2^{169}$ bytes, a time complexity of $2^{172.6-5} + 2^{151.7+7} \approx 2^{167.6}$ 14-round Camellia-192 encryptions and a success probability of 98%. The MitM attack on 16-round Camellia-256 without $\mathbf{FL}/\mathbf{FL}^{-1}$ functions has a data complexity of at most $2^{112+8} = 2^{120}$ chosen plaintexts, a memory complexity of $64 \times 2^{30 \times 8} = 2^{246}$ bytes and a time complexity of $64 \times 2^{30 \times 8} \times 5 \times \frac{1}{16} + 64 \times 2^{120+128} \times \frac{8+8+5+1}{8 \times 16} \approx 2^{252}$ 16-round Camellia-256 encryptions. These MitM attacks are effective but less efficient than the HO-MitM attacks described earlier.

6   Conclusions

In this paper, we have proposed an extension of the meet-in-the-middle (MitM) attack, called the higher-order meet-in-the-middle (HO-MitM) attack; it is based on using multiple plaintexts to cancel some key-dependent component(s) or parameter(s) when constructing a basic unit of value-in-the-middle. We have described a novel approach, which combines integral cryptanalysis with the MitM attack, to construct HO-MitM attacks on 10-round Camellia-128 with FL/ FL^{-1} functions, 11-round Camellia-192 with FL/FL^{-1} functions and 12-round Camellia-256 with FL/FL^{-1} functions, all of which do not include the whitening operations. The HO-MitM attack obtained by this approach can also be called the integral-meet-in-the-middle attack, and it can work better than either integral cryptanalysis or the MitM attack in certain circumstances. We have used an existing approach to construct HO-MitM attacks on 14-round Camellia-192 without FL/FL^{-1} functions and 16-round Camellia-256 without FL/FL^{-1} functions, both of which do not include the whitening operations.

The HO-MitM attack is a general cryptanalytic technique, and can potentially be used to cryptanalyse other block ciphers, in particular the integral-meet-in-the-middle attack is applicable to Camellia-like Feistel ciphers (i.e. Feistel ciphers with some function inserted after some round). An interesting direction for future research is to investigate new approaches to construct HO-MitM attacks.

Acknowledgments. The authors would like to thank several anonymous referees for their comments on earlier versions of the paper.

References

1. Aoki, K., Ichikawa, T., Kanda, M., Matsui, M., Moriai, S., Nakajima, J., Tokita, T.: *Camellia*: A 128-Bit Block Cipher Suitable for Multiple Platforms - Design and Analysis. In: Stinson, D.R., Tavares, S. (eds.) SAC 2000. LNCS, vol. 2012, pp. 39–56. Springer, Heidelberg (2001)
2. Bai, D., Li, L.: New Impossible Differential Attacks on Camellia. In: Ryan, M.D., Smyth, B., Wang, G. (eds.) ISPEC 2012. LNCS, vol. 7232, pp. 80–96. Springer, Heidelberg (2012)
3. Biham, E., Biryukov, A., Shamir, A.: Cryptanalysis of Skipjack Reduced to 31 Rounds Using Impossible Differentials. In: Stern, J. (ed.) EUROCRYPT 1999. LNCS, vol. 1592, pp. 12–23. Springer, Heidelberg (1999)
4. Biham, E., Dunkelman, O., Keller, N.: The Rectangle Attack - Rectangling the Serpent. In: Pfitzmann, B. (ed.) EUROCRYPT 2001. LNCS, vol. 2045, pp. 340–357. Springer, Heidelberg (2001)
5. Biham, E., Shamir, A.: Differential cryptanalysis of DES-like cryptosystems. Journal of Cryptology 4(1), 3–72 (1991)
6. Biryukov, A., Shamir, A.: Structural cryptanalysis of SASAS. Journal of Cryptology 23(4), 505–518 (2010)
7. Chen, J., Jia, K., Yu, H., Wang, X.: New Impossible Differential Attacks of Reduced-Round Camellia-192 and Camellia-256. In: Parampalli, U., Hawkes, P. (eds.) ACISP 2011. LNCS, vol. 6812, pp. 16–33. Springer, Heidelberg (2011)
8. CRYPTREC — Cryptography Research and Evaluatin Committees, report 2002 (2003)
9. Daemen, J., Knudsen, L.R., Rijmen, V.: The Block Cipher SQUARE. In: Biham, E. (ed.) FSE 1997. LNCS, vol. 1267, pp. 149–165. Springer, Heidelberg (1997)
10. Demirci, H., Selçuk, A.A.: A Meet-in-the-Middle Attack on 8-Round AES. In: Nyberg, K. (ed.) FSE 2008. LNCS, vol. 5086, pp. 116–126. Springer, Heidelberg (2008)
11. Demirci, H., Taşkın, İ., Çoban, M., Baysal, A.: Improved Meet-in-the-Middle Attacks on AES. In: Roy, B., Sendrier, N. (eds.) INDOCRYPT 2009. LNCS, vol. 5922, pp. 144–156. Springer, Heidelberg (2009)
12. Diffie, W., Hellman, M.: Exhaustive cryptanalysis of the NBS data encryption standard. Computer 10(6), 74–84 (1977)
13. Dunkelman, O., Keller, N., Shamir, A.: Improved Single-Key Attacks on 8-Round AES-192 and AES-256. In: Abe, M. (ed.) ASIACRYPT 2010. LNCS, vol. 6477, pp. 158–176. Springer, Heidelberg (2010)
14. Lei, D., Chao, L., Feng, K.: New Observation on Camellia. In: Preneel, B., Tavares, S. (eds.) SAC 2005. LNCS, vol. 3897, pp. 51–64. Springer, Heidelberg (2006)
15. Gilbert, H., Minier, M.: A collision attack on 7 rounds of Rijndael. In: Proceedings of the Third Advanced Encryption Standard Candidate Conference, pp. 230–241. NIST (2000)

16. Hatano, Y., Sekine, H., Kaneko, T.: Higher Order Differential Attack of Camellia(II). In: Nyberg, K., Heys, H.M. (eds.) SAC 2002. LNCS, vol. 2595, pp. 39–56. Springer, Heidelberg (2003)
17. Hellman, M.E.: A cryptanalytic time–memory trade-off. IEEE Transcations on Information Theory 26(4), 401–406 (1980)
18. Hu, Y., Zhang, Y., Xiao, G.: Integral cryptanalysis of SAFER+. Electronics Letters 35(17), 1458–1459 (1999)
19. International Standardization of Organization (ISO), International Standard – ISO/IEC 18033-3, Information technology – Security techniques – Encryption algorithms – Part 3: Block ciphers (2005)
20. Knudsen, L.R.: Truncated and Higher Order Differentials. In: Preneel, B. (ed.) FSE 1994. LNCS, vol. 1008, pp. 196–211. Springer, Heidelberg (1995)
21. Knudsen, L.R.: DEAL — a 128-bit block cipher. Technical report, Department of Informatics, University of Bergen, Norway (1998)
22. Knudsen, L.R., Wagner, D.: Integral Cryptanalysis. In: Daemen, J., Rijmen, V. (eds.) FSE 2002. LNCS, vol. 2365, pp. 112–127. Springer, Heidelberg (2002)
23. Lai, X.: Higher order derivatives and differential cryptanalysis. In: Communications and Cryptography, pp. 227–233. Academic Publishers (1994)
24. Liu, Y., Li, L., Gu, D., Wang, X., Liu, Z., Chen, J., Li, W.: New Observations on Impossible Differential Cryptanalysis of Reduced-Round Camellia. In: Canteaut, A. (ed.) FSE 2012. LNCS, vol. 7549, pp. 90–109. Springer, Heidelberg (2012)
25. Lu, J.: Cryptanalysis of block ciphers. PhD thesis, University of London, UK (2008)
26. Lu, J., Wei, Y., Kim, J., Fouque, P.-A.: Cryptanalysis of reduced versions of the Camellia block cipher. In: Miri, A., Vaudenay, S. (eds.) Pre-proceedings of SAC 2011 (2011), http://sac2011.ryerson.ca/SAC2011/LWKF.pdf, An editorially revised version is to appear in IET Information Security
27. Lu, J., Wei, Y., Pasalic, E., Fouque, P.-A.: Meet-in-the-Middle Attack on Reduced Versions of the Camellia Block Cipher. In: Hanaoka, G., Yamauchi, T. (eds.) IWSEC 2012. LNCS, vol. 7631, pp. 197–215. Springer, Heidelberg (2012)
28. Mala, H., Shakiba, M., Dakhilalian, M., Bagherikaram, G.: New Results on Impossible Differential Cryptanalysis of Reduced–Round Camellia–128. In: Jacobson Jr., M.J., Rijmen, V., Safavi-Naini, R. (eds.) SAC 2009. LNCS, vol. 5867, pp. 281–294. Springer, Heidelberg (2009)
29. Mala, H., Dakhilalian, M., Shakiba, M.: Impossible differential cryptanalysis of reduced-round Camellia-256. IET Information Security 5(3), 129–134 (2011)
30. Matsui, M.: Linear Cryptanalysis Method for DES Cipher. In: Helleseth, T. (ed.) EUROCRYPT 1993. LNCS, vol. 765, pp. 386–397. Springer, Heidelberg (1994)
31. NESSIE — New European Schemes for Signatures, Integrity, and Encryption, Final report of European project IST-1999-12324 (2004)
32. Wei, Y., Lu, J., Hu, Y.: Meet-in-the-Middle Attack on 8 Rounds of the AES Block Cipher under 192 Key Bits. In: Bao, F., Weng, J. (eds.) ISPEC 2011. LNCS, vol. 6672, pp. 222–232. Springer, Heidelberg (2011)
33. Wu, W., Feng, D., Chen, H.: Collision Attack and Pseudorandomness of Reduced-Round Camellia. In: Handschuh, H., Hasan, M.A. (eds.) SAC 2004. LNCS, vol. 3357, pp. 252–266. Springer, Heidelberg (2004)
34. Wagner, D.: The Boomerang Attack. In: Knudsen, L.R. (ed.) FSE 1999. LNCS, vol. 1636, pp. 156–170. Springer, Heidelberg (1999)
35. Yeom, Y., Park, S., Kim, I.: A study of integral type cryptanalysis on Camellia. In: Proceedings of the 2003 Symposium on Cryptography and Information Security, pp. 453–456. IEICE (2003)

Double-SP Is Weaker Than Single-SP: Rebound Attacks on Feistel Ciphers with Several Rounds

Yu Sasaki

NTT Secure Platform Laboratories, NTT Corporation
3-9-11 Midori-cho, Musashino-shi, Tokyo 180-8585 Japan
sasaki.yu@lab.ntt.co.jp

Abstract. The current paper presents rebound attacks on generalized Feistel network (GFN) with double-SP functions, and show that double-SP functions are weaker than single-SP functions when a number of rounds is small. In 2011, Bogdanov and Shibutani showed that double-SP functions for R rounds could generate more active bytes than single-SP functions for $2R$ rounds, when R approaches to infinity. Hence, double-SP functions resist the differential and linear attacks more efficiently than single-SP functions. However, in practice, R is relatively small, and thus a comparison with dedicated attacks is also important. For 4-branch type-2 GFN with single-SP functions, the current best attack is up to 11 rounds (22 SP-layers) while no result exists for double-SP functions. In this paper, we present the first cryptanalysis for 4-branch type-2 GFN with double-SP functions. Up to 6 rounds (24 SP-layers), we can find near-collisions when such functions are instantiated in compression function modes, e.g. Davies-Meyer mode. The attack is extended to 7 rounds (28 SP-layers) with respect to a non-ideal property. The important knowledge provided with this paper is that including more active bytes does not immediately indicate stronger security. This is because attackers may control behaviors of several active S-boxes and mount efficient attacks.

Keywords: rebound attack, generalized Feistel network, double-SP, single-SP, near-collision, known-key distinguisher, (controlled) active S-box.

1 Introduction

Designing good block-ciphers and hash functions has been a challenging topic for a long time. Various designs have been considered to achieve high security and good performance. Feistel network and generalized Feistel network (GFN), which are shown in Fig. 1, are widely used structures to build such primitives. Specifically, GFN is known to be suitable for light-weight designs rather than the standard Feistel network.

The core of these constructions is the design of the round function. One of the most popular methods is combining an S-box transformation (S-layer) and a linear transformation (P-layer). The round function that consists of a subkey XOR, S-layer, and P-layer, is called a *single-SP* function.

S. Galbraith and M. Nandi (Eds.): INDOCRYPT 2012, LNCS 7668, pp. 265–282, 2012.

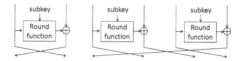

Fig. 1. Left: Feistel network **Right:** 4-branch type-2 GFN

In 2011, Bogdanov and Shibutani [1] showed that 4-branch GFN with double-SP functions was stronger than the one with single-SP functions against differential and linear cryptanalysis in terms of the number of active S-boxes, where the *double-SP* function consists of a subkey XOR and two iterations of the S-layer, and P-layer. Considering the fact that several designs adopt single-SP functions but no design adopts double-SP functions, the result in [1] is a breakthrough toward good block-cipher and hash function designs. [1] compares the security of the single-SP and double-SP functions only from the viewpoint of the number of active S-boxes. Therefore, comparing the single-SP and double-SP functions from other perspectives is useful to enhance the design theory.

Note that good designs for Feistel network or its variant has been discussed actively in recent years. In 2010, Bogdanov analyzed round functions consisting of one SP-layer and additional S-layer (SPS functions) [2]. Actually, SPS functions are adopted by several designs, e.g., E2 [3] and Piccolo [4]. In 2011, Bogdanov and Shibutani showed another result of double-SP functions for 3-branch GFN [5]. Regarding the network, Generalized GFN was proposed in 2010 [6], and its philosophy can be seen in several block ciphers, e.g., LBlock [7] and TWINE [8]. The security of source-heavy and target-heavy GFN is also discussed [9]. From this background, the comparison of single-SP and double-SP functions would be useful. Note that the hash function SHAvite-3 [10] adopts Feistel network or 4-branch type-2 GFN using 3 AES rounds as a round function, which can be viewed as a triple-SP function. Its security discussion, especially the rebound attack by Minier *et al.* [11], is also useful to understand the security of the Feistel scheme.

Recently cryptographers have paid attention to the security of block ciphers when the key value is known to the attackers. Such an approach is called known-key distinguisher, which was firstly discussed by Knudsen and Rijmen [12] and partially formalized by Minier *et al.* [13]. If the analysis target is a compression function such as the Davies-Meyer (DM), Matyas-Meyer-Oseas (MMO) and Miyaguchi-Preneel modes instantiating an internal block cipher, the known-key distinguisher on the internal block cipher can be directly converted into a distinguisher on the compression function. For example, a known-key distinguisher proposed by Mendel *et al.* [14] on 7-round AES can also be viewed as a distinguisher on AES-based compression function.

In 2011, Sasaki and Yasuda applied the rebound attack for the standard (2-branch) Feistel ciphers with single-SP round functions in the known-key setting [15]. They successfully mounted a distinguisher up to 11 rounds (including 11 SP-layers). Note that their results can be trivially extended to 4-branch type-2

GFN with single-SP functions by simply running the same procedure for two round functions in parallel. Therefore, for single-SP functions, up to 11 rounds of 4-branch type-2 GFN (including 22 SP-layers) can be distinguished. In 2012, Sasaki *et al.* improved the complexity of this attack [16]. However, the number of attacked round is still 11.

The number of total S-boxes in R rounds with double-SP functions is the same as the one in $2R$ rounds with single-SP functions. It is proved by [1] that, the minimum number of active S-boxes in every 6 rounds for single-SP and double-SP functions are $2(r+1)+2$ and $6(r+1)$, respectively, where r is the number of S-boxes included in a single S-layer, and r is bigger than one for the SP structure. From this result, for any choice of r, 6-round double-SP functions always achieve a bigger minimum number of active S-boxes than 12-round single-SP functions. In [1], the security of single-SP and double-SP functions are compared regarding the ratio of the number active S-boxes when the number of rounds, R, approaches to infinity. While it is a reasonable approach in some sense, such a metric ignores a constant factor of the number of active S-boxes. Because R is relatively small in practice to obtain a good performance, we cannot ignore the constant factor for evaluating specific designs. Therefore, comparing the security with dedicated attacks is also important. For single-SP functions, the current best attack is up to 11 rounds (22 SP-layers) while no result exists for double-SP functions.

Our Contributions

In this paper, we present the first cryptanalytic results for 4-branch type-2 GFN with double-SP functions in order to compare the security of single-SP and double-SP functions with respect to the number of attacked rounds. Our approach is a rebound attack. As a result of the analysis, we show that up to 6 rounds (24 SP-layers), we can find paired values whose input and output differences are identical for a half of the state. The attack complexity is 2^c round function computations and a memory to store 2^c internal state. The attack can be used to find near-collisions when such functions are instantiated in compression function modes. The attack can also be regarded as a known-key distinguisher for the block cipher. We then extended the attack to 7 rounds (28 SP-layers) with some artificial distinguished property. Compared to the current best attack for single-SP functions, our attacks work for more SP-layers. This gives a different view of the security of single-SP and double-SP functions compared to the previous analysis by Bogdanov and Shibutani [1] discussing the security when the number of rounds approaches to infinity.

The important knowledge provided with this paper is that including more active bytes does not immediately indicate stronger security. This is because attackers can control the behavior of several active S-boxes and mount efficient attacks.

Note that our results do not contradict the claim [1]. It is obvious that the number of rounds (or active S-boxes) which the attackers can control is limited. Hence, as long as infinite rounds are considered, the impact of the attacker's

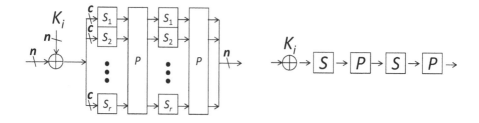

Fig. 2. Left: Double-SP round function **Right:** Simplified description

ability to control active S-boxes can be ignored. However, in practice, R is relatively small to achieve a good performance. Then, the attacker's ability cannot be ignored. If one may design GFN with double-SP functions, we recommend taking such impact into consideration.

Paper Outline

The paper is organized as follows. In Sect. 2, we give basic notions and review previous work. We present our 6-round attacks on GFN with double-SP functions in Sect. 3 We then extend the attack to 7 rounds in Sect. 4. Finally, we conclude the paper in Sect. 5.

2   Preliminaries

2.1   Description of the Double-SP Round Function

We specify the double-SP round function used in the 4-branch generalized Feistel network. Hereafter, we use the following notations.

N: The block length of the cipher (in bits),
n: The word size in bits, equal to the size of the input and output of the round function, so that $n = N/4$,
c: The byte size in bits (In this paper the byte size is not fixed), equal to the size of an S-box,
r: The number of S-boxes in a word, so that $r = n/c$.

Note that the design of the CLEFIA block-cipher [17] is close to the 4-branch type-2 GFN with single-SP functions. CLEFIA uses the parameter

$$(N, n, c, r) = (128, 32, 8, 4). \tag{1}$$

Eq. (1) is useful to consider the impact of the attack. In this paper, we often use these parameters to demonstrate the complexities of our attacks.

The double-SP round function is depicted in Fig. 2, consisting of the five operations: subkey XOR, S-box layer, permutation layer, S-box layer, and permutation layer. Details of each operation are as follows.

Subkey XOR: This layer computes the XOR of a round-function input and a round key K_i.

S-box layer: This layer substitutes each byte value by using one or several S-boxes; the S-boxes $S_1, S_2, \ldots, S_r$ may differ from each other. To simplify the explanation, we explain our attacks based on the S-box which is designed to be resistant to differential and linear cryptanalyses, like the ones used in AES [18,19]. Hence, given a pair of randomly chosen input and output differences, there exist paired values following the given input/output differences with a probability of approximately 2^{-1}. If exist, then the number of such paired values is approximately two.

Permutation layer: This layer mixes values by multiplying the word value and an $r \times r$ matrix P over $\mathbb{F}(2^c)$ together. To simplify the explanation, we make the assumption that the branch number of P is $r+1$, so that the total number of active bytes in the input and output of P is always greater than or equal to $r+1$, as long as there is at least one active byte.

To simplify the analysis, we assume that all round functions are the same.

2.2 Comparison of Double-SP and Single-SP Functions in 4-Branch Type-2 GFN

Bogdanov and Shibutani showed that instantiating double-SP functions in 4-branch GFN is significantly more efficient with respect to differential and linear cryptanalysis than instantiating single-SP functions in terms of the proportion of the active S-boxes [1]. The metric to measure the resistance, E_r, is defined as $E_r = \lim_{R \to \infty} \frac{A_{r,R}}{S_{r,R}}$, where r is the number of S-boxes in a word, $S_{r,R}$ is the total number of S-boxes over R rounds, and $A_{r,R}$ is the number of active S-boxes over R rounds.

Therefore intuitively, the security of single-SP and double-SP functions can be compared by counting the number of active S-boxes in $2R$ rounds for single-SP functions and R rounds for double-SP functions. Forcing more active bytes results in the stronger security. It is shown in [1] that $A_{r,R}$ is $2(r+1)+2$ for single-SP functions in every 6 rounds and $6(r+1)$ for double-SP functions in every 6 rounds. Hence, the security is compared by $4(r+1)+4$ and $6(r+1)$, which leads to the conclusion that instantiating double-SP functions is more efficient than instantiating single-SP functions.

Note that the value of $\frac{A_{r,R}}{S_{r,R}}$ for double-SP and single-SP functions are irrespective of the round number R as long as R is a multiple of 6. Hence, comparing $\frac{A_{r,12}}{S_{r,12}}$ for single-SP functions and $\frac{A_{r,6}}{S_{r,6}}$ for double-SP functions derives the same results as the case $R \to \infty$. Also note that $6(r+1)$ is always bigger than $4(r+1)+4$ as long as $r > 1$, which is usually satisfied.

2.3 Rebound-Attack Technique

The rebound attack was introduced by Mendel et al. [20]. An attacker tries to find efficiently a pair of values that follows a pre-determined truncated differen-

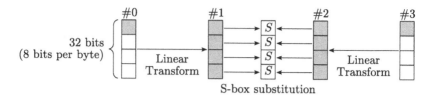

Fig. 3. Example of the inbound phase with 4-byte state (gray bytes are active)

tial path. The search procedure is divided into two phases: The *inbound phase* and the *outbound phase*. The attacker generates sufficiently many paired values that satisfy the truncated differential path of the inbound phase. This can be done with a small complexity on average. These paired values are called *starting points*. Then, the attacker computes the outbound phase with the generated starting points and checks whether or not any of the starting points satisfies the truncated differential path of the outbound phase. The attacker succeeds if he finds a starting point conforming to the path of the outbound phase.

Let us explain the basic procedure of the inbound phase by using a 4-byte (1 byte = 8 bits) state with an 8-bit S-box as an example. In this example, the goal of an attacker is to find a pair of values (M, M') that satisfies the truncated differential path $1 \rightarrow 4 \rightarrow 1$ (one active byte diverging to four active bytes and then converging to one active byte again) which is illustrated in Fig. 3.

The rebound attack can generate 2^8 pairs satisfying the differential path with a complexity of approximately 2^8—in other words, each pair is generated with a complexity of 1 on average. The detailed attack procedure is as follows:

1. For all 2^8 (more precisely, $2^8 - 1$) possible differences of state #0, compute the corresponding 4-byte differences of state #1 and store them in a table T.
2. Choose a difference of state #3 and compute the corresponding 4-byte difference of state #2. For each 4-byte difference in T, check whether or not the computed 4-byte difference of state #2 can be output through the S-boxes by looking up the differential distribution table (DDT). If the differences match, output such paired values.

For a pair of randomly determined input difference ΔS_{in} and output difference ΔS_{out}, an equation $S(x) \oplus S(x \oplus \Delta S_{\text{in}}) = \Delta S_{\text{out}}$ has approximately one solution with a probability of approximately 2^{-1}. If we find a solution x, then we will automatically obtain *two* paired values $(x, x \oplus \Delta S_{\text{in}})$ and $(x \oplus \Delta S_{\text{in}}, x)$ that satisfy the differential propagation through the S-box. In this example there are four S-boxes between state #1 and state #2. Therefore, if we have 2^4 pairs of Δ#1 and Δ#2, then one of these pairs can be expected to have a solution for each of the four S-boxes. Hence we obtain 2^4 paired values that satisfy the truncated differential path of the inbound phase.

Consequently, for a difference of state #3 in the above procedure, there are $2^8 \cdot 2^{-4} = 2^4$ differences in T that have approximately one solution for each of the four S-boxes. So we obtain $2^4 \cdot 2^4 = 2^8$ paired values that satisfy the truncated differential path (starting points). In other words, we can obtain one starting point with a complexity of 1 on average.

We also need to count the number of starting points obtained. The above procedure can be iterated for all 2^8 differences of state #3 in Step 2. As a result, we obtain $2^8 \cdot 2^8 = 2^{16}$ starting points at maximum.

2.4 Previous Rebound Attack on Feistel-SP Ciphers

Sasaki and Yasuda applied the rebound attack for Feistel (2-branch) ciphers with single-SP round function in the known-key setting [15]. They successfully mounted a distinguisher up to 11 rounds with most of practical parameters of (N, c). Their results can be trivially extended to 4-branch type-2 GFN ciphers with single-SP round function by simply running the same procedure for two round functions in parallel. Sasaki et al. later improved the complexity [16]. However, the number of attacked rounds is still 11.

In this paper, we use the results in [16,15] to compare the effect of attacks against 4-branch type-2 GFN ciphers with single-SP round function and with double-SP round function. Given the result that the current best attack does not reach 12 rounds for single-SP round functions, the first goal of this paper is attacking 6 rounds for double-SP functions.

3 6R Attack on Type-2 GFN with Double-SP Functions

In this section, we present a rebound attack for six rounds of 4-branch type-2 GFN with double-SP functions (including 24 SP-layers). In all attacks in this paper, we assume that the key value is fixed to a randomly determined value without any control of the attacker. Note that the current best attack for single-SP functions only works for 11 rounds (including 22 SP-layers). Although the differential path used in this attack activates many S-boxes, the attacker can control their behavior, and successfully distinguish 6 rounds. The result indicates that activating more S-boxes does not immediately result in the stronger security against differential cryptanalysis because the attacker may be able to control the behavior of active S-boxes.

3.1 Overview

Hereafter, we fix a system of notations as follows:

X^j: The j-th byte of a word X, where $1 \leq j \leq r$ and the size of X^j is c bits,

  0: A word where all bytes are non-active,

  1: A word where only one byte of the predetermined (j-th) position is active,

  F: A word where all bytes are active.

Our attacks, for a randomly determined key value, amount to finding a pair of values whose input difference is of the form $(\mathbf{1}, \mathbf{F}, \mathbf{1}, \mathbf{F})$ and whose output difference is also $(\mathbf{1}, \mathbf{F}, \mathbf{1}, \mathbf{F})$. The point is that our attacks can find such a pair more efficiently than one could for a random permutation.

Recall that in rebound attacks, an attacker needs to construct a differential path and divide it into inbound and outbound phases. Here the truncated differential path that we use is

$$(\mathbf{1}, \mathbf{F}, \mathbf{1}, \mathbf{F}) \xrightarrow{1^{st}R} (\mathbf{0}, \mathbf{1}, \mathbf{0}, \mathbf{1}) \xrightarrow{2^{nd}R}$$
$$(\mathbf{1}, \mathbf{0}, \mathbf{1}, \mathbf{0}) \xrightarrow{3^{rd}R} (\mathbf{1}, \mathbf{1}, \mathbf{1}, \mathbf{1}) \xrightarrow{4^{th}R} (\mathbf{0}, \mathbf{1}, \mathbf{0}, \mathbf{1})$$
$$\xrightarrow{5^{th}R} (\mathbf{1}, \mathbf{0}, \mathbf{1}, \mathbf{0}) \xrightarrow{6^{th}R} (\mathbf{1}, \mathbf{F}, \mathbf{1}, \mathbf{F})$$

which is shown in Fig. 11 in Appendix.

In the 6R attack, we use a 2-round truncated differential path for the inbound phase. The difference propagates from $(\mathbf{1}, \mathbf{0}, \mathbf{1}, \mathbf{0})$ to $(\mathbf{0}, \mathbf{1}, \mathbf{0}, \mathbf{1})$ through the two rounds ($3^{rd}R$ – $4^{th}R$). Our attack can find a pair of values following the 2-round truncated differential path with a complexity of 2^c in time and 2^c in memory.

The outbound phase consists of two rounds in backward direction ($2^{nd}R$ – $1^{st}R$) and two rounds in forward direction ($5^{th}R$ – $6^{th}R$), in total four rounds. In both directions, the differences propagate to $(\mathbf{1}, \mathbf{F}, \mathbf{1}, \mathbf{F})$, i.e., only one byte is active in the first and third words.

Given any paired values satisfying the truncated differential path of the inbound phase, the truncated differential path of the outbound phase is satisfied with a probability of 1. Hence, we need only one starting point from the inbound phase. Finally, we can find the pair that satisfies the entire path with a complexity of 2^c in time and 2^c in memory.

This means that our attack works effectively, because for a random permutation such a pair cannot be found with that complexity. Namely, let us consider the complexity to find a pair of values that has the differential form of $(\mathbf{1}, \mathbf{F}, \mathbf{1}, \mathbf{F})$ for both of the input and output states in a random permutation. Attackers have an access to both encryption and decryption oracles. For such attackers, this problem is regarded as finding a $2^{2(n-c)}$-bit collision. Because enough freedom degrees are available to mount the birthday attack, this requires a complexity of 2^{n-c}.

Two-Round Inbound Phase. In this phase, our goal is to find a pair of values whose difference will propagate as $(\mathbf{1}, \mathbf{0}, \mathbf{1}, \mathbf{0}) \xrightarrow{3^{rd}R} (\mathbf{1}, \mathbf{1}, \mathbf{1}, \mathbf{1}) \xrightarrow{4^{th}R} (\mathbf{0}, \mathbf{1}, \mathbf{0}, \mathbf{1})$. We use the differential path depicted in Fig. 4.

In this analysis, we only need to analyze one round function. Then, we can copy the result to the other three round functions. We set differences of the form $\mathbf{1}$ for the word just before the first P-layer and immediately after the second P-layer, i.e., ($\#A$ and $\#A'$) in Fig. 4. The analysis becomes almost the same as the one explained in Fig. 3. We propagate these differences through the linear operations with actual word values undetermined. We then search for a matched set of differences at the S-box operation in the middle. To connect the results of four round functions, we need to ensure that the difference before

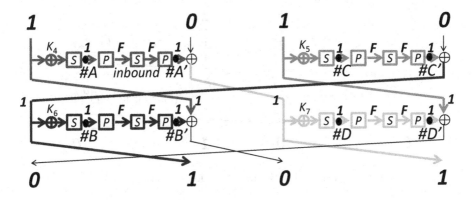

Fig. 4. Two-round inbound phase

the first S-layer is the same as the difference after the second P-layer. This is satisfied probabilistically. Finally, we can find word values that follow the desired differential path of the two-round inbound phase. More specifically, our attack procedure is as follows:

0. Choose a byte-position j ($1 \leq j \leq r$) in a word to be activated in the differential form of **1**.
1. For all 2^c possible differences in word $\#A$, compute the corresponding full-byte differences after applying the P-layer and store the results in a table T.
2. For each of the 2^c possible differences in word $\#A'$, compute the corresponding full-byte difference after applying the inverse permutation. For the middle S-layer, check whether or not we can match the full-byte difference evolved from $\Delta\#A'$ with 2^c differences stored in T. This can be done by looking up the DDTs.
3. We expect to find such 2^{2c-r} matched sets of differences, and thus we find $2^{2c-r} \cdot 2^r = 2^{2c}$ paired values that satisfy the truncated differential path between $\#A$ and $\#A'$. We then choose one of 2^{2c} pairs and fix word values in accordance with the chosen differences. Also, the word values colored by red in Fig. 4 are all fixed.
4. We check that the difference before the first S-layer is the same as the one after the second P-layer. Namely, check the following:

$$\Delta S^{-1}(\#A) = \Delta\#A'. \tag{2}$$

If Eq. (2) is not satisfied, we go back to Step 3, and choose another values from 2^{2c} possibilities.
5. After obtain one solution for (2), we copy the solution to the other three round functions. Namely, we fix the word values between $\#B$ and $\#B'$, $\#C$ and $\#C'$, and $\#D$ and $\#D'$ in Fig. 4 to the same as the one between $\#A$ and $\#A'$. Because the difference among different round functions is only the subkey value, the input and output differences for each round function are

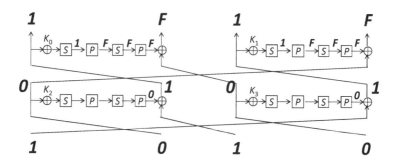

Fig. 5. Backward outbound phase for six-round attack

identical, i.e., $\Delta S^{-1}(\#A) = \Delta S^{-1}(\#B) = \Delta S^{-1}(\#C) = \Delta S^{-1}(\#D)$ and $\Delta\#A' = \Delta\#B' = \Delta\#C' = \Delta\#D'$. Because $\Delta S^{-1}(\#A) = \Delta\#A'$ due to Eq. (2), the difference of these 8 words are identical.

6. Now all word values are fixed with satisfying the desired differences for the two-round inbound phase. Moreover, all differences of the form **1** for the input and output of the inbound phase are identical.

Let us estimate the time and memory complexities necessary for each of the above steps. We also verify that the success probability of the inbound phase is sufficiently high.

- Step 1 requires 2^c computations and 2^c memory.
- In Step 2, with a complexity of 2^c, we can check the match of 2^{2c} pairs at maximum. Because each match succeeds with a probability of 2^{-r}, we can expect to find 2^{2c-r} matched pairs as long as $2c \geq r$. Note that the parameter in Eq. (1) satisfies $2c \geq r$. Also note that the match can be identified with the meet-in-the-middle manner, thus the complexity of this step is 2^c rather than 2^{2c}.
- Eq. (2) is satisfied with probability 2^{-c}. Therefore, Step 3 and 4 require 2^c 1-round computations and negligible memory.
- Step 5 requires one computation for each round function.
- The total complexity is 2^c round-function computations for Steps 1, 2, and 4 which are much lower than 2^c 6-round 4-branch GFN computations. The amount of required memory is 2^c words for Step 1.

Outbound Phase. The outbound phase for the first and last two rounds are shown in Fig. 5 and 6.

We explain the last two rounds in details. As a result of the inbound phase, we obtain the difference $(\mathbf{0}, \mathbf{1}, \mathbf{0}, \mathbf{1})$ as an input to the 5th round. This propagates as $(\mathbf{0}, \mathbf{1}, \mathbf{0}, \mathbf{1}) \xrightarrow{5^{\text{th}}\text{R}} (\mathbf{1}, \mathbf{0}, \mathbf{1}, \mathbf{0}) \xrightarrow{6^{\text{th}}\text{R}} (\mathbf{1}, \mathbf{F}, \mathbf{1}, \mathbf{F})$ with a probability of 1. Hence, $n - c$ bits in the first and third output words do not have any difference. Moreover, the active-word differences of these two words are identical.

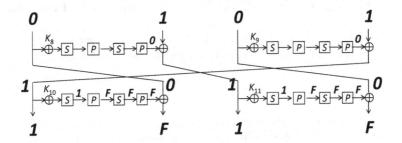

Fig. 6. Forward outbound phase for six-round attack

The same applies to the differential propagation in backward direction for the first 2 rounds. The difference reaches $(\mathbf{1}, \mathbf{F}, \mathbf{1}, \mathbf{F})$ after the 2 rounds with a probability of 1.

Attack Summary and Its Impact. The inbound phase finds a starting point with a time complexity of 2^c using 2^c memory. Given any solution of the inbound phase, the outbound phase, with a probability of 1, generates a pair of values that has a differential form of $(\mathbf{1}, \mathbf{F}, \mathbf{1}, \mathbf{F})$ for both plaintext and ciphertext.

The above attack can be viewed as a valid known-key distinguisher for block ciphers. Let us compare the above complexity with the generic birthday bound (2^{n-c}). Then, the condition which our attacks can work more effectively than the generic attack is derived as $2^c < 2^{n-c}$. This gives us a condition

$$c < \frac{n}{2}, \tag{3}$$

which is usually satisfied by taking into account current designs of SP-ciphers.

The attack can be used to generate near-collisions on half of the state when double-SP functions are instantiated in some compression function modes. Matyas-Meyer-Oseas (MMO) and Miyaguchi-Preneel modes provide efficient ways to construct a compression function from a block cipher. They are among the 12 secure schemes [21] of PGV style [22]. Let E be a block cipher, and let E_K denote its encryption algorithm with a key K. The MMO compression function outputs H_i by computing

$$H_i = E_{H_{i-1}}(M_{i-1}) \oplus M_{i-1}$$

for a message block M_{i-1} and a previous chaining value H_{i-1}. Similarly, the Miyaguchi-Preneel mode computes H_i by

$$H_i = E_{H_{i-1}}(M_{i-1}) \oplus M_{i-1} \oplus H_{i-1},$$

given M_{i-1} and H_{i-1}. In both modes, the XOR of the plaintext and ciphertext of the internal cipher is computed to generate the output. In our attack, the differential form of the plaintext and ciphertext are $(\mathbf{1}, \mathbf{F}, \mathbf{1}, \mathbf{F})$, and more over, the difference of $\mathbf{1}$ in the first and third words are identical between plaintext and ciphertext. Therefore, these are canceled each other in the final output, which results in a near-collision on a half of the state.

4   7R Attack on Type-2 GFN with Double-SP Functions

For a deeper understanding of the double-SP functions, we extend the six-round known-key distinguisher in the previous section by one more round.

4.1   Overview

In this distinguisher, the distinguished property is somehow artificial. The attacker, for a randomly determined key value, aims to find a pair of values where the XOR of the first (resp. third) word in the plaintext and the third (resp. first) word in the ciphertext is limited to the 2^c pre-specified patterns.

The truncated differential path that we use for seven rounds is as follows;

$$(\mathbf{Y}, \mathbf{F}, \mathbf{X}, \mathbf{F}) \xrightarrow{1^{\text{st}}\text{R}} (0, \mathbf{Y}, 0, \mathbf{X}) \xrightarrow{2^{\text{nd}}\text{R}} (\mathbf{X}, 0, \mathbf{Y}, 0) \xrightarrow{3^{\text{rd}}\text{R}}$$

$$(\mathbf{F}, \mathbf{X}, \mathbf{F}, \mathbf{Y}) \xrightarrow{4^{\text{th}}\text{R}} (\mathbf{Y} \oplus P(1), \mathbf{F}, \mathbf{X} \oplus P(1), \mathbf{F}) \xrightarrow{5^{\text{th}}\text{R}} (0, \mathbf{Y} \oplus P(1), 0, \mathbf{X} \oplus P(1))$$

$$\xrightarrow{6^{\text{th}}\text{R}} (\mathbf{X} \oplus P(1), 0, \mathbf{Y} \oplus P(1), 0) \xrightarrow{7^{\text{th}}\text{R}} (\mathbf{X} \oplus P(1), \mathbf{F}, \mathbf{Y} \oplus P(1), \mathbf{F}).$$

Note that $\mathbf{X}$ and $\mathbf{Y}$ are full-active differences determined in the middle of the attack, and $P(1)$ is a difference where 1 is processed through the P-layer. The entire differential path is shown in Fig. 12 in Appendix.

In the 7R attack, the inbound phase covers three middle rounds and the outbound phases covers the first and last two rounds, in total four rounds. Our attack can find a pair of values following the three-round inbound phase with a complexity of 2^{3c-r} in time and 2^c in memory. The outbound phase is satisfied with a probability of 1. Finally, we can find the pair that satisfies the entire path with a complexity of 2^{3c-r} in time and 2^c in memory. Note that, if the above path is satisfied, XOR of the first (resp. third) word in the plaintext and the third (resp. first) word in the ciphertext is form $P(1)$ irrespective of the difference $\mathbf{X}$ and $\mathbf{Y}$, which only takes 2^c pre-specified patterns.

Similarly to the six-round attack, finding such a pair for a random permutation requires 2^{n-c} queries. With the parameter in Eq. (1), 2^{3c-r} is $2^{3 \cdot 8-4} = 2^{20}$ and 2^{n-c} is $2^{32-8} = 2^{24}$. Hence, our attack is a valid distinguisher.

Three-Round Inbound Phase. The differential path of the three-round inbound phase is

$$(\mathbf{X}, 0, \mathbf{Y}, 0) \xrightarrow{3^{\text{rd}}\text{R}} (\mathbf{F}, \mathbf{X}, \mathbf{F}, \mathbf{Y}) \xrightarrow{4^{\text{th}}\text{R}} (\mathbf{Y} \oplus P(1), \mathbf{F}, \mathbf{X} \oplus P(1), \mathbf{F})$$

$$\xrightarrow{5^{\text{th}}\text{R}} (0, \mathbf{Y} \oplus P(1), 0, \mathbf{X} \oplus P(1))$$

which is depicted in Fig. 7. We start with the difference of the form 1 in words #A and #A' in Fig. 7, and then try to find matched sets of differences at the S-boxes in the right-hand side of the 5th round. This computation is colored by red in Fig. 7. Note that the difference in word #A'' is always the same as the

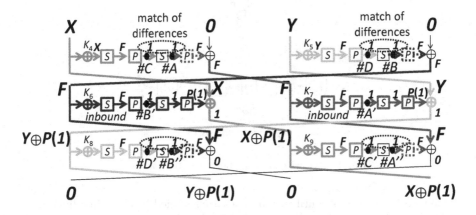

Fig. 7. 3-round inbound phase

one in $\#A$, which is the form **1**. We then do the same in words $\#B$ and $\#B'$. So far, paired values for red and blue bold lines in Fig. 7 are fixed and difference for dotted lines are fixed. We then search for paired values between $\#C$ and $\#C'$ (green) and $\#D$ and $\#D'$ (yellow). This part can be analyzed in bytewise, and thus efficient. Finally, we check the match of differences between $\#A$ and $\#C$, $\#B$ and $\#D$, $\#A''$ and $\#C'$, and $\#B''$ and $\#D'$. These matches are satisfied probabilistically. The detailed attack procedure is as follows:

0. Choose a byte-position j ($1 \le j \le r$) in a word to be activated in the differential form of **1**.
1. Randomly choose differences of the active byte in words $\#A$ and $\#A'$. Then, check if the solution exists for the sets of differences at the S-boxes in the right-hand side of the 5th round. Repeat this procedure until a solution is obtained. After solutions are obtained, choose one of them and fix words depicted by bold red lines in Fig. 7 to the solution. Note that differences of red dotted lines in Fig. 7 are also fixed though values are not determined yet.
2. Do the same as step 1 for words $\#B$ and $\#B'$. After this step, values and differences for blue bold lines, and differences for blue dotted lines in Fig. 7 are fixed.
3. Randomly choose differences of the active byte in words $\#C$ and $\#C'$. The detailed analysis is shown in Fig. 8. Propagate these differences to the words $\#C1$ and $\#C2$ described in Fig. 8.
   (a) For each byte in $\#C1$, try all possible 2^c values and process the computation until $\#C2$. Check if the corresponding difference at $\#C2$ matches the one propagated from $\#C'$. Store all solutions for each byte.
   (b) For each combination of the solutions for each byte, compute the paired value until the words $\#C$ and $\#C'$. Then, check if both of $\Delta S(\#C) = \Delta\#A$ and $\Delta S(\#C') = \Delta\#A''$.

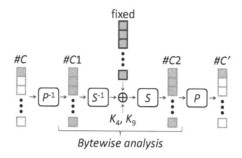

Fig. 8. Bytewise analysis between $\#C$ and $\#C'$

4. Repeat Steps 3 to 3b until two matches are satisfied by changing differences of the active byte in words $\#C$ and $\#C'$.
5. Do the same as Steps 3 to 4 for the words $\#D$ and $\#D'$. After this step, all words are fixed and we obtain the solution for the 3-round inbound phase.

Let us evaluate the time and memory complexities necessary for the above procedure:

- Step 1 requires at most 2^c computations and 2^c memory. As explained in Sect. 3, we can expect to find matched pairs as long as $2c \geq r$,
- The solution of Step 1 can be copied in order to satisfy Step 2. Hence, the complexity for Step 2 is negligible.
- For each fixed difference of Step 3, Step 3a requires 2^c computations and produces 2^r solutions. Then, Step 3b requires 2^r computations. The probability of the match with 2^r trials in Step 3b is 2^{-2c+r}.
- Step 3 is repeated 2^{2c-r} times, and each iteration requires $2^c + 2^r$ computations. Thus, the complexity is $2^{2c-r}(2^c + 2^r)$.
- Step 5 is the same as Steps 3 to 4, which is $2^{2c-r}(2^c + 2^r)$ computations.

To sum up, we can find a starting point for the 3-round inbound phase with a complexity of $2^{2c-r}(2^c + 2^r) \approx 2^{3c-r}$ computations and at most 2^c memory.

With the parameter in Eq. (1), 2^2 computations and 2^2 memory are enough to find the match over 4 S-boxes in Step 1. The bottle-neck of the complexity is for Steps 3 to 4 and Step 5, which is $2^{16-4} \cdot (2^8 + 2^4) \approx 2^{20}$ half-round computations.

Outbound Phase. The outbound phase is a deterministic differential propagation, which is shown in Fig. 9 and 10. Because it is straight-forward, we omit the explanation. After the outbound phase, the plaintext difference becomes $(\mathbf{Y}, \mathbf{F}, \mathbf{X}, \mathbf{F})$ and the ciphertext difference becomes $(\mathbf{X} \oplus P(\mathbf{1}), \mathbf{F}, \mathbf{Y} \oplus P(\mathbf{1}), \mathbf{F})$, where $\mathbf{X}$ and $\mathbf{Y}$ are full-active differences determined in the middle of the inbound phase.

Attack Summary and Its Impact. The inbound phase finds a starting point with a time complexity of $2^{2c-r}(2^c + 2^r)$ using 2^c memory. Given any solution

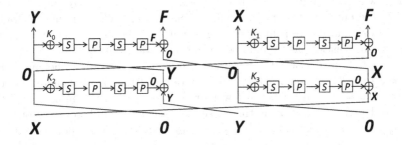

Fig. 9. Backward outbound phase for seven-round attack

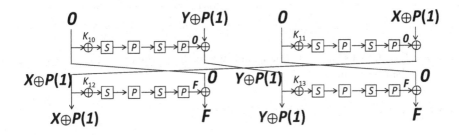

Fig. 10. Forward outbound phase for seven-round attack

of the inbound phase, the outbound phase, with a probability of 1, generates a pair of values whose plaintext difference is the form of $(\mathbf{Y}, \mathbf{F}, \mathbf{X}, \mathbf{F})$ and ciphertext difference is the form of $(\mathbf{X} \oplus P(1), \mathbf{F}, \mathbf{Y} \oplus P(1), \mathbf{F})$. Therefore, with that complexity, we can find a pair of values where the XOR of the first (resp. third) word in the plaintext and the third (resp. first) word in the ciphertext is form $P(1)$, which only takes 2^c pre-specified patterns.

Similarly to the six-round attack, finding such a pair for a random permutation requires 2^{n-c} queries. With the parameter in Eq. (1), 2^{3c-r} is $2^{3 \cdot 8 - 4} = 2^{20}$ and 2^{n-c} is $2^{32-8} = 2^{24}$. Hence, our attack is a valid distinguisher.

5   Conclusion

In this paper, we have presented the first cryptanalytic results for 4-branch type-2 GFN with double-SP functions. Our attack generates near-collision up to 6 rounds and forms a valid distinguisher up to 7 rounds of compression functions consisting of double-SP functions. Compared to the current best attack (up to 11 rounds) for single-SP functions, our attack works for more SP-layers. The attack exploits the fact that the attacker can control the behavior of several active S-boxes, and can mount efficient attacks for a small number of rounds. This gives a different view of the security of single-SP and double-SP functions compared to the previous analysis by Bogdanov and Shibutani [1] discussing the security when the number of rounds approaches to infinity. We hope that the analysis presented in this paper leads to the deeper understanding of double-SP functions and future primitive designs will consider the impact of controlled active S-boxes as well as the number of total active S-boxes.

One of the future possible research directions is extending the number of attacked rounds. Although the complexity of our attack is very small, it seems hard to extend the attack by a few more rounds due to the limited available freedom degrees. Investigating the chosen-key scenario could solve the problem.

Another possible direction is implementing the attack to see the actual impact to ciphers. However, there does not exist any concrete example of Feistel double-SP functions even for 2-branch and generalized Feistel network. Hence, the best way seems to borrow the parameters (N, n, c, r) and the MDS matrix from those designs and to construct the imaginary designs in order to see the actual behavior of the attacks.

Acknowledgments. The author would like to thank the anonymous reviewers for many helpful comments.

References

1. Bogdanov, A., Shibutani, K.: Double SP-Functions: Enhanced Generalized Feistel Networks. In: Parampalli, U., Hawkes, P. (eds.) ACISP 2011. LNCS, vol. 6812, pp. 106–119. Springer, Heidelberg (2011)
2. Bogdanov, A.: Bounds for balanced and generalized feistel constructions. In: ECRYPT II Symmetric Techniques Virtual Lab (2011)
3. Kanda, M., Moriai, S., Aoki, K., Ueda, H., Miyako Ohkubo, Y.T., Ohta, K., Matsumoto, T.: A new 128-bit block cipher E2. Technical Report ISEC98-12, The Institute of Electronics, Information and Communication Engineers (1998)
4. Shibutani, K., Isobe, T., Hiwatari, H., Mitsuda, A., Akishita, T., Shirai, T.: *Piccolo*: An Ultra-Lightweight Blockcipher. In: Preneel, B., Takagi, T. (eds.) CHES 2011. LNCS, vol. 6917, pp. 342–357. Springer, Heidelberg (2011)
5. Bogdanov, A., Shibutani, K.: Analysis of 3-line generalized feistel networks with double sd functions. Inf. Process. Lett. 111(13), 656–660 (2011)
6. Suzaki, T., Minematsu, K.: Improving the Generalized Feistel. In: Hong, S., Iwata, T. (eds.) FSE 2010. LNCS, vol. 6147, pp. 19–39. Springer, Heidelberg (2010)
7. Wu, W., Zhang, L.: LBlock: A Lightweight Block Cipher. In: Lopez, J., Tsudik, G. (eds.) ACNS 2011. LNCS, vol. 6715, pp. 327–344. Springer, Heidelberg (2011)
8. Suzaki, T., Minematsu, K., Morioka, S., Kobayashi, E.: TWINE: A lightweight block cipher for multiple platforms. In: Knudsen, L.R., Wu, H. (eds.) Selected Areas in Cryptography SAC 2012. LNCS, Springer, Heidelberg (2012)
9. Yanagihara, S., Iwata, T.: On Permutation Layer of Type 1, Source-Heavy, and Target-Heavy Generalized Feistel Structures. In: Lin, D., Tsudik, G., Wang, X. (eds.) CANS 2011. LNCS, vol. 7092, pp. 98–117. Springer, Heidelberg (2011)
10. Biham, E., Dunkelman, O.: The SHAvite-3 hash function. Submission to NIST (Round 2) (2009)
11. Minier, M., Naya-Plasencia, M., Peyrin, T.: Analysis of Reduced-SHAvite-3-256 v2. In: Joux, A. (ed.) FSE 2011. LNCS, vol. 6733, pp. 68–87. Springer, Heidelberg (2011)

12. Knudsen, L.R., Rijmen, V.: Known-Key Distinguishers for Some Block Ciphers. In: Kurosawa, K. (ed.) ASIACRYPT 2007. LNCS, vol. 4833, pp. 315–324. Springer, Heidelberg (2007)

13. Minier, M., Phan, R.C.-W., Pousse, B.: Distinguishers for Ciphers and Known Key Attack against Rijndael with Large Blocks. In: Preneel, B. (ed.) AFRICACRYPT 2009. LNCS, vol. 5580, pp. 60–76. Springer, Heidelberg (2009)

14. Mendel, F., Peyrin, T., Rechberger, C., Schläffer, M.: Improved Cryptanalysis of the Reduced Grøstl Compression Function, ECHO Permutation and AES Block Cipher. In: Jacobson Jr., M.J., Rijmen, V., Safavi-Naini, R. (eds.) SAC 2009. LNCS, vol. 5867, pp. 16–35. Springer, Heidelberg (2009)

15. Sasaki, Y., Yasuda, K.: Known-Key Distinguishers on 11-Round Feistel and Collision Attacks on Its Hashing Modes. In: Joux, A. (ed.) FSE 2011. LNCS, vol. 6733, pp. 397–415. Springer, Heidelberg (2011)

16. Sasaki, Y., Emami, S., Hong, D., Kumar, A.: Improved Known-Key Distinguishers on Feistel-SP Ciphers and Application to Camellia. In: Susilo, W., Mu, Y., Seberry, J. (eds.) ACISP 2012. LNCS, vol. 7372, pp. 87–100. Springer, Heidelberg (2012)

17. Shirai, T., Shibutani, K., Akishita, T., Moriai, S., Iwata, T.: The 128-Bit Blockcipher CLEFIA (Extended Abstract). In: Biryukov, A. (ed.) FSE 2007. LNCS, vol. 4593, pp. 181–195. Springer, Heidelberg (2007)

18. Daemen, J., Rijmen, V.: AES Proposal: Rijndael (1998)

19. U.S. Department of Commerce, National Institute of Standards and Technology: Specification for the ADVANCED ENCRYPTION STANDARD (AES) (Federal Information Processing Standards Publication 197) (2001)

20. Mendel, F., Rechberger, C., Schläffer, M., Thomsen, S.S.: The Rebound Attack: Cryptanalysis of Reduced Whirlpool and Grøstl. In: Dunkelman, O. (ed.) FSE 2009. LNCS, vol. 5665, pp. 260–276. Springer, Heidelberg (2009)

21. Black, J., Rogaway, P., Shrimpton, T.: Black-Box Analysis of the Block-Cipher-Based Hash-Function Constructions from PGV. In: Yung, M. (ed.) CRYPTO 2002. LNCS, vol. 2442, pp. 320–335. Springer, Heidelberg (2002)

22. Preneel, B., Govaerts, R., Vandewalle, J.: Hash Functions Based on Block Ciphers: A Synthetic Approach. In: Stinson, D.R. (ed.) CRYPTO 1993. LNCS, vol. 773, pp. 368–378. Springer, Heidelberg (1994)

A   Entire Differential Characteristics

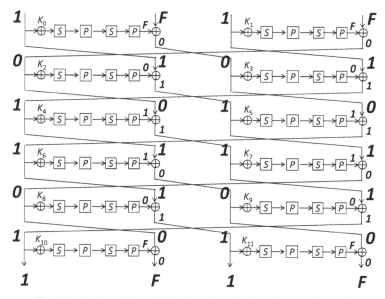

Fig. 11. Entire differential characteristic for six-round attack

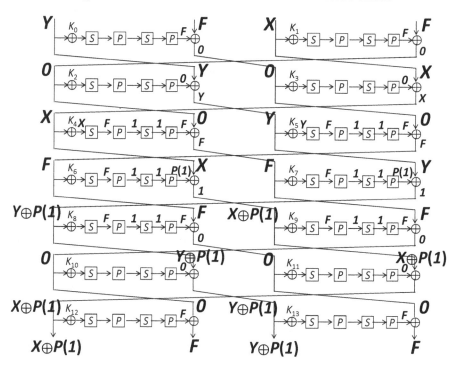

Fig. 12. Entire differential characteristic for seven-round attack

Automatic Search of Truncated Impossible Differentials for Word-Oriented Block Ciphers

Shengbao Wu[1,2] and Mingsheng Wang[3]

[1] Institute of Software, Chinese Academy of Sciences,
Beijing 100190, PO Box 8718, China
[2] Graduate School of Chinese Academy of Sciences, Beijing 100190, China
[3] State Key Laboratory of Information Security, Institute of Information
Engineering, Chinese Academy of Sciences, Beijing, China
wushengbao@is.iscas.ac.cn, mingsheng_wang@yahoo.com.cn

Abstract. Impossible differential cryptanalysis is a powerful technique to recover the secret key of block ciphers by exploiting the fact that in block ciphers specific input and output differences are not compatible. This paper introduces a novel tool to search truncated impossible differentials for word-oriented block ciphers with bijective Sboxes. Our tool generalizes the earlier $\mathcal{U}$-method and the UID-method. It allows to reduce the gap between the best impossible differentials found by these methods and the best known differentials found by ad hoc methods that rely on cryptanalytic insights. The time and space complexities of our tool in judging an r-round truncated impossible differential are about $O(c \cdot l^4 \cdot r^4)$ and $O(c' \cdot l^2 \cdot r^2)$ respectively, where l is the number of words in the plaintext and c, c' are constants depending on the machine and the block cipher. In order to demonstrate the strength of our tool, we show that it does not only allow to automatically rediscover the longest truncated impossible differentials of many word-oriented block ciphers, but also finds new results. It independently rediscovers all 72 known truncated impossible differentials on 9-round CLEFIA. In addition, it finds new truncated impossible differentials for AES, ARIA, Camellia without FL and FL^{-1} layers, E2, LBlock, MIBS and Piccolo. Although our tool does not improve the lengths of impossible differentials for existing block ciphers, it helps to close the gap between the best known results of previous tools and those of manual cryptanalysis.

Keywords: word-oriented block ciphers, truncated impossible differentials, difference propagation system, $\mathcal{U}$-method, UID-method.

1 Introduction

Impossible differential cryptanalysis is one of the most popular cryptanalytic tools for block ciphers. It was firstly proposed by Knudsen to analyze DEAL [13] in 1998 and then extended by Biham *et al.* to attack IDEA [5] and Skipjack [4]. Unlike traditional differential cryptanalysis [7], which uses differential

S. Galbraith and M. Nandi (Eds.): INDOCRYPT 2012, LNCS 7668, pp. 283–302, 2012.

characteristics with high probabilities to recover the right key, impossible differential cryptanalysis is a sieving method which exploits differentials with probability zero to retrieve the right key by filtering out all wrong keys. Until now, impossible differential cryptanalysis has shown its superiority over differential cryptanalysis in many block ciphers such as IDEA, Skipjack, CLEFIA [22] and AES [8].

Impossible differential cryptanalysis mainly consists of two steps. Firstly, an attacker tries to find impossible differentials, that is, differentials that never occur. Then, after gaining a list of plaintext-ciphertext pairs, the attacker guesses some subkey material involved in the outer rounds of the impossible differentials, and then partially encrypts/decrypts each plaintext-ciphertext pair to check whether the corresponding internal differences are identical to the input and output differences of the impossible differentials. Once that happens, the guessed subkey will be discarded. The right key will be recovered if we discard all wrong keys.

Several factors influence the success of impossible differential cryptanalysis, including the length of impossible differentials, specific input/output difference patterns and the strength of one-round encryption/decryption. Among them, the most important factor is the length of an impossible differential. The longer the impossible differential is, the better the attack will be. Another important factor is the input/output difference pattern when two impossible differentials have the same length, because the new impossible differentials may well result in improved attacks [26,16,9]. If we find more impossible differentials, we can perform a successful attack or improve the time/data complexities of known attacks with higher possibilities.

In Indocrypt 2003, Kim et al. [12] proposed the $\mathcal{U}$-method to find impossible differentials for various block cipher structures with bijective round functions. The $\mathcal{U}$-method is based on the miss-in-the-middle approach (see 1-(a) of Fig. 1): it first constructs two differentials with probability one from the encryption and decryption direction and subsequently demonstrates some contradictions by combining them. In the $\mathcal{U}$-method, the propagation of differences in a block cipher (structure) is translated into simple matrix operations, and some inconsistent conditions are used to detect impossible differentials. Luo et al. [17] developed the idea of the $\mathcal{U}$-method and proposed a more general method — the UID-method. The UID-method removed some limitations in the $\mathcal{U}$-method and harnesses more inconsistent conditions to evaluate impossible differentials. So far, the $\mathcal{U}$-method and the UID-method have been employed as tool by some block cipher designers to evaluate the security of their designs against impossible differential cryptanalysis, for instance, LBlock [25] and Piccolo [21].

However, the $\mathcal{U}$-method and the UID-method only focus on finding impossible differentials with the miss-in-the-middle approach, which limits their power. An example is illustrated in 1-(b) of Fig. 1. In this case, we cannot detect a contradiction in the match point of the two probability-one differentials, but instead recover some useful information, which will feed back to the internal rounds to produce a contradiction. The $\mathcal{U}$-method and the UID-method fail to detect this

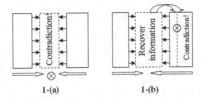

Fig. 1. Basic idea of the miss-in-the-middle approach (1-(a)) and impossible differentials with information feedback (1-(b))

kind of impossible differentials because they do not fully use the information in the match point.

Indeed, the longest impossible differentials of many block ciphers known so far are not found by the $\mathcal{U}$-method and the UID-method, but constructed by ad hoc approaches and the experience of cryptanalysts, such as 8-round MIBS [3], 6-round E2 [24], 8-round Camellia without the FL and FL^{-1} layers [26,24] and all 72 9-round impossible differentials of CLEFIA listed in [23]. Almost all impossible differentials of these block ciphers fall under the model of 1-(b) of Fig. 1, which implies that they are beyond the abilities of the $\mathcal{U}$-method and the UID-method. Of course, ad hoc approaches also have some disadvantages. For example, it is very computationally intensive to find even one impossible differential, and the success of finding an impossible differential is highly dependent to the experience of a cryptanalyst. Thus, they are not efficient and systematic in practice. Especially in designing a block cipher, one has to modify his/her design and frequently re-evaluate its security against impossible differential cryptanalysis.

Our Contributions. In this work, we propose a new tool for automatically finding impossible differentials of word-oriented block ciphers with bijective S-boxes. The goals of our tool are to reduce the gap between previous automatic tools (i.e., the $\mathcal{U}$-method and the UID-method) and ad hoc approaches, and to provide an automated method to find impossible differentials with a reduced complexity. The development of attack algorithms to exploit these impossible differentials is outside the scope of this paper; we leave it for further work.

Unlike the miss-in-the-middle approach, which splits a block cipher into two parts factitiously, we treat it as an entirety. The inputs of our tool are some constraints of the plaintext difference and the ciphertext difference, and a system of equations that describes the propagation behavior when differences pass through the inner primitives of a block cipher. Then, our tool predicts information about unknown variables from the known ones iteratively, with probability one in each step. Finally, it outputs a flag indicating whether a truncated differential is impossible under several filtering conditions. The time and space complexities of our tool in judging an r-round truncated impossible differential are about $O(c \cdot l^4 \cdot r^4)$ and $O(c' \cdot l^2 \cdot r^2)$ respectively, where l is the number of words in the plaintext and c, c' are constants depending on the machine and the block cipher.

Although our tool does not improve the lengths of impossible differentials for existing block ciphers, it helps in reducing the gap between previous automatic tools and ad hoc approaches. Experimental results also indicate that our tool is efficient and systematic. It not only rediscovers the longest truncated impossible differentials of many word-oriented block ciphers known so far, but also finds new results. It independently rediscovers all 72 known truncated impossible differentials on 9-round CLEFIA [23]. In addition, besides the best results known so far, our tool finds new truncated impossible differentials for many other word-oriented block ciphers, such as AES, Camellia [1] without FL and FL^{-1} layers, MIBS [10], LBlock [25], ARIA [14], E2 [11] and Piccolo [21]. The number of new truncated impossible differentials obtained by our tool is summarized in Table 1.

Table 1. Summary of new truncated impossible differentials (ID) obtained by our tool. Camellia* is a variant of Camellia without FL and FL^{-1} layers.

Block Cipher	Word unit	Previous results		In this paper		
		Round	No. of IDs	Round	No. of IDs	New IDs
AES	byte	4 ([6,19,20,2,18])	269,554	4	3,608,100	3,338,546
ARIA	byte	4 ([26,16,9,15])	156	4	94,416	94,260
Camellia*	byte	8 ([26,24])	3	8	4	1
E2	byte	6 ([24])	1	6	56	55
MIBS	nibble	8 ([3])	2	8	8	6
LBlock	nibble	14 ([25])	64	14	80	16
Piccolo	nibble	7 ([21])	1	7	450	449

An interesting observation is that the $\mathcal{U}$-method and the UID-method are specific cases of our tool, and our tool is more powerful than them. A new impossible differential of 8-round MIBS obtained by our tool is given to indicate that our tool can find longer impossible differentials than the $\mathcal{U}$-method and the UID-method. Thus, we expect that our tool is useful in evaluating the security of block ciphers against impossible differential cryptanalysis, especially when one tries to design a word-oriented block cipher with bijective Sboxes.

Outline of This Paper. In Sect. 2, we discuss how to build difference propagation systems, which describe the propagation behavior when differences pass through the inner primitives of block ciphers. In Sect. 3, we discuss our idea to find new impossible differentials. Then, a tool for automatically searching truncated impossible differentials is proposed in Sect. 4. Experimental results are also provided in this section. Finally, we compare our tool with the $\mathcal{U}$-method and the UID-method in Sect. 5 and conclude this paper in Sect. 6.

2   Difference Propagation System

Throughout this paper, we consider the exclusive-or difference, and we assume that: (1) $\mathcal{E}$ is an r-round word-oriented block cipher with block length $l \cdot s$ bits (where s is the bit length of a word), that is, the plaintext and the ciphertext of $\mathcal{E}$ are vectors in $\mathbb{F}_{2^s}^l$, and all inner operations of $\mathcal{E}$ consist only of calculations over $\mathbb{F}_{2^s}$; (2) bijective Sboxes over $\mathbb{F}_{2^s}$ are the only nonlinear primitives of $\mathcal{E}$; (3) all subkeys are exclusive-ored to the internal state. Thus, we do not consider the subkey addition operation since it does not influence the propagation of differences. Although some block ciphers do not satisfy all the above conditions, e.g., IDEA, we believe that similar ideas can also be applied, with some modifications.

In this section, we discuss how to build a system of equations that describes the propagation behavior when differences pass through the inner primitives of a word-oriented block cipher. This system will be called *difference propagation system* in the subsequent discussions.

2.1   Difference Propagation of Basic Primitives

Before studying block ciphers, we first investigate the difference propagation of four basic primitives which are often employed as parts of a word-oriented block cipher, namely the branching operation, the XOR-operation, the bijective Sbox layer and the linear permutation layer. These primitives are illustrated in 1-(a), 1-(b) and 1-(c) of Fig. 2.

Suppose $\Delta X = (\Delta x_i)_{1 \leq i \leq n}$, $\Delta Y = (\Delta y_i)_{1 \leq i \leq n}$ and $\Delta Z = (\Delta z_i)_{1 \leq i \leq n}$ are row vectors in $\mathbb{F}_{2^s}^n$, the difference propagation of basic primitives can be described as follows.

Lemma 1. *(The branching operation.) For a branching operation (see 1-(a) of Fig. 2), we have $\Delta X = \Delta Y = \Delta Z$. This equation can be written as $2n$ linear equations $\Delta x_i \oplus \Delta y_i = 0$ and $\Delta x_i \oplus \Delta z_i = 0$.*

Lemma 2. *(The XOR-operation.) For an XOR-operation (see 1-(b) of Fig. 2), we have $\Delta X \oplus \Delta Y = \Delta Z$. This equation can be written as n linear equations $\Delta x_i \oplus \Delta y_i \oplus \Delta z_i = 0$.*

Lemma 3. *(The linear permutation layer.) A linear permutation (see 1-(c) of Fig. 2) has matrix representation $P = (p_{i,j})_{1 \leq i,j \leq n}$ over $\mathbb{F}_{2^s}$, that is, $\Delta Y^T = P \cdot \Delta X^T$, where ΔX^T is the transposed vector of ΔX. This equation can be written as n linear equations $\Delta y_i \oplus \bigoplus_{j=1}^{n} p_{i,j} \cdot \Delta x_j = 0$.*

Lemma 4. *(The Sbox layer.) For an Sbox layer consisting of n bijective Sboxes $S_i : \mathbb{F}_{2^s} \to \mathbb{F}_{2^s}$ (see 1-(c) of Fig. 2), we build n formal equations $\overline{S}_i(\Delta x_i, \Delta y_i) = 0$.*

Remark 1. Notice that $\overline{S}_i(\cdot, \cdot)$ is inherently a nonlinear map if we try to write its concrete expression, since S_i is a nonlinear bijective Sbox. Each pair $(\Delta x_i, \Delta y_i)$ with $Pr(\Delta x_i \to \Delta y_i) \neq 0$ in the Difference Distributed Table [7] of S_i is a solution of $\overline{S}_i(\Delta x_i, \Delta y_i) = 0$, which means that an input difference Δx_i may propagate to the output difference Δy_i. However, in Lemma 4, we build a formal equation for an Sbox without considering its concrete expression because the only property used in our tool is that it is bijective.

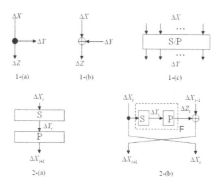

Fig. 2. Basic primitives of a block cipher: the branching operation (1-(a)), the XOR-operation (1-(b)) and the bijective Sbox/linear permutation layer (1-(c)), and the two most important classes of block ciphers: SPN ciphers (2-(a)) and Feistel ciphers with SPN round function (2-(b))

2.2 Build Difference Propagation Systems

The two most important classes of block ciphers are SPN ciphers and Feistel ciphers with SPN round functions (see 2-(a) and 2-(b) of Fig. 2). In this section, we choose them as examples to display how to build difference propagation systems.

SPN Cipher. One round of an SPN cipher typically has three layers (see 2-(a) of Fig. 2): the SubkeyAddition layer, the Sbox layer and the linear permutation layer. As mentioned above, we omit the SubkeyAddition layer since it does not influence the propagation of differences. Additionally, we can omit the last linear permutation layer since it does not influence the length of an impossible differential.

We denote by $\Delta X_i = (\Delta X_{i,j})_{1 \le j \le l}$ and $\Delta Y_i = (\Delta Y_{i,j})_{1 \le j \le l}$ the differences before and after the Sbox layer of round i, respectively. Then, for an r-round SPN cipher, we build a difference propagation system as follows:

$$\begin{cases} \overline{S}_j(\Delta X_{i,j}, \Delta Y_{i,j}) = 0 & \text{for } 1 \le i \le r \text{ and } 1 \le j \le l \ , \\ \Delta X_{i+1}^T \oplus P \cdot \Delta Y_i^T = 0 & \text{for } 1 \le i \le r-1 \ , \end{cases} \qquad (1)$$

where P is the matrix of linear permutation layer. This system contains $2 \cdot l \cdot r$ unknown variables, $l \cdot r$ formal (and nonlinear) equations from Sbox layers and $l \cdot (r-1)$ linear equations (using Lemma 3). ΔX_1 is the plaintext difference and ΔY_r is the ciphertext difference.

Feistel Cipher with SPN Round Functions. For an r-round Feistel cipher (see 2-(b) of Fig. 2), we denote by $\Delta X_{i-1} = (\Delta X_{i-1,j})_{1 \le j \le \frac{l}{2}}$ and $\Delta X_i = (\Delta X_{i,j})_{1 \le j \le \frac{l}{2}}$ the differences of the right branch and the left branch of round

i, respectively. Note that we assume that l is even. Thus, $(\Delta X_1, \Delta X_0)$ is the plaintext difference and $(\Delta X_{r+1}, \Delta X_r)$ is the ciphertext difference.

The SPN round function typically has three layers: the SubkeyAddition layer, the Sbox layer and the linear permutation layer. We introduce $\Delta Y_i = (\Delta Y_{i,j})_{1 \leq j \leq \frac{l}{2}}$ to denote the difference after the Sbox layer of round i and $\Delta Z_i = (\Delta Z_{i,j})_{1 \leq j \leq \frac{l}{2}}$ to represent the output difference of F function in round i. Then, we build a difference propagation system as follows:

$$\begin{cases} \overline{S}_j(\Delta X_{i,j}, \Delta Y_{i,j}) = 0 & \text{for } 1 \leq i \leq r \text{ and } 1 \leq j \leq \frac{l}{2} , \\ \Delta Z_i^T \oplus P \cdot \Delta Y_i^T = 0 & \text{for } 1 \leq i \leq r , \\ \Delta X_{i-1} \oplus \Delta X_{i+1} \oplus \Delta Z_i = 0 & \text{for } 1 \leq i \leq r , \end{cases} \quad (2)$$

where P is the matrix of the linear permutation layer in F. This system contains $\frac{l}{2} \cdot (3r + 2)$ unknown variables, $\frac{l}{2} \cdot r$ formal (and nonlinear) equations and $l \cdot r$ linear equations (using Lemma 2 and Lemma 3).

Other Block Ciphers. From Lemma 1 to Lemma 4, we know that a basic primitive costs $3l$ variables and provides $2l$ equations at most. Thus, for one round block cipher consisting of m basic primitives, we can build a difference propagation system with $3l \cdot m$ variables and $2l \cdot m$ equations in the worst case. In general, m is a small constant. For example, m is 2 in SPN ciphers (see 2-(a) of Fig. 2) and m is 4 in Feistel ciphers with SPN round functions (see 2-(b) of Fig. 2). Hence, for an r-round block cipher, we can build a difference propagation system with $O(c_1 \cdot r \cdot l)$ variables and $O(c_2 \cdot r \cdot l)$ equations, where c_1 and c_2 are constants depending on specific block ciphers.

3   Finding Impossible Differentials

In this section, we first introduce the basic idea of finding impossible differentials. Then, we discuss how to predict information from a given difference propagation system and how to detect contradictions. Finally, we briefly review the $\mathcal{U}$-method and the UID-method.

3.1   Basic Idea

The idea of finding impossible differentials in this paper is simple: given some information of the plaintext difference and the ciphertext difference, we may predict the information of new variables according to the difference propagation system, yielding a new set of "known" variables. Then, new information may again be predicted from these "known" variables. This process will continue until we find a contradiction or we can no longer obtain any new information. Notice that every prediction we made is deterministic (i.e., with probability one), which implies that the system does not have any solution if a contradiction is found. In other words, we obtain an impossible differential if a contradiction is detected in the process of predicting information, under given plaintext difference and ciphertext difference.

3.2 Predict Information and Detect Contradictions

We can divide a difference propagation system into two subsystems — $\mathcal{L}$ and $\mathcal{NL}$. $\mathcal{L}$ includes all linear equations while $\mathcal{NL}$ contains all formal (and nonlinear) equations from bijective Sboxes. Then, the information can be obtained with probability one in the following two ways:

(i) Predict information from the linear system $\mathcal{L}$. If system $\mathcal{L}$ has solutions, then they can be solved by the *Gauss-Jordan Elimination* algorithm, which gets solutions by firstly reducing the augmented matrix of $\mathcal{L}$ to row echelon form using elementary row operations and then back-substituting until the entire solution is found. The reduced augmented matrix after the back-substituting step represents a linear system that is equivalent to the original. Then, we have the following lemma.

Lemma 5. *Suppose $\mathcal{L}$ has solutions and the reduced augmented matrix of $\mathcal{L}$ is obtained, then*

1) *If an affine equation with only one variable, that is, $\Delta X \oplus c = 0$ (c is a constant), is found in the reduced system of $\mathcal{L}$, we have $\Delta X = 0$ if $c = 0$ and $\Delta X \neq 0$ if $c \neq 0$.*
2) *If a linear equation with two variables, that is, $\Delta X \oplus \Delta Y = 0$, is found in the reduced system of $\mathcal{L}$, we have $\Delta X \neq 0$ if and only if $\Delta Y \neq 0$.*

(ii) Predict information from the nonlinear system $\mathcal{NL}$. We have

Lemma 6. *Suppose S is a bijective Sbox, ΔX is the input difference and ΔY is the output difference. Then, ΔX is zero (respectively, nonzero) if and only if ΔY is zero (respectively, nonzero).*

According to the basic idea of finding impossible differentials, the strategy of predicting information is clear now: predict information from system $\mathcal{L}$ and $\mathcal{NL}$ alternately until a contradiction is found or we can no longer obtain any new information. An impossible differential is detected by the following proposition:

Proposition 1. *We denote by ΔP and ΔC the plaintext difference and the ciphertext difference, respectively. Then, $\Delta P \to \Delta C$ is impossible if one of the following two situations happens:*

- **I.** *The linear system $\mathcal{L}$ does not have any solution. That is, the rank of its coefficient matrix is not equal to the rank of its augmented matrix.*
- **II.** *There exists a variable with both zero and nonzero values.*

A tiny example of the second case is given below.

Example 1. Suppose the equations $\Delta Y \oplus \Delta Z = 0$ and $S(\Delta X, \Delta Y) = 0$ are included in a difference propagation system, and we know that $\Delta X = 0$ and $\Delta Z \neq 0$ from the previous information. Then, in the next prediction, we know that $\Delta Y \neq 0$ from Lemma 5 while $\Delta Y = 0$ from Lemma 6, which is a contradiction.

3.3 Related Work — The $\mathcal{U}$-Method and the UID-Method

In this section, we briefly review the $\mathcal{U}$-method and the UID-method. The specification of these tools can be found in [12] and [17].

Both the $\mathcal{U}$-method and the UID-method mainly have three steps in finding an impossible differential. First, both tools construct a characteristic matrix which describes the propagation of differences in one round encryption/decryption. For example, for one round of Feistel structure (see 2-(b) of Fig. 2), we have $\Delta X_{i+1} = F(\Delta X_i) \oplus \Delta X_{i-1}$ and $\Delta X_i = \Delta X_i$, that is, $(\Delta X_{i+1}, \Delta X_i)^T = \mathcal{E} \cdot (\Delta X_i, \Delta X_{i-1})^T$, where $\mathcal{E} = \left(\begin{smallmatrix} F & 1 \\ 1 & 0 \end{smallmatrix} \right)$ is the characteristic matrix of one round encryption. The characteristic matrix $\mathcal{D}$ of one round decryption can be defined similarly. Secondly, the $\mathcal{U}$-method and the UID-method defined some operations to calculate the multiplications between a characteristic matrix and a vector, because the output difference after r_1-round encryptions (resp., r_2-round decryptions) can be described as $\Delta U = \mathcal{E}^{r_1} \cdot \Delta P^T$ (resp., $\Delta V = \mathcal{D}^{r_2} \cdot \Delta C^T$). Finally, suppose $\Delta U = (\Delta u_j)_{1 \leq j \leq l}$ and $\Delta V = (\Delta v_j)_{1 \leq j \leq l}$ are two vectors which should be combined in the miss-in-the-middle approach, the following filtering conditions are used to detect contradictions.

Definition 1. *(Definition 1 of [17]) Vectors ΔU and ΔV are inconsistent if there exists a subset $I \subseteq \{1, 2, \ldots, l\}$ such that $\oplus_{i \in I}(\Delta u_i \oplus \Delta v_i) \neq 0$, where Δu_i (respectively, Δv_i and $f = \oplus_{i \in I}(\Delta u_i \oplus \Delta v_i)$) is a linear XOR combination of the four types of differences: zero difference, nonzero fixed difference, nonzero unspecified difference and unknown difference. Especially, the $\mathcal{U}$-method always considers subsets that have exactly one index.*

There are two main differences between the $\mathcal{U}$-method and the UID-method. First, the UID-method relaxes the 1-Property (i.e., the number of 1 entries in each column of the characteristic matrix is zero or one) required in the $\mathcal{U}$-method. Secondly, the UID-method exploits a more general filtering condition to detect contradictions than the $\mathcal{U}$-method, which has shown in Definition 1. Thus, the UID-method is more general than the $\mathcal{U}$-method.

4 Algorithm to Find Truncated Impossible Differentials

In this section, we first sketch our algorithm in finding impossible differentials. Then, we discuss some details of our algorithm. Finally, experimental results and some discussions of our tool are introduced.

4.1 Sketch of Our Algorithm

After building a difference propagation system as (1) or (2), we first implement it in a computer. Then, we choose a set of promising $(\Delta P, \Delta C)$ pairs. Finally, for each of these pairs, our algorithm judges whether it is an impossible differential automatically, that is, our algorithm predicts information from the linear system $\mathcal{L}$ and the nonlinear system $\mathcal{NL}$ alternately until a contradiction is found or we can no longer obtain any new information. The outline of our algorithm is shown in Algorithm 1. *flag* indicates whether $\Delta P \to \Delta C$ is impossible, and *index* controls the termination of predicting information.

---

1 Implement the difference propagation system, i.e., $\mathcal{L}$ and $\mathcal{NL}$, on a computer;
2 **for** *every pair of* $(\Delta P, \Delta C)$ *we choose* **do**
3   *flag*:=false; *index*:=true;
4   **while** *index* **do**
5   **if** *System $\mathcal{L}$ does not have any solution* **then**
6   *flag*:=true; *index*:=false;
         else
7   Predict information from the reduced augmented matrix of $\mathcal{L}$;
8   Predict information from the nonlinear system $\mathcal{NL}$;
9   **if** *do not find new information* **then**
10   *index*:=false;
            else
11   **if** *find a variable with both zero and nonzero values* **then**
12   *flag*:=true; *index*:=false;

13   **return** *flag*;

Algorithm 1: The outline of our algorithm

In the subsequent sections, we will discuss some details of Algorithm 1, including how we implement a difference propagation system on a computer, the choices of the plaintext difference and the ciphertext difference, and a specific algorithm for automatically judging a truncated impossible differential. Λ_0 and Λ_1 are sets for storing variables with zero difference and nonzero differences in a difference propagation system.

4.2   Implementation of a Difference Propagation System

Since a difference propagation system is divided into two parts — systems $\mathcal{L}$ and $\mathcal{NL}$ in our tool, we discuss how to implement them on a computer, respectively.

Implement System $\mathcal{L}$ with a Matrix. System $\mathcal{L}$ can be written formally as $A \cdot x = b$, where A, b and $B = [A|b]$ are called the *coefficient matrix, constant matrix* and *augmented matrix* of this system respectively, x is the set of *all* variables (in order) involved in the whole difference propagation system.

In the following example, we show that matrix A can be easily constructed if we choose a proper order of variables.

Example 2. We consider Feistel ciphers with SPN round functions, that is, we need to generate the coefficient matrix of linear equations in (2). Firstly, we may simplify the second and third equations of (2) as

$$I_1 \cdot \Delta X_{i-1}^T \oplus I_2 \cdot \Delta X_{i+1}^T \oplus P \cdot \Delta Y_i^T = 0 \text{ for } 1 \le i \le r , \qquad (3)$$

where $I_1 = I_2 = I$ and I is the identity matrix. Notice that variables ΔZ_is are eliminated in the simplified system. For a further step, the coefficient matrices of ΔX_{i-1}^T and ΔX_{i+1}^T may be substituted by other matrices according to the specification of the block ciphers.

Next, we fix the order of all variables, i.e., x, in the simplified system as

$$[\Delta X_0, \Delta X_1, \Delta X_2, \Delta X_3, \ldots, \Delta X_{r-1}, \Delta X_r, \Delta X_{r+1}, \Delta Y_1, \Delta Y_2, \ldots, \Delta Y_r] \ . \quad (4)$$

Finally, (3) can be represented as $A \cdot x^T = 0$, where

$$A = \begin{pmatrix} I_1 & 0 & I_2 & 0 & \cdots & 0 & 0 & 0 & P & 0 & \cdots & 0 \\ 0 & I_1 & 0 & I_2 & \cdots & 0 & 0 & 0 & 0 & P & \cdots & 0 \\ \vdots & \vdots & \vdots & \vdots & \ddots & \vdots & \vdots & \vdots & \vdots & \vdots & \ddots & \vdots \\ 0 & 0 & 0 & 0 & \cdots & I_1 & 0 & I_2 & 0 & 0 & \cdots & P \end{pmatrix}$$

is a block matrix with r rows and $2r+2$ columns. Now, A is a highly structured matrix. We can easily generate it in a computer if we obtain the coefficient matrices of ΔX_{i-1}, ΔX_{i+1} and ΔY_i in (3).

For iterative block ciphers with other structures, similar techniques can be used to construct highly structured coefficient matrices.

A technical detail in solving a linear system is to dispose a zero variable. In our tool, we keep the variable in the system while setting its coefficient matrix (i.e, a column of A) to a zero vector. This method keeps the solutions of the other variables unchanged and needs an additional space to store the value of this variable. Variables with zero value will be stored in set Λ_0. Notice that for this method, B is a matrix with a fixed number of columns.

Implement System $\mathcal{NL}$ with a Table. Once the order of variables in x is given, the i-th variable in x corresponds to the i-th column of A (and B). Then, we can store equations of system $\mathcal{NL}$ with a simple table using the column indexes of B. First, we initialize an empty table $\mathcal{T}$, then for each equation in $\mathcal{NL}$, we add an element $\{v_1, v_2\}$ to $\mathcal{T}$, where v_1, v_2 are two integers (i.e., the column indexes of B) indicating the two variables involved in the formal equation of an Sbox. For example, if the order of variables in a Feistel cipher is fixed as that in (4), then table $\mathcal{T}$ for formal equations in (2) is

$$\mathcal{T} = [\{\frac{l}{2} + \frac{l}{2} \cdot (i-1) + j, \frac{l}{2} \cdot (r+2) + \frac{l}{2} \cdot (i-1) + j\} : 1 \leq i \leq r, 1 \leq j \leq \frac{l}{2}] \ .$$

4.3 The Choices of the Plaintext Difference and the Ciphertext Difference

According to the property of word-oriented block ciphers, a natural way is to consider truncated differences. That is, for each word of $\Delta P = (\Delta P_i)_{1 \leq i \leq l}$ and $\Delta C = (\Delta C_i)_{1 \leq i \leq l}$, we assign an indicator to indicate the choice of its difference, representing by 0 a word without difference and by 1 a word with a difference.

In such representation, the indicator vectors of ΔP and ΔC are row vectors in $\mathbb{F}_2^l$, and the number of all possible combinations of ΔP and ΔC is $(2^l-1)\cdot(2^l-1)$.

Notice that the value of ΔP_i (or ΔC_i) is zero if its indicator is 0, which implies that, in a difference propagation system, the variable ΔP_i (or ΔC_i) can be evaluated as zero. However, if the indicator of ΔP_i (or ΔC_i) is 1, we cannot evaluate any specific value for ΔP_i (or ΔC_i). In this case, we just leave ΔP_i (or ΔC_i) as an undetermined variable in a difference propagation system while storing its indicator information. In our tool, we will add the variable ΔP_i (or ΔC_i) to set Λ_1.

4.4 Algorithm for Judging a Truncated Impossible Differential

Let $p = [p_1, p_2, \ldots, p_l]$ (respectively, $c = [c_1, c_2, \ldots, c_l]$) be a vector, indicating the variable positions of ΔP (respectively, ΔC) in x. Fox example, $p = [\frac{l}{2} + 1, \frac{l}{2} + 2, \ldots, l, 1, 2, \ldots, \frac{l}{2}]$ if the order of variables in an r-round Feistel cipher is fixed as given in (4), since $[\Delta X_1, \Delta X_0]$ is the plaintext difference. $MulCol(\text{B,i,j})$ is a function that multiplies the j-th column of matrix B with element i, and $ColSubMatrix(B, i, j)$ is the submatrix of B with columns from i to j.

Our algorithm takes the matrix B, table $\mathcal{T}$, vector p, vector c, indicator vectors of ΔP and ΔC as inputs, and then predicts information as described in Algorithm 1 (i.e., from step 3 to step 12) automatically. Finally, it outputs a *flag* indicating whether $\Delta P \to \Delta C$ is impossible, under the filter conditions listed in Proposition 1. The specific description of our algorithm is shown in Algorithm 2, and some sketches are listed as follows.

- In step 1, we introduce a new matrix B' to protect the matrix B against revision, because B can be re-used for different indicator vectors of ΔP and ΔC (we also discuss it below in the remarks of Algorithm 2).
- From step 2 to step 6, our tool scans the indicator vectors of ΔP and ΔC, and stores their information in Λ_0 or Λ_1. As mentioned in Sect. 4.2, if the value of a variable is zero, then the corresponding column of B' will be set to a zero vector.
- From step 7 to step 17, our tool predicts information from system $\mathcal{L}$ and $\mathcal{NL}$ alternately. This process is terminated if *index* is false, which implies that a contradiction is found by using Proposition 1 or we can no longer obtain any new information. Especially,
  - in step 10 and step 11, our tool detects whether there is a type **I** contradiction of Proposition 1;
  - in step 12, it preforms *Gauss-Jordan Elimination* algorithm to obtain the reduced augmented matrix of the system $\mathcal{L}$;
  - in step 13, it predicts information from the reduced augmented matrix of the system $\mathcal{L}$ and the system $\mathcal{NL}$ by calling the subprogram **Predict_Info**;
  - in step 15 and step 16, it detects whether there is a type **II** contradiction of Proposition 1.

- From step 18 to step 36, our tool predicts information from the reduced augmented matrix of the system $\mathcal{L}$ and the system $\mathcal{NL}$. Especially,
  - in step 19, our tool defines a *temp* to represent whether it finally obtain some new information in the information prediction;
  - from step 20 to step 27, our tool scans all linear equations in the system $\mathcal{L}$ and recovering information using Lemma 5;
  - from step 28 to step 33, our tool predicts information from the system $\mathcal{NL}$ using Lemma 6;
  - in step 34 and step 35, the returned *temp* will be true if our tool recovers some new information.

For an r-round word-oriented block cipher, we may obtain all r-round truncated impossible differentials by enumerating all possible nonzero indicator vectors of ΔP and ΔC. To obtain truncated impossible differentials with different lengths, it only needs to try different round numbers.

Some remarks on our algorithm are given below.

1. Since the encryption process of a block cipher is deterministic, for a fixed round number r, we only need to build the difference propagation system once. So, matrix B, table $\mathcal{T}$, vector p and vector c can be reused for different indicator vectors of ΔP and ΔC. Thus, we introduce a new matrix B' in Algorithm 2 to protect the matrix B against revision.
2. Besides indicator vectors of ΔP and ΔC, some linear constraints between nonzero variables in ΔP and ΔC can also be added while selecting the initial constraints. Our tool still works in this case by translating all linear constraints to row matrices firstly and then adding these rows to matrix B'.
3. Our tool only uses the bijective property of an Sbox to predict information, without solving nonlinear systems. The most time-consuming steps are calculating the rank of a matrix and solving a linear system.

Algorithm Complexity. Suppose B' is a matrix with M rows and N columns. The time consumption for computing $rank(A')$, $rank(B')$ and *Gauss-Jordan Elimination* is about $O(M^2 \cdot N)$, and the time consumption for the subprogram **Predict_Info** is about $O(M \cdot N)$. Algorithm 2 terminates if *index* is false. Otherwise, at least one of sets Λ_0 and Λ_1 is updated after each while loop. According to the pigeon-hole principle, there must be a type **II** contradiction mentioned in Proposition 1 when $|\Lambda_0| + |\Lambda_1| > N$. Therefore, while loop runs $N + 1$ times at most. In summary, the time complexity of judging a truncated impossible differential does not exceed $O(M^2 \cdot N^2)$. From Sect. 2.2, we know M is about $O(c_1 \cdot l \cdot r)$ and N is about $O(c_2 \cdot l \cdot r)$. Thus, the time complexity of Algorithm 2 is about $O(c \cdot l^4 \cdot r^4)$, where c is a constant depending on the machine and the block cipher. The space complexity of Algorithm 2 is dominated by storing the matrices B, B' and A'. Thus, the space complexity is about $O(M \cdot N)$, that is, $O(c' \cdot l^2 \cdot r^2)$, where c' is also a constant depending on the machine and the block cipher.

Input: Matrix B, table $\mathcal{T}$, vector p, vector c and indicator vectors of ΔP and ΔC.
Output: A $flag$ indicates whether $\Delta P \to \Delta C$ is impossible.

1 $B' := B$; $\Lambda_0 := \emptyset$; $\Lambda_1 := \emptyset$; $n :=$ NumberOfColumns(B');
   // Initialize the truncated information of ΔP and ΔC ...
   // ... using their indicator vectors SdeltaP and SdeltaC.

2 **for** $i := 1$ *to* l **do**
3    $\Lambda_1 := \Lambda_1 \cup \{p_i\}$ if SdeltaP=1;
4    $\Lambda_0 := \Lambda_0 \cup \{p_i\}$ and $B' :=$ MulCol$(B', 0, p_i)$ if SdeltaP=0;
5    $\Lambda_1 := \Lambda_1 \cup \{c_i\}$ if SdeltaC=1;
6    $\Lambda_0 := \Lambda_0 \cup \{c_i\}$ and $B' :=$ MulCol$(B', 0, c_i)$ if SdeltaC=0;
   // Predict information and find contradictions.

7 $flag :=$ false; $index :=$ true;
8 **while** $index$ **do**
9    $A' :=$ ColSubMatrix$(B', 1, n-1)$;
10   **if** $rank(A') \neq rank(B')$ **then**
11      $flag :=$ true; $index :=$ false;
     else
        // Gauss-Jordan Elimination.
12      $B' :=$ Reduced-row-echelon-form-of(B');
13      $< B', \Lambda_0, \Lambda_1, temp >:=$ **Predict_Info**$\big(B', \mathcal{T}, \Lambda_0, \Lambda_1\big)$;
14      $index := temp$;
15   **if** $\Lambda_0 \cap \Lambda_1 \neq \emptyset$ **then**
16         $flag :=$ true; $index :=$ false;

17 **return** $flag$.

18 **Predict_Info**$\big(B', \mathcal{T}, \Lambda_0, \Lambda_1\big)$;
19 $n_0 := |\Lambda_0|$; $n_1 := |\Lambda_1|$; $temp :=$ false; $n :=$ NumberOfColumns(B');
   // Predict information from matrix B' using Lemma 5.

20 **for** $i := 1$ *to* $NumberOfRows(B')$ **do**
21   $S := \emptyset$;
22   **for** $j := 1$ *to* $n-1$ **do**
23      $S := S \cup \{j\}$ if $B'[i,j] \neq 0$;
     // A linear equation with form $\Delta X = 0$.
24   **if** $|S| = 1$ *and* $B'[i,n] = 0$ **then**
25      $\Lambda_0 := \Lambda_0 \cup S$; $B' :=$ MulCol$(B', 0, j)$ for $j \in S$;
     // An equation with form $\Delta X \oplus c = 0$ ($c \neq 0$) or $\Delta X \oplus \Delta Y = 0$.
26   **if** $(|S| = 1$ *and* $B'[i,n] \neq 0)$ *or* $(|S| = 2, B'[i,n] = 0$ *and* $S \cap \Lambda_1 \neq \emptyset)$ **then**
27      $\Lambda_1 := \Lambda_1 \cup S$;
   // Scan table $\mathcal{T}$ and use Lemma 6 to predict information.

28 **for** $j := 1$ *to* $NumberOfElements(\mathcal{T})$ **do**
29   **if** $\mathcal{T}[j] \cap \Lambda_0 \neq \emptyset$ *and* $\mathcal{T}[j] \setminus \Lambda_0 \neq \emptyset$ **then**
30      $B' :=$ MulCol$(B', 0, e)$ for $e \in \mathcal{T}[j] \setminus \Lambda_0$;
31      $\Lambda_0 := \Lambda_0 \cup \mathcal{T}[j]$;
32   **if** $\mathcal{T}[j] \cap \Lambda_1 \neq \emptyset$ **then**
33      $\Lambda_1 := \Lambda_1 \cup \mathcal{T}[j]$;

34 **if** $|\Lambda_0| > n_0$ *or* $|\Lambda_1| > n_1$ **then**
35   $temp :=$ true;

36 **return** $< B', \Lambda_0, \Lambda_1, temp >$.

Algorithm 2: Evaluation of a truncated impossible differential

4.5   Experimental Results

We apply our tool to find truncated impossible differentials for various byte-(or nibble-)oriented block ciphers, such as the AES, CLEFIA, E2, Camellia without FL and FL^{-1} layers, ARIA, LBlock, MIBS and Piccolo. All of these block ciphers have $l = 16$. We may classify them into three groups according to their underlying structures: SPN ciphers (AES and ARIA), Feistel ciphers (Camellia without FL and FL^{-1} layers, LBlock, MIBS and E2), and generalized Feistel ciphers (CLEFIA and Piccolo).

Our tool independently rediscovers all 72 known truncated impossible differentials on 9-round CLEFIA. In addition, besides the best results known so far, our tool finds new truncated impossible differentials for AES, Camellia without FL and FL^{-1} layers, MIBS, LBlock, ARIA, E2 and Piccolo. The number of new truncated impossible differentials obtained by our tool is summarized in Table 1. Due to the lack of space, the specification of new results obtained by our tool and the source code for searching truncated impossible differentials of some block ciphers will be given in the full version.

4.6   Discussions

For our tool, the only thing we can confirm is that impossible differentials obtained by our tool must be correct if one implements a difference propagation system and our algorithm on a computer correctly, because the conditions of judging an impossible differential are sufficient conditions.

However, our tool still has some limitations, which are also unsolved by the U-method and the UID-method. First, the choice of truncated difference may result in missing some impossible differentials. For example, $\Delta P \to \Delta C$ should be an impossible differential if ΔP and ΔC are evaluated to some specific values, but our tool may miss it if we only know the indicator information of ΔP and ΔC. Secondly, our tool is not able to exploit any properties of the Sboxes beyond the fact that they are bijective. Thus, we may also miss some impossible differentials if we need some specific properties of an Sbox to detect these impossible differentials. Finally, our tool may fail if a block cipher is not word-oriented or uses an Sbox that is not bijective.

5   Comparison of Our Tool with the U-Method and the UID-Method

In this section, we investigate the relationship of our tool with the U-method and the UID-method.

On the one hand, we have

Proposition 2. *The U-method and the UID-method are specific cases of our tool.*

Proof. First, the characteristic matrices defined by the $\mathcal{U}$-method and the UID-method are specific forms of our difference propagation systems.

Secondly, the operations used in these tools (see Table 2 and 3 in [12], Table 1 and Definition 3 in [17]) are included in Lemma 5, Lemma 6 and the difference propagation system. For example, in Table 3 of [12], the operation $1 \cdot 1_F = 1$, i.e., the result of a nonzero difference propagating through a nonlinear bijective function (Sbox) is also a nonzero difference, is described in Lemma 6.

Finally, we show that a contradiction that is detected by the $\mathcal{U}$-method or the UID-method can also be found by our tool. From Definition 1, to detect an impossible differential, we have to show that the final value of $f = \oplus_{i \in I}(\Delta u_i \oplus \Delta v_i)$ is a nonzero difference. However, the value of f is unpredictable if its final expression contains a term with an unknown difference or two terms with nonzero unspecified differences. And f is useless if it is a zero difference, or it contains a term with a nonzero fixed difference and a term with a nonzero unspecified difference. Thus, a useful f only consists of a single term with a nonzero fixed difference or a nonzero unspecified difference. In our tool, we can build l linear equations, i.e., $\Delta u_i \oplus \Delta v_i = 0$ for $1 \leq i \leq l$, in a difference propagation system by introducing proper internal variables. These linear equations are included in the subsystem $\mathcal{L}$, then we have

(1) $f = \Delta c$, where Δc is a nonzero fixed difference. That is, a nonzero fixed difference is in the linear space spanned by $\{\Delta u_i \oplus \Delta v_i : 1 \leq i \leq l\}$. In this case, we conclude that the rank of the coefficient matrix of $\mathcal{L}$ is not equal to the rank of the augmented matrix of $\mathcal{L}$. Thus, an impossible differential is detected by case **I** of Proposition 1.

(2) $f = \Delta a$, where Δa is a nonzero unspecified difference. In this case, we have known that there is a variable with nonzero unspecified difference, i.e., $\Delta a \neq 0$, from previous information. Meanwhile, we obtain $\Delta a = 0$ if linear system $\mathcal{L}$ is solved by the *Gauss-Jordan Elimination* algorithm. Thus, an impossible differential is detected by case **II** of Proposition 1.

In summary, the $\mathcal{U}$-method and the UID-method are specific cases of our tool.
□

On the other hand, our tool is more powerful than the $\mathcal{U}$-method and the UID-method. A new impossible differential of 8-round MIBS obtained by our tool is given to illustrate that our tool finds longer impossible differentials. The $\mathcal{U}$-method and the UID-method fail to detect this impossible differential because they do not fully use the information hiding in the match point of the two probability-one differentials (see Fig. 1).

Example 3. MIBS is a nibble-oriented block cipher following the Feistel structure. It operates on 64-bit blocks, uses keys of 64 or 80 bits, and iterates 32 rounds for both key sizes. Therefore, the plaintext/ciphertext can be represented with a vector with 16 nibbles, i.e., $l = 16$.

The difference propagation system of r-round MIBS is

$$\begin{cases} \overline{S}(\Delta X_{i,j}, \Delta Y_{i,j}) = 0 & \text{for } 1 \leq i \leq r \text{ and } 1 \leq j \leq 8 , \\ I_1 \cdot \Delta X_{i-1}^T \oplus I_2 \cdot \Delta X_{i+1}^T \oplus P \cdot \Delta Y_i^T = 0 & \text{for } 1 \leq i \leq r , \end{cases} \quad (5)$$

where I_1 and I_2 are the identity matrix, and P is the linear permutation layer in round functions. Here,

$$P = \begin{pmatrix} 1&1&0&1&1&0&1&1 \\ 0&1&1&1&1&1&1&0 \\ 1&1&1&0&1&1&0&1 \\ 0&1&1&1&0&0&1&1 \\ 1&0&1&1&1&0&0&1 \\ 1&1&0&1&1&1&0&0 \\ 1&1&1&0&0&1&1&0 \\ 1&0&1&1&0&1&1&1 \end{pmatrix} \text{ and its inverse } P^{-1} = \begin{pmatrix} 0&1&0&1&0&1&1&1 \\ 1&0&0&1&1&0&1&1 \\ 1&0&1&1&1&1&0&0 \\ 0&1&1&0&1&1&1&0 \\ 0&1&1&1&1&0&1&1 \\ 1&1&0&1&1&1&0&1 \\ 1&0&1&0&1&1&1&1 \\ 1&1&1&1&0&1&1&0 \end{pmatrix}.$$

Now, for given $(\Delta X_1 \| \Delta X_0) = (0,0,0,0,0,0,0,0 \| 0,0,a,0,0,0,0,0)$ $(a \neq 0)$ and $(\Delta X_9 \| \Delta X_8) = (0,0,0,0,g,0,0,0 \| 0,0,0,0,0,0,0,0)$ $(g \neq 0)$, we can deduce the internal state of MIBS round by round (see Fig. 3). After 4-round deductions in the forward direction, we obtain that $(\Delta X_5 \| \Delta X_4)$ is

$$(e_1, e_2 \oplus b, e_3 \oplus b, e_4 \oplus b, e_5 \oplus b, e_6, e_7 \oplus b, e_8 \oplus b \| c_1, c_2, c_3 \oplus a, c_4, c_5, c_6, c_7, c_8) \ , \quad (6)$$

where a, b and b_t $(t \in \{2,3,4,5,7,8\})$ are nonzero differences while c_t, d_t and e_t $(1 \leq t \leq 8)$ are unspecific differences. While in the backward direction, after 4-round deductions, we obtain that $(\Delta X_5 \| \Delta X_4)$ is

$$(i_1, i_2, i_3, i_4, i_5 \oplus g, i_6, i_7, i_8 \| k_1 \oplus h, k_2 \oplus h, k_3 \oplus h, k_4, k_5 \oplus h, k_6 \oplus h, k_7, k_8) \ , \quad (7)$$

where g, h and h_t $(t \in \{1,2,3,5,6\})$ are nonzero differences while i_i, j_t and k_t $(1 \leq t \leq 8)$ are unknown differences.

Now, if we combine (6) and (7) together and try to find a contradiction by the filtering conditions of the $\mathcal{U}$-method and the UID-method, we get nothing because e_t and k_t $(1 \leq t \leq 8)$ are unknown differences. Thus, the $\mathcal{U}$-method and the UID-method can not detect this impossible differential.

However, our tool retrieves some important information by solving a system of linear equations deduced from (6) and (7). Notice that from (6), we know that $\Delta X_4^T = \Delta Z_3^T \oplus \Delta X_2^T = P \cdot \Delta Y_3^T \oplus \Delta X_2^T$, and we also have $\Delta X_4^T = \Delta Z_5^T \oplus \Delta X_6^T = P \cdot \Delta Y_5^T \oplus P \cdot \Delta Y_7^T$ from (7). Thus, we get

$$P \cdot \Delta Y_3^T \oplus \Delta X_2^T \oplus P \cdot \Delta Y_5^T \oplus P \cdot \Delta Y_7^T = 0 \ . \quad (8)$$

It is equivalent to solve the following linear system

$$\Delta Y_3^T \oplus P^{-1} \cdot \Delta X_2^T \oplus \Delta Y_5^T \oplus \Delta Y_7^T = 0 \ . \quad (9)$$

Since $\Delta Y_3^T = (0, b_2, b_3, b_4, b_5, 0, b_7, b_8)^T$, $P^{-1} \cdot \Delta X_2^T = (0, 0, a, a, a, 0, a, a)^T$ and $\Delta Y_7^T = (0,0,0,0,h,0,0,0)^T$, we deduce that the first nibble and the six nibble of ΔY_5 are zero, that is, $j_1 = j_6 = 0$. From Lemma 6, we know $i_1 = i_6 = 0$. For a further step, since $\Delta Z_6^T = P \cdot \Delta Y_6^T$, we have $i_1 = h_1 \oplus h_2 \oplus h_5$, $i_6 = h_1 \oplus h_2 \oplus h_5 \oplus h_6$. Now, $i_1 = i_6 = 0$ implies that $h_6 = 0$, which contradicts with $h_6 \neq 0$.

We observe that the $\mathcal{U}$-method and UID-method fail to find any of the 6 impossible differentials of 8-round MIBS.

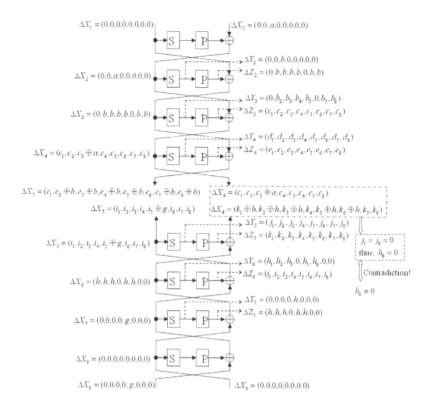

Fig. 3. A truncated impossible differential of 8-round MIBS

6   Conclusions

This paper presents an automated tool for finding truncated impossible differentials of word-oriented block ciphers with bijective S-boxes. The $\mathcal{U}$-method and the UID-method are specific cases of our tool. Although our tool does not improve the lengths of impossible differentials for existing block ciphers, it reduces the gap between previous automated tools (i.e., the $\mathcal{U}$-method and the UID-method) and ad hoc approaches. With the application of our tool, we not only rediscover the longest truncated impossible differentials of many byte-(and nibble-)oriented block ciphers known so far, but also find new results. Although it is not clear whether new results found by our tool are useful to improve known attacks or not, they bring more choices in designing attack algorithms. Hence it may be possible to improve the known attacks.

To obtain a better tool in the future, one may find some more general filtering conditions than those given in Proposition 1 or manage to exploit any properties of the Sboxes beyond the fact that they are bijective.

Acknowledgements. We are grateful to the anonymous reviewers for their valuable comments on this paper. This work was partially supported by the 973 Program (Grant No. 2013CB834203) and the National Natural Science Foundation of China (Grant No. 60970134 and Grant No. 11171323).

References

1. Aoki, K., Ichikawa, T., Kanda, M., Matsui, M., Moriai, S., Nakajima, J., Tokita, T.: *Camellia*: A 128-Bit Block Cipher Suitable for Multiple Platforms - Design and Analysis. In: Stinson, D.R., Tavares, S. (eds.) SAC 2000. LNCS, vol. 2012, pp. 39–56. Springer, Heidelberg (2001)

2. Bahrak, B., Aref, M.R.: Impossible differential attack on seven-round AES-128. IET Information Security 2, 28–32 (2008)

3. Bay, A., Nakahara Jr., J., Vaudenay, S.: Cryptanalysis of Reduced-Round MIBS Block Cipher. In: Heng, S.-H., Wright, R.N., Goi, B.-M. (eds.) CANS 2010. LNCS, vol. 6467, pp. 1–19. Springer, Heidelberg (2010)

4. Biham, E., Biryukov, A., Shamir, A.: Cryptanalysis of Skipjack Reduced to 31 Rounds Using Impossible Differentials. In: Stern, J. (ed.) EUROCRYPT 1999. LNCS, vol. 1592, pp. 12–23. Springer, Heidelberg (1999)

5. Biham, E., Biryukov, A., Shamir, A.: Miss in the Middle Attacks on IDEA and Khufu. In: Knudsen, L.R. (ed.) FSE 1999. LNCS, vol. 1636, pp. 124–138. Springer, Heidelberg (1999)

6. Biham, E., Keller, N.: Cryptanalysis of Reduced Variants of Rijndael. In: The Third AES Candidate Conference (2000)

7. Biham, E., Shamir, A.: Differential Cryptanalysis of DES-like Cryptosystems. Journal of Cryptology 4(1), 3–72 (1991)

8. Daemen, J., Rijmen, V.: The Design of Rijndael: AES – The Advanced Encryption Standard. Springer (2002)

9. Du, C., Chen, J.: Impossible Differential Cryptanalysis of ARIA Reduced to 7 rounds. In: Heng, S.-H., Wright, R.N., Goi, B.-M. (eds.) CANS 2010. LNCS, vol. 6467, pp. 20–30. Springer, Heidelberg (2010)

10. Izadi, M.I., Sadeghiyan, B., Sadeghian, S.S., Khanooki, H.A.: MIBS: A New Lightweight Block Cipher. In: Garay, J.A., Miyaji, A., Otsuka, A. (eds.) CANS 2009. LNCS, vol. 5888, pp. 334–348. Springer, Heidelberg (2009)

11. Kanda, M., Moriai, S., Aoki, K., Ueda, H., Takashima, Y., Ohta, K., Matsumoto, T.: E2 — A new 128-bit block cipher. IEICE Transactions Fundamentals – SpecialSection on Cryptography and Information Security E83–A (1), 48–59 (2000)

12. Kim, J., Hong, S., Sung, J., Lee, S., Lim, J., Sung, S.: Impossible Differential Cryptanalysis for Block Cipher Structures. In: Johansson, T., Maitra, S. (eds.) INDOCRYPT 2003. LNCS, vol. 2904, pp. 82–96. Springer, Heidelberg (2003)

13. Knudsen, L.R.: DEAL—A 128-bit block cipher. Technical Report 151, Department of Informatrics, University of Bergen, Bergen, Norway (1998)

14. Kwon, D., Kim, J., Park, S., et al.: New Block Cipher: ARIA. In: Lim, J.-I., Lee, D.-H. (eds.) ICISC 2003. LNCS, vol. 2971, pp. 432–445. Springer, Heidelberg (2004)

15. Li, R., Sun, B., Li, C.: Impossible differential cryptanalysis of SPN ciphers. IET Information Security 5(2), 111–120 (2011)

16. Li, S., Song, C.: Improved Impossible Differential Cryptanalysis of ARIA. In: ISA 2008, pp. 129–132. IEEE Computer Society, Los Alamitos (2008)

17. Luo, Y., Wu, Z., Lai, X., Gong, G.: A Unified Method for Finding Impossible Differentials of Block Cipher Structures. Cryptology ePrint Archive: Report 2009/627, http://eprint.iacr.org/2009/627

18. Mala, H., Dakhilalian, M., Rijmen, V., Modarres-Hashemi, M.: Improved Impossible Differential Cryptanalysis of 7-Round AES-128. In: Gong, G., Gupta, K.C. (eds.) INDOCRYPT 2010. LNCS, vol. 6498, pp. 282–291. Springer, Heidelberg (2010)

19. Phan, R.C.-W., Siddiqi, M.U.: Generalised Impossible Differentials of Advanced Encryption Standard. Electronics Letters 37(14), 896–898 (2001)

20. Phan, R.C.-W.: Classes of Impossible Differentials of Advanced Encryption Standard. Electronics Letters 38(11), 508–510 (2002)

21. Shibutani, K., Isobe, T., Hiwatari, H., Mitsuda, A., Akishita, T., Shirai, T.: *Piccolo*: An Ultra-Lightweight Blockcipher. In: Preneel, B., Takagi, T. (eds.) CHES 2011. LNCS, vol. 6917, pp. 342–357. Springer, Heidelberg (2011)

22. Shirai, T., Shibutani, K., Akishita, T., Moriai, S., Iwata, T.: The 128-Bit Blockcipher CLEFIA (Extended Abstract). In: Biryukov, A. (ed.) FSE 2007. LNCS, vol. 4593, pp. 181–195. Springer, Heidelberg (2007)

23. Tsunoo, Y., Tsujihara, E., Shigeri, M., Suzaki, T., Kawabata, T.: Cryptanalysis of CLEFIA Using Multiple Impossible Differentials. In: International Symposium on Information Theory and its Applications, ISITA 2008, pp. 1–6 (2008)

24. Wei, Y., Li, P., Sun, B., Li, C.: Impossible Differential Cryptanalysis on Feistel Ciphers with *SP* and *SPS* Round Functions. In: Zhou, J., Yung, M. (eds.) ACNS 2010. LNCS, vol. 6123, pp. 105–122. Springer, Heidelberg (2010)

25. Wu, W., Zhang, L.: LBlock: A Lightweight Block Cipher. In: Lopez, J., Tsudik, G. (eds.) ACNS 2011. LNCS, vol. 6715, pp. 327–344. Springer, Heidelberg (2011)

26. Wu, W., Zhang, W., Deng, D.: Impossible Diffrential Cryptanalysis of ARIA and Camellia. Journal of Computer Science and Technology 22(3), 449–456 (2007)

High-Speed Parallel Implementations of the Rainbow Method in a Heterogeneous System

Jung Woo Kim[1,*], Jungjoo Seo[1,*], Jin Hong[2,**],
Kunsoo Park[1,*,***], and Sung-Ryul Kim[3,*]

[1] Department of Computer Science and Engineering and Institute of Computer
Technology, Seoul National University, Seoul, Korea
{jkim,jjseo,kpark}@theory.snu.ac.kr
[2] Department of Mathematical Sciences and ISaC,
Seoul National University, Seoul, Korea
jinhong@snu.ac.kr
[3] Division of Internet and Media, Konkuk University, Seoul, Korea
kimsr@konkuk.ac.kr

Abstract. The computing power of graphics processing units (GPU) has increased rapidly, and there has been extensive research on general-purpose computing on GPU (GPGPU) for cryptographic algorithms such as RSA, ECC, NTRU, and AES. With the rise of GPGPU, commodity computers have become complex heterogeneous GPU+CPU systems. This new architecture poses new challenges and opportunities in high-performance computing. In this paper, we present high-speed parallel implementations of the rainbow method, which is known as the most efficient time-memory tradeoff, in the heterogeneous GPU+CPU system. We give a complete analysis of the effect of multiple checkpoints on reducing the cost of false alarms, and take advantage of it for load balancing between GPU and CPU. Our implementation with multiple checkpoints requires no more time on average for resolving false alarms and it actually finishes earlier than generating all online chains unlike other implementations on GPU.

Keywords: Cryptanalysis, Cryptanalytic Time-Memory Tradeoff, Rainbow Method, GPGPU, CUDA, Heterogeneous Computing.

1 Introduction

With the GPU's rapid evolution from a graphics processor to a programmable parallel processor, GPU is a many-core multi-threaded multiprocessor that excels

---

* This work was supported by the National Research Foundation of Korea (NRF) grant funded by the Korea government (MEST) (No. 20120006492).
** This work was supported by the Basic Science Research Program through the National Research Foundation of Korea (NRF) funded by the Ministry of Education, Science and Technology (2012003379).
*** Corresponding author.

S. Galbraith and M. Nandi (Eds.): INDOCRYPT 2012, LNCS 7668, pp. 303–316, 2012.
© Springer-Verlag Berlin Heidelberg 2012

at not only graphics but also computing applications. Today's GPUs have hundreds of parallel processor cores executing tens of thousands of parallel threads. Using a large number of processors, GPUs are used for accelerating the performance of mathematical and scientific works. General-purpose computing on GPUs (GPGPU) was first introduced in 2006 by unveiling CUDA by NVIDIA [6]. CUDA enables programmers to easily control GPUs by writing programs similar to C.

Recently, researchers and developers have enthusiastically adopted CUDA and GPU computing for cryptographic algorithms. In 2007, Manavski et al. efficiently implemented the Advanced Encryption Standard (AES) algorithm using CUDA [22]. In 2008, Szerwinski and Güneysu made use of CUDA for GPGPU processing of asymmetric cryptosystems (RSA, DSA, ECC) [27]. In 2009, Bernstein et al. showed that GPU can be used for cryptanalysis as well as implementation of cryptographic algorithms [9]. They implemented the elliptic-curve method for integer factorization on GPUs. In 2010, NTRU cryptosystem was implemented on CUDA by Hermans et al. [16].

One-way functions are fundamental tools for cryptography, and it is a hard problem to invert them. There are three generic approaches to invert them. The simplest approach is an exhaustive search. An attacker tries all possible values until the pre-image is found; however, it needs a lot of time. Another simple approach is a table lookup, in which an attacker precomputes the images of a one-way function for all possible pre-images and stores them in a table. The attack can be carried out quickly, but a large amount of memory is needed to store all precomputed values. Cryptanalytic time-memory tradeoffs [21,11,14,26,20,10,23,8,28,18] are compromise solutions between time and memory. Cryptanalytic time-memory tradeoff was introduced by Hellman in 1980 [15]. Rivest proposed to apply *distinguished points* technique [13] to Hellman's method which reduces the number of table lookup operations. In 2003, a new method, which is referred to as *rainbow method*, was suggested by Oechslin [25]. The rainbow method saves a factor of two in the worst case time complexity compared to Hellman's method. Up until now, the rainbow method is the most efficient time-memory tradeoff. Avoine et al. introduced a technique detecting false alarms, called *checkpoints* [7]. Using the technique, the cost of false alarms is reduced with a minute amount of memory.

With the rise of GPGPU, commodity computers are complex heterogeneous GPU+CPU systems that provide high computational power [24,12]. The GPU and CPU can execute in parallel and have their own independent memory systems connected through the PCIe bus. The GPU+CPU co-processing and data transfers use the bidirectional PCIe bus. This new architecture poses new challenges and opportunities in high-performance computing.

In this paper, we propose high-speed parallel implementations of the rainbow method in the heterogeneous GPU+CPU system through the analysis of the behavior of time-memory tradeoffs. We give a complete analysis of the effect of multiple checkpoints on reducing the cost of false alarms for the non-perfect rainbow table, and take advantage of it for load balancing between GPU and CPU.

The proposed implementation requires no more time on average for resolving false alarms by fully parallelizing the rainbow method on GPU and CPU. Our implementation actually finishes earlier on average than generating all online chains unlike other implementations on GPU. To the best of our knowledge, this is the first work implementing the rainbow method in a heterogeneous system.

The rest of the paper is organized as follows. We begin with an overview of modern GPUs and CUDA in Section 2, followed by a brief review of the rainbow method in Section 3. In Section 4, we describe our fast implementations in a heterogeneous GPU+CPU system. In Section 5, we analyze the checkpoint technique. Finally, Section 6 presents the experimental results.

2   GPGPU and CUDA

While traditional GPUs were used for graphical applications, many modern GPUs can deal with general parallel programs which had been performed normally on CPUs. CUDA [6] is NVIDIA's software and hardware architecture that enables GPUs to be programmed with a variety of high-level programming languages, and it is a parallel computing architecture that is used to improve computing performance by exploiting the power of GPU. NVIDIA has released several improved versions of architectures since its first architecture, G80, and the newest one is called Fermi [4], which was introduced in 2009.

One of the most attractive features of GPUs is that it has a large number of processor cores. Basically, GPUs consist of a number of streaming multiprocessors (SM), and each SM contains multiple processor cores. The clock rate of each core is relatively lower than that of a CPU core. In our experiment, we used the GeForce GTX580 which belongs to the Fermi architecture. The GTX580 accommodates 16 SMs, each of which consists of 32 processor cores operating in the clock rate 1,544 MHz, as presented in Figure 1. Hence, the total number of processor cores is 512.

One can program the GPU with a high-level programming language. We write programs in CUDA C that supports the CUDA programming with a minimal set of extensions to the C language. In the rest of this section, we will describe the key features of the CUDA that we must take into account for programming.

Thread Hierarchy. One of the key abstractions of the CUDA is a hierarchy of threads. By this abstraction, we can divide the whole problem into coarse-grained subproblems, *blocks*, which can be solved independently in parallel. A block can be further partitioned into fine-grained subproblems that can also be solved in parallel within the block. This fine-grained subproblem unit is called a *thread*. CUDA's hierarchy of threads maps to a hierarchy of processors on the GPU. An SM executes one or more blocks, and CUDA cores in the SM execute threads.

Scheduling & Branch. The way threads are scheduled in GPUs is somewhat different from that in CPUs. The unit of thread scheduling in SMs is a *warp* which is a collection of 32 threads.

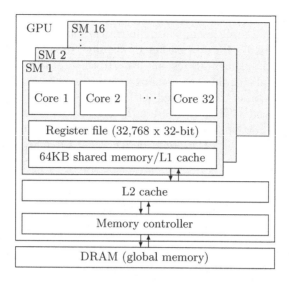

Fig. 1. Fermi architecture

Basically, all the threads within a single warp execute the same instruction at the same time. However, multiple threads of the same warp may execute serially. When they meet any flow control instruction such as if A else B, they could take different execution paths. Then, different execution paths within a warp are serialized. It is called *warp serialization* [5,6], which will slow down the overall performance.

Memory. The physically separated place where CUDA threads are executed is referred to as *device*, which includes the GPU. The *host* is where the C program runs, and this includes the CPU. The host and device have their own memory address space. The data to be processed are firstly loaded on the host memory and then copied to the device memory, so that threads running on the GPU can access the data. The processed data on the device needs to be copied back to the host memory after the execution.

The device memory has a hierarchy and it consists of registers, shared memory, caches and global memory. Registers are the fastest on-chip memory and the GTX580 contains about $32K$ registers for each stream multiprocessor. The global memory resides in the off-chip DRAM on the graphics board. It has the longest access latency but has the largest space.

3   Rainbow Method

In this section, we summarize the rainbow method [25]. Let g be a one-way function from $\mathcal{N}$ to $\mathcal{H}$ and R_i be a reduction function from $\mathcal{H}$ to $\mathcal{N}$. The function f_i, defined by $f_i(x) = R_i(g(x))$, maps $\mathcal{N}$ into $\mathcal{N}$, where $|\mathcal{N}| = N$.

t

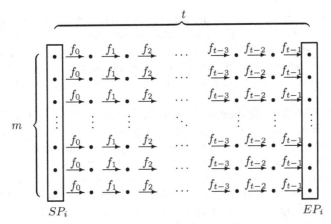

SP_i EP_i

Fig. 2. A rainbow table

The rainbow method consists of two phases: precomputation and online phases. In the precomputation phase, we randomly choose m start points in $\mathcal{N}$, labeled SP_0, SP_1, ..., SP_{m-1}. For each $0 \leq i < m$, we set

$$x_{i,0} = SP_i,$$

and compute

$$x_{i,j} = f_{j-1}(x_{i,j-1}), 1 \leq j \leq t$$

recursively. In other words, m chains of length t are produced starting from SP_i ($0 \leq i < m$) as shown in Figure 2. The last element $x_{i,t}$ for each i-th chain is called an end point (EP_i). The pairs of the start and end points, (SP_i, EP_i), are stored in a table, and they are sorted with respect to the end points. Note that all intermediate points are discarded to reduce memory requirements.

In the online phase, given an image $y_0 = g(x_0)$, we try to invert the one-way function $g(\cdot)$ to find the pre-image x_0, by generating online chains that start from y_0.

At the first iteration, the online chain of length one is generated by computing $y_1 = R_{t-1}(y_0) = f_{t-1}(x_0)$, and we check whether it is an end point on the table by conducting a binary search. If $y_1 = EP_i$ for some i, which is referred to as an *alarm*, it means that x_0 is next to EP_i in Figure 2 or EP_i has more than one inverse images. The latter case is referred to as a *false alarm*. Therefore, we regenerate a chain starting from SP_i to compute $x_{i,t-1}$, and check whether it is a false alarm or not by computing $g(x_{i,t-1}) = y_0$. If $g(x_{i,t-1}) = y_0$, we find the pre-image x_0, which is equal to $x_{i,t-1}$, and the online phase stops. If $y_1 \neq EP_i$ or a false alarm occurred, then we compute $y_2 = f_{t-1}(R_{t-2}(y_0))$, the online chain of length two, and check whether it is an end point. The above process is repeated until x_0 is found or all t online chains fail to invert the given image y_0.

The online phase of the rainbow method can be divided into three parts: *online chain*, *lookup* and *regenerating chain*. The *online chain* procedure generates the online chain of length i at the i-th iteration. The *lookup* procedure checks whether each of these is an end point (alarm) through a binary search in the rainbow table. The *regenerating chain* procedure regenerates the chains of length $(t - i)$, starting from start points for resolving alarms. Because table lookup time through a binary search is negligible in comparison to the one-way function invocation time, the one-way function invocation is the dominant factor in the overall cost of the rainbow method.

Note that the rainbow method is a probabilistic algorithm. That is, success is not guaranteed and the success probability depends on the time and memory allocated for cryptanalysis. If the pre-image x_0 that we want to find exists in the rainbow table, the rainbow method will succeed in finding it; Otherwise, it will fail. The success probability can be computed by the equation presented in [19,25]. In the case of failure, the online phase generates t online chains, and it carries out t lookups. Also, it regenerates some chains starting from start points whenever alarms occur in the lookup procedure. On the other hand, if the rainbow method succeeds in finding the pre-image x_0, it immediately stops in the middle of the online phase.

4   Implementation in a Heterogeneous GPU+CPU System

In this section, we describe our implementations in a heterogeneous GPU+CPU system. Using both GPU and CPU, we implement the rainbow method in parallel. The key factors for achieving good performance are: (i) eliminating the warp serialization by splitting the online phase of the rainbow method, and (ii) load balancing between GPU and CPU using checkpoints.

Before explaining our implementations, we first present the table used in our experiment. Cryptographic hash algorithm SHA-1 was used as the one-way function. We assumed that our table is used for cracking passwords which consist of lowercase, uppercase alphabets (a-z, A-Z) and numbers (0-9), and their lengths are shorter than or equal to 7. That is, $N = 62 + 62^2 + \cdots + 62^7 \approx 3.58 \times 10^{12} \approx 2^{41.7}$. We created a single non-perfect[1] rainbow table with 70% success probability, in which $m = 80,530,636$, $t = 73,403$. For reasons of efficient memory access, a start point of $\lceil \log_2 m \rceil = 27$ bits is stored in a 32-bit data type, uint32_t, and an end point of $\lceil \log_2 N \rceil = 42$ bits[2] is stored in a 64-bit data type, uint64_t. Thus, the total size of the table is about 0.9 GB. We conducted our experiments on an Intel i7 2.8GHz quad-core CPU and a GTX580 1544MHz 512-core GPU. We used Microsoft Visual Studio 2008 environment on Window 7.

The naive implementation of the parallel rainbow method is that each thread generates the corresponding online chain in parallel. That is, the i-th thread $(1 \leq i \leq t)$ generates the online chain of length i (the online chain procedure), and it checks whether an alarm occurs (the lookup procedure). If an alarm occurs, the i-th thread regenerates the chain of length $(t - i)$ and it checks whether the

---

[1] None of the colliding chains in the rainbow table are removed.

[2] For the simple implementation, efficient storage techniques [19] such as the index file and the end point truncation were not considered.

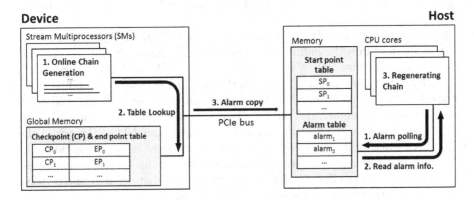

Fig. 3. Implementation in a heterogeneous GPU+CPU system

element in the $(t - i)$-th column is x_0 or a false alarm (the regenerating chain procedure). We created 896 threads per SM, i.e., total $896 \times 16 = 14,336$ threads. Thus, at first, threads generate the online chains whose lengths are between 1 and 14,336, and some of them in which alarms occur regenerate the chains and check whether each of these is a success or a false alarm. If some SM finishes its workload, the next 896 online chains, whose lengths are between 14,337 and 15,232, are assigned to the SM. We call this implementation *the Naive GPU*.

Table 1 shows the execution time when it fails to find a pre-image. The second row represents the time for executing all three procedures, and the third row represents the time for executing the online chain and the lookup procedures excluding the regenerating chain procedure. The third column in the table represents the total length of the chains generated in the online chain and regenerating chain procedures.

Table 1. Time of online phase when it fails

procedures	time	chain length
online chain+lookup+regenerating chain	258 sec	4.2×10^9
online chain+lookup	13 sec	2.7×10^9

Generally, the sum of the chain lengths in the regenerating chain procedure is smaller than that of the lengths in the online chain procedure, because alarms occur only in some of the online chains. [17] As can be seen in Table 1, the sum of chain lengths in the online chain procedure (2.7×10^9) is larger than that in the regenerating chain (1.5×10^9). However, the regenerating chain procedure takes much more time than the online chain procedure in the Naive GPU. This is because of *warp serialization*. Since alarms occur in some of the 32 threads within a warp, only these threads regenerate chains for resolving alarms. Thus, the other threads within a warp should wait until the threads finish the regenerating chain procedure. We should eliminate the warp serialization to improve the performance.

To solve this problem (warp serialization), we split the online phase of the rainbow method into the online chain+lookup procedures (A) and the regenerating chain procedure (B). A is processed in the GPU, and B is processed in the CPU, as in Figure 3. Each thread in the GPU (i) generates the online chain assigned to itself and (ii) checks whether it is an end point (alarm). (iii) If an alarm occurs, the number and the length of the corresponding chain are copied to the alarm table in the host memory. At the same time, (i) the threads in the CPU check whether the values copied from the GPU exist in the alarm table. (ii) If so, they read the copied values and (iii) regenerate chains for resolving alarms. By doing this, we can eliminate the warp serialization that occured in the Naive GPU. We call this implementation *the GPU+CPU*.

The execution time of the GPU+CPU is shown in Table 2. The GPU processes A in 13 seconds, whereas on the CPU it takes 102 seconds to process B. While the workload on the GPU is heavier than that on the CPU, the computing power of the GPU is much better than that of the CPU. Therefore, it is necessary to reduce the workload on the CPU for the efficient GPU+CPU implementation.

Table 2. Time of online phase when it fails

online chain+lookup (GPU)	regenerating chain (CPU)	total
13 sec	102 sec	102 sec

We take advantage of checkpoints [7] for load balancing between GPU and CPU. By decreasing the number of false alarms with checkpoints, we can reduce the workload on the CPU. The more checkpoints we use, the less workload the CPU have to process. In the following section, we analyze the performance improvement using checkpoints and their optimal positions. In Section 6, we present the experimental results when the checkpoints are applied to the GPU+CPU.

5   Checkpoints

By using checkpoints [7], we can reduce the time for the regenerating chain procedure. We store not only the start and end points of the chains in the table but also the information of some intermediate points, i.e., *checkpoints*. The least significant bits of the intermediate points are usually stored. Using the information, we can detect false alarms in advance without regenerating the chains starting from start points. If alarms occur, we compare the information stored in the table with those of the online chain for each checkpoint. If they differ at least for one checkpoint, we know for certain that this is a false alarm. In [7], Avoine *et al.* analyzed the effect of checkpoints for the perfect rainbow table. Analysis for the non-perfect rainbow table was done only for one checkpoint in [17]. In this section, we analyze the performance improvement of checkpoints and

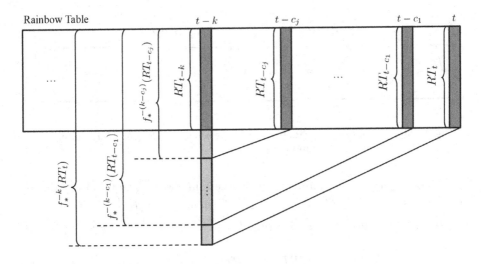

Fig. 4. Sizes of the pre-images at the $(t - k)$-th column

their optimal positions when multiple checkpoints are used for the non-perfect rainbow table.

The set of elements in the k-th column of the rainbow table is denoted by RT_k. Let $c_1, c_2, \ldots, c_n$ ($c_1 < c_2 < \cdots < c_n$) be the positions of n 1-bit checkpoints. That is, n checkpoints are located at $(t - c_j)$-th columns of the table for $j = 1, \ldots, n$.

First, at the k-th iteration (an online chain of length k is generated) for $k \leq c_1$, the checkpoints cannot filter out false alarms. Thus, we assume that an alarm is observed at the k-th iteration such that $c_j < k \leq c_{j+1}$ for $j = 1, \ldots, n$, where $c_{n+1} = t$. This means that the pre-image x_0 is in $f_\star^{-k}(RT_t)$, where $f_\star$ is function f_j whose index j is not explicitly specified and $f_\star^{-k}(RT_t)$ is the set of pre-images under $f_\star^k (= f_\star \circ \cdots \circ f_\star)$ of the end points RT_t. As can be seen in Figure 4, the following relations hold:

$$RT_{t-k} \subset f_\star^{-(k-c_j)}(RT_{t-c_j}) \subset \cdots \subset f_\star^{-(k-c_1)}(RT_{t-c_1}) \subset f_\star^{-k}(RT_t).$$

We compute the probability of false alarms when checkpoints are used. If $x_0 \in RT_{t-k}$, x_0 can be found. If $x_0 \in f_\star^{-(k-c_j)}(RT_{t-c_j}) \setminus RT_{t-k}$ (Figure 5), a false alarm always occurs. It is because the online chain starting from x_0 is merged with an precomputed chain in the rainbow table before the $(t - c_j)$-th column, and j checkpoints are thus useless in detecting false alarms. If $x_0 \in f_\star^{-(k-c_u)}(RT_{t-c_u}) \setminus f_\star^{-(k-c_{u+1})}(RT_{t-c_{u+1}})$ for $1 \leq u \leq j - 1$ (Figure 6), this means that the online chain is merged with an chain in the table between c_u and c_{u+1}. Hence, a false alarm occurs with probability $1/2^{j-u}$ by $(j - u)$ 1-bit checkpoints, i.e., $c_{u+1}, \ldots, c_j$. Finally, if $x_0 \in f_\star^{-k}(RT_t) \setminus f_\star^{-(k-c_1)}(RT_{t-c_1})$ (Figure 7), a false alarm occurs with probability $1/2^j$.

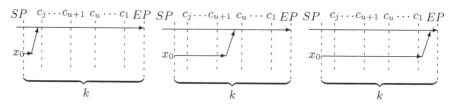

Fig. 5. Merge before c_j **Fig. 6.** Merge between c_u **Fig. 7.** Merge after c_1
and c_{u+1}

We now compute the improvement in the number of $f_\star$ applications due to checkpoints. Let $z_\star = |RT_{t-k}|$, $z_0 = |f_\star^{-k}(RT_t)|$, and $z_j = |f_\star^{-(k-c_j)}(RT_{t-c_j})|$ for $j = 1, \ldots, n$. The probability that $x_0 \in f_\star^{-(k-c_j)}(RT_{t-c_j}) \setminus RT_{t-k}$ is

$$\frac{1}{N}\left|f_\star^{-(k-c_j)}(RT_{t-c_j}) \setminus RT_{t-k}\right| = \frac{1}{N}(z_j - z_\star),$$

where N is the size of $\mathcal{N}$. In this case, a false alarm always occurs. The probability that $x_0 \in f_\star^{-(k-c_u)}(RT_{t-c_u}) \setminus f_\star^{-(k-c_{u+1})}(RT_{t-c_{u+1}})$ is

$$\frac{1}{N}\left|f_\star^{-(k-c_u)}(RT_{t-c_u}) \setminus f_\star^{-(k-c_{u+1})}(RT_{t-c_{u+1}})\right| = \frac{1}{N}(z_u - z_{u+1}).$$

In this case, a false alarm occurs with probability $1/2^{j-u}$. The probability that $x_0 \in f_\star^{-k}(RT_t) \setminus f_\star^{-(k-c_1)}(RT_{t-c_1})$ is

$$\frac{1}{N}\left|f_\star^{-k}(RT_t) \setminus f_\star^{-(k-c_1)}(RT_{t-c_1})\right| = \frac{1}{N}(z_0 - z_1)$$

In this case, a false alarm occurs with probability $1/2^j$. Therefore, the expected number of false alarms at the k-th iteration such that $c_j < k \le c_{j+1}$ ($j = 1, \ldots, n$) can be written as

$$\frac{1}{N}\left\{(z_j - z_\star) + \sum_{u=0}^{j-1}\frac{1}{2^{j-u}}(z_u - z_{u+1})\right\}. \tag{1}$$

Also, the expected number of false alarms at the k-th iteration without checkpoints is

$$\frac{1}{N}(z_0 - z_\star). \tag{2}$$

Hence, the expected decreasing number of false alarms at the k-th iteration due to checkpoints is $(2) - (1)$, which simplifies to

$$\frac{1}{N}\left\{(1 - \frac{1}{2^j})z_0 - \sum_{u=0}^{j-1}\frac{1}{2^{j-u}}z_{u+1}\right\}. \tag{3}$$

According to Propositions 4 and 5 in [17], $z_0 \approx m(1+k), z_u \approx m(1+k-c_u)$. Simplification of (3) using these approximations results in

$$\frac{m}{N} \sum_{u=0}^{j-1} \left(\frac{c_u+1}{2^{j-u}} \right).$$

We shall write $D(j)$ for this. At the k-th iteration such that $c_j < k \leq c_{j+1}$, the expected decreasing number of false alarms due to checkpoints is $D(j)$ and the number of $f_\star$ applications for checking false alarms is $t - k + 1$[3]. The probability that the k-th iteration is processed is equal to the probability to fail until the $(k-1)$-th iteration. This probability is

$$\prod_{i=1}^{k-1} \left(1 - \frac{m_{t-i}}{N} \right),$$

where m_j denotes the distinct number of elements in the j-th column of the rainbow table, i.e., $m_j \approx \frac{N}{N/m+j/2}$ [7]. Therefore, the expected number of $f_\star$ applications that can be removed through n 1-bit checkpoints is

$$\sum_{j=1}^{n} \left\{ \sum_{c_j < k \leq c_{j+1}} (t - k + 1) \cdot D(j) \cdot \prod_{i=1}^{k-1} \left(1 - \frac{m_{t-i}}{N} \right) \right\},$$

where $c_{n+1} = t$.

Table 3 shows the performance improvement due to the checkpoints and the optimal positions of those, where $N = 3.58 \times 10^{12}$, $m = 80,530,636$, and $t = 73,403$. The optimal positions represent the ratio from the rightmost column of the table. We used Maple 12 [3] to obtain these positions. The number of $f_\star$ applications in the regenerating chain procedure without checkpoints can be calculated from Theorem 3 of [17].

We made use of 22 1-bit checkpoints. Because $N = 3.58 \times 10^{12} \approx 2^{41.7}$, we used uint64_t, which is the data type of 64 bits, to store an end point, as mentioned in Section 4. An end point was stored in the lower 42 bits, and 22 1-bit checkpoints were stored in the upper 22 bits which remained empty. Therefore, no additional memory is needed to store the checkpoints. The 22 checkpoints are expected to decrease the number of $f_\star$ applications due to false alarms by about 84%. The optimal positions of 22 checkpoints are 0.0416, 0.0633, 0.0855, 0.1083, 0.1316, 0.1556, 0.1802, 0.2056, 0.2318, 0.2589, 0.2870, 0.3162, 0.3465, 0.3783, 0.4117, 0.4470, 0.4845, 0.5247, 0.5684, 0.6168, 0.6718, and 0.7381.

6   Experimental Results

In this paper, we introduced three different kinds of implementations using the GPU: naive GPU, GPU+CPU and GPU+CPU with checkpoints. Figure 8 also

---

[3] Strictly speaking, one extra g application follows $(t - k)$ number of $f_\star$ applications in order to check false alarms.

Table 3. Expected numbers of f_* applications (unit: t^2) in the regenerating chain procedure and performance improvement due to checkpoints at the optimal positions

# of checkpoints	1	2	3	4	5	6	7	8
# of f_* applications without checkpoints[†] (1)	0.1676							
Reduced # of f_* applications with checkpoints[†] (2)	0.0354	0.0577	0.0732	0.0847	0.0936	0.1008	0.1066	0.1115
Improvement ((2)/(1))	21.1%	34.4%	43.7%	50.5%	55.9%	60.1%	63.6%	66.5%
Optimal positions	0.2792	0.2123	0.1732	0.1470	0.1281	0.1136	0.1022	0.0930
		0.3591	0.2827	0.2356	0.2028	0.1785	0.1597	0.1446
			0.4179	0.3379	0.2863	0.2495	0.2216	0.1996
				0.4637	0.3826	0.3287	0.2894	0.2590
					0.5008	0.4199	0.3649	0.3239
						0.5317	0.4517	0.3962
							0.5579	0.4792
								0.5806

[†] The f_* applications in the online chain procedure are not included.

shows the experimental results using the CPU, as well as those of the three implementations presented in this paper. In the case of the CPU, the i-th thread generates the online chain of length i and regenerates the chain of length $(t - i)$ from a start point if an alarm occurs, as in the naive GPU. We used the i7 quad-core CPU for our experiment. Every experiment was carried out 50 times, and numerical values in the figure represent the average times for searching a pre-image.

There are several GPU-accelerated implementations of the rainbow method: RainbowCrack [2] and Cryptohaze [1]. The overall performance of the rainbow method depends on the implementations of the one-way function and the reduction function. Because their source codes are not publicly available, however, direct comparisons are not possible. Also, our work does not focus on the optimized implementations of these functions. Thus, we show the advantage of ours through an indirect comparison with RainbowCrack and Cryptohaze. Assume that the implementations of the one-way function and the reduction function are the same. Then, the online chain procedures of all implementations will take more or less the same time, since the parallelization of the online chain procedure is straightforward. The time for the lookup procedure is negligible compared to the other two procedures. We regard the time for the online chain procedure as 100%, and measure the total time as its percentage. According to our experiments, RainbowCrack and Cryptohaze take about 56% and 158% time for the regenerating chain procedure, respectively. Therefore, RainbowCrack and Cryptohaze take about 156% and 258% for the total time, since they regenerate chains for resolving false alarms after the online chain and lookup procedures

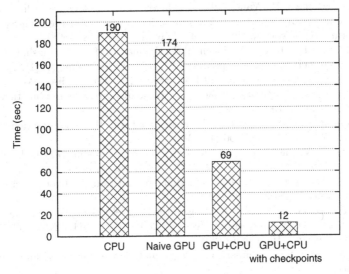

Fig. 8. Timings for searching a pre-image. Each bar represents the average time for the whole 50 experiments.

are finished. However, our method requires no more time on average for the regenerating chain procedure because the online chain+lookup procedures and the regenerating chain procedure are simultaneously executed in GPU and CPU. Our GPU+CPU with checkpoints (12 seconds) actually finishes earlier on average than the worst case of the online chain+lookup procedures (13 seconds). That is, our implementation takes about 92% on average for the total time.

References

1. Cryptohaze gpu rainbow cracker, https://www.cryptohaze.com
2. RainbowCrack Project, http://project-rainbowcrack.com
3. Maplesoft, Maple 12 user manual (2007)
4. Nvidia, Nvidia's next generation CUDA compute architecture: Fermi (2009)
5. Nvidia, CUDA best practices guide (2012)
6. Nvidia, CUDA C programming guide (2012)
7. Avoine, G., Junod, P., Oechslin, P.: Characterization and improvement of time-memory trade-off based on perfect tables. ACM Trans. Inf. Syst. Secur. 11(4) (2008)
8. Barkan, E., Biham, E., Shamir, A.: Rigorous Bounds on Cryptanalytic Time/Memory Tradeoffs. In: Dwork, C. (ed.) CRYPTO 2006. LNCS, vol. 4117, pp. 1–21. Springer, Heidelberg (2006)
9. Bernstein, D.J., Chen, T.-R., Cheng, C.-M., Lange, T., Yang, B.-Y.: ECM on Graphics Cards. In: Joux, A. (ed.) EUROCRYPT 2009. LNCS, vol. 5479, pp. 483–501. Springer, Heidelberg (2009)
10. Biryukov, A., Mukhopadhyay, S., Sarkar, P.: Improved Time-Memory Trade-Offs with Multiple Data. In: Preneel, B., Tavares, S. (eds.) SAC 2005. LNCS, vol. 3897, pp. 110–127. Springer, Heidelberg (2006)

11. Borst, J., Preneel, B., Vandewalle, J.: On the time-memory tradeoff between exhaustive key search and table precomputation. In: Proc. of the 19th Symposium in Information Theory in the Benelux, WIC, pp. 111–118 (1998)

12. Brodtkorb, A.R., Dyken, C., Hagen, T.R., Hjelmervik, J.M., Storaasli, O.O.: State-of-the-art in heterogeneous computing. Scientific Programming 18(1), 1–33 (2010)

13. Denning, D.E.: Cryptography and Data Security, p. 100. Addison-Wesley (1982)

14. Fiat, A., Naor, M.: Rigorous time/space trade-offs for inverting functions. SIAM J. Comput. 29(3), 790–803 (1999)

15. Hellman, M.: A cryptanalytic time-memory trade-off. IEEE Transactions on Information Theory 26(4), 401–406 (1980)

16. Hermans, J., Vercauteren, F., Preneel, B.: Speed Records for NTRU. In: Pieprzyk, J. (ed.) CT-RSA 2010. LNCS, vol. 5985, pp. 73–88. Springer, Heidelberg (2010)

17. Hong, J.: The cost of false alarms in Hellman and rainbow tradeoffs. Des. Codes Cryptography 57(3), 293–327 (2010)

18. Hong, J., Lee, G.W., Ma, D.: Analysis of the Parallel Distinguished Point Tradeoff. In: Bernstein, D.J., Chatterjee, S. (eds.) INDOCRYPT 2011. LNCS, vol. 7107, pp. 161–180. Springer, Heidelberg (2011)

19. Hong, J., Moon, S.: A comparison of cryptanalytic tradeoff algorithms. Journal of Cryptology (to appear)

20. Hong, J., Sarkar, P.: New Applications of Time Memory Data Tradeoffs. In: Roy, B. (ed.) ASIACRYPT 2005. LNCS, vol. 3788, pp. 353–372. Springer, Heidelberg (2005)

21. Kusuda, K., Matsumoto, T.: Optimization of time-memory trade-off cryptanalysis and its application to DES, FEAL-32 and Skipjack. IEICE Transactions on Fundamentals of Electronics, Communications and Computer Sciences E79-A(1), 35–48 (1996)

22. Manavski, S.A.: CUDA compatible GPU as an efficient hardware accelerator for AES cryptography. In: ICSPC (2007)

23. Mukhopadhyay, S., Sarkar, P.: Application of LFSRs in Time/Memory Trade-Off Cryptanalysis. In: Song, J.-S., Kwon, T., Yung, M. (eds.) WISA 2005. LNCS, vol. 3786, pp. 25–37. Springer, Heidelberg (2006)

24. Nickolls, J., Dally, W.J.: The GPU computing era. IEEE Micro 30(2), 56–69 (2010)

25. Oechslin, P.: Making a Faster Cryptanalytic Time-Memory Trade-Off. In: Boneh, D. (ed.) CRYPTO 2003. LNCS, vol. 2729, pp. 617–630. Springer, Heidelberg (2003)

26. Standaert, F.-X., Rouvroy, G., Quisquater, J.-J., Legat, J.-D.: A Time-Memory Tradeoff Using Distinguished Points: New Analysis & FPGA Results. In: Kaliski Jr., B.S., Koç, Ç.K., Paar, C. (eds.) CHES 2002. LNCS, vol. 2523, pp. 593–609. Springer, Heidelberg (2003)

27. Szerwinski, R., Güneysu, T.: Exploiting the Power of GPUs for Asymmetric Cryptography. In: Oswald, E., Rohatgi, P. (eds.) CHES 2008. LNCS, vol. 5154, pp. 79–99. Springer, Heidelberg (2008)

28. Wang, W., Lin, D., Li, Z., Wang, T.: Improvement and Analysis of VDP Method in Time/Memory Tradeoff Applications. In: Qing, S., Susilo, W., Wang, G., Liu, D. (eds.) ICICS 2011. LNCS, vol. 7043, pp. 282–296. Springer, Heidelberg (2011)

Computing Small Discrete Logarithms Faster

Daniel J. Bernstein[1,2] and Tanja Lange[2]

[1] Department of Computer Science
University of Illinois at Chicago, Chicago, IL 60607–7053, USA
djb@cr.yp.to
[2] Department of Mathematics and Computer Science
Technische Universiteit Eindhoven, P.O. Box 513, 5600 MB Eindhoven,
The Netherlands
tanja@hyperelliptic.org

Abstract. Computations of small discrete logarithms are feasible even in "secure" groups, and are used as subroutines in several cryptographic protocols in the literature. For example, the Boneh–Goh–Nissim degree-2-homomorphic public-key encryption system uses generic square-root discrete-logarithm methods for decryption. This paper shows how to use a small group-specific table to accelerate these subroutines. The cost of setting up the table grows with the table size, but the acceleration also grows with the table size. This paper shows experimentally that computing a discrete logarithm in an interval of order ℓ takes only $1.93 \cdot \ell^{1/3}$ multiplications on average using a table of size $\ell^{1/3}$ precomputed with $1.21 \cdot \ell^{2/3}$ multiplications, and computing a discrete logarithm in a group of order ℓ takes only $1.77 \cdot \ell^{1/3}$ multiplications on average using a table of size $\ell^{1/3}$ precomputed with $1.24 \cdot \ell^{2/3}$ multiplications.

Keywords: Discrete logarithms, random walks, precomputation.

1 Introduction

Fully homomorphic encryption is still prohibitively slow, but there are much more efficient schemes achieving more limited forms of homomorphic encryption. We highlight Freeman's variant [11] of the scheme by Boneh, Goh, and Nissim [7]. The Boneh–Goh–Nissim (BGN) scheme can handle adding arbitrary subsets of encrypted data, multiplying the sums, and adding any number of the products. Freeman's variant works in groups typically encountered in pairing-based protocols. The scheme is vastly more efficient than schemes handling unlimited numbers of additions and multiplications. Encryption takes only one exponentiation, as does addition of encrypted messages; multiplication takes a pairing computation.

The limitation to one level of multiplication means that polynomial expressions of degree at most 2 can be evaluated over the encrypted messages, but

This work was supported by the National Science Foundation under grant 1018836, by the Netherlands Organisation for Scientific Research under grant 639.073.005, and by the European Commission under Contract ICT-2007-216676 ECRYPT II. Permanent ID of this document: b83446575069e4e1d5517415fa8a2421. Date: 2012.09.18.

S. Galbraith and M. Nandi (Eds.): INDOCRYPT 2012, LNCS 7668, pp. 317–338, 2012.
© Springer-Verlag Berlin Heidelberg 2012

this is sufficient for a variety of protocols. For example, [7] presented protocols for private information retrieval, elections, and generally universally verifiable computation. There are 395 citations of [7] so far, according to Google Scholar.

The BGN protocol does not have any built-in limit on the number of cipher-texts added, but it does take more time to decrypt as this number grows. The problem is that decryption requires computing a discrete logarithm, where the message is the unknown exponent. If this message is a sum of B products of sums of A input messages from the space $\{0, \ldots, M\}$, then the final message can be essentially anywhere in the interval $[0, (AM)^2 B]$. This means that even if the space for the input messages is limited to bits $\{0, 1\}$, the discrete-logarithm computation needs to be able to handle the interval $[0, A^2 B]$. For "random" messages the result is almost certainly in a much shorter interval, but most applications need to be able to handle non-random messages.

Boneh, Goh, and Nissim suggested using Pollard's kangaroo method for the discrete-logarithm computation. This method runs in time $\Theta(\ell^{1/2})$ for an interval of size ℓ. This bottleneck becomes quite troublesome as A and B grow.

For larger message spaces, Hu, Martin, and Sunar in [18] sped up the discrete-logarithm computation at the cost of expanding the ciphertext length and slowing down encryption and operations on encrypted messages. They suggested representing the initial messages by their residues modulo small coprime numbers $d_1, \ldots, d_j$ with $\prod d_i > (AM)^2 B$, and encrypting these j residues separately. This means that the ciphertexts are j times as long and that each operation on the encrypted messages is replaced by j operations of the same type on the components. The benefit is that each discrete logarithm is limited to $[0, (Ad_i)^2 B]$, which is a somewhat smaller interval. The original messages are reconstructed using the Chinese remainder theorem.

Contributions to BGN. This paper explains (Section 3) how to speed up computations of small discrete logarithms, i.e., discrete logarithms in small intervals. The speedup requires a one-time computation of a small group-specific table. The speedup grows as the table grows; an interesting special case is a table of size $\Theta(\ell^{1/3})$, speeding up the discrete logarithm to $\Theta(\ell^{1/3})$ group operations. The space for the table (and the one-time cost for computing the table) is not a problem for the sizes of ℓ used in these applications.

Our experiments (Section 4) show discrete logarithms in an interval of order ℓ taking only $1.93 \cdot \ell^{1/3}$ multiplications on average using a table of size $\ell^{1/3}$. Precomputation of the table used $1.21 \cdot \ell^{2/3}$ multiplications. This paper also explains (Section 5) how to compress each table entry below $\lg \ell$ bits with negligible overhead.

This algorithm directly benefits the BGN scheme for any message size M. As an illustration, consider the common binary case $M = 1$, and assume $A = B$. The cost of decryption then drops from $\Theta(A^{3/2})$ (superlinear in the number of additions carried out) to just $\Theta(A)$, using a table of size $\Theta(A)$. The same speedup means that [18] can afford to use fewer moduli, saving both space and time.

Further Applications of Discrete Logarithms in Small Intervals. Many protocols use only degree-1-homomorphic encryption: i.e., addition without any multiplications. The pairing used in the BGN protocol is then unnecessary: one can use a faster elliptic curve that does not support pairings. Decryption still requires a discrete-logarithm computation, this time on the elliptic curve rather than in a multiplicative group. These protocols can also use Paillier's homomorphic cryptosystem, but elliptic curves provide faster encryption and smaller ciphertexts.

As an example we mention the basic aggregation protocol proposed by Kursawe, Danezis, and Kohlweiss in [21] to enable privacy for smart-meter power-consumption readings. The power company obtains the aggregated consumptions $\sum c_j$ in the exponent as $g^{\sum c_j}$, and compares this to its own measurement of the total consumption c by checking whether $\log_g(g^{\sum c_j}/g^c)$ lies within a tolerance interval. This is another example of a discrete-logarithm computation in a small fixed interval within a large, secure group; we use a small group-specific table to speed up this computation, allowing larger intervals, more aggregation, and better privacy. In cases of sufficiently severe cheating, the discrete logarithm will be too large, causing any discrete-logarithm computation to fail; one recognizes this case by seeing that the computation is running several times longer than expected.

Applications of Discrete Logarithms in Small Groups. Another interesting category of applications uses "trapdoor discrete-logarithm groups": groups in which computations of discrete logarithms are feasible with some trapdoor information and hard otherwise. These applications include the Maurer–Yacobi ID-based encryption system in [24], for example, and the Henry–Henry–Goldberg privacy-preserving protocol in [16].

Maurer and Yacobi in [24, Section 4] introduced a construction of a trapdoor discrete-logarithm group, with a quadratic gap between the user's cost and the attacker's cost. It is generally regarded as preferable to have constructive applications be polynomial time and cryptanalytic computations exponential time, but this quadratic gap is adequate for practical applications. A different construction uses Weil descent with isogenies as trapdoor; see [27] for credits and further discussion of both constructions.

The Maurer–Yacobi construction works as follows. Choose an RSA modulus $n = pq$, where $p-1$ and $q-1$ have many medium-size factors — distinct primes ℓ_i chosen small enough that a user knowing the factors of $p-1$ and $q-1$ can solve discrete logarithms in each of these subgroups, using $\Theta(\ell_i^{1/2})$ multiplications modulo p or q, but large enough that the $p-1$ method for factoring n, using $\Theta(\ell_i)$ multiplications modulo n, is out of reach. The group $(\mathbf{Z}/n)^*$ is then a trapdoor discrete-logarithm group. The trapdoor information consists of p, q, and the primes ℓ_i. Note that the trapdoor computation here consists of computations of discrete logarithms in small groups, not small intervals inside larger groups; this turns out to make our techniques slightly more efficient.

Henry and Goldberg in [15] presented a fast GPU implementation of the trapdoor computation of discrete logarithms, using Pollard's rho method. A simple GMP-based implementation of our algorithm on a single core of a low-cost AMD

CPU takes an order of magnitude less wall-clock time than the optimized GPU implementation described in [15], for the same DLP sizes considered in [15], even though the GPU is more than 30 times faster than the CPU core at modular multiplications. Specifically, for a group of prime order almost exactly 2^{48}, our experiments show a discrete-logarithm computation taking just $115729 \approx 1.77 \cdot 2^{16}$ multiplications on average, using a table of size $65536 = 2^{16}$. Precomputation of the table used $5333245354 \approx 1.24 \cdot 2^{32}$ multiplications.

Previous Work. Escott, Sager, Selkirk, and Tsapakidis in [9, Section 4.4] showed experimentally that attacking a total of $2, 3, 4, 5$ DLPs with the parallel rho method took, respectively, $1.52, 1.90, 2.22, 2.49$ times longer than solving just one DLP. The basic idea, which [9] said "has also been suggested by Silverman and Stapleton" in 1997, is to compute $\log_g h_1$ with the rho method; compute $\log_g h_2$ with the rho method, reusing the distinguished points produced by h_1; compute $\log_g h_3$ with the rho method, reusing the distinguished points produced by h_1 and h_2; etc.

Kuhn and Struik in [20] analyzed this method and concluded that solving a batch of L discrete logarithms in a group of prime order ℓ reduces the cost of an average discrete logarithm to $\Theta(L^{-1/2}\ell^{1/2})$ multiplications — but only for $L \ll \ell^{1/4}$; see [20, Theorem 1]. Each discrete logarithm here costs at least $\Theta(\ell^{3/8})$; see [20, footnote 5].

Hitchcock, Montague, Carter, and Dawson in [17, Section 3] viewed the computation of many preliminary discrete logarithms $\log_g h_1, \log_g h_2, \ldots$ as a precomputation for the main computation of $\log_g h_k$, and analyzed some tradeoffs between the main computation time and the precomputation time. Two much more recent papers, independent of each other, have instead emphasized tradeoffs between the main computation time and the *space* for a table of precomputed distinguished points. The earlier paper, [22] by Lee, Cheon, and Hong, pointed out that these algorithms are tools not just for the cryptanalyst but for the cryptographer, specifically for trapdoor discrete-logarithm computations. The later paper, our paper [6], pointed out that these algorithms illustrate the gap between the time and space taken by an attack and the difficulty of finding the attack, causing trouble for security definitions in the provable-security literature. Both [22] and [6] clearly break the $\Theta(\ell^{3/8})$-time-per-discrete-logarithm barrier from [20].

In this paper we point out that the same idea, suitably adapted, works not only for discrete logarithms in small groups ("rho") but also for discrete logarithms in small intervals ("kangaroos"). This is critical for BGN-type protocols. We also point out three improvements applicable to both the rho setting and the kangaroo setting: we reduce the number of multiplications by a constant factor by choosing the table entries more carefully; we further reduce the number of multiplications by choosing the iteration function more carefully; and we reduce the space consumed by each table entry. This paper includes several illustrative experiments.

2   Review of Generic Discrete-Logarithm Algorithms

This section reviews several standard "square-root" methods to compute discrete logarithms in a group of prime order ℓ. Throughout this paper we write the group operation multiplicatively, write g for the standard generator of the group, and write h for the DLP input; our objective is thus to compute $\log_g h$, i.e., the unique integer k modulo ℓ such that $h = g^k$.

All of these methods are "generic": they work for any order-ℓ group, given an oracle for multiplication (and assuming sufficient hash randomness, for the methods using a hash function). "Square-root" means that the algorithms take $\Theta(\ell^{1/2})$ multiplications on average over all group elements h.

Shanks's Baby-Step-Giant-Step Method. The baby-step-giant-step method [31] computes $\lceil \ell/W \rceil$ "giant steps" $g^0, g^W, g^{2W}, g^{3W}, \ldots$ and then computes a series of W "baby steps" $h, hg, hg^2, \ldots, hg^{W-1}$. Here W is an algorithm parameter. It is easy to see that there will be a collision $g^{iW} = hg^j$, revealing $\log_g h = iW - j$.

Normally W is chosen as $\Theta(\ell^{1/2})$, so that there are $O(\ell^{1/2})$ multiplications in total; more precisely, as $(1+o(1))\ell^{1/2}$ so that there are $\leq (2+o(1))\ell^{1/2}$ multiplications in total. Interleaving baby steps with giant steps, as suggested by Pollard in [29, page 439, top], obtains a collision after $(4/3+o(1))\ell^{1/2}$ multiplications on average. We have recently introduced a "two grumpy giants and a baby" variant that reduces the constant $4/3$; see [5].

The standard criticism of these methods is that they use a large amount of memory, around $\ell^{1/2}$ group elements. One can reduce the giant-step storage to, e.g., $\Theta(\ell^{1/3})$ group elements by taking W as $\Theta(\ell^{2/3})$, but this also increases the average number of baby steps to $\Theta(\ell^{2/3})$. This criticism is addressed by the rho and kangaroo methods discussed below, which drastically reduce space usage while still using just $\Theta(\ell^{1/2})$ multiplications.

Pollard's Rho Method. Pollard's original rho method [28, Section 1] computes a pseudorandom walk $1, F(1), F(F(1)), \ldots$. Here $F(u)$ is defined as gu or u^2 or hu, depending on whether a hash of u is 0 or 1 or 2. Each iterate $F^n(1)$ then has the form $g^y h^x$ for some easily computed pair $(x, y) \in (\mathbf{Z}/\ell)^2$, and any collision $g^y h^x = g^{y'} h^{x'}$ with $(x, y) \neq (x', y')$ immediately reveals $\log_g h$. One expects a sufficiently random-looking walk on ℓ group elements to collide with itself within $O(\ell^{1/2})$ steps. There are several standard methods to find the collision with negligible memory consumption.

Van Oorschot and Wiener in [35] proposed running many walks in parallel, starting from different points $g^y h^x$ and stopping each walk when it reaches a "distinguished point". Here a fraction $1/W$ of the points are defined (through another hash function) as "distinguished", where W is an algorithm parameter; each walk reaches W points on average. One checks for collisions only among the occasional distinguished points, not among all of the group elements produced. The critical observation is that if two walks reach the same group element then they will eventually reach the same distinguished point — or will enter cycles, but cycles have negligible chance of appearing if W is below the scale of $\ell^{1/2}$.

There are many other reasonable choices of F. One popular choice — when there are many walks as in [35], not when there is a single walk as in [28, Section 1] — is a "base-g r-adding walk": this means that the hash function has r different values, and $F(u)$ is defined as $s_1 u$ or $s_2 u$ or ... or $s_r u$ respectively, where $s_1, s_2, \ldots, s_r$ are precomputed as random powers of g. One then starts each walk at a different power h^x. This approach has several minor advantages (for example, x is constant in each walk and need not be updated) and the major advantage of simulating a random walk quite well as r increases. See, e.g., [30], [33], and [5] for further discussion of the impact of r. The bottom line is that this method finds a discrete logarithm within $(\sqrt{\pi/2} + o(1))\ell^{1/2}$ multiplications on average.

The terminology "r-adding walk" is standard in the literature but the terminology "base-g r-adding walk" is not. We use this terminology to distinguish a base-g r-adding walk from a "base-(g, h) r-adding walk", in which $s_1, s_2, \ldots, s_r$ are precomputed as products of random powers of g and h. This distinction is critical in Section 3.

Pollard's Kangaroo Method. An advantage of baby-step-giant-step, already exploited by Shanks in the paper [31] introducing the method, is that it immediately generalizes from computing discrete logarithms in any *group* of prime order ℓ to computing discrete logarithms in any *interval* of length ℓ inside any group of prime order $p \geq \ell$. The rho method uses $\Theta(p^{1/2})$ group operations, often far beyond $\Theta(\ell^{1/2})$ group operations.

Pollard in [28, Section 3] introduced a "kangaroo" method that combines the advantages of the baby-step-giant-step method and the rho method: it takes only $\Theta(\ell^{1/2})$ group operations to compute discrete logarithms in an interval of length ℓ, while still using negligible memory. This method

- chooses a base-g r-adding iteration function whose steps have average exponents $\Theta(\ell^{1/2})$, instead of exponents chosen uniformly modulo ℓ;
- runs a walk starting from g^y (the "tame kangaroo"), where y is at the right end of the interval;
- records the Wth step in this walk (the "trap"), where W is $\Theta(\ell^{1/2})$; and
- runs a walk (the "wild kangaroo") starting from h, checking at each step whether this walk has fallen into the trap.

van Oorschot and Wiener in [35] proposed a parallel kangaroo method in which tame kangaroos start from g^y for many values of y, all close to the middle of the interval, and a similar number of wild kangaroos start from hg^y for many small values of y. Collisions are detected by distinguished points as in the parallel rho method, but the distinguished-point property is chosen to have probability considerably higher than $1/W$; walks continue past distinguished points. The walks are adjusted to avoid collisions between tame kangaroos and to avoid collisions between wild kangaroos. Several subsequent papers have proposed refinements of the kangaroo method, obtaining constant-factor speedups.

The Nechaev–Shoup Bound. Shoup proved in [32] that all generic discrete-logarithm algorithms have success probability $O(m^2/\ell)$ after m multiplications.

(The same bound had been proven by Nechaev in [25] for a more limited class of algorithms, which one might call "representation oblivious" generic discrete-logarithm algorithms.) All generic discrete-logarithm algorithms therefore need $\Omega(\ell^{1/2})$ multiplications on average; i.e., the usual square-root discrete-logarithm algorithms are optimal up to a constant factor. A closer look shows that the lower bound is $(2\sqrt{2}/3+o(1))\ell^{1/2}$, so both the baby-step-giant-step method and the rho method are within a factor $2 + o(1)$ of optimal.

There are much faster discrete-logarithm algorithms (e.g., index-calculus algorithms) for specific classes of groups. However, the conventional wisdom is that these square-root algorithms are the fastest discrete-logarithm algorithms for "secure" groups: a sensibly chosen elliptic-curve group, for example, or the order-ℓ subgroup of $\mathbf{F}_p^*$ for sufficiently large p.

In the rest of this paper we discuss algorithms that improve upon these square-root algorithms by a non-constant factor. Evidently these improved algorithms do not fit Shoup's model of "generic" algorithms — but these improved algorithms *do* apply to "secure" groups. The algorithms deviate from the "generic" model by requiring an extra input, a small table that depends on the group but not on the particular discrete logarithm being computed. The table is set up by a generic algorithm, and if one views the setup and use of the table as a single unified algorithm then Shoup's bound applies to that algorithm; but if the table is set up once and used enough times to amortize the setup costs then each use of the table evades Shoup's bound.

3   Using a Small Table to Accelerate Generic Discrete-Logarithm Algorithms

This section explains how to use a small table to accelerate Pollard's rho and kangaroo methods. The table depends on the group, and on the base point g, but not on the target h. For intervals the table depends on the length of the interval but not on the position of the interval: dividing h by g^A reduces a discrete logarithm in the interval $\{A, A + 1, \ldots, A + \ell - 1\}$ to a discrete logarithm in the interval $\{0, 1, \ldots, \ell - 1\}$, eliminating the influence of A.

The speedup factor grows as the square root of the table size T. As T grows, the average number of multiplications needed to compute a discrete logarithm drops far below the $\approx\ell^{1/2}$ multiplications used in the previous section.

The cost of setting up the table is larger than $\ell^{1/2}$, also growing with the square root of T. However, this cost is amortized across all of the targets h handled with the same table. Comparing the table-setup cost $(\ell T)^{1/2}$ to the discrete-logarithm cost $(\ell/T)^{1/2}$ shows that the table-setup cost becomes negligible as the number of targets handled grows past T.

The main parameters in this algorithm are the table size T and the walk length W. Sensible parameter choices will satisfy $W \approx \alpha(\ell/T)^{1/2}$, where α is a small constant discussed below. Auxiliary parameters are various decisions used in building the table; these decisions are analyzed below.

For simplicity we begin this section by describing a "basic algorithm" that uses a small table to accelerate the rho method. We then describe speedups to the basic algorithm, and finally a variant that uses a small table to accelerate the kangaroo method.

The Basic Algorithm. To build the table, simply start some walks at g^y for random choices of y. The table entries are the distinct distinguished points produced by these walks, together with their discrete logarithms.

It is critical here for the iteration function used in the walks to be independent of h. A standard base-g r-adding walk satisfies this condition, and for simplicity we focus on the case of a base-g r-adding walk, although we recommend that implementors also try "mixed walks" with some squarings. Sometimes walks collide (this happens frequently when parameters are chosen sensibly), so setting up the table requires more than T walks; see below for quantification of this effect.

To find the discrete logarithm of h using this table, start walks at h^x for random choices of x, producing various distinguished points $h^x g^y$, exactly as in the usual rho method. Check for two of these new distinguished points colliding, but also check for one of these new distinguished points colliding with one of the distinguished points in the precomputed table. Any such collision immediately reveals $\log_g h$.

In effect, the table serves as a free foundation for the list of distinguished points naturally accumulated by the algorithm. If the number of h-dependent walks is small compared to T (this happens when parameters are chosen sensibly) then one can reasonably skip the check for two of the new distinguished points colliding; the algorithm almost always succeeds from collisions with distinguished points in the precomputed table.

Special Cases. The extreme case $T = 0$ of this algorithm is the usual rho method with a base-g r-adding walk (or, more generally, the rho method with any h-independent iteration function). However, our main interest is in the speedups provided by larger values of T.

We also draw attention to the extreme case $r = 1$ with exponent 1, simply stepping from u to gu. In this case the main "rho" computation consists of taking, on average, W baby steps $h^x, h^x g, h^x g^2, \ldots$ and then looking up the resulting distinguished point in a table. What is interesting about this case is its evident similarity to the baby-step-giant-step method, but with the advantage of carrying out a table access only after W baby steps; the usual baby-step-giant-step method checks the table after every baby step. What is bad about this case is that the walk is highly nonrandom, requiring $\Theta(\ell)$ steps to collide with another such walk; larger values of r create collisions within $\Theta(\ell^{1/2})$ steps.

Recall from Section 1 the classic algorithm to solve multiple discrete logarithms: for each k in turn, compute $\log_g h_k$ with the rho method, reusing the distinguished points produced by $h_1, \ldots, h_{k-1}$. The $\log_g h_k$ part of this computation obviously fits the algorithm discussed here, with T implicitly defined as the number of distinguished points produced by $h_1, \ldots, h_{k-1}$. We emphasize, however, that this is a special choice of T, and that the parameter curve

(T, W) used implicitly in this algorithm as k varies does not obey the relationship $W \approx \alpha(\ell/T)^{1/2}$ mentioned above. Treating T and W as explicit parameters allows several optimizations that we discuss below.

Optimizing the Walk Length. Assume that $W \approx \alpha(\ell/T)^{1/2}$, and consider the chance that a *single* walk already encounters one of the T distinguished points in the table, thereby solving the DLP. The T table entries were obtained from walks that, presumably, each covered about W points, for a total of TW points. The new walk also covers about W points and thus has $TW^2 \approx \alpha^2\ell$ collision opportunities. If these collision opportunities were independent then the chance of escaping all of these collisions would be $(1 - 1/\ell)^{\alpha^2\ell} \approx \exp(-\alpha^2)$.

This heuristic analysis suggests that a single walk succeeds with, e.g., probability $1 - \exp(-1/16) \approx 6\%$ for $\alpha = 1/4$, or probability $1 - \exp(-1/4) \approx 22\%$ for $\alpha = 1/2$, or probability $1 - \exp(-1) \approx 63\%$ for $\alpha = 1$, or probability $1 - \exp(-4) \approx 98\%$ for $\alpha = 2$.

The same analysis also suggests that the end of the precomputation, finding the Tth point in the table, will require trying $\exp(1/16) \approx 1.06$ length-W walks for $\alpha = 1/4$, or $\exp(1/4) \approx 1.28$ length-W walks for $\alpha = 1/2$, or $\exp(1) \approx 2.72$ length-W walks for $\alpha = 1$, or $\exp(4) \approx 54.6$ length-W walks for $\alpha = 2$.

The obvious advantage of taking very small α is that one can reasonably carry out several walks in parallel. Taking (e.g.) $\alpha = 1/8$ requires 64 walks on average, and if one carries out (e.g.) 4 walks in parallel then at most 3 walks are wasted. The most common argument for parallelization is that it allows the computation to exploit multiple cores, decreasing latency. Parallelization is helpful even when latency is not a concern: for example, it allows merging inversions in affine elliptic-curve computations (Montgomery's trick), and it often allows effective use of vector units in a single core. Solving many independent discrete-logarithm problems produces the same benefits, but requires the application to have many independent problems ready at the same time.

The obvious disadvantage of taking very small α is that the success probability per walk drops quadratically with α, while the walk length drops only linearly with α. In other words, chopping a small α in half makes each step half as effective, doubling the number of steps expected in the computation. Sometimes this is outweighed by the increase in parallelization (there are now four times as many walks), but clearly there is a limit to how small α can reasonably be taken.

Clearly there is also a limit to how large α can reasonably be taken. Doubling α beyond 1 does not make each step twice as effective: an $\alpha = 1$ walk already succeeds with chance 63%; an $\alpha = 2$ walk succeeds with chance 98% but is twice as expensive.

We actually recommend optimizing α experimentally (and not limiting it to powers of 2), rather than trusting the exact details of the heuristic analysis shown above. A small issue with the heuristic analysis is that the new walk sometimes takes only, say, $W/2$ steps, obtaining collisions with much lower probability than indicated above, and sometimes $2W$ steps; the success probability of a walk is not the same as the success probability of a length-W walk. A larger issue is that TW is only a crude approximation to the table coverage. Discarding previously discovered distinguished points when building the table creates a bias towards short walks, especially for large α; on the other hand, a walk finding a distinguished point will rarely see all of the ancestors of that point, and in a moment we will see that this is a controllable effect, allowing the table coverage to be significantly increased.

Lee, Cheon, and Hong in [22, Lemma 1 and Theorem 1] give a detailed heuristic argument that starting M walks in the precomputation will produce $T \approx M(\sqrt{1+2a}-1)/a$ distinct distinguished points, where $a = MW^2/\ell$ (so our α is $(\sqrt{1+2a}-1)^{1/2}$), and that each walk in the main computation then succeeds with probability $1 - 1/\sqrt{1+2a}$ (i.e., $1 - 1/(\alpha^2+1)$). In [22, page 13] they recommend taking $a = (1+\sqrt{5})/4 \approx 0.809$ (equivalently, $\alpha \approx 0.786$); the heuristics then state that $T \approx 0.764M$ and that each walk in the main computation succeeds with probability $1 - 1/\sqrt{1+2a} \approx 0.382$, so the main computation uses $W/0.382 \approx 2.058(\ell/T)^{1/2}$ multiplications on average. We issue three cautions regarding this recommendation. First, assuming the same heuristics, it is actually better to take $a = 1.5$ (equivalently, $\alpha = 1$); then the main computation uses just $2(\ell/T)^{1/2}$ multiplications on average. Second, our improvements to the table coverage (see below) reduce the number of multiplications, and this reduction is different for different choices of a (see our experimental results in Section 4), rendering the detailed optimization in [22] obsolete. Third, even though we emphasize number of multiplications as a simple algorithm metric, the real goal is to minimize time; the parallelization issues discussed above seem to favor considerably smaller choices of α, depending on the platform.

Choosing the Most Useful Distinguished Points. Instead of randomly generating T distinguished points, we propose generating more distinguished points, say $2T$ or $10T$ or $1000T$, and then keeping the T most useful distinguished points. (This presumably means storing $2T$ or $10T$ or $1000T$ points during the precomputation, but we follow standard practice in distinguishing between the space consumed during the precomputation and the space required for the output of the precomputation. As an illustrative example in support of this practice, consider the precomputed rainbow tables distributed by the A5/1 Cracking Project [26]; the cost of local RAM used temporarily by those computations is much less important than the network cost of distributing these tables to users and the long-term cost of storing these tables.)

The natural definition of "most useful" is "having the largest number of ancestors". By definition the ancestors of a distinguished point are the group elements that walk to this point; the chance of a uniform random group element walking to this point is exactly the number of ancestors divided by ℓ.

Unfortunately, without taking the time to survey all ℓ group elements, one does not know the number of ancestors of a distinguished point. Fortunately, one has a statistical estimate of this number: a distinguished point found by many walks is very likely to be more useful than a distinguished point found by fewer walks. This estimate is unreliable for a distinguished point found by very few walks, especially for distinguished points found by just one walk; we thus propose using the walk length as a secondary estimate. (In our experiments we computed a weight for each distinguished point as the total length of all walks reaching the point, plus $4W$ per walk; we have not yet experimented with modifications to this weighting.) This issue disappears as the number of random walks increases towards larger multiples of T.

This table-generation strategy reduces the number of walks required for the main discrete-logarithm computation. The table still has size T, and each walk still has average length W, but the success probability of each walk increases. The only disadvantage is an increase in the time spent setting up the table.

Interlude: The Penalty for Iteration Functions That Depend on h.
Escott, Sager, Selkirk, and Tsapakidis in [9, Section 4.4] chose an iteration function "that is independent of all the Q_is" (the targets h_i): namely, a base-g r-adding walk, optionally mixed with squarings. Kuhn and Struik in [20] said nothing about this independence condition; instead they chose a base-(g, h_k) r-adding walk. See [20, Section 2.2] ("$g^{a_i} h^{b_i}$") and [20, Section 4] ("all distinguished points $g^{a_j} h_i^{b_j}$ that were calculated in order to find x_i"). No experiments were reported in [20], except for a brief comment in [20, Remark 2] that the running-time estimate in [20, Theorem 1] was "a good approximation of practically observed values".

Hitchcock, Montague, Carter, and Dawson in [17, page 89] pointed out that "the particular random walk recommended by Kuhn and Struik", with the iteration function used for h_k different from the iteration functions used for $h_1, \ldots, h_{k-1}$, fails to detect collisions "from different random walks". They reported experiments showing that a base-(g, h_k) r-adding walk was much less effective for multiple discrete logarithms than a base-g r-adding walk.

To understand this penalty, consider the probability that the main computation succeeds with one walk, i.e., that the resulting distinguished point appears in the table. There are $\approx \ell/W$ distinguished points, and the table contains T of those points, so the obvious first approximation is that the main computation succeeds with probability TW/ℓ. If the table is generated by a random walk independent of the walk used in the main computation then this approximation is quite reasonable. If the table was generated by the *same* walk used in the main computation then the independence argument no longer applies and the approximation turns out to be a severe underestimate.

In effect, the table-generation process in [20] selects the table entries uniformly at random from the set of distinguished points. The table-generation process in [9], [17], and [22] instead starts from random group elements and walks to distinguished points; this produces a highly non-uniform distribution of distinguished points covered by the table, biasing the table entries towards more useful distinguished points. We go further, biasing the table entries even more by selecting them carefully from a larger pool of distinguished points.

Choosing the Most Useful Iteration Function. Another useful way to spend more time on table setup is to try different iteration functions, i.e., different choices of exponents for the r-adding walk.

The following examples are a small illustration of the impact of varying the iteration function. http://cr.yp.to/dlog/20120727-function1.pdf is a directed graph on 1000 nodes obtained as follows. Each node marked itself as distinguished with probability $1/W$ where $W = 10$. (We did not enforce exactly 100 distinguished points; each node made its decision independently.) Each non-distinguished node created an outgoing edge to a uniform random node. We then used the `neato` program, part of the standard `graphviz` package [13], to draw the digraph with short edges. The $T = 10$ most useful distinguished points are black squares; the 593 nontrivial ancestors of those points are black circles; the other 99 distinguished points are white squares; the remaining points are white circles.

http://cr.yp.to/dlog/20120727-function2.pdf is another directed graph obtained in the same way, with the same values of W and T but different distinguished points and a different random walk. For this graph the 10 most useful distinguished points have 687 nontrivial ancestors, for an overall success probability of $697/1000 \approx 70\%$, significantly better than the first graph and also significantly above the 63% heuristic mentioned earlier.

These graphs were not selected as outliers; they were the first two graphs we generated. Evidently the table coverage has rather high variance.

Of course, a larger table coverage by itself does not imply better performance: graphs with larger coverage tend to have longer walks. We thus use the actual performance of the resulting discrete-logarithm computations as a figure of merit for the graph.

For small examples it is easy to calculate the exact average-case performance, rather than just estimate it statistically. Our second sample graph uses, on average, 10.8506 steps to compute a discrete logarithm if walks are limited to 27 steps. (Here 27 is optimal for that graph. The graph has cycles, so some limit or other cycle-detection mechanism is required. One can also take this limit into account in deciding which distinguished points are best.) Our first sample graph uses, on average, 11.2007 steps.

Adapting the Method to a Small Interval. We now explain a small set of tweaks that adapt the basic algorithm stated above to the problem of computing discrete logarithms in an interval of length ℓ. These tweaks trivially combine with the refinements stated above, such as choosing the most useful distinguished points.

As in the standard kangaroo method, choose the steps $s_1, s_2, \ldots, s_r$ as powers of g where the exponents are $\beta\ell/W$ on average. We recommend numerical optimization of the constant β.

Start walks at g^y for random choices of y in the interval. As in the basic algorithm, stop each walk when it reaches a distinguished point, and build a table of discrete logarithms of the resulting distinguished points.

To find the discrete logarithm of h, start a walk at hg^y for a random small integer y; stop at the first distinguished point; and check whether the resulting distinguished point is in the table. In our experiments we defined "small" as "bounded by $\ell/256$", but it would also have been reasonable to start the first walk at h, the second at hg, the third at hg^2, etc.

We are deviating in several ways here from the typical kangaroo methods stated in the literature. Our walks starting from g^y can be viewed as tame kangaroos, but our tame kangaroos are spread through the interval rather than being clustered together. We do not continue walks past distinguished points. We select the most useful distinguished points experimentally, rather than through preconceived notions of how far kangaroos should be allowed to jump.

We do not claim that the details of this approach are optimal. However, this approach has the virtue of being very close to the basic algorithm, and our experiments so far have found discrete logarithms in intervals of length ℓ almost as quickly as discrete logarithms in groups of order ℓ.

4   Experiments

This section reports several experiments with the algorithm described in Section 3, both for small groups and for small intervals inside larger groups. To aid in verification we have posted our software for a typical small-interval experiment at http://cr.yp.to/dlog/cuberoot.html.

Case Study: A Small-Group Experiment. We began with several experiments targeting the discrete-logarithm problem modulo pq described in [15, Table 2, first line]. Here p and q are "768-bit primes" generated so that $p-1$ and $q-1$ are "2^{48}-smooth"; presumably this means that $(p-1)/2$ is a product of 16 primes slightly below 2^{48}, and similarly for $(q-1)/2$. The original discrete-logarithm problem then splits into 16 separate 48-bit DLPs modulo p and 16 separate 48-bit DLPs modulo q.

What [15] reports is that a 448-ALU NVIDIA Tesla M2050 graphics card takes an average of 23 seconds for these 32 discrete-logarithm computations, i.e., 0.72 seconds for each 48-bit discrete-logarithm computation. The discrete-logarithm computations in [15] use standard techniques, using more than 2^{24} modular multiplications; the main accomplishment of [15] is at a lower level, using the graphics card to compute 52 million 768-bit modular multiplications per second.

The Tesla M2050 card is currently advertised for $1300. We do not own one; instead we are using a single core of a 6-core 3.3GHz AMD Phenom II X6 1100T CPU. This CPU is no longer available but it cost only $190 when we purchased it last year.

We generated an integer p as $1 + 2\ell_1\ell_2\cdots\ell_{16}$, where $\ell_1, \ell_2, \ldots, \ell_{16}$ are random primes between $2^{48} - 2^{20}$ and 2^{48}. We repeated this process until p was prime, and then took $\ell = \ell_1$. This ℓ turned out to be $2^{48} - 313487$. We do not claim that this narrow range of 48-bit primes is cryptographically secure in the context of [15]; we stayed very close to 2^{48} to avoid any possibility of our order-ℓ DLP being noticeably smaller than the DLP in [15]. We chose g as $2^{(p-1)/\ell}$ in $\mathbf{F}_p^*$.

For modular multiplications we used the standard C++ interface to the well-known GMP library (version 5.0.2). This interface allows writing readable code such as

```
x = (a * b) % p
```

which turns out to run slightly faster than 1.4 million modular multiplications per second on our single CPU core for our 769-bit p. This understates GMP's internal speeds — it is clear from other benchmarks that we could gain at least a factor of 2 by precomputing an approximate reciprocal of p — but for our experiments we decided to use GMP in the most straightforward way.

We selected $T = 64$ and $W = 1048576$; here $\alpha = 1/2$, i.e., $W \approx (1/2)(\ell/T)^{1/2}$. Precomputing T table entries used a total of $80289876 \approx 1.20TW \approx 0.60(\ell T)^{1/2}$ multiplications; evidently some distinguished points were found more than once. We then carried out a series of 1024 discrete-logarithm experiments, all targeting the same h. Each experiment chose a random y and started a walk from hg^y, hoping that (1) the walk would reach a distinguished point within $8W$ steps and (2) the distinguished point would be in the table. If both conditions were satisfied, the experiment double-checked that it had correctly computed the discrete logarithm of h, and finally declared success.

These experiments used a total of $1040325443 \approx 0.97 \cdot 1024W$ multiplications (not counting the occasional multiplications for the randomization of hg^y and for the double-checks) and succeeded 192 times, on average using $5418361 \approx 2.58(\ell/T)^{1/2}$ multiplications per discrete-logarithm computation. Note that the randomization of hg^y made these speeds independent of h.

More Useful Distinguished Points. We then changed the precomputation, preserving $T = 64$ and $W = 1048576$ but selecting the T table entries as the most useful 64 table entries from a pool of $N = 128$ distinguished points. This increased the precomputation cost to $167040079 \approx 1.24NW \approx 1.24(\ell T)^{1/2}$ multiplications. We ran 4096 new discrete-logarithm experiments, using a total of $3980431381 \approx 0.93 \cdot 4096W$ multiplications and succeeding 1060 times, on average using $3755123 \approx 1.79(\ell/T)^{1/2}$ multiplications per discrete-logarithm computation.

The $T^{1/2}$ Scaling. We then reduced W to 262144, increased T to 1024, and increased N to 2048. This increased the precomputation cost to $626755730 \approx 1.17NW \approx 1.17(\ell T)^{1/2}$ multiplications. We then ran 8192 new experiments, using a total of $2123483139 \approx 0.99 \cdot 8192W$ multiplications and succeeding 2265 times, on average using just $937520 \approx 1.79(\ell/T)^{1/2}$ multiplications per discrete-logarithm computation. As predicted the increase of T by a factor of 16 decreased the number of steps by a factor of 4.

We also checked that these computations were running at more than 1.4 million multiplications per second, i.e., under 0.67 seconds per discrete-logarithm computation — less real time on a single CPU core than [15] needed on an entire GPU. There was no noticeable overhead beyond GMP's modular multiplications. The precomputation for $T = 1024$ took several minutes, but this is not a serious problem for a cryptographic protocol that is going to be run many times.

We then reduced W to 32768, increased T to 65536, and increased N to 131072. This increased the precomputation cost to $5333245354 \approx 1.24NW \approx 1.24(\ell T)^{1/2}$ multiplications, roughly an hour. We then ran 4194304 experiments,

Table 4.1. Observed cost for 15 types of discrete-logarithm computations in a group of order $\ell \approx 2^{48}$. Each discrete-logarithm experiment used T table entries selected from N distinguished points, and used $W = 524288 \approx \alpha(\ell/T)^{1/2}$. Each "main computation" table entry reports, for a series of 2^{20} discrete-logarithm experiments, the average number of multiplications per successful discrete-logarithm computation, scaled by $(\ell/T)^{1/2}$. Each "precomputation" table entry reports the total number of multiplications to build the table, scaled by $(\ell T)^{1/2}$.

T	512	640	768	896	1024
α	0.70711	0.79057	0.86603	0.93541	1.00000
precomputation, $N = T$	0.84506	0.94916	1.11884	1.23070	1.34187
precomputation, $N = 2T$	1.89769	2.33819	2.74627	3.27589	3.66113
precomputation, $N = 8T$	15.7167	20.7087	26.1621	31.2112	36.9350
main computation, $N = T$	2.13856	2.03391	2.01172	1.98725	2.01289
main computation, $N = 2T$	1.62474	1.59358	1.58893	1.59218	1.61922
main computation, $N = 8T$	1.38323	1.40706	1.42941	1.46610	1.49688

using a total of $137426510228 \approx 1.00 \cdot 4194304W$ multiplications and succeeding 1187484 times, on average using just $115729 \approx 1.77(\ell/T)^{1/2}$ multiplications per discrete-logarithm computation — under 0.1 seconds.

Optimizing α. We then carried out a series of experiments with $W = 524288$, varying both T and N/T as shown in Table 4.1. Each table entry is rounded to 6 digits. The smallest "main computation" table entry, 1.38314 for $T = 512$ and $N/T = 8$, means (modulo this rounding) that a series of discrete-logarithm experiments used $1.38314(\ell/T)^{1/2}$ multiplications per successful discrete-logarithm computation. Each table entry involved 2^{20} discrete-logarithm experiments, of which more than 2^{18} were successful, so each table entry is very likely to have an experimental error below 0.02.

This table shows that the optimal choice of α depends on the ratio N/T, but also that rather large variations in α around the optimum make a relatively small difference in performance. Performance is much more heavily affected by increased N/T, i.e., by extra precomputation.

To better understand the tradeoffs between precomputation time and main-computation time, we plotted the 15 pairs of numbers in Table 4.1, obtaining Figure 4.3. For example, Table 4.1 indicates for $T = 512$ and $N = 2T$ that each successful discrete-logarithm computation took $1.62474(\ell/T)^{1/2}$ multiplications on average after $1.89769(\ell T)^{1/2}$ multiplications in the precomputation, so $(1.89769, 1.62474)$ is one of the points plotted in Figure 4.3. Figure 4.3 suggests that optimizing α to minimize main-computation time for fixed N/T does not produce the best tradeoff between main-computation time and precomputation time; one should instead decrease α somewhat and increase N/T. To verify this theory we are performing more computations to fill in more points in Figure 4.3.

Small-Interval Experiments. Starting from the same software, we then made the following tweaks to compute discrete logarithms in a short interval inside a much larger prime-order group:

Table 4.2. Observed cost for 15 types of discrete-logarithm computations in an interval of length $\ell = 2^{48}$ inside a much larger group. Table entries have the same meaning as in Table 4.1.

T	512	640	768	896	1024
α	0.70711	0.79057	0.86603	0.93541	1.00000
precomputation, $N = T$	0.85702	1.00463	1.14077	1.28112	1.41167
precomputation, $N = 2T$	1.99640	2.38469	2.81441	3.17253	3.61816
precomputation, $N = 8T$	15.5307	20.2547	25.2022	30.7112	36.7452
main computation, $N = T$	2.32320	2.21685	2.14892	2.10155	2.09915
main computation, $N = 2T$	1.66106	1.64183	1.63488	1.65603	1.66895
main computation, $N = 8T$	1.44377	1.44808	1.46581	1.49548	1.52502

- We replaced p by a "strong" 256-bit prime, i.e., a prime for which $(p-1)/2$ is also prime. Of course, 256 bits is not adequate for cryptographic security for groups of the form $\mathbf{F}_p^*$, but it is adequate for these experiments.
- We replaced g by a large square modulo p.
- We replaced ℓ by exactly 2^{48}, and removed the reductions of discrete logarithms modulo ℓ.
- We increased r, the number of precomputed steps, from 32 to 128.
- We generated each step as g^y with y chosen uniformly at random between 0 and $\ell/(4W)$, rather than between 0 and ℓ.
- We started each walk from hg^y with y chosen uniformly at random between 0 and $\ell/2^8$, rather than between 0 and ℓ.
- After each successful experiment, we generated a new target h for the following experiments.

For $W = 131072$, $T = 4096$, and $N = 8192$ the precomputation cost was $1337520628 \approx 1.25NW \approx 1.25(\ell T)^{1/2}$ multiplications. We ran 8388608 experiments, using a total of $1100185139821 \approx 1.00 \cdot 8388608W$ multiplications and succeeding 2195416 times, on average using $501128 \approx 1.91(\ell/T)^{1/2}$ multiplications per discrete-logarithm computation.

For $W = 32768$, $T = 65536$, and $N = 131072$ the precomputation cost was $5214755468 \approx 1.21NW \approx 1.21(\ell T)^{1/2}$ multiplications. We ran 33554432 experiments, using a total of $1097731367293 \approx 1.00 \cdot 33554432W$ multiplications and succeeding 8658974 times, on average using just $126773 \approx 1.93(\ell/T)^{1/2}$ multiplications per discrete-logarithm computation.

Table 4.2 reports the results of experiments for $W = 524288$ with various choices of T and N/T, all using the same bounds $\ell/(4W)$ and $\ell/2^8$ stated above. Comparing Table 4.2 to Table 4.1 shows that this approach to computing discrete logarithms in an interval of length ℓ uses — for the same table size, and essentially the same amount of precomputation — only slightly more multiplications than computing discrete logarithms in a group of order ℓ.

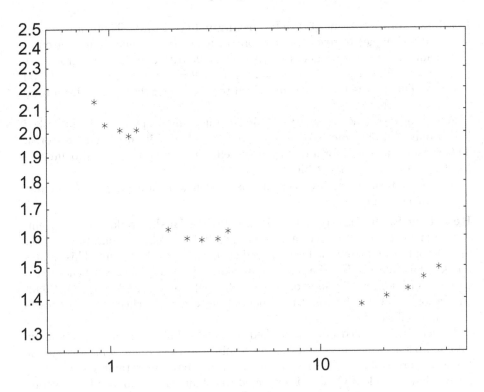

Fig. 4.3. Observed tradeoffs between precomputation time and main-computation time in Table 4.1. Horizontal axis is observed average precomputation time, scaled by $(\ell T)^{1/2}$. Vertical axis is observed average main-computation time, scaled by $(\ell/T)^{1/2}$.

5   Space Optimization

Each table entry described in Section 3 consists of a group element, at least $\lg \ell$ bits, and a discrete logarithm, also $\lg \ell$ bits, for a total of at least $2T \lg \ell$ bits. This section explains several ways to compress the table to a much smaller number of bits.

Many of these compression mechanisms slightly increase the number of multiplications used to compute $\log_g h$. This produces a slightly worse tradeoff between the number of multiplications and the number of table *entries*, but produces a much better tradeoff between the number of multiplications and the number of table *bits*.

For comparison, [**22**, Table 2] took $T = 586463$ and $W = 2^{11}$ for a group of size $\ell \approx 2^{42}$, and reported about $2(\ell/T)^{1/2}$ multiplications per discrete-logarithm computation, using 150 megabytes for the table. Previous sections of this paper explain how to use significantly fewer multiplications for the same T; this section reduces the space consumption by two orders of magnitude for the same T, with only a small increase in the number of multiplications. Equivalently, for the same number of table bits, we use an order of magnitude fewer multiplications.

Lossless Compression of Each Distinguished Point. There are several standard techniques to reversibly compress elements of commonly used groups. For example, nonzero elements of the "Curve25519" elliptic-curve group are pairs $(x, y) \in \mathbf{F}_q \times \mathbf{F}_q$ satisfying $y^2 = x^3 + 486662x^2 + x$; here $q = 2^{255} - 19$ and $\ell \approx 2^{252}$. This pair is trivially compressed to x and a single bit of y, for a total of $256 \approx \lg \ell$ bits.

A typical distinguished-point definition states that a point is distinguished if and only if its bottom $\lg W$ bits are 0. These $\lg W$ bits need not be stored. This reduces the space for a distinguished elliptic-curve point to approximately $\lg(\ell/W)$ bits; e.g., $(2/3) \lg \ell$ bits for $W \approx \ell^{1/3}$.

The other techniques discussed in this section work for any group, not just an elliptic-curve group.

Replacing Each Distinguished Point with a Hash. To do better we simply suppress some additional bits: we hash each distinguished point to a smaller number of bits and store the hash instead of the distinguished point. This creates a risk of false alarms, but the cost of false alarms is merely the cost of checking a bad guess for $\log_g h$. Checking one guess takes only about $(1 + 1/\lg \lg \ell) \lg \ell$ multiplications, and standard multiexponentiation techniques check several guesses even more efficiently.

If each distinguished point is hashed to $\lg(T/\gamma)$ bits then one expects many false alarms as γ increases past 1. Specifically, a distinguished point outside the table has probability γ/T of colliding with any particular table entry (if the hash behaves randomly), so it is expected to collide with γ table entries overall, creating γ bad guesses for $\log_g h$. For a successful walk, the expected number of bad guesses drops approximately in half, or slightly below half if the discrete logarithms with each hash are sorted in decreasing order of utility.

If γ is far below $W/\lg \ell$ then the cost of checking γ bad guesses is far below W multiplications, the average cost of a walk. For example, if T is much smaller than W then one can afford to hash each distinguished point to 0 bits: the table then consists simply of T discrete logarithms, occupying $T \lg \ell$ bits, and one checks the end of each walk against each table entry.

Compressing a Sorted Sequence of Hashes. It is well known that a sorted sequence of n d-bit integers contains far fewer than nd bits of information when n and d are not very small. "Delta compression" takes advantage of this by computing differences between successive integers and using a variable-length encoding of the differences. For random integers the average difference is close to $2^d/n$ and is encoded as slightly more than $d - \lg n$ bits if $d \geq \lg n$, saving nearly $n \lg n$ bits.

Delta compression does not allow fast random access: to search for an integer one must read the sequence from the beginning. This is not visible in this paper's multiplication counts, but it nevertheless becomes a bottleneck as T grows past W. We instead use a simpler approach that allows fast random access: namely, store the sorted sequence $x_1, x_2, \ldots, x_n$ of d-bit integers as

- the sorted sequence $x_1, x_2, \ldots, x_m$ of $(d - 1)$-bit integers where m is the largest index such that $x_m < 2^{d-1}$; and

– the sorted sequence $x_{m+1} - 2^{d-1}, x_{m+2} - 2^{d-1}, \ldots, x_n - 2^{d-1}$ of $(d-1)$-bit integers.

To search for an integer s we search $x_1, \ldots, x_m$ for s if $s < 2^{d-1}$ and search $x_{m+1} - 2^{d-1}, \ldots, x_n - 2^{d-1}$ for $s - 2^{d-1}$ if $s \geq 2^{d-1}$. The second search requires a pointer to the second sorted sequence, i.e., a count of the number of bits used to encode $x_1, x_2, \ldots, x_m$.

This transformation saves 1 bit in each of the n table entries at the expense of a small amount of overhead. This is a sensible transformation if the overhead is below n bits. The transformation is inapplicable if $d = 0$; we encode a sequence of 0-bit integers as simply the number of integers.

Of course, we can and do apply the transformation recursively. The recursion continues for nearly $\lg n$ levels if $d \geq \lg n$, again saving nearly $n \lg n$ bits. For small d the compressed sequence drops to a fraction of n bits.

For example, if each distinguished point is hashed to $d \approx \lg(4T)$ bits, at the expense of $1/4$ bad guesses for each walk, then the hashes are compressed from $Td \approx T \lg(4T)$ bits to just a few bits per table entry. If each distinguished point is hashed to slightly fewer bits, at the expense of more bad guesses for each walk, then the T hashes are compressed to fewer than T bits; in this case one should concatenate the hashes with the discrete logarithms before applying this compression mechanism.

Compressing Each Discrete Logarithm. We finish by considering two mechanisms for compressing discrete logarithms in the table. The first mechanism was introduced in the ongoing ECC2K-130 computation; see [3]. The second mechanism appears to be new.

The first mechanism is as follows. Instead of choosing a random y and starting a walk at g^y, choose a pseudorandom y determined by a short seed. The seed is about $\lg T$ bits, or slightly more if one tries more than T walks; for example, the seed is about 3 times shorter than the discrete logarithm if $T \approx \ell^{1/3}$. Store the seed as a proxy for the discrete logarithm of the resulting distinguished point. Reconstructing the discrete logarithm then takes about W multiplications to recompute the walk starting from g^y. This reconstruction is a bottleneck if distinguished points are hashed to fewer than $\lg T$ bits (creating many bad guesses), and it slows down the main computation by a factor of almost 2 if α is large, but if distinguished points are hashed to more than $\lg T$ bits and α is small then the reconstruction cost is outweighed by the space savings.

The second mechanism is to simply suppress most of the bits of the discrete logarithm. Reconstructing those bits is then a discrete-logarithm problem in a smaller interval; solve these reconstruction problems with the same algorithm recursively, using a smaller table and a smaller number of multiplications. For example, communicating just 9 bits of an ℓ-bit discrete logarithm means reducing an ℓ-bit DLP to an $(\ell - 9)$-bit DLP, which takes $1/8$th as many multiplications using a $T/8$-entry table (or $1/16$th as many multiplications using a $T/2$-entry table); if the number of bad guesses is sufficiently small then this is a good tradeoff.

Note that this second mechanism relies on being able to quickly compute discrete logarithms in small intervals, even if the original goal is to compute discrete logarithms in small groups.

References

[1] — (no editor): 2nd ACM conference on computer and communication security, Fairfax, Virginia, November 1994. Association for Computing Machinery (1994). See [34]

[2] Atallah, M.J., Hopper, N.J. (eds.): Privacy enhancing technologies, 10th interational symposium, PETS 2010, Berlin, Germany, July 21–23, 2010, proceedings. LNCS, vol. 6205. Springer (2010). ISBN 978-3-642-14526-1. See [16]

[3] Bailey, D.V., Batina, L., Bernstein, D.J., Birkner, P., Bos, J.W., Chen, H.-C., Cheng, C.-M., Van Damme, G., de Meulenaer, G., Perez, L.J.D., Fan, J., Güneysu, T., Gürkaynak, F., Kleinjung, T., Lange, T., Mentens, N., Niederhagen, R., Paar, C., Regazzoni, F., Schwabe, P., Uhsadel, L., Van Herrewege, A., Yang, B.-Y.: Breaking ECC2K-130 (2010), http://eprint.iacr.org/2009/541/. Citations in this document: §5

[4] Bao, F., Samarati, P., Zhou, J. (eds.): Applied cryptography and network security, 10th international conference, ACNS 2012, Singapore, June 26–29, 2012, proceedings (industrial track) (2012), http://icsd.i2r.a-star.edu.sg/acns2012/proceedings-industry.pdf. See [18]

[5] Bernstein, D.J., Lange, T.: Two grumpy giants and a baby. In: Proceedings of ANTS 2012, to appear (2012), http://eprint.iacr.org/2012/294. Citations in this document: §2, §2

[6] Bernstein, D.J., Lange, T.: Non-uniform cracks in the concrete: the power of free precomputation (2012), http://eprint.iacr.org/2012/318. Citations in this document: §1, §1

[7] Boneh, D., Goh, E.-J., Nissim, K.: Evaluating 2-DNF formulas on ciphertexts. In: TCC 2005 [19], pp. 325–341 (2005), http://crypto.stanford.edu/~dabo/abstracts/2dnf.html. Citations in this document: §1, §1, §1

[8] Davies, D.W. (ed.): Advances in cryptology — EUROCRYPT '91, workshop on the theory and application of cryptographic techniques, Brighton, UK, April 8–11, 1991, proceedings. LNCS, vol. 547. Springer (1991). See [24]

[9] Escott, A.E., Sager, J.C., Selkirk, A.P.L., Tsapakidis, D.: Attacking elliptic curve cryptosystems using the parallel Pollard rho method. CryptoBytes 4 (1999), ftp://ftp.rsa.com/pub/cryptobytes/crypto4n2.pdf. Citations in this document: §1, §1, §3, §3

[10] Fischer-Hübner, S., Hopper, N. (eds.): Privacy enhancing technologies — 11th international symposium, PETS 2011, Waterloo, ON, Canada, July 27–29, 2011, proceedings. LNCS, vol. 6794. Springer (2011). See [21]

[11] Freeman, D.M.: Converting pairing-based cryptosystems from composite-order groups to prime-order groups. In: Eurocrypt 2010 [14], pp. 44–61 (2010), http://theory.stanford.edu/~dfreeman/papers/subgroups.pdf. Citations in this document: §1

[12] Fumy, W. (ed.): Advances in cryptology — EUROCRYPT '97, international conference on the theory and application of cryptographic techniques, Konstanz, Germany, May 11–15, 1997. LNCS, vol. 1233. Springer (1997). See [32]

[13] Gansner, E.R., North, S.C.: An open graph visualization system and its applications to software engineering. Software: Practice and Experience 30, 1203–1233 (2000). Citations in this document: §3

[14] Gilbert, H. (ed.): Advances in cryptology — EUROCRYPT 2010, 29th annual international conference on the theory and applications of cryptographic techniques, French Riviera, May 30–June 3, 2010, proceedings. LNCS, vol. 6110. Springer (2010). See [11]

[15] Henry, R., Goldberg, I.: Solving discrete logarithms in smooth-order groups with CUDA. In: Workshop Record of SHARCS 2012: Special-purpose Hardware for Attacking Cryptographic Systems, pp. 101–118 (2012), http://2012.sharcs.org/record.pdf. Citations in this document: §1, §1, §1, §4, §4, §4, §4, §4, §4, §4

[16] Henry, R., Henry, K., Goldberg, I.: Making a nymbler Nymble using VERBS. In: PETS 2010 [2], pp. 111–129 (2010), http://www.cypherpunks.ca/~iang/pubs/nymbler-pets.pdf. Citations in this document: §1

[17] Hitchcock, Y., Montague, P., Carter, G., Dawson, E.: The efficiency of solving multiple discrete logarithm problems and the implications for the security of fixed elliptic curves. International Journal of Information Security 3, 86–98 (2004). Citations in this document: §1, §3, §3

[18] Hu, Y., Martin, W.J., Sunar, B.: Enhanced flexibility for homomorphic encryption schemes via CRT. In: ACNS 2012 industrial track [4], pp. 93–110 (2012). Citations in this document: §1, §1

[19] Kilian, J. (ed.): Theory of cryptography, second theory of cryptography conference, TCC 2005, Cambridge, MA, USA, February 10–12, 2005, proceedings. LNCS, vol. 3378. Springer (2005). ISBN 3-540-24573-1. See [7]

[20] Kuhn, F., Struik, R.: Random walks revisited: extensions of Pollard's rho algorithm for computing multiple discrete logarithms. In: SAC 2001 [36], pp. 212–229 (2001), http://www.distcomp.ethz.ch/publications.html. Citations in this document: §1, §1, §1, §1, §3, §3, §3, §3, §3, §3, §3

[21] Kursawe, K., Danezis, G., Kohlweiss, M.: Privacy-friendly aggregation for the smart-grid. In: PETS 2011 [10], pp. 175–191 (2011), http://research.microsoft.com/pubs/146092/main.pdf. Citations in this document: §1

[22] Lee, H.T., Cheon, J.H., Hong, J.: Accelerating ID-based encryption based on trapdoor DL using pre-computation. 11 Jan 2012 (2012), http://eprint.iacr.org/2011/187. Citations in this document: §1, §1, §3, §3, §3, §3, §5

[23] Lewis, D.J. (ed.): 1969 Number Theory Institute: proceedings of the 1969 summer institute on number theory: analytic number theory, Diophantine problems, and algebraic number theory; held at the State University of New York at Stony Brook, Stony Brook, Long Island, New York, July 7–August 1, 1969. Proceedings of Symposia in Pure Mathematics, vol. 20. American Mathematical Society, Providence, Rhode Island (1971). ISBN 0-8218-1420-6. MR 47:3286. See [31]

[24] Maurer, U.M., Yacobi, Y.: Non-interactive public-key cryptography. In: Eurocrypt 1991 [8], pp. 498–507 (1991). Citations in this document: §1, §1

[25] Nechaev, V.I.: Complexity of a determinate algorithm for the discrete logarithm. Mathematical Notes 55, 165–172 (1994). Citations in this document: §2

[26] Nohl, K., Paget, C.: GSM — SRSLY? (2009), http://events.ccc.de/congress/2009/Fahrplan/attachments/1519_26C3.Karsten.Nohl.GSM.pdf. Citations in this document: §3

[27] Paterson, K.G., Srinivasan, S.: On the relations between non-interactive key distribution, identity-based encryption and trapdoor discrete log groups. Designs, Codes and Cryptography 52, 219–241 (2009), http://www.isg.rhul.ac.uk/~prai175/PatersonS09.pdf. Citations in this document: §1

[28] Pollard, J.M.: Monte Carlo methods for index computation (mod p). Mathematics of Computation 32, 918–924 (1978), http://www.ams.org/mcom/1978-32-143/S0025-5718-1978-0491431-9/S0025-5718-1978-0491431-9.pdf. Citations in this document: §2, §2, §2

[29] Pollard, J.M.: Kangaroos, Monopoly and discrete logarithms. Journal of Cryptology 13, 437–447 (2000). Citations in this document: §2

[30] Sattler, J., Schnorr, C.-P.: Generating random walks in groups. Annales Universitatis Scientiarum Budapestinensis de Rolando Eötvös Nominatae. Sectio Computatorica 6, 65–79 (1989). ISSN 0138-9491. MR 89a:68108, http://ac.inf.elte.hu/Vol_006_1985/065.pdf. Citations in this document: §2

[31] Shanks, D.: Class number, a theory of factorization, and genera. In: [23], pp. 415–440 (1971). MR 47:4932. Citations in this document: §2, §2

[32] Shoup, V.: Lower bounds for discrete logarithms and related problems. In: Eurocrypt 1997 [12], pp. 256–266 (1997), http://www.shoup.net/papers/. Citations in this document: §2

[33] Teske, E.: On random walks for Pollard's rho method. Mathematics of Computation 70, 809–825 (2001), http://www.ams.org/journals/mcom/2001-70-234/S0025-5718-00-01213-8/S0025-5718-00-01213-8.pdf. Citations in this document: §2

[34] van Oorschot, P.C., Wiener, M.: Parallel collision search with application to hash functions and discrete logarithms. In: [1], pp. 210–218 (1994); see also newer version [35]

[35] van Oorschot, P.C., Wiener, M.: Parallel collision search with cryptanalytic applications. Journal of Cryptology 12, 1–28 (1999); see also older version [34]. ISSN 0933-2790, http://members.rogers.com/paulv/papers/pubs.html. Citations in this document: §2, §2, §2

[36] Vaudenay, S., Youssef, A.M. (eds.): Selected areas in cryptography: 8th annual international workshop, SAC 2001, Toronto, Ontario, Canada, August 16–17, 2001, revised papers. LNCS, vol. 2259. Springer (2001). ISBN 3-540-43066-0. MR 2004k:94066. See [20]

Embedded Syndrome-Based Hashing*

Ingo von Maurich and Tim Güneysu

Horst Görtz Institute for IT-Security, Ruhr-University Bochum, Germany
{ingo.vonmaurich,tim.gueneysu}@rub.de

Abstract. We present novel implementations of the syndrome-based hash function RFSB on an Atmel ATxmega128A1 microcontroller and a low-cost Xilinx Spartan-6 FPGA. We explore several trade-offs between speed and area/code size on both platforms and show that RFSB is extremely versatile with applications ranging from lightweight to high performance. Our lightweight microcontroller implementation requires just 732 byte of ROM while still achieving a competitive performance with respect to other established hash functions. Our fastest FPGA implementation is based on embedded block memories available in Xilinx Spartan-6 devices and runs at 0.21 cycles/byte, with a throughput of 5.35 Gbit/s. To the best of our knowledge, this is the first time the RFSB hash function is implemented on either of these wide-spread platforms.

Keywords: RFSB, hash function, code-based cryptography, microcontroller, hardware, FPGA.

1 Introduction

Cryptographic hash functions are used in a broad range of applications where mapping an arbitrary amount of data to a fixed length bit string is required. Examples are digital signatures, messages authentication codes, data integrity checks, and password protection. Prominent and widely deployed hash functions such as MD5 [38], SHA-1 [37], and the SHA-2 family [37] are used in various products and implementations whose security depends on the collision resistance of those hash functions. However, over the last years (chosen-prefix) collision attacks have been published for MD5 [42] [43] and SHA-1 [30] and are already exploited in the real-world. Recently, a major attack based on MD5 collisions was performed by the Flame espionage malware which injects itself into the Microsoft Windows operating system. The malware code is signed by a rogue Microsoft certificate and disguises itself as a Microsoft Windows update. The rogue certificate was obtained using a previously unknown chosen-prefix collision attack on a Microsoft Terminal Server Licensing Service certificate which still used the MD5 algorithm [34].

Although the SHA-2 family withstands these attacks so far, its similar structure to SHA-1 raised concerns about its long term security. Therefore, the

---

* This work was partially supported by the European Commission through the ICT programme under contract ICT-2007-216676 ECRYPT II.

S. Galbraith and M. Nandi (Eds.): INDOCRYPT 2012, LNCS 7668, pp. 339–357, 2012.
© Springer-Verlag Berlin Heidelberg 2012

National Institute of Standards and Technology (NIST) announced the public SHA-3 competition in the end of 2007 [36]. A total of 64 candidates entered the competition, of which 14 advanced to the second round, and the last round still has 5 competing candidates. Apart from their security the main selection criteria are hardware and software speed as well as scalability. The SHA-3 competition announcement explicitly demands efficiency in 8-bit microcontrollers and in hardware.

Embedded 8-bit microcontrollers are a common representative of low-cost and energy efficient computation units used in automotive applications, digital signature smart cards, wireless sensor networks and many more. Field-Programmable Gate Arrays (FPGA) on the other hand allow reconfigurable implementations of cryptography in hardware, usually yielding a much better performance than achievable with 8-bit microcontrollers or PCs. FPGA device classes range from low-cost (e.g., Xilinx Spartan family) to high-end high-speed (e.g., Xilinx Virtex family). Since both microcontrollers and FPGAs are used for applications handling sensitive data, efficiently computable cryptographic primitives are essential for successful real-world applications.

Code-based cryptography offers a variety of cryptographic primitives that are built upon the hardness of well-known NP-complete problems in coding theory. The Fast Syndrome Based (FSB) hash function is a code-based hash function that entered the SHA-3 competition but due to its inefficiency compared to other candidates, FSB did not advance to the second round. The Really Fast Syndrome-Based (RFSB) hash function is an improved version of FSB that aims to overcome this problem.

Other cryptographic primitives based on codes are the McEliece [31], the Niederreiter [35] or the Hybrid McEliece (HyMES) [11] asymmetric encryption scheme. Digital signatures based on McEliece can be computed using CFS [14], Parallel-CFS [17], or quasi-dyadic CFS [5] and even one-time signatures are possible using the BMS-OTS scheme [6]. Code-based stream ciphers such as SYND [20] and 2SC [32] also offer security reductions to the syndrome decoding problem.

McEliece and Niederreiter have been reported to be efficiently implementable in 8-bit microcontrollers [12][16][25][26] as well as in reconfigurable hardware [24] [27][41]. Software and hardware implementations of the code-based signature scheme CFS have been published as well [10][29].

1.1 Contribution

With code-based encryption and signature schemes proven to be feasible in hard- and software, it still is an open question how code-based hash functions perform on these platforms. In this paper we set out to answer this question by evaluating the feasibility and achievable performance of RFSB-509 in embedded systems. We explore different design choices for embedded microcontrollers and reconfigurable hardware using the wide-spread 8-bit microcontroller Atmel ATxmega128A1 and a Xilinx Spartan-6 FPGA. We show that RFSB-509 can be efficiently implemented on both platforms and that RFSB can, in contrast to its predecessor FSB, keep up with current SHA-3 candidates and hash standards. Source code for both platforms is made publicly available in order to allow other researchers to use and evaluate our implementations[1]. To the best of our knowledge this is the first work of its kind.

1.2 Organization

This paper is organized as follows. We present related work on code-based hash functions and the history that led to the development of RFSB in Section 2. After shortly recalling general specifications of the RFSB hash function, we detail on the concrete parameter proposal RFSB-509 and give an implementors' view on RFSB-509 in Section 3. Next, design considerations for implementations on embedded microcontrollers and on reconfigurable hardware are evaluated in Section 4 and Section 5. We present our result in Section 6 before we draw a conclusion in Section 7.

2 Related Work on Code-Based Hash Functions

Augot, Finiasz, Gaborit, Manuel, and Sendrier entered the SHA-3 competition with the Fast Syndrome Based (FSB) hash function [2] that relies on the syndrome decoding problem for linear codes. Previous attempts to build such a hash function by Augot, Finiasz, and Sendrier [3][4], and Finiasz, Gaborit, and Sendrier [18] turned out to be flawed and were consequently broken by Coron and Joux [13], Saarinen [40], and Fouque and Leurent [19]. Hence, the FSB parameters were adjusted to withstand these attacks for the SHA-3 submission and to date this parameter set remains unbroken. However, FSB did not advance to the second round of the SHA-3 competition mainly because of its lack in efficiency compared to other submissions.

Meziani, Dagdelen, Cayrel, and Yousfi Alaoui use the ideas of FSB and combine them with a sponge construction instead of the Merkle-Damgård principle to construct the S-FSB hash function [33]. Their main goal is to improve the performance compared to FSB and they report a C implementation of S-FSB-256 on an Intel Core 2 Duo CPU running at 183 cycles/byte. Compared to FSB requiring 264 cycles/byte on the same CPU, S-FSB is about 30% faster but when looking at the overall picture S-FSB is still an order of magnitude slower than the current hash standard SHA-256 which runs at 15.49 cycles/byte on a similar CPU according to eBASH[2] [8].

Bernstein, Lange, Peters, and Schwabe introduce the Really Fast Syndrome-Based (RFSB) hash function as an improved version of FSB and propose concrete parameters (RFSB-509) in [9]. The authors report an implementation of RFSB-509 that outperforms the current hash standard SHA-256 on Intel Core 2

---

[1] See our web page at http://www.sha.rub.de/research/projects/code/

[2] (6fd); 2007 Intel Core 2 Duo E4600; 2 x 2400MHz; cobra, supercop-20111120.

Quad Q9550 CPUs at 13.62 vs. 15.26 cycles/byte. According to the latest measurements on eBASH[3], new implementations allow to compute RFSB-509 even faster at 10.64 cycles/byte while SHA-256 remains at 15.31 cycles/byte on the same CPU.

Furthermore, a PC implementation of RFSB in Java and C is reported by Rothamel and Weiel [39]. In addition to RFSB-509, the authors suggest parameter sets RFSB-227, RFSB-379, and RFSB-1019 and provide performance measurements for all four sets. However, their results do not come anywhere close to the speeds reported in the original RFSB paper (e.g., they report RFSB-509 to run at 120.5 cycles/byte on an Intel i7 CPU).

3   The RFSB Hash Function

The RFSB [9] hash function is constructed very similar to the FSB hash function [2], both are designed to be used inside a collision resistant hash function. A fixed length compression function is combined with the Merkle-Damgård domain extender [15] to enable processing of an arbitrary amount of data. An initialization vector (IV) is compressed together with the first message block. The output is used as chaining value together with the second message block and is again fed into the compression function. This continues until the second to last message block has been processed. The last block is padded by appending a single 1 bit followed by sufficiently many zeros and a 64-bit message length counter. After all input has been processed a final output filter (called final compression function in FSB terms) is applied. In case of FSB Whirlpool is used as final compression function, the authors of RFSB suggests to use SHA-256 or an AES-based output filter. The basic hashing principle of FSB and RFSB is illustrated in Figure 1.

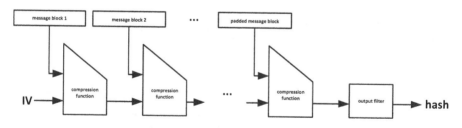

Fig. 1. Illustration of the hashing principle based on the Merkle-Damgård domain extender used by FSB and RFSB. The initialization vector (IV) is set to zero in RFSB and the output filter is defined to be SHA-256.

3.1   The RFSB Compression Function

The RFSB compression function is defined by four parameters: an odd prime r, positive integers b and w, and a compressed matrix of size $2^b \times r$-bit. The

---

[3] (10677); 2008 Intel Core 2 Quad Q9550; 4 x 2833MHz; berlekamp, supercop-20120704.

compression function takes a bw-bit string as input which is interpreted as a sequence of $\lceil bw/8 \rceil$ bytes $(m_1, m_2, \ldots, m_w)$ where each $m_i \in \{0, 1, \ldots, 2^b - 1\}$. The output is a r-bit string that is interpreted as a sequence of $\lceil r/8 \rceil$ bytes. Both input and output are interpreted in little-endian format. The compressed matrix consists of constants $c[0], c[1], \ldots, c[2^b - 1]$ where each constant has a length of r-bit. The uncompressed RFSB matrix is derived from these constants by defining

$$c_i[j] = c[j] x^{128(w-i)}, 1 \leq i \leq w, 0 \leq j \leq 2^b - 1$$

in the ring $\mathbb{F}_2[x]/(x^r - 1)$ which essentially are rotations of the compressed matrix constants.

The input is mapped to the output using the message bytes m_i as indices of the uncompress matrix constants c_i. The constants are summed up by exclusive-or addition to form the output as follows:

$$(m_1, m_2, \ldots, m_w) \mapsto c_1[m_1] \oplus c_2[m_2] \oplus \cdots \oplus c_w[m_w].$$

When using the compressed matrix notation the mapping from input to output is given by:

$$(m_1, m_2, \ldots, m_w) \mapsto c[m_1] x^{128(w-1)} \oplus c[m_2] x^{128(w-2)} \oplus \cdots \oplus c[m_w]$$

in $\mathbb{F}_2[x]/(x^r - 1)$.

3.2   A Concrete Proposal: RFSB-509

RFSB-509 is a concrete parameter proposal by the designers of RFSB which achieves good software speed. In the original paper RFSB-509 is assumed to provide a collision resistance of more than 2^{128}. The proposed parameters are $r = 509, b = 8$, and $w = 112$. Hence, the RFSB-509 input message size is 896 bit (112 byte) and the output size is 509 bit. The compressed matrix is of size $2^b \times r = 256 \times 509$ bit which roughly amounts to 16 kByte. A recent result by Kirchner [28] suggests an improved generalized birthday attack which claims to lower the complexity to about 2^{79}. Hence, the parameters need to be adjusted if a collision resistance of 128-bit is required.

Each element of the compressed matrix is generated using a concatenation of the ciphertexts output by four AES-128 calls with fixed all-zero key and a plaintext which is a function of the index of the constant. We denote the AES calls by $y = \text{AES}_k(x)$ with y being the 16-byte ciphertext, k being the 16-byte key, and x being the 16-byte plaintext. The plaintext is set to zero except for the last two bytes. The second to last byte is set to j which is the index of the constant and $0 \leq j \leq 255$. The last byte of the plaintext is a counter which increases with each AES-128 call from 0 to 3. In total this results in a 512-bit string

$$c'[j] = \text{AES}_0(0, \ldots, 0, j, 0) \| \text{AES}_0(0, \ldots, 0, j, 1) \| \ldots \| \text{AES}_0(0, \ldots, 0, j, 3)$$

which is reduced to

$$c[j] = c'[j] \mod x^{509} - 1$$

to stay in the ring $\mathbb{F}_2[x]/(x^{509}-1)$. The 112-byte input block $(m_1, m_2, \ldots, m_{112})$ with each $m_i \in \{0, 1, \ldots, 255\}$ is mapped to the 509-bit output by computing

$$(m_1, m_2, \ldots, m_{112}) \mapsto c[m_1] x^{128(112-1)} \oplus c[m_2] x^{128(112-2)} \oplus \cdots \oplus c[m_{112}]$$

in $\mathbb{F}_2[x]/(x^{509}-1)$.

3.3 RFSB-509 from an Implementors' Point-of-View

When designing RFSB-509 for embedded systems a few aspects have to be considered beforehand. In the following we detail considerations and optimizations for implementations of RFSB-509 in embedded devices.

At first, the constant matrix, although compressed, still has a size of 16 kByte which poses a challenge in embedded systems where memory usually is scarce. Due to the computability of the constants there are basically two choices that can be made. Either memory is spent to store the table or each constant is, probably multiple times, generated on-the-fly when needed. For the on-the-fly generation one has to keep in mind that each constant requires four calls to the AES-128 encryption function, thus a total of $4 \times 112 = 448$ AES encryptions are required during one compression.

When compressing a message block the rotations applied to each constant depend on the position of the current message byte. For example the first computation in RFSB-509 is $c[m_1] x^{128(112-1)} = c[m_1] x^{14208}$ which requires to rotate $c[m_1]$ by 14208 bit positions. When using 512-bit wide registers the amount of different rotations performed during RFSB compression can be reduced to just four since $128(112-i) \equiv 128i \mod 512 \in \{0, 128, 256, 384\}$. Using this the RFSB compression of the first four messages bytes (m_1, m_2, m_3, m_4) can be rewritten as

$$s_1 = \mathrm{ROL}_{384}(c[m_1]) \oplus \mathrm{ROL}_{256}(c[m_2]) \oplus \mathrm{ROL}_{128}(c[m_3]) \oplus c[m_4]$$

where ROL_j denotes a j-bit rotation to the left (towards the most significant bit) of a 512-bit register. The four different rotations and their exclusive-or sum can be seen as *basic compression unit* of RFSB-509, which can be generalized to

$$s_i = \mathrm{ROL}_{384}(c[m_{4i+1}]) \oplus \mathrm{ROL}_{256}(c[m_{4i+2}]) \oplus \mathrm{ROL}_{128}(c[m_{4i+3}]) \oplus c[m_{4i+4}].$$

In order to process all 112 input message bytes this computation has to be repeated 28 times. Adding up all intermediate results s_i then yields the output of the compression function

$$\mathrm{compress}_{509}(m_1, \ldots, m_{112}) = \sum_{i=0}^{27} s_i \mod x^{509} - 1$$

$$= \sum_{i=0}^{27} \sum_{j=1}^{4} \mathrm{ROL}_{512-128j}(c[m_{4i+j}]) \mod x^{509} - 1$$

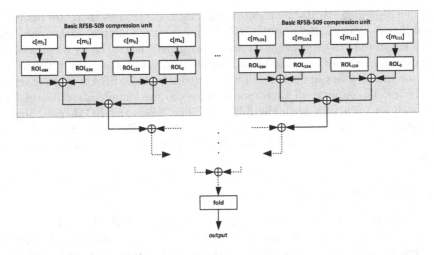

Fig. 2. The basic compression unit in RFSB consists of looking up four constants, rotating them according to their position by either 384, 256, 128, or 0 bits and xoring the result. The *fold* unit represents the reduction modulo $x^{509} - 1$ by folding the three most significant bits onto the three least significant bits.

where the sums are formed using exclusive-or addition. Figure 2 illustrates the tree-like structure of the RFSB-509 compression function and shows the repetitions of the *basic compression unit*.

One further important detail is the computation of the reduction modulo $x^{509} - 1$ for 512-bit registers. It is achieved by folding the three most significant bits onto the three least significant bits and setting the three most significant bits to zero. Such a reduction does not pose a problem on both platforms and can be realized at minimal cost.

4   Designing RFSB-509 for Embedded Microcontrollers

For our evaluations of RFSB-509 on an embedded microcontroller we use the wide-spread 8-bit ATxmega microcontroller family from Atmel. These microcontrollers are low-cost yet powerful enough for a wide range of cryptographic and non-cryptographic applications. Apart from the usual features available in this kind of devices (analog to digital converter, digital to analog converter, timers, counters, several communication interfaces, etc.) the ATxmega offers dedicated hardware accelerators for the encryption standards DES and AES-128.

All following designs are split into three basic functions *init*, *update*, and *final*. During *init* we reset the internal state, output and counter to zero. The *update* function implements the Merkle-Damgård domain extender, processes new message blocks and updates the internal state accordingly until the last message block is reached. The last message block is processed by the finalization function which pads the message, appends the length counter, compresses the message and sets the output when finished.

When designing RFSB-509 for embedded microcontrollers there are basically two different ways of realizing the RFSB compression function. Either the constants are stored in a table or the constants are generated on-the-fly when needed. One can also think of a hybrid mode, where the constants are not stored in the program memory but are generated and stored in volatile SRAM when starting the device. We explore all three possibilities an give details about the design approach for each version in the following. The AES- and ROM-based implementations are done on an Atmel ATxmega128A1 microcontroller while the SRAM-based implementation is done on an Atmel ATxmega384C3 microcontroller.

4.1   On-the-Fly Constant Generation

When designing a small memory footprint version of RFSB-509, on-the-fly constant generation is required since the compressed constant table would consume 16 kByte of program memory which would render a lightweight implementation impossible. Especially the hardware AES-128 offered in ATxmega devices is useful in such a setting. The AES-128 crypto module runs concurrently to the CPU and takes 375 clock cycles after loading the key and the plaintext block into the module to en-/decrypt a 128-bit block. When taking loading and storing of key, plaintext and ciphertext into account, an AES-128 encryption takes about 500 clock cycles or 31.25 cycles/byte. Thus when running at its maximum frequency of 32 MHz the ATxmega is able to achieve a AES-128 encryption throughput of about 8 Mbit/s.

Our small footprint implementation of the RFSB-509 hash function is built around a parameterizable constant generation function that is capable of providing rotation widths of 0, 128, 256, and 384 bit. In each iteration the constant generation function computes four AES encryptions. After each encryption the ciphertext is transferred into 16 general purpose registers and immediately afterwards the next plaintext and key (which is the all-zero key for all encryptions but has to be reloaded before every encryption nevertheless) are loaded into the AES module and the next encryption is triggered. While waiting for the next encryption to finish, we concurrently process the previous ciphertext by accumulating it to the output and reducing the computed constant modulo $x^{509} - 1$. Thus, we make use of otherwise wasted cycles while the next encryption is running in parallel. In order to maintain a reasonable performance, parts of the implementation are unrolled, e.g., storing and loading data to and from the AES crypto module is unrolled since this part is critical for the overall runtime.

If the constants would be generated using DES instead of AES-128, the performance of the on-the-fly constant generation could be further improved. Since the output length of DES is half the output length of AES-128, twice the amount of DES calls would be required. However, at 16 cycles per DES encryption after loading the key and plaintext to the corresponding registers, this would still be an order of magnitude faster than AES-128 encryption on an ATxmega microcontroller. Since the performance of the encryption function is the limiting factor in such an implementation, the overall performance would greatly benefit from such an improvement.

Note, the short key length of DES and its vulnerability to brute-force attacks does not pose a thread to the security of RFSB-509 since all plaintexts and keys are already known by definition. As stated in the original RFSB paper: "The full security of AES is certainly not required for RFSB: all that we need is a function generating a few elements of $\mathbb{F}_2[x]/(x^r - 1)$ without any obvious linear structure" [9].

4.2   ROM-Based Lookup Table

A total of 16 kByte of program memory is required when storing the precomputed constant table in the ROM of the microcontroller. Each of the 256 entries in the table consists of 64 byte, thus we multiply each message byte by 2^6 (shifting the message byte six times to the left) to compute the index of the required constant. Instead of first reading out the constant and then rotating it according to the position of the current message byte, we adjust the table pointer beforehand to directly read out the rotated constant. This is possible since all rotation widths are a multiple of 8 and the basic addressable unit in our 8-bit microcontroller is a byte. For example if a constant has to be rotated by 384 bit, we add $\frac{384}{8} = 48$ to the current index, read out 16 constant bytes, then subtract 64 from the current index and read out the remaining 48 constant bytes. Thus we achieve nearly free rotations by only adjusting the table pointer.

This process is repeated for all message bytes and rotation widths, and after all constants have been read out and accumulated, the result is reduced modulo $x^{509} - 1$.

In our evaluation we explore two different approaches, a rolled and an unrolled version. In the unrolled version we remove every loop inside the computation of the basic compression unit which computes the intermediate output of four consecutive message bytes with four different rotation widths applied to the read out constants (cf. Figure 2).

4.3   RAM-Based Lookup Table

In order to estimate the maximum achievable performance in embedded micro-controllers, we move the constants from the program memory into the faster SRAM. Accessing a byte in the program memory of the ATxmega takes 3 clock cycles while accessing the internal SRAM takes 2 clock cycles. Since $112 \times 64 = 7168$ byte have to be looked up when hashing one message block, this small difference can have a larger impact on the overall runtime than one might expect on first sight. The hashing itself is constructed similar to the previously described setup, with some minor adjustments taking care of the modified memory locations.

For this evaluation we use the Atmel ATxmega384C3 microcontroller since it offers 32 kByte of SRAM. Devices offering 8 or 16 kByte of SRAM do not suffice in this scenario since in addition to the constant table also the current state and the next message block have to be held in the SRAM.

A remaining question is how to place the RFSB-509 constants into the SRAM. Since SRAM is volatile memory, its content has to be reloaded after every power cycle. As designers we are left with two choices. Either we store the constants in the program memory as done in Section 4.2 and copy them into SRAM at every power up, or we generate the constants once at every power up and directly store them in the SRAM. The decision which of the proposed methods to use basically depends on two factors. Firstly, it has to be considered how much time is available after a power cycle before the hash function has to be used for the first time. Generating the constants takes longer then just copying them from program memory to SRAM. Secondly, it depends on the available program memory. The generation function takes up less space in program memory compared to a 16 kByte table. In our implementation, we generate the constants after each power up, thus avoiding redundant tables in RAM and ROM.

Again we explore two approaches: a rolled and an unrolled version similar to the previously described ROM-table design.

5   Designing RFSB-509 for Reconfigurable Hardware

For our evaluation of RFSB-509 in reconfigurable hardware we use the low-cost Xilinx Spartan-6 device family. Spartan-6 devices are powerful, up-to-date FPGAs offering hundreds to (ten-)thousands of slices, where each slice contains four 6-input lookup tables (LUT), eight flip-flops (FF), and surrounding logic. In addition to the general purpose logic, embedded resources such as block memories (BRAM) and digital signal processors (DSP) are available. Yet Spartan-6 devices are about an order of magnitude cheaper than Xilinxs' high-performance devices families Virtex-5 and Virtex-6.

For the design of RFSB-509 in reconfigurable hardware, we again follow two different strategies of implementing the lookup of compressed matrix constants. In one architecture we generate the constants when needed using on-the-fly AES computations, in the other architectures we make use of the available block memories to store the matrix constants.

Since different choices for the constant lookups only affect the compression function of RFSB, all implementations share the same top-level component that takes care of handling the input and output of data through FIFOs and passes the data and control signals to the Merkle-Damgård construction which is also the same for all hardware implementations. Thus we design a modular system in which the compression function can be easily exchanged. We detail on the different compression function designs in the following.

5.1   Implementing RFSB-509 Using Embedded Block Memories

Spartan-6 FPGAs feature dual-ported block memories (BRAM) each capable of storing up to 18 Kbit of data. They can be configured to represent one of five different memory types. For our purpose we choose to configure the BRAMs as dual-port read-only memory (ROM) since we do not need the write capability. In each clock cycle two separate values can be read from two different memory addresses because of the BRAMs' dual-port layout.

Minimal BRAM Consumption. Since the compressed matrix constant table has a size of about 15.9 Kbyte a theoretical minimum of $\frac{15.9 \cdot 8}{18} = 7.07$ BRAMs is required to store the full table. However, a wide-access port of 509 bit for each constant is not natively supported by the BRAM primitives. The maximum natively supported width is 32 bit (36 when using the parity bits) or 64 bit (72 when using the parity bits) when combining both ports. Thus, for achieving a minimal block memory usage, we use eight BRAMs to store the constants as shown in Figure 3.

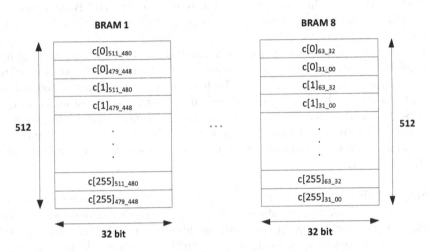

Fig. 3. Our smallest table based FPGA implementation of RFSB-509 requires 8 block memories configured as 512×32 bit dual-port ROM. Every BRAM holds a 64-bit chunk of the 509-bit constants (prepended by three zero bits) split into two 32-bit parts. Since two memory slots of each BRAM can be read out in one clock cycle, one constant can be read out in one clock cycle.

We configure the BRAMs to store 512 values of 32 bit each, which is natively supported. The RFSB constants are divided into eight 64-bit chunks and are distributed to the BRAMs. The 64-bit chunks are again split and stored in two consecutive memory slots. Hence, $BRAM_1$ holds the topmost 64-bit of all 256 RFSB constants, $BRAM_2$ the following 64-bit of all RFSB constants and so forth.

The index into the table is the current message byte m_i appended by a zero and a one bit to address both memory slots. Because of the dual-port layout of the block memories, both 32-bit memory slots can be read out simultaneously. This is done for all block memories at the same time and the results are concatenated and rotated to form the demanded constant $ROL_x(c[m_i])$. Because of this set-up, a complete already rotated RFSB-509 constant is available in one clock cycle.

We sequentially iterate over all input message bytes, accumulate the read out constants and reduce the result after all message byte have been processed.

Due to its tree-like structure, RFSB allows for very scalable designs which allow to process multiple message bytes in one clock cycle since the inputs to the block memories are independent of each other. In the following we make different proposals of how to implement multiple constant lookups in one clock cycle.

Wide-Access Block Memories. Our next architecture uses block memories with wide-access ports to provide the matrix constants. Creating a 256×509-bit table using the Xilinx block memory generator results in 15 occupied 18 Kbit BRAMs. With this architecture it is possible to read out two RFSB-509 constants in one clock cycle, thus reducing the necessary cycles spent for table lookups by a factor of 2 to 56 cycles.

The internal compression module now handles two bytes at once and applies two different rotations to the read out constants depending on the position of the message byte in the input string. In the first mode, one constant is rotated by 384-bit, the second by 256-bit, in the second mode the first constant is rotated by 128-bit and the second is not rotated. Both constants are accumulated to the intermediate result, the rotation mode is switched with every input message pair and after the complete input block has been processed, the result is reduced modulo $x^{509} - 1$.

Multiple Table Instances. For high-performance applications we explore architectures in which we instantiate multiple of the aforementioned wide-access block memories containing the full constant table. We go only so far that we still stay within reasonable (i.e., realizable on current Spartan-6 devices) resource boundaries.

In the first setting we use two tables which allows to process four message bytes in one clock cycle, essentially representing the basic compression unit introduced in Figure 2. Furthermore, it is now possible to hardwire the rotations applied to the constants because the output of each of the block memory ports only handles either $c\,[m_{4i+1}]$, $c\,[m_{4i+2}]$, $c\,[m_{4i+3}]$, or $c\,[m_{4i+4}]$, $0 \le i \le 27$. The two tables require 29 block memories and again halve the required clock cycles to 28 for the constant lookups for one 112-byte input block.

In a second design we use four separate instances of the constant table, which requires 58 BRAMs. It enables us to look up eight message bytes per clock cycle and finish the lookups after 14 clock cycles.

The third design quadruples the amount of block memories to be able to hold eight parallel instances of the constant table. This requires 116 BRAMs and allows to lookup 16 constant at the same time which means all constants are retrieved in just 7 clock cycles.

5.2   Implementing RFSB-509 Using an AES Core

To cover all possible designs, we also include on-the-fly generation of the matrix constants using an hardware AES core. Since the key is always fixed to the all-zero key, the key-schedule does not have to be implemented as the round-keys can be precomputed. This of course is only true if the AES core is not used by other applications that require the key to be adjustable during runtime. The AES core in use is a T-table based implementation that occupies eight block memories for storing the tables.

The constant computation unit is built straightforward. It receives a message byte and starts four consecutive encryptions with the respective input blocks as described in Section 3.2. Each result is xored to an internal output signal and after the fourth encryption is finished, a modular reduction is performed and the constant is output. The higher level unit receives the constant, rotates it according to the position of the current message byte and passes the next message byte to the constant computation unit.

6   Results

All our implementations are verified against testvectors generated using the reference implementation of RFSB-509 by Schwabe which was submitted to the ECRYPT Benchmarking of Cryptographic Systems (eBACS) [7].

The results for embedded microcontrollers are provided by the Atmel AVR Studio 6, and the implementations are additionally tested in real hardware using an AVR XPLAIN board equipped with an ATxmega128A1. In addition, the microcontroller implementations feature a full padding unit.

The FPGA results are achieved using post place-and-route results from Xilinx ISE Design Suite 14.1. As target device we use a Spartan-6 FPGA XC6SLX100, but we stress that for all implementations smaller Spartan-6 FPGAs suffice.

The output filter is currently not implemented because a wide range of SHA-256 implementation is already available in hard- and software. Neglecting the output filter arguably does not effect the performance measurements when hashing long messages since it is only applied once to the output of the RFSB-509 compression function.

In the following we present our microcontroller and FPGA results and compare them to other hash function implementations on similar devices.

6.1   Embedded Microcontrollers

Table 1 shows the results of our implementations of RFSB-509 on the embedded microcontroller ATxmega. The achievable performance is measured in cycles/byte, where the amount of clock cycles required for calling the *update* function is divided by 48 byte although in total 112 byte are hashed. This is due to the fact, that only 48 fresh message bytes enter each compression function because of the Merkle-Damgård construction. Thus, these figures give a realistic performance overview when hashing long messages.

Table 1. Results of RFSB-509 achieved on the embedded microcontroller Atmel ATxmega128A1. *Results for the SRAM table based implementations are measured on an ATxmega384C3.

Design	ROM [byte]	RAM [byte]	Cycles/ Byte	Used ROM	Used RAM
HW-AES	732	232	4753.1	0.5%	2.8%
ROM table	602+16384	232	1573.9	12.2%	2.8%
ROM table unrolled	3100+16384	232	1114.9	14.0%	2.8%
RAM table*	996	232+16384	1424.5	4.2%	50.7%
RAM table unrolled*	3494	232+16384	965.6	4.9%	50.7%

All implementations require 232 byte of RAM, split into 112-byte state, 48-byte input, 64-byte output and an 8-byte counter. An additional 16 kByte of SRAM are used by the SRAM-based table implementations to store the table.

The fastest implementation is running at 965.6 cycles/byte but is so far only realizable in a few 8-bit microcontrollers since only newer devices meet the RAM requirements. The fastest ROM-based implementation computes one RFSB-509 round at 1114.9 cycles/byte. The counterpart to the unrolled version does not seem to be a good choice, since program memory at this size is not a problem for current microcontrollers and spending an additional 2.5 Kbyte of ROM seems to be worth the 460 cycles/byte performance improvement.

Our smallest implementation, the one based on AES encryptions, only requires 732 byte of ROM which falls into the lightweight cryptography category. If ROM memory is scarce, the current version could be implemented even smaller since some loops have been unrolled to improve the performance. Since for every constant the AES encryption is called four times, 448 AES encryptions are needed during compression. When assuming 500 clock cycles for each AES encryption we get a lower bound of 224000 clock cycles or 4666.7 cycles/byte for the encryptions, not counting rotations, modular reductions and the combination of looked-up constants to form the output. Our result of 4753.1 cycles/byte comes very close to this lower bound.

Although RFSB fits well on current embedded microcontrollers and performs at a decent speed, beating implementations of the SHA-3 candidates is not possible due to memory requirements caused by the size of the matrix constant table. When comparing the lightweight AES-based implementation to the results of a ECRYPT initiative that aims to provide a comprehensive collection of lightweight implementations of hash functions [1], RFSB-509 beats well known hash functions such as SHA-256, BLAKE-256, JH-256, and Skein-256 in terms of code size and outperforms JH-256, and sponge-based construction such as PHOTON and SPONGENT. However, it has to be noted that the other implementations do not use the crypto accelerators.

6.2   Reconfigurable Hardware

Table 2 contains our FPGA results taken from the post place-and-route reports which in contrast to post-synthesis figures take actual delays in hardware into account. The different designs using BRAM tables are named RFSB-509_x where x denotes the amount of used block memories.

Table 2. Results of RFSB-509 achieved on a Xilinx Spartan-6 XC6SLX100. We measure the occupied slices, used flip flops(FF), 6-input lookup tables (LUT), and the maximum clock frequency f. From this we compute the cycles/byte, the throughput (Tp), and for comparison the throughput to area ratio (Tp/Area).

Design	Occ. Slices	Slice FFs	Slice LUTs	18 Kbit BRAM	f [MHz]	Cycles/ Byte	Tp [Mbit/s]	Tp/Area [$\frac{\text{Mbit/s}}{\text{Slices}}$]
AES-based	1526	5793	4920	8	260.2	213.8	9.3	0.01
RFSB-509_8	1402	4621	4316	8	259.4	2.46	805.1	0.57
RFSB-509_15	1381	4106	4277	15	234.7	1.25	1,432.8	1.04
RFSB-509_29	1409	4101	4309	29	223.0	0.65	2,633.9	1.87
RFSB-509_58	1447	4070	3709	58	171.1	0.38	3,480.2	2.41
RFSB-509_116	2112	4071	4690	116	146.2	0.21	5,354.0	2.54

To measure the performance of our implementations we count the clock cycles consumed for loading new message bits into the Merkle-Damgård state, compressing the current state and updating it accordingly. We divide the number of clock cycles by 48 instead of 112 byte since due to the Merkle-Damgård construction only 48 new message bytes enter each 112-byte compression. In addition, we compute the achieved throughput of each implementation as Tp = $\frac{\text{clock frequency} \times 8}{\text{cycles/byte}}$. In terms of speed the amount of utilized block memories directly correlates with the performance. When using just 8 BRAMs a throughput of 805.1 Mbit/s can be achieved. Our fastest implementation runs at 5.35 Gbit/s and consumes 116 block memories. A designer is thus left with the decision of how many block memory resources he is willing to spend for the hash function or from a different perspective how many block memories have to be spent for a certain performance goal.

We measure the required area on an FPGA in terms of occupied flip-flops, LUTs, and BRAMs. We also include the number of occupied slices for comparison even though this number has to be considered with care since the slice count itself does not reveal the actual degree of used logic inside the slice and neglects the number of occupied embedded resources (e.g., DSPs and BRAMs). The overall slice count stays on the same level for nearly all of our implementations, only the fastest implementation occupies more slices but the amount of used flip-flops and LUTs does not increase on the same scale. This is due to fact that block memories are spread out over the FPGA and are not located at only one designated area. Usually this leaves more freedom of where to place an implementation on the FPGA but when combining more than just a few BRAMs, the design is spread out leading to partly used slices. It also increases the critical path which explains the decreasing clock frequency for the 58 and 116 BRAM variants.

Note, the performance and size of the AES-based design is inherently depended on the underlying AES core. Nevertheless, using on-the-fly constant generation on an FPGA does not seem to be a good choice since the required resources are nearly the same as in our smallest BRAM implementation plus additional logic for the AES core (393 flip-flops, 326 LUTs, 130 slices, 8 BRAMs, and 21 clock cycles for one encryption). The performance however is two orders of magnitude slower. The only scenario in which an AES-based implementation could be favorable is a setting in which no block memories are present (which of course would also require a non BRAM-based AES implementation).

We compare our results to a recent evaluation of the hardware performance of the five SHA-3 finalists [21] and a recent implementation of the lattice-based hash function SWIFFTX [22] in Table 3. When comparing the plain numbers one has to keep in mind that our implementation results are achieved on low-cost Xilinx Spartan-6 devices while the other results are measured using high-end Xilinx Virtex-5 and Virtex-6 devices. Nevertheless, our implementations keep up with most implementations and get only clearly beaten by the Keccak-256 hardware implementation.

Table 3. In this table our results are compared to other hash functions implemented in hardware. The results of [21] are given for high-end Xilinx Virtex-6 devices, [22] for Xilinx Virtex-5 and our results for the low-cost Xilinx Spartan-6.

Hash Function	Occ. Slices	Tp [Gbit/s]	Tp/Area [$\frac{\text{Mbit/s}}{\text{Slices}}$]	Device [Xilinx]
RFSB-509_58	1,447	3.48	2.41	Spartan-6
RFSB-509_116	2,112	5.34	2.54	Spartan-6
SWIFFTX [22]	16,645	4.85	0.29	Virtex-5
SHA-256 [21]	239	1.63	6.83	Virtex-6
Helion Fast SHA-256 [23]	214	1.5	7.01	Spartan-6
BLAKE-256 [21]	2,530	8.06	3.18	Virtex-6
Grøstl-256 [21]	898	4.20	4.68	Virtex-6
JH-256 [21]	849	5.41	6.37	Virtex-6
Keccak-256 [21]	1,474	18.80	12.76	Virtex-6
Skein-256 [21]	1,628	6.21	3.82	Virtex-6

7   Conclusion

In this work, we presented the first implementations of RFSB-509 for embedded microcontrollers and reconfigurable hardware. Different designs from lightweight to high speed implementations have been evaluated and proven to be feasible on both platforms with competitive results in code size/area and performance. Our result show that code-based hash functions are practical and can be used in real-world application involving embedded systems.

References

1. ECRYPT Benchmarking of Lightweight Hash Functions in Atmel AVR devices (2012), http://perso.uclouvain.be/fstandae/source_codes/hash_atmel/ (accessed July 21, 2012)
2. Augot, D., Finiasz, M., Gaborit, P., Manuel, S., Sendrier, N.: SHA-3 proposal: FSB (2008), http://www.rocq.inria.fr/secret/CBCrypto/fsbdoc.pdf
3. Augot, D., Finiasz, M., Sendrier, N.: A Fast Provably Secure Cryptographic Hash Function. Cryptology ePrint Archive, Report 2003/230 (2003), http://eprint.iacr.org/
4. Augot, D., Finiasz, M., Sendrier, N.: A Family of Fast Syndrome Based Cryptographic Hash Functions. In: Dawson, E., Vaudenay, S. (eds.) Mycrypt 2005. LNCS, vol. 3715, pp. 64–83. Springer, Heidelberg (2005)
5. Barreto, P.S.L.M., Cayrel, P.-L., Misoczki, R., Niebuhr, R.: Quasi-Dyadic CFS Signatures. In: Lai, X., Yung, M., Lin, D. (eds.) Inscrypt 2010. LNCS, vol. 6584, pp. 336–349. Springer, Heidelberg (2011)
6. Barreto, P., Misoczki, R., Simplicio Jr., M.: One-time signature scheme from syndrome decoding over generic error-correcting codes. Journal of Systems and Software 84(2), 198–204 (2011)
7. Bernstein, D., Lange, T.: eBACS: ECRYPT Benchmarking of Cryptographic Systems (2012), http://bench.cr.yp.to (accessed July 21, 2012)
8. Bernstein, D., Lange, T.: eBASH: ECRYPT Benchmarking of All Submitted Hashes (2012), http://bench.cr.yp.to/results-hash.html (accessed July 21, 2012)
9. Bernstein, D.J., Lange, T., Peters, C., Schwabe, P.: Really Fast Syndrome-Based Hashing. In: Nitaj, A., Pointcheval, D. (eds.) AFRICACRYPT 2011. LNCS, vol. 6737, pp. 134–152. Springer, Heidelberg (2011)
10. Beuchat, J., Sendrier, N., Tisserand, A., Villard, G.: FPGA Implementation of a Recently Published Signature Scheme. Rapport de recherche RR LIP 2004-14 (2004)
11. Biswas, B., Sendrier, N.: McEliece Cryptosystem Implementation: Theory and Practice. In: Buchmann, J., Ding, J. (eds.) PQCrypto 2008. LNCS, vol. 5299, pp. 47–62. Springer, Heidelberg (2008)
12. Cayrel, P.-L., Hoffmann, G., Persichetti, E.: Efficient Implementation of a CCA2-Secure Variant of McEliece Using Generalized Srivastava Codes. In: Fischlin, M., Buchmann, J., Manulis, M. (eds.) PKC 2012. LNCS, vol. 7293, pp. 138–155. Springer, Heidelberg (2012)
13. Coron, J.-S., Joux, A.: Cryptanalysis of a Provably Secure Cryptographic Hash Function. Cryptology ePrint Archive, Report 2004/013 (2004), http://eprint.iacr.org/
14. Courtois, N.T., Finiasz, M., Sendrier, N.: How to Achieve a McEliece-Based Digital Signature Scheme. In: Boyd, C. (ed.) ASIACRYPT 2001. LNCS, vol. 2248, pp. 157–174. Springer, Heidelberg (2001)
15. Damgård, I.B.: A Design Principle for Hash Functions. In: Brassard, G. (ed.) CRYPTO 1989. LNCS, vol. 435, pp. 416–427. Springer, Heidelberg (1990)
16. Eisenbarth, T., Güneysu, T., Heyse, S., Paar, C.: MicroEliece: McEliece for Embedded Devices. In: Clavier, C., Gaj, K. (eds.) CHES 2009. LNCS, vol. 5747, pp. 49–64. Springer, Heidelberg (2009)
17. Finiasz, M.: Parallel-CFS: Strengthening the CFS McEliece-Based Signature Scheme. In: Biryukov, A., Gong, G., Stinson, D.R. (eds.) SAC 2010. LNCS, vol. 6544, pp. 159–170. Springer, Heidelberg (2011)

18. Finiasz, M., Gaborit, P., Sendrier, N.: Improved fast syndrome based cryptographic hash functions. In: Proceedings of ECRYPT Hash Workshop, vol. 2007, p. 155 (2007)

19. Fouque, P.-A., Leurent, G.: Cryptanalysis of a Hash Function Based on Quasi-cyclic Codes. In: Malkin, T. (ed.) CT-RSA 2008. LNCS, vol. 4964, pp. 19–35. Springer, Heidelberg (2008)

20. Gaborit, P., Lauradoux, C., Sendrier, N.: SYND: a Fast Code-Based Stream Cipher with a Security Reduction. In: IEEE International Symposium on Information Theory, ISIT 2007, pp. 186–190 (2007)

21. Gaj, K., Homsirikamol, E., Rogawski, M., Shahid, R., Sharif, M.U.: Comprehensive Evaluation of High-Speed and Medium-Speed Implementations of Five SHA-3 Finalists Using Xilinx and Altera FPGAs. Cryptology ePrint Archive, Report 2012/368 (2012), http://eprint.iacr.org/

22. Gyrfi, T., Cre, O., Hanrot, G., Brisebarre, N.: High-Throughput Hardware Architecture for the SWIFFT / SWIFFTX Hash Functions. Cryptology ePrint Archive, Report 2012/343 (2012), http://eprint.iacr.org/

23. Helion: Fast Hash Core Family for Xilinx FPGA (2011), http://heliontech.com/fast_hash.htm (accessed July 21, 2012)

24. Heyse, S.: Code-based cryptography: Implementing the McEliece scheme in reconfigurable hardware. Diploma thesis (2009)

25. Heyse, S.: Low-Reiter: Niederreiter Encryption Scheme for Embedded Microcontrollers. In: Sendrier, N. (ed.) PQCrypto 2010. LNCS, vol. 6061, pp. 165–181. Springer, Heidelberg (2010)

26. Heyse, S.: Implementation of McEliece Based on Quasi-dyadic Goppa Codes for Embedded Devices. In: Yang, B.-Y. (ed.) PQCrypto 2011. LNCS, vol. 7071, pp. 143–162. Springer, Heidelberg (2011)

27. Heyse, S., Güneysu, T.: Towards One Cycle per Bit Asymmetric Encryption: Code-Based Cryptography on Reconfigurable Hardware. In: Prouff, E., Schaumont, P. (eds.) CHES 2012. LNCS, vol. 7428, pp. 340–355. Springer, Heidelberg (2012)

28. Kirchner, P.: Improved Generalized Birthday Attack. Cryptology ePrint Archive, Report 2011/377 (2011), http://eprint.iacr.org/

29. Landais, G., Sendrier, N.: CFS Software Implementation. Cryptology ePrint Archive, Report 2012/132 (2012), http://eprint.iacr.org/

30. Manuel, S.: Classification and Generation of Disturbance Vectors for Collision Attacks against SHA-1. Cryptology ePrint Archive, Report 2008/469 (2008), http://eprint.iacr.org/

31. McEliece, R.: A public-key cryptosystem based on algebraic coding theory. DSN progress report 42(44), 114–116 (1978)

32. Meziani, M., Cayrel, P.-L., El Yousfi Alaoui, S.M.: 2SC: An Efficient Code-Based Stream Cipher. In: Kim, T.-H., Adeli, H., Robles, R.J., Balitanas, M. (eds.) ISA 2011. CCIS, vol. 200, pp. 111–122. Springer, Heidelberg (2011)

33. Meziani, M., Dagdelen, Ö., Cayrel, P.-L., El Yousfi Alaoui, S.M.: S-FSB: An Improved Variant of the FSB Hash Family. In: Kim, T.-H., Adeli, H., Robles, R.J., Balitanas, M. (eds.) ISA 2011. CCIS, vol. 200, pp. 132–145. Springer, Heidelberg (2011)

34. Ness, J.: Microsoft certification authority signing certificates added to the Untrusted Certificate Store. Microsoft Security Research and Defense (2012), http://blogs.technet.com/b/srd/archive/2012/06/03/microsoft-certification-authority-signing-certificates-added-to-the-untrusted-certificate-store.aspx (accessed July 21, 2012)

35. Niederreiter, H.: A Public-Key Cryptosystem Based on Shift Register Sequences. In: Pichler, F. (ed.) EUROCRYPT 1985. LNCS, vol. 219, pp. 35–39. Springer, Heidelberg (1986)
36. NIST. Announcing Request for Candidate Algorithm Nominations for a New Cryptographic Hash Algorithm (SHA3) Family (2007), http://csrc.nist.gov/ groups/ST/hash/documents/FR_Notice_Nov07.pdf (accessed July 21, 2012)
37. U. D. of Commerce. Secure Hash Standard (SHS). Technical report, National Institute of Standards and Technology (2008)
38. Rivest, R.: RFC 1321: The MD5 message-digest algorithm (April 1992)
39. Rothamel, L., Weiel, M.: Report Cryptography Lab SS2011 Implementation of the RFSB hash function (2011), http://www.cayrel.net/IMG/pdf/Report.pdf
40. Saarinen, M.-J.O.: Linearization Attacks Against Syndrome Based Hashes. In: Srinathan, K., Rangan, C.P., Yung, M. (eds.) INDOCRYPT 2007. LNCS, vol. 4859, pp. 1–9. Springer, Heidelberg (2007)
41. Shoufan, A., Wink, T., Molter, G., Huss, S., Strentzke, F.: A novel processor architecture for McEliece cryptosystem and FPGA platforms. In: 20th IEEE International Conference on Application-specific Systems, Architectures and Processors, ASAP 2009, pp. 98–105. IEEE (2009)
42. Stevens, M.: On collisions for MD5. Master's thesis, Eindhoven University of Technology, Department of Mathematics and Computing Science (June 2007)
43. Stevens, M., Lenstra, A., de Weger, B.: Chosen-Prefix Collisions for MD5 and Colliding X.509 Certificates for Different Identities. In: Naor, M. (ed.) EUROCRYPT 2007. LNCS, vol. 4515, pp. 1–22. Springer, Heidelberg (2007)

Compact Hardware Implementations of the Block Ciphers mCrypton, NOEKEON, and SEA

Thomas Plos, Christoph Dobraunig, Markus Hofinger, Alexander Oprisnik,
Christoph Wiesmeier, and Johannes Wiesmeier

Institute for Applied Information Processing and Communications (IAIK),
Graz University of Technology, Inffeldgasse 16a, 8010 Graz, Austria
Thomas.Plos@iaik.tugraz.at
{dobraunig,markus.hofinger,oprisnik,
c.wiesmeier,johannes.wiesmeier}@student.tugraz.at

Abstract. Compact hardware implementations are important for enabling security services on constrained devices like radio-frequency identification (RFID) tags or sensor nodes where chip area is highly limited. In this work we present compact hardware implementations of the block ciphers: mCrypton, NOEKEON, and SEA. Our implementations are significantly smaller in terms of chip area than the results available in related work. In case of NOEKEON, we even provide the first hardware-implementation results of this algorithm at all. Our implementations are designed as stand-alone hardware modules, contain an 8-bit interface for communication, and support encryption as well as decryption operation. We give results for different datapath widths and evaluate also the impact of using shift registers or latch-based memory instead of flip flops. The most-compact implementation of mCrypton requires 2 709 GEs when using a 130 nm CMOS process technology from Faraday. NOEKEON and SEA consume 2 880 and 2 562 GEs, respectively.

Keywords: low-resource hardware implementation, RFID, symmetric cryptography, block cipher, low power consumption, shift register, 8-bit interface.

1 Introduction

Compact hardware implementations of cryptographic algorithms are inevitable for resource-constrained devices where chip area has to be kept low. A prominent example are radio-frequency identification (RFID) tags that will be used in the future Internet of Things (IoT). The vision of the future IoT is to provide communication capabilities to every object by equipping it with an RFID tag. Equipping every object with a tag requires the tags to be produced in high volume and at low price to make them competitive. Since chip area is a significant cost factor, such tags are designed towards low resource usage.

In this work we present compact hardware implementations of the block ciphers: mCrypton, NOEKEON, and SEA. For mCrypton and NOEKEON a key size of 128 bits is used. SEA uses a key size of 96 bits ($SEA_{96,8}$). The implementations are stand-alone hardware modules that already include an interface for communication. With the interface, commands are sent to the hardware modules (e.g., to start an encryption operation)

S. Galbraith and M. Nandi (Eds.): INDOCRYPT 2012, LNCS 7668, pp. 358–377, 2012.
© Springer-Verlag Berlin Heidelberg 2012

and data is exchanged with them (e.g., to load the plaintext or to read the ciphertext). In many publications, costs in terms of additional chip area and clock cycles introduced by the interface are not considered when providing low-area hardware implementations. In this work, we compare the impact of using shift-registers and scan-chain flip flops to lower the overhead costs of the interface.

For each block cipher, we implement different datapath widths and optimization variants. Data and key are completely stored in the hardware modules. All modules support encryption as well es decryption operation. Our most-compact implementations are in the range of 2 562 to 2 880 GEs (one *gate equivalent* (GE) is the area consumed by a two-input NAND gate) when using a 130 nm CMOS process technology, which is significantly smaller than results given in related work.

The remainder of this paper is structured as follows. Section 2 describes the three block ciphers in short. In Section 3 we outline the general optimization techniques to lower the area requirements of our hardware circuits. Detailed implementation results are provided in Section 4, followed by a short summary and comparison with related work in Section 5. Conclusions are drawn in Section 6.

2   Description of the Block Ciphers

In this section we give a short description of the three block ciphers that we have used for our low-resource implementations. The block ciphers are: mCrypton, NOEKEON, and SEA. Block ciphers with different key lengths, state sizes, and structures have been selected to allow a better comparison of the effects of our optimization techniques on them (i.e., to evaluate if a technique suites better, e.g., for a certain block-cipher structure). Besides resource usage, security provided by the block ciphers has also been a selection criteria.

2.1   mCrypton

mCrypton has been designed by Chae Hoon Lim and Tymur Korkishko in 2005 [13]. The authors aimed at developing a very resource-efficient algorithm suitable for constrained devices like RFID tags. The architecture is very similar to Crypton but a redesign and simplification of all components has allowed a smaller implementation. mCrypton is a 64-bit block cipher with support for three different key sizes: 64, 96, and 128 bits. In our implementation we have used a key size of 128 bits (mCrypton-128). The cipher is round based with each round consisting of four subsequent steps called: *Gamma, Pi, Tau,* and *Sigma*. This round function is applied 12 times. In addition, a key-scheduling mechanism derives a round key from the cipher key for each round. mCrypton provides a good security level. The best-known attack applies on 9 rounds of mCrypton and has been published by Mala *et al.* in 2012 [15].

The function *Gamma* used by mCrypton is a nonlinear substitution using four different S-boxes to transform the input. *Pi* is a more complex column-wise bit permutation using constants, AND and XOR operations followed by *Tau*. The function *Tau* is only a permutation (column-to-row transposition) of the state. The last part of a round is the key addition *Sigma*, which is the state XORed with the current round key. In order to derive the round key, mCrypton uses a key-scheduling mechanism. For encryption, the key schedule uses round constants as well as AND and XOR operations to determine the current round key. Afterwards, a permutation of the key register is performed. For decryption an additional permutation *Phi* (consisting of the functions *Tau* and *Pi*) is applied to derive the round key. For further details we refer to [13].

2.2 NOEKEON

The second block cipher that we have implemented is NOEKEON, which has been designed by Joan Daemen, Michaël Peeters, Gilles Van Assche, and Vincent Rijmen in 2000 [4]. The cipher is based on a substitution-linear transformation network where both block size and key size are fixed to 128 bits. NOEKEON mainly consists of a simple round function that involves five operations: XOR with a round constant r_{const}, *Theta*, *Pi1*, *Gamma*, and *Pi2*. All operations only rely on simple bit-wise Boolean operations and cyclic shifts. The operation *Theta* also involves an XOR with the working key, which is either the initial cipher key itself (direct mode) or the cipher key deduced from using a key schedule (indirect mode). The key schedule in indirect mode applies the NOEKEON cipher itself on the cipher key with a null string as working key. The authors of the cipher recommend using the key schedule since it increases the resistance against related-key attacks.

A full encryption or decryption operation of NOEKEON consists of 16 iterations of the round function, followed by an additional application of *Theta*. Due to the self-inverse structure of NOEKEON, implementing the decryption operation causes only little overhead in terms of chip area. The best-known attack on NOEKEON applies on 5 out of 16 rounds and has been published by Z'aba *et al.* in 2008 [19]. Related-key attacks on both modes the direct one and the indirect one have been presented by Knudsen *et al.* in 2001 [11].

2.3 SEA

SEA stands for Scalable Encryption Algorithm and is a round-based block cipher that has been published by Standaert *et al.* in 2006 [17]. The block cipher has been designed for software implementations with small throughput requirements and limited hardware resources. One of the key features of SEA is the ability to adjust the algorithm to a specific processor word size, which allows to run it efficiently on various platforms. Block size and key size are equal and can be freely adjusted to an integer multiple of 6 times the word size. Besides the flexibility the algorithm also provides a very simple structure that uses only operations like XOR, addition, and bit/word rotation. For this work $SEA_{96,8}$ with a block size of 96 bits and a word size of 8 bits has been chosen.

SEA uses a Feistel structure for both data and round-key calculation. Round keys are generated symmetrically which means that the first round key equals the last one. Because of this property there is no difference in the key schedule for encryption and decryption operation (also no need to reload the key when processing multiple data blocks). Data and round-key computation of SEA are quite similar, easing a compact implementation of the block cipher. Four different basic operations are applied by SEA: word/bit rotations, XOR computation, modulo addition, and an S-box transformation.

The word-rotate functions R and R^{-1} rotate a complete branch at word level. The bit rotate function r, which also works on word level, rotates the bits inside the words. The S-box transformation operates on single bits of three different words at the same time.

In addition to software-based implementations, various publications have shown that SEA is also well suited for hardware implementations on FPGAs [12] or ASICs [17]. This results from the fact that only simple logic operations and bit-wise shifts are used, which can be efficiently realized in hardware. So far only little security-analysis results of SEA are available. However, as the authors have reported in [17], several security aspects have already been investigated and incorporated during the development process of the block cipher, making it resistant against state-of-the-art cryptanalysis.

3   General Optimization Techniques

In order to minimize the area requirement of the block-cipher implementations, different optimization techniques have been used and combined. The first and often one of the most powerful techniques concerns optimization on algorithmic level, e.g., to use alternative representations or to rearrange the execution order of operations. Other optimizations that are more general aim at changing the architecture or use advantageous realizations on implementation level. Details about optimizations on algorithmic level are explained individually for each algorithm in the result section. In this section we focus on general optimizations such as reducing the datapath width, and shrinking implementation costs of interface and data memory.

3.1   Datapath Width

The block ciphers presented in the previous section have been implemented with focus on minimum area requirement. For this reason, implementations with different datapath widths have been designed. Reducing the datapath width of an implementation has the advantage that less hardware resources are required for realizing the required operations (e.g., less XOR gates, S-boxes or adders). However, designs with smaller datapath widths have typically higher control complexity (i.e., larger control unit) and need often also extra multiplexers that introduce additional area overhead.

Starting from full-size versions we scaled the datapath width down to minimum-size versions. The full-size versions use the same datapath width as the state/key. The minimum-size versions use a datapath width of 8 bits for NOEKEON/SEA and 4 bits for mCrypton, respectively. Although full-size datapath implementations have already been published for two of the selected block ciphers (mCrypton and SEA), we have decided to reimplement them for allowing a fairer comparison. Typically, it is rather difficult to compare implementations that have been obtained by using different synthesis parameters and/or CMOS process technologies.

3.2   Realization of Interface

When a block-cipher module is used in an embedded device or an RFID tag for encrypting/decrypting data, it is not practical to have an interface with a data width equal

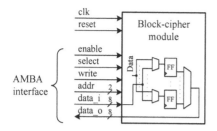
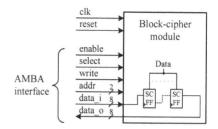

Fig. 1. Schematic overview of the AMBA interface when using standard flip flops and input/output multiplexers

Fig. 2. Schematic overview of the AMBA interface when using a shift-register approach based on scan-chain flip flops

to that of the state/key size. This is an aspect that is often ignored in related work, since depending on the realization of the interface significant overhead costs in terms of area requirements arise. Not only area requirement but also execution time and energy consumption are strongly influenced by the interface. Many microcontrollers (e.g., AVR from Atmel [3] or MSP430 from Texas Instruments [18]) that are available for embedded applications provide interfaces with 8 or 16-bit data width, which requires that data is exchanged byte/word wise with the block-cipher module.

All our block-cipher modules have an 8-bit AMBA APB interface [2] integrated. The interface allows not only data exchange, but also sending of commands to the module (e.g., to start a computation) as well as reading status information from it (e.g., whether computation is finished). In that way, data is written to and read from the hardware modules in a byte-wise manner. A schematic overview of the AMBA interface is provided in Figure 1. A single read or write operation requires two clock cycles. When applying multiple operations on the same address, which is typically the case when writing or reading data in a *burst* way, only the very first operation consumes two clock cycles. Consecutive operations require only a single clock cycle, leading to a lower communication overhead. For handling a single block of data we get a communication overhead of n_{AMBA} clock cycles as stated in Equation 1. This communication overhead consists of loading plaintext and key, starting the encryption/decryption operation, polling for the end of the operation, as well as reading the result (ciphertext). When using for example mCrypton with a 64-bit state and a 128-bit key, this results in a communication overhead of 39 clock cycles.

$$n_{AMBA} = (Byte\_num_{State} + 1) \cdot 2 + (Byte\_num_{Key} + 1) + 4 \qquad (1)$$

A big disadvantage of such an interface are the additional area costs that result from the integration of additional multiplexers that are necessary for loading data into the state/key registers as well as selecting a single byte from the state register for reading the result. Especially for larger datapath widths, additional area costs are significant for two reasons. First, a larger datapath width implies larger additional multiplexers for loading data into the state/key registers. Second, for a full-size datapath version, an output multiplexer that selects a single byte from the state register is typically not required for the block-cipher implementation itself and is therefore only used for the interface.

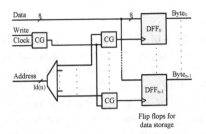 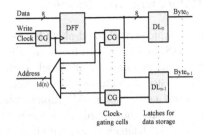

Fig. 3. Data memory with RAM-based structure using standard flip flops   **Fig. 4.** Data memory with RAM-based structure using latches to reduce area requirements

For a minimal-datapath version on the other hand that operates on a single byte of the state at once, such an output multiplexer is inherently used for the computation and consequently causes no overhead costs when reused by the interface.

Instead of using standard flip flops and multiplexers for realizing the AMBA interface as shown in Figure 1, we have also implemented a shift-register based variant with so-called scan-chain flip flops. Using shift registers to lower the area requirement has been proposed, e.g., by Hämäläinen et al. [9] and Moradi et al. [16]. A schematic overview of the AMBA interface with shift registers is given in Figure 2. No output multiplexer is required for reading the result, as data is simply shifted out byte-by-byte. Moreover, using scan-chain flip flops has the advantage that also no extra multiplexers are required for realizing the shift-register functionality. Scan-chain flip flops provide the functionality to handle both shifting of data and loading of data in parallel. Looking at the size of a scan-chain flip flop shows that it has a lower footprint than a standard flip flop with an extra multiplexer, leading to further area savings (about 1 GE per bit for our 130 nm CMOS technology).

3.3 Latch-Based Memory

Another design concept that helps reducing the area requirements of a hardware circuit is using latches instead of flip flops. State and key are typically stored in flip flops (or scan-chain flip flops), consuming around 6 GEs per bit (exact value depends on CMOS technology and driving strength of flip flop). An alternative to flip flops for storing data are latches, which have the advantage that they are smaller in size (1-2 GEs per bit). However, using latches requires careful circuit design to obtain reliable results. One such aspect that has to be considered is, for example, that input data needs to be buffered in an extra register stage. Figure 3 and Figure 4 show the typical realization of data memory with flip flops and latches, respectively (CG in the figures stands for clock-gating cell). Due to the additional register stage for buffering the input data that is written into the memory, the latch-based approach is only attractive when the input data-bit width is rather small (e.g., for RAM-like memory structures). We have used latch-based memory for our minimum-size datapath implementations to further reduce the area requirements.

4   Implementation Results

We have implemented stand-alone hardware modules of the block ciphers mCrypton, NOEKEON, and SEA with focus on minimum area requirement. Encryption and decryption functionality of all block ciphers are realized. Both state and key are completely stored in the modules, requiring therefore no external storage memory or input during computation. All implementations contain an 8-bit AMBA interface for exchanging data and control information. The naming scheme of the implementations consists of the datapath (DP) width in bits, the type of storage elements (i.e., flip flops (FF), scan-chain flip flops (SC-FF) or latches (Lh)) used for state/key memory, the number of cycles per full round, and eventually further properties of the implementation. Tables 1, 3, and 5 summarize the execution times of the implementations and list the cycles consumed by the interface as well as by the computation of the block cipher (algorithm) itself.

The block ciphers have been implemented in VHDL and synthesized for a 130 nm CMOS standard-cell technology from Faraday with 1.2 volts supply voltage [5]. For synthesis, we have used a semi-custom design flow working with Cadence RTL Compiler. Power simulations have been performed with Cadence First Encounter Power Estimator (operates on extracted netlist and VCD file obtained from simulation). Tables 2, 4, and 6 provide an overview of the area values obtained for our implementations after synthesis and after place and route for the 130 nm technology as well as power-consumption values.

4.1   mCrypton

Our first implementation of mCrypton is a one-cycle-per round version that uses a 64-bit datapath for the state and a 128-bit datapath for the key schedule. The implementation of Lim *et al.* [13] served as starting point, which uses a slightly different arrangement of the operations than the original specification. The reason for this is that replacing *Pi* by its transposed version Pi^T in the round function simplifies the overall structure, as shown in Equation 2 (note that *Gamma* and its inverse commute, i.e., *Gamma* ∘ *Tau* is equal to *Tau* ∘ *Gamma*). As suggested by the authors, we also instantiated only a single set of 16 4-bit S-boxes for realizing both *Gamma* and *Gamma*$^{-1}$. The *Gamma*$^{-1}$ operation is implemented by simply rearranging the data (via *Mix* and *Mix*$^{-1}$), saving around 300 GEs. When using standard flip flops for state and key memory, 4 290 GEs are required for this implementation (*64DP_FF_1cycle*) as listed in Table 2. Using scan-chain flip flops connected as 8-bit shift registers reduces the area to 3 835 GEs (*64DP_SC-FF_1cycle*). The variant with standard flip flops leads to a similar size than the implementation of Lim *et al.* , whereas our module also contains already an 8-bit AMBA interface for communication. An overview of the datapath with shift registers is provided in Figure 5.

A first optimization that can be applied concerns the key schedule. Three shift operations are used by the key schedule: *Shift$_0$* derives the initial state of the key register for decryption operation from the cipher key, *Shift$_1$* and *Shift$_2$* are used for updating the key register after each round for encryption and decryption, respectively. Instead of using *Shift$_0$* , the initial state of the key register for decryption (corresponds to the last state of the key register during encryption) can also be derived by applying the shift operation

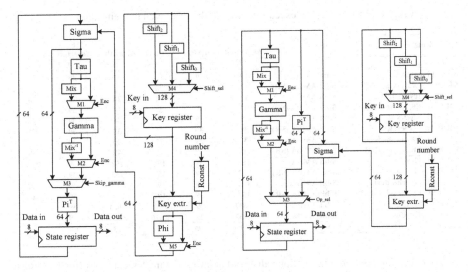

Fig. 5. Architecture of the one-cycle-per-round implementation of mCrypton (*64DP_SC-FF_1cycle*)

Fig. 6. Architecture of mCrypton with datapath split into three parts (*64DP_SC-FF_3cycles*)

Shift$_1$ that is used for updating the key register during encryption multiple times. This reduces the size of multiplexer *M4*, but increases the time for decrypting data from 13 to 24 clock cycles. The impact of this measure on the overall size of the implementation is around 50 to 100 GEs, which as rather small (cf. *64DP_FF_1cycle_drvKey* and *64DP_FF_1cycle_drvKey* in Table 2).

$$
\begin{aligned}
mCrypton &= Sigma \circ \underbrace{Gamma \circ Pi \circ Tau \circ Sigma \circ Tau \circ Pi \circ Tau}_{12\times} \\
&= Sigma \circ \underbrace{Gamma \circ Tau \circ Pi^T \circ Sigma \circ Pi^T}_{12\times} \\
&= Sigma \circ \underbrace{Gamma \circ Tau \circ Pi^T \circ Sigma \circ Pi^T}_{12\times} \\
&= \underbrace{Sigma \circ Tau \circ Gamma \circ Pi^T \circ Sigma \circ Pi^T}_{12\times}
\end{aligned}
\tag{2}
$$

Another concept to lowered the area requirement is splitting the datapath into three parts, one for computing *Sigma*, one for *Pi^T*, and one for *Tau* and *Gamma*. In that way, one round of mCrypton requires three clock cycles instead of one, but execution order of operations can be rearranged as illustrated in Equation 3 when decrypting data. By using this rearranged execution order, *Phi* transformation of the round keys is no longer required (note that *Phi* is equal to *Pi^T*) when decrypting data. Hence, not only *Phi* can be omitted but also multiplexer *M5*, saving around 400 GEs. The resulting implementations consume 3 858 GEs with standard flip flops (*64DP_FF_3cycles*) and 3 425 GEs with scan-chain flip flops (*64DP_SC-FF_3cycles*). Figure 6 sketches the architecture of

Table 1. Overview of the execution times of the mCrypton implementations

Implementation	Interface	Encryption			Decryption		
		Algorithm	Interface	Total	Algorithm	Interface	Total
-	-	[Cycles]	[Cycles]	[Cycles]	[Cycles]	[Cycles]	[Cycles]
64DP_FF_1cycle	8-bit AMBA	13	39	52	13	39	52
64DP_SC-FF_1cycle	8-bit AMBA (SR)	13	39	52	13	39	52
64DP_FF_1cycle_drvKey	8-bit AMBA	13	39	52	24	39	63
64DP_SC-FF_1cycle_drvKey	8-bit AMBA (SR)	13	39	52	24	39	63
64DP_FF_3cycles	8-bit AMBA	38	39	77	38	39	77
64DP_SC-FF_3cycles	8-bit AMBA (SR)	38	39	77	38	39	77
64DP_FF_3cycles_drvKey	8-bit AMBA	38	39	77	49	39	88
64DP_SC-FF_3cycles_drvKey	8-bit AMBA (SR)	38	39	77	49	39	88
16DP_FF_12cycles	8-bit AMBA	152	39	191	152	39	191
16DP_SC-FF_12cycles	8-bit AMBA (SR)	152	39	191	152	39	191
16DP_FF_12cycles_drvKey	8-bit AMBA	152	39	191	163	39	202
16DP_SC-FF_12cycles_drvKey	8-bit AMBA (SR)	152	39	191	163	39	202
4DP_FF_207cycles	8-bit AMBA	2637	39	2676	3021	39	3060
4DP_Lh_207cycles	8-bit AMBA	2637	39	2676	3021	39	3060
Lim et al. [13], 64DP_FF_1cycle	no	13	n/a	13	13	n/a	13

Table 2. Area and power-consumption values of the mCrypton implementations (for 130 nm)

Implementation	Synthesis			Place&route			Power
	Control	Datapath	Total	Control	Datapath	Total	(at 100 kHz)
-	[GEs]	[GEs]	[GEs]	[GEs]	[GEs]	[GEs]	[µW]
64DP_FF_1cycle	97	4042	4139	97	4193	4290	1.08
64DP_SC-FF_1cycle	42	3661	3703	40	3795	3835	1.05
64DP_FF_1cycle_drvKey	125	3943	4068	126	4087	4213	1.15
64DP_SC-FF_1cycle_drvKey	110	3522	3632	71	3630	3701	0.96
64DP_FF_3cycles	128	3619	3747	128	3730	3858	0.72
64DP_SC-FF_3cycles	74	3257	3331	71	3354	3425	0.78
64DP_FF_3cycles_drvKey	161	3531	3692	163	3662	3825	0.76
64DP_SC-FF_3cycles_drvKey	110	3169	3279	109	3275	3384	0.83
16DP_FF_12cycles	216	2983	3199	215	3131	3346	0.79
16DP_SC-FF_12cycles	91	2575	2666	85	2674	2759	0.68
16DP_FF_12cycles_drvKey	261	2882	3143	262	3014	3276	0.77
16DP_SC-FF_12cycles_drvKey	122	2472	2594	118	2591	2709	0.70
4DP_FF_207cycles	616	2127	2743	619	2343	2962	0.68
4DP_Lh_207cycles	585	1982	2567	587	2197	2784	0.58
Lim et al. [13], 64DP_FF_1cycle	n/a	n/a	4108	n/a	n/a	n/a	n/a

the three-cycle-per-round variant. Again, $Shift_0$ can be omitted when deriving the initial state of the key register during decryption operation by applying $Shift_1$ multiple times, saving another 50 to 100 GEs.

$$
\underbrace{Sigma(Phi(K_{12-i})) \circ Tau \circ Gamma^{-1} \circ Pi^{T}}_{i=0\cdots11} \circ Sigma(Phi(K_0)) \circ Pi^{T} =
$$
$$
Pi^{T} \circ Sigma(K_{12}) \circ \underbrace{Pi^{T} \circ Tau \circ Gamma^{-1} \circ Sigma(K_{12-i})}_{i=1\cdots12} \tag{3}
$$

The highly regular structure of mCrypton allows to independently handle a single row of the state (note that we use Pi^{T} that operates row wise). This property makes it rather easy to reduce the datapath width of the round function from 64 bits to 16 bits (i.e.,

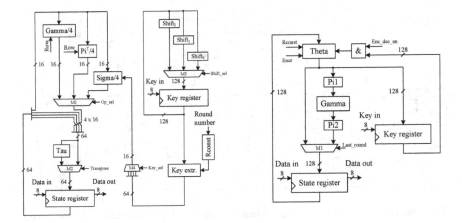

Fig. 7. Architecture of the mCrypton implementation with 16-bit datapath width (*16DP_SC-FF_12cycles*)

Fig. 8. Datapath of NOEKEON with one-cycle-per-round architecture (*128DP_SC-FF_1cycle*)

operating on a single row). Hence, only a quarter of the operations *Sigma*, *Pi^T*, and *Gamma* has to be implemented. Moreover, by implicitly shifting the data 16 bits to the right (via rewiring) within each round, no additional multiplexes are needed for selecting the proper row from the state, as illustrated in the schematic view of the architecture in Figure 7. Only multiplexer *M2* for selecting between transposed and not transposed data and *M4* for selecting the proper row from the round key are required. Computing one round of mCrypton lasts 12 clock cycles, whereas each of the three operations is sequentially applied four times on the state data. This results in 152 clock cycles for encrypting and decrypting a data block, respectively (cf. Table 1). The reduced datapath architecture brings an area gain of about 600 GEs. Hence, with standard flip flops 3 346 GEs are required (*16DP_FF_12cycles*), with scan-chain flip flops connected as 8-bit shift registers only 2 759 GEs (*16DP_SC-FF_12cycles*). Deriving the initial state of the key register during decryption operation (i.e., omitting *Shift_0*) brings another gain of 50 GEs.

The minimum-datapath version operates on 4 bits of the state and on 16 bits of the key at once. Two additional registers for temporary storage are required, the 4-bit *ACC* register and the 16-bit *Temp* register. An overview of the minimal datapath is given in Figure 9. The state is stored in a RAM-like memory of 8×8 bits, and the key in a RAM-like memory of 8×16 bits. The execution time significantly increases to more than 200 cycles per round. Extracting the round key from the key register and applying *Sigma* consumes 57 cycles. *Gamma* and *Pi^T* require 32 and 96 cycles, respectively. Key-register update consumes 22 cycles for encryption and 32 cycles for decryption. For decryption operation another 264 cycles are required to derive the initial state of the key register from the cipher key. The minimum-datapath implementation has a size of 2 962 GEs when using standard flip flops (*4DP_FF_207cycles*). When using a latch-based memory for state and key memory, area requirement is lowered to 2 784 GEs

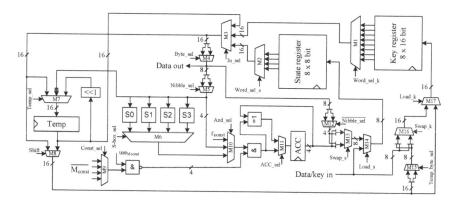

Fig. 9. Schematic overview of the minimal-datapath implementation of mCrytpton (*4DP_Lh_207cycles*)

(*4DP_Lh_207cycles*). However, even with latch-based memory the design is still slightly larger than the 12-cycle-per-round implementation. The main reason for this is the large control unit of the minimum-datapath version, which consumes around 600 GEs, nullifying all the area savings obtained by compacting the datapath.

4.2 NOEKEON

Datapath widths of 128, 32, and 8 bits have been implemented for NOEKEON. Indirect version of NOEKEON is used, but in order to shorten computation time, we have directly loaded the working key that has been derived from the cipher key into the hardware modules. This simplification can be done since the working key is independent of the processed data and constant for all rounds.

The straightforward implementation with 128-bit datapath computes a single round within one clock cycle. Hence, encrypting a data block lasts 17 clock cycles (16 rounds plus a final application of *Theta*). Decrypting data requires an additional cycle, as the decryption key K_D is first derived from the working key K. Deriving K_D has the advantage that *Theta* can be reused for decrypting data without the need for implementing *Theta⁻¹* as well. The relation between *Theta* and *Theta⁻¹* is given in Equation 4. K_D is derived from K (which is loaded first into the state register) by applying *Theta* on K with an all-zero value. The one-cycle-per-round architecture consumes 4 981 GEs with standard flip flops (*128DP_FF_1cycle*) and 4 316 GEs (*128DP_SC-FF_1cycle*) with scanchain flip flops as listed in Table 4. An overview of the datapath structure with scanchain flip flops is given in Figure 8. Around 150 GEs can be saved when externally filling the key register (i.e., via the AMBA interface) with zeros before deriving K_D (cf. *128DP_FF_1cycle_extLoad* and *128DP_SC-FF_1cycle_extLoad* in Table 4), as this allows to remove the AND gate at the output of the key register. However, externally filling the key register increases execution time by 17 clock cycles when decrypting data.

$$Theta^{-1}(K, data) = Theta(K_D, data) = Theta(Theta(0, K), data) \qquad (4)$$

Table 3. Overview of the execution times of the NOEKEON implementations

Implementation	Interface	Encryption			Decryption		
		Algorithm	Interface	Total	Algorithm	Interface	Total
-	-	[Cycles]	[Cycles]	[Cycles]	[Cycles]	[Cycles]	[Cycles]
128DP_FF_1cycle	8-bit AMBA	17	55	72	18	55	73
128DP_SC-FF_1cycle	8-bit AMBA (SR)	17	55	72	18	55	73
128DP_FF_1cycle_extLoad	8-bit AMBA	17	55	72	18	72	90
128DP_SC-FF_1cycle_extLoad	8-bit AMBA (SR)	17	55	72	18	72	90
128DP_FF-Lh_6cycles	8-bit AMBA	99	55	154	99	55	154
128DP_SC-FF_6cycles	8-bit AMBA (SR)	99	55	154	99	55	154
128DP_FF-Lh_9cycles	8-bit AMBA	148	55	203	148	55	203
128DP_SC-FF_9cycles	8-bit AMBA (SR)	148	55	203	148	55	203
32DP_FF-Lh_24cycles	8-bit AMBA	393	55	448	393	55	448
32DP_SC-FF_24cycles	8-bit AMBA (SR)	393	55	448	393	55	448
8DP_FF_225cycles	8-bit AMBA	3 669	55	3 724	3 669	55	3 724
8DP_Lh_225cycles	8-bit AMBA	3 669	55	3 724	3 669	55	3 724

Table 4. Area and power-consumption values of the NOEKEON implementations (for 130 nm)

Implementation	Synthesis			Place&route			Power
	Control	Datapath	Total	Control	Datapath	Total	(at 100 kHz)
-	[GEs]	[GEs]	[GEs]	[GEs]	[GEs]	[GEs]	[µW]
128DP_FF_1cycle	150	4 719	4 869	149	4 832	4 981	0.93
128DP_SC-FF_1cycle	54	4 177	4 231	51	4 265	4 316	1.28
128DP_FF_1cycle_extLoad	151	4 588	4 739	149	4 699	4 848	0.87
128DP_SC-FF_1cycle_extLoad	55	4 046	4 101	51	4 115	4 166	1.19
128DP_FF-Lh_6cycles	167	4 020	4 187	171	4 173	4 344	1.41
128DP_SC-FF_6cycles	98	3 668	3 766	101	3 799	3 900	1.46
128DP_FF-Lh_9cycles	197	3 740	3 937	198	3 964	4 162	1.43
128DP_SC-FF_9cycles	128	3 396	3 524	128	3 517	3 645	1.25
32DP_FF-Lh_24cycles	311	3 106	3 417	313	3 323	3 636	0.94
32DP_SC-FF_24cycles	221	3 040	3 261	224	3 162	3 386	0.96
8DP_FF_225cycles	511	2 386	2 897	513	2 520	3 033	0.78
8DP_Lh_225cycles	487	2 137	2 624	489	2 391	2 880	0.67

Having a detailed look at the functions *Theta* and *Gamma* shows that they consist of three sub parts, which we call $Theta_1$, $Theta_2$, $Theta_3$, $Gamma_1$, $Gamma_2$, and $Gamma_3$. $Gamma_1$ and $Gamma_3$ are equal, $Theta_1$ and $Theta_3$ provide the same functionality but operate on different data (cf. [4]). Hence, by sequentially computing $Theta_i$ and $Gamma_i$, redundances in the datapath can be removed, leading to a more compact implementation. Another advantage of computing *Theta* sequentially is that $Theta^{-1}$ can be easily realized with K instead of K_D (i.e., computing K_D is no longer required) by reversing the execution order of the sub parts as shown in Equation 5. In that way, computing one round of NOEKEON needs 6 clock cycles, reducing the area by about 500 GEs. The version with standard flip flops for the state memory and latches for the key memory (latches can be used as the key is only loaded once at the beginning in 8-bit blocks) consumes 4 344 GEs (*128DP_FF-Lh_6cycles*). When using scan-chain flip flops for both state and key memory, 3 900 GEs are needed (*128DP_SC-FF_6cycles*). The datapath of this version is sketched in Figure 10. Dividing the execution of a single

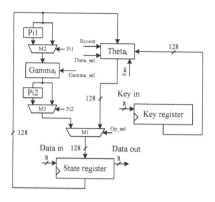

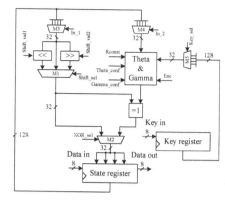

Fig. 10. Datapath of the NOEKEON implementation with *Theta* and *Gamma* split into tree sub parts (*128DP_SC-FF_6cycles*)

Fig. 11. Overview of NOEKEON architecture with 32-bit datapath width (*2DP_SC-FF_24cycles*)

round even further from 6 into 9 parts (i.e., 9 cycles per round are used as stated in Table 3) brings another area gain of more than 200 GEs.

$$Theta_1 \circ Theta_2(K_D) \circ Theta_3 = Theta_3 \circ Theta_2(K) \circ Theta_1 \tag{5}$$

In the next step we have reduced the datapath width from 128 to 32 bits, which increases the execution time to 24 cycles per round (9 for *Theta*, 3 for *Pi1*, 9 for Gamma, and 3 for *Pi2*), but saving again 250 to 500 GEs. *Theta* and *Gamma* have been implemented in a combined function as both use basically XOR operations that can be better reused in this way. Two input multiplexers (*M3* and *M4*) are used to independently select values for the combined *Theta-Gamma* function and shifting of data. With the 32-bit architecture, NOEKEON can be implemented with 3 636 GEs using standard flip flops for the state memory and latches for the key memory (*32DP_FF-Lh_24cycles*). Implementing both memories with scan-chain flip flops connected as 8-bit shift registers leads to an area requirement of only 3 386 GEs (*32DP_SC-FF_24cycles*). An overview of the datapath with scan-chain flip flops for state and key memory is depicted in Figure 11.

The minimum-datapath implementation of NOEKEON operates on 8 bits of the state/key at once. The resulting design can be seen as a very simplified sequencer circuit with a highly-optimized instruction set. Temporary results are placed in an 8-bit accumulator (ACC) register. The sequences that need to be executed are hard coded within the state machine. Writing to the state memory is only done in blocks of 8 bits. This has allowed us to also use a latch-based register file (16×8-bit) for the state memory. Utilizing latches instead of flip flops for both key memory and state memory results in about 150 GEs of area savings. A schematic overview of the minimum datapath is given in Figure 12. Bit shifts that are used within the functions *Pi1* and *Pi2* are only done on a bit basis to avoid additional multiplexers. Other components of the datapath are OR, AND, and XOR operations (each 8-bit wide) as well as several multiplexers. Executing one round requires 225 clock cycles (69 for *Theta*, 40 for *Pi1*, 76 for Gamma, and 40 for *Pi2*). The implementation of NOEKEON with an 8-bit datapath is very compact and

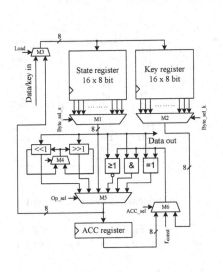

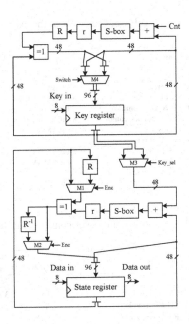

Fig. 12. Schematic overview of the NOEKEON implementation with 8-bit datapath width (*8DP_Lh_225cycles*)

Fig. 13. Datapath of the one-cycle-per-round architecture of SEA$_{96,8}$ (*48DP_SC-FF_1cycle*) according to [14]

requires only 3 033 GEs with standard flip flops (*8DP_FF_225cycles*) and 2 880 GEs with latch-based memory (*8DP_Lh_225cycles*). In contrast to mCrypton, using a minimum datapath brings a significant area gain. The main reason for this is that NOEKEON has no highly regular structure (e.g., columns/rows can be computed independently) that would allow implementations with larger datapath width to implicitly select data by simply rewiring instead of using extra multiplexers.

4.3 SEA

We have used the one-cycle-per-round architecture proposed by Macè *et al.* [14] as starting point for the full-size datapath version. Hence, only 93 clock cycles are required for encrypting or decrypting a data block (cf. Table 5). Additionally, an 8-bit AMBA interface has been added as for all other implementations. Since we have synthesized our design for a lower target clock frequency, the area requirement of our implementation is only 3 445 GEs with standard flip flops (*48DP_FF_1cycle*) and 2 854 GEs when using scan-chain flip flops (*48DP_SC-FF_1cycle*)—compared to 4 313 GEs without any interface in [14]. Using scan-chain flip flops has significantly lowered the area requirement of the implementation (by more then 600 GEs). The datapath of the one-cycle-per-round architecture with scan-chain flip flops is depicted in Figure 13.

The first optimization measure that seems to be obvious when looking at the structure of SEA is using a combined datapath for both data and round-key computation. By sequentially computing data and round key, execution time is roughly doubled, but

Table 5. Overview of the execution times of the SEA implementations

Implementation	Interface	Encryption			Decryption		
		Algorithm	Interface	Total	Algorithm	Interface	Total
-	-	[Cycles]	[Cycles]	[Cycles]	[Cycles]	[Cycles]	[Cycles]
48DP_FF_1cycle	8-bit AMBA	93	43	136	93	43	136
48DP_SC-FF_1cycle	8-bit AMBA (SR)	93	43	136	93	43	136
48DP_FF_2cycles_comb	8-bit AMBA	185	43	228	185	43	228
48DP_SC-FF_2cycles_comb	8-bit AMBA (SR)	185	43	228	185	43	228
24DP_FF_6cycles	8-bit AMBA	554	43	597	554	43	597
24DP_SC-FF_6cycles	8-bit AMBA (SR)	554	43	597	554	43	597
8DP_FF_31cycles	8-bit AMBA	2871	43	2914	2871	43	2914
8DP_Lh_31cycles	8-bit AMBA	2871	43	2914	2871	43	2914
Macè et al. [14], 48DP_FF_1cycle	no	93	n/a	93	93	n/a	93
Macè et al. [14], 8DP_FF_15cycles	8-bit	1395	33	1428	1395	33	1428

Table 6. Area and power-consumption values of the SEA implementations (for 130 nm)

Implementation	Synthesis			Place&route			Power
	Control	Datapath	Total	Control	Datapath	Total	(at 100 kHz)
-	[GEs]	[GEs]	[GEs]	[GEs]	[GEs]	[GEs]	[μW]
48DP_FF_1cycle	125	3182	3307	128	3317	3445	1.35
48DP_SC-FF_1cycle	78	2682	2760	76	2778	2854	1.16
48DP_FF_2cycles_comb	122	3120	3242	119	3304	3423	1.71
48DP_SC-FF_2cycles_comb	77	2886	2963	71	3002	3073	1.62
24DP_FF_6cycles	180	2701	2881	183	2856	3039	1.39
24DP_SC-FF_6cycles	123	2444	2567	125	2539	2664	1.22
8DP_FF_31cycles	587	1960	2547	593	2161	2754	0.85
8DP_Lh_31cycles	564	1776	2340	569	1993	2562	0.64
Macè et al. [14], 48DP_FF_1cycle	n/a	n/a	3758	n/a	n/a	4313	2.04
Macè et al. [14], 8DP_FF_15cycles	n/a	n/a	3925	n/a	n/a	4472	19.24

hardware components for modular addition, S-box operation, and XOR computation can be shared (i.e., implemented only once). However, our implementation results show that the additional multiplexers ($M5$, $M6$, $M7$) that are introduced by combining the datapath of data and round-key computation nearly nullify the area gain or even lead to a larger design. An overview of the combined datapath architecture is given in Figure 14.

Due to the regular structure of SEA, the combined datapath version mentioned above can be adapted to an implementation with a datapath width of only 24 bits instead of 48 bits. A schematic view of the architecture of this adapted version is provided in Figure 15. The implementation requires no additional multiplexers (note that the size of $M7$ is halved compensating the introduction of $M8$). Switching of higher and lower parts of left/right side of state/key is done implicitly by rewiring. Modular addition, S-box operation, bit rotation, and XOR computation can be independently applied on 24-bit blocks of data. This allows reducing the area requirements to 3039 GEs with standard flip flops (24DP_FF_6cycles) and 2664 GEs with scan-chain flip flops (24DP_SC-FF_6cycles) as listed in Table 6. Computing one round of SEA requires 6 clock cycles with this version. In the first cycle, R is applied on the left side of the state, followed by XORing it with the lower part of the computed right side. In the second cycle the higher part of the right side is XORed with the higher part of the left side. The round-key computation consumes 4 cycles as we have to use a small trick for the word rotation. Looking on

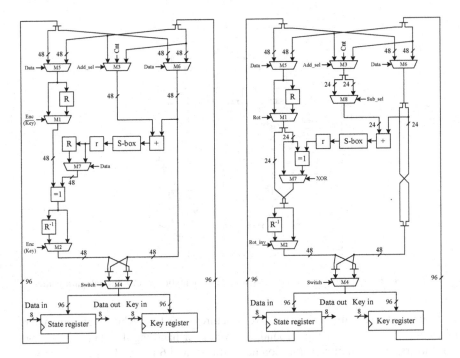

Fig. 14. Architecture of the SEA$_{96,8}$ implementation with combined 48-bit datapath (*48DP_SC-FF_2cycles_comb*)

Fig. 15. Architecture of the SEA$_{96,8}$ implementation with combined 24-bit datapath (*24DP_SC-FF_6cycles*)

the datapath of the round-key computation part in Figure 13 shows that a word rotation is applied on the right side before XORing it with the left side. As the word rotation operates on 48 bits, a simple reduction of the round-key datapath to 24 bits is on first sight not possible. We solve this issue by using the relation given in Equation 6. First, an inverse word rotation is applied on the left side of the key in the third cycle (replaces the word rotate on the right side). Now, lower and higher part of the right part are XORed in a straightforward way in cycles four and five with the left part. In the sixth cycle, a word rotation is applied on left part, finalizing the round-key computation. The change of the command sequence results in an execution time of 185 clock cycles for encrypting or decrypting one block of data (without interface communication).

$$Key_{Left,i+1} = Key_{Left,i} \oplus R(r(Sbox(Key_{Right,i} + Cnt)))$$
$$= R(R^{-1}(Key_{Left,i}) \oplus r(Sbox(Key_{Right,i} + Cnt))) \qquad (6)$$

By further reducing the datapath size to 1 byte we have obtained the minimum-datapath version of SEA. The architecture of the minimum-datapath version is shown in Figure 16 and differs from the proposal presented in [14] (which sketches only the datapath but provides no stand-alone implementation). Three temporary registers T0, T1, and T2 are used to buffer the input data for the 8 S-boxes (since each S-box operates on bits from three different bytes). T0 serves also as accumulator register. State as well

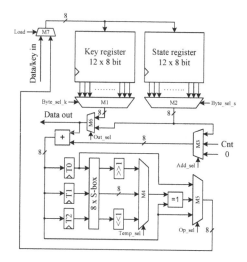

Fig. 16. Schematic overview of the SEA$_{96,8}$ implementation with 8-bit datapath width

as key memory are realized as RAM-like structures with 12×8 bits. Our approach requires only 31 clock cycles per round, 19 cycles for handling the state data (7 for a word/inverse word rotation, 6 for modular addition, and 6 for XORing) and 12 cycles for updating the round key (6 for modular addition, and 6 for XORing). Hence, a full encryption/ecryption process requires 2 871 GEs as stated in Table 5. The round-key update is faster as the word rotation is done implicitly during XOR operation by proper addressing. The minimum-datapath proposal in [14] for comparison requires 50 clock cycles per round, which is significantly slower than our implementation. When using a flip-flop-based memory, a chip size of 2 754 GEs is achieved (*8DP_FF_31cycles*). Interestingly, this version is only 100 GEs smaller than the one-cycle-per-round implementation with shift registers, and still larger than the combined variant with 24-bit datapath. Main reasons are the large output multiplexers (*M1* and *M2*) of state and key memory and the significantly increased size of the controller. By replacing the flip-flop-based memory with latches (*8DP_Lh_31cycles*), the area requirement is further reduced to 2 562 GEs, making it the by far smallest stand-alone hardware implementation of SEA$_{96,8}$ currently available.

5   Summary of Implementation Results

In this section we summarize in short the achieved results of our block-cipher implementations. All hardware modules support encryption and decryption functionality and contain already an 8-bit AMBA interface for communication. In order to lower the area requirements of the implementations, optimization techniques on different levels (e.g., algorithmic or architectural level) are applied. For selected implementation

Table 7. Comparison of performance values of selected implementation variants for different CMOS process technologies

Block cipher	Implementation	CMOS techn.	Area (after P&R)	Energy cons.	Throughput (at 100 kHz)	Energy-per-bit	A-T product
		[nm]	[GEs]	[nJ]	[kbits/s]	[pJ]	[k₁ᵃ]
mCrypton	64DP_SC-FF_1cycle	130	3835	0.54	123.08	8	31
	64DP_SC-FF_3cycles		3425	0.60	83.12	9	41
	16DP_SC-FF_12cycles_drvKey		2709	1.35	33.51	21	81
	4DP_Lh_207cycles		2784	15.55	2.39	243	1164
	64DP_SC-FF_1cycle	180	3979	1.85	123.08	29	32
	64DP_SC-FF_3cycles		3593	2.27	83.12	35	43
	16DP_SC-FF_12cycles_drvKey		2760	4.76	33.51	74	82
	4DP_Lh_207cycles		2769	49.37	2.39	771	1158
	64DP_SC-FF_1cycle	350	3694	11.13	123.08	174	30
	64DP_SC-FF_3cycles		3370	13.12	83.12	205	41
	16DP_SC-FF_12cycles_drvKey		2594	28.38	33.51	443	77
	4DP_Lh_207cycles		2534	330.75	2.39	5168	1060
	Lim et al. [13], 64DP_FF_1cycle	130	4108	-	492.31ᵇ	-	8ᵇ
NOEKEON	128DP_FF_1cycle	130	4981	0.67	177.78	5	28
	128DP_SC-FF_6cycles		3900	2.25	83.12	18	47
	32DP_SC-FF_24cycles		3386	4.30	28.57	34	119
	8DP_Lh_225cycles		2880	25.04	3.44	196	838
	128DP_FF_1cycle	180	5427	2.72	177.78	21	31
	128DP_SC-FF_6cycles		4170	8.74	83.12	68	50
	32DP_SC-FF_24cycles		3692	15.90	28.57	124	129
	8DP_Lh_225cycles		2862	69.19	3.44	541	833
	128DP_FF_1cycle	350	4597	18.42	177.78	144	26
	128DP_SC-FF_6cycles		3813	54.12	83.12	423	46
	32DP_SC-FF_24cycles		3376	98.92	28.57	773	118
	8DP_Lh_225cycles		2604	459.91	3.44	3593	758
SEA	48DP_SC-FF_1cycle	130	2854	1.57	70.59	16	40
	48DP_SC-FF_2cycles		3073	3.69	42.11	38	73
	24DP_SC-FF_6cycles		2664	7.28	16.08	76	166
	8DP_Lh_31cycles		2562	18.74	3.29	195	778
	48DP_SC-FF_1cycle	180	2941	5.77	70.59	60	42
	48DP_SC-FF_2cycles		3151	14.46	42.11	151	75
	24DP_SC-FF_6cycles		2714	26.84	16.08	280	169
	8DP_Lh_31cycles		2569	55.31	3.29	576	780
	48DP_SC-FF_1cycle	350	2679	31.35	70.59	327	38
	48DP_SC-FF_2cycles		2845	82.22	42.11	856	68
	24DP_SC-FF_6cycles		2492	137.49	16.08	1432	155
	8DP_Lh_31cycles		2300	361.34	3.29	3764	698
	Macè et al. [14], 48DP_FF_1cycle	130	4313	-	103.23ᵇ	-	42ᵇ
	Macè et al. [14], 8DP_FF_15cycles		4472	-	6.72	-	665
AES-128	Feldhofer et al. [8]	350	3400	-	12.40	-	274
	Hämäläinen et al. [9]	130	3875ᶜ	-	80.00ᵇ	-	48ᵇ
XTEA	Feldhofer et al. [7]	350	2636	-	9.08	-	290
HIGHT	Hong et al. [10]	250	3048	-	200.00ᵇ	-	15ᵇ

variants we also provide results for a 180 nm CMOS technology from Faraday [6] and a 350 nm CMOS technology from AMS [1]. The results are listed in Table 7. Note that all results are obtained by using a common tool chain, which eases comparability. Additionally, computed performance values such as energy consumption, energy-per-bit values, throughput, and size-time (AT) product are given. As expected, energy consumption significantly decreases when feature size shrinks. Size of the implementations in terms of GEs is also influenced by the CMOS technology. Using the 350 nm technology leads to even more-compact implementations (between 2 300 and 2 603 GEs for

---

ᵃ k_1 is given in GEs × s/kbits (smaller is better).

ᵇ Pure execution time of the algorithm without communication overhead of an interface.

ᶜ Estimated value when supporting encryption and decryption functionality (according to [9]).

the most-compact algorithm implementations), whereas the 180 nm technology results in slightly larger GE values (with respect to the 130 nm technology). We also compare area, throughput and AT product with other low-resource block ciphers that have a similar security level (i.e., key size) and implement encryption and decryption functionality. Energy consumption and energy-per-bit values are omitted for implementations from related work as a fair comparison (e.g., different supply voltage or power-simulation tools) is hardly possible.

Comparing our results with related work clearly illustrates that our area-optimized hardware modules are much more compact than currently published implementations of these algorithms. In case of NOEKEON we even provide the first stand-alone hardware implementations at all. SEA is the algorithm with the lowest area requirement. Although SEA and mCrypton have the same memory footprint (i.e., number of bits needed to store key and state), SEA is *cheaper* in hardware. mCrypton has a higher throughput as SEA since it uses less rounds. Note that the throughput of SEA could be further improved when encrypting/depcrypting multiple data blocks, as reloading the key is not necessary (due to the symmetry property of SEA). NOEKEON on the other hand has the highest throughput as it operates on the largest block size and uses also a rather small number of rounds.

Another important observation is that using a shift-register based interface with scan-chain flip flops can significantly lower the area requirements of a circuit. As we have seen for mCrypton and SEA that have both a rather regular structure, implementations with shift-registers and larger datapath width can even be smaller than minimum-datapath versions with rather long execution times and much higher energy consumption. Especially for embedded devices like sensor-enabled RFID tags that are powered by a battery, energy consumption is an important criterion. Last but not least, using latch-based memory if possible is advantageous as it lowers not only the area requirement, but also the power consumption of the cryptographic module.

6 Conclusions

In this work we presented hardware implementations of the block ciphers mCrypton, NOEKEON, and SEA with focus on low-area design. Implementations with different datapath widths and various optimization levels are provided. All hardware modules contain already an 8-bit interface for communication. We further show that using a shift-register based approach with scan-chain flip flops is advantageous in terms of chip size for realizing the interface. Another observation is that replacing flip-flop-based memory by latches can lower both area requirement and power consumption. Our implementations are the most-compact one of these algorithms that are available so far. When using a 130 nm CMOS technology, SEA requires only 2 562 GEs, followed by mCrypton and NOEKEON with 2 709 and 2 880 GEs, respectively. In case of NOEKEON we even provide the first results of stand-alone hardware implementations.

Acknowledgements. This work has been supported by the European Commission through the ICT program ECRYPT II (European Network of Excellence for Cryptology II) under contract ICT-2007-216676 and through the ICT program TAMPRES (Tamper Resistant Sensor Node) under contract ICT-SEC-2009-5-258754.

References

[1] AMS. Standard Cell Library 0.35μm CMOS (C35), http://asic.ams.com

[2] ARM Ltd. AMBA Advanced Microcontroller Bus Architecture Specification, http://www.arm.com (1997)

[3] Atmel Corporation. 8-bit AVR Microcontroller with 128K Bytes In-System Programmable Flash (August 2007), http://www.atmel.com

[4] Daemen, J., Peeters, M., Assche, G.V., Rijmen, V.: Nessie proposal: NOEKEON (2000), http://gro.noekeon.org/Noekeon-spec.pdf

[5] Faraday Technology Corporation. Faraday FSA0A_C 0.13 μm ASIC Standard Cell Library (2004), http://www.faraday-tech.com

[6] Faraday Technology Corporation. Faraday FSA0A_C 0.18 μm ASIC Standard Cell Library (2004), http://www.faraday-tech.com

[7] Feldhofer, M., Wolkerstorfer, J.: Hardware Implementation of Symmetric Algorithms for RFID Security. In: RFID Security: Techniques, Protocols and System-On-Chip Design, pp. 373–415. Springer (2008)

[8] Feldhofer, M., Wolkerstorfer, J., Rijmen, V.: AES Implementation on a Grain of Sand. IEEE Proceedings on Information Security 152(1), 13–20 (2005)

[9] Hämäläinen, P., Alho, T., Hännikäinen, M., Hämäläinen, T.D.: Design and Implementation of Low-Area and Low-Power AES Encryption Hardware Core. In: Conference on Digital System Design (DSD 2006), pp. 577–583. IEEE (September 2006)

[10] Hong, D., Sung, J., Hong, S., Lim, J., Lee, S., Koo, B.-S., Lee, C., Chang, D., Lee, J., Jeong, K., Kim, H., Kim, J., Chee, S.: HIGHT: A New Block Cipher Suitable for Low-Resource Device. In: Goubin, L., Matsui, M. (eds.) CHES 2006. LNCS, vol. 4249, pp. 46–59. Springer, Heidelberg (2006)

[11] Knudsen, L.R., Raddum, H.: On Noekeon NES/DOC/UIB/WP3/009/1 (2001), https://www.cosic.esat.kuleuven.be/nessie/reports/phase1/uibwp3-009.pdf

[12] Kumar, K.J., Salivahanan, S., Reddy, K.C.K.: Implementation of Low Power Scalable Encryption Algorithm. International Journal of Computer Applications 11(1), 14–18 (2010)

[13] Lim, C.H., Korkishko, T.: mCrypton – A Lightweight Block Cipher for Security of Low-Cost RFID Tags and Sensors. In: Song, J.-S., Kwon, T., Yung, M. (eds.) WISA 2005. LNCS, vol. 3786, pp. 243–258. Springer, Heidelberg (2006), doi:10.1007/11604938

[14] Mace, F., Standaert, F.-X., Quisquater, J.-J.: ASIC Implementations of the Block Cipher SEA for Constrained Applications. In: Workshop on RFID Security (RFIDSec 2007), pp. 103–114 (2007)

[15] Mala, H., Dakhilalian, M., Shakiba, M.: Cryptanalysis of mCrypton - A lightweight block cipher for security of RFID tags and sensors. International Journal of Communication Systems 25(4), 415–426 (2012)

[16] Moradi, A., Poschmann, A., Ling, S., Paar, C., Wang, H.: Pushing the Limits: A Very Compact and a Threshold Implementation of AES. In: Paterson, K.G. (ed.) EUROCRYPT 2011. LNCS, vol. 6632, pp. 69–88. Springer, Heidelberg (2011)

[17] Standaert, F.-X., Piret, G., Gershenfeld, N., Quisquater, J.-J.: SEA: A Scalable Encryption Algorithm for Small Embedded Applications. In: Domingo-Ferrer, J., Posegga, J., Schreckling, D. (eds.) CARDIS 2006. LNCS, vol. 3928, pp. 222–236. Springer, Heidelberg (2006)

[18] Texas Instruments. MSP430C11x1 - Mixed Signal Microcontroller (2008), http://focus.ti.com

[19] Z'aba, M.R., Raddum, H., Henricksen, M., Dawson, E.: Bit-Pattern Based Integral Attack. In: Nyberg, K. (ed.) FSE 2008. LNCS, vol. 5086, pp. 363–381. Springer, Heidelberg (2008)

Efficient Arithmetic on Elliptic Curves in Characteristic 2

David Kohel

Institut de Mathématiques de Luminy
Université d'Aix-Marseille
163, avenue de Luminy, Case 907
13288 Marseille Cedex 9
France

Abstract. We present normal forms for elliptic curves over a field of characteristic 2 analogous to Edwards normal form, and determine bases of addition laws, which provide strikingly simple expressions for the group law. We deduce efficient algorithms for point addition and scalar multiplication on these forms. The resulting algorithms apply to any elliptic curve over a field of characteristic 2 with a 4-torsion point, via an isomorphism with one of the normal forms. We deduce algorithms for duplication in time $2\mathbf{M} + 5\mathbf{S} + 2\mathbf{m}_c$ and for addition of points in time $7\mathbf{M} + 2\mathbf{S}$, where $\mathbf{M}$ is the cost of multiplication, $\mathbf{S}$ the cost of squaring, and $\mathbf{m}_c$ the cost of multiplication by a constant. By a study of the Kummer curves $\mathscr{K} = E/\{[\pm 1]\}$, we develop an algorithm for scalar multiplication with point recovery which computes the multiple of a point P with $4\mathbf{M} + 4\mathbf{S} + 2\mathbf{m}_c + \mathbf{m}_t$ per bit where $\mathbf{m}_t$ is multiplication by a constant that depends on P.

1 Introduction

The last five years have seen significant improvements in the efficiency of known algorithms for arithmetic on elliptic curves, spurred by the introduction of the Edwards model [11] and its analysis [1,2,13]. Previously, it had been recognized that alternative models of elliptic curves could admit efficient arithmetic [8], but the fastest algorithms could be represented in terms of functions on elliptic curves embedded in $\mathbb{P}^2$ as Weierstrass models.

Among the best alternative models one finds a common property of symmetry. They admit a large number of (projective) linear automorphisms, often given by signed or scaled coordinate permutations. An elliptic curve with j-invariant $j \neq 0, 12^3$ admits only $\{[\pm 1]\}$ as automorphism group *fixing the identity element*. However, as a genus 1 curve, it also admits translations by rational points, and a translation morphism $\tau_Q(P) = P + Q$ on E is projectively linear, i.e. induced by a linear transformation of the ambient projective space, if and only if E is a degree n model determined by a complete linear system in $\mathbb{P}^{n-1}$ and Q is in the n-torsion subgroup. As a consequence the principal models of cryptographic interest are elliptic curves in $\mathbb{P}^2$ with rational 3-torsion points (e.g. the Hessian

S. Galbraith and M. Nandi (Eds.): INDOCRYPT 2012, LNCS 7668, pp. 378–398, 2012.

models) and in $\mathbb{P}^3$ with 2-torsion or 4-torsion points (e.g. the Jacobi quadratic intersections and Edwards model), and unfortunately, the latter models do not have good reduction to characteristic 2. The present work aims to fill this gap.

A rough combinatorial explanation for the role of symmetry in efficiency is the following. Suppose that the sum of $x = (x_0 : \cdots : x_r)$ and $y = (y_0 : \cdots : y_r)$ is expressed by polynomials $(p_0(x,y) : \cdots : p_r(x,y))$ of low bidegree, say $(2,2)$, in x_i and y_j. Such polynomials form a finite dimensional space. A translation morphism τ given by scaled coordinate transformation on E determines a new tuple $(p_0(\tau(x), \tau^{-1}(y)) : \ldots, p_r(\tau(x), \tau^{-1}(y)))$. If $(p_0(x,y) : \cdots : p_r(x,y))$ is an eigenvector for this transformation then it tends to have few monomials. In the case of Hessian, Jacobi, Edwards, and similar models, there exist bases of eigenvector polynomial addition laws such that the p_j achieve the minimal value of two terms.

Section 2 recalls several results, observations, and conclusions of Kohel [17] on symmetries of elliptic curves in their embeddings. As illustration, Section 3 recalls the main properties of the Edwards model as introduced by Edwards [11], reformulated by Bernstein and Lange [1] with twists by Bernstein et al. [2], and properties of its arithmetic described in Hisil et al. [13] and Bernstein and Lange [3].

This background motivates the introduction and classification of new models for elliptic curves in Section 4, based on imitation of the desired properties of Edwards curves, and in Section 5 we present new elliptic curve models, the **Z/4Z**-normal form and the split **μ_4**-normal form, which satisfy these properties. In Section 6 we classify all symmetric quartic elliptic curves in $\mathbb{P}^3$ with a rational 4-torsion point, up to projective linear isomorphism.[1] In particular we prove that any such curve is linearly isomorphic to one of these two models. In Section 7 we determine the polynomial addition laws and resulting complexity for arithmetic on these forms. Finally Section 8 develops models for the Kummer curve $\mathscr{K} = E/\{[\pm 1]\}$ and exploits an embedding of E in $\mathscr{K}^2$ in order to develop a Montgomery ladder for scalar multiplication with point recovery. Section 9 summarizes the new complexity results for these models in comparison with previously known models and algorithms. An appendix gives the addition laws for a descended μ_4-normal form that allows us to save on multiplications by constants involved in the curve equation.

Notation

In what follows we use **M** and **S** for the complexity of multiplication and squaring, respectively, in the field k, and $\mathbf{m}_c$ for a multiplication by a fixed (possibly small) constant c (or constants c_i).[2] For the purposes of complexity analysis we ignore field additions.

---

[1] Note that any quartic plane model has a canonical extension to a nonsingular quartic model in $\mathbb{P}^3$ by extending to a complete linear system.

[2] When the small constant is a bounded power of a fixed constant we omit the squarings or products entailed in its construction and continue to consider $c^{O(1)}$ a fixed constant.

When describing a morphism $\varphi : X \to Y$ given by polynomial maps, we write

$$\varphi(x) = \begin{cases} \big(p_{1,0}(x) : \cdots : p_{1,n}(x)\big), \\ \qquad\qquad \vdots \\ \big(p_{m,0}(x) : \cdots : p_{m,n}(x)\big), \end{cases}$$

to indicate that each of the tuples of polynomials $(p_{i,0}(x), \ldots, p_{i,n}(x))$ defines the morphism on an open neighborhood $U_i \subset X$, namely on the complement of the common zeros $p_{i,0}(x) = \cdots p_{i,m}(x) = 0$, that any two agree on the intersections $U_i \cap U_j$, and that the union of the U_i is all of X.

For the projective coordinate functions on $\mathbb{P}^r$, with $r > 3$, we use X_i and so $x = (X_0 : \cdots : X_r)$ represents a generic point. We also use X_i for their restriction to a curve E, in which case the X_i are defined modulo the defining ideal of E. In the product $\mathbb{P}^r \times \mathbb{P}^r$, we continue to write x for the first coordinate and use (x, y) for a generic point in $\mathbb{P}^r \times \mathbb{P}^r$, where $y = (Y_0 : \cdots : Y_r)$.

2 Elliptic Curves with Symmetries

We consider conditions for an elliptic curve embedding in $\mathbb{P}^r$ to admit many projective linear transformations, or symmetries. In what follows, we recall standard definitions and conclusions drawn from Kohel [17] (reformulated here without the language of invertible sheaves). The examples of Hessian curves and Edwards curves[3] play a pivotal role in motivating [17] and further examples (see Bernstein and Lange [3], Joye and Rezaeian Farashahi [14], Kohel [17, Section 8]) suggest that such symmetries go hand-in-hand with efficient forms for their arithmetic.[4]

The automorphism group of an elliptic curve E is a finite group, and if $j(E) \neq 0, 12^3$, this group is $\{[\pm 1]\}$. Inspection of standard projective models for elliptic curves shows that the symmetry group can be much greater. The disparity is explained by the existence of subgroups of rational torsion. The automorphism group of an elliptic curve is defined to be the automorphisms of the curve which fix the identity point, which does not include translations. For any rational torsion point T, the translation-by-T map τ_T is an automorphism of the curve, which may give rise to the additional symmetries.

We restrict to models of elliptic curves given by complete linear systems of a given degree d. Basically, such a curve is defined by $E \subset \mathbb{P}^r$ such that $r = d - 1$, E is not contained in any hyperplane, and any hyperplane H intersects E in exactly d points, counted with multiplicities. For embedding degree 3, such a

---

[3] In particular my discussions of symmetries with Bernstein and Lange motivated a study of symmetries in the unpublished work [5] (see the EFD [6]) on twisted Hessian curves, picked up by Joye and Rezaeian Farashahi [14] after posting to the EFD). This further led the author to develop a general framework for symmetries and to classify the linear action of torsion in [17].

[4] By efficient forms, we mean sparse polynomials expressions with small coefficients. These may or may not yield the most efficient *algorithms*, as seen in comparing the evaluation of similarly sparse addition laws for the Edwards and $\mathbf{Z}/4\mathbf{Z}$-normal forms.

curve is given by a single homogeneous form $F(X, Y, Z)$ of degree 3, and for degree 4 we have an intersection of two quadrics in $\mathbb{P}^3$. Quartic plane models formally lie outside of this scope — they are neither nonsingular nor given by a complete linear system — but determine a unique degree 4 elliptic curve in $\mathbb{P}^3$ after completing the basis of functions. As in the case of the Edwards curve, we always pass to this model to apply the theory.

Definition 1. *Let $E \subset \mathbb{P}^r$ be an elliptic curve embedded with respect to a complete linear system. We say that E is a symmetric model if $[-1]$ is induced by a projective linear transformation of $\mathbb{P}^r$.*

We next recall a classification of symmetric embeddings of elliptic curves (cf. Kohel [17, Lemma 2] for the statement in terms of invertible sheaves).

Lemma 2. *Let $E \subset \mathbb{P}^r$ be an elliptic curve over k embedded with respect to a complete linear system. There exists a point S in $E(k)$ such that for any hyperplane H in $\mathbb{P}^r$ not containing E, the set of points in the intersection $E \cap H = \{P_0, \ldots, P_r\}$, in $E(\bar{k})$, counted with multiplicity, sum to S. The model is symmetric if and only if S is in the subgroup $E[2]$ of 2-torsion points.*

Definition 3. *Let E be a degree d embedding in $\mathbb{P}^r$ with respect to a complete linear system, and let S be the point as in the previous lemma. We define the embedding divisor class of E to be $(d-1)(O) + (S)$.*

We describe here the classification of elliptic curves with projective embedding, up to *linear* isomorphism, rather than isomorphism.[5] The notion of isomorphisms given by linear transformations plays an important role in the addition laws, since such a change of variables gives an isomorphism between the respective spaces of addition laws of fixed bidegree (m, n), as described in Kohel [17, Section 7]. For a point T, we denote the translation-by-T morphism by τ_T, given by $\tau_T(P) = P + T$. We now recall the classification of symmetries which arise from the group law [17, Lemma 5].

Lemma 4. *Let $E \subset \mathbb{P}^r$ be embedded with respect to the complete linear system of degree d and let T be in $E(\bar{k})$. The translation-by-T morphism is induced by a projective linear automorphism of $\mathbb{P}^r$ if and only if $dT = O$.*

Similarly, we recall the classification of projective linear isomorphisms between curves in $\mathbb{P}^r$ (see Kohel [17, Lemma 3] for a slightly stronger formulation).

Lemma 5. *Let E_1 and E_2 be isomorphic elliptic curves embedded in $\mathbb{P}^r$ with respect to complete linear systems of the same degree d. An isomorphism $\varphi : E_1 \to E_2$ is induced by a projective linear transformation if and only if $\varphi(S_1) = S_2$, where $S_i \in E_i(k)$ determine the embedding divisor classes $(d-1)(O) + (S_i)$ of the embeddings.*

---

[5] In recent cryptographic literature, there has been a trend to refer to existence of a *birational equivalence*. In the context of elliptic curves, by definition nonsingular projective curves, this concept coincides with isomorphism, and we want to identity the subclass of isomorphisms which are linear with respect to the coordinate functions of the given embedding.

Remark. By definition, an isomorphism $\varphi : E_1 \to E_2$ of elliptic curves takes the identity of E_1 to the identity of E_2. It may be possible to define a projective linear transformation from E_1 to E_2 which does not respect the group identities (hence is not a group isomorphism).

3   Properties of the Edwards Normal Form

In this section we suppose that k is a field of characteristic different from 2. To illustrate the symmetry properties of the previous section and motivate the analogous construction in characteristic 2, we recall the principal properties of the Edwards normal form, summarizing work of Edwards [11], Hisil et al. [13], and Bernstein and Lange [3]. We follow the definitions and notation of Kohel [17], defining the twisted Edwards normal form E/k in $\mathbb{P}^3$:

$$cX_1^2 + X_2^2 = X_0^2 + dX_3^2, \ X_0X_3 = X_1X_2, \ O = (1:0:1:0).$$

Edwards Model for Elliptic Curves

In 2007, Edwards introduced a new model for elliptic curves [11], defined by the affine model

$$x^2 + y^2 = a^2(1 + z^2), \ z = xy,$$

over any field k of characteristic different from 2. The complete linear system associated to this degree 4 model has basis $\{1, x, y, z\}$ such that the image $(1 : x : y : z)$ is a nonsingular projective model in $\mathbb{P}^3$:

$$X_1^2 + X_2^2 = a^2(X_0^2 + X_3^2), \ X_0X_3 = X_1X_2,$$

with identity $O = (a : 0 : 1 : 0)$, as a family of curves over $k(a)$ We hereafter refer to this model as the split Edwards model. Bernstein and Lange [1] introduced a rescaling to descend to $k(d) = k(a^4)$, and subsequently (with Joye, Birkner, and Peters [2]) a quadratic twist by c, to define the twisted Edwards model with $O = (1 : 0 : 1 : 0)$:

$$cX_1^2 + X_2^2 = X_0^2 + dX_3^2, \ X_0X_3 = X_1X_2.$$

The twisted Edwards model in this form appears in Hisil et al. [13] (as extended Edwards coordinates), which provides the most efficient arithmetic. We next recall the principal properties of the Edwards normal form (with $c = 1$).

Symmetry Properties

1. The embedding divisor class is $3(O) + (S)$ where $S = 2T$.
2. The point $T = (1 : -1 : 0 : 0)$ is a rational 4-torsion point.
3. The translation–by–T and inverse morphisms are given by:

$$\tau_T(X_0 : X_1 : X_2 : X_3) = (X_0 : -X_2 : X_1 : -X_3),$$
$$[-1](X_0 : X_1 : X_2 : X_3) = (X_0 : -X_1 : X_2 : -X_3).$$

4. The model admits a factorization $s \circ (\pi_1 \times \pi_2)$ through $\mathbb{P}^1 \times \mathbb{P}^1$, where

$$\pi_1(X_0 : X_1 : X_2 : X_3) = \begin{cases} (X_0 : X_1), \\ (X_2 : X_3) \end{cases}, \; \pi_2(X_0 : X_1 : X_2 : X_3) = \begin{cases} (X_0 : X_2), \\ (X_1 : X_3). \end{cases}$$

and s is the Segre embedding

$$s((U_0 : U_1), (V_0 : V_1)) = (U_0 V_0 : U_1 V_0 : U_0 V_1 : U_1 V_1).$$

Remark. The linear expression for $[-1]$ implies that the embedding is symmetric. This linearity is a consequence of the form of the embedding divisor $3(O) + (S)$, in view of Lemma 2. In addition the two projections are symmetric, in the sense that they are stable under $[-1]$. This is due to the fact that the divisors $2(O) = \pi_1^*(\infty)$ and $(O) + (T) = \pi_2^*(\infty)$ are symmetric.

A Remarkable Factorization

Hisil et al. [13] discovered amazingly simple bilinear rational expressions for the affine addition laws, which can be described as a factorization of the addition laws through the isomorphic curve in $\mathbb{P}^1 \times \mathbb{P}^1$ (see Bernstein and Lange [3] for further properties). As a consequence of the symmetry of the embedding and its projections, the composition of the addition morphism $\mu : E \times E \longrightarrow E$ with each of the projections $\pi_i : E \to \mathbb{P}^1$ admits a basis of *bilinear* defining polynomials. For $\pi_1 \circ \mu \; \pi_1 \circ \mu$, respectively, we have

$$\begin{cases} (X_0 Y_0 + d X_3 Y_3, \; X_1 Y_1 + X_2 Y_1), \\ (c X_1 Y_1 + X_2 Y_2, \; X_0 Y_3 + X_3 Y_0) \end{cases} \text{ and } \begin{cases} (X_1 Y_2 - X_2 Y_1, \; -X_0 Y_3 + X_3 Y_0), \\ (X_0 Y_0 - d X_3 Y_3, \; -c X_1 Y_1 + X_2 Y_2) \end{cases}.$$

Addition laws given by polynomial maps of bidegree $(2,2)$ are recovered by composing with the Segre embedding. This factorization led the author to prove dimension formulas for these addition law projections and classify the exceptional divisors [17]. In particular, this permits one to prove *a priori* the form of the exceptional divisors described in Bernstein and Lange [3, Section 8], show that these addition laws span all possible addition laws of the given bidegree, and conclude their completeness.

4   Axioms for a D_4-Linear Model

The previous sections motivate the study of symmetric quartic models of elliptic curves with a rational 4-torsion point T. For such a model, we obtain a 4-dimensional linear representation of $D_4 \cong \langle [-1] \rangle \ltimes \langle \tau_T \rangle$, induced by the action on the linear automorphisms of $\mathbb{P}^3$. Here we give characterizations of elliptic curve models for which this representation is given by coordinate permutation.

Suppose that E/k is an elliptic curve with $\operatorname{char}(k) = 2$ and T a rational 4-torsion point. In view of the previous lemmas and the properties of Edwards' normal form, we consider reasonable hypotheses for a characteristic 2 analog. We note that in the Edwards model, τ_T acts by signed coordinate permutation, which we replace with a permutation action in characteristic 2.

1. The embedding of $E \to \mathbb{P}^3$ is a quadratic intersection.
2. E has a rational 4-torsion point T.
3. The group $\langle[-1]\rangle \ltimes \langle\tau_T\rangle \cong D_4$ acts by coordinate permutation, and in particular $\tau_T(X_0 : X_1 : X_2 : X_3) = (X_3 : X_0 : X_1 : X_2)$.
4. There exists a symmetric factorization of E through $\mathbb{P}^1 \times \mathbb{P}^1$.

Combining conditions 3 and 4, we assume that E lies in the skew-Segre image $X_0X_2 = X_1X_3$ of $\mathbb{P}^1 \times \mathbb{P}^1$. In order for the representation of τ_T to stabilize the image of $\mathbb{P}^1 \times \mathbb{P}^1$, we have

$$\mathbb{P}^1 \times \mathbb{P}^1 \longrightarrow \mathbb{P}^3,$$

whose image is $X_0X_2 = X_1X_3$, in isomorphism with $\mathbb{P}^1 \times \mathbb{P}^1$ by the projections

$$\pi_1(X_0 : X_1 : X_2 : X_3) = \begin{cases} (X_0 : X_1), \\ (X_3 : X_2), \end{cases}, \; \pi_2(X_0 : X_1 : X_2 : X_3) = \begin{cases} (X_0 : X_3), \\ (X_1 : X_2). \end{cases}$$

Secondly, up to isomorphism, there are *two* permutation representations of D_4, both having the same image. The two representations are distinguished by the image of $[-1]$, up to coordinate permutation, being one of the two

$$[-1](X_0 : X_1 : X_2 : X_3) = (X_3 : X_2 : X_1 : X_0) \text{ or } (X_0 : X_3 : X_2 : X_1).$$

Considering the form of the projection morphisms π_1 and π_2, we see that only the first of the possible actions of $[-1]$ stabilizes π_1 and π_2, while the second exchanges them. In the next section we are able to write down a normal form with D_4-permutation action associated to each of the possible actions of $[-1]$.

5 Normal Forms

The objective of this section is to introduce elliptic curve models which satisfy the desired axioms of the previous section. After their definition we list their main properties, whose proof is essentially immediate from the symmetry properties of the model. We first present the objects of study over a general field k before passing to k of characteristic 2. Additional details of their construction can be found in the talk notes [18] where they were first introduced.

Definition 6. *An elliptic curve E/k in $\mathbb{P}^3$ is said to be in $\mathbf{Z}/4\mathbf{Z}$-normal form if it is given by the equations*

$$X_0^2 - X_1^2 + X_2^2 - X_3^2 = eX_0X_2 = eX_1X_3,$$

with identity $O = (1 : 0 : 0 : 1)$.

The $\mathbf{Z}/4\mathbf{Z}$-normal form is the unique model, up to linear isomorphism (see Theorem 12), satisfying the complete set of axioms of the previous section. If we drop the condition for the factorization through $\mathbb{P}^1 \times \mathbb{P}^1$ (condition 4), we obtain the following normal form, which admits the alternative action of $[-1]$.

Definition 7. *An elliptic curve C/k in $\mathbb{P}^3$ is said to be in split $\boldsymbol{\mu}_4$-normal form if it is given by the equations*

$$X_0^2 - X_2^2 = c^2 X_1 X_3, \quad X_1^2 - X_3^2 = c^2 X_0 X_2,$$

with identity $O = (c : 1 : 0 : 1)$.

These normal forms both have good reduction in characteristic 2. The $\mathbb{Z}/4\mathbb{Z}$-normal form admits a rational 4-torsion point $T = (1 : 1 : 0 : 0)$, and the isomorphism

$$\langle T \rangle = \{(1 : 0 : 0 : 1), (1 : 1 : 0 : 0), (0 : 1 : 1 : 0), (0 : 0 : 1 : 1)\} \cong \mathbb{Z}/4\mathbb{Z}$$

gives the name to curves in this form.

On the split $\boldsymbol{\mu}_4$-normal form, the point $T = (1 : c : 1 : 0)$ is a rational 4-torsion point, and if $\text{char}(k) \neq 2$ and there exists a primitive 4-th root of unity i in k, then $R = (c : i : 0, -i)$ is a rational 4-torsion point (dual to T under the Weil pairing) such that $\langle T, R \rangle = C[4]$. The subgroup

$$\langle R \rangle = \{(c : 1 : 0 : 1), (c : i : 0 : -i), (c : -1 : 0 : -1), (c : -i : 0 : i)\} \cong \boldsymbol{\mu}_4$$

is a group (scheme) isomorphic to the group (scheme) $\boldsymbol{\mu}_4$ of 4-th roots of unity, which gives the name to this normal form. The nonsplit variant (see Remark following Corollary 21) descends to any subfield containing c^4, does not necessarily have a rational 4-torsion point, but in the application to elliptic curves over finite fields of characteristic 2, every such model can be put in the split form. The action of the respective points T by translation gives the coordinate permutation action which we desire, the dual subgroup $\langle R \rangle$ degenerates in characteristic 2 to the identity group $\{O\} = \{(c : 1 : 0 : 1)\}$, and the embedding divisor $3(O) + (S)$, where $S = 2R$, degenerates to $4(O)$. Hereafter we consider these models only over a field of characteristic 2.

We now formally state and prove the main symmetry properties of the new models over a field of characteristic 2 with analogy to the Edwards model.

Theorem 8. *Let E/k be a curve in $\mathbb{Z}/4\mathbb{Z}$-normal form over a field of characteristic 2.*

1. *The embedding divisor class is $3(O) + (S)$ where $S = (0 : 1 : 1 : 0) = 2T$.*
2. *The point $T = (1 : 1 : 0 : 0)$ is a rational 4-torsion point.*
3. *The translation–by–T and inverse morphisms are given by:*

$$\tau_T(X_0 : X_1 : X_2 : X_3) = (X_3 : X_0 : X_1 : X_2),$$
$$[-1](X_0 : X_1 : X_2 : X_3) = (X_3 : X_2 : X_1 : X_0).$$

4. *E admits a factorization through $\mathbb{P}^1 \times \mathbb{P}^1$, where*

$$\pi_1(X_0 : X_1 : X_2 : X_3) = \begin{cases} (X_0 : X_1), \\ (X_3 : X_2), \end{cases}, \quad \pi_2(X_0 : X_1 : X_2 : X_3) = \begin{cases} (X_0 : X_3), \\ (X_1 : X_2). \end{cases}$$

More precisely, if (U_0, U_1) and (V_0, V_1) are the coordinate functions on $\mathbb{P}^1 \times \mathbb{P}^1$, the product morphism $\pi_1 \times \pi_2$ determines an isomorphism $E \to E_1$, where E_1 is the curve $(U_0 + U_1)^2(V_0 + V_1)^2 = c\, U_0 U_1 V_0 V_1$, whose inverse is the restriction of the skew-Segre embedding $((U_0 : U_1), (V_0 : V_1)) \longrightarrow (U_0 V_0 : U_1 V_0 : U_1 V_1 : U_0 V_1)$.

Proof. The correctness of the forms for $[-1]$ and τ_T follow from the fact that they are automorphisms, that the asserted map for $[-1]$ fixes O and that for τ_T has no fixed point, and that $\tau_T(O) = T$. Since $\tau_T^4 = 1$, it follows that T is 4-torsion. The hypersurface $X_0 + X_1 + X_2 + X_3 = 0$ cuts out the subgroup $\langle T \rangle \cong \mathbf{Z}/4\mathbf{Z}$, which determines the embedding divisor class as $3(O)+(S)$ where $S = O+T+2T+3T = 2T \in E[2]$. The factorization is determined by the automorphism group, and the image curve can be verified by elementary substitution. $\qquad\square$

Lemma 9. *The $\mathbf{Z}/4\mathbf{Z}$-normal form is isomorphic to a curve in Weierstrass form $Y(Y + X)Z = X(X + c^{-1}Z)^2$. The linear map $(X : Y : Z) = (X_1 + X_2 : X_2 : c(X_0 + X_3))$ defines the isomorphism except at O.*

Proof. The existence of a linear map is implied by Kohel [17, Lemma 3], and the exact form of this map can be easily verified. The exceptional divisor of the given rational map follows since $X_1 = X_2 = 0$ only meets the curve at O. $\qquad\square$

Theorem 10. *Let C/k be a curve in $\boldsymbol{\mu}_4$-normal form over a field of characteristic 2.*

1. *The embedding divisor class of C is $4(O)$.*
2. *The point $T = (1 : c : 1 : 0)$ is a rational 4-torsion point.*
3. *The translation–by–T and inverse morphisms are given by:*

$$\tau_T(X_0 : X_1 : X_2 : X_3) = (X_3 : X_0 : X_1 : X_2),$$
$$[-1](X_0 : X_1 : X_2 : X_3) = (X_0 : X_3 : X_2 : X_1).$$

Proof. As in Theorem 8, the correctness of automorphisms is implied by action on the points O and T, and the relation $\tau_T^4 = 1$ shows that T is 4-torsion. Since the hyperplanes $X_i = 0$ cut out the divisors $4(T_{i+2})$ where $T_k = kT$, and T is 4-torsion, this gives the form of the embedding divisor class. $\qquad\square$

Lemma 11. *An elliptic curve in split $\boldsymbol{\mu}_4$-normal form is isomorphic to the curve $Y(Y + X)Z = X(X + c^{-2}Z)^2$ in Weierstrass form. The linear map $(X : Y : Z) = (c(X_1 + X_3) : X_0 + cX_1 + X_2 : c^4 X_2)$ defines the isomorphism except at O.*

Proof. As above, the existence of a linear map is implied by Kohel [17, Lemma 3], and the exact form of this map can be easily verified. The exceptional divisor of the given rational map follows since $X_2 = 0$ only meets the curve at O. $\qquad\square$

Remark. The rational maps of Lemma 9 and 11 extend to isomorphisms, but there is no base-point free linear representative for these isomorphisms.

6   Isomorphisms with Normal Forms

Let E_{c^2} denote an elliptic curve in $\mathbf{Z}/4\mathbf{Z}$-normal form and C_c a curve in μ_4-normal form. By Lemmas 9 and 11, the curves E_{c^2} and C_c are isomorphic, but by classification of their embedding divisor classes in Theorems 8 and 10, it follows from Lemma 4 that there is no linear isomorphism between them. In this section we obtain a classification of curves over with rational 4-torsion point and make the isomorphism explicit for E_{c^2} and C_c.

Theorem 12. *Let X/k be an elliptic curve over a field k of characteristic 2, with identity O and k-rational point T of order 4, and suppose that c is an element of k such that $j(X) = c^8$.*

1. *There exists a unique isomorphism of X over k to a curve E_{c^2} in $\mathbf{Z}/4\mathbf{Z}$-normal form sending O to $(1:0:0:1)$ and T to $(1:1:0:0)$.*
2. *There exists a unique isomorphism of X over k to a curve C_c in split μ_4-normal form sending O to $(c:1:0:1)$ and T to $(1:c:1:0)$.*

If X is embedded as a symmetric quartic model in $\mathbb{P}^3$, then either the isomorphism of X with E_{c^2} or the isomorphism with C_c is induced by a linear automophism of $\mathbb{P}^3$.

Proof. The j-invariants of E_{c^2} and C_c are each c^8 ($\neq 0$ since X is not supersingular by existence of a 2-torsion point), which implies the existence of the isomorphisms over the algebraic closure. The rational 4-torsion point T fixes the quadratic twist, hence the isomorphism is defined over k. Since there is a unique 2-torsion point $S = 2T$, the embedding divisor of X in $\mathbb{P}^3$ is either $3(O) + (S)$ or $4(O)$ by Lemma 2. In the former case, the isomorphism to E_{c^2} is linear, and in the latter case the isomorphism to C_c is linear by Lemma 5.   □

The following theorem classifies the isomorphisms between E_{c^2} and C_c.

Theorem 13. *Let C_c be an elliptic curve in split μ_4-normal form and E_{c^2} an elliptic curve in $\mathbf{Z}/4\mathbf{Z}$-normal form. Then there exists an isomorphism $\iota : C_c \to E_{c^2}$ determined by the projections*

$$\pi_1 \circ \iota((X_0 : X_1 : X_2 : X_3)) = \begin{cases} (cX_0 : X_1 + X_3), \\ (X_1 + X_3 : cX_2), \end{cases}$$

$$\pi_2 \circ \iota((X_0 : X_1 : X_2 : X_3)) = \begin{cases} (X_0 + X_2 : cX_1), \\ (cX_3 : X_0 + X_2). \end{cases}$$

The morphism to E_{c^2} is recovered by composing $\pi_1 \times \pi_2$ with the skew-Segre embedding. The inverse morphism is given by

$$\iota^{-1}(X_0 : X_1 : X_2 : X_3) = \begin{cases} (X_0X_1 + X_2X_3 : cX_2^2 : X_0X_1 + c^2X_1X_2 + X_2X_3 : cX_1^2), \\ (X_0X_3 : (X_2 + X_3)^2 : X_1X_2 : (X_0 + X_1)^2), \\ ((X_0 + X_3)^2 : cX_2X_3 : (X_1 + X_2)^2 : cX_0X_1), \\ (cX_3^2 : X_0X_3 + X_1X_2 + c^2X_2X_3 : cX_2^2 : X_1X_2 + X_0X_3). \end{cases}$$

Neither ι nor its inverse can be represented by a projective linear transformation.

Proof. This correctness of this isomorphism can be verified explicitly (e.g. as implemented in Echidna [19]). The nonexistence of a linear isomorphism is a consequence of Lemma 4 and the classification of the embedding divisor classes in Theorems 8 and 10.   □

7   Addition Law Structure and Efficient Arithmetic

The interest in alternative models of elliptic curves has been driven by the simple form of their *addition laws* — the polynomial maps which define the addition morphism $\mu : E \times E \to E$ as rational maps. In this section we determine bases of simple forms for the addition laws of the $\mathbf{Z}/4\mathbf{Z}$-normal form and of the $\boldsymbol{\mu}_4$-normal form.

The verification that a system of putative addition laws determines a well-defined morphism can be verified symbolically. In particular we refer to the implementations of these models and their addition laws in Echidna [19] (in the Magma [21] language) for a verification that the systems are consistent and define rational maps. The dimensions of the spaces of given bidegree are known *a priori* by Kohel [17], as well as their completeness as morphisms. By the Rigidity Theorem [22, Theorem 2.1], a morphism μ of abelian varieties is the composition of a homomorphism and translation. In order to verify that $\mu : E \times E \to E$ is the addition morphism, it suffices to check that the restrictions of μ to $E \times \{O\}$ and $\{O\} \times E$ agree with the restrictions of π_1 and π_2, respectively. Similarly, for a particular addition law of bidegree $(2,2)$, the exceptional divisors, on which the polynomials of the addition law simultaneously vanish, are known by Lange and Ruppert [20] and the generalizations in Kohel [17] to have components of the form $\Delta_P = \{(P+Q, Q) \mid Q \in E\}$. Consequently, as pointed out in Kohel [17] (Corollary 11 and the Remark following Corollary 12), the exceptional divisors can be computed (usually by hand) by intersecting with $E \times \{O\}$.

Addition Law Structure for the Z/4Z-Normal Form

Theorem 14. *Let E/k, $\mathrm{char}(k) = 2$, be an elliptic curve in $\mathbf{Z}/4\mathbf{Z}$-normal form:*

$$(X_0 + X_1 + X_2 + X_3)^2 = cX_0X_2 = cX_1X_3.$$

Bases for the bilinear addition law projections $\pi_1 \circ \mu$ and $\pi_2 \circ \mu$ are, respectively:

$$\left\{ \begin{array}{l} (X_0Y_3 + X_2Y_1,\ X_1Y_0 + X_3Y_2), \\ (X_1Y_2 + X_3Y_0,\ X_0Y_1 + X_2Y_3) \end{array} \right\} \quad and \quad \left\{ \begin{array}{l} (X_0Y_0 + X_2Y_2,\ X_1Y_1 + X_3Y_3), \\ (X_1Y_3 + X_3Y_1,\ X_0Y_2 + X_2Y_0) \end{array} \right\}.$$

Addition laws of bidegree $(2,2)$ are recovered by composition with the skew-Segre embedding $s((U_0 : U_1), (V_0 : V_1)) = (U_0V_0 : U_1V_0 : U_1V_1 : U_0V_1)$. Each of these basis elements has an exceptional divisor of of the form $2\Delta_{nT}$ for some $0 \le n \le 3$.

Proof. That the addition laws determine a well-defined morphism is verified symbolically.[6] The morphism is the addition morphism since the substitution $(Y_0, Y_1, Y_2, Y_3) = (1, 0, 0, 1)$, gives the projection onto the first factor. By symmetry of the spaces in X_i and Y_i, the same holds for the second factor.

The form of the exceptional divisor is verified by a similar substitution. For example, for the exceptional divisor $X_1Y_2 + X_3Y_0 = X_0Y_1 + X_2Y_3 = 0$, we intersect with $(Y_0, Y_1, Y_2, Y_3) = (1, 0, 0, 1)$ to find $X_3 = X_2 = 0$, which defines the unique point $T = (1, 1, 0, 0)$ with a multiplicity of 2, hence the exceptional divisor is $2\Delta_T$. The other exceptional divisors are determined similarly. □

Remark. We observe that the entire space of addition laws of bidgree $(2, 2)$ is independent of the curve parameters. This is not a feature of the Edwards addition laws.

Corollary 15. *Addition of generic points on E can be carried out in* 12**M**.

Proof. Since each of the pairs is equivalent under a permutation of the input variables it suffices to consider the first, which each require 4**M**. Composition with the skew-Segre embedding requires an additional 4**M**, which yields the bound of 12**M**. □

Evaluation of the addition forms along the diagonal yields the duplication formulas.

Corollary 16. *Let $E = E_c$ be an elliptic curve in* **Z**$/4$**Z**-*normal form. The duplication morphism on E is given by*

$$\pi_1 \circ [2](X_0 : X_1 : X_2 : X_3) = (X_0X_3 + X_1X_2 : X_0X_1 + X_2X_3),$$
$$\pi_2 \circ [2](X_0 : X_1 : X_2 : X_3) = ((X_0 + X_2)^2 : (X_1 + X_3)^2),$$

composed with the skew-Segre embedding.

This immediately gives the following complexity for duplication.

Corollary 17. *Duplication on E can be carried out in* 7**M** + 2**S**.

Proof. The pair $(X_0X_3 + X_1X_2, X_0X_1 + X_2X_3)$ can be computed with 3**M** by exploiting the usual Karatsuba trick using the factorization of their sum:

$$(X_0X_3 + X_1X_2) + (X_0X_1 + X_2X_3) = (X_0 + X_2)(X_1 + X_3).$$

After the two squarings, the remaining 4**M** come from the Segre morphism. □

---

[6] In Echidna [19], the constructor is `EllipticCurve_C4_NormalForm` after which `AdditionMorphism` returns this morphism as a composition.

Addition Law Structure for the Split μ_4-Normal Form

Theorem 18. *Let C be an elliptic curve in split μ_4-normal form:*

$$(X_0 + X_2)^2 = c^2\, X_1 X_3, \quad (X_1 + X_3)^2 = c^2\, X_0 X_2.$$

A basis for the space of addition laws of bidegree $(2,2)$ is given by:

$$\left\{ \begin{array}{l} \left((X_0 Y_0 + X_2 Y_2)^2,\ c(X_0 X_1 Y_0 Y_1 + X_2 X_3 Y_2 Y_3),\ (X_1 Y_1 + X_3 Y_3)^2,\ c(X_0 X_3 Y_0 Y_3 + X_1 X_2 Y_1 Y_2)\right), \\ \left(c(X_0 X_1 Y_0 Y_3 + X_2 X_3 Y_1 Y_2),\ (X_1 Y_0 + X_3 Y_2)^2,\ c(X_0 X_3 Y_2 Y_3 + X_1 X_2 Y_0 Y_1),\ (X_0 Y_3 + X_2 Y_1)^2\right), \\ \left((X_3 Y_1 + X_1 Y_3)^2,\ c(X_0 X_3 Y_1 Y_2 + X_1 X_2 Y_0 Y_3),\ (X_0 Y_2 + X_2 Y_0)^2,\ c(X_0 X_1 Y_2 Y_3 + X_2 X_3 Y_0 Y_1)\right), \\ \left(c(X_0 X_3 Y_0 Y_1 + X_1 X_2 Y_2 Y_3),\ (X_0 Y_1 + X_2 Y_3)^2,\ c(X_0 X_1 Y_1 Y_2 + X_2 X_3 Y_0 Y_3),\ (X_1 Y_2 + X_3 Y_0)^2\right). \end{array} \right\}$$

The exceptional divisor of each addition law is of the form $4\Delta_{nT}$.

Proof. As for the $\mathbf{Z}/4\mathbf{Z}$-normal form the consistency of the addition laws is verified symbolically[7] and the space is known to have dimension four by Kohel [17]. Evaluation of the first addition law at $(Y_0, Y_1, Y_2, Y_3) = (c, 1, 0, 1)$ gives

$$(c^2 X_0^2,\ c^2 X_0 X_1,\ (X_1 + X_3)^2,\ c^2 X_0 X_3).$$

Using $(X_1 + X_3)^2 = c^2 X_0 X_2$, after removing the common factor $c^2 X_0$, this agrees with projection to the first factor, and identifies the exceptional divisor $4\Delta_S$ where S is the 2-torsion point $(0 : 1 : c : 1)$ with $X_0 = 0$. $\square$

Corollary 19. *Addition of generic points on C can be carried out in $7\mathbf{M} + 2\mathbf{S} + 2\mathbf{m}_c$.*

Proof.   – Evaluate $(Z_0, Z_1, Z_2, Z_3) = (X_0 Y_0, X_1 Y_1, X_2 Y_2, X_3 Y_3)$ with $4\mathbf{M}$.
– Evaluate $(X_0 Y_0 + X_2 Y_2)^2 = (Z_0 + Z_2)^2$ with $1\mathbf{S}$.
– Evaluate $(X_1 Y_1 + X_3 Y_3)^2 = (Z_1 + Z_3)^2$ with $1\mathbf{S}$.
– Evaluate $(X_0 Y_0 + X_2 Y_2)(X_1 Y_1 + X_3 Y_3) = (Z_0 + Z_2)(Z_1 + Z_3)$ followed by

$$X_0 X_1 Y_0 Y_1 + X_2 X_3 Y_2 Y_3 = Z_0 Z_1 + Z_2 Z_3$$
$$X_0 X_3 Y_0 Y_3 + X_1 X_2 Y_1 Y_2 = Z_0 Z_3 + Z_1 Z_2$$

using $3\mathbf{M}$, exploiting the linear relation (following Karatsuba):

$$(Z_0 + Z_2)(Z_1 + Z_3) = (Z_0 Z_1 + Z_2 Z_3) + (Z_0 Z_3 + Z_1 Z_2).$$

After two scalar multiplications by c, we obtain $7\mathbf{M} + 2\mathbf{S} + 2\mathbf{m}_c$ for the computation using the first addition law. $\square$

Specializing this to the diagonal we find defining polynomials for duplication.

Corollary 20. *The duplication morphism on an elliptic curve C in split μ_4-normal form is given by*

$$[2](X_0 : X_1 : X_2 : X_3) =$$
$$((X_0 + X_2)^4 : c(X_0 X_1 + X_2 X_3)^2 : (X_1 + X_3)^4 : c(X_0 X_3 + X_1 X_2)^2).$$

---

[7] The Echidna [19] constructor is `EllipticCurve_Split_Mu4_NormalForm` after which `AdditionMorphism` returns this morphism as a composition.

This gives an obvious complexity bound of $3\mathbf{M} + 6\mathbf{S} + 2\mathbf{m}_c$ for duplication, however we note that along the curve we have the following equivalent expressions:

$$c^2(X_0X_1 + X_2X_3)^2 = (X_0 + X_2)^4 + c^{-4}(X_1 + X_3)^4 + F^2,$$
$$c^2(X_0X_3 + X_1X_2)^2 = (X_0 + X_2)^4 + c^{-4}(X_1 + X_3)^4 + G^2,$$

for $F = (X_0 + cX_3)(cX_1 + X_2)$ and $G = (X_0 + cX_1)(X_2 + cX_3)$, and that

$$F + G = c(X_0 + X_2)(X_1 + X_3).$$

This leads to a savings of $1\mathbf{M} + 1\mathbf{S}$ from the naive analysis, at the cost of extra multiplications by c.

Corollary 21. *Duplication on C can be carried out in* $2\mathbf{M} + 5\mathbf{S} + 7\mathbf{m}_c$.

Proof. We describe the evaluation of the forms of Corollary 20, using the equivalent expressions. Setting $(U, V, W) = ((X_0 + X_2)^2, (X_1 + X_3)^2, (X_0 + cX_1)^2)$,

$$G^2 = (U + c^2V + W)WP \text{ and } F^2 = G^2 + c^2UV,$$

from which the duplication formula can be expressed as:

$$(cU^2 : U^2 + c^{-4}V^2 + (U + c^2V + W)W + c^2UV : cV^2 : U^2 + c^{-4}V^2 + (U + c^2V + W)W).$$

We scale by c^4 to have only integral powers of c, which gives the

- Evaluate $(U, V, W) = ((X_0 + X_2)^2, (X_1 + X_3)^2, (X_0 + cX_1)^2)$ with $3\mathbf{S} + 1\mathbf{m}_c$.
- Evaluate $c^5(X_0 + X_2)^4 = c^5U^2$ with $1\mathbf{S} + 1\mathbf{m}_c$, storing U^2.
- Evaluate $c^5(X_1 + X_3)^4 = c^5V^2$ with $1\mathbf{S} + 1\mathbf{m}_c$, storing V^2.
- Evaluate c^2V, c^2UV, $(U + c^2V + W)W$ with $2\mathbf{M} + 1\mathbf{m}_c$, then set

$$c^4(X_0 + X_2)^4 + (X_1 + X_3)^4 = c^4U^2 + V^2,$$
$$c^4G^2 = c^4(U + c^2V + W)W,$$
$$c^4F^2 = c^4G^2 + c^6UV,$$

using $3\mathbf{m}_c$, followed by additions. This gives the asserted complexity.   □

Remark. The triple (U, V, W), up to scalars, can be identified with the variables (A, B, C) of the EFD [6] in the improvement of Bernstein et al. [4] to the duplication algorithm of Kim and Kim [16] in "extended López-Dahab coordinates" with $a_2 = 0$. In brief, the extended López-Dahab coordinates defines a curve $Y^2 = (X^2 + a_6)XZ$, in a $(1, 2, 1, 2)$-weighted projective space with coordinate functions X, Y, Z, Z^2. We embed this in a standard $\mathbb{P}^3$, with embedding divisor class $4(O)$, by the map (X^2, Y, XZ, Z^2). By Lemma 5 this is linearly isomorphic to the curve C in split $\boldsymbol{\mu}_4$-normal form. One derives an equivalent complexity for duplication on this $\mathbb{P}^3$ model, and duplication on C differs only by the cost of scalar multiplications involved in the linear transformation to C.

We remark that this can be interpretted as a factorization of the duplication map as follows. Letting D be the image of C given by (U, V, W) in $\mathbb{P}^2$, the Kim and Kim algorithm can be expressed as a composition $C \xrightarrow{\varphi} D \xrightarrow{\psi} C$ where φ and ψ are each of degree 2, with φ purely inseparable and ψ separable. The curve D is a singular quartic curve in $\mathbb{P}^2$, given by a well-chosen incomplete linear system. The nodal singularities of D are oriented such that the resolved points have the same image under ψ. The omission of a fourth basis element of the complete linear system allows one to save $1\mathbf{S}$ in its computation.

In order to best optimize the multiplications by scalars, we can apply a coordinate scaling. The split $\boldsymbol{\mu}_4$-normal descends to a (non-split) $\boldsymbol{\mu}_4$-normal form over any subfield containing the parameter $s = c^{-4}$, by renormalization of coordinates:

$$s(X_1 + X_3)^2 + X_0 X_2, \ (X_0 + X_2)^2 = X_1 X_3.$$

In this form the duplication polynomials require fewer multiplications by constants:

$$\big((X_0 + X_2)^4 : (X_0 + X_2)^4 + s^2(X_1 + X_3)^4 + (X_0 + X_3)^2(X_1 + X_2)^2 : $$
$$s(X_1 + X_3)^4 : (X_0 + X_2)^4 + s^2(X_1 + X_3)^4 + (X_0 + X_1)^2(X_2 + X_3)^2 \big),$$

yielding $2\mathbf{M} + 5\mathbf{S} + 2\mathbf{m}_s$.

Addition Law Projections for the Split $\boldsymbol{\mu}_4$-Normal Form

Let $C = C_c$ be an elliptic curve in $\boldsymbol{\mu}_4$-normal form and $E = E_{c^2}$ be an elliptic curve in $\mathbf{Z}/4\mathbf{Z}$-normal form. In view of Theorem 13, there is an explicit isomorphism $\iota : C \to E$, determined by the application of the skew-Segre embedding to the pair of projections $\pi_i : C \to \mathbb{P}^1$:

$$\pi_1((X_0 : X_1 : X_2 : X_3)) = \begin{cases} (cX_0 : X_1 + X_3), \\ (X_1 + X_3 : cX_2), \end{cases}$$
$$\pi_2((X_0 : X_1 : X_2 : X_3)) = \begin{cases} (X_0 + X_2 : cX_1), \\ (cX_3 : X_0 + X_2). \end{cases}$$

The first projection π_1 determines a map to $C/\langle[-1]\rangle \cong \mathbb{P}^1$, and the second projection π_2 satisfies $\pi_2 \circ [-1] = \sigma \circ \pi_2$, where $\sigma((U_0 : U_1)) = (U_1 : U_0)$. As a consequence of the addition law structure of Theorem 18, the addition law projections $C \times C \to \mathbb{P}^1$ associated to these projections take a particularly simple form.

Corollary 22. *If $\pi_i : C \to \mathbb{P}^1$ are the projections defined above, the addition law projections $\pi_1 \circ \mu$ and $\pi_2 \circ \mu$ are respectively spanned by*

$$\begin{Bmatrix} (X_0Y_0 + X_2Y_2, X_1Y_1 + X_3Y_3), \\ (X_1Y_3 + X_3Y_1, X_2Y_0 + X_0Y_2) \end{Bmatrix} \ and \ \begin{Bmatrix} (X_0Y_3 + X_2Y_1, X_1Y_0 + X_3Y_2), \\ (X_1Y_2 + X_3Y_0, X_0Y_1 + X_2Y_3) \end{Bmatrix}.$$

Proof. The addition law projections can be verified in Echidna [19]. □

The skew-Segre embedding of $\mathbb{P}^1 \times \mathbb{P}^1$ in $\mathbb{P}^3$ induces a map to the isomorphic curve E in $\mathbf{Z}/4\mathbf{Z}$-normal form, rather than the $\boldsymbol{\mu}_4$-normal form. These addition law projections play a central role in the study of the Kummer arithmetic in Section 8, defined more naturally in terms of E.

8   Kummer Quotients and the Montgomery Ladder

For an abelian variety A, the quotient variety $\mathcal{K}(A) = A/\{[\pm 1]\}$ is called the Kummer variety of A. We investigate explicit models for the Kummer curves $\mathcal{K}(E)$ and $\mathcal{K}(C)$ where $E = E_{c^2}$ and $C = C_c$ are isomorphic elliptic curves in $\mathbf{Z}/4\mathbf{Z}$-normal form and $\boldsymbol{\mu}_4$-normal form, respectively. The objective of this study is to obtain a Montgomery ladder [23] for efficient scalar multiplication on these curves. Such a Montgomery ladder was developed for Kummer curves (or *lines* since they are isomorphic to the projective line $\mathbb{P}^1$) in characteristic 2 by Stam [24]. More recently Gaudry and Lubicz [12] developed efficient pseudo-addition natively on a Kummer line $\mathcal{K} = \mathbb{P}^1$ by means of theta identities. Neither the method of Stam nor Gaudry and Lubicz provides recovery of points on the curve. We show that for fixed P, the morphism $E \to \mathcal{K} \times \mathcal{K}$ sending Q to $(\overline{Q}, \overline{Q - P})$, used for initialization of the Montgomery ladder, is in fact an isomorphism with its image. As a consequence we rederive the equations of Gaudry and Lubicz for pseudo-addition, together with an algorithm for point recovery. In addition, knowledge of the curve equation (in $\mathcal{K} \times \mathcal{K}$) permits the trade-off of a squaring for a multiplication by a constant depending on the base point P (see Corollary 26).

Kummer Curves

We consider the structure of $\mathcal{K}(E) = E/\{[\pm 1]\}$ and $\mathcal{K}(C) = C/\{[\pm 1]\}$ for elliptic curves $E = E_{c^2}$ in $\mathbf{Z}/4\mathbf{Z}$-normal form and $C = C_c$ in split $\boldsymbol{\mu}_4$-normal form, respectively. The former has a natural identification with $\mathbb{P}^1$ equipped with the covering $\pi_1 : E \to \mathcal{K}(E)$, given by

$$(X_0 : X_1 : X_2 : X_3) \mapsto \begin{cases} (X_0 : X_1), \\ (X_3 : X_2). \end{cases}$$

The latter quotient has a plane model $\mathcal{K}(C) : Y^2 = c^2 XZ$ in $\mathbb{P}^2$ obtained by taking the $[-1]$-invariant basis $\{X_0, X_1 + X_3, X_2\}$. For $E = E_{c^2}$ and $C = C_c$ as above, the isomorphism $\iota : C \to E$ of Theorem 13 induces an isomorphism $\iota : \mathcal{K}(C) \to \mathcal{K}(E) = \mathbb{P}^1$ of Kummer curves given by

$$\iota(X : Y : Z) = \begin{cases} (cX : Y), \\ (Y : cZ), \end{cases}$$

with inverse $(U_0 : U_1) \mapsto (U_0^2 : cU_0U_1 : U_1^2)$. Hereafter we fix this isomorphism, and obtain the covering morphism $C \to \mathcal{K}(E)$:

$$(X_0 : X_1 : X_2 : X_3) \mapsto \begin{cases} (cX_0 : X_1 + X_3), \\ (X_1 + X_3 : cX_2). \end{cases}$$

We denote this common Kummer curve by $\mathcal{K}$, to distinguish the curve with induced structure from the elliptic curve covering (by both E and C) from $\mathbb{P}^1$.

Montgomery Endomorphism

The Kummer curve $\mathscr{K}$ (of an arbitrary elliptic curve E) no longer supports an addition morphism, however scalar multiplication $[n]$ is well-defined, since $[-1]$ commutes with $[n]$. We investigate the general construction of the Montgomery ladder for the Kummer quotient. For this purpose we define the *Montgomery endomorphism* $E \times E \to E \times E$:

$$\begin{pmatrix} 2 & 0 \\ 1 & 1 \end{pmatrix} (Q, R) = (2Q, Q + R).$$

In general this endomorphism, denoted φ, is not well-defined on $\mathscr{K} \times \mathscr{K}$. Instead, for fixed $P \in E(k)$ we consider

$$\Delta_P = \{(Q, R) \in E \times E \mid Q - R = P\} \cong E,$$

and let $\mathscr{K}(\Delta_P)$ be the image of Δ_P in $\mathscr{K} \times \mathscr{K}$, which we call a *Kummer-oriented curve*. In what follows we develop algorithmically the following observations (see Theorems 23, 24, and 25):

1. The morphism $\Delta_P \to \mathscr{K}(\Delta_P)$ is an isomorphism for any $P \notin E[2]$.
2. The Montgomery endomorphism is well-defined on $\mathscr{K}(\Delta_P)$.

By means of the elliptic curve structure on Δ_P determined by the isomorphism $E \to \Delta_P$ given by $Q \mapsto (Q, Q - P)$, the Montgomery endomorphism is the duplication morphism (i.e. $\varphi(Q, Q - P) = (2Q, 2Q - P)$). On the other hand, the Montgomery endomorphism allows us to represent scalar multiplication on P symmetrically as a sequence of compositions. Precisely, we let $\varphi_0 = \varphi$, let σ be the involution $\sigma(Q, R) = (-R, -Q)$ of Δ_P, which induces the exchange of factors on $\mathscr{K}(\Delta_P)$, and set $\varphi_1 = \sigma \circ \varphi \circ \sigma$. For an integer n with binary representation $n_r n_{r-1} \ldots n_1 n_0$ we may compute nP by the sequence

$$\varphi_{n_0} \circ \varphi_{n_1} \cdots \circ \varphi_{n_{r-1}}(P, O) = ((n+1)P, nP),$$

returning the second component.

This composition representation for scalar multiplication on $E \times E$ is a double-and-always-add algorithm [9], which provides a symmetry protection against side-channel attacks in cryptography (see Joye and Yen [15, Section 4]), but is inefficient due to insertion of redundant additions. When applied to $\mathscr{K}(\Delta_P)$, on the other hand, this gives a (potentially) efficient algorithm, conjugate duplication, for carrying out scalar multiplication. In view of this, $\mathscr{K}(\Delta_P)$ should be thought of as a model oriented for carrying out efficient scalar multiplication on a fixed point P in $E(k)$.

The Kummer-Oriented Curves $\mathscr{K}(\Delta_P)$

Let $E = E_{c^2}$ be a curve in $\mathbf{Z}/4\mathbf{Z}$-normal form, let $P = (t_0 : t_1 : t_2 : t_3)$ be a fixed point in $E(k)$, and let $\mathscr{K}(\Delta_P)$ be the Kummer-oriented curve in $\mathscr{K}^2$, with coordinate functions $((U_0, U_1), (V_0, V_1))$.

Theorem 23. *The Kummer-oriented curve $\mathscr{K}(\Delta_P)$ in $\mathscr{K}^2$, for $P = (t_0 : t_1 : t_2 : t_3)$, has defining equation*

$$t_0^2(U_0V_1 + U_1V_0)^2 + t_1^2(U_0V_0 + U_1V_1)^2 = c^2 t_0 t_1 U_0 U_1 V_0 V_1.$$

If P is not a 2-torsion point, the morphism $\kappa : E \to \mathscr{K}(\Delta_P)$, defined by $Q \mapsto (\overline{Q}, \overline{Q - P})$, is an isomorphism, given by

$$\pi_1 \circ \kappa(X_0 : X_1 : X_2 : X_3) = (U_0 : U_1) = \begin{cases} (X_0 : X_1), \\ (X_3 : X_2), \end{cases}$$

$$\pi_2 \circ \kappa(X_0 : X_1 : X_2 : X_3) = (V_0 : V_1) = \begin{cases} (t_0X_0 + t_2X_2 : t_3X_1 + t_1X_3), \\ (t_1X_1 + t_3X_3 : t_2X_0 + t_0X_2), \end{cases}$$

with inverse

$$\pi_1 \circ \kappa^{-1}((U_0 : U_1), (V_0 : V_1)) = (U_0 : U_1)$$

$$\pi_2 \circ \kappa^{-1}((U_0 : U_1), (V_0 : V_1)) = \begin{cases} (t_1U_0V_0 + t_2U_1V_1 : t_0U_0V_1 + t_3U_1V_0), \\ (t_3U_0V_1 + t_0U_1V_0 : t_2U_0V_0 + t_1U_1V_1). \end{cases}$$

Proof. The form of κ follows from the definition of the addition law. The equation for the image curve can be computed by taking resultants, and verified symbolically. The composition of κ with projection onto the first factor is the Kummer quotient of degree 2. However, for all P not in $E[2]$, the inverse morphism induces a nontrivial involution

$$(\overline{Q}, \overline{Q - P}) \longmapsto (\overline{-Q}, \overline{-Q - P}) = (\overline{Q}, \overline{Q + P})$$

on $\mathscr{K}(\Delta_P)$. Consequently the map to $\mathscr{K}(\Delta_P)$ has degree one, and being nonsingular, gives an isomorphism. □

Remark. We observe that $\mathscr{K}(\Delta_P) = \mathscr{K}(\Delta_{-P})$ in $\mathscr{K}^2$, but that a change of base point changes κ by $[-1]$.

The isomorphism of E_{c^2} with C_c lets us derive the analogous result for curves in μ_4-normal form.

Theorem 24. *Let $C = C_c$ be an elliptic curve in split μ_4-normal form with rational point $S = (s_0 : s_1 : s_2 : s_3)$. The Kummer-oriented curve $\mathscr{K}(\Delta_S)$ in $\mathscr{K}^2$ is given by the equation*

$$s_0(U_0V_1 + U_1V_0)^2 + s_2(U_0V_0 + U_1V_1)^2 = c(s_1 + s_3)U_0U_1V_0V_1.$$

If S is not a 2-torsion point, the morphism $\lambda : C \to \mathscr{K}(\Delta_S)$ is an isomorphism, and defined by

$$\pi_1 \circ \lambda(X_0 : X_1 : X_2 : X_3) = \begin{cases} (cX_0 : X_1 + X_3), \\ (X_1 + X_3 : cX_2), \end{cases}$$

$$\pi_2 \circ \lambda(X_0 : X_1 : X_2 : X_3) = \begin{cases} (s_0X_0 + s_2X_2 : s_1X_1 + s_3X_3), \\ (s_3X_1 + s_1X_3 : s_2X_0 + s_0X_2), \end{cases}$$

with inverse $\lambda^{-1}((U_0 : U_1), (V_0 : V_1))$ equal to

$$\begin{cases} ((s_1+s_3)U_0^2V_0 : (s_0U_0^2+s_2U_1^2)V_1 + cs_1U_0U_1V_0 : (s_1+s_3)U_1^2V_0 : (s_0U_0^2+s_2U_1^2)V_1 + cs_3U_0U_1V_0), \\ ((s_1+s_3)U_0^2V_1 : (s_2U_0^2+s_0U_1^2)V_0 + cs_3U_0U_1V_1 : (s_1+s_3)U_1^2V_1 : (s_2U_0^2+s_0U_1^2)V_0 + cs_1U_0U_1V_1). \end{cases}$$

Proof. The isomorphism $\iota : E_{c^2} \to C_c$ sending S to $T = (t_0 : t_1 : t_2 : t_3)$ induces the isomorphism $(s_0 : s_1+s_3 : s_2) = (t_0^2 : ct_0t_1 : t_1^2)$, by which we identify $\mathscr{K}(\Delta_P)$ and $\mathscr{K}(\Delta_S)$. The form of the morphism λ follows from the form of projective addition laws of Corollary 22, and its inverse can be verified symbolically.   □

We now give explicit maps and complexity analysis for the Montgomery endomorphism $\varphi(Q, R) = (2Q, Q+R)$, on the Kummer quotient $\mathscr{K}(\Delta_P)$ (or $\mathscr{K}(\Delta_S)$ setting $(t_0 : t_1) = (cs_0 : s_1 + s_3) = (s_1 + s_3 : cs_2)$). In view of the application to scalar multiplication on E or C, this gives an asymptotic complexity per bit of n, for computing $[n]P$.

Theorem 25. *The Montgomery endomorphism φ is defined by:*

$$\pi_1 \circ \varphi((U_0 : U_1), (V_0 : V_1)) = (U_0^4 + U_1^4 : cU_0^2U_1^2),$$
$$\pi_2 \circ \varphi((U_0 : U_1), (V_0 : V_1)) = (t_1(U_0V_0 + U_1V_1)^2 : t_0(U_0V_1 + U_1V_0)^2).$$

The sets of defining polynomials are well-defined everywhere and the following maps are projectively equivalent modulo the defining ideal:

$$(t_1(U_0V_0 + U_1V_1)^2 : t_0(U_0V_1 + U_1V_0)^2)$$
$$= (t_0(U_0V_0 + U_1V_1)^2 : t_1(U_0V_0 + U_1V_1)^2 + ct_0(U_0V_0)(U_1V_1))$$
$$= (t_0(U_0V_1 + U_1V_0)^2 + ct_1(U_0V_1)(U_1V_0) : t_1(U_0V_1 + U_1V_0)^2).$$

Assuming the point normalization with $t_0 = 1$ or $t_1 = 1$, this immediately gives the following corollary.

Corollary 26. *The Montgomery endomorphism on $\mathscr{K}(\Delta_P)$ can be computed with $4\mathbf{M} + 5\mathbf{S} + 1\mathbf{m}_t + 1\mathbf{m}_c$ or with $4\mathbf{M} + 4\mathbf{S} + 1\mathbf{m}_t + 2\mathbf{m}_c$.*

The formulas so obtained agree with those of Gaudry and Lubicz [12]. The first complexity result agrees with theirs and the second obtains a trade-off of one $\mathbf{m}_c$ for one $\mathbf{S}$ using the explicit equation of $\mathscr{K}(\Delta_P)$ in $\mathscr{K}^2$. Finally, the isomorphisms of Theorems 23 and 24 permit point recovery, hence scalar multiplication on the respective elliptic curves.

9   Conclusion

We conclude with a tabulation of the best known complexity results for doubling and addition algorithms on projective curves (taking the best reported algorithm from the EFD [6]). We include the Hessian model, the only cubic curve model, for comparison. It covers only curves with a rational 3-torsion point. Binary Edwards curves [4] cover general ordinary curves, but the best complexity result we give here is for $d_1 = d_2$ which has a rational 4-torsion point. Similarly, the López-Dahab model with $a_2 = 0$ admits a rational 4-torsion point, hence covers the same classes, but the fastest arithmetic is achieved on the quadratic twists with $a_2 = 1$. The results here for addition and duplication on μ_4-normal form report the better result (in terms of constant multiplications $\mathbf{m}$) for the non-split μ_4 model (see the remark after Corollary 21 and Corollary 28 in the appendix).

Curve model	Doubling	Addition
$\mathbf{Z}/4\mathbf{Z}$-normal form	7M + 2S	12M
Hessian	6M + 3S	12M
Binary Edwards	2M + 5S + 2m	16M + 1S + 4m
López-Dahab ($a_2 = 0$)	2M + 5S + 1m	14M + 3S
López-Dahab ($a_2 = 1$)	2M + 4S + 2m	13M + 3S
$\boldsymbol{\mu}_4$-normal form	2M + 5S + 2m	7M + 2S

This provides for the best known addition algorithm combined with essentially optimal doubling. We note that binary Edwards curves with $d_1 = d_2$ and the López-Dahab model with $a_2 = 0$ and have canonical projective embeddings in $\mathbb{P}^3$ such that the transformation to $\boldsymbol{\mu}_4$-normal form is linear, so that, conversely, these models can benefit from the efficient addition of the $\boldsymbol{\mu}_4$-normal form.

References

1. Bernstein, D.J., Lange, T.: Faster Addition and Doubling on Elliptic Curves. In: Kurosawa, K. (ed.) ASIACRYPT 2007. LNCS, vol. 4833, pp. 29–50. Springer, Heidelberg (2007)

2. Bernstein, D.J., Birkner, P., Joye, M., Lange, T., Peters, C.: Twisted Edwards Curves. In: Vaudenay, S. (ed.) AFRICACRYPT 2008. LNCS, vol. 5023, pp. 389–405. Springer, Heidelberg (2008)

3. Bernstein, D.J., Lange, T.: A complete set of addition laws for incomplete Edwards curves. J. Number Theory 131, 858–872 (2011)

4. Bernstein, D.J., Lange, T., Rezaeian Farashahi, R.: Binary Edwards Curves. In: Oswald, E., Rohatgi, P. (eds.) CHES 2008. LNCS, vol. 5154, pp. 244–265. Springer, Heidelberg (2008)

5. Bernstein, D.J., Kohel, D., Lange, T.: Twisted Hessian curves (unpublished 2009)

6. Bernstein, D.J., Lange, T.: Explicit-formulas database (2012), http://www.hyperelliptic.org/EFD/

7. Bosma, W., Lenstra Jr., H.W.: Complete systems of two addition laws for elliptic curves. J. Number Theory 53(2), 229–240 (1995)

8. Chudnovsky, D.V., Chudnovsky, G.V.: Sequences of numbers generated by addition in formal groups and new primality and factorization tests. Adv. in Appl. Math. 7(4), 385–434 (1986)

9. Coron, J.-S.: Resistance against Differential Power Analysis for Elliptic Curve Cryptosystems. In: Koç, Ç.K., Paar, C. (eds.) CHES 1999. LNCS, vol. 1717, pp. 292–302. Springer, Heidelberg (1999)

10. Diao, O.: Quelques aspects de l'arithmtique des courbes hyperelliptiques de genre 2, Ph.D. thesis, Université de Rennes (2011)

11. Edwards, H.: A normal form for elliptic curves. Bulletin of the American Mathematical Society 44, 393–422 (2007)

12. Gaudry, P., Lubicz, D.: The arithmetic of characteristic 2 Kummer surfaces and of elliptic Kummer lines. Finite Fields and Their Applications 15(2), 246–260 (2009)

13. Hisil, H., Wong, K.K.-H., Carter, G., Dawson, E.: Twisted Edwards Curves Revisited. In: Pieprzyk, J. (ed.) ASIACRYPT 2008. LNCS, vol. 5350, pp. 326–343. Springer, Heidelberg (2008)

14. Farashahi, R.R., Joye, M.: Efficient Arithmetic on Hessian Curves. In: Nguyen, P.Q., Pointcheval, D. (eds.) PKC 2010. LNCS, vol. 6056, pp. 243–260. Springer, Heidelberg (2010)

15. Joye, M., Yen, S.-M.: The Montgomery Powering Ladder. In: Kaliski Jr., B.S., Koç, Ç.K., Paar, C. (eds.) CHES 2002. LNCS, vol. 2523, pp. 291–302. Springer, Heidelberg (2003)

16. Kim, K.H., Kim, S.I.: A new method for speeding up arithmetic on elliptic curves over binary fields (2007), http://eprint.iacr.org/2007/181

17. Kohel, D.: Addition law structure of elliptic curves. Journal of Number Theory 131(5), 894–919 (2011)

18. Kohel, D.: A normal form for elliptic curves in characteristic 2. In: Arithmetic, Geometry, Cryptography and Coding Theory (AGCT 2011), Luminy, talk notes (March 15, 2011)

19. Kohel, D., et al.: Echidna algorithms, v.3.0 (2012), http://echidna.maths.usyd.edu.au/echidna/index.html

20. Lange, H., Ruppert, W.: Complete systems of addition laws on abelian varieties. Invent. Math. 79(3), 603–610 (1985)

21. Magma Computational Algebra System, Computational Algebra Group, University of Sydney (2012), http://magma.maths.usyd.edu.au/

22. Milne, J.S.: Abelian Varieties, version 2.00 (2012), http://www.jmilne.org/math/CourseNotes/av.html

23. Montgomery, P.L.: Speeding the Pollard and elliptic curve methods of factorization. Mathematics of Computation 48, 243–264 (1987)

24. Stam, M.: On Montgomery-Like Representationsfor Elliptic Curves over $GF(2^k)$. In: Desmedt, Y.G. (ed.) PKC 2003. LNCS, vol. 2567, pp. 240–253. Springer, Heidelberg (2002)

Appendix

By means of a renormalization of variables, the split μ_4-normal form can be put in μ_4-normal form $(X_0 + X_2)^2 = X_1 X_3$, $s(X_1 + X_3)^2 = X_0 X_2$, where $s = c^{-4}$. This form loses the elementary symmetry given by cyclic permutation of the coordinates, but by the Remark following Corollary 21, we are able to save on multiplications by scalars in duplication. This renormalization gives the following addition laws (as a consequence of Theorem 18), and give an analogous savings for addition.

Theorem 27. *Let C be an elliptic curve in μ_4-normal form: A basis for the space of addition laws of bidegree $(2,2)$ is given by:*

$$\left\{ \begin{array}{l} \left((X_0Y_0 + X_2Y_2)^2,\ X_0X_1Y_0Y_1 + X_2X_3Y_2Y_3,\ s(X_1Y_1 + X_3Y_3)^2,\ X_0X_3Y_0Y_3 + X_1X_2Y_1Y_2\right), \\ \left(X_0X_1Y_0Y_3 + X_2X_3Y_1Y_2,\ (X_1Y_0 + X_3Y_2)^2,\ X_0X_3Y_2Y_3 + X_1X_2Y_0Y_1,\ (X_0Y_3 + X_2Y_1)^2\right), \\ \left(s(X_1Y_3 + X_3Y_1)^2,\ X_0X_3Y_1Y_2 + X_1X_2Y_0Y_3,\ (X_0Y_2 + X_2Y_0)^2,\ X_0X_1Y_2Y_3 + X_2X_3Y_0Y_1\right), \\ \left(X_0X_3Y_0Y_1 + X_1X_2Y_2Y_3,\ (X_0Y_1 + X_2Y_3)^2,\ X_0X_1Y_1Y_2 + X_2X_3Y_0Y_3,\ (X_1Y_2 + X_3Y_0)^2\right). \end{array} \right\}$$

The absence of the constant s in the 2nd and 4th addition laws permits us to save the 2m in the computation of addition.

Corollary 28. *Addition of generic points on C can be carried out in $7\mathbf{M} + 2\mathbf{S}$.*

Proof. After evaluating $(Z_0, Z_1, Z_2, Z_3) = (X_0Y_1, X_1Y_2, X_2Y_3, X_3Y_0)$ in the last addition law, the algorithm follows that of Corollary 19.   □

A New Model of Binary Elliptic Curves

Hongfeng Wu[1,*], Chunming Tang[2], and Rongquan Feng[2,**]

[1] College of Sciences, North China University of Technology, Beijing 100144, China
whfmath@gmail.com
[2] LMAM, School of Mathematical Sciences, Peking University, Beijing 100871, China
tangchunmingmath@163.com, fengrq@math.pku.edu.cn

Abstract. In this paper, we present a new model of elliptic curves over finite fields of characteristic 2. We first describe the group law on this new binary curve. Furthermore, this paper presents the unified addition formulas for new binary elliptic curves, that is the point addition formulas which can be used for almost all doubling and addition. Finally, this paper presents explicit addition formulas for differential addition.

Keywords: Elliptic curve, point multiplication, unified addition law, differential addition, cryptography.

1 Introduction

Elliptic curve cryptography (ECC) was first proposed by Koblitz [8] and Miller [13]. In recent years, ECC has gained widespread exposure and acceptance. One of the key operations of ECC is the scalar multiplication, i.e., to compute kP by giving a point P on the curve and an integer k. In the last twenty years, there are a lot of techniques introduced in the literature to improve the scalar multiplication of ECC. Two types of elliptic curves are mainly used to implement ECC: elliptic curve over large prime fields and elliptic curve over finite fields of characteristic 2.

A well known elliptic curve model is Weierstrass model, and many efficient formulae for this model can be found in [1,6,7,9,11,16,14] etc.. Let $K = \mathbb{F}_{2^m}$ be a finite field of characteristic 2, a non-supersingular elliptic curve over K can be written in Weierstrass form [18,12]

$$E/\mathbb{F}_{2^m} : y^2 + xy = x^3 + ax + b$$

with $a, b \in \mathbb{F}_{2^m}$ and $b \neq 0$. There are alternate models of elliptic curves, such as binary Edwards curves and binary Huff curves, have been proposed for usage in cryptography. Expressing an elliptic curve with these models can lead to more efficient arithmetics. These curves can also lead to improved security because

---

* Supported in part by the National Natural Science Foundation of China (No. 11101002).
** The work of R. Feng is supported by the National Natural Science Foundation of China (No. 10990011).

S. Galbraith and M. Nandi (Eds.): INDOCRYPT 2012, LNCS 7668, pp. 399–411, 2012.
© Springer-Verlag Berlin Heidelberg 2012

they have unified addition and doubling laws, which can reduce information leakage through side channels.

A binary Edwards curve [2] over $\mathbb{F}_{2^m}$ is the affine curve

$$E_{d_1,d_2} : d_1(x + y) + d_2(x^2 + y^2) = xy + xy(x + y) + x^2y^2$$

with $d_1, d_2 \in K$ and $d_1 \neq 0$, $d_2 \neq d_1^2 + d_1$. This curve is symmetric in x and y and thus has the property that if (x_1, y_1) is a point on the curve then so is (y_1, x_1). The dedicated doubling formulas for binary Edwards curves using $2M + 6S + 3D$, where M is the cost of a field multiplication, S is the cost of a field squaring, and D is the cost of multiplying by a curve parameter. The projective addition formulas for binary Edwards curve using $18M + 2S + 7D$. For differential addition, i.e., addition of points with known difference, which is the basic step in a Montgomery ladder, uses $5M + 5S + 1D$ when the known difference is given in affine form.

A binary Huff curve [10] over $\mathbb{F}_{2^m}$ is the affine curve

$$ax(y^2 + y + 1) = by(x^2 + x + 1)$$

with $a, b \in K$ and $a \neq b$. This curve is not symmetric in x and y. The dedicated doubling formulas for binary Huff curves using $6M + 2D$. The projective unified addition formulas for binary Huff curve using $15M + 2D$. The differential addition uses $5M + 4S + 2D$.

In the spirit of [2] and [10], we introduce a new model of elliptic curves

$$S_t : x^2y + xy^2 + txy + x + y = 0$$

over $\mathbb{F}_{2^m}$. This curve is symmetric in x and y. The dedicated doubling formulas, the projective unified addition formulas, and the differential addition and doubling formulas for this new binary curve are described. Our addition formulas and differential-addition formulas are extremely fast: for example, the dedicated doubling formula uses $3M + 6S + 1D$. The projective unified addition formula uses $12M + 4S + 2D$. The basic step in a Montgomery ladder for a differential addition and doubling cost $5M + 5S + 1D$ or $5M + 4S + 2D$.

This paper is organized as follows. In Section 2 a new model of binary curves and their birational equivalence with Weierstrass model are presented. In Section 3, the group law on this new binary curve is described. Explicit addition and doubling formulas are given in Sections 4 and 5 respectively. In Section 6, differential addition and doubling formulas are given. Finally, we conclude in Section 7.

2   A New Model of Binary Edwards Curves

Let K be a field of characteristic 2. Consider the set of projective points $(X : Y : Z) \in \mathbb{P}^2(K)$ satisfying the equation

$$S_t : X^2Y + XY^2 + tXYZ + XZ^2 + YZ^2 = 0 \tag{1}$$

where $t \in K$ and $t \neq 0$. The tangent line at $(1 : 1 : 0)$ is $X + Y + tZ = 0$, which intersects the curve with multiplicity 3, so that $(1 : 1 : 0)$ is an inflection point of S_t. The partial derivatives of the curve equation are $Y^2 + tYZ + Z^2, X^2 + tXZ + Z^2$ and tXY. A singular point $(X_1 : Y_1 : Z_1)$ must have $Y_1^2 + tY_1Z_1 + Z_1^2 = X_1^2 + tX_1Z_1 + Z_1^2 = tX_1Y_1 = 0$, and therefore $X_1 = Y_1 = Z_1 = 0$ since $t \neq 0$. Therefore, S_t is nonsingular. The affine form of the curve is

$$S_t : x^2y + xy^2 + txy + x + y = 0.$$

Denote $S_t(K)$ as

$$S_t(K) = \{(x,y) \in K^2 | x^2y + xy^2 + txy + x + y = 0\} \cup \{(1:0:0), (0:1:0), (1:1:0)\}$$

by a slight abuse notation.

Note that the variant form $x^2y + xy^2 + axy + b(x + y) = 0$ is isomorphic to $x^2y + xy^2 + txy + (x+y) = 0$ via the change of variables $(x, y) \to (ax/\sqrt{b}, ay/\sqrt{b})$ with $t = a/\sqrt{b}$, and the curve $x^2y + xy^2 + xy + b(x + y) = 0$ is isomorphic to $x^2y + xy^2 + txy + (x + y) = 0$ by $(x, y) \to (x/\sqrt{b}, y/\sqrt{b})$ with $t = 1/\sqrt{b}$.

The new curve $x^2y + xy^2 + xy + b(x + y) = 0$ looks similar to the binary Edwards curve $E_{B,b,0} : b(x + y) = xy + xy(x + y) + x^2y^2$ without the quartic term. Hence, we call this new curve a new binary Edwards curve model. The generalized form $S_{a,b} : x^2y + xy^2 + axy + (x + y) + b(x^2 + y^2) = 0$ of S_t is isomorphic to $v^2 + uv = u^3 + (b/a)u^2 + a^{-8}(1 + ab)$. So one can change $S_{a,b}$ to the form $d_1(x + y) + d_2(x^2 + y^2) = xy + xy(x + y)$.

2.1 Birational Equivalence with Weierstrass Model

Let $S_t : x^2y + xy^2 + txy + x + y = 0$ be defined over the finite field $\mathbb{F}_{2^m}$. Then S_t is birational equivalent to the Weierstrass elliptic curve $v^2 + uv = u^3 + \dfrac{1}{t^8}$ over $\mathbb{F}_{2^m}$ via the change of variables $\varphi(x, y) = (u, v)$, where

$$u = \frac{x + y}{t^2(x + y + t)}, \quad \text{and} \quad v = \frac{x + y + t^2x + t}{t^4(x + y + t)}.$$

The inverse map is $\psi(u, v) = (x, y)$, where $x = \frac{t^4v+1}{t^3u+t}$ and $y = \frac{t^4(u+v)+1}{t^3u+t}$. Note that $x + y + t = 0$ lead to $x + y = 0$, thus $t = 0$, hence, the rational map covers all points and defines an isomorphism.

In projective coordinates, the corresponding projective transformations from $X^2Y + XY^2 + tXYZ + XZ^2 + YZ^2 = 0$ to $V^2W + UVW = U^3 + \dfrac{1}{t^8}W^3$ is given by $(X : Y : Z) \mapsto (U : V : W)$, where

$$\begin{cases} U = t^2(X + Y), \\ V = X + Y + t^2X + tZ, \\ W = t^4(X + Y + tZ). \end{cases}$$

The inverse transformation is $(U : V : W) \mapsto (X : Y : Z)$, where

$$\begin{cases} X = t^4 V + W, \\ Y = t^4 (U + V) + W, \\ Z = t^3 U + tW. \end{cases}$$

The above change of variables map the element $(1 : 1 : 0)$ on S_t to the identity element $(0 : 1 : 0)$ on the Weierstrass curve.

Note that the curve $x^2 y + xy^2 + xy + b(x + y) = 0$ is isomorphic to $v^2 + uv = u^3 + b^4$ via the change of variables

$$x = \frac{v + b^2}{u + b}, \quad \text{and} \quad y = \frac{u + v + b^2}{u + b}.$$

Lemma 1. *An elliptic curve E defined over $\mathbb{F}_{2^m}$ satisfies $4 \mid \sharp E(\mathbb{F}_{2^m})$ if and only if E is isomorphic to an elliptic curve of the form $x^2 y + xy^2 + txy + x + y = 0$.*

Proof. For any $a \in \mathbb{F}_{2^m}^*$, there exists a t such that $S_t : x^2 y + xy^2 + txy + x + y = 0$ is isomorphic to $v^2 + uv = u^3 + a$. We only need to prove that an elliptic curve E defined over $\mathbb{F}_{2^m}$ satisfies $4 \mid \sharp E(\mathbb{F}_{2^m})$ if and only if E is isomorphic to a curve of the form $W_a : v^2 + uv = u^3 + a$.

Assume that E is isomorphic to $W_a : v^2 + uv = u^3 + a$. For any point $P = (x, y) \in S_a$ with $P \neq (0 : 1 : 0), (0, \sqrt{a})$, we have $x \neq 0$. Therefore the number of points on W_a is $\sharp W_a(\mathbb{F}_{2^m}) = 2 + 2\sharp\{t \in \mathbb{F}_{2^m} | t^2 + t = x + \frac{a}{x^2}, x \neq 0\}$. The equation $t^2 + t = x + \frac{a}{x^2}$ has a solution if and only if $Tr(x + \frac{a}{x^2}) = 0$, that is, $Tr(x) = Tr(\frac{\sqrt{a}}{x})$. Note that $\sharp\{x \in \mathbb{F}_{2^m}^* | Tr(x) = Tr(\frac{\sqrt{a}}{x})\}$ is an odd number since $x \mapsto \frac{\sqrt{a}}{x}$ is an involution on $\mathbb{F}_{2^m}^*$ with precisely one fixed point. Actually, the point $(\sqrt[4]{a}, \sqrt{a})$ belongs to W_a and has order 4. Hence $4 \mid \sharp E(\mathbb{F}_{2^m})$.

Conversely, if $4 \mid \sharp E(\mathbb{F}_{2^m})$, then E is ordinary, and has an equation $E : y^2 + xy = x^3 + rx^2 + a$ after a suitable choice of coordinates, where $r \in \mathbb{F}_{2^m}$. We can change $v^2 + uv = u^3 + a$ to a standard form $E_a : y^2 + xy = x^3 + bx^2 + a$ with some $b \in \mathbb{F}_{2^m}$. The curve E is isomorphic to E_a if and only if $Tr(r) = Tr(b)$. If E is not isomorphic to E_a, then $Tr(r) \neq Tr(b)$ and $t = a$. Thus E is a quadratic twist of E_a and $\sharp E_a(\mathbb{F}_{2^m}) + \sharp E(\mathbb{F}_{2^m}) = 2^{m+1} + 2 \equiv 2 \pmod 4$. □

3   The Addition Law

Let C be a nonsingular cubic curve defined over a field K, and let O be a point on $C(K)$. For any two points P and Q, the line through P and Q meets the cubic curve C at one more point, denoted by PQ. With a point O as zero element and the chord-tangent composition PQ we can define the group law $P + Q$ by the relation $P + Q = O(PQ)$, this means that $P + Q$ is the third intersection point on the line through O and PQ. This makes $C(K)$ into an abelian group with O as zero element and $-P = P(OO)$. If O be an inflection point then $-P = PO$ and $OO = O$.

Note that $(1 : 1 : 0)$ is an inflection point on the curve S_t. Define $O = (1 : 1 : 0)$ as the neutral element. Let $P = (x_1, y_1)$ be a finite point on the curve S_t. The inverse of P is therefore defined as $-P = PO$, that is $-P = (y_1, x_1)$. For the points at infinity, we obtain $-(0 : 1 : 0) = (1 : 0 : 0)$ and $-(1 : 0 : 0) = (0 : 1 : 0)$. Observe that $(0 : 0 : 1)$ is of order 2.

Let $P = (x_1, y_1) \in S_t$ be a finite point. The third point of intersection of the tangent line to the curve at P is $R := PP$. Then $[2]P = OR$. After some calculation, whenever defined, we get $[2]P = (x_3, y_3)$ with

$$x_3 = \frac{t(1 + x_1^2)}{x_1^2 + y_1^2 + x_1^2 y_1^2 + t^2 x_1^2 + x_1^4}, \quad \text{and} \quad y_3 = \frac{t(1 + y_1^2)}{x_1^2 + y_1^2 + x_1^2 y_1^2 + t^2 y_1^2 + y_1^4}. \tag{2}$$

For the points at infinity, we have $2(0 : 1 : 0) = (0 : 0 : 1)$ and $2(1 : 0 : 0) = (0 : 0 : 1)$.

Let $P = (x_1, y_1)$ and $Q = (x_2, y_2) \in S_t$ be two finite points with $P \neq Q$. As explained above, the addition of P and Q is given by $P + Q = O(PQ)$. Then, whenever defined, we obtain the *dedicated point addition* formula $P + Q = (x_3, y_3)$ as

$$x_3 = \frac{x_1 + y_1 + x_2 + y_2 + x_1 y_2(x_1 + y_1 + t) + x_2 y_1(x_2 + y_2 + t)}{(x_1 + x_2)(x_1 + y_1 + x_2 + y_2)},$$

$$y_3 = \frac{x_1 + y_1 + x_2 + y_2 + x_1 y_2(x_2 + y_2 + t) + x_2 y_1(x_1 + y_1 + t)}{(y_1 + y_2)(x_1 + y_1 + x_2 + y_2)}. \tag{3}$$

For infinite points, we also have $(0 : 1 : 0) + (1 : 0 : 0) = (1 : 1 : 0)$, $(0 : 1 : 0) + (0 : 0 : 1) = (0 : 1 : 0)$ and $(1 : 0 : 0) + (0 : 0 : 1) = (1 : 0 : 0)$.

If $P = (x_1, y_1)$ is a finite point and Q is a point at infinity or Q is $(0 : 0 : 1)$, we have the following formulas whenever defined.

$$\begin{cases} (x_1 : y_1 : 1) + (1 : 0 : 0) = (y_1 : \dfrac{1}{x_1} : 1), \\[2mm] (x_1 : y_1 : 1) + (0 : 1 : 0) = (\dfrac{1}{y_1} : x_1 : 1), \\[2mm] (x_1 : y_1 : 1) + (0 : 0 : 1) = (\dfrac{x_1 y_1 + t y_1}{x_1} : x_1 + t : 1). \end{cases}$$

We can also delete t from the above dedicated addition formula and get the following dedicated addition formula which is independent of the curve parameters.

$$x_3 = \frac{(y_1 + y_2)(y_1 x_2 + y_2 x_1)}{y_1 y_2(x_1 + x_2)(x_1 + y_1 + x_2 + y_2)}, \quad y_3 = \frac{(x_1 + x_2)(y_1 x_2 + y_2 x_1)}{x_1 x_2(y_1 + y_2)(x_1 + y_1 + x_2 + y_2)}. \tag{4}$$

Note that if (x_1, y_1) is on the curve $x^2 y + x y^2 + t x y + x + y = 0$, then so do $(\frac{1}{x_1}, y_1)$, $(x_1, \frac{1}{y_1})$, and $(\frac{1}{x_1}, \frac{1}{y_1})$ whenever defined. We have $(x_1, y_1) + (\frac{1}{x_1}, y_1) = (0 : 1 : 0)$ when $x_1 \neq 0$, and $(x_1, y_1) + (x_1, \frac{1}{y_1}) = (1 : 0 : 0)$ when $y_1 \neq 0$.

The following facts will be useful in the sequal sections.

$$(x_1, y_1) + (\frac{1}{x_1}, \frac{1}{y_1}) = 2(y_1, \frac{1}{x_1}) = \left(\frac{(1+y_1^2)(1+x_1^2 y_1^2)}{tx_1^2(1+y_1^2)}, \frac{(1+x_1^2)(1+x_1^2 y_1^2)}{ty_1^2(1+x_1^2)} \right).$$

and $(x_1, y_1) - (\frac{1}{x_1}, \frac{1}{y_1}) = (0, 0)$.

After some calculation, we can get the following *unified point addition* formula. Let $(x_1, y_1) + (x_2, y_2) = (x_3, y_3)$. Then

$$x_3 = \frac{(x_1 x_2 + y_1 y_2)(y_1 + y_2) + t y_1 y_2 (1 + x_1 x_2)}{(x_1 x_2 + y_1 y_2)(1 + y_1 y_2)},$$

$$y_3 = \frac{(x_1 x_2 + y_1 y_2)(x_1 + x_2) + t x_1 x_2 (1 + y_1 y_2)}{(x_1 x_2 + y_1 y_2)(1 + x_1 x_2)}.$$

(5)

One can prove that the addition law corresponds to the usual addition law on an elliptic curve in Weierstrass form. That is, fix $(x_1, y_1), (x_2, y_2), (x_3, y_3) \in S_t(K)$, if $(x_1, y_1) + (x_2, y_2) = (x_3, y_3)$, then $\varphi(x_1, y_1) + \varphi(x_2, y_2) = \varphi(x_3, y_3)$. The corresponding Sage[17] script is given in **Appendix A**.

Completeness of the Addition Law

Let $P = (x_1, y_1)$ and $Q = (x_2, y_2)$ be points on S_t. From formula (5), the addition law is defined when the denominators $(x_1 x_2 + y_1 y_2)(1 + y_1 y_2)$ and $(x_1 x_2 + y_1 y_2)(1 + x_1 x_2)$ are non-zero.

If $1 + y_1 y_2 = 0$, then $y_2 = \frac{1}{y_1}$, and $Q \in \{(x_1, \frac{1}{y_1}), (\frac{1}{x_1}, \frac{1}{y_1})\}$. If $1 + x_1 x_2 = 0$, then $x_2 = \frac{1}{x_1}$, and $Q \in \{(\frac{1}{x_1}, y_1), (\frac{1}{x_1}, \frac{1}{y_1})\}$.

Lemma 2. *Let $P = (x_1, y_1)$ and $Q = (x_2, y_2)$ on the curve S_t. If $x_1 x_2 + y_1 y_2 = 0$, then $Q = (\frac{1}{x_1}, \frac{1}{y_1})$ or $Q = -P$.*

Proof. If $x_1 x_2 + y_1 y_2 = 0$, then $x_1 x_2 = y_1 y_2$. If $x_1 x_2 = y_1 y_2 = 1$, then $Q = (\frac{1}{x_1}, \frac{1}{y_1})$. If $x_1 x_2 = y_1 y_2 = a \neq 0, 1$, then $x_2 = a/x_1, y_2 = a/y_1$. Since $x_1^2 y_1 + x_1 y_1^2 + t x_1 y_1 + x_1 + y_1 = 0$,

$$\frac{1}{x_1^2 y_1} + \frac{1}{x_1 y_1^2} + \frac{t}{x_1 y_1} + \frac{1}{x_1} + \frac{1}{y_1} = 0 \text{ and } \frac{a^2}{x_1^2 y_1} + \frac{a^2}{x_1 y_1^2} + \frac{ta}{x_1 y_1} + \frac{1}{x_1} + \frac{1}{y_1} = 0.$$

Therefore,

$$\frac{1}{x_1^2 y_1} + \frac{a^2}{x_1^2 y_1} + \frac{1}{x_1 y_1^2} + \frac{a^2}{x_1 y_1^2} + \frac{t}{x_1 y_1} + \frac{ta}{x_1 y_1} = 0.$$

Hence $x_1 + a^2 x_1 + y_1 + a^2 y_1 + t x_1 y_1 + t a x_1 y_1 = 0$ and $x_1 + y_1 = \frac{t x_1 y_1}{1+a}$. Thus $x_1 y_1 = a$ since $x_1 + y_1 = \frac{t x_1 y_1}{x_1 y_1 + 1}$. From $x_1 x_2 = y_1 y_2 = a$ and $x_1 y_1 = a$, we get $x_2 = y_1$ and $y_2 = x_1$. That is, $Q = -P$.   □

Note that if $P = (x_1, y_1)$ and $Q \in \{(\frac{1}{x_1}, y_1), (x_1, \frac{1}{y_1}), (\frac{1}{x_1}, \frac{1}{y_1})\}$, then $P + Q = (0 : 1 : 0), (1 : 0 : 0)$ or $(0 : 0 : 1)$. Therefore, we have the following theorem.

Theorem 3. *Let the curve* $S_t : x^2y + xy^2 + txy + x + y = 0$ *be defined over* $\mathbb{F}_{2^m}$ *and let* $G \subset S_t(\mathbb{F}_{2^m})$ *be a subgroup that does not contain points* $(0 : 1 : 0)$, $(1 : 0 : 0)$ *or* $(0 : 0 : 1)$. *Then the unified addition formula is complete.*

In particular, the addition formula is complete in a subgroup of odd order, since all points $(0 : 1 : 0)$, $(1 : 0 : 0)$ and $(0 : 0 : 1)$ have even order.

Projective Formula. We now present the projective version of the addition formula. For $P = (X_1 : Y_1 : Z_1)$ and $Q = (X_2 : Y_2 : Z_2)$, Let $P + Q = (X_3 : Y_3 : Z_3)$. Then the projective version of formula (3) is

$$\begin{aligned}
X_3 = (Y_1Z_2 + Y_2Z_1) \cdot (Z_1Z_2^2(X_1 + Y_1) + Z_1^2Z_2(X_2 + Y_2) \\
+ X_1Y_2Z_2(X_1 + Y_1 + tZ_1) + X_2Y_1Z_1(X_2 + Y_2 + tZ_2)),
\end{aligned}$$

$$\begin{aligned}
Y_3 = (X_1Z_2 + X_2Z_1) \cdot (Z_1Z_2^2(X_1 + Y_1) + Z_1^2Z_2(X_2 + Y_2) \\
+ X_1Y_2Z_1(X_2 + Y_2 + tZ_2) + X_2Y_1Z_2(X_1 + Y_1 + tZ_1)),
\end{aligned} \quad (6)$$

$$Z_3 = (X_1Z_2 + X_2Z_1)(Y_1Z_2 + Y_2Z_1)(X_1Z_2 + Y_1Z_2 + X_2Z_1 + Y_2Z_1).$$

The projective version of the unified formula (5) is

$$X_3 = (X_1X_2 + Z_1Z_2) \cdot ((X_1X_2 + Y_1Y_2)(Y_1Z_2 + Y_2Z_1) + tY_1Y_2(Z_1Z_2 + X_1X_2)),$$

$$Y_3 = (Y_1Y_2 + Z_1Z_2) \cdot ((X_1X_2 + Y_1Y_2)(X_1Z_2 + X_2Z_1) + tX_1X_2(Z_1Z_2 + Y_1Y_2)),$$

$$Z_3 = (X_1X_2 + Y_1Y_2)(X_1X_2 + Z_1Z_2)(Y_1Y_2 + Z_1Z_2).$$

$$(7)$$

4 Explicit Addition Formulas

In this section, explicit formulas for affine addition, projective addition, and mixed addition on the binary curve S_t are presented.

Affine Addition. Given (x_1, y_1) and (x_2, y_2) on the curve $S_t : x^2y + xy^2 + txy + x + y = 0$, we have the following formulas by using the formula (3) to compute the sum $(x_3, y_3) = (x_1, y_1) + (x_2, y_2)$ if it is defined:

$$\begin{aligned}
w_1 &= x_1 + y_1 + t, \ w_2 = x_2 + y_2 + t, \ A = x_1y_2, \ B = x_2y_1, \\
C &= A \cdot w_1, \ D = B \cdot w_2, \ E = (A + B) \cdot (w_1 + w_2) + C + D, \\
F &= (x_1 + x_2) \cdot (y_1 + y_2), \ G = (x_1 + x_2)^2 + F, \ H = (y_1 + y_2)^2 + F \\
x_3 &= (w_1 + w_2 + C + D)/G, \ y_3 = (w_1 + w_2 + E)/H.
\end{aligned}$$

These formulas cost $2I + 8M + 2S$, where I is the cost of a field inversion, M is the cost of a field multiplication, and S is the cost of a field squaring. We will use D to denote the cost of a multiplication by a curve parameter. One can replace $2I$ with $1I + 3M$ by using Montgomery's inversion trick. Then the affine addition costs $1I + 11M$. Note that the cost of additions and squaring in $\mathbb{F}_{2^m}$ can be neglected.

By the formula (5), we have the following algorithm to compute the sum $(x_3, y_3) = (x_1, y_1) + (x_2, y_2)$ if it is defined:

$$A = x_1 \cdot x_2, \; B = y_1 \cdot y_2, \; C = (A + B) \cdot (y_1 + y_2), \; D = (A + B) \cdot (x_1 + x_2),$$
$$E = A \cdot B, \; F = B + E, \; G = A + E, \; H = A + B + E + B^2,$$
$$J = A + B + E + A^2, \; x_3 = (C + tF)/H, \; y_3 = (D + tG)/J.$$

This algorithm costs $2I + 7M + 2D + 2S$ or $1I + 10M + 2D + 2S$, the $2D$ here is two multiplications by t.

Projective Addition. Given points $(X_1 : Y_1 : Z_1)$ and $(X_2 : Y_2 : Z_2)$ on the curve S_t, the following algorithm can compute the sum $(X_3 : Y_3 : Z_3) = (X_1 : Y_1 : Z_1) + (X_2 : Y_2 : Z_2)$ by using the unified formula (7) if it is defined.

$$A = X_1 \cdot X_2, \; B = Y_1 \cdot Y_2, \; C = Z_1 \cdot Z_2,$$
$$D = (X_1 + Z_1) \cdot (X_2 + Z_2) + A + C,$$
$$E = (Y_1 + Z_1) \cdot (Y_2 + Z_2) + B + C,$$
$$X_3 = (A + C) \cdot ((A + B) \cdot E + tB \cdot (A + C)),$$
$$Y_3 = (B + C) \cdot ((A + B) \cdot D + tA \cdot (B + C)),$$
$$Z_3 = (A + B) \cdot (A + C) \cdot (B + C).$$

This algorithm costs $13M + 2D$, where the $2D$ is multiplications by the curve parameter t.

Since the squaring in $\mathbb{F}_{2^m}$ can be neglected, we have the following algorithm to compute the sum.

$$A = X_1 \cdot X_2, \; B = Y_1 \cdot Y_2, \; C = Z_1 \cdot Z_2, \; D = (X_1 + Z_1) \cdot (X_2 + Z_2) + A + C,$$
$$E = (Y_1 + Z_1) \cdot (Y_2 + Z_2) + B + C, \; F = (A + C)^2, \; G = (B + C)^2, \; H = A \cdot (B + C),$$
$$I = B \cdot C, \; J = A^2, \; K = B^2, \; X_3 = (J + H + I) \cdot E + tB \cdot F,$$
$$Y_3 = (H + K + I) \cdot D + tA \cdot G, \; Z_3 = (J + H + I) \cdot (B + C).$$

This algorithm costs $12M + 4S + 2D$, where the $2D$ is multiplication by the curve parameter t.

Mixed Addition. The mixed addition is to compute $(X_3 : Y_3 : Z_3) = (X_1 : Y_1 : Z_1) + (x_2, y_2)$ by given $(X_1 : Y_1 : Z_1)$ and (x_2, y_2) on the curve S_t. From projective addition algorithm we can get the mixed addition which costs $12M + D$ by using the formula (6) since $Z_2 = 1$. However, by using the formula (7), the mixed addition costs $11M + 2D$.

Comparison with Previous Works. The following comparison shows that our addition formulas are more efficient than binary Edwards curves and Weierstrass curves.

The projective addition formulas of binary Edwards curves in [2] cost $21M + 1S + 4D$, or $18M + 2S + 7D$ when the curve parameters are small. The fastest formulas cost $16M + 1S + 4D$ when the parameters $d_1 = d_2$ for binary Edwards curves. The best operation counts $13M + 4S$ for Weierstrass curves with projective coordinates reported in Explicit-Formulas Database [1]. The projective addition formulas of binary Huff curves in [10] cost $15M + 2S + 2D$. Therefore, our formulas are more faster than the formulas in the literature.

5   Doubling

The doubling formulas on the binary curve S_t will be described in this section.

Affine Doubling. Let $P = (x_1, y_1)$ be a point on S_t, and assume that the sum $2(x_1, y_1)$ is defined. From unified formula (5) we get

$$2P = 2(x_1, y_1) = \left(\frac{ty_1^2(1+x_1)^2}{(x_1^2+y_1^2)(1+y_1^2)}, \frac{tx_1^2(1+y_1)^2}{(x_1^2+y_1^2)(1+x_1^2)} \right).$$

From $(x_1+y_1)(1+x_1)(1+y_1) = x_1(1+y_1^2) + y_1(1+x_1)^2 + x_1^2 + y_1^2$, we have the following algorithm to compute $2P$:

$$A = y_1 \cdot (1+x_1^2), \ B = x_1 \cdot (1+y_1^2), \ D = (A+B+x_1^2+y_1^2)^{-1},$$
$$E = tD^2, \ x_3 = E \cdot A^2, \ y_3 = E \cdot B^2.$$

The algorithm costs $1I + 4M + 5S + D$, where the $1D$ is the multiplication by t.

Projective Doubling. Let $P = (X_1, Y_1, Z_1)$ and $2P = (X_3, Y_3, Z_3)$. From the unified formula (5) we get

$$2P = \left(tY_1^2(X_1^2+Z_1^2)^2, tX_1^2(Y_1^2+Z_1^2)^2, (X_1^2+Y_1^2)(X_1^2+Z_1^2)(Y_1^2+Z_1^2)\right)$$
$$= \left(Y_1^2(X_1^2+Z_1^2)^2, X_1^2(Y_1^2+Z_1^2)^2, (1/t)(X_1^2+Y_1^2)(X_1^2+Z_1^2)(Y_1^2+Z_1^2)\right).$$

Since

$$(X_1^2+Y_1^2)(X_1^2+Z_1^2)(Y_1^2+Z_1^2) = \left(Y_1(X_1^2+Z_1^2) + X_1(Y_1^2+Z_1^2) + Z_1(X_1^2+Y_1^2)\right)^2,$$

This leads to the following doubling algorithm.

$$A = X_1^2, \ B = Y_1^2, \ C = Z_1^2, \ D = Y_1 \cdot (A+C), \ E = X_1 \cdot (B+C)$$
$$X_3 = D^2, \ Y_3 = E^2, \ Z_3 = (1/t)(D+E+Z_1 \cdot (A+B))^2.$$

This algorithm costs $3M + 6S + 1D$, where the $1D$ is the multiplication by $1/t$.

Comparison with Previous Works. We compare our addition formulas of new binary curve with other models of elliptic curves. The following comparison (Table 1) shows that the new binary curve are competitive to binary Edwards curves, binary Huff curves, binary Hessian curves and Weierstrass curves.

We denote Edwards coordinates by $\mathcal{E}$, projective coordinates by $\mathcal{P}$, and extended López-Dahab coordinates [11] by $\mathcal{L}$.

Table 1. Comparisons of points operations in binary fields

Models	Addition	Doubling
$\mathcal{L}$, Weierstrass [1]	13M+4S	2M+4S+2D
$\mathcal{E}$, Binary Edwards [2]	18M+2S+7D	2M+6S+3D
$\mathcal{E}$, Binary Edwards, $d_1 = d_2$ [2]	16M+1S+4D	2M+5S+2D
$\mathcal{E}$, Binary Huff [10]	13M+3S+2D	6M+5S+2D
$\mathcal{P}$, Hessian curve [4]	12M	6M+3S+2D
$\mathcal{P}$, new model in this paper	12M+4S+2D	3M+6S+1D

6 Differential Addition

In this section, fast explicit formulas for w-coordinate differential addition on the curve $S_t : x^2y + xy^2 + txy + x + y = 0$ will be presented. We can define the w-function in two ways, where $w(P) = x + y$ or $w(P) = xy$ for $P = (x, y)$. Note that $w(-P) = w(P)$ since $-(x, y) = (y, x)$. In the following we only present explicit formulas of differential addition and doubling for w-coordinates with $w(P) = xy$.

Differential addition means computing $Q + P$ by giving P, Q, and $Q - P$ or computing $2P$ by giving P. A general differential point addition consists in calculating $w(P + Q)$ from $w(P)$, $w(Q)$ and $w(Q - P)$ for some coordinate function w. Montgomery [15] developed a method, called Montgomery ladder, allowing faster scalar multiplication than usual methods. Montgomery presented fast formulas for u-coordinate differential addition on non-binary elliptic curves $v^2 = u^3 + a_2u^2 + u$. The Montgomery ladder can compute $u(mP), u((m + 1)P)$ efficiently by giving $u(P)$, and is one of the most important methods to compute the scalar multiplication. Bernstein et al. [2] presented a fast w-coordinate differential addition on binary Edwards curves by using the idea of the Montgomery ladder.

More concretely, write $Q - P = (x_1, y_1)$, $P = (x_2, y_2)$, $Q = (x_3, y_3)$, $2P = (x_4, y_4)$ and $Q + P = (x_5, y_5)$, and write $w_i = x_iy_i$ for $i = 1, 2, 3, 4, 5$. We will present fast explicit formulas to compute $w(P+Q)$ and $w(2P)$ from $w(P)$, $w(Q)$ and $w(Q - P)$.

From Section 5, we know that the doubling formula is

$$2P = 2(x, y) = \left(\frac{ty^2(1 + x^2)}{(x^2 + y^2)(1 + y^2)}, \frac{tx^2(1 + y^2)}{(x^2 + y^2)(1 + x^2)} \right).$$

Let $w_1 = w(P)$. Then $w(2P) = \dfrac{t^2x^2y^2}{(x^2 + y^2)^2}$. Thus we have $w_4 = \dfrac{1 + w_2^4}{t^2w_2^2}$.

By a tedious but straightforward calculation, we can get that

$$w_1 + w_5 = \frac{t^2w_2w_3}{w_2^2 + w_3^2} \quad \text{and} \quad w_1w_5 = \frac{1 + w_2^2w_3^2}{w_2^2 + w_3^2}.$$

Cost of Affine w-Coordinate Differential Addition and Doubling. The explicit addition formulas

$$A = w_2^2, \ B = w_3^2, \ C = w_2 w_3, \ D = (A+B)^{-1}, \ \text{and}$$
$$w_5 = w_1 + t^2 C \cdot D$$

cost $1I + 2M + 2S + 1D$, where the $1D$ is a multiplication by the curve parameter t^2. The explicit doubling formulas

$$A = w_2^2, \ B = A^2, \ C = t^2 A, \ D = C^{-1}, \ \text{and}$$
$$w_4 = (1+B) \cdot D$$

cost $1I + 1M + 2S + 1D$, where the $1D$ is a multiplication by the curve parameter t^2. The total cost of a differential addition and doubling is $2I + 3M + 4S + 2D$, or $I + 6M + 4S + 2D$ with Montgomery's inversion trick.

Cost of Projective w-Coordinate Differential Addition. Assume that w_1, w_2, w_3 are given as fractions $W_1/Z_1, W_2/Z_2, W_3/Z_3$ and that w_4, w_5 are to be output as fractions $W_4/Z_4, W_5/Z_5$.

The explicit addition formulas

$$A = W_2 \cdot Z_3, \ B = W_3 \cdot Z_2, \ C = (A+B)^2,$$
$$W_5 = t^2 A \cdot Z_1 \cdot B + W_1 \cdot C, \ Z_5 = Z_1 \cdot C$$

cost $6M + 1S + 1D$, where the $1D$ is a multiplication by the curve parameter t^2. From the $w_1 w_5$ formula, we have alternate formulas

$$A = Z_2 \cdot Z_3, \ B = W_2 \cdot w_3, \ C = (A+B)^2,$$
$$D = (W_2 + Z_2) \cdot (W_3 + Z_3) + A + B,$$
$$W_5 = Z_1 \cdot C, \ Z_5 = W_1 \cdot D^2$$

which cost $5M + 2S$ for differential addition.

Cost of Mixed w-Coordinate Differential Addition and Doubling. Assume that w_1, w_2, w_3 are given as fractions $W_1/Z_1, W_2/Z_2, W_3/Z_3$ and that w_4, w_5 are to be output as fractions $W_4/Z_4, W_5/Z_5$, where $Z_1 = 1$. The explicit mixed addition formulas

$$A = W_2 \cdot Z_3, \ B = W_3 \cdot Z_2, \ C = (A+B)^2,$$
$$W_5 = t^2 A \cdot B + w_1 \cdot C, \ Z_5 = C$$

use $4M + 1S + 1D$, where the $1D$ is multiplication by the curve parameter t^2. From the $w_1 w_5$ formula, we have alternate formulas

$$A = Z_2 \cdot Z_3, \ B = W_2 \cdot w_3, \ C = (A+B)^2,$$
$$D = (W_2 + Z_2) \cdot (W_3 + Z_3) + A + B,$$
$$W_5 = C, \ Z_5 = w_1 \cdot D^2$$

which cost $4M + 2S$ for mixed differential addition.

Moreover, the explicit doubling formulas

$$A = W_2^2, \ C = Z_2^2,$$
$$W_4 = (A + C)^2, \ Z_4 = t^2 A \cdot C$$

cost $1M + 3S + 1D$, where the $1D$ is multiplication by the curve parameter t^2. Thus, the total cost of differential addition and doubling is $5M + 4S + 2D$ or $5M + 5S + 1D$.

Table 2. Comparisons of differential addition over binary fields

Models	differential doubling	differential addition	Total
Weierstrass [1]	1M+3S+1D	4M+1S	5M+4S+1D
Binary Edwards [2]	1M+3S+1D	4M+1S+1D	5M+4S+2D
Binary Huff [10]	1M+3S+1D	4M+2S	5M+5S+1D
Hessian curve [4]	1M+3S+1D	4M+1S+1D	5M+4S+2D
Gaudry and Lubicz [5]	1M+3S+1D	4M+2S	5M+5S+1D
The new model in this paper	1M+3S+1D	4M+2S or 4M+1S+1D	5M+5S+1D or 5M+4S+2D

7   Conclusion

In this paper, we firstly introduce a new model of elliptic curve over binary fields. The new curve can be seen as a new model of binary Edwards curves by canceling the quartic term. Furthermore we propose explicit formulas for the addition, doubling and differential addition. We believe that this new model of elliptic curves is worthy of consideration in elliptic curve cryptography.

Acknowledgements. The authors wish to thank Marc Joye, Julio López, and the anonymous referees for helpful comments and suggestions on earlier versions of this paper.

References

1. Bernstein, D.J., Lange, T.: Explicit-formulas database, http://www.hyperelliptic.org/EFD
2. Bernstein, D.J., Lange, T., Farashahi, R.R.: Binary Edwards Curves. In: Oswald, E., Rohatgi, P. (eds.) CHES 2008. LNCS, vol. 5154, pp. 244–265. Springer, Heidelberg (2008)
3. Brier, E., Joye, M.: Weierstraß Elliptic Curves and Side-Channel Attacks. In: Naccache, D., Paillier, P. (eds.) PKC 2002. LNCS, vol. 2274, pp. 335–345. Springer, Heidelberg (2002)
4. Farashahi, R.R., Joye, M.: Efficient Arithmetic on Hessian Curves. In: Nguyen, P.Q., Pointcheval, D. (eds.) PKC 2010. LNCS, vol. 6056, pp. 243–260. Springer, Heidelberg (2010)
5. Gaudry, P., Lubicz, D.: The arithmetic of characteristic 2 Kummer surfaces and of elliptic Kummer lines. Finite Fields and Their Applications 15(2), 246–260 (2009)
6. Järvinen, K.U., Skytta, J.: Fast Point Multiplication on Koblitz Curves: Parallelization Method and Implementations. Microproc. Microsyst. 33(2), 106–116 (2009)
7. Kim, K.H., Kim, S.I.: A new method for speeding up arithmetic on elliptic curves over binary fields (2007), http://eprint.iacr.org/2007/181

8. Koblitz, N.: Elliptic curve cryptosystems. Mathematics of Computation 48(177), 203–209 (1987)
9. Lange, T.: A note on López-Dahab coordinates. Tatra Mountains Mathematical Publications 33, 75-81 (2006), http://eprint.iacr.org/2004/323
10. Devigne, J., Joye, M.: Binary Huff Curves. In: Kiayias, A. (ed.) CT-RSA 2011. LNCS, vol. 6558, pp. 340–355. Springer, Heidelberg (2011)
11. López, J., Dahab, R.: Fast Multiplication on Elliptic Curves over $GF(2^m)$ without Precomputation. In: Koç, Ç.K., Paar, C. (eds.) CHES 1999. LNCS, vol. 1717, pp. 316–327. Springer, Heidelberg (1999)
12. Menezes, A.J.: Elliptic Curve Public Key Cryptosystems. Kluwer Academic Publishers (1993)
13. Miller, V.S.: Use of Elliptic Curves in Cryptography. In: Williams, H.C. (ed.) CRYPTO 1985. LNCS, vol. 218, pp. 417–426. Springer, Heidelberg (1986)
14. Roy, S.S., Rebeiro, C., Mukhopadhyay, D., Takahashi, J., Fukunaga, T.: Scalar Multiplication on Koblitz Curves using τ^2-NAF. Cryptology ePrint Archive, Report 2011/318, http://eprint.iacr.org/2011/318
15. Montgomery, P.L.: Speeding the Pollard and elliptic curve methods of factorization. Mathematics of Computation 48, 243–264 (1987)
16. Stam, M.: On Montgomery-Like Representationsfor Elliptic Curves over $GF(2^k)$. In: Desmedt, Y.G. (ed.) PKC 2003. LNCS, vol. 2567, pp. 240–253. Springer, Heidelberg (2002)
17. Stein, W.A. (ed.): Sage Mathematics Software (Version 4.6), The Sage Group (2010), http://www.sagemath.org
18. Silverman, J.H.: The Arithmetic of Elliptic Curves. Graduate Texts in Mathematics, vol. 106. Springer (1986)

Appendix A. Sage script to check $P + Q = R$.

```
R.<t,x1,y1,x2,y2>=GF(2)[ ]
S=R.quotient([
x1*y1^2+y1*x1^2+t*x1*y1+x1+y1),
x2*y2^2+y2*x2^2+t*x2*y2+x2+y2),])
x3=(x1*x2*(y1+y2)+y1*y2*(t+y1+y2)+t*(x1*y1*x2*y2)
)/(x1*x2+y1*y2+y1^2*y2^2+x1*y1*x2*y2)
y3=(y1*y2*(x1+x2)+x1*x2*(t+x1+x2)+(a+b)*(x1*y1*x2*y2)
)/(x1*x2+y1*y2+x1^2*x2^2+x1*y1*x2*y2)
u1=(x1+y1)/(t^2*(x1+y1+t))
v1=(x1+y1+t^2*x1+t)/(t^4*(x1+y1+t))
u2=(x2+y2)/(t^2*(x2+y2+t))
v2=(x2+y2+t^2*x2+a+b)/(t^4*(x2+y2+t))
u3=(x3+y3)/(t^3*(x3+y3+t))
v3=(x3+y3+t^3*x3+t)/(t^4*(x3+y3+t))
lam=(v1+v2)/(u1+u2)
u4=lam^2+lam=u1+u2
v4=v1+lam*(u1+u4)+u4
S(numerator(u3-u4))==0
S(numerator(v3-v4))==0
```

Analysis of Optimum Pairing Products at High Security Levels

Xusheng Zhang[1,2] and Dongdai Lin[3]

[1] Institute of Software, Chinese Academy of Sciences, Beijing, 100190, China
[2] University of Chinese Academy of Sciences, Beijing, 100049, China
xszhang.is@gmail.com
[3] Institute of Information Engineering, Chinese Academy of Sciences, Beijing, 100093, China
ddlin@iie.ac.cn

Abstract. In modern pairing implementations, considerable researches target at the optimum pairings at different security levels. However, in many cryptographic protocols, computing products or quotients of pairings is needed instead of computing single pairings. In this paper, we mainly analyze the computations of fast pairings on Kachisa-Schaefer-Scott curves with embedding degree 16 (KSS16) for the 192-bit security and Barreto-Lynn-Scott curves with embedding degree 27 (BLS27) for the 256-bit security, and then compare the cost estimations for implementing products and quotients of pairings at the 192 and 256-bit security levels. Being different from implementing single pairings, our results show that KSS16 curves could be most efficient for computing products or quotients of pairings for the 192-bit security; and for the 256-bit security, BLS27 curves might be more efficient for computing products of no less than 25 pairings, otherwise BLS24 curves are much more efficient. In addition, for the fast pairing computation on BLS27 curves, we propose faster Miller formulas in both affine and projective coordinates on curves with only cubic twist and embedding degree divisible by 3.

Keywords: pairing computation, Miller's algorithm, BLS curve, KSS curve.

1 Introduction

Bilinear pairing has performed an important role in modern public key cryptography and led to so-called pairing-based cryptography. Considerable works focus on constructions and implementations of bilinear pairings, and the speed records for implementing pairings have been greatly improved, e.g. a pairing on Barreto-Naehrig curves (BN, [5]) for the 128-bit security can be implemented under 2 million cycles on 64-bit desktop platforms [2]. So there are real needs to find the optimum pairings at different security levels.

Currently at high levels of security, asymmetric pairings admitting higher embedding degrees are more fit for implementing. For example, the optimal ate pairing on BN curves is perfectly suitable for the 128-bit security, and the speed

S. Galbraith and M. Nandi (Eds.): INDOCRYPT 2012, LNCS 7668, pp. 412–430, 2012.

of its performance has been improved unceasingly [28,7,2]. For the 256-bit security, the ate pairing on Barreto-Lynn-Scott curves [4] with embedding degree 24 (BLS24) are recommended by Scott [30] and studied by Costello, Lauter and Naehrig [11]. In contrast, for the 192-bit security, Scott mentioned that the optimum choice of pairing-friendly curves is not straightforward, and recommended Kachisa-Schaefer-Scott curves [21] with embedding degree 18 or 16. More recently, however, Aranha et al. [1] compared the implementations of single pairings derived from KSS18, BN, BLS12 and BLS24 curves at the 192-bit security level. Surprisingly their result shows that the ate pairing on BLS12 curves has the fastest implementation. Table 1 shows the currently suitable choices of pairings for the 128, 192, 256-bit security.

Table 1. Recommended choices of pairings at high security levels

Security level	Bits of r	Bits of p^k	$b_{p^k}/b_r = \rho k$	Pairing
128-bit	256	3072	12	BN optimal ate [30]
192-bit	384	8192	$21\frac{1}{3}$	BLS12 ate [1]
256-bit	512	15360	30	BLS24 ate [30]

However, as noticed in [30], pairings in protocols are often used as products or quotients of pairings. So there are also needs to analyze optimum pairing products and quotients. As far as we know, for the 128-bit security no other known pairings could be competitive with the optimal ate pairing on BN curves. Thus it still perfectly fits for implementing products and quotients of pairings. But for the 192 and 256-bit security, the situations are relatively complicated.

Our motivation of choosing the optimum pairing products is that products of pairings benefit from more loop reduction than they do from a higher degree twist. Regardless of security levels, the cost for ate-like Miller loop is decided by the loop length and the high-degree twist formulas, so the sextic twist and optimal ate pairing are preferred. But, in ate-like Miller loop, since decreasing the total number of full extension field operations could have a greater influence than decreasing the size of the field where the subfield operations take place with high-degree twist technique, then the cost for ate-like Miller loop may be reduced further when taking a balance between the twisting degree and the reduction of loop length. In this paper, we show that two ignored curves, namely KSS16 and BLS27 curves, could admit fast ate-like pairings with much more efficient Miller loop at the 192 and 256-bit security levels. Thus, the fast ate-like pairings on these curves could be optimum for computing products and quotients of pairings in some situations.

First, in Section 3 we propose a new denominator elimination to derive new Miller formulas for ate-like pairings in both affine and projective coordinates on pairing-friendly curves with cubic twist. Our formulas not only extend the study of affine coordinates for pairing computation by Lauter et al. [24], but also work more efficiently compared with the previous work [10] for ate-like pairings on pairing-friendly curves with cubic twist.

Then, we give the first detailed cost estimations for computing the optimal ate pairing on BLS27 curves for the 256-bit security, and KSS16 curves for the

192-bit security. Concretely, in Section 4 we carefully analyze the choices of field extensions and twists for BLS27 curves, and estimate the costs for computing the ate pairing on BLS27 curves with our new affine Miller formulas. In Section 5 we use a linear algebra method for constructing optimal pairings on a parameterized complete family of pairing-friendly curves, to give an optimal ate pairing on KSS16 curves. Further we provide an efficient computation of the final exponentiation of the optimal ate pairing on KSS16 curves, and estimate the total costs for this pairing computation.

At last, we provide the comparisons for computing products and quotients of pairings at the 192 and 256-bit security levels. We show that for the 192-bit security KSS16 curves are the most efficient choices for implementing products of pairings and quotients of two pairings among other recommended curves: BLS12, KSS18, and BN curves; for the 256-bit security, BLS27 curves might be better than BLS24 curves when implementing products of no less than 25 pairings, but in other cases BLS24 curves are the more suitable choices.

2   Background

Let $p > 3$ be a prime and let E be an elliptic curve defined over $\mathbb{F}_p$ with short Weierstrass equation $E : y^2 = x^3 + ax + b$, and point at infinity $\mathcal{O}$. Let r be a prime factor of $\#E(\mathbb{F}_p) = p - t + 1$ and let $k > 1$ be the smallest integer such that $r|p^k - 1$ named the embedding degree with respect to r. Elliptic curves having small embedding degrees and large prime-order subgroups are known as pairing-friendly elliptic curves. To construct an ordinary elliptic curve E with a given embedding degree k, one finds parameters r, t, p satisfying the CM equation $4p - t^2 = Dy^2$ where D is a square-free integer called the CM discriminant, and then make use of the CM method to solve the curve equation. For efficiency, define the ρ-value $\rho = \log_2(q)/\log_2(r) \geq 1$, and the ideal ρ-value is $\rho = 1$.

In current pairing implementations, asymmetric pairings are usually preferred. When $k > 1$, let $G_1 = E[r] \cap \mathrm{Ker}(\pi_p - 1)$ and $G_2 = E[r] \cap \mathrm{Ker}(\pi_p - p)$ denote two eigenspaces of the p-th Frobenius endomorphism $\pi_p : (x, y) \mapsto (x^p, y^p)$, and let $G_T \subset \mathbb{F}_{p^k}^*$ denote the group of r-th roots of unity. There are many improvements on pairings from $G_1 \times G_2$ to G_T, such as the ate pairing [19]

$$a_T : G_2 \times G_1 \to G_T, \quad (Q, P) \mapsto f_{T,Q}(P)^{(q^k-1)/r},$$

where $T = t - 1$ and $f_{T,R}$ is a $\mathbb{F}_{q^k}$-rational function with divisor $\mathrm{div}(f_{T,Q}) = T(Q) - ([T]Q) - (T - 1)(\mathcal{O})$. Let E' over $\mathbb{F}_{q^{k/d}}$ be a twist of degree d of E given by $y^2 = x^3 + a\omega^4 x + b\omega^6$ for some $\omega \in \mathbb{F}_{q^k}$, and the twisting isomorphism is $\psi : E' \to E$, $(x', y') \mapsto (x'/\omega^2, y'/\omega^3)$ with the inverse $\psi^{-1} : E \to E'$, $(x, y) \mapsto (\omega^2 x, \omega^3 y)$. Thus G_1 and G_2 can be represented by their image under ψ^{-1} as $G_1' \subseteq E'(\mathbb{F}_{q^k})[r]$ and $G_2' = E'(\mathbb{F}_{q^{k/d}})[r]$. Costello *et al.* [10] showed that the ate pairing computation can be moved entirely on the twist by using the twisting isomorphism. So the ate pairing can be transformed as

$$a_T' : G_2' \times G_1' \to G_T, (Q', P') \mapsto f_{T,Q'}(P')^{(q^k-1)/r}.$$

Miller [27] proposed an algorithm to evaluate $f_{m,P}(Q)$ in a double-and-add manner, and nowadays the algorithm is widely used in an extended double-and-add manner. Let $m = m_L 2^L + \cdots + m_1 2 + m_0 > 0$ with $m_i \in \{-1, 0, 1\}$, and initialize $R = P$, $f = 1$, then compute $f_{m,P}(Q)$ as

for j from $L - 1$ downto 0 **do**
    $f \leftarrow f^2 \cdot f_{DBL(R)} = f^2 \cdot l_{R,R}(Q)/v_{2R}(Q)$; $R \leftarrow 2R$;   (DBL)
    if $m_j = 1$ **then**
        $f \leftarrow f \cdot f_{ADD(R,P)} = f \cdot l_{R,P}(Q)/v_{R+P}(Q)$; $R \leftarrow R + P$;   (ADD)
    elif $m_j = -1$ **then**
        $f \leftarrow f/f_{ADD(R,P)} = f \cdot v_R(Q)/l_{-R,P}(Q)$; $R \leftarrow R - P$;
return $f = f_{m,P}(Q)$.

where l_{R_1,R_2} is the line passing through points R_1, R_2 with divisor $\operatorname{div}(l_{R_1,R_2}) = (R_1) + (R_2) + (-(R_1 + R_2)) - 3(\mathcal{O})$, and v_R is the vertical line passing through point R with divisor $\operatorname{div}(v_R) = (R) + (-R) - 2(\mathcal{O})$.

Up to now, the most efficient ate-like pairing is given from Vercauteren's work [32] (also see [18]), namely optimal ate pairing, which can be computed in approximately $\frac{1}{\varphi(k)} \log_2 r$ basic Miller iterations. For $\lambda = mr$ with $r \nmid m$, and let $\lambda = h(p) = \sum_{i=0}^{l} c_i p^i$ $(h(\chi) \in \mathbb{Z}[\chi])$, and let $s_i = \sum_{j=i}^{l} c_j p^i$, if $mkp^{k-1} \not\equiv ((p^k - 1)/r) \cdot \sum_{i=0}^{l} i c_i p^{i-1} \pmod{r}$, then a nondegenerate pairing is defined as

$$a_h : G_2 \times G_1 \to G_T, \ (Q, P) \mapsto \prod_{i=0}^{l} f_{c_i,Q}^{p^i}(P) \cdot \prod_{i=0}^{l-1} \frac{l_{s_{i+1}Q,[c_i p^i]Q}(P)}{v_{s_i Q}(P)}.$$

Using a lattice-based method, one may obtain an optimal $h(\chi)$ such that each coefficient of $h(\chi)$ satisfies $\mid c_i \mid \leq r^{1/\varphi(k)}$. For example, optimal ate pairings for BN and KSS18 curves are given in [32] concretely, and especially the ate pairings on BLS curves are optimal (see [1]).

3   New Miller Formulas Using Cubic Twists

The denominator elimination with quadratic twist [3] greatly simplifies pairing computation so that almost all pairing computations focus on curves with even embedding degrees. Recently, pairing-friendly curves with embedding degree 9 and 15 studied by Lin et al. [26] and El Mrabet et al. [13] attract attention to produce fast Miller formulas using cubic twist. For curves of the from $y^2 = x^3 + b$ admitting embedding degree divisible by 3, Lin et al. first observed that

$$\frac{1}{v_R(S)} = \frac{1}{x_S - x_R} = \frac{x_S^2 + x_S x_R + x_R^2}{y_S^2 - y_R^2},$$

which leads to a denominator elimination with cubic twist since $(y_S^2 - y_R^2)$ lies in a subfield. Later, Costello et al. [10] refined the projective formulas in this case using more efficient optimization.

In order to computing pairing on BLS27 curves in this paper, we will propose new Miller formulas using cubic twist. At high security levels, Lauter *et al.* [24] showed that in the ate-like pairing implementation the formulas in affine coordinates are faster than the best currently known formulas in projective coordinates. However, they only analyzed the case of even embedding degree and quadratic or sextic twist. In this section we present both new faster affine and projective Miller formulas of the ate-like pairing on the curve of the from $y^2 = x^3 + b$ with cubic twist and embedding degree k divisible by 3.

3.1   Affine Miller Formulas

We first analyze the case of k divisible by 6 and then extend the result to the case of odd k divisible by 3. Since $E/\mathbb{F}_p$ has the sextic twist $E'/\mathbb{F}_{p^{k/6}}$, we take $P = (x_P, y_P) \in G_1 = E(\mathbb{F}_p)[r]$ and $Q' = (x'_Q, y'_Q) \in G'_2 \subset E'(\mathbb{F}_{p^{k/6}})[r]$. Then we have $P' = \psi^{-1}(P) = (x_P\omega^2, y_P\omega^3)$ with twisting map $\psi : E' : y^2 = x^3 + b\omega^6 \to E,\ (x', y') \mapsto (x'/\omega^2, y'/\omega^3)$, where $\{1, \omega, \dots, \omega^5\}$ are used as the basis of $\mathbb{F}_{p^k}$ over $\mathbb{F}_{p^{k/6}}$ in favor of the representation of the Miller function.

We give a new denominator elimination method for computing the ate-like pairing on the cubic twist as follows. For $R_1 = (x_1, y_1)$, $R_2 = (x_2, y_2)$, $R_3 = R_1 + R_2 = (x_3, y_3) \in \langle Q' \rangle$, we utilize the equivalent line function $l_{R_1, -R_3}$ instead of l_{R_1, R_2} to compute the Miller linear function $f_{R_1, R_2} = l_{R_1, -R_3}/v_{R_3}$ in affine coordinates. Take the regular addition formulas $x_3 = \lambda^2 - x_1 - x_2$ and $y_3 = \lambda(x_2 - x_3) - y_2$, where $\lambda = 3x_1^2/2y_1$ if $R_1 = R_2$ and $\lambda = (y_2 - y_1)/(x_2 - x_1)$ otherwise. Then we have

$$f_{R_1, R_2} = \frac{l_{R_1, -R_3}}{v_{R_3}} = \frac{y + y_3 - \lambda(x - x_3)}{x - x_3} = \frac{y + y_3}{x - x_3} - \lambda$$

and obtain a new function $f'_{R_1, R_2} = (y - y_3)f_{R_1, R_2}$ without the denominator

$$f'_{R_1, R_2} = \frac{y^2 - y_3^2}{x - x_3} - \lambda(y - y_3) = x^2 + x_3 x + x_3^2 - \lambda(y - y_3).$$

On the cubic twist E', the y-values of P' and Q' belong to the subfield $\mathbb{F}_{p^{k/3}}$. Thus the evaluations of $f'_{R_1, R_2}(P')$ and the original Miller function $f_{R_1, R_2}(P')$ are the same under the final exponentiation. Then the new Miller function $f'_{R_1, R_2}(P')$ can be formalized as

$$f'_{R_1, R_2}(P') = x_3^2 + \lambda y_3 - \lambda y_P \omega^3 + x_3 x_P \omega^2 + x_P^2 \omega^4.$$

Furthermore, we introduce new affine coordinates $(x, y, t) = (x, y, x^2)$ to compute the Miller doubling and addition steps as follows, and let M_i, S_i, and I_i denote multiplication, squaring, and inversion in field $\mathbb{F}_{p^i}$.

Doubling steps in affine coordinates. With the precomputation of x_P^2, the formulas for computing the doubling and the Miller function require $1\mathbf{I}_{k/6} + 3\mathbf{M}_{k/6} + 2\mathbf{S}_{k/6} + \frac{k}{3}\mathbf{M}_1$ by computing the following sequence of operations.

$$A = 3t_1,\ B = 2y_1,\ C = B^{-1},\ D = A \cdot C,\ x_3 = D^2 - 2x_1,\ y_3 = D \cdot (x_1 - x_3) - y_1,$$
$$t_3 = x_3^2,\ E = D \cdot y_3,\ F = t_3 + E,\ G = -D \cdot y_P,\ H = x_3 \cdot x_P.$$

Addition steps in affine coordinates. The point addition and affine function can be computed as follows by costing $1\mathbf{I}_{k/6} + 3\mathbf{M}_{k/6} + 2\mathbf{S}_{k/6} + \frac{k}{3}\mathbf{M}_1$.

$$A = (x_2 - x_1)^{-1}, \; B = A \cdot (y_2 - y_1), \; x_3 = B^2 - x_1 - x_2, \; y_3 = B \cdot (x_2 - x_3) - y_2,$$
$$t_3 = x_3^2, \; C = B \cdot y_3, \; D = t_3 + C, \; E = -B \cdot y_P, \; F = x_3 \cdot x_P.$$

When k is an odd integer divisible by 3, we have $G_2' \subset E'(\mathbb{F}_{p^{k/3}})[r]$ and y_3, $y_P \omega^3 \in \mathbb{F}_{p^{k/3}}$. In both doubling and addition steps, when precomputing $y_P \omega^3$, $\lambda y_3 - \lambda y_P \omega^3 = \lambda(y_3 - y_P \omega^3)$ can be computed with $1\mathbf{M}_{k/3}$ instead of $1\mathbf{M}_{k/6} + \frac{k}{6}\mathbf{M}_1$ in the case of $6|k$. Thus both our Miller doubling and addition steps require $1\mathbf{I}_{k/3} + 3\mathbf{M}_{k/3} + 2\mathbf{S}_{k/3} + \frac{k}{3}\mathbf{M}_1$.[1]

3.2 Projective Miller Formulas

Our new denominator elimination can also be used to derive new efficient projective formulas. First, we transform $f'_{DBL(R_1)}(P')$ and $f'_{ADD(R_1,Q')}(P')$ to the homogeneous forms $F_{DBL(R_1)}(P')$ and $F_{ADD(R_1,Q')}(P')$. Using the projective formulas for scalar multiplication in [10], where $Z_3 = 8Y_1^3 Z_1$ for point doubling, and $Z_3 = -Z_1 Z_2 (Z_1 X_2 - X_1 Z_2)^3$ for point addition, we prefer to compute

$$F'_{DBL(R_1)}(P') = (Z_3^2) F_{DBL(R_1)}(P')$$
$$= X_3^2 + 12X_1^2 Y_1^2 (Y_3 - Z_3 y_P \omega^3) + 2X_3 Z_3 (\frac{x_P}{2}\omega^2) + Z_3^2(x_P^2 \omega^4)$$

$$F'_{ADD(R_1,Q')}(P') = (Z_3^2) F_{ADD(R_1,Q')}(P')$$
$$= X_3^2 - Z_1 Z_2 (Z_1 X_2 - X_1 Z_2)^2 (Z_1 Y_2 - Y_1 Z_2)(Y_3 - Z_3 y_P \omega^3)$$
$$= +2X_3 Z_3 (\frac{x_P}{2}\omega^2) + Z_3^2(x_P^2 \omega^4).$$

We introduce new coordinates $(X, Y, Z, T, U) = (X, Y, Z, X^2, Z^2)$ to compute the point doubling and Miller function $F'_{DBL(R_1)}(P')$ as follows.

$$A = Y_1^2, \; B = 3b \cdot U_1, \; C = (X_1 + Y_1)^2 - T_1 - A, \; D = (Y_1 + Z_1)^2 - A - U_1, \; E = 3B,$$
$$X_3 = C \cdot (A - E), \; Y_3 = (A + E)^2 - 3(2B)^2, \; Z_3 = 4A \cdot D, \; T_3 = X_3^2, \; U_3 = Z_3^2,$$
$$F = (X_3 + Z_3)^2 - T_3 - U_3, \; G = 3C^2, \; H = Z_3 \cdot y_P \omega^3, \; L_0 = G \cdot (Y_3 - H) + T_3,$$
$$L_1 = F \cdot (x_P/2), \; L_2 = U_3 \cdot (x_P^2).$$

Similarly, the point addition and Miller function $F'_{ADD(R_1,Q')}(P')$ can be computed in the new coordinates as the following sequence of operations.

$$A = X_1 \cdot Z_2, \; B = Y_1 \cdot Z_2, \; C = Z_1 \cdot Z_2, \; D = A - Z_1 \cdot X_2, \; E = B - Z_1 \cdot Y_2, \; F = D^2,$$
$$G = E^2, \; H = D \cdot F, \; I = F \cdot A, \; J = H + C \cdot G - 2I, \; K = C \cdot F \cdot E, \; X_3 = D \cdot J,$$
$$Y_3 = E \cdot (I - J) - H \cdot B, \; Z_3 = C \cdot H, \; T_3 = X_3^2, \; U_3 = Z_3^2, \; L = (X_3 + Z_3)^2 - T_3 - U_3,$$
$$M = Z_3 \cdot y_P \omega^3, \; L_0 = T_3 - K \cdot (Y_3 - M), \; L_1 = L \cdot (x_P/2), \; L_2 = U_3 \cdot (x_P^2).$$

---

[1] When $R_1 = P$ and $R_2 = R - P$ and $R_3 = R$, the formulas for $l_{-R,P}/v_R = l_{P,R-P}/v_R$ are similar and require the same cost.

Denote $\alpha = \omega^k$ and since there are field representations $\mathbb{F}_{p^k} = \mathbb{F}_{p^{k/6}}(\omega)$ and $\mathbb{F}_{p^{k/3}} = \mathbb{F}_{p^{k/6}}(\omega^3)$, and $Z_3 \cdot y_P$ is of the form $\sum_{i=0}^{k/3} b_i \omega^{3i}$, then the multiplication of $Z_3 \cdot y_P$ by ω^3 gives the form of $b_{k/3}\alpha + \sum_{i=0}^{k/3-1} b_i \omega^{3(i+1)}$. When α is small, this multiplication by ω^3 is just a rotation of the vector representing the field element with respect to basis ω^3 and nearly free. Then computation of $Z_3 \cdot y_P \omega^3$ in our formulas costs $\frac{k}{3}\mathbf{M}_1$.

With the precomputation of $x_P/2$ and x_P^2, computing the projective Miller doubling formulas requires $3\mathbf{M}_{k/3} + 9\mathbf{S}_{k/3} + k\mathbf{M}_1 + 1\mathbf{M}_{3b}$. And when taking $Z_2 = 1$, then the operations for the projective Miller mixed addition formulas cost $12\mathbf{M}_{k/3} + 5\mathbf{S}_{k/3} + k\mathbf{M}_1$.

Comparison. We give the comparison of operation counts of our affine and projective formulas and the currently fastest projective formulas in [10] in Table 2. From our comparison, our projective formulas for DBL are faster than those in [10]. Although our projective formulas for mADD is a little shower than those in [10],[2] our formulas for both DBL and mADD could be a little faster than those in [10] since the percentage of S is larger for our formulas.

Assume that $\mathbf{S}_{k/3} \geq 0.8\mathbf{M}_{k/3}$ and the inversion-to-multiplication ratio $\mathbf{R}_{k/3} = \mathbf{I}_{k/3}/\mathbf{M}_{k/3} \leq 5.6$, our affine formulas are better than projective ones. This case could happen when $\mathbb{F}_{p^k}$ is a special larger extension field, e.g. in the case of $k = 27$ in Table 3.

Table 2. Operation counts in the ate-like Miller loop with cubic twist

$3\mid k$	coordinates	$\mathbf{M}_1$	$\mathbf{I}_{k/3}$	$\mathbf{M}_{k/3}$	$\mathbf{S}_{k/3}$	$\mathbf{M}_{(\cdot)}$
DBL	new affine	$k/3$	1	3	2	–
DBL	new proj.	k	–	3	9	$1\mathbf{M}_{(3b)}$
DBL	proj. [10]	k	–	6	7	$1\mathbf{M}_{(b)}$
ADD	new affine	$k/3$	1	3	2	–
mADD	new proj.	k	–	12	5	–
mADD	proj. [10]	k	–	13	3	–
DBL+mADD	new proj.	$2k$	–	15	14	$1\mathbf{M}_{(3b)}$
DBL+mADD	proj. [10]	$2k$	–	19	10	$1\mathbf{M}_{(b)}$

4   Pairing Computation on BLS27 Curves

In this section we consider such a family of BLS curves [4] which has embedding degree 27 and ρ-value 10/9, and is parameterized by

$$
\begin{aligned}
r(z) &= \tfrac{1}{3}(z^{18} + z^9 + 1), \\
t(z) &= z + 1, \\
p(z) &= \tfrac{1}{3}(z-1)^2(z^{18} + z^9 + 1) + z.
\end{aligned}
\tag{1}
$$

---

[2] For fast computations, one can not simply use our formulas for DBL and the formulas from [10] for mADD, since in that case the latter formulas need additional $2\mathbf{S}_{k/3}$.

Note that the ate pairing derived from the above curve family has the optimal loop length $[\log_2(t(z) - 1)] = [\log_2 z] \approx [\frac{1}{\varphi(k)} \log_2 r(z)]$. In addition, as the recommended BLS24 curves at the 256-bit security level, the BLS27 curves we considered also exactly match the recommended ratio $b_{p^k}/b_r = 30$ at the 256-bit security level.

4.1   Choices of Finite Field $\mathbb{F}_{p^{27}}$ and Twists

For the parameters (1), when $3|z^{18} + z^9 + 1$, then we have $z \equiv 1 \pmod 3$, i.e. $z = 3\gamma + 1$ for some γ, and transform that $p(\gamma) = 3\gamma^2 \cdot \left((3\gamma + 1)^{18} + (3\gamma + 1)^9 + 1\right) + 3\gamma + 1$. According to the definition given by Benger and Scott [6], $\mathbb{F}_{p^{27}}$ is a towering-friendly field due to $p(\gamma) \equiv 1 \pmod 3$, and can be constructed from Theorem 4 in [6]. First, according to Euclid's algorithm, we find that $p(\gamma) = a(\gamma)^2 + 3b(\gamma)^2$, where $a(\gamma) = \frac{3}{2}\gamma + 1$ and $b(\gamma) = 19683\gamma^{10} + 59049\gamma^9 + 78732\gamma^8 + 61236\gamma^7 + 30618\gamma^6 + 10206\gamma^5 + 2268\gamma^4 + 324\gamma^3 + 27\gamma^2 + \frac{3}{2}\gamma$.

When γ takes an odd value, we define new $\mathbb{Z}$-polynomials $A(\gamma) = 2a(\gamma)$ and $B(\gamma) = \frac{2}{3}b(\gamma)$, and then we have $4p(\gamma) = A(\gamma)^2 + 27B(\gamma)^2$, where $A(\gamma)$ and $B(\gamma)$ are uniquely determined up to sign according to Proposition 8.3.2 in [20]. In this case we can construct the extension of $\mathbb{F}_{p^{27}}$ directly by using the same technique in [11].

Proposition 1. *Let $p(\gamma)$, $A(\gamma)$, and $B(\gamma)$ be defined as above. For a given γ, let p be prime and congruent to 1 modulo 3. If γ is an odd integer, then the binomial $t^{27} - 2$ is irreducible in $\mathbb{F}_p[t]$.*

Proof. From Proposition 9.6.2 in [20], for a prime $p \equiv 1 \pmod 3$, 2 is a cubic residue modulo p if and only if there exist integers C and D so that $p = C^2 + 27D^2$. Since $4p = A^2 + 27B^2$ with uniquely determined A and B, then 2 is a cubic residue modulo p if and only if both A and B are even. Equivalently one only needs to verify A or B to be even. Since $A(\gamma) = 3\gamma + 2$, it follows that A is even if and only if γ is even. Hence, when γ is odd, we conclude that $t^{27} - 2$ is irreducible in $\mathbb{F}_p[t]$ from Theorem 4 in [6]. □

When γ takes an even value, take $\gamma = 2\gamma'$, and define new $\mathbb{Z}$-polynomials $a'(\gamma') = a(\gamma)$ and $b'(\gamma') = b(\gamma)$ satisfying $p(\gamma) = a'(\gamma')^2 + 3b'(\gamma')^2$. Using Euler's criteria for cubic residues of small integers [25], we know that 3 is a cubic residue if and only if $9|b(\gamma')$, or $9|(a(\gamma') \pm b(\gamma'))$. Immediately another irreducible binomial can be found from the following proposition.

Proposition 2. *Assume that $\gamma = 2\gamma'$ takes an even integer and $p = a'(\gamma')^2 + 3b'(\gamma')^2$. If 9 does not divide $b'(\gamma')$ and $a'(\gamma') \pm b'(\gamma')$, the binomial $t^{27} - 3$ is irreducible in $\mathbb{F}_p[t]$.*

Searching even values of γ so that p represents primes with the help of a computer, we find that nearly two-thirds of these γ satisfy the conditions in the above proposition.

On the other hand, the element α used for constructing $\mathbb{F}_{p^k}$ can define the dth twist element $\delta \in \mathbb{F}_{p^{k/d}}$ in favor of the efficient pairing computation. For the

BLS27 curves, we prefer $\alpha = 2$ or 3 in Proposition 1 or Proposition 2 to define $\omega = \alpha^{1/27}$ or $\omega = \alpha^{5/27} \in \mathbb{F}_{p^{27}}$, and then the twisting isomorphism is given as $\psi : E' : y^2 = x^3 + b \cdot \omega^6 \to E : y^2 = x^3 + b$, $(x, y) \mapsto (x/\omega^2, y/\omega^3)$.

4.2 Computation in the Miller Loop

We choose the parameter z as $2^{28} + 2^{27} + 2^{25} + 2^8 - 2^3$ so that $r(z)$ has a prime factor of 516 bits and $p(z)$ is a prime of 573 bits, and the corresponding curve equation is $y^2 = x^3 - 2$. According to Proposition 1, since $(z - 1)/3$ is an odd integer, $\mathbb{F}_{p^{27}}$ can be constructed as $\mathbb{F}_p[t]/(t^{27} - 2)$, and furthermore the cubic twist curve has the equation $y^2 = x^3 - 2\delta$ with $\delta = 2^{2/9} \in \mathbb{F}_{p^9}$.

Note that Lauter *et al.* [24] recommended to use affine formulas at high security levels, and Scott [30] holds the currently fastest speed of pairing computation at 256-bit security level by using affine formulas. Hence, in the Miller loop we can compute the Miller function $f_{z,Q}(P)$ costing 28 doubling steps (affine formulas), 4 mixed addition steps (affine formulas), 27 squarings in $\mathbb{F}_{p^{27}}$, 30 multiplications in $\mathbb{F}_{p^{27}}$, and 1 inversion in $\mathbb{F}_{p^{27}}$.[3]

4.3 Final Exponentiation

For the parameters (1), the final exponentiation of the ate pairing has the factorization $(p^{27} - 1)/r = (p^9 - 1)(p^{18} + p^9 + 1)/r = (p^9 - 1) \cdot d$, where the power of d is called the hard part. Since $p(z) = \frac{1}{3}(z - 1)^2 \cdot r(z) + z = (z - 1)^2 \cdot r + z$, we obtain a recursion relation $p^{m+1} = r \cdot (z - 1)^2 \cdot p^m + z \cdot p^m$, and therefore expand the parameter polynomial of the hard part to the base p as

$$(p^{18} + p^9 + 1)/r = (z - 1)^2 \cdot (p^9 + z^9 + 1) \cdot (p^8 + z \cdot p^7 + \cdots + z^7 \cdot p + z^8) + 3.$$

For the output of the Miller loop $f \in \mathbb{F}_{p^{27}}$, we compute $M = f^{(p^9 - 1)}$ and then

$$M_1 = M^{p^8} \cdot M^{zp^7} \cdot M^{z^2 p^6} \cdot M^{z^3 p^5} \cdot M^{z^4 p^4} \cdot M^{z^5 p^3} \cdot M^{z^6 p^2} \cdot M^{z^7 p} \cdot M^{z^8},$$
$$M_2 = M_1^{(z-1)^2}, \quad M_3 = M_2^{z^9} \cdot M_2^{p^9} \cdot M_2, \quad M_4 = M_3 \cdot M_3.$$

Thus the total calculations of the final exponentiation require 1 inversion in $\mathbb{F}_{p^{27}}$, 17 powers of z, 2 powers of $z - 1$, 11 multiplications in $\mathbb{F}_{p^{27}}$, and p, p^2, p^3, p^4, p^5, p^6, p^7, p^8-Frobenius and 2 p^9-Frobenius maps, where the cost for M^3 is luckily included in computing M^z, and especially for any cyclotomic subgroup element $M_0 \in G_{\Phi_{27}(p)}$ we can compute M_0^{-1} as $M_0^{p^{18}} + M_0^{p^9}$ to avoid a full inversion.

4.4 Cost Estimation

Here the Karatsuba method is used for our estimation in Table 3. Since p is a prime of 573 bits, the costs of the base field $\mathbb{F}_p$ can be in terms of a 573 bits

---

[3] Note that z has one negative coefficient -1, we need a full inversion to compute f/f_{ADD} in the case of odd embedding degree.

multiplication ($\mathbf{m}_{573}$), squaring ($\mathbf{s}_{573}$), and inversion ($\mathbf{i}_{573}$), Let $\mathbf{M}, \mathbf{S}, \mathbf{I}$ denote multiplication, squaring, inversion in $\mathbb{F}_{p^{27}}$, and $\widetilde{\mathbf{m}}, \widetilde{\mathbf{s}}, \widetilde{\mathbf{i}}$ denote multiplication, squaring, inversion in $\mathbb{F}_{p^9}$

Table 3. Operation costs for computing the ate pairing on the BLS27 curve

Operation	Costs (Karatsuba)
$\widetilde{\mathbf{m}}$ Multiplication in $\mathbb{F}_{p^9}$	$36\mathbf{m}_{573}$
$\widetilde{\mathbf{s}}$ Squaring in $\mathbb{F}_{p^9}$	$36\mathbf{s}_{573}$
Inversion in $\mathbb{F}_{p^3}$ [24]	$1\mathbf{i}_{573} + 9\mathbf{m}_{573} + 2\mathbf{s}_{573}$
$\widetilde{\mathbf{i}}$ Inversion in $\mathbb{F}_{p^9}$	$1\mathbf{i}_{573} + 63\mathbf{m}_{573} + 14\mathbf{s}_{573}$
$\mathbf{I}$ Inversion in $\mathbb{F}_{p^{27}}$	$1\mathbf{i}_{573} + 387\mathbf{m}_{573} + 86\mathbf{s}_{573}$
Formulas for DBL or mADD	$1\widetilde{\mathbf{i}} + 3\widetilde{\mathbf{m}} + 2\widetilde{\mathbf{s}} + 9\mathbf{m}_{573}$
Exponentiation by z	$28\mathbf{S} + 5\mathbf{M} + 36\mathbf{m}_{573}$
Exponentiation by $z - 1$	$28\mathbf{S} + 6\mathbf{M} + 36\mathbf{m}_{573}$
$p/p^2/p^4/p^5/p^7/p^8$-Frobenius	$26\mathbf{m}_{573}$
$p^3/p^6/p^9$-Frobenius	$18\mathbf{m}_{573}$

The Miller loop cost is $28(3\widetilde{\mathbf{m}} + 2\widetilde{\mathbf{s}} + 1\widetilde{\mathbf{i}} + 9\mathbf{m}_{573}) + 4(3\widetilde{\mathbf{m}} + 2\widetilde{\mathbf{s}} + 1\widetilde{\mathbf{i}} + 9\mathbf{m}_{573}) + 27(6\widetilde{\mathbf{s}}) + 30(6\widetilde{\mathbf{m}}) + 1\mathbf{i}_{573} + 387\mathbf{m}_{573} + 86\mathbf{s}_{573} = 12627\mathbf{m}_{573} + 8670\mathbf{s}_{573} + 33\mathbf{i}_{573}$. And the final exponentiation cost is $1\mathbf{I} + 17(28 \cdot 6\widetilde{\mathbf{s}} + 5 \cdot 6\widetilde{\mathbf{m}} + 36\mathbf{m}_{573}) + 2(28 \cdot 6\widetilde{\mathbf{s}} + 6 \cdot 6\widetilde{\mathbf{m}} + 36\mathbf{m}_{573}) + 11(6\widetilde{\mathbf{m}}) + 228\mathbf{m}_{573} = 24627\mathbf{m}_{573} + 114998\mathbf{s}_{573} + 1\mathbf{i}_{573}$. Thus the total cost is $37254\mathbf{m}_{573} + 123668\mathbf{s}_{573} + 34\mathbf{i}_{573}$.

5   Pairing Computation on KSS16 Curves

In this section we give a detailed analysis of computing the optimal ate pairing on KSS16 curve recommended by Scott [30] at the 192-bit security level, which has embedding degree 16 and ρ-value 5/4 and is parameterized by

$$
\begin{aligned}
r(z) &= z^8 + 48z^4 + 625, \\
t(z) &= \frac{1}{35}(2z^5 + 41z + 35), \\
p(z) &= \frac{1}{980}(z^{10} + 2z^9 + 5z^8 + 48z^6 + 152z^5 + 240z^4 + 625z^2 + 2398z \\
&\quad + 3125).
\end{aligned}
\tag{2}
$$

5.1   Optimal Ate Pairing

We first note that the optimal pairing on a parameterized complete family of pairing-friendly curves defined in [14] could be constructed by using the linear algebra method, which is much easier than the general lattice-based method proposed by Vercauteren [32].

Lemma 1. *Let $t(z), r(z), p(z)$ parameterize a complete family of pairing-friendly curves with embedding degree $k > 1$. Then, there exist $m(z) \in \mathbb{Q}[z]$ and $c_i(z) \in$*

$\mathbb{Z}[z]$ for $0 \le i < \varphi(k)$ so that $m(z)r(z) = \sum_{i=0}^{\varphi(k)-1} c_i(z)p(z)^i$, where $\deg c_0(z) = 1$ and $\deg c_i(z) = 0$ for $0 < i < \varphi(k)$.

Proof. Note that $\deg r(z) = \varphi(k)$, and let N denote the number of all nonzero coefficients of $\{c_i(z)\}_{0 \le i < \varphi(k)}$, then from $\sum_{i=0}^{\varphi(k)-1} c_i(z)p(z)^i \equiv 0 \pmod{r(z)}$, we can derive a homogeneous system of $\varphi(k)$ linear equations of N variables as the coefficients of $\{c_i(z)\}_{0 \le i < \varphi(k)}$, which has nonzero solutions in $\mathbb{Q}^N$ when N is greater than $\varphi(k)$. Thus we can take a single $c_0(z)$ of degree one and others of degree zero so that the homogeneous system has a solution in $\mathbb{Z}^{\varphi(k)+1}$, and the leading coefficient of $c_0(z)$ must be nonzero. Otherwise, one has $\sum_{i=0}^{\varphi(k)-1} c_i p(z)^i \equiv 0 \pmod{r(z)}$ with $c_i \in \mathbb{Z}$, which contradicts the lower bound of the shortest vector in the corresponding lattice in [32] (Theorem 7) when z is sufficiently large. Thus the coefficient matrix of this system is full row rank, and this system has a unique nonzero solution in $\mathbb{Z}^{\varphi(k)+1}$ up to a unit in $\mathbb{Q}$. $\qquad\square$

From Lemma 1, the polynomials $c_i(z)$ for $0 \le i < \varphi(k)$ can be solved easily for any complete family of pairing-friendly curves, and therefore lead to nondegenerate optimal pairing on this family. Concretely, we solve an optimal polynomial $h(\chi) = \sum_{i=0}^{\varphi(k)-1} c_i(z)\chi^i = z + \chi - 2\chi^5 \in \mathbb{Z}[\chi]$ for KSS16 curves. According to [32], the optimal ate pairing $a_h : G_2 \times G_1 \to G_T$ is given as

$$a_h(Q, P) = \left(f_{z,Q}(P) \cdot f_{-2,Q}^{p^5}(P) \cdot l_{[z]Q,[p]Q}(P) \right)^{(p^{16}-1)/r}$$

$$= \left(f_{z,Q}(P) \cdot l_{Q,Q}^{-p^5}(P) \cdot l_{[z]Q,[p]Q}(P) \right)^{(p^{16}-1)/r}.$$

Since $p^8 + 1 \equiv 0 \pmod{r}$, then $a_{opt}(Q, P) = a_h(Q, P)^{p^3}$ also defines a nondegenerate pairing as

$$a_{opt}(Q, P) = \left(\left(f_{z,Q}(P) \cdot l_{[z]Q,[p]Q}(P) \right)^{p^3} \cdot l_{Q,Q}(P) \right)^{(p^{16}-1)/r}. \qquad (3)$$

5.2   Computation in the Miller Loop

Here we choose the parameter z as $2^{49} + 2^{26} + 2^{15} - 2^7 - 1$ so that $r(z)$ has a prime factor of 377 bits and $p(z)$ is a prime of 481 bits, and the corresponding curve equation is $y^2 = x^3 - 3x$ which fits for fast scalar multiplication. From the law of Quadratic Reciprocity (Theorem 1, §5.2 of [20]), 3 is a quadratic non-residue modulo p if $p \equiv 5, 7 \pmod{12}$. Since $p(z) \equiv 5 \pmod{12}$, we can construct $\mathbb{F}_{p^{16}}$ as $\mathbb{F}_p[z]/(z^{16} - 3)$ according to [6]. Furthermore, we can choose the quartic twist $y^2 = x^3 - 3\delta x$ with $\delta = 3^{1/4} \in \mathbb{F}_{p^4}$.

The Miller loop consists of computing $f_{z,Q}(P)$, $l_{[z]Q,[p]Q}(P)$, $l_{Q,Q}(P)$, two sparse multiplications in $\mathbb{F}_{p^{16}}$, and one p^3-Frobenius map in $\mathbb{F}_{p^{16}}$. First we can compute $f_{z,Q}(P)$ costing 49 doubling steps, 4 mixed addition steps, 48 squarings in $\mathbb{F}_{p^{16}}$, and 52 sparse multiplications in $\mathbb{F}_{p^{16}}$. Then, we obtain $[z]Q$ and only need extra 2 p-Frobenius maps in $\mathbb{F}_{p^{16}}$ for computing $[p]Q = \psi(Q)$ by using the skew-Frobenius map. Thus we require 8 multiplications in $\mathbb{F}_p$, 4 multiplications in $\mathbb{F}_{p^4}$, and 2 p-Frobenius maps in $\mathbb{F}_{p^{16}}$ to compute $l_{[z]Q,[p]Q}(P)$. And we also require 8 multiplications in $\mathbb{F}_p$, 1 multiplication in $\mathbb{F}_{p^4}$, and 1 squaring in $\mathbb{F}_{p^4}$ to compute $l_{Q,Q}(P)$.

Unlike the case of the 256-bit security level, Aranha et al. [1] recommended projective formulas for computing pairing at the 192-bit security level. So we follow their strategies. The total costs for the Miller loop are 49 doubling steps (projective formulas), 4 mixed addition steps (projective formulas), 48 squarings in $\mathbb{F}_{p^{16}}$, 54 sparse multiplications in $\mathbb{F}_{p^{16}}$, 16 multiplications in $\mathbb{F}_p$, 5 multiplications and 1 squaring in $\mathbb{F}_{p^4}$, 2 p-Frobenius and 1 p^3-Frobenius maps in $\mathbb{F}_{p^{16}}$.

5.3   Final Exponentiation

In the case of KSS16 curves, the final exponentiation of the optimal ate pairing can be divisible into $(p^{16} - 1)/r = (p^8 - 1) \cdot (p^8 + 1)/r = (p^8 - 1) \cdot d$. For simplicity, we consider a multiple of the hard part d as $857500 \cdot d$ and expand it as $857500 \cdot d = \sum_{i=0}^{7} c_i p^i$, where taking $A = z^3 \cdot B + 56$ and $B = (z+1)^2 + 4$,

$$c_0 = -11z^9 - 22z^8 - 55z^7 - 278z^5 - 1172z^4 - 1390z^3 + 1372,$$
$$= -11(z^4 \cdot A + 27z^3 \cdot B + 28) + 19A,$$
$$c_1 = 15z^8 + 30z^7 + 75z^6 + 220z^4 + 1280z^3 + 1100z^2,$$
$$= 5(3z^3 \cdot A + 44z^2 \cdot B) = 5c_1',$$
$$c_2 = 25z^7 + 50z^6 + 125z^5 + 950z^3 + 3300z^2 + 4750z,$$
$$= 25(z^2 \cdot A + 38z \cdot B) = 25c_2',$$
$$c_3 = -125z^6 - 250z^5 - 625z^4 - 3000z^2 - 1300z - 15000,$$
$$= -125(z \cdot A + 24B) = -125c_3',$$
$$c_4 = -2z^9 - 4z^8 - 10z^7 + 29z^5 - 54z^4 + 145z^3 + 4704,$$
$$= -(2z^4 \cdot A + 55z^3 \cdot B) + 84A,$$
$$c_5 = -20z^8 - 40z^7 - 100z^6 - 585z^4 - 2290z^3 - 2925z^2,$$
$$= -5(4z^3 \cdot A + 117z^2 \cdot B) = -5c_5',$$
$$c_6 = 50z^7 + 100z^6 + 250z^5 + 1025z^3 + 4850z^2 + 5125z,$$
$$= 25(2z^2 \cdot A + 41z \cdot B) = 25c_6',$$
$$c_7 = 875z^2 + 1750z + 4375 = 125 \cdot 7B = 125c_7'.$$

For the output of the Miller loop $f \in \mathbb{F}_{p^k}$, we compute $M = f^{(p^8-1)}$ as an element in the cyclotomic subgroup $G_{\Phi_2(p^8)}$ [15]. To compute M^{c_0}, $M^{c_1'}$, $M^{c_2'}$, $M^{c_3'}$, M^{c_4}, $M^{c_5'}$, $M^{c_6'}$, $M^{c_7'}$, we note that $c_2' = z(c_3' + 2c_7')$, $c_6' = z(2c_3' - c_7')$, $c_5' = z^2(4c_3' + 3c_7')$, and $c_1' = z^2(3c_3' - 4c_7')$. Thus they can be computed roughly in the following diagram, and their concrete formulas are given in Appendix A.

$$M \longrightarrow M^B \longrightarrow M^A \nearrow \begin{matrix} M^{c_3'} \\ \searrow M^{c_7'} \end{matrix} \quad M^{zc_3'} \longrightarrow M^{z^2 c_3'} \longrightarrow M^{z^3 c_3'} \begin{matrix} M^{c_2'} & M^{c_1'} & M^{c_0} \\ M^{c_6'} & M^{c_5'} & M^{c_4} \end{matrix}$$

As we show in Appendix A, the computation of the final exponentiation requires 1 inversion in $\mathbb{F}_{p^{16}}$, 7 exponentiations by z, 2 exponentiations by $z + 1$, 16 cyclotomic squarings in $G_{\Phi_2(p^8)}$, 34 multiplications in $\mathbb{F}_{p^{16}}$, 2 cyclotomic cubings in $\mathbb{F}_{p^{16}}$, and $p, p^2, p^3, p^4, p^5, p^6, p^7, p^8$-Frobenius maps.

5.4   Cost Estimation

We estimate the extension field multiplication using the Karatsuba method, and list the costs for computing the optimal ate pairing on the KSS16 curve in Table 4.[4] We denote $\mathbf{m}_{481}$ multiplication, $\mathbf{s}_{481}$ squaring, $\mathbf{i}_{481}$ inversion in $\mathbb{F}_p$ with p a 481 bits prime; and $\mathbf{M}$ multiplication, $\mathbf{S}$ squaring, and $\mathbf{I}$ inversion in $\mathbb{F}_{p^{16}}$; $\widetilde{\mathbf{m}}$ multiplication and $\widetilde{\mathbf{s}}$ squaring in $\mathbb{F}_{p^4}$.

Table 4. Operation costs for computing the optimal ate pairing on the KSS16 curve

Operation	Costs (Karatsuba)
$\widetilde{\mathbf{m}}$ Multi. in $\mathbb{F}_{p^4}$	$9\mathbf{m}_{481}$
$\widetilde{\mathbf{s}}$ Squaring in $\mathbb{F}_{p^4}$	$6\mathbf{m}_{481}$
$\mathbf{M}_s$ Sparse Multi. in $\mathbb{F}_{p^{16}}$	$7\widetilde{\mathbf{m}}$
$\mathbf{S}_c$ Squaring in $G_{\Phi_2(p^8)}$ [15]	$4\widetilde{\mathbf{m}}$
$\mathbf{C}_c$ Cubing in $G_{\Phi_2(p^8)}$	$8\widetilde{\mathbf{m}}$
p^8-Frobenius / $\mathbf{I}_c$ Inversion in $G_{\Phi_2(p^8)}$	Conjugation
$\mathbf{I}$ Inversion in $\mathbb{F}_{p^{16}}$ [24]	$1\mathbf{i}_{481} + 132\mathbf{m}_{481} + 2\mathbf{s}_{481}$
Formulas for DBL [10]	$8\mathbf{m}_{481} + 2\widetilde{\mathbf{m}} + 8\widetilde{\mathbf{s}}$
Formulas for mADD [10]	$8\mathbf{m}_{481} + 9\widetilde{\mathbf{m}} + 5\widetilde{\mathbf{s}}$
Exponentiation by z (or $z + 1$)	$49\mathbf{S}_c + 4\mathbf{M}$ (or $49\mathbf{S}_c + 3\mathbf{M}$)
$p/p^3/p^5/p^7$-Frobenius	$15\mathbf{m}_{481}$
p^2/p^6-Frobenius	$12\mathbf{m}_{481}$
p^4-Frobenius	$8\mathbf{m}_{481}$

In the Miller loop for computing the function (3), the cost requires $49(8\mathbf{m}_{481} + 2\widetilde{\mathbf{m}} + 8\widetilde{\mathbf{s}}) + 4(8\mathbf{m}_{481} + 9\widetilde{\mathbf{m}} + 5\widetilde{\mathbf{s}}) + 48(9\widetilde{\mathbf{s}}) + 54(7\widetilde{\mathbf{m}}) + 16\mathbf{m}_{481} + 5\widetilde{\mathbf{m}} + \widetilde{\mathbf{s}} + 45\mathbf{m}_{481} = 10208\mathbf{m}_{481}$. Assume $\mathbf{m}_{481} = \mathbf{s}_{481}$, the final exponentiation cost is $1\mathbf{i}_{481} + 134\mathbf{m}_{481} + 7(49(4\widetilde{\mathbf{m}} + 4(9\widetilde{\mathbf{m}})) + 2(49(4\widetilde{\mathbf{m}} + 3(9\widetilde{\mathbf{m}})) + 16(4\widetilde{\mathbf{m}}) + 34(9\widetilde{\mathbf{m}}) + 2(8\widetilde{\mathbf{m}}) + 92\mathbf{m}_{481} = 22330\mathbf{m}_{481} + 1\mathbf{i}_{481}$. Hence, the total cost for the optimal ate pairing on the KSS16 curve is $32538\mathbf{m}_{481} + 1\mathbf{i}_{481}$.

6   Comparison

As noticed by Scott [30], many protocols involve the computation of products or quotients of pairings, such as Boneh-Boyen IBE [8] needs to compute a quotient

---

[4] Since $p(z) \equiv 1 \pmod 4$ and -1 is a 8-th residue modulo $p(z)$, then the costs for $p/p^3/p^5/p^7$, and p^2/p^6, and p^4 Frobenius maps in $\mathbb{F}_{p^{16}}$ are $15\mathbf{m}_{481}$, $12\mathbf{m}_{481}$, $8\mathbf{m}_{481}$.

of two pairings, and ABE scheme due to Waters [33] and the non-interactive proof systems proposed by Groth and Sahai [17] need to check pairing product equations. Fast computation of products of pairings has already been investigated by Scott [29] and Granger and Smart [16]. In this section, we give the estimated comparisons of computing products and quotients of pairings at the 192 and 256-bit security levels, by using the strategies of combining full field squarings in the Miller loop and the final exponentiation [16].

6.1   192-Bit Security Level

When the embedding degree is even, since the pairing-friendly field $\mathbb{F}_{p^k}$ can be constructed as the quadratic extension of $\mathbb{F}_{p^{k/2}}$, then dividing by a pairing equals multiplying the conjugate of it under the action of the final exponentiation.[5] So we can regard quotients of pairings as products of pairings for KSS16, KSS18, BLS12, and BN curves that we considered here. Therefore, we only compare the cost estimations for computing the products of n optimal ate pairings on these curves using the pairing product techniques in Table 5.

Table 5. Costs comparison of products of n pairings at the 192-bit security level

Costs	n	KSS16	BLS12 [1]	BN [1]	KSS18 [1]
Full Squarings for DBL ·	1	$2592m_{481}$	$\approx 5892m_{512}$	$\approx 8837m_{512}$	$4158m_{512}$
Others in Miller loop	1	$7616m_{481}$	$\approx 10883m_{512}$	$\approx 16720m_{512}$	$9544m_{512}$
Final exponentiation	1	$22330m_{481}$ $+1i_{481}$	$\approx 13068m_{512}$ $+9i_{512}$	$\approx 11145m_{512}$ $+6i_{512}$	$23821m_{512}$ $+8i_{512}$
Total costs	1	$32538m_{481}$ $+1i_{481}$	$30023m_{512}$ $+9i_{512}$	$36702m_{512}$ $+6i_{512}$	$37523m_{512}$ $+8i_{512}$
	2	$40154m_{481}$ $+1i_{481}$	$40726m_{512}$ $+9i_{512}$	$53422m_{512}$ $+6i_{512}$	$47067m_{512}$ $+8i_{512}$
	7	$78234m_{481}$ $+1i_{481}$	$95141m_{512}$ $+9i_{512}$	$137022m_{512}$ $+6i_{512}$	$94787m_{512}$ $+8i_{512}$
	20	$177242m_{481}$ $+1i_{481}$	$236620m_{512}$ $+9i_{512}$	$354382m_{512}$ $+6i_{512}$	$218859m_{512}$ $+8i_{512}$

From Table 5 we conclude that the pairing having fewer costs in Miller loop may be more suitable for implementing pairing products. Our comparison shows that the KSS16 curve can derive the fastest product or quotient of n pairings when $n \geq 2$, and moreover a product or quotient of n pairings derived from the KSS18 curve is even faster than the BLS12 curve when $n \geq 7$.

---

[5] In the Miller algorithm of dividing by a pairing, we need to replace the element in $\mathbb{F}_{p^k}$ by its conjugate over $\mathbb{F}_{p^{k/2}}$.

6.2 256-Bit Security Level

We take the first comparison of the pairing computations at the 256-bit security level, by using the affine formulas in [24] for the BLS24 curve and our new ones in this paper for the BLS27 curve to estimate their costs for computing the ate pairings.

For computing the ate pairing on BLS24 curves, we choose the parameter $z = -2^{64} - 2^{60} - 2^{21} - 2^{12}$ so that $r(z)$ has a prime factor of 513 bits and $q(z)$ is a prime of 640 bits. As with the estimation in [1], we reestimate the cost for implementing the ate pairing on the BLS24 curve over a 640-bit field. The cost for the Miller loop is $64(3(9\mathbf{m}_{640}) + 2(6\mathbf{m}_{640}) + 4\mathbf{m}_{640} + (14\mathbf{m}_{640} + \mathbf{i}_{640})) + 3(3(9\mathbf{m}_{640}) + 6\mathbf{m}_{640} + 4\mathbf{m}_{640} + (14\mathbf{m}_{640} + \mathbf{i}_{640})) + 63(108\mathbf{m}_{640}) + 66(162\mathbf{m}_{640}) = 21297\mathbf{m}_{640} + 67\mathbf{i}_{640}$; and the cost for the final exponentiation (using cyclotomic squaring) is $83(3\mathbf{m}_{640}) + 11(2\mathbf{m}_{640}) + (4\mathbf{m}_{640} + \mathbf{i}_{640}) + 9(64(36\mathbf{m}_{640}) + 3(162\mathbf{m}_{640}) + 89(3\mathbf{m}_{640}) + 2(2\mathbf{m}_{640}) + 4\mathbf{m}_{640} + \mathbf{i}_{640}) + 14(162\mathbf{m}_{640}) + 2(54\mathbf{m}_{640}) + 360\mathbf{m}_{640} = 30596\mathbf{m}_{640} + 10\mathbf{i}_{640}$. As noticed in [1], in software implementations in 64-bit platforms, a $\mathbb{F}_p$ field element is represented with $l = 1 + \log_2(p)$ binary coefficients packed in $n_{64} = \lceil \frac{l}{64} \rceil$ 64-bit processor words, and a $\mathbb{F}_p$ field multiplication can be implemented with complexity $O(2n_{64}^2 + n_{64})$. So we can estimate that $\mathbf{m}_{640} \approx 1.228\mathbf{m}_{576}$, and list the cost comparison for computing pairing products on the BLS24 and BLS27 curves in Table 6.

Table 6. Costs comparison of the product of n pairings at the 256-bit security level

Costs	n	BLS24	BLS27
Full Squarings for DBL	1	$6804\mathbf{m}_{640}$ $\approx 8355\mathbf{m}_{576}$	$5832\mathbf{s}_{573}$
Others in Miller loop	1	$14493\mathbf{m}_{640} + 67\mathbf{i}_{640}$ $\approx 17797\mathbf{m}_{576} + 82\mathbf{i}_{576}$	$12627\mathbf{m}_{573} + 2838\mathbf{s}_{573}$ $+33\mathbf{i}_{573}$
Final exponentiation	1	$30596\mathbf{m}_{640} + 10\mathbf{i}_{640}$ $\approx 37572\mathbf{m}_{576} + 12\mathbf{i}_{576}$	$24627\mathbf{m}_{573} + 114998\mathbf{s}_{573}$ $+1\mathbf{i}_{573}$
Total costs ($\mathbf{s}_{576} \approx 0.8\mathbf{m}_{576}$)	1	$63724\mathbf{m}_{576} + 94\mathbf{i}_{576}$	$136188\mathbf{m}_{573} + 34\mathbf{i}_{573}$
	25	$490852\mathbf{m}_{576} + 2062\mathbf{i}_{576}$	$493716\mathbf{m}_{573} + 826\mathbf{i}_{573}$
	26	$508649\mathbf{m}_{576} + 2144\mathbf{i}_{576}$	$508613\mathbf{m}_{573} + 859\mathbf{i}_{573}$
	30	$579837\mathbf{m}_{576} + 2472\mathbf{i}_{576}$	$568201\mathbf{m}_{573} + 991\mathbf{i}_{573}$

Note that for a single pairing or quotients of pairings implementation the BLS24 curves have significantly faster pairings compared with the BLS27 curve. However, when implementing a product of no less than 25 pairings, the pairing on the BLS27 curve might be a better choice than the pairing on the BLS24 curve.

Remark 1. Compared with our estimated result at the 192-bit security level, the estimated superiority for computing the pairing products on the BLS27 curve is tiny, since the lack of the compressed squaring method [15] makes the final exponentiation for the BLS27 curve much more costly. So how to apply a compressed squaring method for curves with cubic twist and embedding degree divisible by 3 would be a interesting open question.

Remark 2. Scott [30] also mentioned that at the 256-bit level of security the KSS $k = 32$ curve [21] might turn out to be a better choice than the BLS24 curve if the fixed argument optimization applies. According to Lemma 1 we solve an optimal polynomial $h(\chi) = z - 3\chi + 2\chi^9 \in \mathbb{Z}[\chi]$ for KSS32 curves, which can derive an optimal ate pairing. However, since the parameters $(p(z), r(z), t(z))$ of KSS32 curves are relatively complicated, we find that the bit size of prime p is no less than 600 bits when assuming r has a prime factor of no less than 500 bits. So when targeting implementation of pairings on KSS32 curves for the 256-bit security, the base field is at least of 600 bits and the field multiplication is estimated as $\mathbf{m}_{640}$ for a 64-bit processer. Unluckily, the best z we can find is $z = 2^{35} + 2^{33} - 2^{22} + 2^{19} + 2^{10} - 2^7 + 1$ ensuring p a prime of 615 bits and r having a prime factor of 519 bits, and then the cost for its Miller loop $(19889\mathbf{m}_{640} + 41\mathbf{i}_{640})$ is larger than the cost for the BLS24 curve. Besides, the complicated final exponentiation for the KSS32 curve is still much more costly than the case of the BLS24 curve. This is the reason why we didn't add the optimal ate pairing on KSS32 curves in our comparison.

7 Conclusion

This paper has studied the optimum computations of products and quotients of pairings widely used in protocols at the 192 and 256-bit security levels. We have provided a detailed analysis of computing the optimal ate pairings on the BLS27 and KSS16 curves, and estimated their computing costs at the 256 and 192-bit security levels, respectively. From our comparisons of computing pairings derived from KSS16, BLS12, BN, KSS18 curves for the 192-bit security, and BLS24 and BLS27 curves for the 256-bit security, we show that for computing products or quotients of pairings at the 192-bit security level, the optimal ate pairing on the KSS16 curve could be the best choice; and at the 256-bit security level, the ate pairing on the BLS27 curve might be better when computing products of no less than 25 pairings, otherwise the ate pairing on the BLS24 curve is much better. In addition, we have presented new faster affine and projective formulas for the ate-like pairing on pairing-friendly curves with embedding degree divisible by 3 and cubic twist.

Being limited to the general estimated results, the thresholds where our proposed curves become faster than the ones previously suggested might vary a little when different target platforms are used. In future, we'd like to develop further improvements and give comparative timings for implementing products of pairings for concrete protocols.

Acknowledgments. We are heartily grateful to the anonymous reviewers for their helpful and insightful comments and corrections, they help us to remedy the shortcomings and improve the readability. This work was supported by the National 973 Program of China under Grant 2011CB302400, the National Natural Science Foundation of China under Grant 60970152, the Strategic Priority Research Program of the Chinese Academy of Sciences under Grant XDA06010701.

References

1. Aranha, D.F., Fuentes-Castañeda, L., Knapp, E., Menezes, A., Rodríguez-Henríquez, F.: Implementing Pairings at the 192-bit Security Level. In: Pairing 2012 (to appear, 2012); online Version: Cryptology ePrint Archive, Report 2012/232
2. Aranha, D.F., Karabina, K., Longa, P., Gebotys, C.H., López, J.: Faster Explicit Formulas for Computing Pairings over Ordinary Curves. In: Paterson, K.G. (ed.) EUROCRYPT 2011. LNCS, vol. 6632, pp. 48–68. Springer, Heidelberg (2011)
3. Barreto, P.S.L.M., Kim, H.Y., Lynn, B., Scott, M.: Efficient Algorithms for Pairing-Based Cryptosystems. In: Yung, M. (ed.) CRYPTO 2002. LNCS, vol. 2442, pp. 354–369. Springer, Heidelberg (2002)
4. Barreto, P.S.L.M., Lynn, B., Scott, M.: Constructing Elliptic Curves with Prescribed Embedding Degrees. In: Cimato, S., Galdi, C., Persiano, G. (eds.) SCN 2002. LNCS, vol. 2576, pp. 257–267. Springer, Heidelberg (2003)
5. Barreto, P.S.L.M., Naehrig, M.: Pairing-Friendly Elliptic Curves of Prime Order. In: Preneel, B., Tavares, S. (eds.) SAC 2005. LNCS, vol. 3897, pp. 319–331. Springer, Heidelberg (2006)
6. Benger, N., Scott, M.: Constructing Tower Extensions of Finite Fields for Implementation of Pairing-Based Cryptography. In: Hasan, M.A., Helleseth, T. (eds.) WAIFI 2010. LNCS, vol. 6087, pp. 180–195. Springer, Heidelberg (2010)
7. Beuchat, J.-L., González-Díaz, J.E., Mitsunari, S., Okamoto, E., Rodríguez-Henríquez, F., Teruya, T.: High-Speed Software Implementation of the Optimal Ate Pairing over Barreto–Naehrig Curves. In: Joye, M., Miyaji, A., Otsuka, A. (eds.) Pairing 2010. LNCS, vol. 6487, pp. 21–39. Springer, Heidelberg (2010)
8. Boneh, D., Boyen, X.: Efficient Selective-ID Secure Identity-Based Encryption Without Random Oracles. In: Cachin, C., Camenisch, J.L. (eds.) EUROCRYPT 2004. LNCS, vol. 3027, pp. 223–238. Springer, Heidelberg (2004)
9. Cook, S.A.: On the Minimum Computation Time of Functions. PhD Thesis, Harvard University Department of Mathematics (1966)
10. Costello, C., Lange, T., Naehrig, M.: Faster Pairing Computations on Curves with High-Degree Twists. In: Nguyen, P.Q., Pointcheval, D. (eds.) PKC 2010. LNCS, vol. 6056, pp. 224–242. Springer, Heidelberg (2010)
11. Costello, C., Lauter, K., Naehrig, M.: Attractive Subfamilies of BLS Curves for Implementing High-Security Pairings. In: Bernstein, D.J., Chatterjee, S. (eds.) INDOCRYPT 2011. LNCS, vol. 7107, pp. 320–342. Springer, Heidelberg (2011)
12. Devegili, A.J., Ó hÉigeartaigh, C., Scott, M., Dahab, R.: Multiplication and squaring on pairing-friendly fields. Cryptography ePrint Archive, Report 2006/471
13. El Mrabet, N., Guillermin, N., Ionica, S.: A study of pairing computation for elliptic curves with embedding degree 15. Cryptology ePrint Archive: Report 2009/370
14. Freeman, D., Scott, M., Teske, E.: A Taxonomy of Pairing-Friendly Elliptic Curves. Journal of Cryptology 23(2), 224–280 (2010)
15. Granger, R., Scott, M.: Faster Squaring in the Cyclotomic Subgroup of Sixth Degree Extensions. In: Nguyen, P.Q., Pointcheval, D. (eds.) PKC 2010. LNCS, vol. 6056, pp. 209–223. Springer, Heidelberg (2010)
16. Granger, R., Smart, N.P.: On computing products of pairings. Cryptology ePrint Archive Report 2006/172
17. Groth, J., Sahai, A.: Efficient Non-interactive Proof Systems for Bilinear Groups. In: Smart, N.P. (ed.) EUROCRYPT 2008. LNCS, vol. 4965, pp. 415–432. Springer, Heidelberg (2008)
18. Hess, F.: Pairing Lattices. In: Galbraith, S.D., Paterson, K.G. (eds.) Pairing 2008. LNCS, vol. 5209, pp. 18–38. Springer, Heidelberg (2008)

19. Hess, F., Smart, N.P., Vercauteren, F.: The Eta Pairing Revisited. IEEE Transactions on Information Theory 52(10), 4595–4602 (2006)

20. Ireland, K.F., Rosen, M.I.: A classical introduction to modern number theory. Graduate Texts in Mathematics, vol. 84. Springer (1990)

21. Kachisa, E.J., Schaefer, E.F., Scott, M.: Constructing Brezing-Weng Pairing-Friendly Elliptic Curves Using Elements in the Cyclotomic Field. In: Galbraith, S.D., Paterson, K.G. (eds.) Pairing 2008. LNCS, vol. 5209, pp. 126–135. Springer, Heidelberg (2008)

22. Karatsuba, A.A., Ofman, Y.: Multiplication of Multidigit Numbers on Automata. Soviet Physics Doklady 7, 595–596 (1963)

23. Koblitz, N., Menezes, A.: Pairing-Based Cryptography at High Security Levels. In: Smart, N.P. (ed.) Cryptography and Coding 2005. LNCS, vol. 3796, pp. 13–36. Springer, Heidelberg (2005)

24. Lauter, K., Montgomery, P.L., Naehrig, M.: An Analysis of Affine Coordinates for Pairing Computation. In: Joye, M., Miyaji, A., Otsuka, A. (eds.) Pairing 2010. LNCS, vol. 6487, pp. 1–20. Springer, Heidelberg (2010)

25. Lemmermeyer, F.: Reciprocity Laws: from Euler to Eisenstein. Springer (2000)

26. Lin, X., Zhao, C.A., Zhang, F., Wang, Y.: Computing the ate pairing on elliptic curves with embedding degree k= 9. IEICE Transactions on Fundamentals of Electronics, Communications and Computer Sciences 91(9), 2387–2393 (2008)

27. Miller, V.: The Weil pairing, and its efficient calculation. Journal of Cryptology 17(4), 235–261 (2004)

28. Naehrig, M., Niederhagen, R., Schwabe, P.: New Software Speed Records for Cryptographic Pairings. In: Abdalla, M., Barreto, P.S.L.M. (eds.) LATINCRYPT 2010. LNCS, vol. 6212, pp. 109–123. Springer, Heidelberg (2010)

29. Scott, M.: Computing the Tate Pairing. In: Menezes, A. (ed.) CT-RSA 2005. LNCS, vol. 3376, pp. 293–304. Springer, Heidelberg (2005)

30. Scott, M.: On the Efficient Implementation of Pairing-Based Protocols. In: Chen, L. (ed.) Cryptography and Coding 2011. LNCS, vol. 7089, pp. 296–308. Springer, Heidelberg (2011)

31. Toom, A.L.: The Complexity of a Scheme of Functional Elements realizing the Multiplication of Integers. Soviet Mathematics 4(3), 714–716 (1963)

32. Vercauteren, F.: Optimal Pairings. IEEE Transactions on Information Theory 56(1), 455–461 (2010)

33. Waters, B.: Ciphertext-Policy Attribute-Based Encryption: An Expressive, Efficient, and Provably Secure Realization. In: Catalano, D., Fazio, N., Gennaro, R., Nicolosi, A. (eds.) PKC 2011. LNCS, vol. 6571, pp. 53–70. Springer, Heidelberg (2011)

A   Computing Final Exponentiation for KSS16 Curve

Here we give the concrete formulas for the final exponentiation of the optimal ate pairing on the KSS16 curve. Luckily we can use an idea of choosing "better" parameter z, which could reduce the cost $39S_c$ to $16S_c$ by increasing a few storages, and all the cost savings occur in the formulas marked with star ($*$). Our idea of choosing parameter is to search for such $z = \pm 2^{n_1} \pm 2^{n_2} \pm \cdots \pm 2^{n_s}$ satisfying $n_1 - n_2 > 6$, and therefore the calculations marked with star could be included in the next calculations of the exponentiations by z (or $z + 1$). Thus one can compute the final exponentiation costing 1 inversion in $\mathbb{F}_{p^{16}}$, 7 exponentiations by z, 2 exponentiations by $z + 1$, 16 cyclotomic squarings in $G_{\Phi_2(p^8)}$, 34 multiplications in $\mathbb{F}_{p^{16}}$, 2 cyclotomic cubings in $\mathbb{F}_{p^{16}}$, and p, p^2, p^3, p^4, p^5, p^6, p^7, p^8-Frobenius maps.

Easy part (Input $f \in \mathbb{F}_{p^{16}}$)	Cost	Value
$E1 = f^{p^8}$, $E2 = E1 \cdot f^{-1}$	1**I**	$E2 = f^{p^8-1} = M$
Hard part	Cost	Value
$T1 = E2^4$, $T2 = T1^8$, $T3 = T2^2$ $*$	(6$\mathbf{S}_c$)	
$F1 = T2 \cdot T1^{-1}$, $F2 = F1^2$	1$\mathbf{S}_c$ + 1**M**	
$F3 = E2^{z+1}$, $F4 = F3^{z+1}$	2 Exp. by $z+1$	
$F5 = F4 \cdot T1$	1**M**	
$T4 = F5^8$ $*$	(3$\mathbf{S}_c$)	
$F6 = F5^z$	1 Exp. by z	
$F7 = \overline{F5} \cdot T6$, $F8 = T4^3$	1**M** + 1$\mathbf{C}_c$	$F7 = M^{c'_7}$
$T5 = F6^8$ $*$	(3$\mathbf{S}_c$)	
$F9 = F6^z$	1 Exp. by z	
$F10 = T5 \cdot \overline{F6}$, $F11 = F10^2$	1$\mathbf{S}_c$ + 1**M**	
$T6 = F9^8$ $*$	(3$\mathbf{S}_c$)	
$F12 = F9^z$	1 Exp. by z	
$F13 = T6 \cdot \overline{\overline{F9}}$, $F14 = F13^3$, $F15 = F12 \cdot F2$	2**M** + 1$\mathbf{C}_c$	
$T7 = F15^2$, $T8 = T7^4$, $T9 = T8^4$ $*$	(5$\mathbf{S}_c$)	
$F16 = F15^z$	1 Exp. by z	
$F17 = T9 \cdot T7$, $F18 = F17 \cdot F15$, $F19 = F18^2$	1$\mathbf{S}_c$ + 2**M**	
$F20 = F19^2$, $F21 = F20 \cdot T8$, $F22 = F16 \cdot F8$	1$\mathbf{S}_c$ + 2**M**	$F22 = M^{c'_3}$
$F23 = F22^z$	1 Exp. by z	
$F24 = F23 \cdot F11$	1**M**	$F24 = M^{c'_2}$
$T10 = F23^2$ $*$	(1$\mathbf{S}_c$)	
$F25 = F23^z$	1 Exp. by z	
$F26 = T10 \cdot \overline{F10}$	1**M**	$F26 = M^{c'_6}$
$T11 = F25^4$ $*$	(2$\mathbf{S}_c$)	
$F27 = F25^z$	1 Exp. by z	
$F28 = T11 \cdot \overline{F25}$, $F29 = T11 \cdot F25$	2**M**	$F29 = M^{c'_5}$
$F30 = F13 \cdot F14$, $F31 = F28 \cdot \overline{F30}$, $F32 = F12^2$	1$\mathbf{S}_c$ + 2**M**	$F31 = M^{c'_1}$
$F33 = F32 \cdot F12$, $F34 = \overline{F27} \cdot F33$, $F35 = F34^2$	1$\mathbf{S}_c$ + 2**M**	
$F36 = F35 \cdot F12$, $F37 = \overline{F36} \cdot F21$, $F38 = F34 \cdot F1$	3**M**	$F37 = M^{c_4}$
$F39 = F38^2$, $F40 = F39^2$, $F41 = F40^2$	3$\mathbf{S}_c$	
$F42 = F39 \cdot F38$, $F43 = F41 \cdot F42$, $F44 = \overline{F43} \cdot F18$	3**M**	$F44 = M^{c_0}$
$H1 = F7^{p^7}$, $H2 = F22^{p^3}$, $H3 = F24^{p^2}$, $H4 = F26^{p^6}$		
$H5 = F29^{p^5}$, $H6 = F31^p$, $H7 = F37^{p^4}$		
$H8 = H1 \cdot \overline{H2}$, $H9 = H8^2$, $H10 = H9^2$	2$\mathbf{S}_c$ + 1**M**	
$H11 = H10 \cdot H8$, $H12 = H11 \cdot H3$, $H13 = H12 \cdot H4$	3**M**	
$H14 = H13^2$, $H15 = H14^2$, $H16 = H15 \cdot H13$	2$\mathbf{S}_c$ + 1**M**	
$H17 = H16 \cdot H6$, $H18 = H17 \cdot \overline{H5}$, $H19 = H18^2$	1$\mathbf{S}_c$ + 2**M**	
$H20 = H19^2$, $H21 = H20 \cdot H18$, $H22 = H21 \cdot H7$	1$\mathbf{S}_c$ + 2**M**	
$H23 = H22 \cdot F44$ (Final output)	1**M**	

Constructing Pairing-Friendly Genus 2 Curves with Split Jacobian

Robert Dryło

Institute of Mathematics, Polish Academy of Sciences,
ul. Śniadeckich 8, 00-950 Warszawa, Poland
Warsaw School of Economics,
al. Niepodległości 162, 02-554 Warszawa, Poland
r.drylo@impan.gov.pl

Abstract. Using genus 2 curves with simple but not absolutely simple Jacobians one can obtain pairing-based cryptosystems more efficient than for a generic genus 2 curve. We describe a new framework to construct pairing-friendly abelian surfaces, which are simple but not absolutely simple. The main contribution is the generalization of the notion of complete, complete with variable discriminant, and sparse families of elliptic curves introduced by Freeman, Scott and Teske [13]. We give algorithms to construct families of abelian surfaces of each type, which generalize the Brezing-Weng method. To realize these abelian surfaces as Jacobians we use curves of the form $y^2 = x^5 + ax^3 + bx$ or $y^2 = x^6 + ax^3 + b$, and apply the method of Freeman and Satoh [12]. As applications we give variable-discriminant families with best ρ-values. We also give some families with record ρ-value.

Keywords: Pairing-friendly hyperelliptic curves, abelian varieties, Weil numbers, CM method.

1 Introduction

Since pairings have been introduced to design cryptographic protocols (see, e.g., [2,3,22,37]), one of the main problems is to construct abelian varieties suitable for these applications. Let $A/\mathbb{F}_q$ be an abelian variety containing an $\mathbb{F}_q$-rational subgroup of prime order r with the embedding degree $k = \min\{l : r \mid (q^l - 1)\}$. To implement pairing-based cryptosystems k should be suitably small so that pairings of r-torsion points with values in the field $\mathbb{F}_{q^k}$ could be efficiently computed, but the discrete logarithm problem in $\mathbb{F}_{q^k}$ remains intractable. Furthermore, in order to reduce the key length for a given security level, we would like the bit size of r to be close to the size of $\#A(\mathbb{F}_q)$. Since $\log(\#A(\mathbb{F}_q)) \approx \dim A \log(q)$, we would like the parameter $\rho = \dim A \log q / \log r$ to be close to one. We can achieve $\rho \approx 1$ using supersingular abelian varieties, which in each dimension have bounded embedding degrees (e.g., $k \leq 6$ or 12 for supersingular elliptic curves or abelian surfaces (see [16,31,33])). For higher security levels we use ordinary varieties, which are unlikely to be found by a random choice and require

S. Galbraith and M. Nandi (Eds.): INDOCRYPT 2012, LNCS 7668, pp. 431–453, 2012.

specific constructions. In practice, we mainly use elliptic curves or Jacobians of hyperelliptic curves of low genus.

Pairing-friendly elliptic curves. In general, to construct an ordinary elliptic curve E with an embedding degree k we first find parameters (r, t, q) of E, where t is the trace of E, q is the size of the field of definition, and r is the order of a subgroup with the embedding degree k. Then we use the Complex Multiplication (CM) method to find the equation of E, which requires that the CM discriminant d of E is sufficiently small, where d is the square-free part of the non-negative integer $4q - t^2$. Parameters (r, t, q) of pairing-friendly elliptic curves are generated either directly, like in the Cocks-Pinch method (see [13, Theorem 4.1]), or are obtained as values of suitable polynomials $(r(x), t(x), q(x))$ called parametric families. The former method is very flexible and allows one to obtain subgroup orders r and discriminants d of almost arbitrary size, however with ρ-value only around 2. Using parametric families we can considerably improve ρ-values for more restricted subgroup orders and discriminants.

Miyaji, Nakabayashi and Takano [27] were the first researchers to use parametric families to characterize elliptic curves of prime orders with embedding degrees $k = 3, 4, 6$. Scott and Barreto [34], and Galbraith et al. [17] generalized their idea to describe elliptic curves with prescribed cofactors for $k = 3, 4, 6$. Currently constructions of families with $\rho = 1$ are also known for $k = 10$ and 12, and were discovered by Freeman [9] and Barreto-Naehrig [1], respectively. Most families used in practice are so-called complete families, which are constructed by the Brezing-Weng method [4]. We now recall the general definition and the classification of families introduced by Freeman, Scott and Teske [13].

Definition 1. ([13, Definition 2.7]) Let k and d be positive integers such that d is square-free. We say that a triple of polynomials $(r(x), t(x), q(x))$ in $\mathbb{Q}[x]$ *parametrizes a family of elliptic curves with embedding degree k and discriminant d* if the following conditions are satisfied:

1. $q(x) = p(x)^s$ for some $s \geq 1$ and $p(x)$ that represents primes.
2. $r(x)$ is irreducible, non-constant, integer valued, and has positive leading coefficient.
3. $r(x)$ divides $q(x) + 1 - t(x)$.
4. $r(x)$ divides $\Phi_k(t(x) - 1)$, where $\Phi_k(x)$ is the kth cyclotomic polynomial.
5. The CM equation $4q(x) - t(x)^2 = dy^2$ has infinitely many integer solutions (x, y).

Properties of the CM equation lead us to the classification of families. It is clear that we can write $4q(x) - t(x)^2 = f(x)g(x)^2$, where $f(x) \in \mathbb{Z}[x]$ is square-free and $g(x) \in \mathbb{Q}[x]$. Then condition (5) implies by Siegel's theorem that $\deg f(x) \leq 2$ (see [9, Proposition 2.10] or Lemma 16). We say that a family is *complete* if f is constant, so $f = d$; then the CM equation is satisfied for any $x \in \mathbb{Z}$. We say that a family is *complete with variable discriminant* if $\deg f = 1$; then substituting $x \leftarrow (dx^2 - b)/a$, where $f(x) = ax + b$, yields a complete family with discriminant d if conditions (1) and (2) of Definition 1 are satisfied. A family is called *sparse* if $\deg f = 2$; then the CM equation can be transformed to the generalized Pell equation, whose solutions grow exponentially. Let us note that the Brezing-Weng method can be generalized to construct families of the latter two types (see [8]). These families can be used to generate elliptic curves with larger discriminant, which may be desired for larger variety of cryptosystems.

Pairing-friendly genus 2 curves. Freeman, Stevenhagen and Streng [14] (see also Freeman [10]) gave a general method to generate pairs (r, π) such that π is a Weil q-number corresponding by the Honda-Tate theory to a simple ordinary abelian variety with embedding degree k with respect to a prime r. In order to use the CM method to realize these varieties as Jacobians, π have to generate a suitable CM field K, where Weil numbers in question are characterized by the condition

$$N_{K/\mathbb{Q}}(\pi - 1) \equiv \Phi_k(\pi\bar{\pi}) \equiv 0 \pmod{r}.$$

If $[K : \mathbb{Q}] = 2g$, then the corresponding varieties are of dimension g with ρ-value around $2g^2$. Freeman [11] generalized this method to construct parametric families of abelian varieties. In order to obtain pairing-friendly ordinary abelian surfaces, which generically have ρ-value around 4, or less than 4 for parametric families, we use genus 2 curves, whose Jacobian is simple but not absolutely simple. Kawazoe and Takahashi [25] use curves of the form $y^2 = x^5 + ax$ and a closed formula for their order [15] (see also Kachisa [23]). Freeman and Satoh [12] give a general method to construct an elliptic curve, whose Weil restriction over some extension contains an abelian surface with a given embedding degree. To realize that surface as a Jacobian, they use curves of the form $y^2 = x^5 + ax^3 + bx$ or $y^2 = x^6 + ax^3 + b$. Recently Guillevic and Vergnaud [19] extended their method using closed formulas for the order of these curves.

Contribution. In this paper we describe a framework to determine Weil numbers of simple but not absolutely simple pairing-friendly abelian varieties, which is based on the following idea (see also [7]). Let K be a CM field of degree $2g$, and suppose that we have a polynomial $\pi(x, y) \in K[x, y]$ such that $q(x, y) = \pi(x, y)\bar{\pi}(x, y) \in \mathbb{Q}[x, y]$ and the image $\pi(\mathbb{Z}^2)$ contains "sufficiently many" Weil numbers in K. Then we can use $\pi(x, y)$ to generate pairing-friendly Weil numbers analogously to the Cocks-Pinch method. If r is a prime such that the system

$$N_{K(x,y)/\mathbb{Q}(x,y)}(\pi(x, y) - 1) = \Phi_k(q(x, y)) = 0 \tag{1}$$

has solutions over $\mathbb{F}_r$, then we check whether $\pi(x, y)$ is a Weil number for lifts $x, y \in \mathbb{Z}$ of these solutions. Under the heuristic assumption that generically solutions over $\mathbb{F}_r$ are of the similar size as r, the resulting varieties have ρ-value $\rho = g \log q(x, y) / \log r \approx 2g \deg \pi(x, y)$. Thus to obtain ρ-value around $2g$, we need suitable polynomials $\pi(x, y)$ of degree one. If K contains an imaginary quadratic subfield $K_0 = \mathbb{Q}(\sqrt{-d})$, then for any $u \in K$ such that $c = u\bar{u} \in \mathbb{Q}$, the polynomial $\pi(x, y) = u(x + y\sqrt{-d})$ satisfies $q(x, y) = \pi(x, y)\bar{\pi}(x, y) = c(x^2 + dy^2) \in \mathbb{Q}[x, y]$, however, if $c \neq 1$, then the image $\pi(\mathbb{Z}^2)$ does not contain sufficiently many primes. Therefore we will use $\pi(x, y) = \zeta_s(x + y\sqrt{-d})$ to generate Weil numbers in the CM field $K = \mathbb{Q}(\zeta_s, \sqrt{-d})$, where ζ_s is an sth primitive root of unity and $d > 0$ is a square-free integer. We note that Weil q-numbers of the form $\pi = \zeta_s \pi_0$ with $\pi_0 \in \mathbb{Q}(\sqrt{-d})$ correspond to simple abelian varieties which are isogenous over $\mathbb{F}_{q^s}$ to a power of an elliptic curve $E/\mathbb{F}_q$ with the Weil q-number π_0 (Corollary 4).

To generalize the Cocks-Pinch and the Brezing-Weng methods we describe in Section 3 prime finite fields and number fields, where system (1) has solutions for $\pi(x, y) = \zeta_s(x + y\sqrt{-d})$, and we give explicit formulas for solutions. In Section 4 we focus on constructing genus 2 curves, whose Jacobian corresponds to Weil numbers $\pi = \zeta_s \pi_0$ in a quartic CM field $K = \mathbb{Q}(\zeta_s, \sqrt{-d})$. We give an algorithm to construct curves of the form $y^2 = x^6 + ax^3 + b$ and $y^2 = x^5 + ax^3 + bx$, which is based on the method of Freeman and Satoh (see [12, Algorithm 5.11]). In Section 5 we generalize Definition 1 and the classification of families of elliptic curves to abelian varieties. In Sections 6, 7, 8 we generalize the Brezing-Weng method to construct families of each type.

As applications we give complete families with variable discriminant of abelian surfaces $(r(x), \pi(x))$ with best ρ-values such that $\deg r(x) < 25$. Some complete families with variable discriminant are given in [12, Section 7], where they are obtained from complete families satisfying certain conditions, however no general method to construct such families is given. We also find some families with record ρ-value $\rho = 2$ for $k = 3, 4, 6, 12$, or $\rho \approx 2.1$ for $k = 27, 54$ (see Examples 19, 24, 27).

2   Background on Abelian Varieties

In this section we gather basic facts on abelian varieties, which will be needed in the sequel (for details see [28,39,40,41,42]).

Let $A/\mathbb{F}_q$ be a g-dimensional abelian variety with qth Frobenius endomorphism π_A, and its characteristic polynomial f_A. Then we have $f_A(\pi_A) = 0$, and $\#A(\mathbb{F}_q) = f_A(1)$. Furthermore, all roots of f_A are Weil q-numbers. Recall that an algebraic integer π is called *a Weil q-number* if $|\alpha(\pi)| = \sqrt{q}$ for every embedding $\alpha : \mathbb{Q}(\pi) \to \mathbb{C}$.

We say that A is *simple* if it is not isogenous over $\mathbb{F}_q$ to a product of two positive dimensional abelian varieties. By the Honda-Tate theorem the map which associates the Frobenius endomorphism π_A to a simple abelian variety $A/\mathbb{F}_q$ induces a one-to-one correspondence between isogeny classes of simple abelian varieties over $\mathbb{F}_q$ and conjugacy classes of Weil q-numbers. Recall also that A is called *ordinary* if it has the maximum number p^g of all p-torsion points over $\overline{\mathbb{F}}_q$, where $p = \mathrm{char}\,\mathbb{F}_q$. We have the following.

Theorem 2. ([42]) *Let $A/\mathbb{F}_q$ be a simple abelian variety of dimension g with the endomorphism algebra $K = \mathrm{End}_{\mathbb{F}_q}(A) \otimes \mathbb{Q}$. Then A is ordinary if and only if K is a CM field of degree $2g$, $K = \mathbb{Q}(\pi_A)$, and $\pi_A, \overline{\pi}_A$ are relatively prime in $\mathcal{O}_K$. Furthermore, if A is ordinary, then f_A is the minimal polynomial of π_A, and*

$$\#A(\mathbb{F}_q) = f_A(1) = \mathrm{N}_{K/\mathbb{Q}}(\pi_A - 1). \tag{2}$$

Recall that a number field K is called a *CM field* if it is an imaginary quadratic extension of a totally real field. Then K has an automorphism, denoted by a bar, which commutes with every embedding $K \to \mathbb{C}$ and the complex conjugation in $\mathbb{C}$. Note that two integers in $\mathcal{O}_K$ are said relatively prime if they generate the unit ideal.

In this paper we are interested in simple abelian varieties, which are not absolutely simple (i.e., they split over some extension of the base field).

Proposition 3. *A simple ordinary abelian variety $A/\mathbb{F}_q$ with a Weil q-number π splits over $\mathbb{F}_{q^s}$ if and only if $\mathbb{Q}(\pi^s) \subsetneq \mathbb{Q}(\pi)$. Then A is isogenous to B^n over $\mathbb{F}_{q^s}$, where $B/\mathbb{F}_{q^s}$ is a simple abelian variety with the Weil q^s-number π^s.*

Proof. For the sake of completeness we give a proof (see also [20, Lemma 4]). Recall that if $f_{A,q}(x) = \prod_{i=1}^{2g}(x - \pi_i)$, then $f_{A,q^s}(x) = \prod_{i=1}^{2g}(x - \pi_i^s)$. Since A is simple and ordinary, $f_{A,q}(x)$ is irreducible, and hence all π_i are conjugated. If $\mathbb{Q}(\pi^s) \subsetneq \mathbb{Q}(\pi)$, then f_{A,q^s} is not the minimal polynomial of π^s, so A splits over $\mathbb{F}_{q^s}$. Conversely, if $A \sim B_1 \times \cdots \times B_m$ for simple abelian varieties $B_i/\mathbb{F}_{q^s}$, then $f_{A,q^s} = f_{B_1} \cdots f_{B_m}$. Since each B_i is ordinary, f_{B_i} is irreducible. Furthermore, since all $\pi_1^s, \ldots, \pi_{2g}^s$ are conjugated, it follows that they are exactly roots of each f_{B_i}. Hence all f_{B_i} are equal, and from Tate's theorem [40] it follows that all B_i are isogenous over $\mathbb{F}_{q^s}$, so $A \sim B_1^n$.

Corollary 4. *Let $A/\mathbb{F}_q$ be an ordinary simple abelian variety with a Weil q-number π, and $E/\mathbb{F}_q$ be an ordinary elliptic curve with a Weil q-number π_0.*

(i) Then A is isogenous to E^g over $\mathbb{F}_{q^n}$ if and only if $\pi = \zeta_s \pi_0$, where ζ_s is an sth primitive root of unity and $s \mid n$.

(ii) If s is even and $\pi = \zeta_s \pi_0$, then A is isogenous to E'^g over $\mathbb{F}_{q^{s/2}}$, where E' is the quadratic twist of E.

(iii) If $\pi = \zeta_s \pi_0$, then $\mathbb{Q}(\pi) = \mathbb{Q}(\zeta_s, \sqrt{-d})$, where $\pi_0 \in \mathbb{Q}(\sqrt{-d})$ and d is a positive square-free integer.

Proof. (i) By Proposition 3 we have $A \sim E^g$ over $\mathbb{F}_{q^n}$ if and only if $\pi^n = \pi_0^n$. So, if s is the minimal integer such that $\pi^s = \pi_0^s$, then $\pi = \zeta_s \pi_0$, and obviously $s \mid n$.

(ii) Since $-\pi_0$ is the Weil q-number of the quadratic twists E' of E, and $\pi = \zeta_{s/2}(-\pi_0)$, it follows from (i) that $A \sim E'^g$ over $\mathbb{F}_{q^{s/2}}$.

(iii) Since E is ordinary, π_0^s and $\bar{\pi}_0^s$ are relatively prime, so $\pi_0^s \notin \mathbb{Z}$. Hence $\pi^s = \pi_0^s$ generates $\mathbb{Q}(\sqrt{-d})$, which implies that $\zeta_s, \sqrt{-d} \in \mathbb{Q}(\pi)$, so $\mathbb{Q}(\pi) = \mathbb{Q}(\zeta_s, \sqrt{-d})$.

2.1   Weil Numbers of Pairing-Friendly Varieties

Recall that *the embedding degree* of an abelian variety $A/\mathbb{F}_q$ with respect to a prime $r \mid \#A(\mathbb{F}_q)$, $r \neq \operatorname{char} \mathbb{F}_q$, is the minimal integer k such that $r \mid (q^k - 1)$. In other words, $q \pmod r$ is a kth primitive root of unity, or equivalently, if $r \nmid k$, it is a root of the kth cyclotomic polynomial $\Phi_k(x)$. By Theorem 2 we have the following.

Lemma 5. ([14, Proposition 2.1]) *Let $K = \mathbb{Q}(\pi)$ be a CM field of degree $2g$, where π is a Weil q-number corresponding to an ordinary abelian variety A. Let k be a positive integer and r be a prime such that $r \nmid kq$. Then A has the embedding degree k with respect to r if and only if*

(1) $r \mid \Phi_k(q)$,
(2) $r \mid N_{K/\mathbb{Q}}(\pi - 1)$.

3   The Generalized Cocks-Pinch Method

Let $K = \mathbb{Q}(\zeta_s, \sqrt{-d})$ be a CM field of degree $2g$, where ζ_s is an sth primitive root of unity and $d > 0$ is a square-free integer. To generate as in the Cooks-Pinch method pairing-friendly Weil numbers of the form $\pi = \zeta_s \pi_0$ with $\pi_0 \in \mathbb{Q}(\sqrt{-d})$, we need to find a prime finite field $\mathbb{F}_r$ where the system

$$N_{K(x,y)/\mathbb{Q}(x,y)} \left(\zeta_s(x + y\sqrt{-d}) - 1 \right) = \Phi_k\left(x^2 + dy^2\right) = 0, \tag{3}$$

has solutions, and check whether $\pi(x,y) = \zeta_s(x + y\sqrt{-d})$ is a Weil number for lifts $x, y \in \mathbb{Z}$ of these solutions. We describe below such prime fields $\mathbb{F}_r$, and give explicit formulas on solutions. We also give an analogous result for number fields in order to further generalize the Brezing-Weng.

Lemma 6. *Let $R = \mathbb{Z}$ or $\mathbb{Q}[x]$, and $r \in R$ be a prime such that the residue field $R/(r)$ contains primitive roots of unity ζ_k, ζ_s and $\sqrt{-d}$ (if $R = \mathbb{Z}$, we assume that $r \nmid 2dks$). If $\sqrt{-d} \notin \mathbb{Q}(\zeta_s)$, then solutions in $R/(r)$ of system (3) are of the form*

$$x = \frac{\zeta_s^{-1} + \zeta_k \zeta_s}{2}, \quad y = \pm\frac{\zeta_s^{-1} - \zeta_k \zeta_s}{2\sqrt{-d}}. \tag{4}$$

If $\sqrt{-d} \in \mathbb{Q}(\zeta_s)$, then one of these pairs is a solution of (3).

Proof. We have

$$N_{K(x,y)/\mathbb{Q}(x,y)} \left(\zeta_s(x + y\sqrt{-d}) - 1 \right) = \prod_{\sigma \in \mathrm{Aut}(K)} \left(\sigma(\zeta_s)(x + y\sigma(\sqrt{-d})) - 1 \right),$$

and

$$x^2 + dy^2 = \sigma(\zeta_s)(x + y\sigma(\sqrt{-d}))\sigma(\zeta_s^{-1})(x - y\sigma(\sqrt{-d})).$$

Thus (3) has the same solutions over $\mathbb{Q}(\zeta_k, \zeta_s, \sqrt{-d})$ as systems

$$\sigma(\zeta_s)(x + y\sigma(\sqrt{-d})) = 1,$$

$$\sigma(\zeta_s^{-1})(x - y\sigma(\sqrt{-d})) = \zeta_k,$$

for each ζ_k and $\sigma \in \mathrm{Aut}(K)$. Hence

$$x = \frac{\sigma(\zeta_s^{-1}) + \zeta_k \sigma(\zeta_s)}{2}, \quad y = \frac{\sigma(\zeta_s^{-1}) - \zeta_k \sigma(\zeta_s)}{2\sigma(\sqrt{-d})}. \tag{5}$$

If $\sqrt{-d} \notin \mathbb{Q}(\zeta_s)$, then the above solutions are of the form (4), since each automorphism of $\mathbb{Q}(\zeta_s)$ has two extensions on K. If $\sqrt{-d} \in \mathbb{Q}(\zeta_s)$, then one of the pairs (4) is equal to such a solution. Now let P be a prime ideal over r in $S = R[\zeta_s, \zeta_k, \sqrt{-d}]$, and S_P be the localization of S at P. It follows from the assumption that $R/(r) = S/P = S_P/PS_P$. Reducing solutions (5) mod PS_P we get solutions in $R/(r)$ of the desired form, sine reduction mod P induces an isomorphism between sth and kth roots of unity in S and $R/(r)$ by the following fact.

Lemma 7. Let $R = \mathbb{Z}$ or $\mathbb{Q}[x]$, and $r \in R$ be a prime such that the residue field $R/(r)$ contains sth primitive roots of unity (if $R = \mathbb{Z}$, we assume that $r \nmid s$). If P is a prime ideal in $R[\zeta_s]$ over r, then $R/(r) = R[\zeta_s]/P$ and reduction mod P induces an isomorphism between sth roots of unity in $R[\zeta_s]$ and $R/(r)$.

Proof. We note that $S = R[\zeta_s]$ is the integral closure of R in the field of fractions of S. This is well-known for $R = \mathbb{Z}$. If $R = \mathbb{Q}[x]$, it follows from the fact that $F[x]$ is integrally closed in $F(x)$ for any field F; in particular for $F = \mathbb{Q}(\zeta_s)$. We also note that the sth cyclotomic polynomial $\Phi_s(x)$ is irreducible over $\mathbb{Q}(x)$, because it is irreducible over $\mathbb{Q}$ and coefficients of monic factors of polynomials in $\mathbb{Q}[x]$ are algebraic over $\mathbb{Q}$. Since $R \subset S$ is an integral extension of Dedekind domains, we have $rS = \prod_{i=1}^{n} P_i^e$, where P_i are prime ideals in S. Let r_i mod r for $r_i \in R$ be different sth primitive roots of unity in $R/(r)$ for $i = 1, \ldots, \varphi(s)$. Since r_i mod r are roots of $\Phi_s(x)$, after rearranging we have $P_i = (r, \zeta_s - r_i)$ (see [26, Proposition I.8.25]). Thus $\zeta_s^j \equiv r_i^j$ mod P_i yields an isomorphism between sth roots of unity.

From Lemma 6 we obtain the following generalization of the Cocks-Pinch algorithm.

Algorithm 8. Input: A CM field $K = \mathbb{Q}(\zeta_s, \sqrt{-d})$ of degree $2g$, and an integer $k > 0$.
Output: $\emptyset$, or a pair (r, π) such that r is a prime and $\pi = \zeta_s \pi_0$ with $\pi_0 \in \mathbb{Q}(\sqrt{-d})$ is a Weil q-number corresponding to a g-dimensional ordinary abelian variety $A/\mathbb{F}_q$ with the embedding degree k with respect to r.

1. Choose a prime r such that $\mathrm{lcm}(s, k) | (r-1)$ and $\sqrt{-d} \in \mathbb{F}_r$.
2. Let $x = \frac{\zeta_s^{-1} + \zeta_k \zeta_s}{2}$ and $y = \frac{\zeta_s^{-1} - \zeta_k \zeta_s}{2\sqrt{-d}}$ for all primitive roots of unity $\zeta_s, \zeta_k \in \mathbb{F}_r$.
3. If $\sqrt{-d} \in \mathbb{Q}(\zeta_s)$ and x, y in the previous step do not satisfy system (3), put $y := -y$.
4. Let $x_1, y_1 \in [0, r)$ be lifts of x, y.
5. Let $\pi = \zeta_s(x_1 + ir + (y_1 + jr)\sqrt{-d})$ for $i, j \in [-m, m]$, where m is a small integer.
6. Return (r, π) if $q = \pi\bar{\pi}$ is prime and $x_1 + ir \neq 0$.

Since usually solutions x or y in step 2 are of the similar size as r, we obtain ρ-value

$$\rho = \frac{g \log((x_1 + ir)^2 + d(y_1 + jr)^2)}{\log r} \approx 2g.$$

Remark. If $d \equiv 3 \bmod 4$, we obtain Weil numbers $\pi = \zeta_s \pi_0$ such that π_0 is in the proper suborder $\mathbb{Z}[\sqrt{-d}]$. If we want to generate Weil numbers without this restriction, we can modify the above method using $\pi(x, y) = \zeta_s(x + y(1 + \sqrt{-d})/2)$.

4   Freeman-Satoh Curves

In this section we focus on constructing genus 2 curves, whose Jacobian corresponds to a given Weil number $\pi = \zeta_s \pi_0$ in a quartic CM field $K = \mathbb{Q}(\zeta_s, \sqrt{-d})$, where $\pi_0 \in \mathbb{Q}(\sqrt{-d})$. Since $\varphi(s) = 2$ or 4, we have $s = 3, 4, 6, 8, 12$ (the quartic CM field $\mathbb{Q}(\zeta_5)$ contains no imaginary quadratic subfield). Note that a simple abelian surface which is not absolutely simple, may be not isogenous to the Jacobian of any curve (see [30]). Since abelian surfaces corresponding to Weil numbers in question have automorphisms of order s, so of order 3 or 4, first it is natural to consider genus 2 curves which have automorphisms of order 3 or 4. We will use the following families of curves

$$y^2 = x^6 + ax^3 + b, \tag{6}$$

$$y^2 = x^5 + ax^3 + bx, \tag{7}$$

which have automorphisms of order 3 and 4 given by $(x, y) \mapsto (\zeta_3 x, y)$ and $(-x, iy)$, respectively (for more details on genus 2 curves with additional automorphisms see [6,18,21,35]). We will need the following result due to Freeman and Satoh [12].

Lemma 9. ([12, Propositions 4.1 and 4.2]) *A curve C given by (6) or (7) is isomorphic to the curve $y^2 = x^6 + cx^3 + 1$ or $y^2 = x^5 + cx^3 + x$, respectively, where $c = a/\sqrt{b}$. Furthermore, $\mathrm{Jac}(C)$ is isogenous over some extension to E^2, where E is an elliptic curve with the j-invariant*

$$j(E) = 2^8 3^3 \frac{(2c - 5)^3}{(c - 2)(c + 2)^3}, \tag{8}$$

$$j(E) = 2^6 \frac{(3c - 10)^3}{(c - 2)(c + 2)^2}, \tag{9}$$

respectively.

We now describe a method based on [12, Algorithm 5.11]. Suppose that an abelian surface $A/\mathbb{F}_q$ corresponding to a Weil q-number $\pi = \zeta_s \pi_0$ is isogenous to the Jacobian of a genus 2 curve C given by (6) or (7). Then A is isogenous over some extension to E^2, where E is an elliptic curve with the j-invariant given by (8) or (9), respectively. By Corollary 4, A is also isogenous to E_0^2 over $\mathbb{F}_{q^s}$, where E_0 is an elliptic curve with the Weil q-number π_0. Hence E and E_0 are isogenous, and so $\mathrm{End}(E)$ is an order in $K_0 = \mathbb{Q}(\sqrt{-d})$. In particular, if $\mathrm{End}(E) = \mathcal{O}_{K_0}$ is the maximal order, then $j(E)$ is a root of the Hilbert class polynomial $H_{K_0}(x)$. Conversely, if $j \in \mathbb{F}_q$ is a root of $H_{K_0}(x)$, and there exists $c \in \mathbb{F}_q$ satisfying equations (8) or (9) with $j(E) = j$, then we determine isomorphism classes over $\mathbb{F}_q$ of curves $y^2 = x^6 + ax^3 + b$ or $y^2 = x^5 + ax^3 + bx$ with $a, b \in \mathbb{F}_q$ satisfying $c = a/\sqrt{b}$, and verify if their Jacobians correspond to π. Recall that to check with high probability if the Jacobian of a curve C corresponds to a Weil number π we pick a random point $P \in \mathrm{Jac}(C)$ and check if $nP = 0$, where $n = \mathrm{N}(\pi - 1)$. The above procedure we give below as an algorithm. The only improvement is that we admit *all* twists of the above curves. The following examples show that this improvement is essential.

Example 10. Let $\pi = \zeta_3(3 + 2\sqrt{-5})$ be a Weil q-number with $q = \pi\bar{\pi} = 29$ and $n = \mathrm{N}_{K/\mathbb{Q}}(1 - \pi) = 1029$. Using Algorithm 11 below we find that π corresponds to the Jacobian of the curve

$$y^2 = 4x^6 + 26x^5 + 7x^4 + 11x^3 + 24x^2 + 27x + 4,$$

which is a twist of the curve $y^2 = x^6 + 5x^3 + 1$. However, checking all $a, b, c \in \mathbb{F}_{29}$, we find that there are no curves $y^2 = ax^6 + bx^3 + c$, whose Jacobian corresponds to π.

Algorithm 11. Input: A square-free positive integer d, $s = 3, 4$, and a Weil q-number $\pi = \zeta_s\pi_0$ with π_0 in $K_0 = \mathbb{Q}(\sqrt{-d})$.
Output: A genus 2 curve over $\mathbb{F}_q$, whose Jacobian corresponds to π, or $\emptyset$.

1. Compute the Hilbert class polynomial $H_{K_0}(x)$.
2. For each root $j \in \mathbb{F}_q$ of $H_{K_0}(x)$ let S_1 and S_2 be the sets of solutions $c \in \bar{\mathbb{F}}_q$ of equations (8) and (9), respectively.
3. For $i = 1, 2$ and for $c \in S_i$ if $i = 1$ let $C_i : y^2 = x^6 + cx^3 + 1$; else let $C_i : y^2 = x^5 + cx^3 + x$. Put $C := C_i$. Remove C if it is not hyperelliptic.
4. If $c \notin \mathbb{F}_q$ and all absolute invariants of C lie in $\mathbb{F}_q$, determine a model $C_0/\mathbb{F}_q$ of C and put $C := C_0$.
5. Determine all twists of C over $\mathbb{F}_q$.
6. For each twist C' choose a random point $P \in \mathrm{Jac}(C')(\mathbb{F}_q)$ and compute nP, where $n = \mathrm{N}_{K/\mathbb{Q}}(\pi - 1)$.
7. Return C' if $nP = 0$.

In this algorithm we need to compute the Hilbert class polynomial $H_{K_0}(x)$, which requires that the discriminant d is sufficiently small (see [38]). Note that if a genus 2 curve $C/\mathbb{F}_q$ has a model over $\mathbb{F}_q$, then all its absolute invariants lie in $\mathbb{F}_q$. The converse property is not always true, but it does hold if C has automorphisms other than the identity and the hyperelliptic involution. Then a model of C over $\mathbb{F}_q$ can be computed using the generalization of the Mestre algorithm [32] due to Cardona and Quer [5].

Remark 12. (i) In the above algorithm it usually suffices to use curves (6) or (7) if $s = 3$ or 4, respectively. However, it may happen for the CM field $K = \mathbb{Q}(\zeta_{12})$ that we need to use curves (6) to realize Weil numbers of the form $i\pi_0$ with $\pi_0 \in \mathbb{Q}(\sqrt{-3})$ (see Example 19).

(ii) For Weil numbers in the CM field $\mathbb{Q}(\zeta_8)$ we can usually use curves $y^2 = x^5 + ax$, which have automorphisms of order 8, $(x, y) \mapsto (\zeta_8^2 x, \zeta_8 y)$. The first construction of pairing-friendly curves of this form due to Kawazoe and Takahashi [25] was based on the closed formula on the order of their Jacobian (see [15]).

Example 13. Let $K = \mathbb{Q}(\zeta_3, \sqrt{-5})$ and $k = 16$. Using Algorithms 8 and 11, we find the following parameters of an abelian surface with $\rho = 4.011$, and the corresponding genus 2 curve:

$r = 48(10^{53} + 2085) + 1$ (181-bits prime),

$\pi = \zeta_3(4305259600539301889028270527319533759867814882609214984 + 5715080678959385503541554725176417909523782410181520093\sqrt{-5})$,

$q = 20168367586386572810015424271002249732267166683454467732594522539415397151727\,6154391831469\,84296058295131523501$,

$y^2 = x^6 + x^3 + 9815329172717304742646668250744383765757406174971515824402826019633848306457589362\,3291386054363203804560511872.$

Example 14. For $K = \mathbb{Q}(i, \sqrt{-7})$ the following abelian surface has embedding degree $k = 31$ and $\rho = 4.016$:

$r = 124(10^{75} + 3) + 1$ (256-bits prime),

$\pi = i(96180181687130548086884708381078859138617038963689573425053970665226825986272 + 9155802799235777905099734889345646175817336249315015553451165600417834527285 3\sqrt{-7})$,

$q = 6793073477831503692897516662864494713051738716943224205560703028558318233544942073946212754109292645543302062880660629280273105153102439936019342663775247$,

$y^2 = 3x^6 + a_4 x^4 + a_3 x^3 + a_2 x^2 + a_1 x + a_0,$

$a_4 = 335988342649190323926068735120533358427441528266958420096170525566380715018406179883690940929053258439399481704008550987597610050426614764135822532343\,4953,$

$a_3 = 535683751760447447062675707174834384291245147597573398640679000616111934596851330165088783087061366317277372428707734988655246788213498779014462967185974 38,$

$a_2 = 208838240340684568805803626096119587252931641073084827363322442046504372669849616916356457171176635445125147016524676328092739235264591714565408500671781 8,$

$a_1 = 577207780651835015003941601259696381939122233621579280719636466755389607774280290500632745899228933245642415050622885662424087983658862392974998163041550 7,$

$a_0 = 406087452863714226140869428855414790591510025332839103372621451075134801243196830732474038065130317947510090708332558470288074276452441302830946228200399 8.$

5   Parametric Families

Here we generalize Definition 1 and the classification of families of elliptic curves introduced by Freeman, Scott, and Teske [13] to simple abelian varieties over $\mathbb{F}_q$, which are isogenous over some extension to a power of an elliptic curve defined over $\mathbb{F}_q$. Recall that by Corollary 4 Weil q-numbers of such abelian varieties are of the form $\pi = \zeta_s \pi_0$, where π_0 is a Weil q-number of an elliptic curve.

Definition 15. Let $K = \mathbb{Q}(\zeta_s, \sqrt{-d})$ be a CM field of degree $2g$, where ζ_s is an sth primitive root of unity and $d > 0$ is a square-free integer. Let $r(x) \in \mathbb{Q}[x]$ and $\pi(x) = \zeta_s(f_1(x) + f_2(x)\sqrt{-f(x)})$, where $f_1(x), f_2(x), f(x) \in \mathbb{Q}[x]$. We say that the pair $(r(x), \pi(x))$ *parametrizes a family of g-dimensional ordinary abelian varieties with embedding degree k and discriminant d* if the following conditions are satisfied:

1. $q(x) = f_1^2(x) + f_2^2(x)f(x)$ is a power of a polynomial in $\mathbb{Q}[x]$ that represents primes, and $\gcd(f_1(x), q(x)) = 1$.
2. $r(x)$ is irreducible, non-constant, integer valued, and has positive leading coefficient.
3. $r(x)$ divides $N_{K_1/\mathbb{Q}(x)}(\pi(x) - 1)$, where $K_1 = \mathbb{Q}(x, \zeta_s, \sqrt{-f})$.
4. $r(x)$ divides $\Phi_k(q(x))$.
5. The CM equation $f(x) = dy^2$ has infinitely many integer solutions (x, y).

We note that the ρ-values $g \log q(x)/\log r(x)$ of parametrized abelian varieties tend to the ρ-value of the family

$$\rho = \frac{g \deg q(x)}{\deg r(x)}.$$

The assumption $\gcd(f_1(x), q(x)) = 1$ is necessary to obtain ordinary varieties. It follows from the fact that an abelian variety with a Weil q-number $\pi = \zeta_s \pi_0$ is ordinary if and only if so is the corresponding elliptic curve with the Weil q-number π_0, which means that its trace $\pi_0 + \bar{\pi}_0$ is relatively prime to q. In the examples below $q(x)$ will always represent primes, then we can assume that $f_1 \neq 0$. As for elliptic curves to obtain parameters of an abelian variety with the endomorphism algebra $K = \mathbb{Q}(\zeta_s, \sqrt{-d})$ we find integer solutions (x_0, y_0) to the CM equation $f(x) = dy^2$, and check whether $\pi(x_0)$ is a Weil number, and $r(x_0)$ is prime, or almost prime. If this is the case, then $N_{K_1/\mathbb{Q}(x)}(\pi(x) - 1)(x_0)$ is the order of an abelian variety corresponding to $\pi(x_0)$, and it is divisible by large prime factors of $r(x_0)$. To generalize the classification of families we will need the following fact (see also [9, Proposition 2.10]).

Lemma 16. *In Definition 15 we can assume that $f \in \mathbb{Z}[x]$ is square-free, $\deg f \leq 2$, and the leading coefficient of f is positive.*

Proof. Obviously, condition (5) in Definition 15 implies that the leading coefficient of f is positive. We can write $f = g_1 g_2^2$, where $g_1 \in \mathbb{Z}[x]$ is square-free and $g_2 \in \mathbb{Q}[x]$. By Siegel's theorem (see [36, Theorem IX.4.3]) the curve $dy^2 = f(x)$ contains finitely many integer points if $f \in \mathbb{Q}[x]$ is square-free of degree $\deg f \geq 3$. Thus replacing f by g_1 and f_2 by $f_2 g_2$ we have $\deg f \leq 2$.

Definition 17. Let $(r(x), \pi(x))$ be a family satisfying Definition 15 with $f(x)$ as in Lemma 16. We say that the family is

1. *complete with discriminant d* if $f = d$,
2. *complete with variable discriminant* if $\deg f = 1$,
3. *sparse* if $\deg f = 2$.

The above conditions have the same interpretation as for elliptic curves, and are useful to obtain algorithms to generate families of each type, which generalize the Brezing-Weng method [4].

6   Complete Families

First we generalize the Brezing-Weng method [4] to construct complete families of abelian varieties. Let $K = \mathbb{Q}(\zeta_s, \sqrt{-d})$ be a CM field of degree $2g$. To construct a complete family $(r(x), \pi(x))$ with $\pi(x) = \zeta_s(f_1(x) + f_2(x)\sqrt{-d})$, we need to find a number field $L = \mathbb{Q}[x]/(r(x))$ where the system

$$N_{K(x,y)/\mathbb{Q}(x,y)}\left(\zeta_s(x + y\sqrt{-d}) - 1\right) = \Phi_k(x^2 + dy^2) = 0 \qquad (10)$$

has solutions, and take $f_1, f_2 \in \mathbb{Q}[x]$ to be lifts of these solutions. Such number fields and formulas on solutions have been described in Lemma 6. Hence we have the following algorithm.

Algorithm 18. Input: A CM field $K = \mathbb{Q}(\zeta_s, \sqrt{-d})$ of degree $2g$, a positive integer k, and a number field L containing $\zeta_s, \zeta_k, \sqrt{-d}$.
Output: A complete family $(r(x), \pi(x))$ of g-dimensional ordinary abelian varieties with embedding degree k, or $\emptyset$.

1. Find a polynomial $r(x) \in \mathbb{Q}[x]$ such that $L = \mathbb{Q}[x]/(r(x))$.
2. Let $x_1 = \frac{\zeta_s^{-1} + \zeta_k \zeta_s}{2}$ and $y_1 = \frac{\zeta_s^{-1} - \zeta_k \zeta_s}{2\sqrt{-d}}$ for all primitive roots of unity $\zeta_s, \zeta_k \in L$.
3. If $\sqrt{-d} \in \mathbb{Q}(\zeta_s)$ and x_1, y_1 do not satisfy system (10), put $y_1 = -y_1$.
4. Let $f_1, f_2 \in \mathbb{Q}[x]$ be lifts of x_1, y_1 with $\deg f_i < \deg r$, $i = 1, 2$.
5. Let $\pi(x) = \zeta_s(f_1(x) + f_2(x)\sqrt{-d})$.
6. Return $(r(x), \pi(x))$ if $f_1 \neq 0$, $2f_1(x) \in \mathbb{Z}$ for some $x \in \mathbb{Z}$, and $q(x) = f_1(x)^2 + df_2^2(x)$ represents primes.

We note that resulting families have ρ-value

$$\rho = \frac{2g\max\{\deg f_1, \deg f_2\}}{\deg r} \leq \frac{2g(\deg r - 1)}{\deg r} < 2g.$$

In the above algorithm we can take as L the cyclotomic field $L = \mathbb{Q}(\zeta_s, \zeta_m, \zeta_k) = \mathbb{Q}(\zeta_l)$, where m is the smallest integer such that $\sqrt{-d} \in \mathbb{Q}(\zeta_m)$ and $l = \mathrm{lcm}(s, m, k)$. We note that such m exists, because $\sqrt{(-1)^{\frac{p-1}{2}}p} \in \mathbb{Q}(\zeta_p)$ for each prime $p > 2$ and $\sqrt{-2} \in \mathbb{Q}(\zeta_8)$ (see [29, Lemma 2.2]). Now we give a few examples; more complete families with variable discriminant we give in Section 8.

Example 19. Let $s = 4$, $d = 3$, and $K = \mathbb{Q}(\zeta_{12}) = \mathbb{Q}(i, \sqrt{-3})$. Let $k = 12$ and $L = K = \mathbb{Q}[x]/(r_0(x))$, where $r_0(x) = x^4 + 2x^3 + 6x^2 - 4x + 4$ is the minimal polynomial of $\zeta_{12} - \zeta_{12}^2 + \zeta_{12}^3$. Using $\pi(x, y) = i(x + y\sqrt{-3})$ we find the following family of simple ordinary abelian surfaces with embedding degree $k = 12$ and $\rho = 2$:

$$r(x) = \tfrac{1}{36}(x^4 + 2x^3 + 6x^2 - 4x + 4),$$
$$\pi(x) = \tfrac{i}{12}\big(x^2(-\sqrt{-3} + 1) - 2x(\sqrt{-3} + 1) - 6\sqrt{-3} - 2\big).$$

We note that this construction is analogous to the Barreto-Naehrig family of elliptic curves with $k = 12$ and $\rho = 1$ (see [1]). For example, we generate the following parameters of abelian surfaces and the corresponding genus 2 curves using Algorithm 11.

$x = 87960930234340,$

$r = 1662864086068056644824292237437174114512687909008301229$ (180-bits prime),

$\pi = \tfrac{i}{2}(1289520874615042134242461153 - 1289520874615100774862617381\sqrt{-3}),$

$q = 1662864086068056644824292237266949891278180041004180996723,$

$y^2 = 3x^6 + 390908738065633359475783013729440679739067632100343921 4x^3$

$\qquad + 840318388709976017122087137087102952585808061504841608$

$x = 4611686018427434730,$

$r = 12564245793980132208559035774981645041883741038087452602908341544711727 0861649$ (256-bits prime),

$\pi = \tfrac{i}{2}(354460798875984764473015759359659256913 - 354460798875984764503760332815842155121\sqrt{-3}),$

$q = 12564245793980132208559035774981645041919187117975051079360254806666120 4465873,$

$y^2 = 10x^6 + 10051396635184105766847228619985316033535349694380040863488203845332 89635572700x^3$

$\qquad + 729329339848958714442434908666134531393324973824215765054704071017354 95968350$

Example 20. Let $s = 8$, $d = 2$, and $K = \mathbb{Q}(\zeta_8)$; we have $\sqrt{-2} = \zeta_8^3 + \zeta_8$. Using $\pi(x, y) = \zeta_8(x + y\sqrt{-2})$ we obtain Kawazoe-Takahashi families [25]. We have the following family with $k = 32$ and $\rho = 3.25$:

$$r(x) = \Phi_{32}(x),$$

$$\pi(x) = \tfrac{\zeta_8}{4}\left(-2x^{13} + 2x^{12} - \sqrt{-2}(x^9 + x^8 + x + 1)\right).$$

For example, we generate the following pairing-friendly curve:

$x = 1011203,$

$r = r(x)/2 = 5975628564030163993716466034887402480498700578175608698339694936788456317153$
10283215375141190561

(318-bits prime),

$\pi = -27636661717843096901242245558493120316710924191467536 2\, \zeta_8^2$

$-5779205224565086112079790018495549298014230975$

$899478553489296182963594766978 41\, \zeta_8 +27636661717843096901242245558493120316710924191467536 2,$

$q = 3339921302764038739820206627349052325432923549582694711656519663209495074190$
$7470195554624141450877072423263\, 78287845324089990264085171394677883056733137233 69,$

$y^2 = x^5 + 21x.$

Example 21. We can also give some families of 3-dimensional varieties with $\rho < 6$. We leave as an open problem constructing the corresponding genus 3 curves. The only sextic CM fields of the form $K = \mathbb{Q}(\zeta_s, \sqrt{-d})$ are the cyclotomic fields $\mathbb{Q}(\zeta_7)$ and $\mathbb{Q}(\zeta_9)$, which contain $\sqrt{-7}$ and $\sqrt{-3}$, respectively.

(i) Let $K = \mathbb{Q}(\zeta_7)$ and $\alpha = \sqrt{-7} = 2\zeta_7^4 + 2\zeta_7^2 + 2\zeta_7 + 1$.

$k = 7, \quad \rho = 4,$

$r(x) = \Phi_7(x),$

$\pi(x) = \tfrac{\zeta_7}{14}(-2\,\alpha\,x^4 + (\alpha+7)x^3 + 2\,\alpha\,x^2 + (\alpha+7)x - 2\,\alpha),$

$k = 21, \quad \rho = 4,$

$r(x) = \Phi_{21}(x),$

$\pi(x) = \tfrac{\zeta_{21}}{14}((-\alpha-7)x^8 + (\alpha-7)x^7 - 2\,\alpha\,x^6 + 2\,\alpha\,x^4 - 2\,\alpha\,x^2 + (\alpha-7)x - \alpha-7).$

(ii) Let $K = \mathbb{Q}(\zeta_9)$ and $\alpha = \sqrt{-3} = 2\zeta_9^3 + 1$.

$k = 9, \quad \rho = 4,$

$r(x) = \Phi_9(x),$

$\pi(x) = \tfrac{\zeta_9}{6}((-\alpha-3)x^4 + (\alpha+3)x^3 + (\alpha-3)x + 2\,\alpha).$

7   Sparse Families

In this section we generalize Algorithm 18 to construct sparse families in an analogous way as the Brezing-Weng method was generalized in [8] to construct such families of elliptic curves. If $(r(x), \pi(x))$ is a family of abelian varieties with $\pi(x) = \zeta_s(f_1(x) + f_2(x)\sqrt{f(x)})$, then $(f_1(x), f_2(x)) \bmod r(x)$ is a solution of the system

$$N_{K_1/\mathbb{Q}(x)}(\zeta_s(X + Y\sqrt{-f}) - 1) = \Phi_k(X^2 + fY^2) = 0, \tag{11}$$

where $K_1 = \mathbb{Q}(x, \zeta_s, \sqrt{-f})$. Hence to construct sparse families we should find polynomials $r(x) \in \mathbb{Q}[x]$ and $f(x) \in \mathbb{Z}[x]$, where $r(x)$ is irreducible and $f(x)$ satisfies Lemma 16, such that system (11) has solutions in the number field $L = \mathbb{Q}[x]/(r(x))$, and take f_1, f_2 to be lifts of these solutions. Such number fields are described in the following lemma, which generalizes Lemma 3.

Lemma 22. *Let $f \in \mathbb{Z}[x]$ satisfy Lemma 16 and $\deg f = 1, 2$. Let $r(x) \in \mathbb{Q}[x]$ be irreducible such that $\zeta_s, \zeta_k, \sqrt{-\bar{f}} \in L = \mathbb{Q}[x]/(r(x))$, where a bar denotes reduction mod $r(x)$. Then system (11) has solutions in L of the form*

$$X = \frac{\zeta_s^{-1} + \zeta_k\zeta_s}{2}, \quad Y = \pm\frac{\zeta_s^{-1} - \zeta_k\zeta_s}{2\sqrt{-\bar{f}}}. \tag{12}$$

Proof. As in the proof of Lemma 6 we first show that solutions in the field of fractions of $S = \mathbb{Q}[x, \zeta_s, \zeta_k, \sqrt{-\bar{f}}]$ are of the above form. Then for a prime ideal P in S over r, the reduction S_P mod PS_P yields the desired result by Lemma 7.

Hence we have the following algorithm; in the next section we give a simplified version to construct complete families with variable discriminant.

Algorithm 23. Input: A number field L containing primitive roots of unity ζ_s, ζ_k.
Output: A sparse family $(r(x), \pi(x))$ of $\varphi(s)$-dimensional ordinary abelian varieties with embedding degree k, or $\emptyset$.

1. Find $r(x) \in \mathbb{Q}[x]$ such that $L = \mathbb{Q}[x]/(r(x))$.
2. Let $f_1 \in \mathbb{Q}[x]$ be the lift of $X = \frac{\zeta_s^{-1} + \zeta_s\zeta_k}{2}$ with $\deg f_1 < \deg r$ for all primitive roots of unity $\zeta_k, \zeta_s \in L$.
3. If $f_1 \neq 0$ and $2f_1(x) \in \mathbb{Z}$ for some $x \in \mathbb{Z}$, let $f(x) = a_2x^2 + a_1x + a_0$ for integers $a_0, a_1, a_2 \in [-m, m]$, where $a_2 > 0$ and $m \in \mathbb{Z}$.
4. If f is square-free and $\sqrt{-\bar{f}} \in L$, let $f_2 \in \mathbb{Q}[x]$ be the lift of $Y = \frac{\zeta_s^{-1} - \zeta_s\zeta_k}{2\sqrt{-\bar{f}}}$
   with $\deg f_2 < \deg r$.
5. Let $\pi(x) = \zeta_s(f_1(x) + f_2(x)\sqrt{-f(x)})$.
6. Return $(r(x), \pi(x))$ if $q(x) = f_1^2(x) + f_2^2(x)f(x)$ represents primes.

Note that the resulting families have ρ-value

$$\rho = \frac{2g \max\{\deg f_1, \deg f_2 + 1\}}{\deg r} \le 2g.$$

We now show how to construct sparse families of ordinary abelian surfaces with $k = 3, 4, 6$ and $\rho = 2$. These families are analogous to constructions for elliptic curves with $k = 3, 4, 6$ and $\rho = 1$ due to Miyaji et al. [27], Scott and Barreto [34], and Galbraith et al. [17].

Example 24. Let $s = 3, 4$, and $K = \mathbb{Q}(\zeta_s)$. Let $k = 3, 4, 6$, and $\zeta_k \in L = K = \mathbb{Q}[x]/(r(x))$ for $r(x) \in \mathbb{Q}[x]$. In order to construct a family $(r(x), \pi(x))$ with $\pi(x) = \zeta_s(f_1(x) + f_2(x)\sqrt{-f(x)})$ and $\rho = 2$, we have to find a polynomial $f(x) \in \mathbb{Z}[x]$ as in step 4 of Algorithm 23 such that f_2 is constant. Since f_2 is the lift of $Y = (\zeta_s^{-1} - \zeta_s\zeta_k)/2\sqrt{-\overline{f}}$, we have $Y \in \mathbb{Q}$. We can assume $Y = 1$, since $c^2 f$ and Y/c yield the same family for each $c \in \mathbb{Q}^\times$. Then for fixed $\zeta_s, \zeta_k \in L$, $\overline{f}$ is uniquely determined by $\overline{f} = -(\zeta_s^{-1} - \zeta_s\zeta_k)^2/4 = a\bar{x} + b$ for some $a, b \in \mathbb{Q}$. So we can take $f = ax + b + cr(x)$ for $c \in \mathbb{Q}$, $c > 0$. As f_1 we take the lift of $X = (\zeta_s^{-1} - \zeta_s\zeta_k)/2$. If $f_1 \ne 0$, $2f_1(x) \in \mathbb{Z}$ for some $x \in \mathbb{Z}$, and $q(x)$ represents primes, we obtain the desired family. For example, we have the following families with $\rho = 2$:

$$k = 3,$$
$$r(x) = 4x^2 + 2x + 1,$$
$$\pi(x) = \tfrac{\zeta_3}{6}\left(6x + 3 + \sqrt{-(12x^2 + 60x + 3)}\right)),$$

$$k = 4,$$
$$r(x) = 4x^2 + 1,$$
$$\pi(x) = \tfrac{i}{2}\left(-2x - 1 + \sqrt{-(12x^2 + 4x + 3)}\right),$$

$$k = 6,$$
$$r(x) = 4x^2 - 2x + 1,$$
$$\pi(x) = \tfrac{\zeta_3}{2}\left(-2x - 1 + \sqrt{-(12x^2 - 4x + 3)}\right).$$

Example 25. Let $k = 8$, $s = 4$, and $L = \mathbb{Q}(\zeta_8)$. For $f(x) = 7x^2 - 10x + 7$ we have $f(x) \bmod \Phi_8(x) = -(-2\zeta_8^3 + 2\zeta_8^2 - \zeta_8 - 1)^2$. We obtain the following family with $\rho = 3$:

$$r(x) = \Phi_8(x),$$
$$\pi(x) = \tfrac{i}{2}\left(-x^2 + x + (2x^2 + 3x + 2)\sqrt{-(7x^2 - 10x + 7)}\right).$$

8   Complete Families with Variable Discriminant

In this section we modify Algorithm 23 to construct complete families with variable discriminant $(r(x), \pi(x))$, where $\pi(x) = \zeta_s(f_1(x) + f_2(x)\sqrt{-f(x)})$ and $f(x) = ax + b$. Substituting $x \leftarrow (x - b)/a$, we can assume that $f = x$. Then by Lemma 22, $L = \mathbb{Q}[x]/(r(x))$ is a number field containing ζ_s, ζ_k, and $\sqrt{-\bar{x}}$. Let us note that a polynomial $r(x) \in \mathbb{Q}[x]$ such that $L = \mathbb{Q}[x]/(r(x))$ and $\sqrt{-\bar{x}} \in L$ can be obtained as the minimal polynomial of a primitive element $z \in L$ such that $\sqrt{-z} \in L$. Hence we have the following variant of Algorithm 23.

Algorithm 26. Input: A number field L such that $\zeta_s, \zeta_k \in L$.
Output: A complete family with variable discriminant $(r(x), \pi(x))$ of $\varphi(s)$-dimensional ordinary abelian varieties with embedding degree k, or $\emptyset$.

1. Find a primitive element $z \in L$ such that $\sqrt{-z} \in L$.
2. Let $r(x)$ be the minimal polynomial of z and $L = \mathbb{Q}[x]/(r(x))$.
3. Let $X = \frac{\zeta_s^{-1} + \zeta_s \zeta_k}{2}$ and $Y = \frac{\zeta_s^{-1} - \zeta_s \zeta_k}{2\sqrt{-\bar{x}}}$ for all primitive roots of unity $\zeta_s, \zeta_k \in L$.
4. Let $f_1(x), f_2(x) \in \mathbb{Q}[x]$ be lifts of X, Y with $\deg f_i < \deg r$, $i = 1, 2$.
5. Let $\pi(x) = \zeta_s(f_1(x) + f_2(x)\sqrt{-x})$.
6. Return $(r(x), \pi(x))$ if $f_1 \neq 0$, $2f_1(x) \in \mathbb{Z}$ for some $x \in \mathbb{Z}$, and $q(x) = f_1^2(x) + xf_2^2(x)$ represents primes.

The resulting families have ρ-value

$$\rho = \frac{g \max\{2 \deg f_1, 1 + 2 \deg f_2\}}{\deg r} \leq \frac{g(2 \deg r - 1)}{\deg r} < 2g.$$

In the examples below we take as L the cyclotomic field $L = \mathbb{Q}(\zeta_s, \zeta_k) = \mathbb{Q}(\zeta_l)$, where $l = \mathrm{lcm}(s, k)$. A crucial step in the above algorithm is to find a primitive element $z \in L$ such that $\sqrt{-z} \in L$, which can be chosen in the following ways.

- If l is odd, then $\sqrt{\zeta_l} = \pm\zeta_l^{(l+1)/2}$, so we can take $z = \zeta_{2l} = -\zeta_l$ and $r(x) = \Phi_{2l}(x)$. Similarly, if $l/2$ is odd, we can take $r(x) = \Phi_l(x)$.
- If $4|l$, then $\sqrt{\pm\zeta_l} \notin \mathbb{Q}(\zeta_l)$, but there may exist $a \in \mathbb{Z}$ such that $\sqrt{-\zeta_l/a} \in \mathbb{Q}(\zeta_l)$. Then we can take $z = \zeta_l/a$ and $r(x) = \Phi_l(ax)$.
- As in the method of Kachisa, Schaefer, Scott [24] we can vary elements $z_0 = a_0 + a_1\zeta_l + \cdots + a_{\varphi(l)-1}\zeta_l^{\varphi(l)-1}$, which have small integer coefficients a_i in the cyclotomic basis, and check if we get a suitable family for $z = -z_0^2$.

In the examples below we will also give necessary conditions on discriminant d so that $q(dx^2)$ could represent primes.

Example 27. (i) Let $k = 27, s = 3$, and $L = \mathbb{Q}(\zeta_{27})$. We obtain a complete family with variable discriminant $d \equiv 3 \pmod 8$ and $\rho = 2.11$

$$r(x) = \Phi_{54}(x),$$
$$\pi(x) = \tfrac{\zeta_3}{2}\left(x^9 - x^5 - 1 - (x^9 - x^4 - 1)\sqrt{-x}\right).$$

For example, we can generate the following parameters:

$d = 987$

$x = 1$

$r = 7901485510647346009300993128257685424898845511187609503$ (179-bits prime)

$\pi = \tfrac{\zeta_3}{2}(8889030043053456721875555919 - 8889030043062813913547490065\sqrt{-987})$

$q = 1951666921129886138225820158709016804569015692496468823659$

$y^2 = x^6 + x^3 + 15110590774962264611862121651343216710922763477745485452 0$

$\rho = 2.078$

$d = 2091$

$x = 3$

$r = 8764714229254862286681699927556088961544289415331105128820662737010542521 5$ 463 (255-bits prime)

$\pi = \tfrac{\zeta_3}{2}(296052600550220841104719607209577744879 - 8881578016506625303949353470223830835 71\sqrt{-2091})$

$q = 41237980448644127058767518369085457198019262788918408381604555266465615677 8750593$

$y^2 = x^6 + x^3 + 56578159329796760688848304124543683168097550241972892000909998577765239565 174952$

$\rho = 2.094$

(ii) Similarly, for $k = 54$, $s = 3$, and $L = \mathbb{Q}(\zeta_{54})$, we obtain a complete family with variable discriminant $d \equiv 3 \pmod 8$ and $\rho = 2.11$

$$r(x) = \Phi_{54}(x),$$
$$\pi(x) = \tfrac{\zeta_3}{2}\left(x^9 + x^5 - 1 + (x^9 + x^4 - 1)\sqrt{-x}\right).$$

Example 28. (i) Let $s = 3$, $k = 12$, and $L = \mathbb{Q}(\zeta_{12})$; then $\sqrt{-\zeta_{12}/2} \in L$. We have the following family with discriminant $d \equiv 3 \pmod 8$ and $\rho = 3.5$:

$$r(x) = \Phi_{12}(2x),$$
$$\pi(x) = \tfrac{\zeta_3}{2}\left(-8x^3 + 4x^2 - 1 + (8x^3 - 4x - 1)\sqrt{-x}\right)$$

Example 29. Let $k = 8$, $s = 4$, and $L = \mathbb{Q}(\zeta_8)$. Let $r(x)$ be the minimal polynomial of $z = -(\zeta_8 - 1)^2$. We have the following family with discriminant $d = 1, 7 \pmod 8$ and $\rho = 3.5$:

$$r(x) = x^4 + 4x^3 + 8x^2 - 8x + 4,$$
$$\pi(x) = \tfrac{i}{24}\left(-3x^3 - 15x^2 - 36x + 6 + (x^3 + 5x^2 + 16x + 2)\sqrt{-x}\right).$$

Table 1. Best ρ-values of complete families with variable discriminant $((r(x), \pi(x))$ given below such that $\deg r(x) < 25$

k	ρ	d	$\deg r(x)$	k	ρ	d	$\deg r(x)$
2	3.00	3 mod 8	2	22	2.70	3 mod 8	10
3	3.00	$1, 3, 7, 9$ mod 10	2	24	3.75	$2, 10, 11, 19$ mod 24	8
4	3.00	3 mod 4	2	26	2.25	3 mod 8	24
5	3.00	1 mod 4	8	27	2.11	3 mod 8	18
6	3.00	any	2	28	3.08	3 mod 8	24
7	2.50	3 mod 8	12	30	2.75	3 mod 8	8
8	3.50	$1, 7$ mod 8	4	33	2.30	3 mod 8	20
9	2.33	3 mod 8	6	36	3.50	3 mod 8	12
10	3.50	any	8	39	2.33	1 mod 4	24
11	2.40	1 mod 4	20	42	2.83	3 mod 8	12
12	3.50	3 mod 8	4	45	2.58	3 mod 8	24
13	2.25	3 mod 8	24	54	2.11	3 mod 8	18
14	2.50	3 mod 8	12	60	3.75	3 mod 8	14
15	2.75	3 mod 8	8	66	2.30	3 mod 8	20
16	3.75	some	8	78	2.42	3 mod 4	24
18	2.33	3 mod 8	6	84	3.75	3 mod 8	24
20	3.75	3 mod 8	8	90	2.58	3 mod 8	24
21	2.66	1 mod 4	12				

$k = 2, \rho = 3, r(x) = \Phi_6(x), \pi(x) = \frac{\zeta_3}{2}\left(2x - 1 + x\sqrt{-x}\right)$

$k = 3, \rho = 3, r(x) = x^2 + 11x + 49, \pi(x) = \frac{\zeta_3}{70}\left(7x + 56 + (x - 17)\sqrt{-x}\right)$

$k = 4, \rho = 3, r(x) = x^2 - 6x + 25, \pi(x) = \frac{i}{40}\left(5x + 5 + (x + 9)\sqrt{-x}\right)$

$k = 5, \rho = 3, r(x) = \Phi_{30}(x), \pi(x) = \frac{\zeta_3}{2}\left(-x^6 + x^5 + x - 1 - (x^3 + x^2)\sqrt{-x}\right)$

$k = 6, \rho = 3, r(x) = \Phi_6(x), \pi(x) = \frac{\zeta_3}{2}\left(x - 2 + (x - 1)\sqrt{-x}\right)$

$k = 7, \rho = 2.5, r(x) = \Phi_{42}(x), \pi(x) = \frac{\zeta_3}{2}\left(x^7 + x^4 - 1 + (x^7 + x^3 - 1)\sqrt{-x}\right)$

$k = 8, \rho = 3.5, r(x) = x^4 + 4x^3 + 8x^2 - 8x + 4, \pi(x) = \frac{i}{24}\left(-3x^3 - 15x^2 - 36x + 6 + (x^3 + 5x^2 + 16x + 2)\sqrt{-x}\right)$

$k = 9, \rho = 2.33, r(x) = \Phi_{18}(x), \pi(x) = \frac{\zeta_3}{2}\left(x^6 - x^3 + 1 + (x^3 + x^2 - 1)\sqrt{-x}\right)$

$k = 10, \rho = 3.5, r(x) = \Phi_{30}(5x), \pi(x) = \frac{\zeta_3}{2}\left(-78125x^7 - 15625x^6 + 3125x^5 + 625x^4 + 125x^3 + 25x^2 - 2 + (15625x^6 - 6250x^5 - 1250x^4 - 250x^3 + 25x^2)\sqrt{-x}\right)$

$k = 11, \rho = 2.4, r(x) = \Phi_{66}(x), \pi(x) = \frac{\zeta_3}{2}\left(-x^{12} + x^{11} + x - 1 + (x^6 + x^5)\sqrt{-x}\right)$

$k = 12, \rho = 3.5, r(x) = \Phi_{12}(2x), \pi(x) = \frac{\zeta_3}{2}\left(-8x^3 + 4x^2 - 1 + (8x^3 - 4x - 1)\sqrt{-x}\right)$

$k = 13, \rho = 2.25, r(x) = \Phi_{78}(x), \pi(x) = \frac{\zeta_3}{2}\left(x^{13} - x^7 - 1 + (x^{13} - x^6 - 1)\sqrt{-x}\right)$

$k = 14, \rho = 2.5, r(x) = \Phi_{42}(x), \pi(x) = \frac{\zeta_3}{2}\left(x^7 - x^4 - 1 + (-x^7 + x^3 + 1)\sqrt{-x}\right)$

$k = 15, \rho = 2.75, r(x) = \Phi_{30}(x), \pi(x) = \frac{\zeta_3}{2}\left(x^5 - x^3 - 1 + (x^5 - x^2 - 1)\sqrt{-x}\right)$

$k = 16, \rho = 3.75, r(x) = x^8 + 76x^6 + 678x^4 + 332x^2 + 1, \pi(x) = \frac{i}{30464}\left(29x^7 - 29x^6 + 2173x^5 - 2173x^4 + 17175x^3 - 17175x^2 - 21009x + 5777 + (5777x^7 - 229x^6 + 439081x^5 - 17389x^4 + 3918979x^3 - 154335x^2 + 1935139x - 71215)\sqrt{-x}\right)$

$k = 18, \rho = 2.33, r(x) = \Phi_{18}(x), \pi(x) = \frac{\zeta_3}{2}\left(x^3 - x^2 - 1 + (-x^3 + x + 1)\sqrt{-x}\right)$

$k = 20, \rho = 3.75, r(x) = \Phi_{20}(2x), \pi(x) = \frac{i}{2}\left(-64x^6 + 32x^5 + 16x^4 - 4x^2 + 1 + (128x^7 - 32x^5 - 4x^2 - 1)\sqrt{-x}\right)$

$k = 21, \rho = 2.66, r(x) = \Phi_{42}(x), \pi(x) = \frac{\zeta_3}{2}\left(-x^8 + x^7 + x - 1 + (x^4 + x^3)\sqrt{-x}\right)$

$k = 22, \rho = 2.7, r(x) = \Phi_{22}(x), \pi(x) = \frac{\zeta_3}{2}\left(x^{11} - x^8 - 1 + (-x^{13} + x^5 + x^2)\sqrt{-x}\right)$

$k = 24, \rho = 3.75, r(x) = x^8 + 80x^6 + 456x^4 + 320x^2 + 16, \pi(x) = \frac{\zeta_3}{10752}\left(-176x^7 + 28x^6 - 14040x^5 + 2240x^4 - 77088x^3 + 12656x^2 - 40480x + 1792 + (177x^7 + 34x^6 + 14150x^5 + 2704x^4 + 79892x^3 + 14248x^2 + 50552x + 9024)\sqrt{-x}\right)$

$k = 26, \rho = 2.25, r(x) = \Phi_{78}(x), \pi(x) = \frac{\zeta_3}{2}\left(x^{13} + x^7 - 1 - (x^{13} + x^6 - 1)\sqrt{-x}\right)$

$k = 27, \rho = 2.11, r(x) = \Phi_{54}(x), \pi(x) = \frac{\zeta_3}{2}\left(x^9 - x^5 - 1 + (-x^9 + x^4 + 1)\sqrt{-x}\right)$

$k = 28, \rho = 3.08, r(x) = \Phi_{84}(2x), \pi(x) = \frac{\zeta_3}{2}\left(16384x^{14} - 32x^5 - 1 + (262144x^{18} - 131072x^{17} + 65536x^{16} + 32768x^{15} - 4096x^{12} - 1024x^{10} + 64x^6 - 4x^2 - 1)\sqrt{-x}\right)$

$k = 30, \rho = 2.75, r(x) = \Phi_{30}(x), \pi(x) = \frac{\zeta_3}{2}\left(x^5 + x^3 - 1 + (-x^5 - x^2 + 1)\sqrt{-x}\right)$

$k = 33, \rho = 2.3, r(x) = \Phi_{33}(-x), \pi(x) = \frac{\zeta_3}{2}\left(x^{11} + x^6 - 1 + (x^{11} + x^5 - 1)\sqrt{-x}\right)$

$k = 36, \rho = 3.5, r(x) = \Phi_{36}(2x), \pi(x) = \frac{\zeta_3}{2}\left(64x^6 - 32x^5 - 1 + (1024x^{10} + 512x^9 - 128x^7 - 16x^4 - 1)\sqrt{-x}\right)$

$k = 39, \rho = 2.33, r(x) = \Phi_{78}(x), \pi(x) = \frac{\zeta_3}{2}\left(-x^{14} + x^{13} + x - 1 - (x^7 + x^6)\sqrt{-x}\right)$

$k = 42, \rho = 2.83, r(x) = \Phi_{42}(x), \pi(x) = \frac{\zeta_3}{2}\left(x^7 + x^5 - 1x + (x^8 + x^3 - x)\sqrt{-x}\right)$

$k = 45, \rho = 2.58, r(x) = \Phi_{90}(x), \pi(x) = \frac{\zeta_3}{2}\left(x^{15} + x^8 - 1(x^{15} + x^7 - 1)\sqrt{-x}\right)$

$k = 54, \rho = 2.11, r(x) = \Phi_{54}(x), \pi(x) = \frac{\zeta_3}{2}\left(x^9 + x^5 - 1 + (x^9 + x^4 - 1)\sqrt{-x}\right)$

$k = 60, \rho = 3.75, r(x) = \Phi_{60}(2x), \pi(x) = \frac{\zeta_3}{2}\left(32768x^{15} + 16384x^{14} + 4096x^{12} - 256x^8 - 64x^6 - 16x^4 + 1 + (4096x^{12} + 1024x^{10} - 128x^7 + 32x^5 - 4x^2 - 1)\sqrt{-x}\right)$

$k = 66, \rho = 2.3, r(x) = \Phi_{66}(x), \pi(x) = \frac{\zeta_3}{2}\left(x^{11} - x^6 - 1 + (-x^{11} + x^5 + 1)\sqrt{-x}\right)$

$k = 78, \rho = 2.42, r(x) = \Phi_{78}(x), \pi(x) = \frac{\zeta_3}{2}\left(x^{13} - x^8 - 1 + (x^{14} - x^6 - x)\sqrt{-x}\right)$

$k = 84, \rho = 3.75, r(x) = \Phi_{84}(2x), \pi(x) = \frac{\zeta_3}{2}\left(16384x^{14} + 2x - 1 + (-4194304x^{22} - 131072x^{17} + 65536x^{16} - 4096x^{12} - 2048x^{11} - 1024x^{10} + 256x^8 + 64x^6 + 16x^4 - 4x^2 - 1)\sqrt{-x}\right)$

$k = 90, \rho = 2.58, r(x) = \Phi_{90}(x), \pi(x) = \frac{\zeta_3}{2}\left(x^{15} - x^8 - 1 + (-x^{15} + x^7 + 1)\sqrt{-x}\right)$

References

1. Barreto, P.S.L.M., Naehrig, M.: Pairing-Friendly Elliptic Curves of Prime Order. In: Preneel, B., Tavares, S. (eds.) SAC 2005. LNCS, vol. 3897, pp. 319–331. Springer, Heidelberg (2006)
2. Boneh, D., Franklin, M.: Identity-Based Encryption from the Weil Pairing. In: Kilian, J. (ed.) CRYPTO 2001. LNCS, vol. 2139, pp. 213–229. Springer, Heidelberg (2001); full version: SIAM J. Comput. 32(3), 586–615 (2003)
3. Boneh, D., Lynn, B., Shacham, H.: Short Signatures from the Weil Pairing. In: Boyd, C. (ed.) ASIACRYPT 2001. LNCS, vol. 2248, pp. 514–532. Springer, Heidelberg (2001); full version: J. Cryptol. 17, 297–319 (2004)
4. Brezing, F., Weng, A.: Elliptic curves suitable for pairing based cryptography. Des. Codes Cryptogr. 37, 133–141 (2005)
5. Cardona, G., Quer, J.: Field of moduli and field of definition for curves of genus 2, http://arxiv.org/abs/math/0207015
6. Cardona, G., Quer, J.: Curves of genus 2 with group of automorphisms isomorphic to D_8 or D_{12}. Trans. Amer. Math. Soc. 359, 2831–2849 (2007)
7. Dryło, R.: A New Method for Constructing Pairing-Friendly Abelian Surfaces. In: Joye, M., Miyaji, A., Otsuka, A. (eds.) Pairing 2010. LNCS, vol. 6487, pp. 298–311. Springer, Heidelberg (2010)
8. Dryło, R.: On Constructing Families of Pairing-Friendly Elliptic Curves with Variable Discriminant. In: Bernstein, D.J., Chatterjee, S. (eds.) INDOCRYPT 2011. LNCS, vol. 7107, pp. 310–319. Springer, Heidelberg (2011)
9. Freeman, D.: Constructing Pairing-Friendly Elliptic Curves with Embedding Degree 10. In: Hess, F., Pauli, S., Pohst, M. (eds.) ANTS 2006. LNCS, vol. 4076, pp. 452–465. Springer, Heidelberg (2006)
10. Freeman, D.: Constructing Pairing-Friendly Genus 2 Curves with Ordinary Jacobians. In: Takagi, T., Okamoto, T., Okamoto, E., Okamoto, T. (eds.) Pairing 2007. LNCS, vol. 4575, pp. 152–176. Springer, Heidelberg (2007)
11. Freeman, D.: A Generalized Brezing-Weng Algorithm for Constructing Pairing-Friendly Ordinary Abelian Varieties. In: Galbraith, S.D., Paterson, K.G. (eds.) Pairing 2008. LNCS, vol. 5209, pp. 146–163. Springer, Heidelberg (2008)
12. Freeman, D., Satoh, T.: Constructing pairing-friendly hyperelliptic curves using Weil restriction. J. Number Theory 131, 959–983 (2011)
13. Freeman, D., Scott, M., Teske, E.: A taxonomy of pairing-friendly elliptic curves. J. Cryptol. 23, 224–280 (2010)
14. Freeman, D., Stevenhagen, P., Streng, M.: Abelian Varieties with Prescribed Embedding Degree. In: van der Poorten, A.J., Stein, A. (eds.) ANTS-VIII 2008. LNCS, vol. 5011, pp. 60–73. Springer, Heidelberg (2008)
15. Furukawa, E., Kawazoe, M., Takahashi, T.: Counting Points for Hyperelliptic Curves of Type $y^2 = x^5 + ax$ Over Finite Prime Fields. In: Matsui, M., Zuccherato, R.J. (eds.) SAC 2003. LNCS, vol. 3006, pp. 26–41. Springer, Heidelberg (2004)
16. Galbraith, S.D.: Supersingular Curves in Cryptography. In: Boyd, C. (ed.) ASIACRYPT 2001. LNCS, vol. 2248, pp. 495–513. Springer, Heidelberg (2001)
17. Galbraith, S., McKee, J., Valença, P.: Ordinary abelian varieties having small embedding degree. Finite Fields Appl. 13, 800–814 (2007)

18. Gaudry, P., Schost, É.: On the Invariants of the Quotients of the Jacobian of a Curve of Genus 2. In: Bozta, S., Sphparlinski, I. (eds.) AAECC- 14. LNCS, vol. 2227, pp. 373–386. Springer, Heidelberg (2001)
19. Guillevic, A., Vergnaud, D.: Genus 2 Hyperelliptic Curve Families with Explicit Jacobian Order Evaluation and Pairing-Friendly Constructions. To appear in Pairing-Based Cryptography – Pairing 2012. LNCS (2012)
20. Howe, E., Zhu, H.: On the existence of absolutely simple abelian varieties of a given dimension over an arbitrary field. J. Number Theory 92, 139–163 (2002)
21. Igusa, J.: Arithmetic Variety of Moduli for Genus Two. Ann. Math. 72, 612–649 (1960)
22. Joux, A.: A One Round Protocol for Tripartite Diffie–Hellman. In: Bosma, W. (ed.) ANTS-IV. LNCS, vol. 1838, pp. 385–393. Springer, Heidelberg (2000); full version: J. Cryptol. 17, 263–276 (2004)
23. Kachisa, E.J.: Generating More Kawazoe-Takahashi Genus 2 Pairing-Friendly Hyperelliptic Curves. In: Joye, M., Miyaji, A., Otsuka, A. (eds.) Pairing 2010. LNCS, vol. 6487, pp. 312–326. Springer, Heidelberg (2010)
24. Kachisa, E.J., Schaefer, E.F., Scott, M.: Constructing Brezing-Weng Pairing-Friendly Elliptic Curves Using Elements in the Cyclotomic Field. In: Galbraith, S.D., Paterson, K.G. (eds.) Pairing 2008. LNCS, vol. 5209, pp. 126–135. Springer, Heidelberg (2008)
25. Kawazoe, M., Takahashi, T.: Pairing-Friendly Hyperelliptic Curves with Ordinary Jacobians of Type $y^2 = x^5 + ax$. In: Galbraith, S.D., Paterson, K.G. (eds.) Pairing 2008. LNCS, vol. 5209, pp. 164–177. Springer, Heidelberg (2008)
26. Lang, S.: Algebraic Number Theory. Graduate Texts in Mathematics, vol. 110. Springer, Berlin (1994)
27. Miyaji, A., Nakabayashi, M., Takano, S.: New explicit conditions of elliptic curve traces for FR-reduction. IEICE Trans. Fundam. E84-A(5), 1234–1243 (2001)
28. Milne, J.S.: Abelian varieties. In: Cornell, G., Silverman, J. (eds.) Arithmetic Geometry, pp. 103–150. Springer, New York (1986)
29. Murphy, A., Fitzpatrick, N.: Elliptic curves for pairing applications, http://eprint.iacr.org/2005/302
30. Maisner, D., Nart, E.: Abelian surfaces over finite fields as Jacobians. Experimental Mathematics 11, 321–337 (2002); With an appendix by Everett W. Howe
31. Menezes, A., Okamoto, T., Vanstone, S.: Reducing elliptic curve logarithms to logarithms in a finite field. IEEE Trans. Inf. Theory 39, 1639–1646 (1993)
32. Mestre, J.F.: Construction de courbes de genre 2 à partir de leurs modules. In: Effective Methods in Algebraic Geometry (Castiglioncello, 1990), pp. 313–334. Birkhäuser, Boston (1991)
33. Rubin, K., Silverberg, A.: Using abelian varieties to improve pairing-based cryptography. J. Cryptol. 22, 330–364 (2009)
34. Scott, M., Barreto, P.S.L.M.: Generating more MNT elliptic curves. Des. Codes Cryptogr. 38, 209–217 (2006)
35. Shaska, T., Voelklein, H.: Elliptic subfields and automorphisms of genus 2 function fields. In: Algebra, Arithmetic and Geometry with Applications (West Lafayette, IN, 2000), 703–723. Springer, Heidelberg (2004)
36. Silverman, J.: The Arithmetic of Elliptic Curves. Springer, Berlin (1986)
37. Sakai, R., Ohgishi, K., Kasahara, M.: Cryptosystems based on pairings. In: 2000 Symposium on Cryptography and Information Security – SCIS 2000, Okinawa, Japan (2000)
38. Sutherland, A.: Computing Hilbert class polynomials with the Chinese remainder theorem. Math. Comp. 80, 501–538 (2011)

39. Tate, J.: Classes d'isogénie des variétés abéliennes sur un corps fini (d'aprés T. Honda.) Séminarie Bourbaki 1968/69, exposé 352. Lect. Notes in Math, vol. 179, pp. 95–110. Springer (1971)
40. Tate, J.: Endomorphisms of abelian varieties over finite fields. Inventiones Mathematicae 2 (1966)
41. Waterhouse, W.C.: Abelian varieties over finite fields. Ann. Sci. École Norm. Sup. 2, 521–560 (1969)
42. Waterhouse, W.C., Milne, J.S.: Abelian varieties over finite fields. Proc. Symp. Pure Math. 20, 53–64 (1971)

Faster Batch Forgery Identification

Daniel J. Bernstein[1,3], Jeroen Doumen[2],
Tanja Lange[3], and Jan-Jaap Oosterwijk[3]

[1] Department of Computer Science
University of Illinois at Chicago, Chicago, IL 60607-7053, USA
djb@cr.yp.to
[2] Irdeto, CTO Research Group, Taurus Avenue 105, 2132 LS, Hoofddorp,
The Netherlands
jdoumen@irdeto.com
[3] Department of Mathematics and Computer Science
Technische Universiteit Eindhoven, P.O. Box 513, 5600 MB Eindhoven,
The Netherlands
tanja@hyperelliptic.org, j.oosterwijk@tue.nl

Abstract. Batch signature verification detects whether a batch of signatures contains any forgeries. Batch forgery identification pinpoints the location of each forgery. Existing forgery-identification schemes vary in their strategies for selecting subbatches to verify (individual checks, binary search, combinatorial designs, etc.) and in their strategies for verifying subbatches. This paper exploits synergies between these two levels of strategies, reducing the cost of batch forgery identification for elliptic-curve signatures.

Keywords: Signatures, batch verification, elliptic curves, scalar multiplication.

1 Introduction

Our goal in this paper is to minimize the cost of elliptic-curve signature verification. As an illustration of our results, one of our algorithms verifies a sequence of 64 elliptic-curve signatures (from 64 different signers) at a 2^{128} security level using

- a total of $0.9 \cdot 64 \cdot 128$ additions if all signatures turn out to be valid,
- a total of $1.3 \cdot 64 \cdot 128$ additions if 2 signatures turn out to be invalid,
- a total of $2.3 \cdot 64 \cdot 128$ additions if 10 signatures turn out to be invalid, and
- a total of $3.6 \cdot 64 \cdot 128$ additions if all 64 signatures turn out to be invalid.

This work was supported by the National Science Foundation under grant 1018836, by the Netherlands Organisation for Scientific Research (NWO) under grant 639.073.005, by the Dutch Technology Foundation STW (which is part of NWO, and which is partly funded by the Ministry of Economic Affairs, Agriculture and Innovation) under grant 10518, and by the European Commission under Contract ICT-2007-216676 ECRYPT II. Permanent ID of this document: 3bde3ab884b9aa2995cb5589e3037232. Date: 2012.09.18.

S. Galbraith and M. Nandi (Eds.): INDOCRYPT 2012, LNCS 7668, pp. 454–473, 2012.
© Springer-Verlag Berlin Heidelberg 2012

For comparison, we use a total of $2.8 \cdot 64 \cdot 128$ additions to separately verify the same 64 signatures.

We emphasize that our algorithms pinpoint the forgeries. These algorithms are not merely "batch signature verification" algorithms, saying yes if and only if all of the signatures are valid; these algorithms are "batch forgery identification" algorithms, telling the user separately for each signature whether that signature is valid. The main challenge we address is to locate each forgery as efficiently as possible.

Cost Metric. We systematically report the costs of our algorithms in group operations: the total number of elliptic-curve doublings, additions, and subtractions. For conciseness we write "additions" rather than "group operations", but readers evaluating costs in more detail should be aware that doublings are less expensive than additions in typical elliptic-curve coordinate systems, that "mixed additions" save time, etc.

We also caution the reader that elliptic-curve computations often involve significant overhead beyond group operations. For example, the CHES 2011 elliptic-curve-signatures paper [4] by Bernstein, Duif, Lange, Schwabe, and Yang reports quite noticeable time, even after various speedups, for decompressing points and for manipulating a priority queue of scalars. We would expect our algorithms to use the same amount of time for decompression and less time for manipulating scalars, but properly verifying these predictions would require an optimized assembly-language implementation at the level of [4].

Our verification algorithms are randomized. Performance depends somewhat on these random choices, but our experiments indicate that the variance in performance (for any particular number of forgeries) is quite small.

The total cost of separately verifying n signatures at a 2^b security level scales linearly in n and almost linearly in b: it has the form $\alpha n b$ where α is independent of n and nearly independent of b. This paper's batch-forgery-identification algorithms use $\alpha n b$ additions where α is a more complicated function of n, b, the number of forgeries, and various algorithm parameters. We systematically report the number of additions in the form $\alpha n b$, as illustrated by the $0.9nb$ example above with $n = 64$ and $b = 128$.

Choice of Signature System. We focus on the EdDSA signature system proposed in [4]. This system is a tweaked version of the classic Schnorr signature system [35]; one of the tweaks allows much faster batch verification.

In EdDSA, verifying a signature (R, S) on a message M under a public key A means verifying an equation of the form $SB = R + hA$. Here B is a standard elliptic-curve point, R and A are elliptic-curve points, S is a scalar, and the scalar h is a hash of R, A, and M.

For comparison, in Schnorr's system, the signature is (h, S) rather than (R, S). The verifier recomputes $R = SB - hA$ and then checks that the hash matches h. This is not compatible with our verification algorithms: our algorithms require R as input.

An analogous tweak for DSA (and the general idea of sending R instead of h) was introduced much earlier by Naccache, M'Raïhi, Vaudenay, and Raphaeli in [23]. We prefer Schnorr to ECDSA (and prefer EdDSA to tweaked ECDSA) for several reasons: Schnorr eliminates inversions, for example, and is resilient to hash-function collisions.

For elliptic-curve signatures at a 2^b security level it is standard practice to use about $2b$ bits for hashes, scalars, and field elements, and to compress points to single coordinates. EdDSA and Schnorr's system then have the same signature size, about $4b$ bits. Additions require uncompressed points, so the standard way to verify a signature in Schnorr's system is to decompress the public key A, compute $SB - hA$, compress the result to obtain R, compute the hash, and check for a match with h. We emphasize that the same operations, in a different order, verify a signature in EdDSA: compute the hash h, decompress the public key A, compute $SB - hA$, compress the result, and check for a match with R. The advantage of EdDSA is that it allows further choices for the verifier: fast batch verification, as discussed in [4], and fast batch forgery identification, as discussed in this paper. These algorithms require decompression of both A and R for each signature, but amply compensate for the extra decompression (an extra square-root computation) by eliminating a large fraction of the subsequent elliptic-curve operations.

One can merge EdDSA with Schnorr's system, simultaneously allowing signatures of the form (h, S) and signatures of the form (R, S). The first step in verifying an EdDSA signature computes, as a side effect, a Schnorr signature for the same message; similarly, one of the (later) steps in verifying a Schnorr signature computes, as a side effect, an EdDSA signature. It is not commonly appreciated that Schnorr's system actually allows hashes as short as b bits (as pointed out by Schnorr), reducing a signature to about $3b$ bits; users then have the flexibility to convert signatures from EdDSA format to Schnorr format to save space, and to convert signatures from Schnorr format to EdDSA format for fast batch forgery identification. One can of course also save decompression time by transmitting uncompressed signatures and uncompressed public keys.

Pairing-based signatures allow shorter signatures, about $2b$ bits, but pairing-based verification is an order of magnitude slower than elliptic-curve verification. Consider, for example, [21, Figures 1(a), 2(a), 3(a), 4(a)]: batch verification of pairing-based signatures with $b = 80$ costs about 2^{14} field multiplications per signature, i.e., about $200nb$ field multiplications. This is the cost in the best case, when there are no forgeries; the cost increases rapidly with the number of forgeries. For comparison, Hisil et al. showed in [12] how to reduce the cost of an elliptic-curve addition to at most 8 field multiplications; we never use more than $4nb$ additions, i.e., $32nb$ field multiplications.

Previous Work on Elliptic-Curve Signature Verification. There is an extensive literature analyzing and optimizing various techniques to verify *one* elliptic-curve signature. The main bottleneck here is double-scalar multiplication, computing an expression of the form $\ell P + mQ$ where ℓ and m are scalars (typically 256 bits) and P and Q are elliptic-curve points. Typical speedups include signed digits, windows, sliding windows, fractional windows, and merged doublings; combining these speedups typically reduces the number of additions by a factor between 2 and 3 compared to the simplest binary methods of computing $\ell P + mQ$. There are also many lower-level speedups inside elliptic-curve additions, field arithmetic, etc., but these speedups have no effect on the number-of-additions metric used for the rest of this paper.

There are, as mentioned above, some papers proposing batch verification of elliptic-curve signatures. The central idea is to check that several quantities $V_1 = R_1 + h_1 A_1 - S_1 B$, $V_2 = R_2 + h_2 A_2 - S_2 B$, etc. are all 0 by checking whether a random linear combination

$$V = z_1 R_1 + z_2 R_2 + \cdots + (z_1 h_1) A_1 + (z_2 h_2) A_2 + \cdots - (z_1 S_1 + z_2 S_2 + \cdots) B$$

is 0. If the verifier chooses the "randomizers" $z_1, z_2, \ldots$ as independent uniform random 128-bit integers then this test cannot be fooled with probability above 2^{-128}. We emphasize the importance of including these randomizers; in Section 2 we explain how to break the non-randomized batch-verification system from a very recent paper.

This linear-combination idea was proposed in [23] for (tweaked) DSA, in the simpler (and faster but obviously less useful) case of verifying multiple signatures of the same user, i.e. $A_1 = A_2 = \cdots$. The speedup in [23] was only a small constant for high security levels, because [23] computed V using only very simple techniques for multi-scalar multiplication, but [4] showed that the Bos–Coster multi-scalar multiplication method produced a much larger speedup. It is easy to see that the speedup here is asymptotically $\Theta(\lg n)$ for a batch of n signatures. The first paper to point out a non-constant speedup was [2] by Bellare, Garay, and Rabin, using a different technique that does not appear to be competitive with advanced multi-scalar multiplication methods.

What is missing from all of these papers is an efficient way to handle forgeries. Consider, for example, the following quote from [4]:

> If verification fails then there must be at least one invalid signature. We then fall back to verifying each signature separately. There are several techniques to identify a *small* number of invalid signatures in a batch, but all known techniques become slower than separate verification as the number of invalid signatures increases; separate verification provides the best defense against denial-of-service attacks.

This strategy means that an attacker sending a low volume of forgeries, enough to have one forgery in each batch, causes a severe slowdown in the software from [4]: each signature ends up being verified separately. It is of course desirable to reduce this damage, if that can be done without compromising performance under heavier denial-of-service floods; what is most desirable is to simultaneously reduce the cost of handling a few forgeries, the cost of handling many forgeries, and every case in between.

Previous Work on Forgery Identification. Pastuszak, Michalek, Pieprzyk, and Seberry in [25] proposed a binary-splitting method of identifying forgeries: if a batch is bad (i.e., fails verification), split it into two halves and apply the same algorithm to each half separately. It is easy to see that this algorithm rapidly becomes slower than separate verification as the number of forgeries increases; however, this algorithm is the foundation for several improved algorithms discussed below.

If one measures algorithm speed by simply counting the *number* of batch verifications then the binary-splitting method seems quite fast, identifying each forgery in $\lg n$ batch verifications where n is the batch size; this is optimal for a single forgery, and diverges only slowly from optimality as the number of forgeries grows. However, the number of batch verifications is not a good measure for the actual amount of time needed to identify the forgeries. Not all verifications require the same amount of time: a larger batch takes longer. Counting additions is a much more realistic cost measure and shows that the binary-splitting method of [25] is actually quite slow.

Pastuszak, Pieprzyk, and Seberry in [26] considered the possibility of *non-adaptively* choosing subbatches to verify. All available evidence suggests that this non-adaptivity restriction compromises performance even when the number of forgeries is somehow known in advance, and it certainly does not improve performance. Furthermore, non-adaptivity is clearly a disaster when the approximate number of forgeries is not known in advance. We therefore focus on the more flexible adaptive case.

Zaverucha and Stinson in [39] pointed out that there was already a long literature on the number of tests required by adaptive and non-adaptive "group testing" algorithms. Aside from terminology, a "group testing" algorithm is precisely a forgery-identification algorithm built on top of batch verification; in particular, both [25] and [26] fit into this framework. However, the following papers (some of which predate [39]) do not fit into this framework.

Law and Matt in [18] were the first to point out, in the context of pairing-based signatures, that batch verification is providing more information than a simple "yes" or "no". The most important idea, transported to the elliptic-curve case discussed in this paper, is that one can reuse the randomizers $z_1, \ldots, z_n$ from $V = z_1 V_1 + \cdots + z_n V_n$. If $V \neq 0$ then the binary-splitting method begins with a half-size multi-scalar multiplication to compute a left-half sum $z_1 V_1 + \cdots + z_{n/2} V_{n/2}$; and then the right-half sum $z_{n/2+1} V_{n/2+1} + \cdots + z_n V_n$ is trivially computed with a single subtraction, rather than another half-size multi-scalar multiplication.

Law and Matt also suggested computing $V' = z_1 V_1 + 2z_2 V_2 + \cdots + n z_n V_n$. If there is just one invalid signature, say $V_i \neq 0$, then $V' = iV$, and one can compute i in $O(\sqrt{n})$ additions by the baby-step-giant-step method. Further development of this approach appears in [18], [20], and [21].

We start from the same ideas, move from pairing-based signatures to elliptic-curve signatures for extra speed, and then point out additional speedups. For example, we introduce two ways to drastically reduce the cost of computing the *left-half* sum described above, without penalizing other parts of the algorithm. To simplify verifiability and reuse of our results we have posted public-domain implementations of our main algorithms at `http://cr.yp.to/badbatch.html`.

2 On the Importance of Being Random

The paper [16] by Karati, Das, Roychowdhury, Bellur, Bhattacharya, and Iyer, appearing at Africacrypt 2012 earlier this year, proposed a scheme for batch verification of ECDSA signatures. This section shows that the scheme is insecure. The main problem is that the scheme does not randomize the linear combination being verified.

ECDSA. The basic ECDSA signature scheme works as follows. The system parameters are a prime ℓ, a generator B of an order-ℓ group $\langle B \rangle$, and a cryptographic hash function H. The secret key of a user is a random integer a in $[1, \ell]$; the user's public key is $A = aB$. The group is a subgroup of the set of $\mathbf{F}_p$-rational points on an elliptic curve given in Weierstrass form $y^2 = x^3 + c_4 x + c_6$ for $c_4, c_6 \in \mathbf{F}_p$. An affine point is a tuple $P = (x(P), y(P))$ satisfying the curve equation; the negative of this point is $-P = (x(P), -y(P))$. The curve consists of the affine points and the point at infinity P_∞, which is the neutral element of the group of points.

A signature on message M under public key A is a tuple (r, s) such that the x-coordinate of $(H(M)/s)B + (r/s)A$ is congruent to r modulo ℓ. The standard approach to verification is to compute $R = (H(M)/s)B + (r/s)A$ and to check that $x(R)$ is congruent to r modulo ℓ.

The Scheme from [16] for Batch ECDSA Verification. The batch verification scheme described in [16] verifies signatures (r_i, s_i) on messages M_i and public keys A_i for $1 \leq i \leq n$ by reconstructing R_i from r_i and checking whether $\sum_{i=1}^{n} R_i$ equals $\left(\sum_{i=1}^{n} H(M_i)/s_i\right)B + \sum_{i=1}^{n}(r_i/s_i)A_i$.

The obvious approach to reconstructing R_i from r_i is to first compute $x(R_i)$ from $x(R_i) \bmod \ell = r_i$ and then compute $y(R_i)$ from the curve equation. The first step is straightforward in the common case that $\ell \approx p$: there is almost always a unique integer $x(R_i) \in \{0, 1, \ldots, p-1\}$ satisfying $x(R_i) \bmod \ell = r_i$. The second step is more difficult: it seems to require a square-root computation, and furthermore can at best determine $\pm y(R_i)$; in a batch of n signatures one needs to guess as many as 2^n combinations of signs. This implies that the batches need to be chosen small; in [16] the maximum batch size considered is 8. The paper puts the main effort into developing new techniques for computing $\sum R_i$ from the x-coordinates in a more efficient manner and reports a good speed-up factor compared to individual verification.

First Attack. A batch signature system is broken if invalid signatures pass as valid. The easiest way to break the above scheme is to submit (r, s) as a signature on a target message M under a target public key A and also $(r, -s)$ as a signature on the same message under the same public key, where r is any x-coordinate of a curve point. The verification algorithm reconstructs two points $R, -R$ having x-coordinate r, and then the contributions of these signatures cancel out in both sums:

$$R + (-R) = P_\infty = (H(M)/s)B + (r/s)A + (H(M)/(-s))B + (r/(-s))A.$$

This attack relies on the fact that r does not pinpoint a unique R: it can be expanded to R and to $-R$.

These forgeries are easy to detect once the system is altered to check for them. Excluding a sum of P_∞ is not adequate if the batch includes other signatures along with these two forgeries, but checking for repeated r values is adequate. However, as we will see in a moment, there are other attacks on the scheme that are much more difficult to detect.

Second Attack. Assume that the attacker knows the secret key a_2 for a public key A_2. The following attack convinces the verifier to accept a signature on any target message M_1 under any target public key A_1, along with a signature on M_2 under A_2.

The attacker picks a random k_1, and computes $R_1 = k_1B$ and $r_1 = x(R_1)$ as in proper signature generation. He then picks a random s_1 and computes $R_2 = (r_1/s_1)A_1$, $r_2 = x(R_2)$, and $s_2 = (H(M_2) + r_2a_2)/(k_1 - H(M_1)/s_1)$; the denominators are nonzero with overwhelming probability. The attacker then submits (r_1, s_1) as signature on M_1 from A_1 and (r_2, s_2) as signature on M_2 from A_2 to the batch system.

The verifier now reconstructs the same R_1 and R_2, and computes $R_1 + R_2$ and $(H(M_1)/s_1 + H(M_2)/s_2)B + (r_1/s_1)A_1 + (r_2/s_2)A_2$, both of which equal $k_1B + (r_1/s_1)A_1$. These forgeries thus pass verification, even though neither of them is valid individually and the attacker does not know the secret key for A_1. The forgeries also work if they are batched together with other signatures in the same verification.

As far as we can tell, the most efficient way to distinguish (r_1, s_1) and (r_2, s_2) from properly formed signatures is to verify them separately. This trivial batch-verification scheme is obviously secure but also sacrifices all of the speedup reported in [16].

Consequences. These attacks show that the scheme considered in [16] is insecure. The second attack would work even if the ECDSA signature system were replaced by a signature system such as EdDSA that transmits R instead of r, removing the $\pm R$ ambiguity. The second attack shows that it is important to use randomness in the tests: to introduce n sufficiently random integers z_i to scale the equations and verify $\sum_{i=1}^{n} z_iR_i = \left(\sum_{i=1}^{n} z_iH(M_i)/s_i\right)B + \sum_{i=1}^{n}(z_ir_i/s_i)A_i$ instead.

Randomizers were used in the original batch signature scheme introduced by Naccache, M'Raïhi, Vaudenay, and Raphaeli in [23]. There is no discussion of randomizers in [16], and in particular no explanation of why the randomizers were omitted in [16], but it is clear that computing $\sum_{i=1}^{n} z_iR_i$ would take much longer than computing $\sum_{i=1}^{n} R_i$, and it is even harder to compute its x-coordinate from the r_i without square-root computations to recover each point R_i first.

3   High Level: Binary Search

This section presents a family of algorithms for verifying a batch of n EdDSA signatures. We begin with a simple binary-search algorithm and then discuss several variants of the algorithm.

These algorithms rely on multi-scalar multiplication as a lower-level subroutine. Section 4 presents several multi-scalar multiplication algorithms usable in this context, pointing out new synergies between these two levels of algorithms. Section 5 analyzes the overall algorithm cost and reports the results of computer experiments with particular algorithm parameters.

For simplicity we assume that the batch size n is a power of 2. Other batch sizes can be split into power-of-2 batch sizes, or handled directly by straightforward generalizations of the algorithms here.

We also assume for simplicity that B has prime order ℓ, and that all input points R_i, A_i are known in advance to be in the group generated by B. For elliptic-curve groups with small cofactors the usual way to ensure this is to multiply all input points by the cofactor, such as the cofactor 8 in [3] and [4]. A closer look shows that this multiplication can safely be suppressed in the context of signature verification, but since the multiplication has very low cost we skip further discussion.

Randomizers. All of our algorithms use the randomizers z_i discussed in Sections 1 and 2. As precomputation we choose $z_1, z_2, \ldots, z_n$ independently and uniformly at random from the set $\{1, 2, 3, \ldots, 2^b\}$, where b is the security level. There are several reasonable ways to do this: for example, generate a uniform random b-bit integer and add 1, or generate a uniform random b-bit integer and replace 0 with 2^b.

Of course, it is also safe to simply generate z_i as a uniform random b-bit integer, disregarding the negligible chance that $z_i = 0$; but this requires minor technical modifications to the security guarantees stated below, so we prefer to require $z_i \neq 0$. It is also safe to simulate random numbers as outputs of a strong stream cipher using a long-term random secret key; this is helpful on platforms where generating randomness is expensive. Rather than maintaining stream-cipher state (e.g., the counter in the AES-CTR stream cipher) one can safely encrypt a collision-resistant hash of the input batch.

We also precompute integers $h_1, h_2, \ldots, h_n$ as the standard (system-specified) hashes of $(R_1, A_1, M_1), (R_2, A_2, M_2), \ldots, (R_n, A_n, M_n)$ respectively. By definition the ith signature is valid if $S_i B = R_i + h_i A_i$, and a forgery if $S_i B \neq R_i + h_i A_i$.

Leaf Randomizers. In this section we define $V_i = z_i(R_i + h_i A_i - S_i B)$. Note the inclusion of z_i here, deviating from Section 1. This is not merely a change of notation: to verify a single signature (when this is required), our algorithm computes this V_i, whereas the standard verification approach from [4] is to compute $S_i B - h_i A_i$. Note that signature i is valid if and only if $V_i = 0$.

The standard approach would seem at first glance to be more efficient: computing $S_i B - h_i A_i$ involves two full-size ($2b$-bit) scalars S_i, h_i, while computing V_i as $z_i R_i + z_i h_i A_i - z_i S_i B$ involves two full-size scalars $z_i h_i \bmod \ell, z_i S_i \bmod \ell$

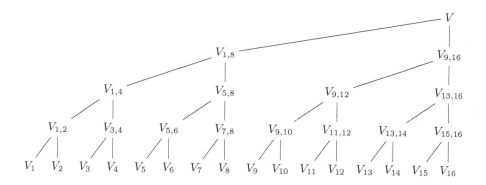

Fig. 3.1. Tree of sums of randomized leaves $V_1, V_2, \ldots, V_n$ for $n = 16$

and a half-size scalar z_i, for a total of 25% more scalar bits. However, the cost of multi-scalar multiplication (see Section 4) is affected much more by the *maximum* number of scalar bits than by the *total* number of scalar bits; the cost of computing V_i turns out to be only slightly higher than the cost of computing $S_i B - h_i A_i$. This slight extra expense pays off in subsequent steps of the batch algorithm, as discussed below.

Shared Randomizers. Starting from these randomized quantities $V_1, \ldots, V_n$ we draw a binary tree as illustrated in Figure 3.1, with $V_{1,2} = V_1 + V_2$ and $V_{3,4} = V_3 + V_4$ and so on at the second level, $V_{1,4} = V_{1,2} + V_{3,4}$ and so on at the third level, etc. In general we write $V_{j,k}$ for the sum $\sum_{j \le i \le k} V_i$ of leaf nodes. If all of the signatures at positions $j, j+1, \ldots, k$ are valid then $V_{j,k} = 0$, while if any of the signatures are invalid then with overwhelming probability $V_{j,k} \ne 0$. The root node $V_{1,n}$ at the top represents the randomized signature verification of the entire batch; we denote this sum by V as a shorthand.

The set of tree nodes actually computed by the algorithm is determined adaptively; see below.

We emphasize that one sequence of randomizers is shared across all levels of the tree, including the leaf nodes. This reuse does not compromise the security of the algorithm: if signature i is invalid then with overwhelming probability *all* of the ancestor tree nodes $V_{j,k}$ with $j \le i \le k$ are nonzero. More precisely, fix a batch of signatures, and define a randomizer sequence $(z_1, \ldots, z_n)$ as "bad" if it produces any zeros among ancestor tree nodes of any invalid signature; then the probability of a randomizer sequence being bad is at most $(n-1)/2^b$. The point is that if signature i is invalid (i.e., $R_i + h_i A_i - S_i B \ne 0$), then any equation $V_{j,k} = 0$ for $j \le i \le k$ is equivalent to a linear equation $\cdots + z_i(R_i + h_i A_i - S_i B) + \cdots = 0$. For each choice of $z_1, \ldots, z_{i-1}, z_{i+1}, \ldots, z_n$ this equation is satisfied by exactly one integer z_i modulo ℓ, and therefore at most one out of the 2^b permitted choices of z_i. A randomizer sequence is therefore "(j, k)-bad" with probability at most $1/2^b$ for $j < k$ (i.e., non-leaf nodes), and with probability 0 for $j = k$ (i.e., leaf nodes). There are $n - 1$ non-leaf nodes, so a randomizer sequence is bad with probability at most $(n - 1)/2^b$.

The Basic Batch-Forgery-Identification Algorithm. The following algorithm takes as input public keys $A_1, A_2, \ldots, A_n$, signatures $(R_1, S_1), \ldots, (R_n, S_n)$, precomputed hashes $h_1, h_2, \ldots, h_n$, and precomputed randomizers $z_1, z_2, \ldots, z_n$. The algorithm also takes an optional input V; this is used when the algorithm calls itself recursively in Step 5.

The algorithm provides two outputs: first, V, whether or not V was provided as input; second, an n-bit string $(b_1, b_2, \ldots, b_n)$. With overwhelming probability $b_i = 1$ if and only if signature i is valid.

The algorithm has six steps:

1. **Batch verification:** Compute $V = \sum_i z_i(R_i + h_i A_i - S_i B)$, if V was not provided as input. Output V. If $V = 0$, output n bits $(1, 1, \ldots, 1)$ and stop.
2. **Forgery rejection:** If $n = 1$, output (0) and stop. (At this point $V \neq 0$, so the signature is invalid.)
3. **Left subtree:** Apply the same algorithm recursively to $A_1, A_2, \ldots, A_{n/2}$; $(R_1, S_1), \ldots, (R_{n/2}, S_{n/2})$; $h_1, \ldots, h_{n/2}$; and $z_1, \ldots, z_{n/2}$; obtaining outputs $V_{1,n/2}$ and $(b_1, \ldots, b_{n/2})$.
4. **Right root:** If $V_{1,n/2} = 0$, set $V_{n/2+1,n} = V$. If $V_{1,n/2} = V$, set $V_{n/2+1,n} = 0$. Otherwise compute $V_{n/2+1,n} = V - V_{1,n/2}$.
5. **Right subtree:** Apply the same algorithm recursively to $A_{n/2+1}, \ldots, A_n$; $(R_{n/2+1}, S_{n/2+1}), \ldots, (R_n, S_n)$; $h_{n/2+1}, \ldots, h_n$; $z_{n/2+1}, \ldots, z_n$; and $V_{n/2+1,n}$; obtaining outputs $V_{n/2+1,n}$ and $(b_{n/2+1}, \ldots, b_n)$.
6. **Final output:** Output $(b_1, \ldots, b_n)$.

This algorithm is optimistic, hoping that there are no forgeries: Step 1 finishes the algorithm as quickly as possible in this case. See Section 4 for details of the computation in this step. The overall binary-splitting structure of this algorithm is taken from [25]. The fast computation of $V_{n/2+1,n}$ in Step 4, using at most one subtraction, is taken from [18]; this is also the reason for treating V as an output and an optional input. This fast computation means that at most n nodes require a multi-scalar multiplication in Step 1; Figure 3.2 illustrates the worst case.

Another way to organize essentially the same computation is to record a partial tree of known $V_{j,k}$ values, and to very quickly update the tree whenever a forgery is discovered, in effect retroactively removing the forgery from the batch. Start the computation at the root; after computing a zero node, deduce without further computation that all descendants of the node are also zero; after computing a nonzero leaf node $V_i \neq 0$, replace all ancestors $V_{j,k}$ by $V_{j,k} - V_i$, skipping the subtraction in the common case that $V_{j,k} = V_i$; after computing a nonzero non-leaf node, compute the left child node (and all of its descendants in order), and then simply copy this (possibly updated) node to the right child node.

Leaf Randomizers, Continued. In the case $n = 1$ this algorithm computes $V_1 = z_1(R_1 + h_1 A_1 - S_1 B)$. As discussed above, this is only slightly more

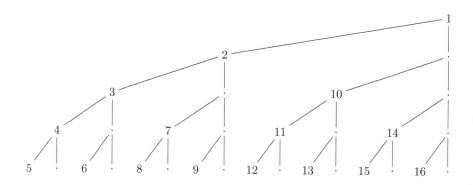

Fig. 3.2. Tests in worst case are depicted in order

expensive than computing $S_1 B - h_1 A_1$. We now explain the compensating advantage of computing V_1.

Consider a batch of two signatures that fails batch verification. i.e., $V_{1,2} \neq 0$. This algorithm computes V_1 (showing whether the first signature is valid), and then deduces V_2 (showing whether the second signature is valid) with at most one subtraction. For comparison, one could instead compare $S_1 B - h_1 A_1$ to R_1 to see whether the first signature is valid, but one then still needs to check whether the second signature is valid. One could check the second signature separately, or multiply $R_1 + h_1 A_1 - S_1 B$ by z_1 to obtain V_1 and thus V_2, but simply starting with V_1 is less expensive.

Early Abort. This algorithm is faster than separate verification when there are not many forgeries, but as discussed in subsequent sections it becomes noticeably slower than separate verification when there are many forgeries. The gap is not very large, but we would still like to minimize it.

We thus propose (1) using the fraction of invalid signatures found so far as an estimate for the expected fraction of invalid signatures in the rest of the tree, and (2) deciding on this basis whether it is best to abort the tree structure and check individual signatures.

An attacker might try to spoil the estimate by, e.g., placing several invalid signatures at the beginning of a large batch. After those signatures the algorithm will confidently, but incorrectly, estimate that the entire batch is invalid. To prevent such attacks one can simply apply a random permutation to the sequence of signatures before applying the algorithm. (One can also imagine tracking forgery percentages long term from one batch to another, but for simplicity we handle each batch separately.)

There is, furthermore, no need for aborts to be permanent: one can return to binary search for the next part of the tree if the fraction of invalid signatures has become small enough again. We actually propose making a new decision

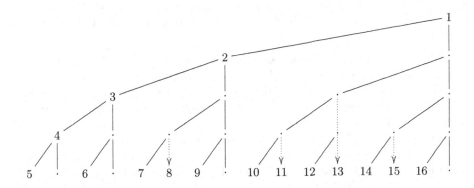

Fig. 3.3. Tests performed for $n = 16$ when all signatures are invalid, using the early abort. Arrows denote the test replacements and savings.

whenever a node is about to be computed. In the notation of the basic algorithm above, we dynamically choose between

- optimism: computing V, and then, if $V \neq 0$, computing $V_{1,n/2}$ and deducing $V_{n/2+1,n} = V - V_{1,n/2}$; or
- pessimism: computing $V_{1,n/2}$ and $V_{n/2+1,n}$, and then deducing $V = V_{1,n/2} + V_{n/2+1,n}$.

If V is provided as input then optimism is better. If V is not provided as input then we use $(1 - p)^n$ as an estimate of the chance that $V = 0$, where p is the fraction of invalid signatures found so far (or 0 at the beginning of the algorithm), and then compare the expected costs of optimism and pessimism, using straightforward models of the costs of computing $V, V_{1,n/2}, V_{n/2+1,n}$.

When there are few forgeries, this approach performs the same computations as the basic algorithm. When there are many forgeries, this approach rapidly converges on checking each signature separately, as shown in Figure 3.3. Compared to the previous worst case, where we computed the top node of each vertical branch, we now only need to compute the top nodes of the main left diagonal branch. In all other vertical branches, the leaf node is computed directly. (One can do marginally better in this extreme case by immediately updating p after discovering $V_{1,16} \neq 0$: there must be a forgery somewhere, even though it has not been located yet.)

When there is a medium fraction of forgeries, this approach skips roots of large subtrees (since those roots are likely to fail verification and require computations of descendant nodes), but computes roots of small subtrees. For example, assume that we identified exactly 2 forgeries out of the first 16 signatures. We expect the same fraction of $1/8$ invalid signatures in the next group of 16, so we estimate that $V_{17,32} = 0$ with probability only 11%, that $V_{17,24} = 0$ with probability 34%, and that $V_{17,20} = 0$ with probability 59%. The next step depends on scalar-multiplication costs; we might decide to skip $V_{17,32}$ and $V_{17,24}$, and proceed directly to $V_{17,20}$. If the fraction of invalid signatures remains stable then we will check these 16 signatures as 4 batches of 4 signatures each. We then decide anew how to check the next 32 signatures.

Smaller Randomizers. Large randomizers z_i are critical for detecting multi-forgeries, as discussed in Section 2, but this does not mean that large randomizers are required at each step of the tree. An alternative approach is to use one sequence of large randomizers at the root, and to use a second sequence of much smaller randomizers, say 20 bits each, for the subsequent levels of the tree.

This approach slightly speeds up multi-scalar multiplication at non-root nodes. However, this approach also has several costs. First, the right child of the root node is no longer obtained for free. Second, the sharing described in Section 4 begins only at the children of the root node, not at the root node itself. Third, an attacker can fool the smaller randomizers with noticeable probability, on the scale of 2^{-20}, so after identifying forgeries using the smaller randomizers one must recompute the corresponding portion of the root node. If this root-node update shows that any forgeries remain then one must choose a new sequence of smaller randomizers and try the computation again on the remaining signatures.

4   Low Level: Trees of Optional Multi-scalar Multiplications

This section looks more closely at the first step of the algorithm of Section 3: namely, batch verification, i.e., computing a linear combination

$$V = z_1 R_1 + \cdots + z_n R_n + (z_1 h_1) A_1 + \cdots + (z_n h_n) A_n - (z_1 S_1 + \cdots + z_n S_n) B$$

of known elliptic-curve points $R_1, \ldots, R_n, A_1, \ldots, A_n, B$. If $V \neq 0$ then the algorithm calls itself recursively and computes a smaller linear combination

$$V_{1,m} = z_1 R_1 + \cdots + z_m R_m + (z_1 h_1) A_1 + \cdots + (z_m h_m) A_m - (z_1 S_1 + \cdots + z_m S_m) B$$

with $m = n/2$.

The computation of V by itself is a standard $(2n + 1)$-scalar-multiplication problem. The only mildly uncommon feature of this problem is that the scalars have variable size, typically n 128-bit scalars (the z_i's) and $n + 1$ 256-bit scalars; but typical scalar-multiplication algorithms can trivially take advantage of the shorter scalars. Similarly, the computation of $V_{1,m}$ by itself is a standard $(2m+1)$-scalar-multiplication problem.

Quite nonstandard, however, is the multi-scalar-multiplication problem that we actually face: computing V and then *perhaps* computing $V_{1,m}$. If we *knew* that we wanted to compute both V and $V_{1,m}$ then the obvious approach would be two separate half-size computations, one for $V_{1,m}$ and one for $V_{m+1,n} = V - V_{1,m}$; but we do not know this in advance. If V turns out to be 0 then we will not need $V_{1,m}$ and $V_{m+1,n}$, and a single full-size computation of V will be more efficient than two separate half-size computations.

The point of this section is that some — although certainly not all — state-of-the-art algorithms to compute V can be modified at negligible cost to remember many intermediate results useful for computing $V_{1,m}$. The same idea can easily be pushed to further levels: for example, computing V, then optionally $V_{1,m}$, then optionally $V_{1,\lfloor m/2 \rfloor}$ and optionally $V_{m+1,m+1+\lfloor (n-m)/2 \rfloor}$.

Overlap in the Bos–Coster Approach. As an illustration of what does *not* seem to work very well in this context, consider the Bos–Coster algorithm reported in [8, Section 4]. This algorithm computes $a_1 P_1 + a_2 P_2 + a_3 P_3 + \cdots$, where $a_1 \geq a_2 \geq a_3 \geq \cdots$, by recursively computing $(a_1 - a_2)P_1 + a_2(P_1 + P_2) + a_3 P_3 + \cdots$. This algorithm was used in [4] to compute V.

The first few additions performed in the Bos–Coster algorithm depend only on the largest scalars. If we permute signatures so that $z_1 h_1 \geq z_2 h_2 \geq \cdots$, and handle $z_1 S_1 + \cdots + z_n S_n$ separately, then the first $\approx m$ additions in the algorithm will involve only $A_1, \ldots, A_m$, and will thus be the same as the first additions involved in computing $V_{1,m}$. However, this is only a slight speedup.

Overlap in the Straus Approach. As a better example, consider the Straus algorithm [37], often miscredited to Shamir. This algorithm computes $a_1 P_1 + a_2 P_2 + \cdots + a_n P_n$ by recursively computing $\lfloor a_1/2^c \rfloor P_1 + \lfloor a_2/2^c \rfloor P_2 + \cdots + \lfloor a_n/2^c \rfloor P_n$, doubling c times, and then adding the precomputed quantity $(a_1 \bmod 2^c)P_1 + (a_2 \bmod 2^c)P_2 + \cdots + (a_n \bmod 2^c)P_n$. Here 2^c is a radix chosen by the algorithm; for example, it is reasonable to take $c = 5$ for 256-bit scalars. We skip discussion of standard speedups such as signed digits.

This algorithm scales poorly to large values of n (because it involves too much precomputation, even for $c = 1$), but a standard variant scales well to large values of n: at the last step one instead adds the separate precomputed quantities $(a_1 \bmod 2^c)P_1$, $(a_2 \bmod 2^c)P_2$, etc.

Evidently one can reuse these precomputed quantities for a subsequent multiscalar multiplication involving $P_1, \ldots, P_m$ with the same choice of c. Furthermore, if the precomputed quantities are added from left to right in each step, then one of the intermediate results is exactly $(a_1 \bmod 2^c)P_1 + \cdots + (a_m \bmod 2^c)P_m$. This drastically reduces the cost of computing $a_1 P_1 + \cdots + a_m P_m$ when m is large: each step of the recursion drops from cost $c + m$ (c doublings and m additions) down to just $c + 1$.

The same overlap applies immediately to $a_1 P_1 + \cdots + a_{\lfloor m/2 \rfloor} P_{\lfloor m/2 \rfloor}$. Even better, if we change the order to add precomputed quantities, recursively adding the $P_1, \ldots, P_m$ part and the $P_{m+1}, \ldots, P_n$ part, then the same overlap applies not just to left descendants but to arbitrary descendants.

Overlap in the Pippenger Approach. As a more advanced example, consider Pippenger's multi-scalar-multiplication method. This method was published in [28] almost forty years ago; various special cases of the method were subsequently reinvented and published in the papers [6] and [19] and continue to be frequently miscredited to those papers. We comment that the patent accompanying [6] (U.S. patent 5299262) expired this year.

Pippenger's method is not as simple as the Bos–Coster method or the Straus method, but it is considerably faster when there are many large scalars. It is almost twice as fast in some cases, and it is within $1 + o(1)$ of optimal for essentially all sequences of scalars; see generally [30]. Of course, this does not imply that Pippenger's method is optimal for the problem of computing V and then perhaps $V_{1,m}$, but inspecting the details shows that Pippenger's approach does allow considerable savings in computing $V_{1,m}$.

The following special case of Pippenger's algorithm has similar performance to the Bos–Coster method and is adequate to illustrate the idea. Choose a radix 2^c as above, and proceed as in Straus's algorithm, but replace the last step with the following computation. Sort the points $P_1, P_2, \ldots, P_n$ into 2^c buckets according to the values $a_1 \bmod 2^c, a_2 \bmod 2^c, \ldots, a_n \bmod 2^c$. Discard bucket 0 and add the points in the remaining buckets, obtaining sums $S_1, \ldots, S_{2^c-1}$. Now compute

$$(a_1 \bmod 2^c)P_1 + \cdots + (a_n \bmod 2^c)P_n = S_1 + 2S_2 + \cdots + (2^c - 1)S_{2^c-1}$$

as the sum of the intermediate quantities S_{2^c-1}, $S_{2^c-1} + S_{2^c-2}$, $\ldots$, $S_{2^c-1} + S_{2^c-2} + \cdots + S_1$.

Observe that computing $a_1 P_1 + \cdots + a_m P_m$ in the same way, using the same value of c, puts $P_1, P_2, \ldots, P_m$ into exactly the same buckets. If for the $a_1 P_1 + \cdots + a_n P_n$ computation we are careful to add points in each bucket from left to right then the intermediate result after $P_1, P_2, \ldots, P_m$ will be exactly the sum relevant to $a_1 P_1 + \cdots + a_m P_m$. For typical parameters there are several points in each bucket, so this approach is several times faster than a standard computation of $a_1 P_1 + \cdots + a_m P_m$. As before, it is even better to change the order to add points in each bucket, recursively adding the points that come from $P_1, P_2, \ldots, P_m$ and the points that come from $P_{m+1}, \ldots, P_n$.

Handling the Base Point. These modified versions of the Straus and Pippenger methods apply directly to

$$z_1 R_1 + (z_1 h_1)A_1 + \cdots + z_n R_n + (z_n h_n)A_n$$

but do not apply directly to $(z_1 S_1 + \cdots + z_n S_n)B$, the last component of V.

The simplest way to handle these multiples of B is to compute them separately. Because B is a fixed base point, one can afford a precomputed table of, e.g., $B, 2B, 3B, \ldots, (2^c - 1)B$ and $2^c B, 2 \cdot 2^c B, 3 \cdot 2^c B, \ldots, (2^c - 1) \cdot 2^c B$ and so on. Computing any desired multiple of B then takes fewer than $1/c$ additions for each bit of the scalar, a very small cost compared to the other computations discussed here.

5   Analysis

This section analyzes the cost of identifying all of the forgeries among n elliptic-curve signatures at a 2^b security level. Full-size scalars such as $h_i, S_i, z_i h_i \bmod \ell, z_i S_i \bmod \ell$ then have $2b$ bits as discussed in Section 1, while the randomizers z_i have b bits.

Our web page http://cr.yp.to/badbatch.html includes all of the software mentioned in this section.

Separate Signature Verification. Solinas' widely used Joint Sparse Form [36] handles a double-scalar multiplication $h_i A_i - S_i B$ using $2b$ doublings and on average b additions, for a total cost of $3nb$ to handle n signatures.

Straus's method is asymptotically more efficient, handling n signatures at cost $(2 + o(1))nb$ as $b \to \infty$. Straus's method involves approximately $2b$ doublings; every c doublings are followed by 2 additions, and on average a fraction $1/2^c$ of the additions are skippable additions of 0. The additions rely on an initial computation of $2A_1, 3A_1, \ldots, (2^c-1)A_1$, which costs $2^c - 2$, and a free precomputation of $2B, 3B, \ldots, (2^c - 1)B$. The total cost for n signatures is approximately $(2b + (1 - 1/2^c)(2/c)2b + 2^c - 2)n$. One can balance the terms $(1 - 1/2^c)(4/c)b$ and $2^c - 2$ by taking c close to $2 + \lg b - \lg\lg b$; the total cost is then roughly $(2 + 8/\lg b)nb$.

Our `separate3.py` software uses Straus's method with $c = 4$ and with two standard speedups, namely signed digits and sliding windows. This software uses, on average, fewer than $2.8nb$ additions for $b = 128$. There is a small variance: $2.75nb$ and $2.82nb$ are not unusual. We would expect more detailed optimization here, in particular using more precomputed multiples of B, to beat $2.7nb$.

Batch Verification. All of our batch-forgery-identification algorithms start with batch verification, computing V. If there are no forgeries — no attackers attempting to fool the receiver or deny service — then this is the end of the computation.

Straus's algorithm computes V with $2b$ doublings as above, approximately $n(b/c)$ additions for parts of $z_i R_i$, approximately $n(2b/c)$ additions for parts of $(z_i h_i)A_i$, and negligible cost for B. The additions rely on initial computations costing $2n(2^c - 2)$. The total cost is approximately $(2/n + 3/c + 2(2^c - 2)/b)nb$. If c is chosen close to $\lg(1.5b) - \lg\lg b$ then this cost is roughly $(2/n + 6/\lg b)nb$.

Our `straus6.py` software, with $b = 128$ and $c = 5$, uses $1.15nb$ additions for $n = 8$; $0.98nb$ additions for $n = 16$; $0.90nb$ additions for $n = 32$; and $0.86nb$ additions for $n = 64$.

We also experimented with the Bos–Coster algorithm (`boscoster2.py`) and did some preliminary analysis of Pippenger's algorithm. Compared to Straus's algorithm, we obtained better batch-verification speeds with the Bos–Coster algorithm (e.g., cost $0.55nb$ for $n = 64$ and $b = 128$) and we expect to obtain better batch-verification speeds with Pippenger's algorithm. Asymptotically the Bos–Coster algorithm costs $O(nb/\lg n)$ and Pippenger's algorithm costs $O(nb/\lg nb)$. However, we decided to focus on Straus's algorithm for our experiments because Straus's algorithm allows much better reuse of intermediate results inside batch forgery identification.

Batch Forgery Identification. For concreteness we focus on the overlap inside Straus's algorithm inside binary search using shared randomizers (including leaf randomizers), without early aborts. After the root node (i.e., the batch verification discussed above), reuse of intermediate results reduces each subsequent multi-scalar multiplication to approximately $2b$ doublings and $4b/c$ additions.

We emphasize that, no matter how many forgeries there are, this strategy is within $1 + o(1)$ of separate signature verification as $b \to \infty$. At most n tree nodes

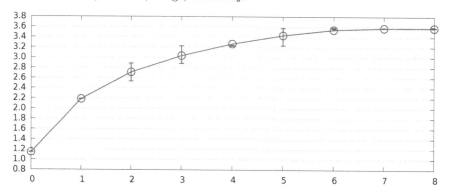

Fig. 5.1. Observed cost $\alpha n b$ of identifying forgeries among $n = 8$ signatures for $b = 128$. Horizontal axis is number of forgeries. Vertical axis is α. Each circle indicates average cost over 101 experiments; error bars indicate quartiles.

require multi-scalar multiplication, and each multi-scalar multiplication costs $(2 + o(1))b$ after $O(nb/\lg b)$ for the root, so the total cost is at most $(2 + o(1))nb$, just like separate signature verification. If a positive constant fraction of the signatures are valid then the number of nodes required is a constant factor below n and this strategy is a constant factor faster than separate signature verification; if the number of forgeries drops then this strategy becomes a logarithmic factor faster than separate signature verification.

For constant b, such as $b = 128$, the picture is more complicated. Each computed non-root node has similar cost to a separate signature verification (in fact slightly lower cost), but the root node adds a significant extra cost, so this algorithm becomes noticeably slower than separate signature verification as the number of forgeries increases. Our `straus6.py` computer experiments indicate that the cutoff is around $n/3$ forgeries for $b = 128$. See Figures 5.1, 5.2, and 5.3.

References

[1] — (no editor): 17th annual symposium on foundations of computer science. IEEE Computer Society, Long Beach, California (1976). MR 56:1766. See [28]
[2] Bellare, M., Garay, J.A., Rabin, T.: Fast batch verification for modular exponentiation and digital signatures. In: Eurocrypt '98 [24], pp. 236–250 (1998), http://cseweb.ucsd.edu/~mihir/papers/batch.html. Citations in this document: §1
[3] Bernstein, D.J.: Curve25519: new Diffie-Hellman speed records. In: PKC 2006 [38], pp. 207–228 (2006), http://cr.yp.to/papers.html#curve25519. Citations in this document: §3
[4] Bernstein, D.J., Duif, N., Lange, T., Schwabe, P., Yang, B.-Y.: High-speed high-security signatures. In: CHES 2011 [31] (2011), http://eprint.iacr.org/2011/368. Citations in this document: §1, §1, §1, §1, §1, §1, §1, §3, §3, §4
[5] Brassard, G. (ed.): Advances in cryptolgy — CRYPTO '89, 9th annual international cryptology conference, Santa Barbara, California, USA, August 20–24, 1989, proceedings. LNCS, vol. 435. Springer (1990). ISBN 3-540-97317-6. MR 91b:94002. See [34]

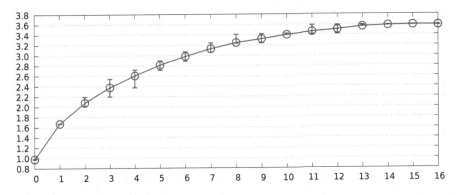

Fig. 5.2. Same as Figure~5.1 but for $n = 16$

[6] Brickell, E.F., Gordon, D.M., McCurley, K.S., Wilson, D.B.: Fast exponentiation with precomputation (extended abstract). In: Eurocrypt '92 [33], pp. 200–207 (1993); see also newer version [7]. Citations in this document: §4, §4

[7] Brickell, E.F., Gordon, D.M., McCurley, K.S., Wilson, D.B.: Fast exponentiation with precomputation: algorithms and lower bounds (1995); see also older version [6], http://research.microsoft.com/~dbwilson/bgmw/

[8] de Rooij, P.: Efficient exponentiation using precomputation and vector addition chains. In: Eurocrypt '94 [9], pp. 389–399 (1995). MR 1479665. Citations in this document: §4

[9] De Santis, A. (ed.): Advances in cryptology—EUROCRYPT '94, workshop on the theory and application of cryptographic techniques, Perugia, Italy, May 9–12, 1994, proceedings. LNCS, vol. 950. Springer (1995). ISBN 3-540-60176-7. MR 98h:94001. See [8], [23]

[10] Desmedt, Y. (ed.): Advances in cryptology—CRYPTO '94, 14th annual international cryptology conference, Santa Barbara, California, USA, August 21–25, 1994, proceedings. LNCS, vol. 839. Springer (1994). ISBN 3-540-58333-5. See [19]

[11] Galbraith, S.D. (ed.): Cryptography and coding, 11th IMA international conference, Cirencester, UK, December 18–20, 2007, proceedings. LNCS, vol. 4887. Springer (2007). ISBN 978-3-540-77271-2. See [18]

[12] Hisil, H., Wong, K.K.-H., Carter, G., Dawson, E.: Twisted Edwards curves revisited. In: Asiacrypt 2008 [27], pp. 326–343 (2008), http://eprint.iacr.org/2008/522. Citations in this document: §1

[13] Imai, H., Zheng, Y. (eds.): Public key cryptography, third international workshop on practice and theory in public key cryptography, PKC 2000, Melbourne, Victoria, Australia, January 18–20, 2000, proceedings. LNCS, vol. 1751. Springer (2000). ISBN 3-540-66967-1. See [25]

[14] Jarecki, S., Tsudik, G. (eds.): Public key cryptography—PKC 2009, 12th international conference on practice and theory in public key cryptography, Irvine, CA, USA, March 18–20, 2009, proceedings. LNCS, vol. 5443. Springer (2009). ISBN 978-3-642-00467-4. See [20]

[15] Joye, M., Miyaji, A., Otsuka, A. (eds.): Pairing-based cryptography—Pairing 2010—4th international conference, Yamanaka Hot Spring, Japan, December 2010, proceedings. LNCS, vol. 6487. Springer (2010). ISBN 978-3-642-17454-4. See [21]

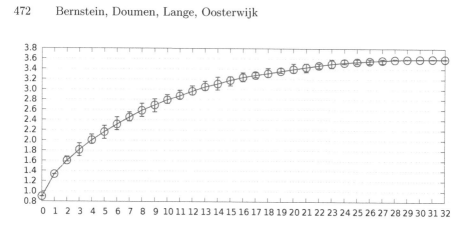

Fig. 5.3. Same as Figure˜5.1 but for $n = 32$

[16] Karati, S., Das, A., Roychowdhury, D., Bellur, B., Bhattacharya, D., Iyer, A.: Batch verification of ECDSA signatures. In: Africacrypt 2012 [22], pp. 1–18 (2012). Citations in this document: §2, §2, §2, §2, §2, §2, §2, §2

[17] Kurosawa, K. (ed.): Information theoretic security, 4th international conference, ICITS 2009, Shizuoka, Japan, December 3–6, 2009, revised selected papers. LNCS, vol. 5973. Springer (2010). ISBN 978-3-642-14495-0. See [39]

[18] Law, L., Matt, B.J.: Finding invalid signatures in pairing-based batches. In: Cirencester 2007 [11], pp. 34–53 (2007). Citations in this document: §1, §1, §3

[19] Lim, C.H., Lee, P.J.: More flexible exponentiation with precomputation. In: Crypto '94 [10], pp. 95–107 (1994). Citations in this document: §4

[20] Matt, B.J.: Identification of multiple invalid signatures in pairing-based batched signatures. In: PKC 2009 [14], pp. 337–356 (2009). Citations in this document: §1

[21] Matt, B.J.: Identification of multiple invalid pairing-based signatures in constrained batches. In: Pairing 2010 [15], pp. 78–95 (2010). Citations in this document: §1, §1

[22] Mitrokotsa, A., Vaudenay, S. (eds.): Progress in cryptology — AFRICACRYPT 2012, 5th international conference on cryptology in Africa, Ifrane, Morocco, July 10-12, 2012, proceedings. LNCS, vol. 7374. Springer (2012). See [16]

[23] Naccache, D., M'Raïhi, D., Vaudenay, S., Raphaeli, D.: Can D.S.A. be improved? Complexity trade-offs with the digital signature standard. In: Eurocrypt '94 [9] (1994). Citations in this document: §1, §1, §1, §1, §2

[24] Nyberg, K. (ed.): Advances in cryptology — EUROCRYPT '98, international conference on the theory and application of cryptographic techniques, Espoo, Finland, May 31–June 4, 1998, proceedings. LNCS, vol. 1403. Springer (1998). ISBN 3-540-64518-7. See [2]

[25] Pastuszak, J., Michalek, D., Pieprzyk, J., Seberry, J.: Identification of bad signatures in batches. In: PKC 2000 [13], pp. 28–45 (2000). Citations in this document: §1, §1, §1, §3

[26] Pastuszak, J., Pieprzyk, J., Seberry, J.: Codes identifying bad signature in batches. In: Indocrypt 2000 [32], pp. 143–154 (2000). Citations in this document: §1, §1

[27] Pieprzyk, J. (ed.): Advances in cryptology—ASIACRYPT 2008, 14th international conference on the theory and application of cryptology and information security, Melbourne, Australia, December 7–11, 2008. LNCS, vol. 5350 (2008). ISBN 978-3-540-89254-0. See [12]

[28] Pippenger, N.: On the evaluation of powers and related problems (preliminary version). In: FOCS '76 [1], pp. 258–263 (1976); newer version split into [29] and [30]. MR 58:3682. Citations in this document: §4

[29] Pippenger, N.: The minimum number of edges in graphs with prescribed paths. Mathematical Systems Theory 12, 325–346 (1979); see also older version [28]. ISSN 0025-5661. MR 81e:05079

[30] Pippenger, N.: On the evaluation of powers and monomials. SIAM Journal on Computing 9, 230–250 (1980); see also older version [28]. ISSN 0097-5397. MR 82c:10064. Citations in this document: §4

[31] Preneel, B., Takagi, T. (eds.): Cryptographic hardware and embedded systems—CHES 2011, 13th international workshop, Nara, Japan, September 28–October 1, 2011, proceedings. LNCS, vol. 6917. Springer (2011). ISBN 978-3-642-23950-2. See [4]

[32] Roy, B.K., Okamoto, E. (eds.): Progress in cryptology—INDOCRYPT 2000, first international conference in cryptology in India, Calcutta, India, December 10–13, 2000, proceedings. LNCS, vol. 1977. Springer (2000). ISBN 3-540-41452-5. See [26]

[33] Rueppel, R.A. (ed.): Advances in cryptology—EUROCRYPT '92, workshop on the theory and application of cryptographic techniques, Balatonfüred, Hungary, May 24–28, 1992, proceedings. LNCS, vol. 658. Springer (1993). ISBN 3-540-56413-6. MR 94e:94002. See [6]

[34] Schnorr, C.P.: Efficient identification and signatures for smart cards. In: Crypto '89 [5], pp. 239–252 (1990); see also newer version [35]

[35] Schnorr, C.P.: Efficient signature generation by smart cards. Journal of Cryptology 4, 161–174 (1991); see also older version [34], http://www.mi.informatik.uni-frankfurt.de/research/papers.html. Citations in this document: §1

[36] Solinas, J.A.: Low-weight binary representations for pairs of integers CORR 2001-41 (2001), http://www.cacr.math.uwaterloo.ca/techreports/2001/corr2001-41.ps. Citations in this document: §5

[37] Straus, E.G.: Addition chains of vectors (problem 5125). American Mathematical Monthly 70, 806–808 (1964). Citations in this document: §4

[38] Yung, M., Dodis, Y., Kiayias, A., Malkin, T. (eds.): Public key cryptography—9th international conference on theory and practice in public-key cryptography, New York, NY, USA, April 24–26, 2006, proceedings. LNCS, vol. 3958. Springer (2006). ISBN 978-3-540-33851-2. See [3]

[39] Zaverucha, G.M., Stinson, D.M.: Group testing and batch verification. In: ICITS 2009 [17], pp. 140–157 (2010). Citations in this document: §1, §1

Implementing CFS

Gregory Landais and Nicolas Sendrier

INRIA Paris-Rocquencourt, Project-Team SECRET
{gregory.landais,nicolas.sendrier}@inria.fr

Abstract. CFS is the first practical code-based signature scheme. In the present paper, we present the initial scheme and its evolutions, the attacks it had to face and the countermeasures applied. We compare the different algorithmic choices involved during the implementation of the scheme and aim to provide guidelines to this task. We will show that all things considered the system remains practical. Finally, we present a state-of-the-art software implementation of the signing primitive to prove our claim. For eighty bits of security our implementation produces a signature in 1.3 seconds on a single core of Intel Xeon W3670 at 3.20 GHz. Moreover the computation is easy to distribute and we can take full profit of multi-core processors reducing the signature time to a fraction of second in software.

Keywords: CFS, digital signature scheme, software implementation.

1   Introduction

CFS [1] is a digital signature scheme based on the Niederreiter cryptosystem [2]. It was published in 2001 and relies on the hardness of the syndrome decoding problem and on the undistinguishability of binary Goppa code.

There are relatively few instances of public-key digital signatures available and most of them are based on number theory. Even though, implementation issues related to CFS have received little attention. This can be explained by an (apparent) lack of practicality and also by some cryptanalytic results that have slightly weakened the scheme.

First, the practicality of the scheme is questionable because of a large public key and a long signing time. The large key size might be a problem for some applications, but a storage requirement of a few megabytes on the verifier's side is not always an issue. The second drawback is the long signing time. In fact each signature requires a large number (several hundred of thousands) of algebraic decoding attempts. The first reported signing time in software [1] was about one minute but was only meant as a proof of concept. An FPGA implementation was reported in [3] with a signing time under one second.

Second, the system has been weakened in several ways. It was recently proven in [4] that the public key of CFS could be distinguished in polynomial time from a random binary matrix. This property certainly needs to be investigated further, but at this time does not seem to lead to an effective attack. More threatening is the Bleichenbacher's attack (unpublished) which essentially reduces the cost

S. Galbraith and M. Nandi (Eds.): INDOCRYPT 2012, LNCS 7668, pp. 474–488, 2012.

of the best decoding attack from $2^{r/2}$ to $2^{r/3}$ ($r = 144$ in [1]) and necessitates an increase of parameters leading to a rather cumbersome scheme (gigabytes of key and minutes of signing time). Fortunately an efficient countermeasure, Parallel-CFS, was proposed in [5].

The purpose of this work is to (try to) clarify the situation, to show that the system is practical and secure when parameters are properly chosen.

Our Contribution: We propose here a study of the implementation of the CFS (and the parallel-CFS) schemes. Since it is the bottleneck of the implementation we will concentrate on the signing algorithm and thus on the decoding of Goppa codes. Other features of the system such as verification, signature size, hash function choice... are certainly significant but are not in the scope of this work. There exist alternatives for implementing finite field arithmetic (we need here extensions of the binary field of degree 16 to 24, which are not common) and for decoding (we will see that the usual choice, Patterson's algorithm, is not necessarily the best in this context). We explore these alternatives and extract guidelines for efficient implementation. To illustrate this study, we have designed a basic state-of-the-art software implementation of parallel-CFS (our target platform is a standard PC). In addition to timings, we have performed a precise field operation count which suggests that a faster field arithmetic may have a spectacular effect on the signing time. We also mean our algorithmic study to be the theoretical basis for dedicated coprocessors for signing with CFS.

We will first review the CFS scheme, its attacks and the Parallel-CFS variant. This will allow us to propose some sets of secure parameters. Next we will describe the software implementation. More specifically, we will explain the various algorithmic options for decoding binary Goppa and compare them. We will conclude with some timings and detailed measurements of our implementation.

2 Background

In this paper we will consider only binary linear codes. Most of the statements are easily generalized to a larger alphabet, but no practical CFS-like signature scheme has ever been proposed so far with non binary codes.

2.1 Syndrome Decoding

We consider the following problem:

Computational Syndrome Decoding Problem: Given a matrix $H \in \{0,1\}^{r \times n}$, a word $s \in \{0,1\}^r$, and an integer $w > 0$, find $e \in \{0,1\}^n$ of Hamming weight $\leq w$ such that $He^T = s$.

We denote $\mathrm{CSD}(H, s, w)$ this problem as well as the set of its solutions for a given instance. This problem is NP-hard [6]. For suitable parameters n, r and w it is conjectured difficult on average which is the basis for the security of many code-based cryptosystems.

2.2 Binary Goppa Codes

Let $\mathbf{F}_{2^m}$ denote a finite field with 2^m elements. Let $n \leq 2^m$, let the *support* $L = (\alpha_0, \ldots, \alpha_{n-1})$ be an ordered sequence of distinct elements of $\mathbf{F}_{2^m}$, and let the *generator* $g(z) \in \mathbf{F}_{2^m}$ be a monic irreducible polynomial of degree t. The binary Goppa code of support L and generator g is defined as

$$\Gamma(L, g) = \{(a_0, \ldots, a_{n-1}) \in \{0,1\}^n \mid \sum_{j=0}^{n-1} \frac{a_j}{z - \alpha_j} \bmod g(z) = 0\}.$$

This code has length $n \leq 2^m$ dimension $\geq n - mt$ and has an algebraic t-error correcting procedure. For the signature we will always take $n = 2^m$, a smaller value would increase the signing cost. For parameters of interest the dimension will always be exactly $k = n - mt$, we will denote $r = mt$ the codimension.

In a code-based cryptosystem using Goppa codes the system parameters are (m, t) and are known to everyone, the secret key is the pair (L, g) and the public key is a parity check matrix $H \in \{0,1\}^{r \times n}$.

Density of Decodable Syndromes for a Goppa Code: A syndrome $s \in \{0,1\}^r$ is decodable (relatively to H) with the algebraic decoder if and only if it is of the form $s = eH^T$ with e of Hamming weight t or less. There are $\sum_{i=0}^{t} \binom{n}{i} \approx \binom{n}{t}$ such syndromes. The total number of syndrome is 2^r and thus the proportion of syndrome decodable with the binary Goppa code algebraic decoder is close to

$$\frac{\binom{n}{t}}{2^r} = \frac{\binom{2^m}{t}}{2^{mt}} = \frac{2^m(2^m - 1) \cdots (2^m - t + 1)}{t! 2^{mt}} \approx \frac{1}{t!}. \tag{1}$$

2.3 Complete Decoding

A complete decoder for a binary linear code defined by some parity check matrix $H \in \{0,1\}^{r \times n}$ is a procedure that will return for any syndrome $s \in \{0,1\}^r$ an error pattern e of minimal weight such that $eH^T = s$. The expected weight w of e will be the integer just above the Gilbert-Varshamov radius τ_{gv}, which we define as the real number[1] such that $\binom{n}{\tau_{\mathrm{gv}}} = 2^r$. The threshold effect can be observed on two examples in Table 1.

In practice we will relax things a little bit and when we mention a complete decoder we mean a w-bounded decoder (with $w \geq \tau_{\mathrm{gv}}$), that is a procedure $\psi : \{0,1\}^r \rightarrow \{0,1\}^n$ returning an error pattern matching with the input of weight $\leq w$ every time there exists one. A w-bounded decoder may return an error pattern of weight $< w$. Also, a w-bounded decoder may fail even if $w \geq \tau_{\mathrm{gv}}$ (see Table 1 for $(m, t) = (20, 8)$ and $w = 9 > \tau_{\mathrm{gv}} = 8.91$), in that case we may either choose a decoding bound larger than $\tau_{\mathrm{gv}} + 1$ (loosing some security) or handle somehow the decoding failure (see §5.4). In the sequel, whenever we refer to complete decoding we implicitly define a decoding bound, an integer larger than τ_{gv}, denoted w.

---

[1] The mapping $x \mapsto \binom{n}{x}$ is easily extended continuously for the positive real numbers making the definition of τ_{gv} sound.

Table 1. Failure probability of a w-bounded decoder for a code of length $n = 2^m$ and codimension $r = mt$

(m, t)	τ_{gv}	$w = 8$	$w = 9$	$w = 10$	$w = 11$
$(20, 8)$	8.91	$1 - 2^{-15}$	0.055	$2^{-131583}$	$2^{-10^{10}}$
$(18, 9)$	10.26	$1 - 2^{-33}$	$1 - 2^{-18}$	0.93	2^{-2484}

2.4 Original CFS

A CFS instance is defined by a binary Goppa code Γ of length n correcting up to t errors; of parity check matrix H; over the finite field $\mathbf{F}_2$. We will denote the decoding function of Γ by *decode*. This function takes a binary syndrome as input and returns a tuple of t error positions matching with the input syndrome or fails if no such error pattern exists. The matrix H is public and the procedure *decode* is secret. Signing a document is done like this :

1. Hash the document.
2. Suppose the hash value is a syndrome and try to decode it using Γ.
3. Use the resulting error pattern as a signature.

Since the hash value of the document is very unlikely to be a decodable syndrome (i.e. syndrome of a word at Hamming distance t or less from a codeword), step 2 is a little more complicated. CFS comes with two workarounds for this step :

- *Complete decoding* [Algorithm 1] adds a certain amount of columns of H to the syndrome until it becomes decodable (*i.e.* guess a few errors).
- *Counter-appending* alters the message with a counter and rehashes it until it becomes decodable.

The two methods require $t!$ decoding in average (a consequence of (1), see [1]), but the *counter-appending* method includes the hash function inside the decoding thus forcing to implement it on the target architecture, which might be an inconvenience on a dedicated coprocessor. Moreover, with this method the size of the signature is variable because the counter has a high standard variation. Finally, the Parallel-CFS countermeasure (see §2.6) does not work with the *counter-appending* method.

2.5 Attacks

There exists key-distinguishing attacks on CFS [4]: it is possible to efficiently distinguish a CFS public key (a binary Goppa parity check matrix) from a random matrix of same size. However this does not lead, for the moment, to any efficient key recovery attack. In practice, the best known techniques for forging a signature are based on generic decoding of linear codes, that is solving the computational syndrome decoding problem (CSD).

The two main techniques for solving CSD are Information Set Decoding (ISD) and the Generalized Birthday Algorithm (GBA). ISD was introduced by Prange

---

Algorithm 1. Signing with complete decoding

function SIGN(M, w, h)   ▷ input: message M; integer $w > t$; hash function h
    $s \longleftarrow h(M)$
    loop
        $(i_{t+1}, \ldots, i_w) \xleftarrow{R} \{0, \ldots, n-1\}^{w-t}$
        $(i_1, \ldots, i_t) \longleftarrow \text{decode}(s + H_{i_{t+1}} + \ldots + H_{i_w}, t)$
        if $(i_1, \ldots, i_t) \neq$ **fail then return** $(i_1, \ldots, i_w)$
        end if
    end loop
end function

---

in 1962 [7]. All practical variant derive from Stern's algorithm [8], the most recent improvements are [9,10]. GBA was introduced by Wagner in 2002 [11], but an order-2 variant was already proposed in [12]. Its first use for decoding came later [13].

Decoding One Out of Many (DOOM): In the signature forgery domain, an attacker can create any number of messages suiting him and be satisfied with one on them being signed. The benefits of having access to several syndromes has been mentioned by Bleichenbacher for GBA[2]. For ISD a proposal was made in [15] and was later generalized and analyzed in [16]. It shows that if N target syndromes are given and if decoding anyone of them is enough, the time complexity is reduced by a factor almost $\sqrt{N}$ compared to the situation where a single specific syndrome has to be decoded. There is a upper limit for N after which there is no gain, it depends on the type of algorithm (ISD or GBA) and on the code parameters. In practice this would mean, for 80 bits of security, multiplying the key size by 400 with a similar signing time, or multiplying the key size by 100 and the signing time by 10.

2.6   Parallel-CFS: A Countermeasure to DOOM

Parallel CFS is a countermeasure proposed by M. Finiasz in 2010 [5], aiming at cancelling the benefits an attacker could have with multiple target syndromes. The idea is to produce λ different hash values from the document to be signed (typically two to four) and to sign (that is decode) each of them separately. The final signature will be the collection of the signatures of all those hash values, see Algorithm 2 for the description using complete decoding. This way, if the attacker forges a signature for the first hash value of one of his multiple messages, he also has to forge a signature for the remaining hash values of this specific message, thus he is back to the initial single target decoding problem. As mentioned in [5], signing with the *counter-appending* method is impossible in this countermeasure since it is necessary to decode several hashes of the exact same message and the counter alters the message. This countermeasure increases by a factor λ the signature time, signature size and verification time.

---

[2] This attack was presented in 2004 but was never published, it is described in [14].

Algorithm 2. Parallel-signing with complete decoding

---

function SIGN_MULT(M, w, λ, H)   ▷ input: message M; integers $w > t$, $\lambda > 0$,
   for $1 \leq i \leq \lambda$ **do**   ▷ set of hash functions H
      $s_i \longleftarrow$ SIGN(M, w, H_i)
   end for
   return $(s_i)_{1 \leq i \leq \lambda}$
end function

In [5], Bleichenbacher's attack is generalized for attacking several hash values. The analysis shows that for most parameters, three hash values, sometimes only two, will cancel the benefits of the attack. For ISD, it is shown in [16] that the benefit of DOOM is not as high as for GBA. There was no generalization as in [5] for several hash values, but it is not likely to change the situation and if the number of hash values is large enough to cancel DOOM-GBA it will probably also cancel DOOM-ISD.

2.7   Previous Implementations

We are not aware of any publicly available software implementation of CFS. There is one FPGA implementation, described in [3], for the original parameters $n = 2^{16}$, $t = 9$, and $w = 11$. It reports an average signing time of 0.86 seconds and implements the Berlekamp-Massey decoding algorithm.

3   Parameter Selection

For single instances ISD is more efficient than GBA, and for multiple instances GBA-DOOM (i.e. generalized Bleichenbacher's attack) is more efficient than ISD-DOOM. To select secure parameters will look for parameters such that we are above the security requirements for the cost of the following attacks:

- ISD-MMT [10], the best known variant of ISD, for solving λ distinct single instances. In [10], only the asymptotic decoding exponent is given. We provide in appendix §A a non asymptotic analysis which we used for Table 2. We also mention the cost of a previous variant ISD-Dum [17] which is more flexible and may have an advantage in some cases (not here though). The numbers for ISD-Dum are derived from [18].
- GBA-DOOM [5], that is the generalized Bleichenbacher's attack, for Parallel-CFS of multiplicity λ.

The Table 2 gives the main features (including security) for some sets of parameters. The original parameters are given for reference but they are a bit undersized. We propose two main families of Goppa codes: 9-error correcting of length 2^{18} and 8-error correcting of length 2^{20}. The latter is faster but also has a larger public key size. All proposed parameter sets achieve 80 bits of security, our main targets are those where the hash multiplicity is $\lambda = 3$. We also give some sets of parameters with higher security which were not implemented.

Table 2. Some parameter sets for Parallel-CFS using full length binary Goppa codes

m	t	τ_{gv}	w	λ	failure prob.	public key size	security bits ($\log_2$ of binary ops.) ISD-MMT	ISD-Dum	GBA-DOOM
16	9	10.46	11	3	~ 0	1 MB	77.4	78.7	74.9
18	9	10.26	11	3	~ 0	5 MB	87.1	87.1	83.4
18	9	10.26	11	4	~ 0	5 MB	87.5	87.5	87.0
20	8	8.91	10	3	~ 0	20 MB	82.6	85.7	82.5
20	8	8.91	9	5	5.5%	20 MB	87.9	91.0	87.3
24	10	11.05	12	3	~ 0	500 MB	126.4	126.9	120.4
26	9	9.82	10	4	10^{-8}	2 GB	125.4	127.5	122.0

To be thorough, there is a very recent improvement of ISD [19]. From what we understand of this variant of ISD-MMT it is not likely to provide a significant non-asymptotic improvement when the target weight is small compared with the length as it is the case for CFS signatures.

4   Algebraic Decoding of Goppa Codes

The secret is a binary Goppa code $\Gamma(L, g)$ of length $n = 2^m$ of dimension r of generator polynomial $g(z) \in \mathbf{F}_{2^m}[z]$, monic irreducible of degree t, and support $L = (\alpha_0, \dots, \alpha_{n-1})$, consisting of (all) distinct elements of $\mathbf{F}_{2^m}$ in a specific order. The public key H is a systematic parity check matrix of $\Gamma(L, g)$. We denote $L_S = (\beta_0, \dots, \beta_{r-1})$ the support elements corresponding to the identity part of H (for instance the first or last r coordinates of L). An algebraic decoder for Goppa codes takes as input a binary syndrome $s = (s_0, \dots, s_{r-1}) \in \{0, 1\}^r$ and returns, if it exists, an error pattern $e \in \{0, 1\}^n$ of weight t such that $eH^T = s$. There are several algorithms (described later in this section) which all have the same three steps:

1. Compute from s a new polynomial syndrome with coefficients in $\mathbf{F}_{2^m}$.
2. Solve a key equation relating this syndrome to the error locator polynomial.
3. Extract the roots of the locator polynomial to recover the error positions.

4.1   Goppa Key Equation

The algebraic syndrome $R(z) = \sum_{0 \leq j < r} s_j f_{\beta_j}(z)$ corresponding to s is computed as a sum of elementary syndromes $f_\beta(z)$ defined for any $\beta \in \mathbf{F}_{2^m}$ as

$$f_\beta(z) = \frac{1}{z - \beta} \bmod g(z) = \frac{1}{g(\beta)} \frac{g(z) - g(\beta)}{z - \beta}. \tag{2}$$

Note that the only elementary syndromes needed are the r elements of L_S. The corresponding key equation is

$$\sigma(z)R(z) = \frac{d}{dz}\sigma(z) \bmod g(z), \deg \sigma \leq t \tag{3}$$

which has a unique solution $\sigma(z) \in \mathbf{F}_{2^m}[z]$ up to a scalar multiplicative constant. If there exists an error pattern $e \in \{0,1\}^n$ of weight $\leq t$ such that $eH^T = s$, then any solution to (3) is a scalar multiple of $\sigma(z) = \prod_{\beta \in \text{supp}(e)}(z - \beta)$ the locator polynomial of e (supp(e) is the subset of L corresponding to the non-zero coordinates of e). Equation (3) is solved with the Patterson algorithm [20].

4.2 Alternant Key Equation

A binary Goppa $\Gamma(L,g)$ can also be viewed as an alternant code. We use the fact that $\Gamma(L,g) = \Gamma(L,g^2)$ when g is square-free. We still have $R(z) = \sum_{0 \leq j < r} s_j f_{\beta_j}(z)$ but the elementary syndrome $f_\beta(z)$ has degree $2t - 1$ instead of $t - 1$ and is now defined for any $\beta \in \mathbf{F}_{2^m}$ as

$$f_\beta(z) = \frac{1}{g(\beta)^2} \frac{1}{1 - \beta z} \bmod z^{2t} = \sum_{i=0}^{2t-1} \frac{\beta^i z^i}{g(\beta)^2}. \tag{4}$$

The corresponding key equation is

$$\sigma_{inv}(z)R(z) = \omega(z) \bmod z^{2t}, \deg \omega < t, \deg \sigma_{inv} \leq t, \tag{5}$$

which has a unique solution $(\sigma_{inv}(z), \omega(z)) \in \mathbf{F}_{2^m}[z]^2$ up to a scalar multiplicative constant. If there exists an error pattern $e \in \{0,1\}^n$ of weight exactly t such that $eH^T = s$ and if $(\sigma_{inv}(z), \omega(z))$ is a solution to (5) then $\sigma(z) = z^t \sigma_{inv}(z^{-1}) = \prod_{\beta \in \text{supp}(e)}(z - \beta)$ up to a scalar multiple. To remain consistent with the Goppa key equation we will speak of $\sigma(z) = z^t \sigma_{inv}(z^{-1})$ as the solution of the equation. The resolution of (5) is achieved either with the Berlekamp-Massey algorithm [21] or with the extended Euclidean algorithm.

4.3 Root Finding

The state-of-the-art for root finding is the Berlekamp trace algorithm [22]. Its complexity is $O((m+t)t^2)$ and is advantageous compared with exhaustive techniques like Chien search or Horner's polynomial evaluation whose complexity is linear in the length n and thus exponential in m.

5 Implementation

5.1 Finite Field Arithmetic

We need to implement extensions of the binary field $\mathbf{F}_2$ of degree $m = 18$ and $m = 20$. For fields of small size, the best approach is to tabulate the logarithm and the exponentiation in base α, a primitive element of $\mathbf{F}_{2^m}$. This is efficient as long as the table fits into the processor cache. This is not the case here and we chose to implement those fields as an extension of degree 2 of $\mathbf{F}_{2^{m/2}}$. We used

$$\mathbf{F}_{2^{20}} = \mathbf{F}_{2^{10}}[x]/(x^2 + x + \alpha), \mathbf{F}_{2^{10}} = \mathbf{F}_2[x]/(x^{10} + x^9 + x^7 + x^6 + 1)$$

with α a primitive element of $\mathbf{F}_{2^{10}}$ such that $\alpha^{10} + \alpha^9 + \alpha^7 + \alpha^6 + 1 = 0$, and

$$\mathbf{F}_{2^{18}} = \mathbf{F}_{2^9}[x]/(x^2 + x + 1), \mathbf{F}_{2^9} = \mathbf{F}_2[x]/(x^9 + x^5 + 1).$$

The field $\mathbf{F}_{2^9}$ and $\mathbf{F}_{2^{10}}$ are small enough to be tabulated (with no cache miss on our target platform) and with Karatsuba's speedup a multiplication in the extension field requires three multiplications in the base field. For constrained architecture higher extension towers might be more effective. Also, bit slicing might offer an interesting alternative which, furthermore, is available also for prime extension degrees like $m = 19$ which have no subfield except $\mathbf{F}_2$.

5.2   Decoding

When signing a message M, we compute a hash value $s = h(M)$ considered as a syndrome according to the public key H. The word e of minimal weight such that $s = eH^T$ has weight $w > t$ and thus s cannot be decoded with the algebraic decoder which is limited to t errors. If, as described in Algorithm 1, we correctly guess $\delta = w - t$ error positions we will be able to successfully apply the algebraic decoder on a modified syndrome. It was proven in [1] that this succeeds on average after $t!$ guesses. We describe in Algorithm 3 a variant where the syndrome is modified in polynomial form. Also, the complete root finding procedure is applied once only.

Algorithm 3. Signing with binary Goppa codes

---

   function SIGN(M, h) $\triangleright$ input: message M; hash function h
      $s \longleftarrow h(M)$
      $R_0(z) \longleftarrow \sum_{0 \le j < r} s_j f_{\beta_j}(z)$ $\triangleright$ once only, either (2) or (4)
      for all $B \subset L$ of cardinality $\delta = w - t$ **do**
         $R(z) \longleftarrow R_0(z) + \sum_{\beta \in B} f_\beta(z)$ $\triangleright$ syndrome adjustment, either (2) or (4)
         $\sigma(z) \longleftarrow$ solve_key_eq($R(z)$) $\triangleright$ key equation solving, either (3) or (5)
         if $z^{2^m} = z \bmod \sigma(z)$ **then** $\triangleright$ split checking
            $A \longleftarrow$ roots_of($\sigma(z)$) $\triangleright$ once only
            return indices of the elements of $A \cup B$ in L
         end if
      end for
      return fail
   end function

Computing the Polynomial Syndrome: The first polynomial syndrome $R_0(z)$ is computed once only from s. Then, as many times as necessary, $R_0(z)$ is adjusted by computing and adding $\delta = w - t$ elementary syndromes $f_\beta(z)$. This adjustment has a cost proportional to δt field operations which is negligible in practice.

Solving the Key Equation: As mentioned above, there are various syndromes and key equations and sometimes several ways to solve them. In all cases this resolution has to be done completely and produces the same locator polynomial $\sigma(z)$. The cost is proportional to t^2 field operations.

Root Finding: The key equation always has a solution $\sigma(z)$ regardless of the existence of a suitable error pattern of weight t. The error has weight t or less if and only if the polynomial $\sigma(z)$ has all its roots in the field $\mathbf{F}_{2^m}$ that is if $\sigma(z) \mid z^{2^m} - z$. In practice we check whether $z^{2^m} = z \bmod \sigma(z)$ and we only compute the roots once. This requires m polynomial squaring modulo $\sigma(z)$ for a cost proportional to mt^2 field operations. This will be the dominant cost for the signature.

5.3   Discarding Degenerate Instances of Decoding

Several syndromes and key equations may be used for implementing the algebraic decoding of Goppa codes. In all cases, there is some amount of control required; at some point a coefficient is checked (leading coefficient in the extended Euclidean algorithm, or the discrepancy in Berlekamp-Massey Algorithm) and if it is zero the sequencing of operations is affected. Ignoring completely this test (*i.e.* assuming the coefficient is non zero) will provide a significant speedup in software (loops are easier to unroll) and a welcome simplicity in constrained devices. The counterpart is that a (small) proportion of decoding attempts produce inconsistent results and will fail. This is not a big deal in the signature context where almost all decoding attempts fail anyway. This was already remarked in [3] to simplify the control in an FPGA implementation.

5.4   How to Handle Decoding Failure

For $(m, t) = (20, 8)$ and $w = 9$ there is a probability of failure of $\nu = 5.5\%$. This means that some messages cannot be signed. When we use Parallel-CFS with multiplicity $\lambda = 5$, this percentage is equal to $\mu = 1 - (1 - \nu)^{\lambda}$ that is almost 25%, which is hardly acceptable. The workaround is to *add a counter:*

> Define a family $\mathcal{F} = \{f_i, 0 \le i < 2^b\}$ of 2^b one-to-one transformations on the syndromes (for instance adding distinct predefined random constant vectors). Let $s_1, \ldots, s_{\lambda}$ denote the hash values of Parallel-CFS of multiplicity λ. We try to decode the tuple $f(s_1), \ldots, f(s_{\lambda})$ for all $f \in \mathcal{F}$. The verifier will also try all $f \in \mathcal{F}$, in the same order as the signer. Optionally, the b bits index i of the transformation can be added to the signature to speed-up the verification.

The decoding fails for all $f \in \mathcal{F}$ with probability μ^{2^b}. For instance with $b = 5$ in our example the probability of failure drops from 0.249 to 2^{-64} and each time we increment b this probability is squared. The security is unchanged because we applied the same transformation on all hash values. Note that, had we allowed different transformations for the λ hash values, the attacker would have been able to apply a 2^b-instances ISD-DOOM for on each hash value, gaining a factor $\sqrt{2^b}$ to the attack.

5.5   Signature Size, Verification and Key Generation

Various interesting tradeoffs are possible between signature size and verification time. They are described already in [1] and we propose no novelty here. For $(m,t,w,\lambda) = (20,8,10,3)$ the signature size ranges from 292 to 535 bits and for $(m,t,w,\lambda) = (18,9,11,3)$ it ranges from 286 to 519 bits. This part of the scheme (as well as the key generation procedure) is out of the scope of this work and is not detailed further.

6   Timings

6.1   Computation Time

We measured with two decoders and the various sets of parameters the average number of algebraic decoding and the running time for producing one signature running on our target platform (a single core of an Intel Xeon W3670 at 3.20 GHz, about 400 000 signatures per parameter set were computed). The finite

Table 3. Average number of algebraic decoding and running time per signature

	(m,t,w,λ)			
	(18,9,11,3)	(18,9,11,4)	(20,8,10,3)	(20,8,9,5)
number of decodings	1 117 008	1 489 344	121 262	360 216
running time (BM)	14.70 s	19.61 s	1.32 s	3.75 s
running time (Pat.)	15.26 s	20.34 s	1.55 s	4.26 s

field arithmetic primitives use 75% of the total amount of CPU time, most of that (66%) for the sole multiplication. We observe in Table 3 that the number of

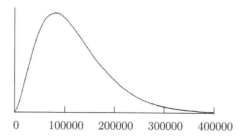

Fig. 1. Distribution of the number of decodings per signature $(m,t,w,\lambda) = (20,8,10,3)$

decodings is very close (but slightly above) the expected value of $\lambda t!$. The only exception is for $(m, t, w, \lambda) = (20, 8, 9, 5)$. In that case each complete decoding has a probability of 5.5% of failure, but when it fails, the number of decoding attempts is equal to the maximum allowed. Experimentally the best tradeoff when $\lambda = 5$ is to allow only 200 000 decoding per binary syndrome (instead of 2^{20}), this raises the probability of failure to 8.4% and with a counter of 6 bits (see §5.4) we fail to sign with probability 2^{-95}. In practice we observe that the signing cost almost doubles.

6.2 Comparing Decoders

We provide here the number of elementary field operations needed for one algebraic decoding attempt. Those numbers were obtained by running the software. Statistics are summarized in Table 4. The "critical" steps are those called inside the loop of Algorithm 3. The "non critical" ones are outside the loop and thus are called only once per complete decoding, that is $\lambda = 3$ times per signature. A field operation is a multiplication, a squaring, a division, an inversion, or a square root. We do not count additions which are implemented with a XOR. In practice, this gives an accurate measure of the complexity and allows and easy comparison of the decoders and an indication about the relative costs of the various steps. All numbers are constant for all steps except the root finding algorithm (Berlekamp trace algorithm). If we consider the non critical parts, it appears that the syn-

Table 4. Number of field operations (excluding additions) per decoding

(m,t)	type	(1)	(2)	(3)	(1)+(2)+(3)	(4)	(5)
(18,9)	BM	58	180	840	1078	2184	3079.1
(18,9)	Pat.	38	329	840	1207	1482	3079.1
(20,8)	BM	52	144	747	943	1950	3024.6
(20,8)	Pat.	34	258	747	1039	1326	3024.6

(1) syndrome adjustment   (4) initial syndrome
(2) key equation solving   (5) root finding
(3) split checking

drome computation and the root finding algorithms are the dominant cost and thus the Patterson algorithm is more efficient than the Berlekamp-Massey algorithm which requires a double sized syndrome. The situation is reversed when we consider only the critical parts because the Berlekamp-Massey key equation solving is more efficient.

7 Conclusion

For a proper choice of parameters we have shown that CFS, in fact Parallel-CFS, is practical, though cumbersome to achieve a reasonable security. The fastest of our instances needs a bit more than one second of CPU time to produce a signature, which is slow but acceptable. The corresponding public key has a size of 20 megabytes, which may disqualify the scheme for some applications. Note that the public key is not needed for signing but only the secret key which consists of a pair (L, g). The generator g has a size of mt bits (160 or 162 bits here) and the support is a permutation of 2^m elements which can be generated on the fly from a seed. The implementation of the signing primitive we describe requires only a relatively small amount of storage[3] and memory, making it suitable for massively parallel architecture (like GPUs) or "hardware-oriented" devices (like FPGAs or even smart cards).

References

1. Courtois, N.T., Finiasz, M., Sendrier, N.: How to Achieve a McEliece-Based Digital Signature Scheme. In: Boyd, C. (ed.) ASIACRYPT 2001. LNCS, vol. 2248, pp. 157–174. Springer, Heidelberg (2001)
2. McEliece, R.: A public-key cryptosystem based on algebraic coding theory. DSN Prog. Rep., Jet Prop. Lab., California Inst. Technol., Pasadena, CA, 114–116 (January 1978)
3. Beuchat, J.L., Sendrier, N., Tisserand, A., Villard, G.: FPGA implementation of a recently published signature scheme. Rapport de recherche 5158, INRIA (2004)
4. Faugère, J.C., Gauthier, V., Otmani, A., Perret, L., Tillich, J.P.: A distinguisher for high rate McEliece cryptosystems. In: ITW 2011, Paraty, Brazil, pp. 282–286 (October 2011)
5. Finiasz, M.: Parallel-CFS: Strengthening the CFS McEliece-Based Signature Scheme. In: Biryukov, A., Gong, G., Stinson, D.R. (eds.) SAC 2010. LNCS, vol. 6544, pp. 159–170. Springer, Heidelberg (2011)
6. Berlekamp, E., McEliece, R., van Tilborg, H.: On the inherent intractability of certain coding problems. IEEE Transactions on Information Theory 24(3) (May 1978)
7. Prange, E.: The use of information sets in decoding cyclic codes. IRE Transactions IT-8, S5–S9 (1962)
8. Stern, J.: A Method for Finding Codewords of Small Weight. In: Cohen, G., Wolfmann, J. (eds.) Coding Theory 1988. LNCS, vol. 388, pp. 106–113. Springer, Heidelberg (1989)
9. Bernstein, D.J., Lange, T., Peters, C.: Smaller Decoding Exponents: Ball-Collision Decoding. In: Rogaway, P. (ed.) CRYPTO 2011. LNCS, vol. 6841, pp. 743–760. Springer, Heidelberg (2011)
10. May, A., Meurer, A., Thomae, E.: Decoding Random Linear Codes in $\tilde{\mathcal{O}}(2^{0.054n})$. In: Lee, D., Wang, X. (eds.) ASIACRYPT 2011. LNCS, vol. 7073, pp. 107–124. Springer, Heidelberg (2011)
11. Wagner, D.: A Generalized Birthday Problem. In: Yung, M. (ed.) CRYPTO 2002. LNCS, vol. 2442, pp. 288–303. Springer, Heidelberg (2002)
12. Camion, P., Patarin, J.: The Knapsack Hash Function Proposed at Crypto 1989 Can Be Broken. In: Davies, D.W. (ed.) EUROCRYPT 1991. LNCS, vol. 547, pp. 39–53. Springer, Heidelberg (1991)

---

[3] Mostly for the finite field for which there are other options.

13. Coron, J.S., Joux, A.: Cryptanalysis of a provably secure cryptographic hash function. Cryptology ePrint Archive, Report 2004/013 (2004),
    http://eprint.iacr.org/
14. Overbeck, R., Sendrier, N.: Code-based cryptography. In: Bernstein, D., Buchmann, J., Dahmen, E. (eds.) Post-Quantum Cryptography, pp. 95–145. Springer (2009)
15. Johansson, T., Jönsson, F.: On the complexity of some cryptographic problems based on the general decoding problem. IEEE-IT 48(10), 2669–2678 (2002)
16. Sendrier, N.: Decoding One Out of Many. In: Yang, B.-Y. (ed.) PQCrypto 2011. LNCS, vol. 7071, pp. 51–67. Springer, Heidelberg (2011)
17. Dumer, I.: On minimum distance decoding of linear codes. In: Proc. 5th Joint Soviet-Swedish Int. Workshop Inform. Theory, Moscow, pp. 50–52 (1991)
18. Finiasz, M., Sendrier, N.: Security Bounds for the Design of Code-Based Cryptosystems. In: Matsui, M. (ed.) ASIACRYPT 2009. LNCS, vol. 5912, pp. 88–105. Springer, Heidelberg (2009)
19. Becker, A., Joux, A., May, A., Meurer, A.: Decoding Random Binary Linear Codes in $2^{n/20}$: How $1 + 1 = 0$ Improves Information Set Decoding. In: Pointcheval, D., Johansson, T. (eds.) EUROCRYPT 2012. LNCS, vol. 7237, pp. 520–536. Springer, Heidelberg (2012)
20. Patterson, N.: The algebraic decoding of Goppa codes. IEEE Transactions on Information Theory 21(2), 203–207 (1975)
21. Massey, J.: Shift-register synthesis and BCH decoding. IEEE Transactions on Information Theory 15(1), 122–127 (1969)
22. Berlekamp, E.: Algebraic Coding Theory. Aegen Park Press (1968)

A   A Non Asymptotic Analysis of ISD-MMT

We refer to the algorithm described in [10] to solve $CSD(H, s, w)$ with $H \in \{0,1\}^{r \times n}$ and $s \in \{0,1\}^r$ (we denote $k = n - r$ the dimension), it uses as parameters three integers p, ℓ_2, and ℓ. First, a Gaussian elimination on H is performed and then three levels of lists are built and successively merged. With a certain probability $\mathcal{P}(p, \ell, \ell_2)$ a solution to $CSD(H, s, w)$ lies in the last of those lists. The whole process has to be repeated $1/\mathcal{P}(p, \ell, \ell_2)$ times on average to find a solution.

- There are 4 lists at the first level, each of size $L_0 = \binom{(k+\ell)/2}{p/4}$.
- There are 2 lists at the second level, obtained by merging the first level lists pairwise, both have size $L_1 = L_0^2 2^{-\ell_2}$ on average.
- The final list at third level is obtained by merging the two second level lists and has size $L_2 = L_1^2 2^{-\ell+\ell_2} = L_0^4 2^{-\ell-\ell_2}$ on average.

To give an expression of the success probability, we cannot use [10] which assumes a unique solution while for signature parameters we may have several[4]. Instead

---

[4] If the weight w is not above the Gilbert-Varshamov radius by more than a constant, the expected number of solutions is polynomial and do not affect the asymptotic analysis, for a non-asymptotic analysis the difference is significant.

we claim (following the analysis of [18,16]) that any particular element of the final list will provide a solution with probability $\approx 2^\ell \varepsilon(p, \ell)$ where

$$\varepsilon(p, \ell) = \frac{\binom{r-\ell}{w-p}}{\min\left(2^r, \binom{n}{w}\right)}.$$

The min in the above expression takes into account the possibility of several solutions. In practice for the signature 2^r will be smaller than $\binom{n}{w}$. We claim that, for practical code parameters and when p and ℓ are near their optimal values

1. if ℓ_2 is not too small the proportion of duplicates in the final list is negligible,
2. if ℓ_2 is not too large the costs for building the first level lists and for the Gaussian eliminations are negligible.

Assuming the first claim is true, the probability of success is

$$\mathcal{P}(p, \ell, \ell_2) = 1 - \left(1 - 2^\ell \varepsilon(p, \ell)\right)^{L_2} \approx L_2 2^\ell \varepsilon(p, \ell) = L_0^4 2^{-\ell_2} \varepsilon(p, \ell).$$

Assuming the second claim is true, the cost for building the final list and checking whether it contains a solution is (crudely) lower bounded by $2L_1 + 2L_2$ elementary operations[5]. The factor 2 in front of L_1 is because there are two lists at the second level and the factor 2 in front of L_2 is because each element of the final list has to be constructed (an addition at least) then checked (a Hamming weight computation). Finally assuming each elementary operation costs at least ℓ binary operations the cost of ISD-MMT is lower bounded by

$$\mathrm{WF}_{\mathrm{MMT}} = \min_{p,\ell} \frac{2\ell}{\varepsilon(p, \ell)} \left(\frac{1}{2^\ell} + \frac{1}{L_0^2}\right). \tag{6}$$

Note that, interestingly, ℓ_2 does not appear in the above expression. It means that, as long as it is neither too small or too large, the choice of ℓ_2 has no impact on the complexity of ISD-MMT. In practice the proper ranges is (roughly) $p/2 \leq \ell_2 \leq \log_2 L_0$. It is best to avoid the extreme values in that range and large values are better because they reduce memory requirements. Finally note that in [10] it is suggested that $\ell_2 \leq p - 2$. This is marginally inside the acceptable range but this has no consequence on the asymptotic exponent analysis.

---

[5] Here an elementary operation is an operation on a column of H, either addition or Hamming weight, possibly with a memory store or read.

SipHash: A Fast Short-Input PRF

Jean-Philippe Aumasson[1] and Daniel J. Bernstein[2]

[1] NAGRA
Switzerland
`jeanphilippe.aumasson@gmail.com`
[2] Department of Computer Science
University of Illinois at Chicago, Chicago, IL 60607–7045, USA
`djb@cr.yp.to`

Abstract. SipHash is a family of pseudorandom functions optimized for short inputs. Target applications include network traffic authentication and hash-table lookups protected against hash-flooding denial-of-service attacks. SipHash is simpler than MACs based on universal hashing, and faster on short inputs. Compared to dedicated designs for hash-table lookup, SipHash has well-defined security goals and competitive performance. For example, SipHash processes a 16-byte input with a fresh key in 140 cycles on an AMD FX-8150 processor, which is much faster than state-of-the-art MACs. We propose that hash tables switch to SipHash as a hash function.

1 Introduction

A message-authentication code (MAC) produces a tag t from a message m and a secret key k. The security goal for a MAC is for an attacker, even after seeing tags for many messages (perhaps selected by the attacker), to be unable to guess tags for any other messages.

Internet traffic is split into short packets that require authentication. A 2000 note by Black, Halevi, Krawczyk, Krovetz, and Rogaway [11] reports that "a fair rule-of-thumb for the distribution on message-sizes on an Internet backbone is that roughly one-third of messages are 43 bytes (TCP ACKs), one-third are about 256 bytes (common PPP dialup MTU), and one-third are 1500 bytes (common Ethernet MTU)."

However, essentially all standardized MACs and state-of-the-art MACs are optimized for long messages, not for short messages. Measuring long-message performance hides the overheads caused by large MAC keys, MAC initialization, large MAC block sizes, and MAC finalization. These overheads are usually quite severe, as illustrated by the examples in the following paragraphs. Applications can compensate for these overheads by authenticating a concatenation of several packets instead of authenticating each packet separately, but then a single forged

This work was supported by the National Science Foundation under grant 1018836. Permanent ID of this document: `b9a943a805fbfc6fde808af9fc0ecdfa`. Date: 2012.09.17.

packet forces several packets to be retransmitted, increasing the damage caused by denial-of-service attacks.

Our first example is HMAC-SHA-1, where overhead effectively adds between 73 and 136 bytes to the length of a message: for example, HMAC-SHA-1 requires two 64-byte compression-function computations to authenticate a short message. Even for long messages, HMAC-SHA-1 is not particularly fast: for example, the OpenSSL implementation takes 7.8 cycles per byte on Sandy Bridge, and 11.2 cycles per byte on Bulldozer. In general, building a MAC from a general-purpose cryptographic hash function appears to be a highly suboptimal approach: general-purpose cryptographic hash functions perform many extra computations for the goal of collision resistance on public inputs, while MACs have secret keys and do not need collision resistance.

Much more efficient MACs combine a large-input universal hash function with a short-input encryption function. A universal hash function h maps a long message m to a short hash $h(k_1, m)$ under a key k_1. "Universal" means that any two different messages almost never produce the same output when k_1 is chosen randomly; a typical universal hash function exploits fast 64-bit multipliers to evaluate a polynomial over a prime field. This short hash is then strongly encrypted under a second key k_2 to produce the authentication tag t. The original Wegman–Carter MACs [34] used a one-time pad for encryption, but of course this requires a very long key. Modern proposals such as UMAC version 2 [11], Poly1305-AES [5], and VMAC(AES) [25] [14] replace the one-time pad with outputs of AES-128: i.e., $t = h(k_1, m) \oplus \text{AES}(k_2, n)$ where n is a nonce. UMAC version 1 argued that "using universal hashing to reduce a very long message to a fixed-length one can be complex, require long keys, or reduce the quantitative security" [10, Section 1.2] and instead defined $t = \text{HMAC-SHA-1}(h(k, m), n)$ where $h(k, m)$ is somewhat shorter than m.

All of these MACs are optimized for long-message performance, and suffer severe overheads for short messages. For example, the short-message performance of UMAC version 1 is obviously even worse than the short-message performance of HMAC-SHA-1. All versions of UMAC and VMAC expand k_1 into a very long key (for example, 4160 bytes in one proposal), and are timed under the questionable assumptions that the very long key has been precomputed and preloaded into L1 cache. Poly1305-AES does not expand its key but still requires padding and finalization in h, plus the overhead of an AES call.

(We comment that, even for applications that emphasize long-message performance, the structure of these MACs often significantly complicates deployment. Typical universal MACs have lengthy specifications, are not easy to implement efficiently, and are not self-contained: they rely on extra primitives such as AES. Short nonces typically consume 8 bytes of data with each tag, and force applications to be stateful to ensure uniqueness; longer nonces consume even more space and require either state or random-number generation. There have been proposals of nonceless universal MACs, but those proposals are significantly slower than other universal MACs at the same security level; see, e.g., [4, Theorem 9.2].)

The short-input performance problems of high-security MACs are even more clear in another context. As motivation we point to the recent rediscovery of "hash flooding" denial-of-service attacks on Internet servers that store data in hash tables. These servers normally use public non-cryptographic hash functions, and these attacks exploit multicollisions in the hash functions to enforce worst-case lookup time. See Section 7 of this paper for further discussion.

Replacing the public non-cryptographic hash functions with strong small-output secret-key MACs would solve this problem. However, to compete with existing non-cryptographic hash functions, the MACs must be extremely fast for very short inputs, even shorter than the shortest common Internet packets. For example, Ruby on Rails applications are reported to hash strings shorter than 10 bytes on average. Recent hash-table proposals such as Google's CityHash [18] and Jenkins' SpookyHash [21] provide very fast hashing of short strings, but these functions were designed to have a close-to-uniform distribution, not to meet any particular cryptographic goals. For example, collisions were found in an initial version of CityHash128 [22], and the current version is vulnerable to a practical key-recovery attack when 64-bit keys are used.

This paper introduces the SipHash family of hash functions to address the needs for high-security short-input MACs. SipHash features include:

- **High security.** Our concrete proposal SipHash-2-4 was designed and evaluated to be a cryptographically strong PRF (pseudorandom function), i.e., indistinguishable from a uniform random function. This implies its strength as a MAC.
- **High speed.** SipHash-2-4 is much faster for short inputs than previous strong MACs (and PRFs), and is competitive in speed with popular non-cryptographic hash functions.
- **Key agility.** SipHash uses a 128-bit key. There is no key expansion in setting up a new key or hashing a message, and there is no hidden cost of loading precomputed expanded keys from DRAM into L1 cache.
- **Simplicity.** SipHash iterates a simple round function consisting of four additions, four xors, and six rotations, interleaved with xors of message blocks.
- **Autonomy.** No external primitive is required.
- **Small state.** The SipHash state consists of four 64-bit variables. This small state size allows SipHash to perform well on a wide range of CPUs and to fit into small hardware.
- **No state between messages.** Hashing is deterministic and doesn't use nonces.
- **No software side channels.** Many cryptographic functions, notably AES, encourage implementors to use secret load/store addresses or secret branch conditions, often allowing timing attacks. SipHash avoids this problem.
- **Minimal overhead.** Authenticated messages are just 8 bytes longer than original messages.

§2 presents a complete definition of SipHash; §3 makes security claims; §4 explains some design choices; §5 reports on our preliminary security analysis; §6 evaluates the efficiency of SipHash in software and hardware; §7 discusses the benefits of switching to SipHash for hash-table lookups.

2   Specification of SipHash

SipHash is a family of PRFs SipHash-c-d where the integer parameters c and d are the number of compression rounds and the number of finalization rounds. A compression round is identical to a finalization round and this round function is called SipRound. Given a 128-bit key k and a (possibly empty) byte string m, SipHash-c-d returns a 64-bit value SipHash-c-$d(k, m)$ computed as follows:

1. **Initialization:** Four 64-bit words of internal state v_0, v_1, v_2, v_3 are initialized as

$$v_0 = k_0 \oplus \texttt{736f6d6570736575}$$
$$v_1 = k_1 \oplus \texttt{646f72616e646f6d}$$
$$v_2 = k_0 \oplus \texttt{6c7967656e657261}$$
$$v_3 = k_1 \oplus \texttt{7465646279746573}$$

   where k_0 and k_1 are the little-endian 64-bit words encoding the key k.

2. **Compression:** SipHash-c-d processes the b-byte string m by parsing it as $w = \lceil (b+1)/8 \rceil > 0$ 64-bit little-endian words $m_0, \ldots, m_{w-1}$ where m_{w-1} includes the last 0 through 7 bytes of m followed by null bytes and ending with a byte encoding the positive integer $b \bmod 256$. For example, the one-byte input string $m = \texttt{ab}$ is parsed as $m_0 = \texttt{01000000000000ab}$. The m_i's are iteratively processed by doing

$$v_3 \oplus = m_i$$

   and then c iterations of SipRound, followed by

$$v_0 \oplus = m_i$$

3. **Finalization:** After all the message words have been processed, SipHash-c-d xors the constant $\texttt{ff}$ to the state:

$$v_2 \oplus = \texttt{ff}$$

   then does d iterations of SipRound, and returns the 64-bit value

$$v_0 \oplus v_1 \oplus v_2 \oplus v_3 \ .$$

Fig. 2.1 shows SipHash-2-4 hashing a 15-byte m.

   The function SipRound transforms the internal state as follows (see also Fig.2.2):

$$
\begin{aligned}
v_0 &+= v_1   &\qquad v_2 &+= v_3 \\
v_1 &\lll= 13   &\qquad v_3 &\lll= 16 \\
v_1 &\oplus = v_0 &\qquad v_3 &\oplus = v_2 \\
v_0 &\lll= 32   & & \\
v_2 &+= v_1   &\qquad v_0 &+= v_3 \\
v_1 &\lll= 17   &\qquad v_3 &\lll= 21 \\
v_1 &\oplus = v_2 &\qquad v_3 &\oplus = v_0 \\
v_2 &\lll= 32   & &
\end{aligned}
$$

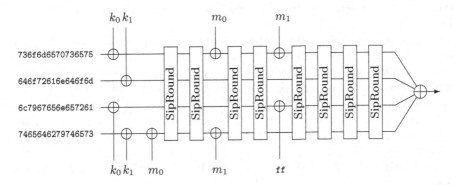

Fig. 2.1. SipHash-2-4 processing a 15-byte message. SipHash-2-4(k, m) is the output from the final ⊕ on the right.

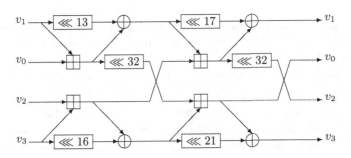

Fig. 2.2. The ARX network of SipRound

3   Expected Strength

SipHash-c-d with $c \geq 2$ and $d \geq 4$ is expected to provide the maximum PRF security possible (and therefore also the maximum MAC security possible) for any function with the same key size and output size. Our fast proposal is thus SipHash-2-4. We define SipHash-c-d for larger c and d to provide a higher security margin: our conservative proposal is SipHash-4-8, which is about half the speed of SipHash-2-4. We define SipHash-c-d for smaller c and d to provide targets for cryptanalysis. Cryptanalysts are thus invited to break

- SipHash-1-0, SipHash-2-0, SipHash-3-0, SipHash-4-0, etc.;
- SipHash-1-1, SipHash-2-1, SipHash-3-1, SipHash-4-1, etc.;
- SipHash-1-2, SipHash-2-2, SipHash-3-2, SipHash-4-2, etc.;

and so on.

Note that the standard PRF and MAC security goals allow the attacker access to the output of SipHash on messages chosen adaptively by the attacker. However, they do not allow access to any "leaked" information such as bits of the key or the internal state. They also do not allow "related keys", "known keys", "chosen keys", etc.

Of course, security is limited by the key size (128 bits). In particular, attackers searching 2^s keys have chance 2^{s-128} of finding the SipHash key. This search is accelerated in standard ways by speedups in evaluation and partial evaluation of SipHash, for one key or for a batch of keys; by attacks against multiple targets; and by quantum computers.

Security is also limited by the output size (64 bits). In particular, when SipHash is used as a MAC, an attacker who blindly tries 2^s tags will succeed with probability 2^{s-64}.

We comment that SipHash is not meant to be, and (obviously) is not, collision-resistant.

4   Rationale

SipHash is an ARX algorithm, like the SHA-3 finalists BLAKE [3] and Skein [16]. SipHash follows BLAKE's minimalism (small code, small state) but borrows the two-input MIX from Skein, with two extra rotations to improve diffusion. SipHash's input injection is inspired by another SHA-3 finalist, JH [36].

Choice of Constants. The initial state constant corresponds to the ASCII string "somepseudorandomlygeneratedbytes", big-endian encoded. There is nothing special about this value; the only requirement was some asymmetry so that the initial v_0 and v_1 differ from v_2 and v_3. This constant may be set to a "personalization string" but we have not evaluated whether it can safely be chosen as a "tweak". Note that two nonzero words of initialization constants would have been as safe as four.

The other constant in SipHash is ff, as xored to v_2 in finalization. We could have chosen any other non-zero value. Without this constant, one can reach the internal state after finalization by just absorbing null words. We found no way to exploit this property, but we felt it prudent to avoid it given the low cost of the defense.

Choice of Rotation Counts. Finding really bad rotation counts for ARX algorithms turns out to be difficult. For example, randomly setting all rotations in BLAKE-512 or Skein to a value in $\{8, 16, 24, \ldots, 56\}$ may allow known attacks to reach slightly more rounds, but no dramatic improvement is expected.

The advantage of choosing such "aligned" rotation counts is that aligned rotation counts are much faster than unaligned rotation counts on many non-64-bit architectures. Many 8-bit microcontrollers have only 1-bit shifts of bytes, so rotation by (e.g.) 3 bits is particularly expensive; implementing a rotation by a mere permutation of bytes greatly speeds up ARX algorithms. Even 64-bit systems can benefit from alignment, when a sequence of shift-shift-xor can be replaced by SSSE3's pshufb byte-shuffling instruction. For comparison, implementing BLAKE-256's 16- and 8-bit rotations with pshufb led to a 20% speedup on Intel's Nehalem microarchitecture.

For SipHash, the rotation distances were chosen as a tradeoff between security and performance, with emphasis on the latter. We ran an automated search that picks random rotation counts, estimates the number of significant statistical biases on three SipRounds with respect to a specific significance threshold, and finally sorts the sets of rotation counts according to that metric. We then manually shortlisted a few sets, by choosing the ones with rotation counts the closest to multiples of eight. We changed some of those values to the closest multiple of eight and benchmarked them against our original security metric, and repeated this process several times until finding a satisfying set of rotation counts.

We chose counts 13, 16, 17, and 21 for the rotations in the two MIX layers: 13 and 21 are three bits away from a multiple of 8, whereas 17 is just one bit away, and 16 can be realized by byte permutation only. We aggressively set the two "asymmetric" rotation counts to 32 to minimize the performance penalty—it is just a swap of words on 32-bit systems. The 32-bit rotations significantly improve diffusion, and their position on the ARX network allows for an efficient scheduling of instructions.

Choice of Injection Structure. Like JH, SipHash injects input before and after each block, with the difference that SipHash leaves less freedom to attackers: whereas JH xors the message block to the two halves of the state before and after the permutation, SipHash xors a block to two quarters of the state. Any attack on the SipHash injection structure can be applied to the JH injection structure, so security proofs for the JH injection structure [30] also apply to the SipHash injection structure.

A basic advantage of the JH/SipHash injection structure compared to the sponge/Keccak [7] injection structure is that message blocks of arbitrary length (up to half the state) can be absorbed without reducing preimage security. A disadvantage is that each message block must be retained while the state is being processed, but for SipHash this extra storage is only a quarter of the state.

Choice of Padding Rule. SipHash's padding appends a byte encoding the message length modulo 256. We could have chosen a slightly simpler padding rule, such as appending a 80 byte followed by zeroes. However, our choice forces messages of different lengths modulo 256 to have different last blocks, which may complicate attacks on SipHash; the extra cost is negligible.

5   Preliminary Cryptanalysis

We first consider attacks that are independent of the SipRound algorithm, and thus that are independent of the c and d parameters. We then consider attacks on SipRound iterations, with a focus on our proposal SipHash-2-4.

Key-Recovery. Brute force will recover a key after on average 2^{127} evaluations of SipHash, given two input/output pairs (one being insufficient to uniquely identify the key). The optimal strategy is to work with 1-word padded messages, so that evaluating SipHash-c-d takes $c + d$ SipRounds.

State-Recovery. A simple strategy to attack SipHash is to choose three input strings identical except for their last word, query for their respective SipHash outputs, and then "guess" the state that produced the output $v_0 \oplus v_1 \oplus v_2 \oplus v_3$ for one of the two strings. The attacker checks the 192-bit guessed value against the two other strings, and eventually recovers the key. On average $d2^{191}$ evaluations of SipRound are computed.

Internal Collisions. As for any MAC with 256-bit internal state, internal collisions can be exploited to forge valid tags with complexity of the order of 2^{128} queries to SipHash. The padding of the message length forces attackers to search for collisions at the same position modulo 256 bytes.

Truncated Differentials. To assess the strength of SipRound, we applied the same techniques that were used [2] to attack Salsa20, namely a search for statistical biases in one or more bits of output given one or more differences in the input. We considered input differences in v_3 and sought biases in $v_0 \oplus v_1 \oplus v_2 \oplus v_3$ after iterating SipRound.

The best results were obtained by setting a 1-bit difference in the most significant bit of v_3. After three iterations of SipRound many biases are found. But after four or more iterations we did not detect any bias after experimenting with sets of 2^{30} samples.

To attempt to distinguish our fast proposal SipHash-2-4 by exploiting such statistical biases, one needs to find a bias on six rounds such that no input difference lies in the most significant byte of the last word (as this encodes the message length).

XOR-Linearized Characteristics. We considered an attacker who injects a difference in the first message word processed by SipHash-2-4, and then that guesses the difference in v_3 every two SipRounds in order to cancel it with the new message word processed. This ensures that at least a quarter of the internal state is free of difference when entering a new absorption phase. Note that such an omniscient attacker would require the leakage of v_3 every two SipRounds, and thus is not covered by our security claims in §3.

We used Leurent's ARX toolkit [27] to verify that our characteristics contain no obvious contradiction, and to obtain refined probability estimates. Table 5.1 shows the best characteristic we found: after two rounds there are 20 bit differences in the internal state, with differences in all four words. The message injection reduces this to 15 bit differences (with no difference in v_3), and after two more rounds there are 96 bit differences. The probability to follow this differential characteristic is estimated to be 2^{-134}. For comparison, Table 5.2 shows the characteristic obtained with the same input difference, but for an attacker who does not guess the difference in v_3: the probability to follow four rounds of the characteristic is estimated to be 2^{-159}.

Better characteristics may exist. However we expect that finding (collections of) characteristics that both have a high probability and are useful to attack SipHash is extremely difficult. SipRound has as many additions as xors, so

Table 5.1. For each SipRound, differences in v_0, v_1, v_2, v_3 before each half-round in the xor-linear model. Every two rounds a message word is injected that cancels the difference in v_3; the difference used is then xored to v_0 after the two subsequent rounds. The probability estimate is given for each round, with the cumulative value in parentheses.

Round	Differences	Prob.
1	`...............  ...............  ...............  8...............` `...............  ...............  8...............  8...........8...`	1 (1)
2	`8...........8... 8............... ........1.8....... 8.....1...1.8...` `....8........... ...........9... 8.....1.8.1.8... 8.1........1.....`	13 (14)
3	`..1.8.....1..... 8....11a.1.1... 8.1.1...8.....1. ...............` `a...1...8.1.8.11 8.12b413a2...... 8.1.1...8.....1. 8.1.1...8.....1.`	33 (47)
4	`2.1.......1.8..1 6825e.1322.1..35 22....1....2a413 2........2..82.3` `22118.344835e.13 f4378453.2172d3. .2....1..2.2261. .2...21.8..1.61.`	87 (134)
5	`a..1..24c834e4.3 fe918.6d5a74e34f ..15.b2.f6378443 ...............` `924..74c5e9.8.49 6e9d2b.7.e29f89e ..15.b2.f6378443 ..15.b2.f6378443`	145 (279)
6	`9255.c6ca8a7.4.a 38863c74.922a1e7 f81e7cdd6e882.27 f64bca9c2.c7.6ab` `a185a5edaad33.18 6d5db13cf5b942fd .e55b6414e4f268c c4c9968648e4d.c7`	160 (439)

linearization with respect to integer addition seems unlikely to give much better characteristics than xor-linearization.

Vanishing Characteristics. A particularly useful class of differential characteristics is that of vanishing characteristics: those start from a non-zero difference and yield an internal state with no difference, that is, an internal collision. Vanishing characteristics obviously do not exist for any iteration of SipRound; one has to consider characteristics for the function consisting of SipRound iterations followed by $v_0 \oplus = \Delta$, with an input difference Δ in v_3.

No vanishing characteristic exists for one SipRound, as a non-zero difference always propagates to v_2. We ensured that no vanishing xor-linear characteristic exists for iterations of two, three, or four SipRounds, by attempting to solve the corresponding linear system. For sequences of two words, we ensured that no sparse vanishing characteristic exists.

Other Attacks. We briefly examine the applicability of other attacks to attack SipHash:

- Rotational attacks are differential attacks with respect to the rotation operator; see, e.g., [6, Section 4] and [23]. Due to the asymmetry in the initial state—at most half of the initial state can be rotation-invariant—rotational attacks are ineffective against SipHash.
- Cube attacks [26] exploit a low algebraic degree in the primitive attacked. Due to the rapid growth of the degree in SipHash, as in other ARX primitives, cube attacks are unlikely to succeed.
- Rebound attacks [28] are not known to be relevant for keyed primitives.

Table 5.2. For each SipRound, differences in v_0, v_1, v_2, v_3 before each half-round in the xor-linear model. Every two rounds a message with no difference is injected. The probability estimate is given for each half-round, with the cumulative value in parentheses.

Round	Differences	Prob.
1	`................ ................ ................ 8...............` `................ ................ 8............... 8...........8...`	1 (1)
2	`8...........8... 8............... ........8....... 8.....1...1.8...` `....8........... ............9... 8.....1.8.1.8... 8.1.......1.....`	13 (14)
3	`..1.8.....1..... 8.....11a.1.1... 8.1.1...8.....1. 8.1.82.......2..` `a...1...8.1.8.11 8.12b413a2...... ....92..8....21. 82..92..82..82..`	42 (56)
4	`22..82...21..211 e835621322.1.235 22...21.8.122613 621.c21.42..42.3` `2.11..24ca35e.13 66778453..57bd22 4.1.c...c212641. 82..82..8.11.6..`	103 (159)
5	`a21182244a24e613 2ec144fcb8.115dd c245d93226674453 e2.18..48a34a6.3` `f225f3ce8cd.c6d8 a44f51d8d.9e5616 2.445936ac53e25. a.4.d3.2.a5...51`	152 (311)
6	`52652.cc868.c689 27baa9d2d.e.fcd8 7ccdb44684.b.8ee 32246acc8cb4ce93` `566.3a5175df891e 2.e5d3.249fb3ea6 4ee9de8a.8bfc67d 2425523ec62cf459`	187 (498)

Fixed Point. Any iteration of SipRound admits a trivial distinguisher: the zero-to-zero fixed-point. This may make theoretical arguments based on the "ideal permutation" assumption irrelevant. But exploiting this property to attack SipHash seems very hard, for

1. Hitting the all-zero state, although easy to verify, is expected to be as hard as hitting any other predefined state;
2. The ability to hit a predefined state implies the ability to recover the key, that is, to completely break SipHash.

That is, the zero-to-zero fixed point cannot be a significant problem for SipHash, for if it were, SipHash would have much bigger problems.

6   Performance

Lower Bounds for a 64-Bit Implementation. SipRound involves 14 64-bit operations, so SipHash-2-4 involves 30 64-bit operations for each 8 bytes of input, i.e., 3.75 operations per byte. A CPU core with 2 64-bit arithmetic units needs at least 1.875 cycles per byte for SipHash-2-4, and a CPU core with 3 64-bit arithmetic units needs at least 1.25 cycles per byte for SipHash-2-4. A CPU core with 4 64-bit arithmetic units needs at least 1 cycle per byte, since SipRound does not always have 4 operations to perform in parallel.

The cost of finalization cannot be ignored for short messages. For example, for an input of length between 16 and 23 bytes, a CPU core with 3 64-bit arithmetic units needs at least 49 cycles for SipHash-2-4.

Table 6.1. Speed measurements of SipHash-2-4 for short messages

Data byte length		8	16	24	32	40	48	56	64
"bulldozer"	Cycles	124	141	156	171	188	203	218	234
	Cycles per byte	15.50	8.81	6.50	5.34	4.70	4.23	3.89	3.66
"ishmael"	Cycles	123	134	145	158	170	182	192	204
	Cycles per byte	15.38	8.38	6.00	4.94	4.25	3.79	3.43	3.19
"latour"	Cycles	135	144	162	171	189	207	216	225
	Cycles per byte	16.88	10.29	6.75	5.34	4.50	4.31	3.86	3.52

Lower Bounds for a 32-Bit Implementation. 32-bit architectures are common in embedded systems, with for example processors of the ARM11 family implementing the ARMv6 architecture. To estimate SipHash's efficiency on ARM11, we can directly adapt the analysis of Skein's performance by Schwabe, Yang, and Yang [31, §7], which observes that six 32-bit instructions are sufficient to perform a MIX transform. Since SipRound consists of four MIX transforms—the 32-bit rotate is transparent—we obtain 24 instructions per SipRound, that is, a lower bound of $3c$ cycles per byte for SipHash on long messages. This is 6 cycles per byte for SipHash-2-4. An input of length between 16 and 23 bytes needs at least 240 cycles.

Implementation Results. We wrote a portable C implementation of SipHash, and ran preliminary benchmarks on three machines:

- "bulldozer", a Linux desktop equipped with a processor from AMD's last generation (FX-8150, 4 × 3600 MHz, "Zambezi" core), using gcc 4.5.2;
- "ishmael", a Linux laptop equipped with a processor from AMD's previous generation (Athlon II Neo Mobile, 1700 MHz, "Geneva" core), using gcc 4.6.3.
- "latour", a Linux desktop equipped with an older Intel processor (Core 2 Quad Q6600, 2394 MHz, "Kentsfield" core), using gcc 4.4.3.

We used compiler options -O3 -fomit-frame-pointer -funroll-loops.

On "bulldozer", our C implementation of SipHash-2-4 processes long messages at a speed of 1.96 cycles per byte. On "ishmael", SipHash-2-4 reaches 1.44 cycles per byte; this is due to the Athlon II's K10 microarchitecture having three ALUs, against only two for the more recent Bulldozer. Similar comments apply to "latour". These speeds are close to the lower bounds reported in §6, with respective gaps of approximately 0.10 and 0.20 cycles per byte.

Table 6.1 reports speeds on short messages. For comparison, the fastest SHA-3 finalist on "bulldozer" (BLAKE-512) takes approximately 1072 cycles to process 8 bytes, and 1280 cycles to process 64 bytes.

Figure 6.1 compares our implementation of SipHash on "bulldozer" with the optimized C++ and C implementations of CityHash (version CityHash64) and SpookyHash (version ShortHash) on short messages, as well as with OpenSSL's

MD5 implementation. Similar relative performance is observed on the other machines considered.

One can see from these tables that SipHash-2-4 is extremely fast, and competitive with non-cryptographic hashes. For example, hashing 16 bytes takes 141 Bulldozer cycles with SipHash-2-4, against 82 and 126 for CityHash and Spooky-Hash, and 600 for MD5. Our conservative proposal SipHash-4-8 is still twice as fast as MD5.

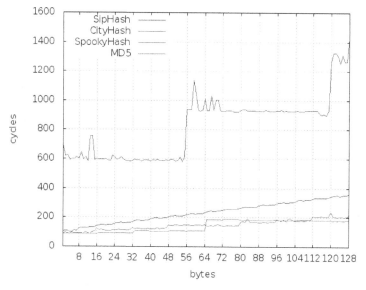

Fig. 6.1. Performance of SipHash-2-4 compared to non-cryptographic hash functions CityHash and SpookyHash and to MD5 on "bulldozer" (AMD FX-8150), for messages of $1, 2, \ldots, 128$ bytes. Curves on the right, from top to bottom, are MD5, SipHash, SpookyHash, and CityHash.

Automated Benchmarks. After the initial publication of SipHash, third-party applications were written in various programming languages, including C, C#, Javascript, Ruby, etc. In particular, Samuel Neves wrote optimized C implementations of SipHash compliant with the `crypto_auth` interface of the SUPERCOP benchmarking software. These implementations (`little`, `mmx`, `sse2-1`, `sse41`) as well as our reference implementation (`ref_le`) were added to SUPERCOP and benchmarked on various machines. A subset of the results are reported in Table 6.2.

Hardware Efficiency. ASICs can integrate SipHash with various degrees of area/throughput tradeoffs, with the following as extreme choices:

- **Compact architecture** with a circuit for a half-SipRound only, that is, two 64-bit full adders, 128 xors, and two rotation selectors. For SipHash-c-d this corresponds a latency of $c/4$ cycles per byte plus $2d$ cycles for the finalization.

Table 6.2. Performance of SipHash-2-4 on processors based on the `amd64` 64-bit architecture, in cycles per byte

Processor	Microarchitecture (core)	Long	64	8
AMD FX-8120	Bulldozer (Zambezi)	1.95	3.75	16.25
AMD E-450	Bobcat (Ontario)	2.03	4.88	22.88
AMD A8-3850	K10 (Llano)	1.44	3.61	26.50
AMD Athlon 64 X2	K8 (Windsor)	1.50	2.91	11.12
Intel Core i3-2310M	Sandy Bridge (206a7)	2.98	6.12	20.50
Intel Atom N435	Bonnell (Pineview)	2.19	4.50	20.00
Intel Xeon E5620	Nehalem (Westmere-EP)	1.63	2.81	11.50
Intel Core 2 Duo E8400	Core (Wolfdale)	1.69	3.38	13.50
VIA Nano U3500	Isaiah	2.38	4.53	17.50

- **High-speed architecture** with a circuit for $e = \max(c, d)$ rounds, that is, $4e$ 64-bit full adders and $256e$ xors. For SipHash-c-d this corresponds to a latency of $1/8$ cycle per byte plus one cycle for finalization.

Both architectures require 256 D-flip-flops to store the internal state, plus 64 for the message blocks. For a technology with 8 gate-equivalents (GE) per full adder, 3 per xor, and 7 per D-flip-flop, this is a total of approximately 3700 GE for the compact architecture of SipHash-2-4, and 13500 GE for the high-speed architecture. With the compact architecture a 20-byte message is hashed by SipHash-2-4 in 20 cycles, against 4 cycles with the high-speed architecture. An architecture implementing $c = 2$ rounds of SipHash-2-4 would take approximately 7900 GE to achieve a latency of $1/8$ cycles per byte plus two cycles for finalization, thus 5 cycles to process 20 bytes.

7   Application: Defense against Hash Flooding

We propose that hash tables switch to SipHash as a hash function. On startup a program reads a secret SipHash key from the operating system's cryptographic random-number generator; the program then uses SipHash for all of its hash tables. This section explains the security benefits of SipHash in this context.

The small state of SipHash also allows each hash table to have its own key with negligible space overhead, if that is more convenient. Any attacks must then be carried out separately for each hash table.

Review of Hash Tables. Storing n strings in a linked list usually takes a total of $\Theta(n^2)$ operations, and retrieving one of the n strings usually takes $\Theta(n)$ operations. This can be a crippling performance problem when n is large.

Hash tables are advertised as providing much better performance. The simplest type of hash table contains ℓ separate linked lists $L[0], L[1], \ldots, L[\ell - 1]$ and stores each string m inside the linked list $L[H(m) \bmod \ell]$, where H is a hash function and ℓ is a power of 2. Each linked list then has, on average, only n/ℓ strings. Normally this improves performance by a factor close to n if ℓ is chosen to be on the same scale as n: storing n strings usually takes only $\Theta(n)$ operations and retrieving a string usually takes $\Theta(1)$ operations.

There are other data structures that guarantee, e.g., $O(n \lg n)$ operations to store n strings and $O(\lg n)$ operations to retrieve one string. These data structures avoid all of the security problems discussed below. However, hash tables are perceived as being simpler and faster, and as a result are used pervasively throughout current programming languages, libraries, and applications.

Review of Hash Flooding. Hash flooding is a denial-of-service attack against hash tables. The attacker provides n strings m that have the same hash value $H(m)$, or at least the same $H(m) \bmod \ell$. The hash-table performance then deteriorates to the performance of one linked list.

The name "hash flooding" for this attack appeared in 1999, in the source code for the first release of the dnscache software from the second author of this paper:

```
if (++loop > 100) return 0; /* to protect against hash flooding */
```

This line of code protects dnscache against the attack by limiting each linked list to 100 entries. However, this is obviously not a general-purpose solution to hash flooding. Caches can afford to throw away unusual types of data, but most applications need to store all incoming data.

Crosby and Wallach reintroduced the same attack in 2003 under the name "algorithmic complexity attack" [13] and explored its applicability to the Squid web cache, the Perl programming language, etc. Hash flooding made headlines again in December 2011, when Klink and Wälde [24] demonstrated its continued applicability to several commonly used web applications. For example, Klink and Wälde reported 500 KB of carefully chosen POST data occupying a PHP5 server for a full minute of CPU time.

Advanced Hash Flooding. Crosby and Wallach recommended replacing public functions H with secret functions, specifically universal hash functions, specifically the hash function $H(m_0, m_1, \ldots) = m_0 \cdot k_0 + m_1 \cdot k_1 + \cdots$ using a secret key $(k_0, k_1, \ldots)$. The idea is that an attacker has no way to guess which strings will collide.

We question the security of this approach. Consider, for example, a hash table containing one string m, where m is known to the attacker. Looking up another string m' will, with standard implementations, take longer if $H(m') \equiv H(m) \pmod{\ell}$ than if $H(m') \not\equiv H(m) \pmod{\ell}$. This timing information will often be visible to an attacker, and can be amplified beyond any level of noise if the application allows the attacker to repeatedly query m'. By guessing ℓ choices of strings $m' \neq m$ the attacker finds one with $H(m') \equiv H(m) \pmod{\ell}$. The linearity of the Crosby–Wallach choice of H then implies that adding any multiple of $m' - m$ to m will produce another colliding string. With twice as many guesses the attacker finds an independent string m'' with $H(m'') \equiv H(m) \pmod{\ell}$; then adding any combination of multiples of $m' - m$ and $m'' - m$

to m will produce even more collisions. With a moderate number of guesses the attacker finds enough information to solve for $(k_0 \bmod \ell, k_1 \bmod \ell, \ldots)$ by Gaussian elimination, and easily computes any number of strings with the same hash value.

One can blame the hash-table implementation for leaking information through timing; but it is not easy to build an efficient constant-time hash table. Even worse, typical languages and libraries allow applications to see all hash-table entries in order of hash value, and applications often expose this information to attackers. One could imagine changing languages and libraries to sort hash-table entries before enumerating them, but this would draw objections from applications that need the beginning of the enumeration to start quickly. One could also imagine changing applications to sort hash-table entries before exposing them to attackers, but ensuring this would require reviewing code in a huge number of applications.

We comment that many of the hash-flooding defenses proposed since December 2011 are vulnerable to the same attack. The most common public hash functions are of the form $m_0 \cdot k_0 + m_1 \cdot k_1 + \cdots$ where $k_0, k_1, \ldots$ are public, and many of the proposed defenses simply add some entropy to $k_0, k_1, \ldots$; but the attack works no matter how $k_0, k_1, \ldots$ are chosen. Many more of the proposed defenses are minor variations of this linear pattern and are broken by easy variants of the same attack.

We do not claim novelty for observing how much damage a single equation $H(m') \equiv H(m) \pmod{\ell}$ does to the unpredictability of this type of hash function; see, e.g., the attacks in [9] and [19] against related MACs. However, the fact that hash tables leak such equations through side channels does not seem to be widely appreciated.

Stopping Advanced Hash Flooding. The worst possible exposure of hash-table indices would simply show the attacker $H(m) \bmod \ell$ for any attacker-selected string m. We advocate protecting against this maximum possible exposure, so that applications do not have to worry about how much exposure they actually provide. The attacker's goal, given this exposure, is to find many strings m having a single value $H(m) \bmod \ell$.

We propose choosing H to be a cryptographically strong PRF. If H is a strong PRF then the truncation $H \bmod \ell$ is also a strong PRF (recall that ℓ is a power of 2), and therefore a strong MAC: even after seeing $H(m) \bmod \ell$ for selected strings m, the attacker cannot predict $H(m) \bmod \ell$ for any other string m. The strength of H as a PRF implies the same unpredictability even if the attacker is given hash values $H(m)$, rather than just hash-table indices $H(m) \bmod \ell$. Achieving this level of unpredictability does not appear to be significantly easier than achieving the full strong-PRF property.

Typical hash-table applications hash a large number of short strings, so the performance of H on short inputs is critical. We therefore propose choosing SipHash as H: we believe that SipHash is a strong PRF, and it provides excellent performance on short inputs. There are previous hash functions with competitive performance, and there are previous functions that have been proposed and evaluated for the same security standards, but as far as we know SipHash is the first function to have both of these features.

Of course, the attacker's inability to predict new hash values does not stop the attacker from exploiting old hash values. No matter how strong H is, the attacker will find two colliding strings after (on average) about $\sqrt{\ell}$ guesses, and then further strings with the same hash value for (on average) ℓ guesses per collision. However, finding n colliding strings in this way requires the attacker to communicate about $n\ell \approx n^2$ strings, so n—the CPU amplification factor of the denial-of-service attack—is limited to the square root of the volume of attacker communication. For comparison, weak secret hash functions and (weak or strong) public hash functions allow n to grow linearly with the volume of attacker communication. A strong secret hash function thus greatly reduces the damage caused by the attack.

The Python Hash Function. Versions 2.7.3 and 3.2.3 of the Python programming language (released in April 2012) introduced an option `-R` with the goal of protecting against hash flooding. According to the Python manual, this option "[turns] on 'hash randomization', so that the hash() values of str, bytes and datetime objects are 'salted' with an unpredictable pseudo-random value. ... This is intended to provide protection against a denial of service caused by carefully-chosen inputs ...". It is therefore expected that outputs of this hash function are unpredictable to parties who were not given the secret key (the salt); obviously, if this key is known, outputs are no longer unpredictable.

We point out that the keyed hashing introduced in Python 2.7.3 and 3.2.3 does not behave as an unpredictable function: the 128-bit key can be recovered efficiently given only two outputs of the keyed hash. The internal state contains only 64 bits, so multicollisions can be found efficiently with a meet-in-the-middle strategy once the key is known.

A proof-of-concept Python script is given in Appendix B. We verified our attack on Python 2.7.3 and 3.2.3, in each case successfully recovering the per-process key.

References

[1] — (no editor): 20th annual symposium on foundations of computer science. IEEE Computer Society, New York (1979). MR 82a:68004. See [33]

[2] Aumasson, J.-P., Fischer, S., Khazaei, S., Meier, W., Rechberger, C.: New features of Latin dances: analysis of Salsa, ChaCha, and Rumba. In: FSE 2008 [**29**], pp. 470–488 (2008), `http://eprint.iacr.org/2007/472`. Citations in this document: §5

[3] Aumasson, J.-P., Henzen, L., Meier, W., Phan, R.C.-W.: SHA-3 proposal BLAKE (version 1.3) (2010), `https://www.131002.net/blake/blake.pdf`. Citations in this document: §4

[4] Bernstein, D.J.: Floating-point arithmetic and message authentication (2004), `http://cr.yp.to/papers.html#hash127`. Citations in this document: §1

[5] Bernstein, D.J.: The Poly1305-AES message-authentication code. In: [**17**], pp. 32–49 (2005), `http://cr.yp.to/papers.html#poly1305`. Citations in this document: §1

[6] Bernstein, D.J.: Salsa20 security. eSTREAM report 2005/025 (2005), `http://cr.yp.to/snuffle/security.pdf`. Citations in this document: §5

[7] Bertoni, G., Daemen, J., Peeters, M., Van Assche, G.: The Keccak reference (version 3.0) (2011), `http://keccak.noekeon.org/Keccak-reference-3.0.pdf`. Citations in this document: §4

[8] Biham, E., Youssef, A.M. (eds.): Selected areas in cryptography, 13th international workshop, SAC 2006, Montreal, Canada, August 17–18, 2006, revised selected papers. LNCS, vol. 4356. Springer (2007). ISBN 978-3-540-74461-0. See [25]

[9] Black, J., Cochran, M.: MAC reforgeability. In: FSE 2009 [15], pp. 345–362 (2009), `http://eprint.iacr.org/2006/095`. Citations in this document: §7

[10] Black, J., Halevi, S., Krawczyk, H., Krovetz, T., Rogaway, P.: UMAC: fast and secure message authentication. In: Crypto '99 [35], pp. 216–233 (1999), `http://fastcrypto.org/umac/umac_proc.pdf`. Citations in this document: §1

[11] Black, J., Halevi, S., Krawczyk, H., Krovetz, T., Rogaway, P.: Update on UMAC fast message authentication (2000), `http://fastcrypto.org/umac/update.pdf`. Citations in this document: §1, §1

[12] Blahut, R.E., Costello Jr., D.J., Maurer, U., Mittelholzer, T. (eds.): Communications and cryptography: two sides of one tapestry. Springer (1994). See [26]

[13] Crosby, S.A., Wallach, D.S.: Denial of service via algorithmic complexity attacks. 12th USENIX Security Symposium (2003), `http://www.cs.rice.edu/~scrosby/hash/CrosbyWallach_UsenixSec2003.pdf`. Citations in this document: §7

[14] Dai, W., Krovetz, T.: VHASH security (2007), `http://eprint.iacr.org/2007/338`. Citations in this document: §1

[15] Dunkelman, O. (ed.): Fast software encryption, 16th international workshop, FSE 2009, Leuven, Belgium, February 22–25, 2009, revised selected papers. LNCS, vol. 5665. Springer (2009). ISBN 978-3-642-03316-2. See [9], [28]

[16] Ferguson, N., Lucks, S., Schneier, B., Whiting, D., Bellare, M., Kohno, T., Callas, J., Walker, J.: The Skein hash function family (version 1.1) (2008), `http://www.skein-hash.info/sites/default/files/skein1.1.pdf`. Citations in this document: §4

[17] Gilbert, H., Handschuh, H. (eds.): Fast software encryption: 12th international workshop, FSE 2005, Paris, France, February 21–23, 2005, revised selected papers. LNCS, vol. 3557. Springer (2005). ISBN 3-540-26541-4. See [5]

[18] Google: The CityHash family of hash functions (2011), `https://code.google.com/p/cityhash/`. Citations in this document: §1

[19] Handschuh, H., Preneel, B.: Key-recovery attacks on universal hash function based MAC algorithms. In: CRYPTO 2008 [32], pp. 144–161 (2008), `http://www.cosic.esat.kuleuven.be/publications/article-1150.pdf`. Citations in this document: §7

[20] Hong, S., Iwata, T.: Fast software encryption, 17th international workshop, FSE 2010, Seoul, Korea, February 7–10, 2010, revised selected papers. LNCS, vol. 6147. Springer (2010). ISBN 978-3-642-13857-7. See [23]

[21] Jenkins, B.: SpookyHash: a 128-bit noncryptographic hash (2010), `http://burtleburtle.net/bob/hash/spooky.html`. Citations in this document: §1

[22] Jenkins, B.: Issue 4: CityHash128 isn't thorough enough (2011), `https://code.google.com/p/cityhash/issues/detail?id=4&can=1`. Citations in this document: §1

[23] Khovratovich, D., Nikolic, I.: Rotational cryptanalysis of ARX. In: FSE 2010 [20], pp. 333–346 (2010), `http://www.skein-hash.info/sites/default/files/axr.pdf`. Citations in this document: §5

[24] Klink, A., Wälde, J.: Efficient denial of service attacks on web application platforms (2011), http://events.ccc.de/congress/2011/Fahrplan/events/4680.en.html. Citations in this document: §7

[25] Krovetz, T.: Message authentication on 64-bit architectures. In: [8], pp. 327–341 (2007), http://eprint.iacr.org/2006/037. Citations in this document: §1

[26] Lai, X.: Higher order derivatives and differential cryptanalysis. In: [12], pp. 227–233 (1994). Citations in this document: §5

[27] Leurent, G.: The ARX toolkit (2012), http://www.di.ens.fr/~leurent/arxtools.html. Citations in this document: §5

[28] Mendel, F., Rechberger, C., Schläffer, M., Thomsen, S.S.: The rebound attack: cryptanalysis of reduced Whirlpool and Grøstl. In: FSE 2009 [15], pp. 260–276 (2009), http://www2.mat.dtu.dk/people/S.Thomsen/MendelRST-fse09.pdf. Citations in this document: §5

[29] Nyberg, K. (ed.): Fast software encryption, 15th international workshop, FSE 2008, Lausanne, Switzerland, February 10–13, 2008, revised selected papers. LNCS, vol. 5086. Springer (2008). ISBN 978-3-540-71038-7. See [2]

[30] Paul, S.: Improved indifferentiability security bound for the JH mode. Third SHA-3 Conference (2012), http://csrc.nist.gov/groups/ST/hash/sha-3/Round3/March2012/documents/papers/PAUL_paper.pdf. Citations in this document: §4

[31] Schwabe, P., Yang, B.-Y., Yang, S.-Y.: SHA-3 on ARM11 processors. In: Proceedings of Africacrypt 2012, to appear (2012), http://cryptojedi.org/papers/sha3arm-20120422.pdf. Citations in this document: §6

[32] Wagner, D. (ed.): Advances in cryptology—CRYPTO 2008, 28th annual international cryptology conference, Santa Barbara, CA, USA, August 17–21, 2008, proceedings. LNCS, vol. 5157. Springer (2008). ISBN 978-3-540-85173-8. See [19]

[33] Wegman, M.N., Lawrence Carter, J.: New classes and applications of hash functions. In: [1], pp. 175–182 (1979); see also newer version [34], http://cr.yp.to/bib/entries.html#1979/wegman

[34] Wegman, M.N., Lawrence Carter, J.: New hash functions and their use in authentication and set equality. Journal of Computer and System Sciences 22, 265–279 (1981); see also older version [33]. ISSN 0022-0000. MR 82i:68017, http://cr.yp.to/bib/entries.html#1981/wegman. Citations in this document: §1

[35] Wiener, M. (ed.): Advances in cryptology—CRYPTO '99. LNCS, vol. 1666. Springer (1999). ISBN 3-5540-66347-9. MR 2000h:94003. See [10]

[36] Wu, H.: The hash function JH (2011), http://www3.ntu.edu.sg/home/wuhj/research/jh/jh_round3.pdf. Citations in this document: §4

A   Test Values

This appendix shows intermediate values of SipHash-2-4 hashing the 15-byte string 000102$\cdots$0c0d0e with the 16-byte key 000102$\cdots$0d0e0f.

Initialization little-endian reads the key as

$$k_0 = 0706050403020100$$
$$k_1 = 0f0e0d0c0b0a0908$$

The key is then xored to the four constants to produces the following initial state (v_0 to v_3, left to right):

7469686173716475 6b617f6d656e6665 6b7f62616d677361 7b6b696e727e6c7b

The first message block 0706050403020100 is xored to v_3 to give

7469686173716475 6b617f6d656e6665 6b7f62616d677361 7c6d6c6a717c6d7b

and after two SipRounds the internal state is:

4d07749cdd0858e0 0d52f6f62a4f59a4 634cb3577b01fd3d a5224d6f55c7d9c8

Xoring the first message block to v_0 concludes the compression phase:

4a017198de0a59e0 0d52f6f62a4f59a4 634cb3577b01fd3d a5224d6f55c7d9c8

The second and last block is the last seven message bytes followed by the message's length, that is, 0f0e0d0c0b0a0908. After xoring this block to v_3, doing two SipRounds, xoring it to v_0 and xoring 00000000000000ff to v_2, the internal state is

3c85b3ab6f55be51 414fc3fb98efe374 ccf13ea527b9f442 5293f5da84008f82

After the four iterations of SipRound, the internal state is

f6bcd53893fecff1 54b9964c7ea0d937 1b38329c099bb55a 1814bb89ad7be679

and the four words are xored together to return a129ca6149be45e5.

B   Computing the Key of the Python Hash Function

The Python script below can be used to compute the key used in the function hash() of a Python process. An example of usage is

```
$ python3.2 -R poc.py
128 candidate solutions
verified solution: 58df0aca50e7f48b 141f57f820cbfefe
verified solution: d8df0aca50e7f48b 941f57f820cbfefe
```

Note the equivalent keys, and the -R option.

```
solutions = []
mask = 0xffffffffffffffff

def bytes_hash( p, prefix, suffix ):
  if len(p) == 0: return 0
  x = prefix ^ (ord( p[0] )<<7)
  for i in range( len(p) ):
    x = ( ( x * 1000003 ) ^ ord(p[i]) ) & mask
  x ^= len(p) ^ suffix
  if x == -1: x = -2
  return x

def solvebit( h1, h2, prefix, bits ):
  f1 = 1000003
  f2 = f1*f1
  target = h1^h2^3
  if bits == 64:
    if ((f1*prefix)^(f2*prefix)^target) & mask: return
    suffix = h1^1^(f1*prefix)
    suffix&= mask
    solutions.append( (prefix,suffix) )
  else:
    if ((f1*prefix)^(f2*prefix)^target) & ((1<<bits)-1):
      return
    solvebit(h1,h2,prefix,bits + 1)
    solvebit(h1,h2,prefix + (1 << bits),bits + 1)
  pass

h1 = hash("\0")      & mask
h2 = hash("\0\0")    & mask
h3 = hash("python")  & mask

solvebit( h1, h2, 0, 0 )

print("%d candidate solutions" % (len(solutions)))
for s in solutions:
  if bytes_hash("python",s[0],s[1]) == hash("python") & mask:
    ok=1
    for i in range(10)[1:]:
      if bytes_hash("\2"*i,s[0],s[1]) != hash("\2"*i) & mask:
        ok=0
    if ok: print("solution: %016x %016x" % (s[0],s[1]))
```

A Novel Permutation-Based Hash Mode of Operation FP and the Hash Function SAMOSA

Souradyuti Paul[1], Ekawat Homsirikamol[2], and Kris Gaj[2]

[1] University of Waterloo, Canada, and K.U. Leuven, Belgium
souradyuti.paul@esat.kuleuven.be
[2] George Mason University, USA
ekawat@gmail.com, kgaj@gmu.edu

Abstract. The contribution of the paper is two-fold. First, we design a novel permutation-based hash mode of operation FP, and analyze its security. We show that any n-bit hash function that uses the FP mode is indifferentiable from a random oracle up to $2^{n/2}$ queries (up to a constant factor), if the underlying $2n$-bit permutation is free from any structural weaknesses. Based on our further analysis and experiments, we conjecture that the FP mode is resistant to all non-trivial generic attacks with work less than the brute force, mainly due to its large internal state. We compare the FP mode with other permutation-based hash modes.

To put this into perspective, we propose a concrete hash function SAMOSA using the new mode and the P-permutations of the SHA-3 finalist Grøstl. Based on our analysis we claim that the SAMOSA family cannot be attacked with work significantly less than the brute force. We also provide hardware implementation (FPGA) results for SAMOSA to compare it with the SHA-3 finalists. In our implementations, SAMOSA family consistently beats Grøstl, Blake and Skein in the throughput to area ratio. With more efficient underlying permutation, it seems possible to design a hash function based on the FP mode that can achieve even higher performances.

1 Introduction

HASH MODE OF OPERATION. Iterative hash functions are generally built from two components: (1) a basic primitive C with finite domain and range, and (2) an iterative mode of operation H to extend the domain of the hash function; the symbol H^C denotes the hash function based on the mode H which invokes C iteratively to compute the hash digest. Therefore, to design an efficient hash function one has to be innovative with both the mode H and the basic primitive C. Merkle-Damgård mode used with a secure block cipher was the most attractive choice to build a practical hash function; some examples are SHA-family [22], and MD5 [26]. The security of a hash function based on the Merkle-Damgård

S. Galbraith and M. Nandi (Eds.): INDOCRYPT 2012, LNCS 7668, pp. 509–527, 2012.
© Springer-Verlag Berlin Heidelberg 2012

mode crucially relies on the fact that C is collision and pre-image resistant. The compression function C achieves these properties when it is constructed using a secure block cipher [8]. However, several security issues changed this popular design paradigm in the last decade. The first concern is that the security of Merkle-Damgård mode of operation – irrespective of the strength of the primitive C – came under a host of generic attacks; the length-extension attack [10], expandable message attack [18], multi-collision attacks [15] and herding attack [9,17] are some of them. Several strategies were discovered to thwart the above attacks. Lucks came up with the proposal of making the output of the primitive C at least twice as large as the hash output; this proposal is outstanding since, apart from rescuing the security of the Merkle-Damgård mode, it is also simple and easy to implement. Another interesting proposal was HAIFA that includes a counter injected into the compression function C to rule out many of the aforementioned attacks. Using the results of [8], it is easy to see that the Wide pipe and the HAIFA constructions are secure when the underlying primitive is a secure block cipher.

Despite the aforementioned foolproof design strategies, it turns out that using a block cipher as the basic primitive of a hash function may not be the best alternative, for several reasons. (1) A hash function does not need both the encryption and the decryption function of a block cipher; one of them could be avoided. (2) The key schedule of a block cipher often turns out to be weak [3]. (3) Furthermore, the key schedule weaknesses of a block cipher render invalid the very common ideal cipher assumption under which the security of block-cipher-based hash functions is usually based; note that an ideal cipher assumption is stronger than an ideal permutation assumption since, in the former case, an extra assumption is that a huge number of ideal permutations need to be *independent* too. (3) The amount of memory needed to implement a wide block cipher is larger due to the 'extra' key schedule than needed for an equally sized permutation.

PERMUTATION-BASED HASH FUNCTIONS. The popularity of permutation-based hash functions has been on the rise since the discovery of weaknesses on the Merkle-Damgård mode. Sponge [4], Grøstl [14], JH [28], Luffa [11] and the Parazoa family [1] are some of them. We note that 9 out of 14 semi-finalist algorithms (3 out of 5 finalist algorithms) of the NIST SHA-3 hash function competition are based on permutations. Other notable example is MD6 [27]. In Table 1, we compare generic security and performance (measured in terms of rate) of various well known permutation-based hash modes.

OUR CONTRIBUTION

FP mode. Our first contribution is to give a proposal for a new hash mode of operation FP based on a single wide pipe permutation (see Fig. 1). The FP mode is derived from the FWP (or Fast-wide pipe) mode designed by Nandi and Paul at Indocrypt 2010 [23]. The difference between the FWP and the FP mode is simple: the FP mode is obtained when the underlying hard-to-invert function

Table 1. Indifferentiability bounds of permutation-based hash modes, where the hash size is n-bit in each case. FP^{Ext1} is a natural variant of FP with parameters shown in the row. The ϵ is a small fraction due to the preimage attack on JH presented in [7].

Mode of operation	Mesg-blk (ℓ)	Size of π (a)	Rate (ℓ/a)	Indiff. bound lower	Indiff. bound upper	# of independent permutations
Hamsi [16]	$n/8$	$2n$	0.07	$n/2$	n	1
Luffa [6]	$n/3$	n	0.33	$n/4$	n	3
Sponge [5]	n	$3n$	0.33	n	n	1
Sponge [5]	n	$2n$	0.5	$n/2$	$n/2$	1
JH [21]	n	$2n$	0.5	$n/2$	$n(1-\epsilon)$	1
Grøstl [14]	n	$2n$	0.5	$n/2$	n	2
FP	n	$2n$	**0.5**	$n/2$	n	1
MD6 [12]	$6n$	$8n$	0.75	n	n	1
FP^{Ext1}	$6n$	$7n$	0.85	$n/2$	n	1

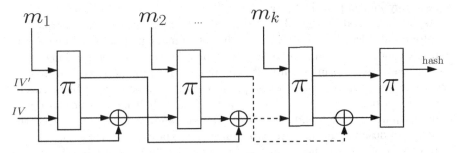

Fig. 1. Diagram of the FP mode. The π is a permutation; all wires are n bits. See Fig. 3(a) for the description.

$f : \{0,1\}^{m+n} \to \{0,1\}^{2n}$ of the FWP mode is replaced by an easy-to-invert permutation $\pi : \{0,1\}^{2n} \to \{0,1\}^{2n}$.[1] There are a number of practical reasons for switching from FWP to FP: (1) Easy-to-invert permutations are usually efficient, and such permutations with strong cryptographic properties are abundant in the literature (*e.g.* JH, Grøstl and Keccak permutations); (2) hard-to-invert functions either turn out to be weak [19], or they are inefficient.

On the other hand, easy-to-invert permutations – even though they are faster – have some drawbacks; the most crucial of them is that they allow the attacker to use reverse queries in addition to forward queries, and, as a result, make the adversary inherently more powerful. Therefore, a good deal of caution is required to design a hash mode of operation that uses permutations. We show that the FP mode based on an ideal permutation is indifferentiable from a random oracle up to approximately $2^{n/2}$ queries (forward and reverse together); this means that the FP mode is secure against *all* generic attacks – including (multi) collision, 2nd preimage, herding attacks – up to approximately $2^{n/2}$ queries, under the assumption that the underlying permutation π is structurally strong. Moving further, a large number of experiments show that the indifferentiability security

---

[1] FP is the shorthand for 'FWP with a permutation'.

of the FP mode could be improved *optimally* to n bits. Another important feature of our work is that the security guarantee is based on only one assumption – like the Sponge and JH – that the underlying permutation should not display any structural weaknesses; note that the security of many permutation-based hash functions (*e.g.* Grøstl and Luffa) requires additional assumptions such as *independence* of several ideal permutations. In Fig. 1, we compare the FP mode and a natural extension of it FP[Ext1] with other permutation-based hash functions. It is noteworthy that the FP mode exhibits the best Rate-Security trade-off when the internal permutation size is fixed.

Design and hardware implementation of the hash function SAMOSA. Our second contribution is establishing the practical usefulness of the FP mode. As an example, we design a concrete hash function family SAMOSA based on the FP mode, where the internal primitives are the P-permutations of the Grøstl hash function.[2] We provide security analysis of SAMOSA, demonstrating its resistance against any known practical attacks.

As demonstrated by the AES and the SHA-3 competitions, the security of a cryptographic algorithm alone is not sufficient to make it stand out from among multiple candidates competing to become a new American or international standard. Excellent performance in software and hardware is necessary to make a cryptographic protocol usable across platforms and commercial products. Assuring good performance in hardware is typically more challenging, since hardware design requires involved and specialized training, and, as it turns out, that the majority of designer groups lack experience and expertise in that area.

In case of SAMOSA, the algorithm design and hardware evaluation have been performed side by side, leading to full understanding of all design decisions and their influence on hardware efficiency. In this paper, we present efficient high-speed architecture of SAMOSA, and show that this architecture outperforms the best known architecture of Grøstl in terms of the throughput to area ratio by a factor ranging between 24 and 51%. These results have been demonstrated using two representative FPGA families from two major vendors, Xilinx and Altera. As shown in [13], these results are also very likely to carry to any future implementations based on ASICs (Application Specific Integrated Circuits). Additionally, we demonstrate that SAMOSA consistently ranks above BLAKE, Skein and Grøstl in our FPGA implementations. Although it still loses to Keccak and JH, nevertheless, a relative distance to these algorithms substantially decreases compared to Grøstl, despite using the same underlying operations. This performance gain is accomplished without any known degradation of the security strength.

Additionally, SAMOSA's dependence on many AES operations makes it suitable for software implementations that use general-purpose processors with AES instruction sets, such as AES-NI. Finally, in both software and hardware, SAMOSA seems to be an attractive choice for applications where both confidentiality and authentication are required to share AES components. One such example is IPSec,

---

[2] SAMOSA is the name of an Indian food.

protocol used for establishing Virtual Private Networks, which is one of the fundamental building blocks of secure electronic transactions over the Internet.

Although SAMOSA comes too late for the current SHA-3 competition, it still has a chance to contribute to better understanding of the security and performance bottlenecks of modern hash functions, and to find niche platforms and applications in which it may outperform the existing and upcoming standards.

NOTATION AND CONVENTION. This part has been moved to the full version of the paper [25].

ORGANIZATION. The remainder of the paper is organized as follows. In Sect. 2 we define the new hash mode FP. In Sect. 4 we present our main indifferentiability theorem for the FP mode, and give a sketch for the proof. The proof is then elaborated in Sects. 5, 6 and 7. In Sect. 8, we propose a new concrete hash function named SAMOSA, and provide its security analysis. Finally, in Sect. 9, we give hardware implementation results for SAMOSA.

2   Definition of the FP Mode

Suppose $n \geq 1$. Let $\pi : \{0,1\}^{2n} \to \{0,1\}^{2n}$ be the $2n$-bit permutation used by the FP mode. The hash function FP^π is a mapping from $\{0,1\}^*$ to $\{0,1\}^n$. The diagram and description of the FP transform are given in Figs. 1 and 3(a), where π is modeled as an ideal permutation. Below we define the padding function $\mathsf{pad}_n(\cdot)$.

Padding function $\mathsf{pad}_n(\cdot)$. It is an injective mapping from $\{0,1\}^*$ to $\cup_{i \geq 1}\{0,1\}^{ni}$, where the message $M \in \{0,1\}^*$ is mapped into a string $\mathsf{pad}_n(M) = m_1 \cdots m_{k-1} m_k$, such that $|m_i| = n$ for $1 \leq i \leq k$. The function $\mathsf{pad}_n(M) = M||1||0^t$ satisfies the above properties (t is the least non-negative integer such that $|M| + 1 + t = 0 \bmod n$). Note that $k = \left\lceil \frac{|M|+1}{n} \right\rceil$.

In addition to the injectivity of $\mathsf{pad}_n(\cdot)$, we will also require that there exists a function $\mathsf{dePad}_n(\cdot)$ that can efficiently compute M, given $\mathsf{pad}_n(M)$. Formally, the function $\mathsf{dePad}_n : \cup_{i \geq 1}\{0,1\}^{in} \to \{\bot\} \cup \{0,1\}^*$ computes $\mathsf{dePad}_n(\mathsf{pad}_n(M)) = M$, for all $M \in \{0,1\}^*$, and otherwise $\mathsf{dePad}_n(\cdot)$ returns $\bot$. The padding rule described above satisfies this property also.

3   Preliminaries: Introduction to the Indifferentiability Framework

Definition 1 (Indifferentiability framework). *[20,10] An interactive Turing machine (ITM) T with oracle access to an ideal primitive $\mathcal{F}$ is said to be $(t_\mathcal{A}, t_\mathcal{S}, \sigma, \varepsilon)$-indifferentiable from an ideal primitive $\mathcal{G}$ if there exists a simulator $\mathcal{S}$ such that, for any distinguisher $\mathcal{A}$, the following equation is satisfied:*

$$\mathbf{Adv}^{T,\mathcal{F}}_{\mathcal{G},\mathcal{S}}(\mathcal{A}) \stackrel{def}{=} \left| \Pr[\mathcal{A}^{T,\mathcal{F}} \Rightarrow 1] - \Pr[\mathcal{A}^{\mathcal{G},\mathcal{S}} \Rightarrow 1] \right| \leq \varepsilon.$$

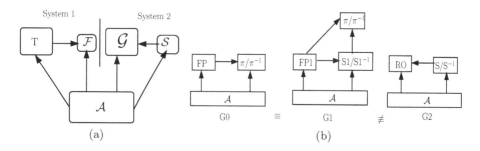

Fig. 2. (a) Indifferentiability framework defined in Def. 1; (b) Schematic diagram for the security games used in the indifferentiability framework for FP (see Sect. 5). The arrows show the directions in which the queries are submitted. System 1 = $(T, \mathcal{F})$ = G0 = $(\mathsf{FP}, \pi, \pi^{-1})$, and System 2 = $(\mathcal{G}, \mathcal{S})$ = G2 = $(\mathsf{RO}, \mathsf{S}, \mathsf{S}^{-1})$.

The simulator $\mathcal{S}$ is an ITM which has oracle access to $\mathcal{G}$ and runs in time at most $t_\mathcal{S}$. The distinguisher $\mathcal{A}$ runs in time at most $t_\mathcal{A}$. The number of queries used by $\mathcal{A}$ is at most σ. Here ε is a negligible function in the security parameter of T. See Fig. 2(a) for a pictorial representation.

Due to page limitation, discussion on the indifferentiability framework has been moved to the full version of the paper [25].

4   Main Theorem: $n/2$-bit Indifferentiability Security of the FP Mode

Let $\mathsf{RO}: \{0,1\}^* \to \{0,1\}^n$ and $\pi: \{0,1\}^{2n} \to \{0,1\}^{2n}$ are a random oracle and an ideal permutation. Our indifferentiability framework uses three systems G0 = $(\mathsf{FP}, \pi, \pi^{-1})$, G1 = $(\mathsf{FP1}, \mathsf{S1}, \mathsf{S1}^{-1})$, and G2 = $(\mathsf{RO}, \mathsf{S}, \mathsf{S}^{-1})$ (see Fig. 2(b)). The correspondence between the entities of Figs. 2(a) and 2(b) are as follows: $\mathcal{G} = \mathsf{RO}$, $T = \mathsf{FP}$ and $\mathcal{F} = (\pi, \pi^{-1})$. The description of FP1, S, S^{-1}, S1, and $\mathsf{S1}^{-1}$ will be provided in Sect. 5. Now we state our main theorem using Def. 1.

Theorem 1 (Main Theorem). *The hash function FP^π (or $\mathsf{FP}^{\pi, \pi^{-1}}$) is $(t_\mathcal{A}, t_\mathcal{S}, \sigma, \varepsilon)$-indifferentiable from RO, where $t_\mathcal{A} = \infty$, $t_\mathcal{S} = \mathcal{O}(\sigma^5)$, and $\sigma \leq K2^{n/2}$, where K is a fixed constant derived from ε.*

In the next few sections, we will prove Theorem 1 by breaking it into several components. First, we briefly describe what the theorem means: it says that no adversary with unbounded running time can mount a *nontrivial generic* attack on the hash function FP^π using at most $K2^{n/2}$ queries. The parameter K is an increasing function in ε, and is constant for all $n > 0$. To reduce the notation complexity, we shall derive the indifferentiability bound assuming $\varepsilon = 0.5$ for which, we shall derive, $K = 1/\sqrt{56}$.

Outline of the Proof. Proof of Theorem 1 consists of the following two components (see Def. 1): (1) Construction of a simulator $\mathcal{S} = (\mathsf{S}, \mathsf{S}^{-1})$ with the worst-case running time $t_S = \mathcal{O}(\sigma^5)$. This is done in Sect. 5. (2) Showing that, for any adversary $\mathcal{A}$ with unbounded running time,

$$\left| \Pr\left[\mathcal{A}^{G0} \Rightarrow 1\right] - \Pr\left[\mathcal{A}^{G2} \Rightarrow 1\right] \right| \leq \frac{28\sigma^2}{2^n}, \tag{1}$$

where the systems $G0 = (\mathsf{FP}, \pi, \pi^{-1})$ and $G2 = (\mathsf{RO}, \mathsf{S}, \mathsf{S}^{-1})$.[3] Proof of (1) is, again, composed of proofs of the following three (in)equations:

- In Sect. 5, we will concretely define the simulator pair (S, S^{-1}) and a new system G1. Using them we will show

$$\Pr\left[\mathcal{A}^{G0} \Rightarrow 1\right] = \Pr\left[\mathcal{A}^{G1} \Rightarrow 1\right]. \tag{2}$$

- In Sect. 6, we will appropriately define a set of events BAD_i and GOOD_i in the system G1, and will establish that

$$\left| \Pr\left[\mathcal{A}^{G1} \Rightarrow 1\right] - \Pr\left[\mathcal{A}^{G2} \Rightarrow 1\right] \right| \leq \sum_{i=1}^{\sigma} \Pr\left[\mathsf{BAD}_i \mid \mathsf{GOOD}_{i-1}\right]. \tag{3}$$

- In Sect. 7, we complete proof of (1) by establishing that

$$\sum_{i=1}^{\sigma} \Pr\left[\mathsf{BAD}_i \mid \mathsf{GOOD}_{i-1}\right] \leq \frac{28\sigma^2}{2^n} \tag{4}$$

where

$$\sum_{i=1}^{\sigma} \Pr\left[\mathsf{BAD}_i \mid \mathsf{GOOD}_{i-1}\right] \leq \varepsilon = 0.5.$$

The next three sections are devoted to proving (2), (3), and (4).

5   Proof of (2)

This section has two parts.

- In the first part – Sects. 5.1, 5.2, and 5.3 – we define the simulator pair (S, S^{-1}) of the indifferentiability framework for FP. This completes the description of systems G0 and G2 that are partially specified in Sect. 4. Then we describe a new system G1.
- In the second part – Proposition 1 – using the definitions of G0 and G1, we prove (2).

In addition, using the definitions of G1 and G2, in Sect. 6, we will prove (3).

The pseudocode for all the systems is given in Figs. 3 and 5. In the remainder of the section we explain the systems G0, G1, and G2 in detail with special emphasis on their usefulness in proving our main theorem.

---

[3] Setting $\frac{28\sigma^2}{2^n} \leq \varepsilon = 1/2$, we get $\sigma \leq \frac{1}{\sqrt{56}} 2^{n/2}$.

5.1   Data Structures

G0, G1, and G2 use several data structures. Due to limited space the definitions of oracle, query, round, global (and local) variable, reconstruction graphs T_π and T_s (built by G1 and G2 resp.), reconstructible message, and view have been moved to the full version of the paper [25].

5.2   Description of the Main Systems G0 and G2

We recall from Sect. 4 that our main task, informally, is to estimate how difficult it is for an arbitrary adversary $\mathcal{A}$ to distinguish between the systems G0 and G2. Essentially, we need to design a simulator pair $(\mathsf{S}, \mathsf{S}^{-1})$ for G2, such that it is 'hard' to tell the systems apart. Although, G0 and G2 are partially defined in Sect. 4, we give the complete description here. The pseudocode is given in Fig. 3.

Description of G0. Following the definition provided in Sect. 2, the system G0 implements the FP hash function using the ideal permutations π and π^{-1}.

Description of G2. The random oracle RO defined in Sect. 4 is implemented through lazy sampling. The only remaining part is to construct the simulator pair (S, S^{-1}). Our design strategy for $(\mathsf{S}, \mathsf{S}^{-1})$ is fairly straightforward and simple. Before going into the details, we first provide a high level intuition.

Intuition behind the design of simulator pair $(\mathsf{S}, \mathsf{S}^{-1})$. The purpose of the simulator pair $(\mathsf{S}, \mathsf{S}^{-1})$ is two-fold: (1) to output values that are indistinguishable from the output from the ideal permutation (π, π^{-1}), and (2) to respond in such a way that $\mathsf{FP}^\pi(M)$ and $\mathsf{RO}(M)$ are identically distributed. It will easily follow that as long as the simulator pair $(\mathsf{S}, \mathsf{S}^{-1})$ is able to output values satisfying the above conditions, no adversary can distinguish between G0 and G2.

In order to serve the above purposes, the primary requirement is that the simulator S, for a distinct input x, be assigned a 'random' value such that the distributions of $\mathsf{S}(x)$ and $\pi(x)$ are close. Similarly, the simulator S^{-1} for a distinct input r should ensure that the random variables $\mathsf{S}^{-1}(r)$ and $\pi^{-1}(r)$ follow statistically close distributions.

Secondly, the simulator S maintains a *reconstruction* graph T_s for D_s; this helps the simulator keep track of all 'FP-mode-compatible' messages (more formally *reconstructible* messages for D_s) that can be formed using the s-queries and responses. This is done by a special subroutine FullGraph. Informally, a reconstruction graph helps the simulator assess the power of the adversary, since any adversary can construct the graph too. Intuitively, at the time of formation of a reconstructible message, the simulator needs to do some adjustment in its database, so that the outputs from G0 and G2 'look' the same to the adversary, no matter what query she submits. The pictorial representation of the reconstruction graph T_s is given in Fig. 4.

Given the current s-query x, the subroutine MessageRecon determines all the new reconstructible messages M by searching through all the branches between

---

FP(M)

01. **If** $M \in Dom(D_l)$ **then**
  return $D_l[M]$;
02. $m_1 m_2 \ldots m_k := \mathsf{pad}_n(M)$;
03. $y_0 := IV, y_0' := IV'$;
04. **for** $(i := 1, 2, \ldots k)$
  $y_i y_i' := \pi(y_{i-1} \| m_i) \oplus (y_{i-1}' \| 0)$;
05. $r := \pi(y_k \| y_k')$;
06. $D_l[M] := r[n, 2n - 1]$;
07. **return** $D_l[M]$;

$\pi(x)$

11. **If** $x \notin Dom(D_\pi)$
  then $D_\pi[x] \xleftarrow{\$} \{0,1\}^{2n} \setminus Rng(D_\pi)$;
12. **return** $D_\pi[x]$;

$\pi^{-1}(r)$

21. **If** $r \notin Rng(D_\pi)$
  then $D_\pi^{-1}[r] \xleftarrow{\$} \{0,1\}^{2n} \setminus Dom(D_\pi)$;
22. **return** $D_\pi^{-1}[r]$;

(a) System G0 = (FP, π, π^{-1}). For all i, $|m_i| = |y_i| = |y_i'| = |r/2| = n$.

RO(M)

001. **If** $M \in Dom(D_l)$ **then**
  return $D_l[M]$;
002. $D_l[M] \xleftarrow{\$} \{0,1\}^n$;
003. **return** $D_l[M]$;

$S^{-1}(r)$

300. **If** $\exists x_1, x_2 \in Dom(D_s)$
  if $D_s[x_1] = D_s[x_2] = r$
  then return $\perp$;
301. if $r \in Rng(D_s)$ then return $D_s^{-1}[r]$;
302. if $r \notin Rng(D_s)$ then $x \xleftarrow{\$} \{0,1\}^{2n}$;
303. if $x \notin Dom(D_s)$ then $D_s[x] := r$;
304. **return** x;

$S(x)$

101. $r \xleftarrow{\$} \{0,1\}^{2n}$;
102. $\mathcal{M} := \mathsf{MessageRecon}(x)$;
103. if $|\mathcal{M}| = 1$ then
  $r[n, 2n - 1] := \mathsf{RO}(M)$;
104. $D_s[x] := r$;
105. FullGraph;
106. **return** r;

MessageRecon(x)

201. $\mathcal{M}' := \mathsf{FindBranch}(x)$;
202. $\mathcal{M} := \{\mathsf{dePad}(X) \mid X \in \mathcal{M}'\}$;
203. **return** $\mathcal{M}$;

(b) System G2 = (RO, S, S^{-1}).

Fig. 3. The main systems G0 and G2

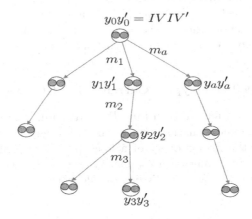

$$y_0 y_0' = IV IV'$$

m_1 $\quad m_a$

$y_1 y_1'$ $\quad y_a y_a'$

m_2

$y_2 y_2'$

m_3

$y_3 y_3'$

Fig. 4. The reconstruction graph T_s (or T_π) updated by FullGraph of G2 (or PartialGraph of G1)

the nodes $IVIV'$ and x in the graph T_s. Then the simulator makes this crucial adjustment: it assigns $\mathsf{FP}^S(M) := \mathsf{RO}(M)$. It is fairly intuitive that, if S and π produce outputs according to statistically close distributions, then the distributions of $\mathsf{FP}^S(M)$ and $\mathsf{FP}^\pi(M)$ are also close. Since $\mathsf{FP}^S(M) = \mathsf{RO}(M)$, the distributions of $\mathsf{RO}(M)$ and $\mathsf{FP}^\pi(M)$ are also close.

Detailed description of the simulator pair (S, S^{-1}). This part has been moved to the full version [25].

5.3   Description of the Intermediate System G1

The pseudocode is provided in Fig. 5. Detailed description will appear in the full version [25].

Now we are well equipped to prove (2).

Proposition 1. *For any distinguishing adversary* $\mathcal{A}$,

$$\Pr[\mathcal{A}^{G0} \Rightarrow 1] = \Pr[\mathcal{A}^{G1} \Rightarrow 1].$$

Proof. From the description of S1 and S1^{-1}, we observe that, for all $x \in \{0,1\}^{2n}$, $\mathsf{S1}(x) = \pi(x)$ and $\mathsf{S1}^{-1}(x) = \pi^{-1}(x)$. Likewise, from the descriptions of FP1 and FP, for all $M \in \{0,1\}^*$, $\mathsf{FP1}(M) = \mathsf{FP}(M)$.

6   Proof of (3)

This section is broadly divided into three parts. In the first two parts we concretely define the Type0, Type1, Type2, Type3 and Type4 events of the system G1 (see Fig. 5), based on whether the current π/π^{-1} is a fresh or an old query. In the last part we prove (3), using the Type0-4 events.

6.1   Events Type0 and Type1: Current π/π^{-1}-Query Is Fresh

Event Type0: Distance of Random Permutation from the Uniform. Consider three different scenarios: when a fresh π-query is an s-query; when a fresh π-query is the final π-query of a long query; when an s^{-1}-query is a fresh π^{-1}-query. Type0 event occurs when, in each of the above scenarios, the output of a fresh π/π^{-1}-query is *distinguishable* from the uniform distribution $\mathcal{U}[0, 2^{2n} - 1]$.

Event Type1: Collision on T_π. Due to limited space in this version we only provide the diagram for these events (Fig. 6). A discussion will be given in the full version [25].

FP1(M)

001. $m_1 m_2 \cdots m_{k-1} m_k := \mathsf{pad}_n(M)$;
002. $y_0 := IV$, $y_0' := IV'$;
003. for $(i := 1, \cdots, k)$ {
004. $r := \pi(y_{i-1} m_i)$;
005. $y_i y_i' := r \oplus (y_{i-1}' \| 0)$;
006.   if $y_{i-1} m_i$ is *fresh* then
              PartialGraph$(y_{i-1} m_i, r)$;}
007. if Type3 then $\boxed{\text{BAD} := \text{True}}$;
008. $r := \pi(y_k y_k')$;
009. if Type0-b then $\boxed{\text{BAD} := \text{True}}$;
010. if $y_k y_k'$ is *fresh* then
              PartialGraph$(y_k y_k', r)$;
011. $D_l[M] := r[n, 2n-1]$;
012. return $D_l[M]$;

MessageRecon(x, T_s)

201. $\mathcal{M}' := \mathsf{FindBranch}(x)$;
202. $\mathcal{M} := \{\mathsf{dePad}(X) \mid X \in \mathcal{M}'\}$;
203. return $\mathcal{M}$;

$\pi(x)$

301. if $x \notin Dom(D_\pi)$ then
         $D_\pi[x] \xleftarrow{\$} \{0,1\}^{2n} \setminus Rng(D_\pi)$;
302. return $D_\pi[x]$;

$\pi^{-1}(r)$

501. If $r \notin Rng(D_\pi)$ then
         $D_\pi^{-1}[r] \xleftarrow{\$} \{0,1\}^{2n} \setminus Dom(D_\pi)$;
502. return $D_\pi^{-1}[r]$;

S1(x)

100. If Type2 then $\boxed{\text{BAD} := \text{True}}$;
101. $r := \pi(x)$;
102. if Type0-a then $\boxed{\text{BAD} := \text{True}}$;
103. $\mathcal{M} := \mathsf{MessageRecon}(x, T_s)$;
104. if $|\mathcal{M}| = 1 \wedge M \notin Dom(D_l)$ then
         $D_l[M] := r[n, 2n-1]$;
105. $D_s[x] := r$;
106. if x is *fresh* then PartialGraph(x, r);
107. return r;

S1$^{-1}(r)$

601. if Type4 then $\boxed{\text{BAD} := \text{True}}$;
602. $x := \pi^{-1}(r)$;
603. if Type0-c then $\boxed{\text{BAD} := \text{True}}$;
604. if Type1-c then $\boxed{\text{BAD} := \text{True}}$;
605. $D_s[x] := r$;
606. return x;

PartialGraph(x, r)

401. $x \xrightarrow{parse} y_c m$; $r \xrightarrow{parse} y^* y'$;
402. $C := \mathsf{CreateCoset}(y_c)$;
403. $E := \{(y_c y_c', m, yy') \mid y := y^* \oplus y_c', y_c y_c' \in C\}$;
404. for $\forall e \in E$ {
         AddEdge(e);
         if Type1-a $\vee$ Type1-b then $\boxed{\text{BAD} := \text{True}}$;}

Fig. 5. System G1. $|m_i| = |m| = |y_i| = |y_i'| = |y_c| = |y_c'| = |y| = |y'| = |y^*| = |r/2| = n$, for all i.

6.2 Events Type2, 3 and 4: Current π/π^{-1}-Query Is Old

Classification of Old Queries and the Branches of T_π. Before we define the Type2, 3 and 4 events, we first classify all the old query-response pairs for the oracles π/π^{-1} stored in D_π, according to its known and unknown parts. The known part of a query-response pair is the part that is present in the view of the system G1, or it can be derived from the view with probability 1; the unknown part is *not* present in the view, and it cannot be derived from the view with probability 1. We observe that there are six types of such a pair, and we denote them by Q0, Q1, Q2, Q3, Q4 and Q5 in Fig. 7(a); the head and tail nodes in each type denote the input and output, each $2n$ bits. Two-sided arrowhead indicates that the corresponding input-output pair is generated from either a π-query or a π^{-1}-query. The *red* and *green* circles denote the unknown and known parts, n bits each.

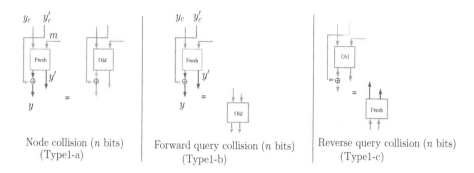

Fig. 6. Type1 events of G1. All arrows are n bits each. Red arrow denotes fresh n bits of output from the ideal permutation π/π^{-1}. The symbol "=" denotes n-bit equality.

In a similar way, the branches of T_π can be classified into four types, as shown in Fig 7(b).

Event Type2: Current s-query is π-query of a Previous Long Query. Discussion on this event has been moved to the full version [25].

Event Type3: Current π-query of a *red* Branch is the Final Query for the Current Long Query. Several types of *red* branch – (I), (II), and (III) – are shown in Fig. 7(b)(I) to (III).

There are three types of Type3 event: (Type3-a) if the current long query M with $m_1 m_2 \cdots m_k = \mathsf{pad}_n(M)$ forms a *red* branch of type (I).[4] (Type3-b) if M is a *red* branch of type (II), and if the most significant n bits of output can be distinguished from the uniform distribution $\mathcal{U}[0, 2^n - 1]$. (Type3-c) if M is a *red* branch of type (III).

Event Type4: Current s^{-1}-query Input Matches Output of a π-query of a Previous Long Query. The Type4 event occurs, if the current s^{-1}-query is equal to an old query of type Q1, Q2, Q3, Q4 or Q5.

6.3   Proof of (3)

With the help of the Type0 to 4 events described before, we are equipped to prove (3). First, we fix a few definitions in order for (3) to make sense.

---

[4] Observe that this case implies a node collision in T_π, since the $y_k y'_k$ is the final π-query for two distinct long queries, the current M and also an old one. Therefore, if Type1 event did not occur in the previous rounds, this event is impossible in the current round.

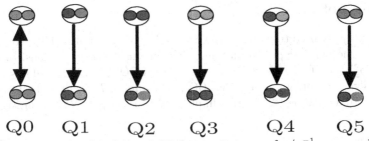

(a) Q0, Q1, Q2, Q3, Q4, and Q5 denote six types of π/π^{-1}-query and response.

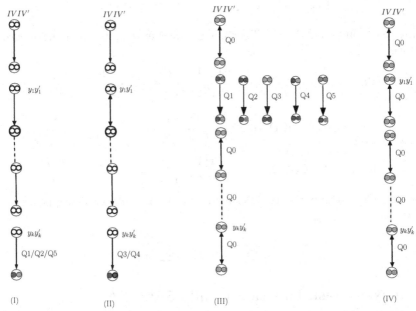

(b) Several types of branch in T_π. (I), (II) and (II) are called *red* branches.

Fig. 7. Several types of old π/π^{-1}-queries and branches in T_π

Events GOOD$_i$ and BAD$_i$. BAD$_i$ denotes the event when the variable BAD is set during round i of G1, that is, when Type0, Type1, Type2, Type3 or Type4 event occurs. Let the symbol GOOD$_i$ denote the event $\neg \bigvee_{j=1}^{i}$ BAD$_i$. The symbol GOOD$_0$ denotes the event when no queries are submitted. From a high level, the intuition behind the construction of the BAD$_i$ event is straightforward: we will show that if BAD$_i$ does not occur, and if GOOD$_{i-1}$ did occur, then the views of G1 and G2 (after i rounds) are identically distributed for *any* attacker $\mathcal{A}$.

We move the entire proof, which is based on induction on the number of rounds, to the full version [25].

7   Proof of (4)

To prove (4), we need to compute the probability of occurrence of Type0, Type1, Type2, Type3 and Type4 events described in the previous sections. Since we assume $\sum_{i=1}^{\sigma} \Pr\left[\mathsf{BAD}_i \mid \mathsf{GOOD}_{i-1}\right] \le \varepsilon = 1/2$, $\mathsf{GOOD}_i \ge 1/2$ for all $0 \le i \le \sigma$.

The Type1 events guarantee that GOOD_{i-1} implies T_π has i nodes after $i-1$ rounds. We assume $i \le 2^{n/2}$; this implies $(2^n - i) \ge \frac{1}{2}2^n$. Now from Fig. 6 we obtain,

$$\Pr\left[\mathsf{Type0}_i \mid \mathsf{GOOD}_{i-1}\right] \le 3/(2^{2n} - i) \le \frac{1}{2^n},$$

$$\Pr\left[\mathsf{Type1}_i \mid \mathsf{GOOD}_{i-1}\right] \le 3i/(2^n - i) \le \frac{6i}{2^n}.$$

Using the definition of Type2, Type3, and Type4 events, it is straightforward to deduce:

$$\Pr\left[\mathsf{Type2}_i \mid \mathsf{GOOD}_{i-1}\right] \le \frac{\Pr\left[\mathsf{Type2}_i\right]}{\Pr\left[\mathsf{GOOD}_{i-1}\right]} \le 2 \cdot \frac{5i}{2^n - i} \le \frac{20i}{2^n},$$

$$\Pr\left[\mathsf{Type3}_i \mid \mathsf{GOOD}_{i-1}\right] \le \frac{\Pr\left[\mathsf{Type3}_i\right]}{\Pr\left[\mathsf{GOOD}_{i-1}\right]} \le 2 \cdot \frac{2}{2^n - i} \le \frac{8}{2^n},$$

$$\Pr\left[\mathsf{Type4}_i \mid \mathsf{GOOD}_{i-1}\right] \le \frac{\Pr\left[\mathsf{Type4}_i\right]}{\Pr\left[\mathsf{GOOD}_{i-1}\right]} \le 2 \cdot \frac{5i}{2^n - i} \le \frac{20i}{2^n}.$$

We conclude by combining the above bounds into the following inequality which holds for $1 \le i \le \sigma$:

$$\sum_{i=1}^{\sigma} \Pr\left[\mathsf{BAD}_i \mid \mathsf{GOOD}_{i-1}\right] \le \sum_{i=1}^{\sigma} \frac{55i}{2^n} \le \frac{28\sigma^2}{2^n}.$$

8   A New Hash Function Family SAMOSA

Now we design a concrete hash function family SAMOSA based on the FP mode defined in Sect. 2. In the subsequent sections, we also provide security analysis and hardware implementation results of SAMOSA.

8.1   Description of SAMOSA

SAMOSA hash family is based on the FP mode and P-permutation of the Grøstl hash function family. Letting n denote the length of hash in bits ($n = 256$ and 512 bits), the complete description of the hash function SAMOSA-n is provided in Fig. 8. SAMOSA is composed of three components: (1) The FP mode and the padding rule $\mathsf{pad}_n(\cdot)$ (see Section 2), (2) $IV IV' = \langle 0 \rangle_n || \langle n \rangle_n$, and (3) the Grøstl permutation P_{2n} (see [14]). The pseudocode for P_{512} and P_{1024} is given in the full version [25].

Security Analysis of the SAMOSA Family. Discussion on this section will appear in the full version of the paper [25].

SAMOSA-256(M)	SAMOSA-512(M)
01. $m_1 m_2 \dots m_{k-1} m_k := \mathsf{pad}_{256}(M);$	11. $m_1 m_2 \dots m_{k-1} m_k := \mathsf{pad}_{512}(M);$
02. $y_0 \| y_0' := \langle 0 \rangle_{256} \| \langle 256 \rangle_{256};$	12. $y_0 \| y_0' := \langle 0 \rangle_{512} \| \langle 512 \rangle_{512};$
03. for$(i := 1, 2, \dots k)$	13. for$(i := 1, 2, \dots k)$
$\quad y_i \| y_i' := P_{512}(y_{i-1} \| m_i) \oplus (y_{i-1}' \| \langle 0 \rangle_{256});$	$\quad y_i \| y_i' := P_{1024}(y_{i-1} \| m_i) \oplus (y_{i-1}' \| \langle 0 \rangle_{512});$
04. $r := P_{512}(y_k \| y_k');$	14. $r := P_{1024}(y_k \| y_k');$
05. return $r[256, 511];$	15. return $r[512, 1023];$

Fig. 8. SAMOSA-256 and SAMOSA-512

9 FPGA Implementations of **SAMOSA**-256 and **SAMOSA**-512

9.1 Motivation and Previous Work

In case the security of two competing cryptographic algorithms is the same or comparable, their performance in software and hardware decides which one of them get selected for use by standardization organizations and industry.

In this section, we will analyze how SAMOSA compares to Grøstl, one of the five final SHA-3 candidates, from the point of view of performance in hardware. This comparison makes sense, because both algorithms share a very significant part, permutation P, but differ in terms of the mode of operation. The FP mode requires only a single permutation, so using permutation P is sufficient. The Grøstl mode relies on the use of two related permutations, P and Q, which can be executed in parallel. Our goal is to determine how much savings in terms of hardware area are introduced by replacing the Grøstl construction for hash function with the FP mode. We also would like to know whether these savings come at the expense of any significant throughput drop. Finally, we would like to analyze how significant is the improvement in terms of the throughput to area ratio, a primary metric used to evaluate the efficiency of hardware implementations in terms of a trade-off between speed and cost of the implementation.

Among many implementations of Grøstl, the one by Latif et al. [24] is currently the most efficient on Virtex 5, this implementation relies on the use of low-level Xilinx FPGA primitives, and as a result is not portable to FPGAs of other vendors, such as Altera. Since our implementation of SAMOSA presented in this paper is fully portable, and does not use any low-level primitives, we compare it with the second best design of Grøstl reported earlier in the literature [13], which has the same features. This design is based on the quasi-pipelined basic iterative architecture denoted as x1 (P/Q). This way, we will be also able to provide comparison for an alternative FPGA family, Stratix III from Altera.

9.2 High-Speed Architectures of **SAMOSA** and Grøstl

Due to the space constraint we move this useful section entirely to the full version of the paper [25].

9.3   Comparison of **SAMOSA** and Grøstl in Terms of the Hardware Performance

Below, we compare SAMOSA and Grøstl in terms of three major hardware performance metrics: Area, Throughput, and Throughput to Area Ratio. The exact results of the comparison generated using Xilinx ISE v13.1 are shown in Table 2. In the full version of the paper [25], we also provide results generated using Altera Quartus II v11.1. Automated Tool for Hardware EvaluatioN (ATHENa) [2] was used to automate the optimization and result extraction process. No low-level primitives and no embedded resources (such as Block Memories or DSP units) were used in our implementations, which makes them fully portable among multiple FPGA families from various vendors. Each design has been implemented in two different versions: with and without padding unit. The designs with padding unit are more complete, while the designs without padding units are more suitable for comparison with hardware implementations presented in earlier academic papers on Grøstl and other SHA-3 candidates (as these implementations typically did not contain padding units).

Comparison in Terms of Area. As shown in Table 2, for comparable hardware architectures, SAMOSA has significantly lower area requirements than Grøstl. For Xilinx FPGAs, the area reduction is between 27 and 35%; for Altera FPGAs, it is between 31 and 34%. This reduction is explained below. First, P round is simpler than P/Q round, as the relevant logic does not need to be switched from implementing P permutation to implementing Q permutation of Grøstl. Although both permutations are quite similar, they still differ in two out of four major operations: AddRoundConstant and ShiftBytes. Additional area requirements may result from inserting a pipeline register between two stages of the P/Q round. The total width of the multiplexers, outside of the P round in SAMOSA is $4h$. The width of similar multiplexers outside of the P/Q round in Grøstl is $5b = 10h$. Finally, the number of the 2-input XOR gates in SAMOSA is h, while in Grøstl it is $3b = 6h$. Additionally, in the designs with padding unit, SAMOSA benefits from eliminating Block Counter from the padding logic. All these differences amount to a significant advantage of SAMOSA over Grøstl in terms of the circuit area. This advantage is particularly important considering one of the major weakness of Grøstl is its inherently large area in any high-speed hardware implementations.

Comparison in Terms of Throughput. We move this material to the full version of the paper [25].

9.4   Comparison of **SAMOSA** with the SHA-3 Finalists

Table 3 presents the comparison between SAMOSA and the SHA-3 finalists using the best single-message architecture, *i.e.*, architecture capable of processing only one message at a time (more results are available in the full version of the paper). All algorithms have been implemented without padding units, in two variants,

Table 2. Implementation results of Grøstl and SAMOSA for Xilinx Virtex 5. CLB stands for Configurable Logic Block.

	Grøstl	Samosa	Percentage Difference [%]	Grøstl	Samosa	Percentage Difference [%]
			Without Padding Unit			*With Padding Unit*
256-bit						
Frequency (MHz)	250.9	215.5	-14.1	269.5	217.0	-19.5
Throughput (Mbit/s)	6117	5516	-9.8	6572	5556	-15.5
Area (CLB slices)	1795	1305	-27.3	2020	1318	-34.8
Throughput/Area ((Mbit/s)/CLB slices)	3.41	4.23	24.0	3.25	4.22	29.6
512-bit						
Frequency (MHz)	217.7	195.0	-10.4	211.3	199.0	-5.9
Throughput (Mbit/s)	7686	7133	-7.2	7462	7276	-2.5
Area (CLB slices)	3853	2559	-33.6	3895	2732	-29.9
Throughput/Area ((Mbit/s)/CLB slices)	1.99	2.79	39.7	1.92	2.66	39.0

Table 3. SAMOSA and the best single message architectures of the SHA-3 finalists for the 256-bit variants of hash functions

Xilinx Virtex 5					*Altera Stratix III*			
Ranking	Architecture	Throughput (Mbits/s)	Area (CLB slices)	TP/Area	Ranking	Architecture	Throughput (Mbits/s)	Area (ALUTs)	TP/Area
Keccak	x1	13337	1369	9.74	Keccak	x1	15493	3531	4.39
JH	x1	4955	982	5.05	JH	x1	5276	3221	1.64
SAMOSA	x1	5516	1305	4.23	SAMOSA	x1	5969	4851	1.23
Grøstl	x1 (P/Q)	6117	1795	3.41	Grøstl	/2(v) (P/Q)	3818	3914	0.98
Skein	x4	3023	1218	2.48	Skein	x4	2475	3943	0.63
BLAKE	/4(v)/4(h)	389	231	1.68	BLAKE	/2(h)	2158	3553	0.61

with 256-bit and 512-bit output, in Xilinx Virtex 5 and Altera Stratix III FPGAs. The primary metric used for comparison is throughput to area ratio.

10   Conclusion and Open Problems

This paper gives proposal for a novel permutation based hash mode of operation named FP. Our indifferentiability security analysis establishes that the new mode is secure against all generic attacks up to approximately $2^{n/2}$ queries; more interestingly, our experimental results suggest that the security bound can be improved to nearly 2^n queries (n is the hash size in bits). We leave the proof of this improved result as an open problem.

We also design a concrete hash function family SAMOSA based on the FP mode and the P permutations of the SHA-3 finalist Grøstl; we claim it is hard to attack SAMOSA with complexities significantly less than the brute force. Our FPGA hardware implementations of SAMOSA show remarkable improvement in the throughput to area ratio compared to the SHA-3 finalists Grøstl, BLAKE and Skein. It is still not known how efficient SAMOSA is in software. We leave the software implementations of SAMOSA as future work.

Acknowledgment. The authors like to thank Dustin Moody and Daniel Smith-Tone for numerous helpful discussions.

References

1. Andreeva, E., Mennink, B., Preneel, B.: The parazoa family: generalizing the sponge hash functions. Int. J. Inf. Sec. 11(3), 149–165 (2012) (Cited on page 510.)
2. ATHENa Project Website, `http://cryptography.gmu.edu/athena` (Cited on page 524.)
3. Biryukov, A., Dunkelman, O., Keller, N., Khovratovich, D., Shamir, A.: Key Recovery Attacks of Practical Complexity on AES-256 Variants with up to 10 Rounds. In: Gilbert, H. (ed.) EUROCRYPT 2010. LNCS, vol. 6110, pp. 299–319. Springer, Heidelberg (2010) (Cited on page 510.)
4. Bertoni, G., Daemen, J., Peeters, M., Van Assche, G.: Sponge Functions. In: ECRYPT 2007 (2007), `http://sponge.noekeon.org/SpongeFunctions.pdf` (accessed March 2012) (Cited on page 510.)
5. Bertoni, G., Daemen, J., Peeters, M., Van Assche, G.: On the Indifferentiability of the Sponge Construction. In: Smart, N.P. (ed.) EUROCRYPT 2008. LNCS, vol. 4965, pp. 181–197. Springer, Heidelberg (2008) (Cited on page 511.)
6. Bhattacharyya, R., Mandal, A.: On the Indifferentiability of Fugue and Luffa. In: Lopez, J., Tsudik, G. (eds.) ACNS 2011. LNCS, vol. 6715, pp. 479–497. Springer, Heidelberg (2011) (Cited on page 511.)
7. Bhattacharyya, R., Mandal, A., Nandi, M.: Security Analysis of the Mode of JH Hash Function. In: Hong, S., Iwata, T. (eds.) FSE 2010. LNCS, vol. 6147, pp. 168–191. Springer, Heidelberg (2010) (Cited on page 511.)
8. Black, J., Rogaway, P., Shrimpton, T.: Black-Box Analysis of the Block-Cipher-Based Hash-Function Constructions from PGV. In: Yung, M. (ed.) CRYPTO 2002. LNCS, vol. 2442, pp. 320–335. Springer, Heidelberg (2002) (Cited on page 510.)
9. Blackburn, S.R., Stinson, D.R., Upadhyay, J.: On the complexity of the herding attack and some related attacks on hash functions. Des. Codes Cryptography 64(1-2), 171–193 (2012) (Cited on page 510.)
10. Coron, J.-S., Dodis, Y., Malinaud, C., Puniya, P.: Merkle-Damgård Revisited: How to Construct a Hash Function. In: Shoup, V. (ed.) CRYPTO 2005. LNCS, vol. 3621, pp. 430–448. Springer, Heidelberg (2005) (Cited on pages 510 and 513.)
11. De Cannière, C., Sato, H., Watanabe, D.: The Luffa Hash Function. In: The 1st SHA-3 Candidate Conference (Cited on page 510.)
12. Dodis, Y., Reyzin, L., Rivest, R.L., Shen, E.: Indifferentiability of Permutation-Based Compression Functions and Tree-Based Modes of Operation, with Applications to MD6. In: Dunkelman, O. (ed.) FSE 2009. LNCS, vol. 5665, pp. 104–121. Springer, Heidelberg (2009) (Cited on page 511.)
13. Gaj, K., Homsirikamol, E., Rogawski, M., Shahid, R., Sharif, M.U.: Comprehensive Evaluation of High-Speed and Medium-Speed Implementations of Five SHA-3 Finalists Using Xilinx and Altera FPGAs. Cryptology ePrint Archive, Report 2012/368 (2012), `http://eprint.iacr.org/2012/368.pdf` (Cited on pages 512 and 523.)
14. Gauravaram, P., Knudsen, L., Matusiewicz, K., Mendel, F., Rechberger, C., Schlaffer, M., Thomsen, S.: Groestl - a SHA-3 candidate. In: The 1st SHA-3 Candidate Conference (Cited on pages 510, 511 and 522.)

15. Joux, A.: Multicollisions in Iterated Hash Functions. Application to Cascaded Constructions. In: Franklin, M. (ed.) CRYPTO 2004. LNCS, vol. 3152, pp. 306–316. Springer, Heidelberg (2004) (Cited on page 510.)
16. Küçük, Ö.: Design and Analysis of Cryptographic Hash Functions. PhD thesis, KU Leuven (2012), http://www.iacr.org/phds/?p=detail&entry=777 (Cited on page 511.)
17. Kelsey, J., Kohno, T.: Herding Hash Functions and the Nostradamus Attack. In: Vaudenay, S. (ed.) EUROCRYPT 2006. LNCS, vol. 4004, pp. 183–200. Springer, Heidelberg (2006) (Cited on page 510.)
18. Kelsey, J., Schneier, B.: Second Preimages on n-Bit Hash Functions for Much Less than 2^n Work. In: Cramer, R. (ed.) EUROCRYPT 2005. LNCS, vol. 3494, pp. 474–490. Springer, Heidelberg (2005) (Cited on page 510.)
19. Matusiewicz, K., Naya-Plasencia, M., Nikolić, I., Sasaki, Y., Schläffer, M.: Rebound Attack on the Full LANE Compression Function. In: Matsui, M. (ed.) ASIACRYPT 2009. LNCS, vol. 5912, pp. 106–125. Springer, Heidelberg (2009) (Cited on page 511.)
20. Maurer, U.M., Renner, R., Holenstein, C.: Indifferentiability, Impossibility Results on Reductions, and Applications to the Random Oracle Methodology. In: Naor, M. (ed.) TCC 2004. LNCS, vol. 2951, pp. 21–39. Springer, Heidelberg (2004) (Cited on page 513.)
21. Moody, D., Paul, S., Smith-Tone, D.: Improved Indifferentiability Security Bound for the JH Mode. In: 3rd SHA-3 Candidate Conference (2012) (Cited on page 511.)
22. NIST. Secure hash standard. In: Federal Information Processing Standard, FIPS 180-2 (April 1995) (Cited on page 509.)
23. Nandi, M., Paul, S.: Speeding Up the Wide-Pipe: Secure and Fast Hashing. In: Gong, G., Gupta, K.C. (eds.) INDOCRYPT 2010. LNCS, vol. 6498, pp. 144–162. Springer, Heidelberg (2010) (Cited on page 510.)
24. Latif, K., Rao, M.M., Aziz, A., Mahboob, A.: Efficient Hardware Implementations and Hardware Performance Evaluation of SHA-3 Finalists. In: The Third SHA-3 Candidate Conference, Washington D.C., March 22-23 (2012) (Cited on page 523.)
25. Paul, S., Homsirikamol, E., Gaj, K.: A Novel Permutation-based Hash Mode of Operation FP and The Hash Function SAMOSA. Full version will appear in IACR ePrint Archive (Cited on pages 513, 514, 516, 518, 520, 521, 522, 523 and 524.)
26. Rivest, R.: The MD5 message-digest algorithm. IETF RFC 1321 (1992) (Cited on page 509.)
27. Rivest, R.: The MD6 Hash Function (Cited on page 510.)
28. Wu, H.: The JH Hash Function. In: The 1st SHA-3 Candidate Conference (2009) (Cited on page 510.)

Resistance against Adaptive
Plaintext-Ciphertext Iterated Distinguishers [*],[**]

Aslı Bay[1], Atefeh Mashatan[2], and Serge Vaudenay[1]

[1] EPFL, Switzerland
{asli.bay,serge.vaudenay}@epfl.ch
[2] Security Engineering, Canadian Imperial Bank of Commerce (CIBC), Canada
Atefeh.Mashatan@cibc.com

Abstract. Decorrelation Theory deals with general adversaries who are mounting iterated attacks, i.e., attacks in which an adversary is allowed to make d queries in each iteration with the aim of distinguishing a random cipher C from the ideal random cipher C^*. A bound for a *non-adaptive* iterated distinguisher of order d, who is making *plaintext (resp. ciphertext)* queries, against a $2d$-decorrelated cipher has already been derived by Vaudenay at EUROCRYPT '99. He showed that a $2d$-decorrelated cipher resists against iterated non-adaptive distinguishers of order d when iterations have almost no common queries. More recently, Bay et al. settled two open problems arising from Vaudenay's work at CRYPTO '12, yet they only consider non-adaptive iterated attacks.

Hence, a bound for an *adaptive* iterated adversary of order d, who can make both *plaintext* and *ciphertext* queries, against a $2d$-decorrelated cipher has not been studied yet. In this work, we study the resistance against this distinguisher and we prove the bound for an adversary who is making adaptive plaintext and ciphertext queries depending on the previous queries to an oracle.

1 Introduction

Attempting to provide provable security to block cipher cryptanalysis, Nyberg [Nyb91] pioneered a new direction where the notion of strength against differential cryptanalysis is formally examined. Similarly, Chabaud and Vaudenay [CV94] examined the notion of strength against linear cryptanalysis. Luby and Rackoff [LR85, LR86] have also considered a Feistel scheme with a random round function and defined the notion of k-wise independent hash function families. The caveat with their approach is that very long secret keys are required. Carter and Wegman [CW79, CW81], however, require smaller key when measuring the effects of pseudorandomness against the adversaries.

---

[*] This work was supported in part by the European Commission through the ICT program under contract ICT-2007-216646 ECRYPT II.

[**] This work was supported by the National Competence Center in Research on Mobile Information and Communication Systems (NCCR-MICS), a center of the SNF under grant number 5005-67322 when the second author was at EPFL.

S. Galbraith and M. Nandi (Eds.): INDOCRYPT 2012, LNCS 7668, pp. 528–544, 2012.

Inspired by the notion of k-wise independence of Luby and Rackoff and the derandomization techniques of Carter and Wegman in sampling pairwise independent numbers, Vaudenay defined and formalized Decorrelation Theory [Vau99c, Vau03] to provide provable security for block ciphers against a wide range of statistical attacks. Indeed perfect decorrelation of order d is equivalent to the d-wise independence of Luby and Rackoff while appropriate norms and measures are defined for imperfect decorrelation in [Vau98a, Vau99a]. Moreover, Decorrelation Theory covers a variety of statistical attacks such as Differential and Linear Attacks, Boomerang Attacks, Truncated Differential Attacks, and Impossible Differential Attacks. However, the attacks covered in Decorrelation Theory are generic attacks complying a certain broad criteria in the Luby and Rackoff model.

Decorrelation Theory considers computationally unbounded attackers who can make d queries in each iteration. When these d queries are random and independent from one another, the attacker is a d-limited *non-adaptive* adversary. In contrast, one can consider *adaptive* adversaries who choose their queries depending on the previous ones. Then, a distinguisher of order d is trying to distinguish between a random cipher C and the ideal random cipher C^* using the aforementioned adversary.

Non-adaptive iterated distinguishers, making plaintext (resp. ciphertext) queries, have been studied in [Vau98b, Vau99b, Vau99c, Vau98a, BV05] extensively, and the security of many block ciphers has been proven by decorrelation techniques, see for example [PV98, Vau03, BF06a, BF06b]. In particular, Vaudenay [Vau99c, Vau03] finds an upper bound for the advantage of a non-adaptive iterated distinguisher of order d, who is making plaintext (resp. ciphertext) queries against a $2d$-decorrelated cipher. He shows that a $2d$-decorrelated cipher resists against iterated non-adaptive attacks of order d when iterations have almost no common queries. His work has been followed by Bay et al. [BMV12] who address two open problems arising from Vaudenay [Vau99c, Vau03] on non-adaptive iterated attacks. When considering resistance against non-adaptive iterated adversaries of order d who are making only plaintext (resp. ciphertext) queries, Bay et al. showed that not only it is sufficient for a cipher to have decorrelation of order $2d$, but this decorrelation order is also necessary. Moreover, they proved that repeating a plaintext query in different iterations may provide a significant advantage to a non-adaptive adversary.

However, a bound for the advantage of an *adaptive* iterated distinguisher of order d, who can make both *plaintext* and *ciphertext* queries has not been computed yet. The significance of studying general distinguishers who can make adaptive queries is not hidden to anyone. Hence, it is important to study adaptive distinguishers. Allowing the adversary to make both plaintext and ciphertext queries strengthens the security results and has already appeared in the literature. Indeed, the Boomerang attack [Wag99] is an example of such an adversary. Studying these general distinguishers making adaptive plaintext-ciphertext queries allows us to, for example, interpret Wagner's Boomerang attack [Wag99] on COCONUT98 [Vau98b, Vau03], a perfect 2-decorrelated block cipher and *provably secure* against differential and linear cryptanalyses and iterated attacks of order 1. Indeed, it could have resisted to Wagner's attack with a decorrelation of order 8.

In this paper, we are going to focus on adaptive iterated distinguishers who can make plaintext and ciphertext queries. We first define a generic adaptive plaintext-ciphertext d-limited distinguisher with an adversary who is making adaptive plaintext queries and ciphertext queries to the oracle depending on the previous queries. We, then, extend this definition to a generic adaptive plaintext-ciphertext *iterated* distinguisher of order d. We prove the bound for the advantage of *adaptive iterated* distinguisher of order d against a $2d$-decorrelated cipher. The appropriate metric for computing the advantage of this kind of adversary was defined by Vaudenay in [Vau99a]. It comes with no surprise that using this metric, we get a looser, i.e., higher, upper bound for adaptive distinguishers than that for non-adaptive distinguishers.

The rest of this paper is organized as follows. Some background results, notations, and definitions are summarized in Section 2. Section 3 defines generic adaptive plaintext-ciphertext iterated distinguishers of order d and Section 4 computes the bound for such adversaries, encapsulating the main contribution of the paper. Appendix A and Appendix B give the details the proof of Theorem 7. Appendix C reminds linear and differential distinguishers.

2 Preliminaries

Vaudenay defines Decorrelation Theory based on the *Luby-Rackoff Model* [LR85] in which the adversary is unbounded in terms of *computational power*, but bounded in the number of d plaintext-ciphertext queries that he can make. In this model, there is an oracle Ω implementing either an instance of a random function (resp. permutation) drawn from all considered functions (resp. permutations) or an instance of a random function (resp. permutation) drawn uniformly at random from all random functions (resp. permutations). The aim of the adversary $\mathcal{A}$ is to guess which of two distributions the oracle Ω selects. There are two main types of adversaries: when the adversary makes his d queries at the same time and this is called a d-limited *non-adaptive* distinguisher; when the adversary makes queries depending on answers to previous queries and this is called a d-limited *adaptive* distinguisher.

Throughout the paper, F denotes a random function (or equivalently a function set up with a random key) from a set $\mathcal{M}_1$ to a set $\mathcal{M}_2$ while F^* denotes an ideal random function from $\mathcal{M}_1$ to $\mathcal{M}_2$ drawn uniformly at random from all $|\mathcal{M}_2|^{|\mathcal{M}_1|}$ random functions. In addition, C denotes a random cipher (or equivalently the encryption function set up with a random key) over a message space $\mathcal{M}$ and C^* denotes an ideal random cipher over $\mathcal{M}$ drawn uniformly at random from all $|\mathcal{M}|!$ permutations of $\mathcal{M}$. Note that F^* and C^* are also denoted as a *perfect function* and a *perfect cipher*, respectively. In Table 1, we provide some notations to be used throughout the paper.

Table 1. Notations

$\lvert S\rvert$: number of elements in S
$\mathcal{M}^d$: set of all sequences of d tuples over the set $\mathcal{M}$
$[F]^d$: d-wise distribution matrix of a random function F
$\mathrm{Adv}_{\mathcal{A}_{\mathsf{NA}(d)}}$: advantage of the d-limited non-adaptive distinguisher $\mathcal{A}_{\mathsf{NA}(d)}$
$\mathrm{Adv}_{\mathcal{A}_{\mathsf{A}(d)}}$: advantage of the d-limited adaptive distinguisher $\mathcal{A}_{\mathsf{A}(d)}$
$\mathrm{Adv}_{\mathcal{A}_{\mathsf{NAI}(d)}}$: advantage of the non-adaptive iterated distinguisher $\mathcal{A}_{\mathsf{NAI}(d)}$ of order d
$\mathrm{Adv}_{\mathcal{A}_{\mathsf{AI}(d)}}$: advantage of the adaptive iterated distinguisher $\mathcal{A}_{\mathsf{AI}(d)}$ of order d
$\mathbb{E}(X)$: expected value of a random variable X
$V(X)$: variance of a random variable X
$\oplus$: addition modulo 2

Decorrelation Theory has a link with Linear and Differential Cryptanalyses
(see Appendix C) which are the essential cryptanalysis methods of both block
ciphers and pseudorandom functions. Both methods have iterative analysis of
an instance of a block cipher and refer to the set of attacks called *iterated
attacks*. More explicitly, iterated attacks are defined as iterations of d-limited
distinguishers. When d-limited *non-adaptive* distinguishers are iterated, we obtain *non-adaptive* iterated distinguishers of order d. When d-limited *adaptive*
distinguishers are iterated, we get *adaptive* iterated distinguishers of order d. A
generic non-adaptive iterated distinguisher of order d is illustrated in Figure 1.
Briefly, a test $\mathcal{T}$ generates the binary output T_i of each iteration i, and then the
acceptance set $\mathcal{Acc}$ produces the decision of the distinguisher based on the tuple
$(T_1, \ldots, T_n)$.

Input: an integer n, a set X, a distribution $\mathcal{X}$ on X, a test $\mathcal{T}$, a set $\mathcal{Acc}$
Oracle: the oracle Ω implementing a permutation c
for $i = 1$ to n **do**
    pick $x = (x_1, \ldots, x_d)$ at random from $\mathcal{X}$
    get $y = (\Omega(x_1), \ldots, \Omega(x_d))$
    set $T_i = 0$ or 1 such that $T_i = \mathcal{T}(x, y)$
end for
if $(T_1, \ldots, T_n) \in \mathcal{Acc}$ **then**
    output 1
else
    output 0
end if

Fig. 1. A generic non-adaptive iterated distinguisher of order d

The success of an adversary is often estimated by a measure called *advantage*
defined as follows.

Definition 1. *Let F_0 and F_1 be two random functions. The advantage of an
adversary $\mathcal{A}$ distinguishing F_0 from F_1 is defined by*

$$\mathrm{Adv}_{\mathcal{A}}(F_0, F_1) = \bigl\lvert \Pr[\mathcal{A}(F_0) = 1] - \Pr[\mathcal{A}(F_1) = 1]\bigr\rvert.$$

When we consider all adversaries distinguishing between F_0 and F_1 and take the maximum of the advantage over all these adversaries in a class ζ, we get the *best advantage* of the distinguisher which is formulated as follows.

$$\text{BestAdv}_\zeta(F_0, F_1) = \max_{\mathcal{A} \in \zeta} \text{Adv}_\mathcal{A}.$$

For example, ζ can consist of all non-adaptive adversaries or adaptive adversaries. Note that in the rest of the paper, when we mention the advantage of an adversary, we mean his best advantage. We now recall Decorrelation Theory by first giving the definition of the *d-wise distribution matrix*.

Definition 2. *[Vau03] Let F be a random function from $\mathcal{M}_1$ to $\mathcal{M}_2$. The d-wise distribution matrix $[F]^d$ of F is a $|\mathcal{M}_1|^d \times |\mathcal{M}_2|^d$-matrix which is defined by $[F]^d_{(x_1,\ldots,x_d),(y_1,\ldots,y_d)} = \Pr_F[F(x_1) = y_1, \ldots, F(x_d) = y_d]$, where $x = (x_1,\ldots,x_d) \in \mathcal{M}_1^d$ and $y = (y_1,\ldots,y_d) \in \mathcal{M}_2^d$.*

There are two main notions of matrix-norms used in this theory and recalled in the following definition.

Definition 3. *[Vau03] Let $M \in \mathbb{R}^{|\mathcal{M}_1|^d \times |\mathcal{M}_2|^d}$ be a matrix. Then, two matrix-norms are defined by*

$$\|M\|_\infty = \max_{x_1,\ldots,x_d} \sum_{y_1,\ldots,y_d} |M_{(x_1,\ldots,x_d),(y_1,\ldots,y_d)}|,$$

$$\|M\|_A = \max_{x_1} \sum_{y_1} \cdots \max_{x_d} \sum_{y_d} |M_{(x_1,\ldots,x_d),(y_1,\ldots,y_d)}|.$$

Vaudenay [Vau03] defines the *decorrelation of order d for a random function F* as the distance between its d-wise distribution matrix and the d-wise distribution matrix of the ideal random function F^*, namely $\mathcal{D}([F]^d, [F^*]^d)$, where $\mathcal{D}$ denotes one of the measures of distance given above. Deciding which matrix-norm to use depends on the type of distinguisher envisaged. While $\|\cdot\|_\infty$ is used for non-adaptive distinguishers, $\|\cdot\|_A$ is used for adaptive distinguishers. When $\mathcal{D}([F]^d, [F^*]^d) = 0$, F is called a *perfect d-decorrelated function*. Now, the following lemma relates the best advantage of a distinguisher with the decorrelation distance.

Theorem 4 (Theorems 10 and 11 in [Vau03]). *Let F and F^* be a random function and the ideal random function, respectively. The respective advantages of the best d-limited non-adaptive and adaptive distinguishers, $\mathcal{A}_{\text{NA}(d)}$ and $\mathcal{A}_{\text{A}(d)}$, are*

$$\text{Adv}_{\mathcal{A}_{\text{NA}(d)}}(F, F^*) = \frac{1}{2}\|[F]^d - [F^*]^d\|_\infty$$

and,

$$\text{Adv}_{\mathcal{A}_{\text{A}(d)}}(F, F^*) = \frac{1}{2}\|[F]^d - [F^*]^d\|_A.$$

We recall one of the main theorems of this theory proving that if a cipher has decorrelation of order $2d$, then it is secure against a non-adaptive iterated attack of order d.

Theorem 5 (Theorem 18 in [Vau03]). *Let C be a random cipher on a message space $\mathcal{M}$ of size M such that $\|[C]^{2d} - [C^*]^{2d}\|_\infty \leq \varepsilon$, for some given $d \leq M/2$, where C^* is the ideal random cipher. Let us consider a non-adaptive iterated distinguisher of order d between C and C^* with n iterations. We assume that the distinguisher generates sets of d plaintexts of independent and identically distributed in all iterations. Then, we can bound the advantage of the adversary as*

$$\mathrm{Adv}_{\mathcal{A}_{\mathsf{NAI}(d)}} \leq 5\sqrt[3]{\left(2\delta + \frac{5d^2}{2M} + \frac{3\varepsilon}{2}\right)n^2} + n\varepsilon,$$

where δ is the probability that any two different iterations send at least one query in common.

Lastly, we will remind the notion of *indicator function*.

Definition 6. *Let S be the sample space and $E \subseteq S$ be an event. The indicator function of the event E, denoted by $\mathbf{1}_E$, is a random variable defined as*

$$\mathbf{1}_E(s) = \begin{cases} 1, & \text{if } s \in E, \\ 0, & \text{if } s \notin E. \end{cases}$$

The indicator function can shortly be denoted as $\mathbf{1}_E$ instead of $\mathbf{1}_E(s)$. In the sequel, we define more general distinguishers, namely *adaptive* plaintext-ciphertext iterated distinguishers of order d.

3   Adaptive Plaintext-Ciphertext Iterated Distinguishers of Order d

In this section, we recall two generic distinguishers, namely an *adaptive* plaintext-ciphertext d-limited distinguisher (see Figure 2) and an *adaptive* plaintext-ciphertext iterated distinguisher of order d (see Figure 3). Both distinguishers are adaptive in a way that the adversary adaptively asks for both encryption and decryption of the queries. Herein we formalize these distinguishers.

We first define a *compact* function G to be distinguished. The goal of defining this function is to specify the input to the oracle to be either encrypted or decrypted (as the adversary makes either the plaintext queries or the ciphertext queries in a specific order depending on his type of attack).

Let $\mathcal{G}$ be the set of functions G such that $G : \mathcal{M} \times \{0,1\} \to \mathcal{M}$ satisfying $G(G(x,0),1) = x$ and $G(G(x,1),0) = x$, for all x. We denote $G_0(x) = G(x,0)$ and $G_1(x) = G(x,1)$ and point out $G_1^{-1} = G_0$ and $G_0^{-1} = G_1$. In what follows, G denotes a random element of $\mathcal{G}$ and G^* is a uniformly distributed element of $\mathcal{G}$.

Input: a function $\mathcal{F}$, a test $\mathcal{T}$, a distribution $\mathcal{R}$ on $\{0,1\}^*$
Oracle: the oracle Ω implementing either an instance of G or an instance of G^*

> Pick $r \in \{0,1\}^*$ at random from $\mathcal{R}$
> Set $u_1 = (a_1, b_1) \leftarrow \mathcal{F}(\cdot; r)$
> Set $v_1 = \Omega(u_1)$
> Set $u_2 = (a_2, b_2) \leftarrow \mathcal{F}(v_1; r)$
> Set $v_2 = \Omega(u_2)$
> $\ldots$
> Set $u_d = (a_d, b_d) \leftarrow \mathcal{F}(v_1, \ldots, v_{d-1}; r)$
> Set $v_d = \Omega(u_d)$

Output $\mathcal{T}(v_1, \ldots, v_d; r)$

Fig. 2. A generic adaptive plaintext-ciphertext d-limited distinguisher

An adaptive d-limited distinguisher. The adversary $\mathcal{A}_{A(d)}$ detailed in Figure 2 has access to an oracle Ω which implements either an instance of G or an instance of G^*, such that G_0 and G_1 perform encryption and decryption, respectively. He picks a random coin r from $\{0,1\}^*$ according to a given distribution $\mathcal{R}$ and queries a function $\mathcal{F}$ which is fed with r and the output of the previous queries $(v_1, v_2, \ldots, v_{i-1})$, where $v_k = \Omega(u_k)$ for all $k \in \{1, 2, \ldots, i-1\}$, and $1 \leq i \leq d$. He then receives a new query u_i. He sends this input u_i to the oracle to receive the output v_i, where –as explained– $v_i = \Omega(u_i)$. Finally, using a test $\mathcal{T}$, he outputs a decision bit "1" if he guesses that Ω implements an instance of the random function G or "0" if he guesses that Ω implements an instance of the ideal random function G^*.

Input: an integer n, a function $\mathcal{F}$, a test $\mathcal{T}$, a set $\mathcal{Acc}$, a distribution $\mathcal{R}$ on $\{0,1\}^*$
Oracle: the oracle Ω implementing a function G or G^*

> **for** $k = 1$ **to** n
>
>> Set T_k (with independent coins) $\leftarrow$ output of Distinguisher in Figure 2
>
> **end for**
> **Output** $1_{\mathcal{Acc}}(T_1, \ldots, T_n)$

Fig. 3. A generic adaptive plaintext-ciphertext iterated distinguisher of order d

An adaptive iterated distinguisher of order d. The iterated distinguisher given in Figure 3 is simply the iteration of the d-limited distinguisher (see Figure 2) in a way that the adversary $\mathcal{A}_{AI(d)}$ repeats the distinguisher n times, then he checks whether the output of n iterations are accepted or not with respect to a set $\mathcal{Acc}$. This gives his final decision.

The Boomerang Attack [Wag99] defined in Figure 4 is an example for an adaptive plaintext-ciphertext iterated distinguisher of order d (see Figure 3) for the case $d = 4$. The adversary queries two (chosen) plaintexts and receives their corresponding ciphertexts, he then constructs two ciphertexts depending on the

Input: an integer n, a set X, differences Δ and ∇
Oracle: the oracle Ω implementing a permutation c

    for $k = 1$ **to** n
      Pick x_1 uniformly at random from the set X
      Set $x_2 = x_1 \oplus \Delta$
      Set $y_1 = c(x_1)$, $y_2 = c(x_2)$
      Set $y_3 = y_1 \oplus \nabla$, $y_4 = y_2 \oplus \nabla$
      Set $x_3 = c^{-1}(y_3)$, $x_4 = c^{-1}(y_4)$
      Set $T_k = \mathbf{1}_{x_3 \oplus x_4 = \Delta}$

    end for
    if $T_1 + \cdots + T_n \neq 0$ **then**
      Output 1
    else
      Output 0

Fig. 4. Boomerang Distinguisher

previous ciphertexts and asks for their decryption. The adaptively chosen queries to the oracle in each iteration of the Boomerang Attack [Wag99] can be written as $(u_1, u_2, u_3, u_4) = ((x_1, 0), (x_1 \oplus \Delta, 0), (c(x_1) \oplus \nabla, 1), (c(x_1 \oplus \Delta) \oplus \nabla, 1))$, where x_1 is selected uniformly at random over the set X, and Δ and ∇ denote non-zero differences.

4 Advantage of Adaptive Plaintext-Ciphertext Iterated Distinguishers of Order d

Vaudenay [Vau03] found a bound for the advantage of non-adaptive iterated distinguishers of order d, which is not apposite for the adaptive adversaries. We extend his result and provide a bound for the advantage of *adaptive* plaintext-ciphertext iterated distinguishers of order d. Strictly speaking, we compute the maximum success of the adversary who is making d adaptive queries to the oracle in each iteration to distinguish a random cipher $2d$-decorrelated upon using the $\| \cdot \|_A$ norm.

Theorem 7. *Let $G \in \mathcal{G}$ be a random function from $\mathcal{M} \times \{0, 1\}$ to $\mathcal{M}$ such that $\||[G]^{2d} - [G^*]^{2d}\||_A \leq \varepsilon$, for some given $d \leq M/2$, where G^* is the ideal random cipher and $|\mathcal{M}| = M$. Let us consider an* adaptive *iterated distinguisher of order d $\mathcal{A}_{\mathsf{AI}(d)}$ who is trying to distinguish G from G^* by performing n iterations (see Figure 3). Then, the advantage $\mathrm{Adv}_{\mathcal{A}_{\mathsf{AI}(d)}}$ of $\mathcal{A}_{\mathsf{AI}(d)}$ is bounded as*

$$\mathrm{Adv}_{\mathcal{A}_{\mathsf{AI}(d)}} \leq 5 \sqrt[3]{\left(2\theta + e^{8d^2/M} + \frac{2d^2}{M} + \frac{3\varepsilon}{2} - 1 \right) n^2 + n\varepsilon},$$

where θ is the expected value of the probability that any two different iterations send at least one query in common for a given G.

Proof. Let one iteration consist of the input queries $u = (u_1, u_2, \ldots, u_d)$ and the output queries $v = (v_1, v_2, \ldots, v_d)$, where $u_i = (a_i, b_i)$ and $v_i = \Omega(u_i)$, for $1 \leq i \leq d$.

We first make two *observations* about the adaptive adversary.

Observation 1: *Inner-collisions* in input queries, i.e., $u_i \neq u_j$, are not allowed, since calling the same query twice in the same iteration will not give any advantage to the adversary.

Observation 2: Let $(u_i = (a_i, b_i), v_i)$ and $(u_j = (a_j, b_j), v_j)$ be two queries in the same iteration. *Cross inner-collisions* are not allowed, that is, we *never* have $a_i = v_j$ and $b_i \neq b_j$. Getting the same information will not give any advantage to the adversary.

Notice that these aforementioned observations do not hold between *different* iterations.

We begin similarly to the proof of Theorem 5 provided in [Vau03]. We first define $T(g)$ to be the probability that the test function $\mathcal{T}$ outputs 1 when $G = g$ (resp. $G^* = g$), i.e.,

$$T(g) = \mathbb{E}_r[\mathcal{T}(v_1, \ldots, v_d; r)|G = g].$$

We let p (resp. p^*) be the probability of the distinguisher outputting 1, let $\mathcal{Acc}$ be the acceptance set, and $T_k(G)$ (resp. $T_k(G^*)$) be the output of iteration k. Then we have

$$p = \Pr_G[(T_1(G), \ldots, T_n(G)) \in \mathcal{Acc}].$$

Notice that all $T_k(G)$'s are *pairwise independent* except that all are only dependent on G, and $T_k(G) = T(G)$. Hence, we obtain

$$p = \mathbb{E}_G\left[\sum_{(t_1, \ldots, t_n) \in \mathcal{Acc}} T(G)^{t_1 + \cdots + t_n}(1 - T(G))^{n - (t_1 + \cdots + t_n)} \right].$$

Then, p can be rewritten as

$$p = \sum_{k=0}^{n} a_k \mathbb{E}_G[T(G)^k(1 - T(G))^{n-k}],$$

for some integers a_k such that $0 \leq a_k \leq \binom{n}{k}$. Similarly, we have the same argument for p^*, i.e., $p^* = \sum_{k=0}^{n} a_k \mathbb{E}_{G^*}[T(G^*)^k(1 - T(G^*))^{n-k}]$.

The advantage of the distinguisher, $|p - p^*|$, is maximal when all a_k's are either 0 or $\binom{n}{k}$ depending on the distributions $T(G)$ and $T(G^*)$. Hence, we assume that $\mathcal{Acc}$ of the best distinguisher is of the form

$$\mathcal{Acc} = \left\{ (t_1, \ldots, t_n) \middle| \sum_{k=1}^{n} t_k \in \mathcal{B} \right\},$$

for some set $\mathcal{B} \subseteq \{0, \ldots, n\}$. Thus, we rewrite $p = \mathbb{E}_G[s(T(G))]$, where $s(x) = \sum_{k \in \mathcal{B}} \binom{n}{k} x^k (1 - x)^{n-k}$.

Now, consider the derivative of s which can be written as

$$s'(x) = \sum_{k \in B} \binom{n}{k} \frac{k - nx}{x(1-x)} x^k (1-x)^{n-k}.$$

Notice that since the sum over all k, such that $0 \leq k \leq n$, is the derivative of $(x + (1-x))^n$, then the total sum is zero. Hence, we obtain

$$|s'(x)| \leq \sum_{nx \leq k \leq n} \binom{n}{k} \frac{k - nx}{x(1-x)} x^k (1-x)^{n-k} \leq \frac{n}{x} \sum_{nx \leq k \leq n} \binom{n}{k} x^k (1-x)^{n-k},$$

since $nx \leq k \leq n$. We note that when $x \geq 1/2$, we have $|s'(x)| \leq 2n$. Similarly, when $x < 1/2$, we have $|s'(x)| \leq 2n$. Hence, we get $|s'(x)| \leq 2n$, for every x. So, according to the Mean Value Theorem, we have

$$|s(T(G)) - s(T(G^*))| \leq 2n|T(G) - T(G^*)|.$$

Furthermore, Theorem 4 gives the exact advantage for the best *adaptive d-limited* distinguisher. Hence, $|\mathbb{E}_G[T(G)] - \mathbb{E}_{G^*}[T(G^*)]| \leq \varepsilon/2$ is obtained. We here notice that in Vaudenay's proof for Theorem 5, the non-adaptive case was considered which leads the same result.

We now define a new random variable $T^2(G)$ which is the output of another test with $2d$ entries, that is,

$$\mathcal{T}(v_1, \ldots, v_d; r) \times \mathcal{T}(v_1', \ldots, v_d'; r').$$

Thanks to Theorem 4, we have $|\mathbb{E}_G[T^2(G)] - \mathbb{E}_{G^*}[T^2(G^*)]| \leq \varepsilon/2$. Hence, we get $|V(T(G)) - V(T(G^*))| \leq 3\varepsilon/2$ (obtained by combining $|\mathbb{E}_G[T(G)] - \mathbb{E}_{G^*}[T(G^*)]| \leq \varepsilon/2$ and $|\mathbb{E}_G[T^2(G)] - \mathbb{E}_{G^*}[T^2(G^*)]| \leq \varepsilon/2$). More precisely, we have

$$
\begin{aligned}
&|V(T(G)) - V(T(G^*))| \\
&= |\mathbb{E}_G[T^2(G)] - \mathbb{E}_G^2[T(G)] - \mathbb{E}_{G^*}[T^2(G^*)] + \mathbb{E}_{G^*}^2[T(G^*)]| \\
&\leq |\mathbb{E}_G[T^2(G)] - \mathbb{E}_{G^*}[T^2(G^*)]| + |\mathbb{E}_G^2[T(G)] - \mathbb{E}_{G^*}^2[T(G^*)]| \\
&\leq \frac{3\varepsilon}{2}.
\end{aligned}
\tag{1}
$$

In 1, we use $|\mathbb{E}_G[T(G)] + \mathbb{E}_{G^*}[T(G^*)]| \leq 2$, since $0 \leq T(G), T(G^*) \leq 1$.

Afterwards, the advantage of the distinguisher is

$$|p - p^*| = |\mathbb{E}_G[s(T(G))] - \mathbb{E}_{G^*}[s(T(G^*))]| \leq \mathbb{E}_{G,G^*}[|s(T(G)) - s(T(G^*))|].$$

By using Tchebichev's inequality, i.e., $\Pr[|T(G) - \mathbb{E}_G[T(G)]| > \lambda] \leq V(T(G))/\lambda^2$ and $\Pr[|T(G^*) - \mathbb{E}_{G^*}[T(G^*)]| > \lambda] \leq V(T(G^*))/\lambda^2$ for any $\lambda > 0$, we have

$$|p - p^*| \leq 5 \sqrt[3]{\left(2V(T(G^*)) + \frac{3\varepsilon}{2}\right) n^2} + n\varepsilon,
\tag{2}$$

when $\lambda = \sqrt[3]{(2V(T(G^*)) + (3\varepsilon/2))/n}$.

So far, everything works similarly to [Vau03]. However, the rest is different since the function implemented in the oracle has new properties. For further details of the proof up to now, refer to [Vau03]. Now, it is left to bound $V(T(G^*))$.

Bounding $V(T(G^*))$. We now bound $V(T(G^*))$ by expanding it as

$$V(T(G^*)) =$$
$$\sum_S \Pr_R[r] \Pr_R[r'] \Big(\Pr_{G^*} [(u, u') \xrightarrow{G^*} (v, v')] - \Pr_{G^*}[u \xrightarrow{G^*} v] \Pr_{G^*}[u' \xrightarrow{G^*} v'] \Big), \quad (3)$$

where $S = \{(v, r), (v', r') \in \mathcal{T}\}$ and u (resp. u') is defined by both r and v (resp. r' and v'). For the sake of simplicity, we denote the expression
$$\Pr_R[r] \Pr_R[r'] \Big(\Pr_{G^*} [(u, u') \xrightarrow{G^*} (v, v')] - \Pr_{G^*}[u \xrightarrow{G^*} v] \Pr_{G^*}[u' \xrightarrow{G^*} v'] \Big) \text{ as } P.$$

In order to find an upper bound for $V(T(G^*))$, we first divide Expression (3) into two *disjoint* sums depending on whether or not u and u' are colliding, i.e., if there exist i and j such that $u_i = u'_j$. In detail, we have $S = S_1 \cup S_2$ such that $S_1 = \{(v, r), (v', r') \in \mathcal{T} | \exists i, j \text{ s.t. } u_i = u'_j\}$ and $S_2 = \{(v, r), (v', r') \in \mathcal{T} | \forall i, j \text{ s.t. } u_i \neq u'_j\}$. Thus, we write

$$\sum_S P = \sum_{S_1} P + \sum_{S_2} P.$$

We now bound each sum separately.

The sum over S_1, $\sum_{S_1} P$, is bounded as

$$\sum_{S_1} P \leq \sum_{v,v'} \sum_{r,r'} \Pr_R[r] \Pr_R[r'] \Pr_{G^*} [(u, u') \xrightarrow{G^*} (v, v')] \mathbf{1}_{S_1}$$
$$= \sum_g \Pr[G^* = g] \sum_{v,v'} \sum_{r,r'} \Pr_R[r] \Pr_R[r'] \mathbf{1}_{(u,u') \xrightarrow{g} (v,v')} \mathbf{1}_{S_1}$$
$$= \mathbb{E}_{G^*}[\Pr_{r,r'}[\exists i, j \text{ s.t. } u_i = u'_j | G]]$$
$$\stackrel{def}{=} \theta,$$

where we denote $\mathbb{E}_{G^*}[\Pr_{r,r'}[\exists i, j \text{ s.t. } u_i = u'_j | G]]$ by θ. This can be interpreted as the expected value of the probability that any two iterations have at least one query in common for given G.

Now, we provide a bound for the sum over S_2, $\sum_{S_2} P$, which is for non-colliding inputs u and u'. We first note that since both G_0^* and G_1^* are from $\mathcal{M}$ to $\mathcal{M}$, and, hence, bijective, they are indeed the ideal cipher C^*, i.e., $G_0^* = G_1^* = C^*$. Therefore, their distribution matrices will be the same as the distribution matrix of the ideal cipher C^*. We define $x = (x_1, x_2, \ldots, x_d)$ and $y = (y_1, y_2, \ldots, y_d)$ as

$$x_i = \begin{cases} a_i, & \text{if } b_i = 0, \\ v_i, & \text{if } b_i = 1, \end{cases} \quad \text{and} \quad y_i = \begin{cases} v_i, & \text{if } b_i = 0, \\ a_i, & \text{if } b_i = 1, \end{cases}$$

where $u = ((a_1, b_1), (a_2, b_2), \ldots, (a_d, b_d))$, with $b_i \in \{0, 1\}$, is the input tuple and $v = (v_1, v_2, \ldots, v_d)$ is its corresponding output tuple. This is basically collecting the plaintexts and ciphertexts into two separate tuples. Now, the sum over S_2 can be rewritten into three *disjoint* sums as

$$\sum_{S_2} A = \sum_{S_3} A + \sum_{S_4} A + \sum_{S_5} A.$$

Here, S_3, S_4 and S_5 are the three partitions of S_2, i.e., $S_2 = S_3 \cup S_4 \cup S_5$,
$S_3 = \left\{ (v, r), (v', r') \in \mathcal{T} \mid \forall i, j, k, m, e, f \ u_i \neq u'_j, x_k \neq x'_m, y_e \neq y'_f \right\}$,
$S_4 = \left\{ (v, r), (v', r') \in \mathcal{T} \mid (\forall i, j, k, m \ u_i \neq u'_j, x_k \neq x'_m) \wedge (\exists e, f \ y_e = y'_f) \right\}$, $S_5 = \left\{ (v, r), (v', r') \in \mathcal{T} \mid (\forall i, j \ u_i \neq u'_j) \wedge (\exists k, m \ x_k = x'_m) \right\}$, and A is
$\Pr_R[r] \Pr_R[r'] \left(\Pr_{G_0^*} [(x, x') \xrightarrow{G_0^*} (y, y')] - \Pr_{G_0^*}[x \xrightarrow{G_0^*} y] \Pr_{G_0^*}[x' \xrightarrow{G_0^*} y'] \right)$. We now deal with these three sums.

The sum over S_3 (all non-colliding u's and u''s, all non-colliding x's and x''s, and all non-colliding y's and y''s), $\sum_{S_3} A$, can be rewritten as

$$\sum_{S_3} A \leq \frac{1}{2} \sum_{v, v'} \sum_{r, r'} A \times 1_{S_3} =$$
$$\frac{1}{2} \left| \Pr_{G_0^*} [(x, x') \xrightarrow{G_0^*} (y, y')] - \Pr_{G_0^*}[x \xrightarrow{G_0^*} y] \Pr_{G_0^*}[x' \xrightarrow{G_0^*} y'] \right| \sum_{v, v'} \sum_{r, r'} \Pr_R[r] \Pr_R[r'] 1_{S_3}.$$
$$(4)$$

Here, since $\left| \Pr_{G_0^*} [(x, x') \xrightarrow{G_0^*} (y, y')] - \Pr_{G_0^*}[x \xrightarrow{G_0^*} y] \Pr_{G_0^*}[x' \xrightarrow{G_0^*} y'] \right|$ is constant when there is no collision between x and x' and between y and y', in Equality (4), we take it out from the sum. Afterwards, since we never have $a_i = v_j$ and $b_i \neq b_j$ according to Observation 2, there will not be any inner-collisions in x.

Now, we bound Equality (4) as

$$\frac{1}{2} \left| \Pr_{G_0^*} [(x, x') \xrightarrow{G_0^*} (y, y')] - \Pr_{G_0^*}[x \xrightarrow{G_0^*} y] \Pr_{G_0^*}[x' \xrightarrow{G_0^*} y'] \right| \sum_{v, v'} \sum_{r, r'} \Pr_R[r] \Pr_R[r'] 1_{S_3}$$
$$\leq \frac{1}{2} \left(\frac{1}{M(M-1) \cdots (M-2d+1)} - \frac{1}{M^2(M-1)^2 \cdots (M-d+1)^2} \right) M^{2d} \quad (5)$$
$$\leq \frac{e^{8d^2/M}}{2} - \frac{d(d-1)}{2M} - \frac{1}{2}. \quad (6)$$

Note that Inequality (5) is due to fact that the sum in (4) is bounded by the total number of v and v' which is M^{2d} and $P_1 \geq P_2^2$. The way to obtain Inequality (6) is shown in Appendix A.

On the other hand, the sum over S_4, $\sum_{S_4} A$, will be the sum over all colliding y's and y''s, all non-colliding x's and x''s, and all non-colliding u's and u''s. When x and x' are non-colliding, it is not possible to have colliding y and y'. Hence, we have $\Pr_{G_0^*}\left[(x, x') \xrightarrow{G_0^*} (y, y')\right] = 0$. Therefore, the sum over S_4 will be negative, i.e., $\sum_{S_4} A \leq 0$.

Finally, we provide a bound for the sum S_5, $\sum_{S_5} A$, as

$$\sum_{S_5} A \leq \sum_{v,v'} \sum_{r,r'} \Pr_R[r] \Pr_R[r'] \Pr_{G^*}\left[(u, u') \xrightarrow{G^*} (v, v')\right] \mathbf{1}_{S_5}$$

$$= \sum_g \Pr[G^* = g] \sum_{r,r'} \Pr_R[r] \Pr_R[r'] \mathbf{1}_{S_5} \tag{7}$$

$$= \mathbb{E}_{G^*}\left[\Pr_{r,r'}[\exists i, j \text{ s.t. } x_i = x'_j \mid \forall k, m \text{ s.t. } u_k \neq u'_m \text{ and } G]\right]$$

$$\overset{def}{=} \gamma$$

$$\leq \frac{d^2}{M}.$$

Here, we define $\gamma = \mathbb{E}_{G^*}[\Pr_{r,r'}[\exists i, j \text{ s.t. } x_i = x'_j \mid \forall k, m \text{ s.t. } u_k \neq u'_m \text{ and } G]]$ as the expected value of the probability that x and x' collide when G is given and there is no collision between u and u'. We get $\gamma \leq d^2/M$ which is proved in Appendix B. Notice that Equality (7) gives the probability γ explicitly.

Now, if we sum up all the results, then we have

$$V(T(G^*)) \leq \theta + \frac{e^{8d^2/M}}{2} + \frac{d^2}{M} - \frac{1}{2}$$

by setting $d/2M \leq d^2/2M$.

When we substitute $V(T(G^*))$ in (2), then we have

$$|p - p^*| \leq 5 \sqrt[3]{\left(2\theta + e^{8d^2/M} + \frac{2d^2}{M} + \frac{3\varepsilon}{2} - 1\right)n^2} + n\varepsilon. \qquad \square$$

Allowing $\theta \approx \delta$ to compare Theorem 5 with Theorem 7, we observe that the bound for adaptive attacks is *higher* than the bound for non-adaptive attacks. This fact comes with no surprise. Adaptive adversaries are stronger than non-adaptive adversaries, in general, and adaptive queries can provide the adversary with some advantage.

5   Conclusion and Final Remarks

In this work, we study the resistance against adaptive plaintext-ciphertext iterated distinguishers of order d which has not been explored before. We prove the bound for this distinguisher in which the adversary is making adaptive plaintext and ciphertext queries to the oracle depending on the previous queries.

This work contributes to proving the security of previous and future designs based on Decorrelation Theory since previously there was no clue with adaptive iterated adversaries in this context.

It is worth mentioning that Theorem 7, provided in this paper, poses two questions. The theorem proves that decorrelation of order $2d$ is sufficient for a cipher to resist an iterated attack of order d. The first question asks whether or not this condition is necessary. The second question is as follows: can the probability θ of having the same query in different iterations increase the advantage of our adaptive adversary? Not surprisingly, similar questions were posed by Theorem 5. Bay et al. [BMV12] have recently answered these questions by providing two counterexamples that are not intuitive. Namely, Bay et al. proceeded as follows for the questions in Theorem 5.

- The first question is answered by showing that the decorrelation of order $2d$ is necessary. They provide a 3-round Feistel construction decorrelated to the order $2d - 1$, that is $\|[C]^{2d-1} - [C^*]^{2d-1}\|_A \leq 2(2d-1)^2/q$, where q is the cardinality of the finite field $\mathsf{GF}(q)$. They then perform a successful non-adaptive iterated distinguisher of order d against this cipher.
- The second one is answered by providing again a 3-round Feistel construction decorrelated to the order $2d$ such that $\|[C]^{2d} - [C^*]^{2d}\|_A \leq 8d^2/2^k$, where 2^k is the number of elements in $\mathsf{GF}(2^k)$. They construct even an iterated distinguisher of order 1 on this cipher, when δ is high.

These counter-intuitive examples can also be applied to our case since the Feistel ciphers used in the solution to both questions are decorrelated by the adaptive norm, and non-adaptive attacks are a subset of adaptive attacks. To conclude, thanks to [BMV12], our two questions for Theorem 7 are immediately answered.

References

[BF06a] Baignères, T., Finiasz, M.: Dial C for Cipher. In: Biham, E., Youssef, A.M. (eds.) SAC 2006. LNCS, vol. 4356, pp. 76–95. Springer, Heidelberg (2007)

[BF06b] Baignères, T., Finiasz, M.: KFC - The Krazy Feistel Cipher. In: Lai, X., Chen, K. (eds.) ASIACRYPT 2006. LNCS, vol. 4284, pp. 380–395. Springer, Heidelberg (2006)

[BMV12] Bay, A., Mashatan, A., Vaudenay, S.: Resistance against Iterated Attacks by Decorrelation Revisited. In: Safavi-Naini, R., Canetti, R. (eds.) CRYPTO 2012. LNCS, vol. 7417, pp. 741–757. Springer, Heidelberg (2012)

[BV05] Baignères, T., Vaudenay, S.: Proving the Security of AES Substitution-Permutation Network. In: Preneel, B., Tavares, S. (eds.) SAC 2005. LNCS, vol. 3897, pp. 65–81. Springer, Heidelberg (2006)

[CV94] Chabaud, F., Vaudenay, S.: Links between Differential and Linear Cryptanalysis. In: De Santis, A. (ed.) EUROCRYPT 1994. LNCS, vol. 950, pp. 356–365. Springer, Heidelberg (1995)

[CW79] Carter, L., Wegman, M.N.: Universal Classes of Hash Functions. Journal of Computer and System Sciences 18(2), 143–154 (1979)

[CW81] Carter, L., Wegman, M.N.: New Hash Functions and Their Use in Authentication and Set Equality. Journal of Computer and System Sciences 22(3), 265–279 (1981)

[LR85] Luby, M., Rackoff, C.: How to Construct Pseudo-random Permutations from Pseudo-random Functions. In: Williams, H.C. (ed.) CRYPTO 1985. LNCS, vol. 218, pp. 447–447. Springer, Heidelberg (1986)

[LR86] Luby, M., Rackoff, C.: Pseudo-random Permutation Generators and Cryptographic Composition. In: Hartmanis, J. (ed.) STOC, pp. 356–363. ACM (1986)

[Nyb91] Nyberg, K.: Perfect Nonlinear S-Boxes. In: Davies, D.W. (ed.) EUROCRYPT 1991. LNCS, vol. 547, pp. 378–386. Springer, Heidelberg (1991)

[PV98] Poupard, G., Vaudenay, S.: Decorrelated Fast Cipher: An AES Candidate Well Suited for Low Cost Smart Card Applications. In: Quisquater, J.-J., Schneier, B. (eds.) CARDIS 1998. LNCS, vol. 1820, pp. 254–264. Springer, Heidelberg (2000)

[Vau98a] Vaudenay, S.: Feistel Ciphers with L_2-Decorrelation. In: Tavares, S., Meijer, H. (eds.) SAC 1998. LNCS, vol. 1556, pp. 1–14. Springer, Heidelberg (1999)

[Vau98b] Vaudenay, S.: Provable Security for Block Ciphers by Decorrelation. In: Morvan, M., Meinel, C., Krob, D. (eds.) STACS 1998. LNCS, vol. 1373, pp. 249–275. Springer, Heidelberg (1998)

[Vau99a] Vaudenay, S.: Adaptive-Attack Norm for Decorrelation and Super-Pseudorandomness. In: Heys, H.M., Adams, C.M. (eds.) SAC 1999. LNCS, vol. 1758, pp. 49–61. Springer, Heidelberg (2000)

[Vau99b] Vaudenay, S.: On Probable Security for Conventional Cryptography. In: Song, J.S. (ed.) ICISC 1999. LNCS, vol. 1787, pp. 1–16. Springer, Heidelberg (2000)

[Vau99c] Vaudenay, S.: Resistance Against General Iterated Attacks. In: Stern, J. (ed.) EUROCRYPT 1999. LNCS, vol. 1592, pp. 255–271. Springer, Heidelberg (1999)

[Vau03] Vaudenay, S.: Decorrelation: A Theory for Block Cipher Security. J. Cryptology 16(4), 249–286 (2003)

[Wag99] Wagner, D.: The Boomerang Attack. In: Knudsen, L.R. (ed.) FSE 1999. LNCS, vol. 1636, pp. 156–170. Springer, Heidelberg (1999)

A   Some Details of Bounding Expression 6

Hence, we will give the detailed upper bounding of the following expression

$$\frac{1}{2}\left(\frac{1}{M}\frac{1}{M-1}\cdots\frac{1}{M-2d+1} - \frac{1}{M^2}\frac{1}{(M-1)^2}\cdots\frac{1}{(M-d+1)^2}\right)M^{2d},$$

or equivalently,

$$\frac{1}{2}\left(\frac{1}{1-\frac{1}{M}}\frac{1}{1-\frac{2}{M}}\cdots\frac{1}{1-\frac{2d-1}{M}}\right) - \frac{1}{2}\left(\frac{1}{(1-\frac{1}{M})^2}\frac{1}{(1-\frac{2}{M})^2}\cdots\frac{1}{(1-\frac{d-1}{M})^2}\right).$$

In order to find an upper bound for Expression 6, we need to maximize $(1-1/M)^{-1}(1-2/M)^{-1}\cdots(1-(2d-1/M))^{-1}$. Hence, we use two inequalities such that $(1-1/x)^{-1} \le 1+2/x$ when $|x| \ge 2$, which holds for $x = M$ since $M \ge 2$ according to Theorem 7 and $(1+r/k)^k \le e^r$, when $1+r/k \ge 0$, then, the upper bound is

$$\frac{1}{1-\frac{1}{M}}\frac{1}{1-\frac{2}{M}}\cdots\frac{1}{1-\frac{2d-1}{M}} \le e^{8d^2/M}.$$

In addition, we get

$$\frac{1}{(1-\frac{1}{M})^2}\frac{1}{(1-\frac{2}{M})^2}\cdots\frac{1}{(1-\frac{d-1}{M})^2} \ge 1+\frac{d(d-1)}{M}.$$

by using *geometric series formula*, i.e., $(1-x)^{-1} = \sum_{n=0}^{\infty} x^n$ for $|x| < 1$, which implies that $(1-1/x)^{-1} \ge 1+1/x$ for $|x| > 1$. Hence, we get the desired upper bound for Expression (6).

B   Bounding the Probability γ

We find an upper bound for γ which is the expected value of the probability that x and x' collide when G is given and there is no collision between u and u'. There is only one way for x and x' to collide when there is no collision between u and u'. This happens when one common query is from u (respectively u') and the other is from v' (respectively v). In detail, let $u_i = (a_i, b_i)$ and $u'_j = (a'_j, b'_j)$ be two respective entries from u and u', and v_i and v'_j be their corresponding output. When $b_i = 0$, $b'_j = 1$ and $a_i = v'_j$, then there is a collision in x and x' such that $x_i = x'_j$. Since u and v' are independent, the probability that u and v' collide is less than $d^2/2M$. Similarly, we have the same result for u' and v. Thus, we bound γ as

$$\gamma \le \frac{d^2}{2M} + \frac{d^2}{2M} = \frac{d^2}{M}.$$

C   Linear and Differential Distinguishers

Input: an integer n, a set X, a distribution $\mathcal{X}$ on X, a set I, masks a and b
Oracle: an oracle Ω implementing a permutation c

> **for** $i = 1$ **to** n
>   Pick x_1 at random from $\mathcal{X}$
>   Set $y_1 = c(x_1)$
>   Set $T_i = a \cdot x_1 \oplus b \cdot y_1$
> **end for**
>
> **if** $T_1 + \cdots + T_n \in I$ **then**
>   Output 1
> **else**
>   Output 0

Fig. 5. Linear Distinguisher

Input: an integer n, a set X, a distribution $\mathcal{X}$ on X, differences α and β
Oracle: an oracle Ω implementing a permutation c

> **for** $i = 1$ **to** n
>   Pick x_1 at random from $\mathcal{X}$
>   Set $x_2 = x_1 \oplus \alpha$
>   Set $y_1 = c(x_1)$, $y_2 = c(x_2)$
>   Set $T_i = \mathbf{1}_{y_1 \oplus y_2 = \beta}$
> **end for**
>
> **if** $T_1 + \cdots + T_n \neq 0$ **then**
>   Output 1
> **else**
>   Output 0

Fig. 6. Differential Distinguisher

Sufficient Conditions on Padding Schemes of Sponge Construction and Sponge-Based Authenticated-Encryption Scheme

Donghoon Chang

IIIT-Delhi, New Delhi, India
donghoon@iiitd.ac.in

Abstract. The sponge construction, designed by Bertoni, Daemen, Peeters, and Van Assche, is the hash domain extension, which allows any hash-output size, and it was also adopted as the hash mode for several concrete hash algorithms. For its security reason, they showed that its padding scheme is required to be injective, reversible, and the last block of a padded message is non-zero. However, firstly we will show that if the output size is less than or equal to the one-block size, then any injective and reversible padding scheme is sufficient. In particular, only for any message whose size is a multiple of block-length, we can take the identity function (which is also injective and reversible) as its padding scheme. Next, we take a look at the padding scheme of SPONGEWRAP which is a sponge-based authenticated encryption scheme and designed by the same authors. Since the padding scheme of SPONGEWRAP is inspired by that of the sponge construction, it requires that the padding scheme of SPONGEWRAP calls its underlying padding scheme for every message block, where the underlying padding scheme is also required to be injective, reversible, and the last block of a padded message is non-zero. In addition, the padding scheme of SPONGEWRAP includes additional frame bits for the privacy and authenticity of SPONGEWRAP. So, the padding scheme of SPONGEWRAP consists of its underlying padding scheme and frame bits. However, secondly, we will show that the non-zero condition on the underlying padding scheme is redundant, in other words, any injective and reversible padding scheme is sufficient for the underlying padding scheme.

Keywords: Sponge Construction, Indifferentiability, Authenticated-Encryption.

1 Introduction

The sponge construction [3], designed by Bertoni, Daemen, Peeters, and Van Assche, is the hash domain extension and it was also adopted as the hash mode for several concrete hash algorithms such as Keccak [6], PHOTON [12], Quark [2] and SPONGENT [8]. In the same paper, they showed that the sponge construction is indifferentiable from a random oracle, which means that the construction

S. Galbraith and M. Nandi (Eds.): INDOCRYPT 2012, LNCS 7668, pp. 545–563, 2012.

behaves as a random oracle. For the security of the sponge construction, it is required that it padding scheme is injective and reversible as well as the last block of a padded message is non-zero. For example, the 10* padding scheme satisfies such requirement. Very recently, Andreeva, Mennink, and Preneel [1] proposed the Parazoa family which generalizes the sponge construction and they proved its indifferentiable security. The Parazoa family also allows any hash-output size and includes more complex ones than the sponge construction so the padding scheme for the Parazoa family requires a complex and stronger condition than the sponge construction.

A keyed sponge construction [4] was also proposed by Bertoni, Daemen, Peeters, and Van Assche and it was proved in the ideal permutation model. Since the proof was done in the ideal permutation model, not the standard model, later, Chang, Dworkin, Hong, Kelsey, and Nandi [10] proposed a new and efficient keyed sponge construction, called the E-M keyed sponge construction (EMKSC), and they gave a proof of its pseudorandomness in the standard model.

Bertoni, Daemen, Peeters, and Van Assche [5] proposed an authenticated-encryption (AE) scheme, called SPONGEWRAP, which is related to the sponge construction. SPONGEWRAP is based on a function, the duplex construction, and it calls the duplex construction for every padded message block. So, the efficiency of SPONGEWRAP is determined by that of the duplex construction. Due to its security reason, the duplex construction needs frame bits as well as a padding scheme, which is the injective and reversible underlying padding scheme of SPONGEWRAP and the last block of its padded message should be non-zero like the condition of the sponge construction.

Our Contribution. Our contribution has two folds as follows.

1. `Padding scheme of the sponge construction`
   When the size of a block is big enough, we may need only the one-block hash-output or its truncation even if the sponge construction is designed to allow any hash-output size. In such case, it is interesting to know if we still need a condition that the last block of the padded message is non-zero. In this paper, we will show that if the size of the output of the sponge construction is less than or equal to the size of one-block, then injective and reversible padding scheme is sufficient. In particular, if a message of a multiple-block is only allowed as input, then we can use the identity function as its padding scheme. In other words, no padding scheme is required for that.

2. `Padding scheme of SPONGEWRAP`
   Basically, the padding scheme of SPONGEWRAP was inspired by that of the sponge construction so SPONGEWRAP needs to repeatedly call a padding scheme such as the 10* padding scheme for every block, not one time for an entire message, to ensure that every padded block is non-zero. In addition, unlike the sponge construction, SPONGEWRAP requires frame bits to ensure that the padding scheme is injective and reversible, and it supports a domain separation to independently generate a ciphertext and a tag. In this paper,

we will show that it is not necessary to make every padded block non-zero for the security of SPONGEWRAP and it is sufficient that the padding scheme of SPONGEWRAP is injective, reversible, and prefix-free. For this purpose, we propose a new AE scheme, called Sponge-AE, with an injective, reversible, and prefix-free padding scheme. And we propose a concrete padding scheme for SPONGEWRAP which is simpler and slightly efficient than the original padding scheme of SPONGEWRAP.

2   Definition and Security Notions

In this section, we explain definitions and security notions we use in this paper.

Authenticated-Encryption with Associate Data in the Ideal Model.
We follow the AE security notion used in [14] but in the ideal model. Our approach is similar to the definition of a pseudo-random function or permutation in the ideal model, which was already used in [4,9]. An authenticated-encryption (AE) scheme with associate data consists of three efficient algorithms with access to an ideal primitive f such as a random oracle, an ideal function, an ideal permutation, etc.: $\Pi^f = (\mathcal{K}, \mathcal{E}^f, \mathcal{D}^f)$. In case that f is an ideal permutation, its inverse oracle f^{-1} is also given to an adversary. The key-generation algorithm $\mathcal{K}(1^k)$ randomly generates a key K from $\{0,1\}^k$. The deterministic encryption algorithm $\mathcal{E}^f$ has four inputs, a secret key K, a nonce N, associate data A, and a message M, and returns the ciphertext and the tag $(C,T) = \mathcal{E}_K^f(N, A, M)$, where the size of C is determined by the size of M as $|C| = g(|M|)$ for a function g and $|T| = t$. The deterministic decryption algorithm $\mathcal{D}^f$ has five inputs, a secret key K, a nonce N, associate data A, a ciphertext C and a tag T, and return M if there exists M such that $(C,T) = \mathcal{E}_K^f(N, A, M)$, otherwise return INVALID. Let $\$(\cdot, \cdot, \cdot)$ be an oracle that, on inputs N, A, and M, returns a random string of length $g(|M|) + t$ for a function g. An adversary is *nonce-respecting* if it never repeats the first component, N, to its oracle, regardless of oracle responses for encryption queries not decryption queries.

PRIVACY. We say that Π^f is indistinguishable from random bits under a chosen-plaintext attack, if for any PPT nonce-respecting adversary $\mathcal{A}$ there exists a PPT simulator S *with no access to* $\$$ such that the advantage below is negligible. Note that, in [4,9], the underlying ideal primitive f is used for the last oracle for both worlds, instead of using a simulator. However, our new definition is same with them because here we only consider a simulator with no access to $\$$, which should perfectly simulate f without access to $\$$.

$$\mathrm{Adv}_{\Pi}^{\mathrm{priv}}(\mathcal{A}) = |\Pr[K \xleftarrow{\$} \{0,1\}^k : \mathcal{A}^{\mathcal{E}_K(\cdot,\cdot,\cdot), f(\cdot)} = 1] - \Pr[\mathcal{A}^{\$(\cdot,\cdot,\cdot), S(\cdot)} = 1]| = \mathsf{negl}.$$

AUTHENCITY. Given an (authenticated) encryption scheme $\Pi^f = (\mathcal{K}, \mathcal{E}^f, \mathcal{D}^f)$, there are two oracles which can be accessed by an adversary, one is encryption oracle $\mathcal{E}_K(\cdot, \cdot, \cdot)$ and the other is decryption oracle $\mathcal{D}_K^f(\cdot, \cdot, \cdot, \cdot)$, where K

is a randomly chosen secret key. We say that an adversary $\mathcal{A}$ forges if $\mathcal{A}$ outputs (N, A, C, T), where $\mathcal{D}_K^f(N, A, C, T)$ is not INVALID, and $\mathcal{A}$ made no earlier query (N, A, M) which resulted in a response (C, T). Note that any forgery adversary does not need to be nonce-respecting for the decryption queries. Nonce-respecting condition is applied only to the encryption queries. More precisely, for an forgery adversary $\mathcal{A}$, the experiment is defined as follows.

Experiment $\mathbf{Exp}_{\Pi^f}^{auth}(\mathcal{A})$
    $K \xleftarrow{\$} \mathcal{K}$
    Run $\mathcal{A}^{\mathcal{E}_K(\cdot,\cdot,\cdot), \mathcal{D}_K(\cdot,\cdot,\cdot,\cdot), f(\cdot)}$
    If $\mathcal{A}$ made a decryption query (N, A, C, T) such that the following are true
    - The decryption oracle's output is not INVALID
    - $\mathcal{A}$ did not, prior to making no earlier query (N, A, M) which resulted
      in a response (C, T)
    Then return 1 else 0

We say that Π^f is unforgeable, if for any PPT nonce-respecting adversary $\mathcal{A}$ the advantage below is negligible.

$$Adv_{\Pi^f}^{auth}(\mathcal{A}) = \Pr[\mathbf{Exp}_{\Pi^f}^{auth}(\mathcal{A}) = 1] = \mathsf{negl}$$

Indifferentiability(Concrete Version). The security notion of indifferentiability was introduced by Maurer et al. in TCC 2004 [13]. In Crypto 2005, Coron et al. were the first to adopt it as a security notion for hash functions [11]. Here, we only consider the security notion i F be a hash function based on an ideal primitive f and R be a VIL random oracle, and S be a simulator with access to R and the upperbound of its query, memory, time complexity, and the maximum length of query and the total block length of all the queries is defined by $(q_S, m_S, t_S, l_S, \sigma_S)$. Let D be an adversary with access to either (R, S) or (F^f, f). Then, we say that F^f is $(q_S, m_S, t_S, l_S, \sigma_S, q, l, \sigma, \epsilon)$-indifferentiable from R if for any adversary D whose the upperbound of the query complexity is q, the maximum length of query is l, and the total block length of all the queries is at most σ, there exists a simulator S such that:

$$\mathrm{Adv}_{F^f, S^R}^{\mathrm{indiff}}(D) = |\Pr[D^{F,f} = 1] - \Pr[D^{R,S} = 1]| \le \epsilon.$$

We say F^f is indifferentiable from R when ϵ is negligible and the simulator S is efficient.

3 Constructions

In this section, we describe all the constructions related to the works of this paper.

The Sponge Construction [3]. The sponge construction, denoted $Sponge_{f,\mathrm{pad},\ell}$ here, is a domain-codomain extension for a hash function that is based on a permutation or function f, with a fixed input and output length, $r + c$. The construction has two other parameters that we omit from our notation: a positive integer r, called the bitrate, and an injective and reversible padding scheme, denoted pad. For any input string M, called the message, the length of pad(M) is a multiple of r. The quantity c is called the capacity. ℓ is the size of output. When the last block of a padded message is non-zero by the padding scheme pad, it was proven in [3] that $Sponge_{f,\text{pad},\ell}$ is indifferentiable from a random oracle. Such non-zero condition on the padding scheme was required for its security when any hash-output size is allowed. This is because, for example, when $\ell > r$ and $\ell' > \ell - r$, the last $(\ell - r)$-bit of the ℓ-bit hash output for a padded message x is always same as the first $(\ell - r)$-bit of the ℓ'-bit hash output for a padded message $x\|0^r$.

The Sponge Construction : $Sponge_{f,\text{pad},\ell}(M)$
Let pad(M) = $(M_0\|....\|M_t)$, for some positive t where each $\|M_i\| = r$.
100 $s_a = 0^r$ and $s_b = 0^c$
200 for $i = 0$ to t,
201    $(s_a\|s_b) = f((s_a \oplus M_i)\|s_b)$.
300 for $i = 0$ to $\lceil\frac{\ell}{r}\rceil$,
301    $z_i = s_a$.
302    $c = f(s_a\|s_b)$.
400 return the first ℓ bits of z.

Fig. 1. The Sponge Construction

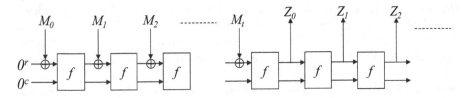

Fig. 2. The Sponge Construction: $Sponge_{f,\text{pad},\ell}(M) = first_\ell(Z_0\|Z_1\|Z_2...)$, where pad($M$) = $M_0\|...\|M_t$

The Duplex Construction [5]. In [5], the duplex construction was introduced in order to define an AE scheme, called SPONGEWRAP. The duplex construction is defined as shown in Fig. 3. The construction consists of three components, (1) an underlying permutation f over b bits, (2) a padding scheme pad, where for any input message, pad returns its padded message of a multiple-block and the last block of the padded message should be non-zero, and (3) r is the bit-size of a block. In Fig. 3, $\rho_{\max}(\text{pad}, r) = max\{x : x + \|\text{pad}[r](x)\| \le r\}$, where pad[$r$]($x$) is the $(r - \|x\|)$-bit padded string to x. For example, if the padding scheme is 10*, which is the simplest one, then $\rho_{\max}(\text{pad}, r) = r - 1$ and for $\|x\| = r - 1$, $\|\text{pad}[r](x)=1$. The function D.$duplexing$ has two inputs, σ and ℓ, where $\sigma \in \bigcup_{n=0}^{\rho_{\max}(\text{pad},r)} \mathbb{Z}_2^n$ and $\ell \le r$. For example, in case of the 10* padding scheme, $0 \le \|\sigma\| \le r - 1$.

Algorithm. The duplex construction DUPLEX$[f, \text{pad}, r]$

Require: $r < b$
Require: $\rho_{\max}(\text{pad}, r) > 0$

   Interface: D.initialize()
   $s = 0^b$

   Interface: $Z = D$.duplexing(σ, ℓ) with $\ell \leq r$, $\sigma \in \bigcup_{n=0}^{\rho_{\max}(\text{pad}, r)} \mathbb{Z}_2^n$, and $Z \in \mathbb{Z}_2^\ell$
   $P = \sigma \| \text{pad}[r](|\sigma|)$
   $s = s \oplus (P \| 0^{b-r})$
   $s = f(s)$
   return $\lfloor s \rfloor_\ell$

Fig. 3. The duplex construction DUPLEX$[f, \text{pad}, r]$ [5]: $\rho_{\max}(\text{pad}, r) = max\{x : x + |\text{pad}[r](x)| \leq r\}$, where $\text{pad}[r](x)$ is the $(r - |x|)$-bit padded string to x

SpongeWrap [5]. The authenticated-encryption scheme SPONGEWRAP, as shown in Fig. 4, is based on the duplex construction in Fig. 3. More precisely, SPONGEWRAP is defined in Fig. 7. As we can see, its underlying padding scheme pad is applied every block, because the duplex construction has to call the padding scheme for every block. Note that each block also has one frame-bit 0 or 1 to separate a key, associate data, and a message and the last frame-bit should be different from the second-last frame-bit to separately compute a ciphertext and a tag. Therefore, the padding scheme of SPONGEWRAP is constructed by the padding scheme of the duplex construction and frame-bits. More in detail, we describe the padding scheme of SPONGEWRAP in Fig. 5. The padding scheme of SPONGEWRAP (here, we call this PAD) has three inputs, K, N, M and returns an output $\widetilde{K} \| \widetilde{N} \| \widetilde{M}$ which is a multiple of r, where r is a block length.

Sponge-AE. We propose a new authenticated-encryption scheme, called Sponge-AE, defined in Fig. 8. Sponge-AE is based on the sponge construction with *any* injective, reversible, and prefix-free four-input padding scheme pad. We say that

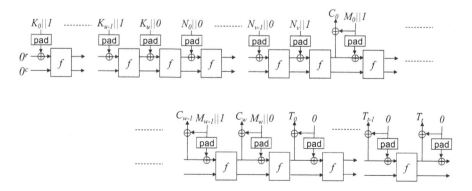

Fig. 4. SPONGEWRAP$[f, \text{pad}, r, \rho]$ [5], where $K_i, N_i, M_i \leq \rho$

The Padding Scheme of SPONGEWRAP$[f, \text{pad}, r, \rho]$: $\text{PAD}(K, N, M) = \widetilde{K}||\widetilde{N}||\widetilde{M}$
Input : K, N, M
Output: $\widetilde{K}||\widetilde{N}||\widetilde{M}$.
Requirement : pad is the 10^* padding.
Requirement : $\rho = r - 2$.
Let $K=K_0||K_1||...||K_u$ with $|K_i|=\rho$ for $i < u$, $|K_u| \le \rho$ and $|K_u| > 0$ if $u > 0$
Let $N=N_0||N_1||...||N_v$ with $|N_i|=\rho$ for $i < v$, $|N_v| \le \rho$ and $|N_v| > 0$ if $v > 0$
Let $M=M_0||M_1||...||M_w$ with $|M_i|=\rho$ for $i < w$, $|M_w| \le \rho$ and $|M_w| > 0$ if $w > 0$
Let $\widetilde{K}=K_0||1||\text{pad}[r](K_0||1)||...||K_{u-1}||1||\text{pad}[r](K_{u-1}||1)||K_u||0||\text{pad}[r](K_u||0)$
Let $\widetilde{N}=N_0||0||\text{pad}[r](N_0||0)||...||N_{v-1}||0||\text{pad}[r](N_{v-1}||0)||N_v||1||\text{pad}[r](N_v||1)$
Let $\widetilde{M}=M_0||1||\text{pad}[r](M_0||1)||...||N_{w-1}||1||\text{pad}[r](M_{w-1}||1)||M_w||0||\text{pad}[r](M_w||0)$

Fig. 5. The Padding Scheme of SPONGEWRAP$[f, \text{pad}, r, \rho]$ [5]: $\text{pad}[r](x)$ is the $(r-|x|)$-bit padded string to x

A Concrete Padding Scheme $\text{mPAD}(K, N, M) = \widetilde{K}||\widetilde{N}||\widetilde{M}$
Input : K, N, M
Output: $\widetilde{K}||\widetilde{N}||\widetilde{M}$.
Requirement : pad is the 10^* padding such that the padded message is a multiple of $r - 1$.
Requirement : $\rho = r - 1$.
Let $K'=\text{pad}(K)=K_0||K_1||...||K_u$ with $|K_i|=r-1$ for each i
Let $N'=\text{pad}(N)=N_0||N_1||...||N_v$ with $|N_i|=r-1$ for each i
Let $M'=\text{pad}(M)=M_0||M_1||...||M_w$ with $|M_i|=r-1$ for each i
Let $\widetilde{K}=K_0||1||...||K_{u-1}||1||K_u||0$
Let $\widetilde{N}=N_0||0||...||N_{v-1}||0||N_v||1$
Let $\widetilde{M}=M_0||1||...||M_{w-1}||1||M_w||0$

Fig. 6. A Concrete Padding Scheme $\text{mPAD}(K, N, M)$ for Sponge-AE

a padding scheme pad is prefix-free if for any pair of $M = (x, y, z, w)$ and $M' = (x', y', z', w')$ ($M \ne M'$) $\text{pad}(M)$ is not a prefix of $\text{pad}(M')$. Note that Sponge-AE allows any tag-size. For example, we can define a concrete padding scheme mPAD in Fig. 6, which is easily shown to be injective, reversible, and prefix-free. Later, we will show that in the ideal permutation model, Sponge-AE is secure as long as its padding scheme is injective, reversible, and prefix-free.

Comparison between Proposed Padding Schemes by SpongeWrap and Sponge-AE. Compared to SPONGEWRAP, the proposed padding scheme of Sponge-AE has at least one-bit gain for every block, because it doesn't need to call the 10^* padding scheme for every block unlike SPONGEWRAP. Also, SPONGEWRAP They both use a frame bit for each block in order to make sure that the padding schemes PAD (defined in Fig. 5) and mPAD (defined in Fig. 6) of SPONGEWRAP and Sponge-AE, respectively, are injective, reversible, and prefix-free.

Algorithm. SPONGEWRAP[f, pad, r, ρ]

Require: $\rho \leq \rho_{\max}(\text{pad}, r)$ - 1
Require: $D = \text{DUPLEX}[f, \text{pad}, r]$

Interface: W.initialize(K) with $K \in \mathbb{Z}_2^*$
Let $K = K_0 || K_1 || ... || K_u$ with $|K_i| = \rho$ for $i < u$, $|K_u| \leq \rho$ and $|K_u| > 0$ if $u > 0$
D.initialize()
for $i = 0$ to $u - 1$ **do**
   D.duplexing($K_i || 1, 0$)
end for
D.duplexing($K_u || 0, 0$)

Interface: $(C, T) = $ W.wrap(N, M, ℓ) with $N, M \in \mathbb{Z}_2^*$, $\ell \geq 0$, $C \in \mathbb{Z}_2^{|B|}$,
and $T \in \mathbb{Z}_2^\ell$
Let $N = N_0 || N_1 || ... || N_v$ with $|N_i| = \rho$ for $i < v$, $|N_v| \leq \rho$ and $|N_v| > 0$ if $v > 0$
Let $M = M_0 || M_1 || ... || M_w$ with $|M_i| = \rho$ for $i < w$, $|M_w| \leq \rho$ and $|M_w| > 0$ if $w > 0$
for $i = 0$ to $v - 1$ **do**
   D.duplexing($N_i || 0, 0$)
end for
$Z = D$.duplexing($N_v || 1, |M_0|$)
$C = M_0 \oplus Z$
for $i = 0$ to $w - 1$ **do**
   $Z = D$.duplexing($M_i || 1, |M_{i+1}|$)
   $C = C || (M_{i+1} \oplus Z)$
end for
$Z = D$.duplexing($M_w || 0, \rho$)
while $|Z| < \ell$ **do**
   $Z = Z || D$.duplexing($0, \rho$)
end while
$T = \lfloor Z \rfloor_\ell$
return (C, T)

Interface: $M = $ W.unwrap(N, C, T) with $N, C, T \in \mathbb{Z}_2^*$, $M \in \mathbb{Z}_2^{|C|} \cup \{\text{error}\}$
Let $N = N_0 || N_1 || ... || N_v$ with $|N_i| = \rho$ for $i < v$, $|N_v| \leq \rho$ and $|N_v| > 0$ if $v > 0$
Let $C = C_0 || C_1 || ... || C_w$ with $|C_i| = \rho$ for $i < w$, $|C_w| \leq \rho$ and $|C_w| > 0$ if $w > 0$
Let $T = T_0 || T_1 || ... || T_x$ with $|T_i| = \rho$ for $i < x$, $|T_x| \leq \rho$ and $|T_x| > 0$ if $x > 0$
for $i = 0$ to $v - 1$ **do**
   D.duplexing($N_i || 0, 0$)
end for
$Z = D$.duplexing($N_v || 1, |C_0|$)
$M_0 = C_0 \oplus Z$
for $i = 0$ to $w - 1$ **do**
   $Z = D$.duplexing($M_i || 1, |C_{i+1}|$)
   $M_{i+1} = C_{i+1} \oplus Z$
end for
$Z = D$.duplexing($M_w || 0, \rho$)
while $|Z| < \ell$ **do**
   $Z = Z || D$.duplexing($0, \rho$)
end while
if $T = \lfloor Z \rfloor_\ell$ **then**
   return $M_0 || M_1 || ... || M_w$
else
   return Error
end if

Fig. 7. SPONGEWRAP[f, pad, r, ρ] [5]

Algorithm. Sponge-AE$[f, \text{pad}, r, k, n, \ell]$

Requirement: $\text{pad}(\cdot, \cdot, \cdot, \cdot) = (pad1(\cdot, \cdot, \cdot) \| pad2(\cdot))$ is an injective, reversible, and prefix-free four-input padding scheme.

Key Generation Algorithm: $\mathsf{SAE.KeyGenerate}(1^k)$

100    $K \leftarrow_r \{0, 1\}^k$

Encryption Algorithm: $C = \mathsf{SAE.encrypt}(K, N, A, M)$ with $N \in \mathbb{Z}_2^n$, $A, M \in \mathbb{Z}_2^*$

200   Let $pad1(K, N, A) = (P_0\|....\|P_v)$, for some positive v where each $|P_i| = r$.

210   Let $pad2(M) = (M_0\|....\|M_w)$, where $|M_i| = r$ for $0 \le i \le w$.

220    $s_a = 0^r$ and $s_b = 0^c$

230   for $i = 0$ to v,

231      $(s_a\|s_b) = f((s_a \oplus P_i)\|s_b)$.

240   for $i = 0$ to w,

241      $C_i = s_a \oplus M_i$

242      $s_a = C_i$

243      $(s_a\|s_b) = f(s_a\|s_b)$.

250   for $i = 0$ to $\lceil \frac{\ell}{r} \rceil$,

251      $z_i = s_a$.

252      $(s_a\|s_b) = f(s_a\|s_b)$.

260   Let T be the first ℓ bits of z.

270   return $(C(= C_0\|...\|C_w), T)$

Decryption Algorithm: $M = \mathsf{SAE.decrypt}(K, N, A, C, T)$ with $N \in \mathbb{Z}_2^n$, $A, C \in \mathbb{Z}_2^*$, $T \in \mathbb{Z}_2^\ell$ and $M \in \mathbb{Z}_2^* \cup \{error\}$

300   Let $pad1(K, N, A) = (P_0\|....\|P_v)$, for some positive v where each $|P_i| = r$.

310   Let $C = (C_0\|....\|C_w)$, where $|C_i| = r$ for $0 \le i \le w$.

320    $s_a = 0^r$ and $s_b = 0^c$

330   for $i = 0$ to v,

331      $(s_a\|s_b) = f((s_a \oplus P_i)\|s_b)$.

340   for $i = 0$ to w,

341      $M_i = s_a \oplus C_i$

342      $s_a = C_i$

343      $(s_a\|s_b) = f(s_a\|s_b)$.

350   for $i = 0$ to $\lceil \frac{\ell}{r} \rceil$,

351      $z_i = s_a$.

352      $(s_a\|s_b) = f(s_a\|s_b)$.

360   if $T = T'$ return $M(= pad2^{-1}(M_0\|...\|M_w))$ otherwise return error

Fig. 8. Sponge-AE$[f, \text{pad}, r, k, n, \ell]$

4 Indifferentiable Security Proof of the Sponge Construction with a Short Output-Size

In [3], it was proved that the sponge construction with any output-size ℓ is indifferentiable from a random oracle with any output size [3] under a condition that the last block of any padded message is non-zero. What if the output-size ℓ is equal to or less than the block size r? In this section, we provide the indifferentiable security of the sponge construction with $\ell \le r$ as follows.

Theorem 1. *Let $Sponge_{f,pad,\ell}$ be the sponge construction with $\ell \leq r$ and any injective and reversible padding scheme pad and an $(r+c)$-bit ideal permutation f. Then, $Sponge_{f,pad,\ell}$ is $(O(q),\ O(q),\ O(lq),\ l,\ O(lq),\ q,\ l,\ \sigma,\ O(\frac{(2\sigma+1)^2}{2^{c+1}}))$-indifferentiable from a VIL random oracle R with ℓ-bit output, where the simulator S is defined in Fig. 10.*

Proof. In Fig. 9 and Fig. 10, we exactly describe what $(Sponge_{f,pad,\ell}, f, f^{-1})$ and (R, S, S^{-1}) are. Now, we want to prove how closely they are to each other by using indifferentiable security notion. In this paper, we follow the code-based game-playing Proof technique [7]. In Fig. 9 and Fig. 10, we exactly describe what $(Sponge_{f,pad,\ell}, f, f^{-1})$ and (R, S, S^{-1}) are. Now, we want to prove how closely they are to each other by using indifferentiable security notion. In this paper, we follow the code-based game-playing Proof technique [7].

From the following differences between Games $G0$ to $G5$, we can know that for any adversary D with at most q queries such that the maximum-bit length of a query is at most l and the bit-length of total queries is at most σ, the following inequality holds:

$$\mathrm{Adv}^{\mathrm{indiff}}_{Sponge_{f,pad,\ell},S}(D) =$$
$$|\Pr[D^{Sponge_{f,pad,\ell},f,f^{-1}} = 1] - \Pr[D^{R,S,S^{-1}} = 1]| \leq \tfrac{\sigma^2}{2^{n+1}} + \tfrac{(2\sigma+1)^2}{2^{c+1}}.$$

Complexity of the Simulator S. Now we want to show that the complexity of S is defined by $(O(q),\ O(q),\ O(lq),\ l,\ O(lq))$ for for any adversary D with at most q queries such that the maximum-bit length of a query is at most l and the bit-length of total queries is at most σ. As shown in Fig. 10, the simulator makes a query to the VIL random oracle R only when the S-query is requested. So, the maximum number of queries of any adversary $\mathcal{A}$ is q, so that of the simulator is also $O(q)$. In case of memory size, the simulator S should keep the graph which has the maximum number of edges $O(q)$. In case of time complexity of the simulator S, S need to backwardly track all the way to the initial value 0^{r+c} in worse case. Since the maximum length of queries is l, so the time complexity is bounded by $O(lq)$. The remaining values l and $O(lq)$ for the maximum length of each query and the total length of queries made by S are clear.

$G0$ **perfectly simulates** $(Sponge_{f,pad,\ell}, f, f^{-1})$. This part is clear so we omit the proof.

$G0$ **and** $G1$ **are identical.** In case of Game $G0$, for O_2 and O_3 queries, $G0$ chooses its response randomly from the outside of previous responses in order to keep the property of a permutation. In case of Game $G1$, for O_2 and O_3 queries, $G0$ firstly chooses its response randomly without considering the previous responses and then if there is a collision among the response, $G0$ again chooses its response randomly from the outside of previous responses in order to keep the property of a permutation. Therefore, two games $G0$ and $G1$ are identical.

G1 and G2 are identical-until-bad. This is clear because all the codes implemented by Games $G1$ and $G2$ are same unless bad events occur. Since all the
bad events are collision events, it is clear that $\Pr[bad] \leq \frac{\sigma^2}{2^{n+1}}$, where σ is the
total block length of queries made by any indifferentiable adversary.

G2 and G3 are identical. This is clear because two games behave identically.

G3 and G4 are identical-until-bad. This is clear because all the codes implemented by Games $G3$ and $G4$ are same unless bad events occur. Since all
the bad events are input and output collision events on the last c-bit, it is clear
that $\Pr[bad] \leq \frac{(2\sigma+1)^2}{2^{c+1}}$, where σ is the total block length of queries made by any
indifferentiable adversary.

**G4 and G5 are identical as long as the number of O_2- or O_3-queries is
less than 2^{c-1}.** This is the most important part of the proof. In case of Game
$G4$, the oracle O_1 internally depends on O_2 in line 1310. In case of Game $G5$,
the oracle O_1 no longer depends on O_2 in line 1110 but compute the response by
itself using a random oracle R. Because of this big change, the attacker will try
to find any inconsistency between the oracles O_1 and O_2 and O_3. However, the
responses of O_2 and O_3 queries are determined in a way that there is no input
and output collision on the last c-bit as shown in lines 2220 and 3300 in Fig. 13
and lines 2211, 2220, and 3200 in Fig. 14, and all the distribution of each re-
sponse is random. And $M_1||...||M_i$ in line 2200 in Fig. 14 is uniquely determined
if it exists, because of there is no input and output collision on the last c-bit.
Therefore, as long as the number of O_2 and O_3 queries is less than 2^{c-1} so that
O_2 and O_3 can generate a response for any query, there is no way to find any
difference between $G4$ and $G5$.

G5 perfectly simulates (R, S, S^{-1}). This part is clear so we omit the proof. ∎

In case that the sponge construction uses the identity function (which is injective
and reversible) as the padding scheme and it accepts only a message of a multiple
block-length, the following corollary comes out from Theorem .

Corollary 1. *Let $Sponge_{f,\ell}$ be the sponge construction with $\ell \leq r$ and an $(r+c)$-
bit ideal permutation f without any padding scheme and it accepts only a message
of multiple block-length. Then, $Sponge_{f,\ell}$ is $(O(q), O(q), O(lq), l, O(lq), q, l, \sigma,$
$O(\frac{(2\sigma+1)^2}{2^{c+1}}))$-indifferentiable from a VIL random oracle R, which accepts only a
message of multiple block-length, with ℓ-bit output, where the simulator S without
a padding scheme is similarly defined in Fig. 10.*

5   The Privacy and Authencity of **Sponge-AE**

The condition of the padding scheme of Sponge-AE is that the padding scheme is
injective, reversible, and prefix-free. Now, in this section, we provide the security
of privacy and authencity of Sponge-AE.

$(Sponge_{f,pad,\ell}, f, f^{-1})$

100 On $Sponge_{f,pad,\ell}$-query M,
110     $pad(M) = M_1\|....\|M_t$ where $|M_i| = r$.
110     $s_a = 0^r$ and $s_b = 0^c$
120   for i=1 to t
121         $(s_a\|s_b) = f((s_a \oplus M_i)\|s_b)$
121     $Z = s_a$
130   return the first ℓ bits of Z.
200 On f-query m,
210     $v = f(m)$.
220   return v.
300 On f^{-1}-query v,
310     $m = f^{-1}(v)$.
320   return m.

Fig. 9. $(Sponge_{f,pad,\ell}, f, f^{-1})$

(R, S, S^{-1})

Initialize : $X = \emptyset$ nd $Y = \{0^c\}$. R is everywhere undefined. the directed graph G is initialized as $\{0^{r+c} \to_\varepsilon 0^{r+c}\}$, where 0^{r+c} is the initial value of the sponge construction and ε is the empty string.
1000 On R-query M,
1100   if $z = R(M)$, then return z.
1110 $z \xleftarrow{\$} \{0,1\}^\ell$ and define $R(M) := z$.
1200   return z.
2000 On S-query $m = (s_a\|s_b)$,   // where $s_a \in \{0,1\}^r$ and $s_b \in \{0,1\}^c$
2100   if $(m,v) \in X$, then return v.
2200   if $\exists M_1\|...\|M_i$ such that $0^{r+c} \to_{M_1\|...\|M_{i-1}} ((s_a \oplus M_i)\|s_b)$ in G,
2210   then if $\exists M$ s.t. $pad(M) = M_1\|...\|M_i$, then $z = R(M)$ else $z \xleftarrow{\$} \{0,1\}^\ell$.
2211 $w \xleftarrow{\$} \{0,1\}^{r-\ell}$, $v_2 \xleftarrow{\$} \{0,1\}^c \backslash Y \cup \{s_b\}$, and add $((s_a \oplus M_i)\|s_b) \to_{M_i}$ $(z\|w\|v_2)$ to G.
2220   else $v_1 \xleftarrow{\$} \{0,1\}^r$ and $v_2 \xleftarrow{\$} \{0,1\}^c \backslash Y \cup \{s_b\}$.
2300 $X = X \cup \{(m, v_1\|v_2)\}$, $Y = Y \cup \{s_b, v_2\}$ and return $v = v_1\|v_2$.
3000 On S^{-1}-query $v = (v_1\|v_2)$,   // where $v_1 \in \{0,1\}^r$ and $v_2 \in \{0,1\}^c$
3100   if $(m,v) \in X$, then return m.
3200 $s_a \xleftarrow{\$} \{0,1\}^r$ and $s_b \xleftarrow{\$} \{0,1\}^c \backslash Y \cup \{v_2\}$.
3300 $X = X \cup \{(s_a\|s_b, v)\}$, $Y = Y \cup \{s_b, v_2\}$ and return $m = s_a\|s_b$.

Fig. 10. (R, S, S^{-1}): G is the directed graph maintained by the simulator S

Theorem 2. *(**Privacy**) Let Π^f be Sponge-AE$[f, pad, r, k, n, \ell]$ defined in Fig. 8. Let a four-input padding scheme pad be injective, reversible, and prefix-free. Let f be an ideal permutation over $\{0,1\}^{r+c}$. Then, for any PPT nonce-respecting adversary $\mathcal{A}$ the following inequality holds.*

Game $G0$
Initialize : $X = \emptyset$.
1000 On O_1-query M,
1100 $pad(M) = M_1\|\ldots\|M_t$ where $\|M_i\| = r$.
1200 $s_a = 0^r$ and $s_b = 0^c$
1300   for $i=1$ to t
1310 $(s_a\|s_b) = O_2((s_a \oplus M_i)\|s_b)$
1400 $Z = s_a$
1500   return the first ℓ bits of Z.
2000 On O_2-query m,   // This oracle implements the ideal permutation query f.
2100   if $(m, v) \in X$, then return v.   // $(m, v) \in X$ means m is a repeated query.
2200 $v \xleftarrow{\$} \{0,1\}^{r+c}$.
2300   if $\exists\ v'$ s.t. $(m, v') \in X$, then $v \xleftarrow{\$} \{0,1\}^{r+c} \setminus \{v' : (m, v') \in X\}$.
2400 $X = X \cup \{(m, v)\}$ and return v.
3000 On O_3-query v,   // this oracle implements the ideal permutation inverse query f^{-1}.
3100   if $(m, v) \in X$, then return m.   // $(m, v) \in X$ means v is a repeated query.
3200 $m \xleftarrow{\$} \{0,1\}^{r+c}$.
3300   if $\exists\ m'$ s.t. $(m', v) \in X$, then $m \xleftarrow{\$} \{0,1\}^{r+c} \setminus \{m' : (m', v) \in X\}$.
3400 $X = X \cup \{(m, v)\}$, and return m.

Fig. 11. $G0$ perfectly simulates $(Sponge_{f,pad,\ell}, f, f^{-1})$

$$\mathrm{Adv}_\Pi^{\mathrm{priv}}(\mathcal{A}) = |\Pr[K \xleftarrow{\$} \{0,1\}^k : \mathcal{A}^{\mathcal{E}_K(\cdot,\cdot,\cdot),f(\cdot),f^{-1}(\cdot)} = 1]$$
$$- \Pr[\mathcal{A}^{\$(\cdot,\cdot,\cdot),S(\cdot),S^{-1}(\cdot)} = 1]| \leq \frac{(2\sigma+1)^2}{2^{c+1}} + \frac{q}{2^k},$$

where q is the number of queries made by $\mathcal{A}$, and σ is the total block length of all the padded messages queried by an adversary $\mathcal{A}$. Let $\mathcal{E}_K(\cdot,\cdot,\cdot)$ be SAE.encrypt$(K, \cdot, \cdot, \cdot)$ which is the encryption algorithm of Sponge-AE. $Sponge_{f,r}$ is the sponge construction accepting only a message of a multiple of the block-size r and returning a hash output of the size r. The simulator S is defined in the same way with the simulator (with no padding scheme) defined in Fig. 10 and S simulates the random oracle R by itself instead of having access to the random oracle R, where R is the random oracle whose input size is a multiple of the block-size r and output size is also the block-size r.

Proof. Firstly, we define pf$R(\mathrm{pad}(K, \cdot, \cdot, \cdot))$, where $\mathrm{pad}(\cdot, \cdot, \cdot, \cdot) = (pad1(\cdot, \cdot, \cdot)\|pad2(\cdot))$ is same as the padding function of Sponge-AE, and R is the random oracle whose input size is a multiple of the block-size r and output size is also the block-size r.

Function pf$R(\mathrm{pad}(K, \cdot, \cdot, \cdot))$
Input: a nonce N, an associate data A, and a message M.

Game $G1$ and $G2$
Initialize : $X = \emptyset$.
1000 On O_1-query M,
1100 $pad(M) = M_1\|....\|M_t$ where $\|M_i\| = r$.
1200 $s_a = 0^r$ and $s_b = 0^c$
1300   for $i=1$ to t
1310 $(s_a\|s_b) = O_2((s_a \oplus M_i)\|s_b)$
1400 $z = s_a$
1500   return the first ℓ bits of Z.
2000 On O_2-query m,
2100   if $(m, v) \in X$, then return v.
2200 $v \xleftarrow{\$} \{0,1\}^{r+c}$.
2220   if $\exists\, v'$ s.t. $(m, v') \in X$, then $\mathsf{bad} \leftarrow \mathsf{true}$,
$\boxed{\text{and } v \xleftarrow{\$} \{0,1\}^{r+c} \setminus \{v' : (m, v') \in X\}}$.
2230 $X = X \cup \{(m, v)\}$ and return v.
3000 On O_3-query v,
3100   if $(m, v) \in X$, then return m.
3200 $m \xleftarrow{\$} \{0,1\}^{r+c}$.
3300   if $\exists\, m'$ s.t. $(m', v) \in X$, then $\mathsf{bad} \leftarrow \mathsf{true}$,
$\boxed{\text{and } m \xleftarrow{\$} \{0,1\}^{r+c} \setminus \{m' : (m', v) \in X\}}$.
3400 $X = X \cup \{(m, v)\}$, and return m.

Fig. 12. $G1$ executes with boxed statements whereas $G2$ executes without these. $G1$ and $G2$ are identical-until-bad. $G1$ perfectly simulates $G0$.

Output: a ciphertext C and a tag T.
100   Let $pad1(K, N, A) = (P_0\|....\|P_v)$, for some positive v where each $|P_i| = r$.
110   Let $pad2(M) = (M_0\|....\|M_w)$, where $|M_i| = r$ for $0 \le i \le w$.
120   for $i = 0$ to w,
121       $C_i = R(P_0\|....\|P_v\|M_0\|...\|M_{i-1}) \oplus M_i$, where R is the random oracle.
130   Let z be the empty string.
140   for $i = 0$ to $\lceil \frac{\ell}{r} \rceil$,
141       $z = z\|R(P_0\|....\|P_v\|M_0\|...\|M_w\|0^{(i+1)r})$
150   Let T be the first ℓ bits of z.
160   return $(C(= C_0\|...\|C_w), T)$

Then, the following inequalities prove the Theorem. Let S^R be a simulator which is same as the simulator R except that S^R has access to the random oracle R instead of simulating R.

$\mathrm{Adv}_{\Pi}^{\mathrm{priv}}(\mathcal{A})$
$= |\Pr[K \xleftarrow{\$} \{0,1\}^k : \mathcal{A}^{\mathcal{E}_K(\cdot,\cdot,\cdot),f,f^{-1}} = 1] - \Pr[\mathcal{A}^{\$(\cdot,\cdot,\cdot),S,S^{-1}} = 1]|$
$\le |\Pr[K \xleftarrow{\$} \{0,1\}^k : B_{\mathcal{A}}^{Sponge_{f,r}(\cdot),\mathcal{E}_K(\cdot,\cdot,\cdot),f,f^{-1}} = 1] - \Pr[B_{\mathcal{A}}^{R,\$(\cdot,\cdot,\cdot),S^R,S^{-1R}} = 1]|$

Game $G3$ and $\boxed{G4}$
Initialize : $X = \emptyset$ nd $Y = \{0^c\}$.
1000 On O_1-query M,
1100 $pad(M) = M_1\|\|....\|\|M_t$ where $\|M_i\| = r$.
1200 $s_a = 0^r$ and $s_b = 0^c$
1300   for i=1 to t
1310 $(s_a\|\|s_b) = O_2((s_a \oplus M_i)\|\|s_b)$
1400 $z = s_a$
1500   return the first ℓ bits of z.
2000 On O_2-query $m = (s_a\|\|s_b)$,   // where $s_a \in \{0,1\}^r$ and $s_b \in \{0,1\}^c$
2100   if $(m, v) \in X$, then return v.
2200 $v_1\|\|v_2 \xleftarrow{\$} \{0,1\}^{r+c}$, where $v_1 \in \{0,1\}^r$ and $v_2 \in \{0,1\}^c$.
2220   if $v_2 \in Y \cup \{s_b\}$, then bad $\leftarrow$ true, $\boxed{\text{and } v_2 \xleftarrow{\$} \{0,1\}^c \setminus Y \cup \{s_b\}}$.
2230 $X = X \cup \{(m, v_1\|\|v_2)\}$, $Y = Y \cup \{s_b, v_2\}$ and return $v = v_1\|\|v_2$.
3000 On O_3-query $v = (v_1\|\|v_2)$,   // where $v_1 \in \{0,1\}^r$ and $v_2 \in \{0,1\}^c$
3100   if $(m, v) \in X$, then return m.
3200 $s_a\|\|s_b \xleftarrow{\$} \{0,1\}^{r+c}$, where $s_a \in \{0,1\}^r$ and $s_b \in \{0,1\}^c$.
3300   if $s_b \in Y \cup \{v_2\}$, then bad $\leftarrow$ true, $\boxed{\text{and } s_b \xleftarrow{\$} \{0,1\}^c \setminus Y \cup \{v_2\}}$.
3400 $X = X \cup \{(s_a\|\|s_b, v)\}$, $Y = Y \cup \{s_b, v_2\}$ and return $m = s_a\|\|s_b$.

Fig. 13. $G4$ executes with boxed statements whereas $G3$ executes without these. $G3$ and $G4$ are identical-until-bad. $G3$ perfectly simulates $G2$.

$$\cdots\cdots \textbf{ by Claim 1}$$

$$\leq |\Pr[K \xleftarrow{\$} \{0,1\}^k : B_{\mathcal{A}}^{Sponge_{f,r}(\cdot),\mathcal{E}_K(\cdot,\cdot,\cdot),f,f^{-1}} = 1]$$
$$-\Pr[K \xleftarrow{\$} \{0,1\}^k : B_{\mathcal{A}}^{R,\text{pfR}(pad(K,\cdot,\cdot,\cdot)),S^R,S^{-1^R}} = 1]|$$
$$+|\Pr[K \xleftarrow{\$} \{0,1\}^k : B_{\mathcal{A}}^{R,\text{pfR}(pad(K,\cdot,\cdot,\cdot)),S,S^{-1}} = 1] - \Pr[B_{\mathcal{A}}^{R,\$(\cdot,\cdot,\cdot),S^R,S^{-1^R}} = 1]|$$
$$= \text{Adv}_{Sponge_{f,r},S}^{indiff}(C_{B_{\mathcal{A}}}) \qquad\qquad \cdots\cdots \textbf{ by Claim 2}$$
$$+|\Pr[K \xleftarrow{\$} \{0,1\}^k : B_{\mathcal{A}}^{R,\text{pfR}(pad(K,\cdot,\cdot,\cdot)),S,S^{-1}} = 1] - \Pr[B_{\mathcal{A}}^{R,\$(\cdot,\cdot,\cdot),S^R,S^{-1^R}} = 1]|$$
$$\leq \text{Adv}_{Sponge_{f,r},S^R}^{indiff}(C_{B_{\mathcal{A}}}) + \frac{q}{2^k} \leq \frac{(2\sigma+1)^2}{2^{c+1}} + \frac{q}{2^k} \cdots \textbf{ by Claim 3 \& Corollary 1}$$

Claim 1. $|\Pr[K \xleftarrow{\$} \{0,1\}^k : \mathcal{A}^{\mathcal{E}_K(\cdot,\cdot,\cdot),f(\cdot),f^{-1}(\cdot)} = 1] - \Pr[\mathcal{A}^{\$(\cdot,\cdot,\cdot),S(\cdot),S^{-1}(\cdot)} = 1]|$
$\leq |\Pr[K \xleftarrow{\$} \{0,1\}^k : B_{\mathcal{A}}^{Sponge_{f,r}(\cdot),\mathcal{E}_K(\cdot,\cdot,\cdot),f,f^{-1}} = 1] - \Pr[B_{\mathcal{A}}^{R,\$(\cdot,\cdot,\cdot),S^R,S^{-1^R}} = 1]|$.

Proof. The only difference between S and S^R (or, S^{-1} and S^{-1^R}) is that S simulates the random oracle R but S^R has access to R. Since $\$$ has no interaction with R, S and S^R have no difference from the adversarial point of view. Moreover, we can easily construct an adversary $B_{\mathcal{A}}$ to perfectly simulate the three oracles of $\mathcal{A}$ because the last three oracles of $B_{\mathcal{A}}$ identically works with the three oracles of $\mathcal{A}$. Therefore, Claim 1 holds. ∎

Game $G5$
Initialize : $X = \emptyset$ nd $Y = \{0^c\}$. $R : \{0,1\}^* \to \{0,1\}^\ell$ be a random oracle. the directed graph G is initialized as $\{0^{r+c} \to_\varepsilon 0^{r+c}\}$, where 0^{r+c} is the initial value of the sponge construction and ε is the empty string.
1000 On O_1-query M,
1100 $z = R(M)$.
1200   return z.
2000 On O_2-query $m = (s_a\|s_b)$,   // where $s_a \in \{0,1\}^r$ and $s_b \in \{0,1\}^c$
2100   if $(m,v) \in X$, then return v.
2200   if $\exists\, M_1\|...\|M_i$ such that $0^{r+c} \to_{M_1\|...\|M_{i-1}} ((s_a \oplus M_i)\|s_b)$ in G,
2210   then if $\exists\, M$ s.t. $pad(M) = M_1\|...\|M_i$, then $z = R(M)$ else $z \xleftarrow{\$} \{0,1\}^\ell$.
2211 $w \xleftarrow{\$} \{0,1\}^{r-\ell}$, $v_2 \xleftarrow{\$} \{0,1\}^c \setminus Y \cup \{s_b\}$, and add $((s_a \oplus M_i)\|s_b) \to_{M_i} (z\|w\|v_2)$ to G.
2220   else $v_1 \xleftarrow{\$} \{0,1\}^r$ and $v_2 \xleftarrow{\$} \{0,1\}^c \setminus Y \cup \{s_b\}$.
2300 $X = X \cup \{(m, v_1\|v_2)\}$, $Y = Y \cup \{s_b, v_2\}$ and return $v = v_1\|v_2$.
3000 On O_3-query $v = (v_1\|v_2)$,   // where $v_1 \in \{0,1\}^r$ and $v_2 \in \{0,1\}^c$
3100   if $(m,v) \in X$, then return m.
3200 $s_a \xleftarrow{\$} \{0,1\}^r$ and $s_b \xleftarrow{\$} \{0,1\}^c \setminus Y \cup \{v_2\}$.
3300 $X = X \cup \{(s_a\|s_b, v)\}$, $Y = Y \cup \{s_b, v_2\}$ and return $m = s_a\|s_b$.

Fig. 14. $G5$ perfectly simulates (R, S, S^{-1}). And $G5$ perfectly simulates $G4$ as long as the number of O_2- or O_3-queries is less than 2^{c-1}.

Claim 2. $|\Pr[K \xleftarrow{\$} \{0,1\}^k : \mathcal{A}^{Sponge_{f,r}(\cdot), \mathcal{E}_K(\cdot,\cdot,\cdot), f, f^{-1}} = 1] - \Pr[K \xleftarrow{\$} \{0,1\}^k : \mathcal{A}^{R, \mathsf{pfR}(pad(K,\cdot,\cdot,\cdot)), S^R, S^{-1}{}^R} = 1]| = \mathbf{Adv}^{\text{indiff}}_{Sponge_{f,r}, S^R}(B_\mathcal{A})$.

Proof. Let $B_\mathcal{A}$ be an adversary which distinguishes $(Sponge_{f,r}(\cdot), \mathcal{E}_K(\cdot,\cdot,\cdot), f, f^{-1})$ and $(R, \mathsf{pfR}(pad(K,\cdot,\cdot,\cdot)), S^R, S^{-1}{}^R)$. Now, we want to construct an indifferentiability adversary $C_{B_\mathcal{A}}$ for $Sponge_{f,r}$ which simulate $B_\mathcal{A}$'s oracles, where $C_{B_\mathcal{A}}$ has oracle-access to $(Sponge_{f,r}(\cdot), f, f^{-1})$ or $(R, S^R, S^{-1}{}^R)$. Let $(O_1^1, O_2^1, O_3^1, O_4^1)$ be the oracles given to $B_\mathcal{A}$ and let (O_1^2, O_2^2, O_3^2) given to $B_\mathcal{A}$. Since $(O_1^1, O_3^1, O_4^1) = (O_1^2, O_2^2, O_3^2)$, we only need to show how $C_{B_\mathcal{A}}$ simulates the oracle O_2^1.

Adversary $C_{B_\mathcal{A}}^{O_1^2, O_2^2, O_3^2}$

Initialization: $K \xleftarrow{\$} \{0,1\}^k$.
On O_1^1-query M by $B_\mathcal{A}$, return $O_1^2(M)$.
On O_3^1-query m by $B_\mathcal{A}$, return $O_2^2(m)$.
On O_4^1-query v by $B_\mathcal{A}$, return $O_3^2(v)$.
On O_2^1-query (N, A, M) by $B_\mathcal{A}$, do the following procedure.
     Let $pad1(K, N, A) = (P_0\|....\|P_v)$, for some positive v where each $|P_i| = r$.
     Let $pad2(M) = (M_0\|....\|M_w)$, where $|M_i| = r$ for $0 \le i \le w$.
     for $i = 0$ to w,
         $C_i = O_1^2(P_0\|....\|P_v\|M_0\|...\|M_{i-1}) \oplus M_i$, where R is the random oracle.
     Let z be the empty string.
     for $i = 0$ to $\lceil \frac{\ell}{r} \rceil$,

$$z = z||O_1^2(P_0||....||P_v||M_0||...||M_w||0^{(i+1)r})$$

Let T be the first ℓ bits of z.

return $(C(= C_0||...||C_w), T)$.

return the final output of $B_{\mathcal{A}}$.

Now, it is trivial that $C_{B_{\mathcal{A}}}$ perfectly simulates the oracle O_2^1 with using the oracle O_1^2. In case that $O_1^2 = Sponge_{f,r}$, $C_{B_{\mathcal{A}}}$ exactly simulates $\mathcal{E}_K(\cdot,\cdot,\cdot)$ by the definition of $\mathcal{E}_K$. In case that $O_1^2 = R$, $C_{B_{\mathcal{A}}}$ also exactly simulates $\mathsf{pf}R(\mathsf{pad}(K,\cdot,\cdot,\cdot))$ by its definition. Therefore, Claim 2 holds. ∎

Claim 3. $|\Pr[K \xleftarrow{\$} \{0,1\}^k : \mathcal{A}^{R,\mathsf{pf}R(\mathsf{pad}(K,\cdot,\cdot,\cdot)),S^R,S^{-1}{}^R} = 1] - \Pr[\mathcal{A}^{R,\$(\cdot,\cdot,\cdot),S^R,S^{-1}{}^R} = 1]| \leq \frac{q}{2^k}$.

Proof. By the definition of $\mathsf{pf}R(\mathsf{pad}(K,\cdot,\cdot,\cdot))$, it always returns a random string as long as the nonce is not repeated, because all the blocks of ciphertexts and tags are randomly generated by the random oracle R on different inputs. And, as long as the adversary does not make the first oracle-query whose first block is the secret K, there is no way to find any inconsistency between the oracles. Therefore, the above difference is bounded by the probability of guessing the key K correctly, which is $\frac{q}{2^k}$, where the key size is k and the maximum number of queries is q. ∎

Theorem 3. *(Authencity) Let Π^f be Sponge-AE$[f, pad, r, k, n, \ell]$ defined in Fig. 8. Let a four-input padding scheme pad be injective, reversible, and prefix-free. Let f be an ideal permutation over $\{0,1\}^{r+c}$. Then, for any PPT nonce-respecting adversary $\mathcal{A}$ the following inequality holds.*

$$Adv_{\Pi^f}^{auth}(\mathcal{A}) = \Pr[\boldsymbol{Exp}_{\Pi^f}^{auth}(\mathcal{A}) = 1] \leq \frac{(2\sigma+1)^2}{2^{c+1}} + \frac{q}{2^k} + \frac{q}{2^\ell},$$

where q is the number of queries made by $\mathcal{A}$, and σ is the total block length of all the padded messages queried by an adversary $\mathcal{A}$.

Sketch of Proof. In order to forge, any adversary $\mathcal{A}$ should make a decryption query (N', A', C', T') such that the decryption oracle's output is not INVALID and $\mathcal{A}$ did not, prior to making no earlier query (N', A', M') which resulted in a response C. Unlike the privacy security, the adversary $\mathcal{A}$ can use a nonce N' used previously. Note that pad is prefix-free. Also, as we proved in Theorem 2, no adversary can distinguish between the encryption algorithm of Sponge-AE and \$ with probability more than $\frac{(2\sigma+1)^2}{2^{c+1}} + \frac{q}{2^k}$. Also, in case of \$, the probablity that a decryption query (N', A', C', T') is valid is at most $\frac{q}{2^\ell}$, where q is the maximum number of queries and ℓ is the tag-size. Therefore, the theorem holds. ∎

6 Conclusion

In this paper, we studied a sufficient condition for the sponge construction and the sponge-based AE scheme. In case of the AE scheme, we showed that any injective, reversible, and prefix-free padding scheme is sufficient for its AE security. As one application, we proposed a concrete padding scheme for the AE scheme which is slightly improved than the padding scheme of SPONGEWRAP by applying the 10* padding scheme in few places unlike the padding scheme of SPONGEWRAP.

Acknowledgements. We would like to appreciate all the reviewers of Indocrypt 2012 for their valuable comments to improve our paper a lot.

References

1. Andreeva, E., Mennink, B., Preneel, B.: The Parazoa Family: Generalizing the Sponge Hash Functions. International Journal of Information Security 11(3), 149–165 (2012)

2. Aumasson, J.-P., Henzen, L., Meier, W., Naya-Plasencia, M.: Quark: A Lightweight Hash. In: Mangard, S., Standaert, F.-X. (eds.) CHES 2010. LNCS, vol. 6225, pp. 1–15. Springer, Heidelberg (2010)

3. Bertoni, G., Daemen, J., Peeters, M., Van Assche, G.: On the Indifferentiability of the Sponge Construction. In: Smart, N.P. (ed.) EUROCRYPT 2008. LNCS, vol. 4965, pp. 181–197. Springer, Heidelberg (2008)

4. Bertoni, G., Daemen, J., Peeters, M., Assche, G.V.: On the security of the keyed sponge construction. Submission to the NIST Second SHA-3 Workshop (2010), http://csrc.nist.gov/groups/ST/hash/sha-3/Round2/Aug2010/documents/papers/VANASSCHE_SpongeKeyed.pdf

5. Bertoni, G., Daemen, J., Peeters, M., Assche, G.V.: Duplexing the sponge: single-pass authenticated encryption and other applications. Submission to the NIST Second SHA-3 Workshop (2010), http://csrc.nist.gov/groups/ST/hash/sha-3/Round2/Aug2010/documents/papers/DAEMEN_DuplexSponge.pdf

6. Bertoni, G., Daemen, J., Peeters, M., Assche, G.V.: The Keccak sponge function family. Submission to NIST (Round 3) (2011), http://keccak.noekeon.org/Keccak-submission-3.pdf

7. Bellare, M., Rogaway, P.: The Security of Triple Encryption and a Framework for Code-Based Game-Playing Proofs. In: Vaudenay, S. (ed.) EUROCRYPT 2006. LNCS, vol. 4004, pp. 409–426. Springer, Heidelberg (2006)

8. Bogdanov, A., Knežević, M., Leander, G., Toz, D., Varıcı, K., Verbauwhede, I.: spongent: A Lightweight Hash Function. In: Preneel, B., Takagi, T. (eds.) CHES 2011. LNCS, vol. 6917, pp. 312–325. Springer, Heidelberg (2011)

9. Bogdanov, A., Knudsen, L.R., Leander, G., Standaert, F.-X., Steinberger, J., Tischhauser, E.: Key-Alternating Ciphers in a Provable Setting: Encryption Using a Small Number of Public Permutations (Extended Abstract). In: Pointcheval, D., Johansson, T. (eds.) EUROCRYPT 2012. LNCS, vol. 7237, pp. 45–62. Springer, Heidelberg (2012)

10. Chang, D., Dworkin, M., Hong, S., Kelsey, J., Nandi, M.: A Keyed Sponge Construction with Pseudorandomness in the Standard Model. Submission to the NIST Third SHA-3 Workshop (2012)
11. Coron, J.-S., Dodis, Y., Malinaud, C., Puniya, P.: Merkle-Damgård Revisited: How to Construct a Hash Function. In: Shoup, V. (ed.) CRYPTO 2005. LNCS, vol. 3621, pp. 430–448. Springer, Heidelberg (2005)
12. Guo, J., Peyrin, T., Poschmann, A.: The PHOTON Family of Lightweight Hash Functions. In: Rogaway, P. (ed.) CRYPTO 2011. LNCS, vol. 6841, pp. 222–239. Springer, Heidelberg (2011)
13. Maurer, U., Renner, R., Holenstein, C.: Indifferentiability, Impossibility Results on Reductions, and Applications to the Random Oracle Methodology. In: Naor, M. (ed.) TCC 2004. LNCS, vol. 2951, pp. 21–39. Springer, Heidelberg (2004)
14. Rogaway, P., Bellare, M., Black, J.: OCB: A Block-Cipher Mode of Operation for Efficient Authenticated Encryption. ACM Transactions on Information and System Security (TISSEC) 6(3), 365–403 (2003)

Author Index